V International Conference on Safety Problems of Civil Engineering Critical Infrastructures (SAFETY2019)

IOP Conference Series: Materials Science and Engineering
Volume 972

Ekaterinburg, Russia
21 – 22 May 2019

ISBN: 978-1-7138-4791-5
ISSN: 1757-8981

Printed from e-media with permission by:

Curran Associates, Inc.
57 Morehouse Lane
Red Hook, NY 12571

Some format issues inherent in the e-media version may also appear in this print version.

Institute of Physics
Dirac House, Temple Back
Bristol BS1 6BE UK

Phone: 44 1 17 929 7481
Fax: 44 1 17 920 0979

techtracking@iop.org

Additional copies of this publication are available from:

Curran Associates, Inc.
57 Morehouse Lane
Red Hook, NY 12571 USA
Phone: 845-758-0400
Fax: 845-758-2633
Email: curran@proceedings.com
Web: www.proceedings.com

TABLE OF CONTENTS

IOP Conf. Series: Materials Science and Engineering **972** (2020) 011001 doi:10.1088/1757-899X/972/1/011001

PREFACE

This volume contains conference proceedings of the V International Conference on safety problems of civil engineering critical infrastructures «SAFETY2019» held on 21th – 22nd May 2019 at the Ural Federal University in Ekaterinburg, Russia

The main conference purpose is integration and coordination of efforts of scientists, experts and specialists in the field of integrated risk analysis related to building interdependent critical infrastructures throughout their life cycle at the international level. Design, fracture mechanics problems, diagnostics, monitoring, maintenance, survivability and optimal management of infrastructure systems issues is considered in the context of the creation and operation of safe, smart and durable infrastructure systems. Question of the creation of real conditions and a practical mechanism for an innovative scientific, educational and organizational breakthrough is one of the most important problems of the twenty-first century – safety and sustainable development of cities, regions and worldwide taking into account the human factor.

The objectives of SAFETY2019 are to collect and disseminate state-of-the-art research and technology for design, fracture mechanics problems, diagnostics, monitoring, maintenance, survivability a ()nd optimal management of infrastructure systems issues is considered in the context of the creation and operation of safe, smart and durable infrastructure systems.

The conference proceedings contain 83 papers, which are divided into 11 sections:

safety technologies for building critical infrastructure and territories; modeling of loads and impacts on buildings and structures; steel and concrete structures; soil mechanics and foundations problems of viable and smart cities and sustainable development of territories; energy efficiency and resource saving in civil engineering; building information technology (BIM); building technologies and materials of innovative type; education in architecture and civil engineering; expertise and management in construction; building engineering systems.

We thank all authors of submitted papers for their participation. They contributed a great deal of effort and creativity to produce this work. We appreciate that they have chosen the V International Conference on safety problems of civil engineering critical infrastructures to present their work. Many thanks also to all editors and reviewers, who donated substantial time to carefully read and evaluate the submitted papers.

At last but not at least, we would like to appreciate all supporting organizations for their support and encouragement and making the event a success.

In addition, we are pleased to express our thanks to Organizing Committee for its great preparing work.

We look forward to the next VI International Conference on Safety Problems of Civil Engineering Critical Infrastructures «SAFETY2020» that will be held next year in November at the same location. We hope that it will be as interesting and enjoying as all five predecessors.

Ekaterinburg, May 2019

Professor
Vladimir Alekhin

Professor
Lawrence Boswell

Professor
Svyatoslav Timashev

Organizing Committee

Chairman

V.A. Koksharov – UrFU, Rector

Vice-chairmen:

A. V. Krupkin – Institute of Civil Engineering and Architecture UrFU, Director

V. N. Alekhin – Institute of Civil Engineering and Architecture UrFU, Deputy Director for Strategic Development and Infrastructure Projects – First Deputy Director, Head of the Department «CAD Systems in Civil Engineering », Professor; Member of Bureau UTB RAACS; Adviser RAACS, Member of the Administrative Council of the Association European Civil Engineering Education and Training – EUCEET;

S. A. Timashev – Science and Engineering Center "Reliability and Safety of Large Systems and Machines" SEC UB RAS, Academic Leader and Principal Scientist; Institute of Civil Engineering and Architecture UrFU, Department «CAD Systems in Civil Engineering», Professor;

Committee members:

L. Boswell– City University, London, Professor, the President of the Association European Civil Engineering Education and Training – EUCEET;

I. Fontanals – OptiCits, Barselona, Spain;

A. Gheorghe – Old Dominion, USA, Professor;

D. Lo Presti – University of Pisa, Professor; the General Secretary of the Association European Civil Engineering Education and Training – EUCEET;

O.N. Priemets – Kazakh Leading Academy of Architecture and Civil Engineering, As. Professor;

P. Akimov – RAACS, Executive Scientific Secretary, Academician RAACS, Professor;

A. A. Antipin – Institute of Civil Engineering and Architecture UrFU, the Department «CAD Systems in Civil Engineering», As. Professor;

Z. V. Beliaeva – Institute of Civil Engineering and Architecture UrFU, Head of the Department «Building structures and soil mechanics»;

E. S. Guryev – Science and Engineering Center "Reliability and Safety of Large Systems and Machines" SEC UB RAS, Scientific Secretary of the Academic Council; Institute of Civil Engineering and Architecture UrFU, the Department «CAD Systems in Civil Engineering», As. Professor;

G.V. Mazaev – UTB RAACS, Chairman, Academician RAACS;

I. N. Maltseva – Institute of Civil Engineering and Architecture UrFU, Deputy Director for International Relations, the Department "Architecture", As. Professor;

O. V. Mashkin – Institute of Civil Engineering and Architecture UrFU, the Department «Industrial, Civil Engineering and real estate expertise», Senior Lecturer;

O. B. Naschetnikova – Institute of Civil Engineering and Architecture UrFU, the Department «Water management and water technology»;

A. S. Noskov – Institute of Civil Engineering and Architecture UrFU, Head of the Department «Hydraulics», Professor; Adviser RAACS;

L. G. Pastukhova – Institute of Civil Engineering and Architecture UrFU, Deputy Director for Science and Innovation, the Department «Hydraulics», As. Professor;

L. V. Poluyan – Science and Engineering Center "Reliability and Safety of Large Systems and Machines" SEC UB RAS, Director; Institute of Civil Engineering and Architecture UrFU, the Department «CAD Systems in Civil Engineering», As. Professor;

S. V. Pridvizhkin – Institute of Civil Engineering and Architecture UrFU, Head of the Department « Pricing in Construction and Industry », Professor; Adviser RAACS;

D. V. Ulrikh – SUSU, Institute of Architecture and Civil Engineering, Director;

A. V. Khrichenkov – Institute of Civil Engineering and Architecture UrFU, Head of the Department «City Planning and Municipal Engineering»;

N. P. Shiriaeva – Institute of Civil Engineering and Architecture UrFU, Deputy Director for Education, Head of the Department «Heat-, Gas- Supply and ventilation»;

ACKNOWLEDGEMENTS

The conference organizers acknowledge the support towards the organization of the V International Conference «Safety Problems of Civil Engineering Critical Infrastructures» (SPCECI2019) to the following organizations:

Ural Federal University named after the first President of Russia B.N.Yeltsin (UrFU)

Ural Branch of the Russian Academy of Sciences

Science and Engineering Center "Reliability and Safety of Large Systems and Machines" (SEC UB RAS)

Ural Territorial Branch of the Russian Academy of Architecture and Construction Sciences (UTB RAACS)

South-Ural State University (SUSU)

The European Civil Engineering Education and Training EUCEET – Association

IOP Conf. Series: Materials Science and Engineering **972** (2020) 011004 doi:10.1088/1757-899X/972/1/011004

Peer review declaration

All papers published in this volume of *IOP Conference Series: Materials Science and Engineering* have been peer reviewed through processes administered by the Editors. Reviews were conducted by expert referees to the professional and scientific standards expected of a proceedings journal published by IOP Publishing.

- **Type of peer review: Double-blind**

- **Conference submission management system:**

by form of the site of conference https://safety-sti.urfu.ru/en/application-and-registration/
and by e-mail safety-int.conf@yandex.ru

- **Number of submissions received: 95**
- **Number of submissions sent for review: 89**
- **Number of submissions accepted: 83**
- **Acceptance Rate (Number of Submissions Accepted / Number of Submissions Received X 100): 87.4**
- **Average number of reviews per paper: 1.62**
- **Total number of reviewers involved: 30**
- **Any additional info on review process:**

Review process of papers were received to International Conference Safety Problems of Civil Engineering Critical Infrastructures:
The papers review process is formalized. The decision on the acceptance of an article is made on the basis of the sum of estimates for a number of formal indicators. The expert opinion of the reviewer is also taken into account.

- **Contact person for queries:**

Pastukhova Liliya
l.g.pastuhova@urfu.ru

IOP Conf. Series: Materials Science and Engineering **972** (2020) 012001 doi:10.1088/1757-899X/972/1/012001

Black-Swan Type Catastrophes and Antifragility/Supra-resilience of Urban Socio-Technical Infrastructures

S A Timashev[1,2,3]

[1]Science and Engineering Centre «Reliability and Safety of Large Systems and Machines», Ural Branch, Russian Academy of Sciences, 54-A, Studencheskaya, 620049, Yekaterinburg, Russia

[2]Ural Federal University, 620002, 19 Mira, Yekaterinburg, Russia

[3]Old Dominion University, Norfolk, VA USA

E-mail: TimashevS@gmail.com

Abstract. This paper may be one of the first attempts dealing with the problem of creating, providing and maintaining *antifragility* of systems of interdependent urban critical infrastructures (CI) in the wake of black-swan type technological, ecological, economic or social catastrophes occurring in a municipality. A synonym is offered to describe antifragility from a *positive psychology* perspective, formulating the problem as the *supraresilience* problem. A brief description is given of the developed innovative approach for creating a supraresilient city/region using black-swan catastrophe and the antifragility concepts. Resilience metrics are formulated as well as methods of assessing damage, interdependence of infrastructures and convergent technologies and sciences needed for practical regional resilience and risk management of the system of systems (SoS) of interdependent urban critical infrastructures).

1. Introduction

The *black swan* and *antifragility* concepts and their mostly phenomenological theories and empirics introduced to the broad public by Nassim Nicolas Taleb (N.N.T.) [1,2] are based on in-depth and convincing analysis of modern finances and macroeconomics, seamlessly coupled with the wisdom of a plethora of Ancient and modern philosophers and leavened by anecdotal and apocryphal cases, mostly from the Levant, Ancient Greece, Ancient Rome and other Near Mediterranean countries.

These concepts are based on three *whales*: (1) profound nonlinearity of *mysterious unknown* random processes of time and financial transactions that rule the financial world of NYSE, NASDAQ and the like; (2) extremely rare events that now and then shock World macroeconomics, create and bring super expensive chaos into lives of millions if not billions of people; (3) ability to see and use the intrinsic asymmetry of the optionality principle to one's (*non-sucker, in NNT's parlance)* benefit. However, their generalization onto other sectors of human activity, science and engineering (practically, to all walks of life, as energetically offered by N.N. Taleb in [2]), need some serious bricolage (adjusting, calibration, tinkering), for we all know: *the devil is in the details*. For potential users considered in this paper (urban decision makers DMs), without specific tinkering this approach may turn out to be counterproductive or an overkill.

2. The Black Swan catastrophe (BSC) concept

According to N.N. Taleb, the black swan catastrophe (BSC) is an event with following attributes: (1) it lies outside the realm of regular expectations (is a huge surprise not only to broad public, but also to the very specialists who designed the system in consideration; in our case, any urban critical infrastructure); (2) it results in (relatively) massive, catastrophic losses; (3) despite being an outlier, risk analysis specialists every time explain the next BSC occurrence *post factum*, making it look as an explainable and predictable event (although actually, it is *unpredictable*).Originally this kind of unpredictability was attributed by N. Taleb exclusively to macroeconomic and financial events. Later, this concept was applied with mixed results by other authors to some other sectors of technology and sciences [3,4].

It should be noted that the BSC as an event is scalable and is *intrinsically conditional.* One could imagine a BSC of global dimension (say, the unexpected and unpredictable Krakatau earthquake, Indonesia, 1897; global financial catastrophes of 1929 or 2008), as well as regional, local and down to personal scale BSC, when a stakeholder or pensioner loses her savings/assets in a surprise (for her) bankruptcy of a company. These conditions could be unknown to (1) the world scientific community; (2) a group of people who were affected by the BSC; (3) a single person due to lack of needed knowledge and/or false perception of the BSC before it happened.

Taleb further gives a classification of the super-rare events: he recognizes black-, grey- and white-swan (benibn rare) events, claiming that he deals only with black swans in macroeconomics and finance. In this context we will consider two types of BS events: in economics, which we will denote as BSE_e, and in Mother Nature or second, man-made Nature, which will be denoted as BSE_{ph}.

The grey-swan events in the field of economics were taken care of by the famous mathematician, Benoit Mandelbrot, when he introduced his fractal theory [5] and showed that many events (distribution of wealth, revenue from selling books of different authors, etc.) obey not the exponent-based Gaussian probability distribution function (PDF), but the more complex power PDFs with so-called *fat tails.* N.Taleb remarked [1] that due to B. Mandelbrot, who was first to use a more accurate and consistent in macroeconomics PDF (namely, the Pareto PDF, that possesses a fat [or heavy] tail), to describe BSE_e, the latter were transformed into *grey* swan events, as they started to be to some extent more predictable. Here it would be appropriate to recollect the famous adage: *The tail wags the dog.*

The white black-swan events are benign, because they produce (very rarely) great benefits, not losses.

Engineers and scientists from the oil and gas industry produced [4,6,7] their own classification of black- swan type events. They recognize following industrial BSE_{ph}: (1) completely unknown to the scientific/engineering community (*unknown unknowns* up to the moment of their occurrence); (2) unknown to the specific /local scientific/engineering community, but known if considered worldwide (unknown knowns – *unknown events to some (suckers), known to others (non-suckers)*); (3) *well known,* but considered extremely unlikely to happen during the entire lifetime or the timeframe between consecutive inspections or maintenance of the designed structure/system (say, a main pipeline or a sea-shelf platform). The tsunami that destroyed the Fukushima Daiichi nuclear power station is a typical example: in that region of Japan there were previously observed (in the 6-th and 13-th century) more severe tsunami, but the probability of them reoccurring was so minuscule (once in a *million* years), that it was deliberately ignored. If the unfortunate nuclear power plant had been built just several meters higher above ground, or had a levee built around it, the critical pumps wouldn't been flooded and there would be no Fukushima catastrophe, and no belated surprise that it happened.

Other BSCph are of the perfect-storm type (when all the components of such an event are well understood, but the probability of them emerging *simultaneously*, creating a *negative synergy* effect is so miniscule that it is deliberately ignored [8]), like the Guadalajara hail storm (in July 2019) when some parts of this city, according to the Washington Post, were covered by a *five feet high* layer of hail; or the 2019 flooding of New Orleans when a storm wave, wind-induced wave and the Mississippi flow created a synergistically combined wall of water that drowned the city.

There are attempts to make more reliable predictions of industrial and natural catastrophes using some combinations of brainstorming and computer modeling [9–13]. The forecasters now claim that some weak signals could serve as early heralds or precursors of BSE_{ph}, although the time to it will still be an unknown. Some clues from global warming, cosmic events or animal behavior can be effectively used for this matter.

In physics and engineering sciences statistical analysis of rare events is performed, using such PDFs, as Freshet, Gumbel, and Weibull, to name a few. Using the Bayes methodology these PDFs can be continuously updated after each occurrence of a rare event. This permits meaningful assessment of probabilities of stronger accidents and catastrophes in the above-mentioned fields

BS events that take place in modern physics, industry and all types of engineering (BSE_{ph}) do not have the devastating effect that is demonstrated by BS events in macroeconomics and finances (BSE_e). The reason for this is that the forces that play out during an industrial disaster are limited by location, physical laws, size of the installation, and the reached level of sophistication of the technology in containing energy, the only exclusion being nuclear energy and nuclear war. Natural catastrophes can be more violent than any human folly, but they also erupt locally, are contained by the size of planet's continents/oceans and laws of mechanics and physics. Damages from BSE_{ph} always have natural physical limits that could be reasonably assessed. Even in a potential Armageddon-type catastrophe involving the whole Universe we have a number, called *googol*, that cannot be exceeded. Bottom line: All engineering accidents are much less grey than Mandelbrot's grey-swan catastrophes.

Some BSE_e that happen in finances and macroeconomics, on the contrary, deliver a global blow to virtually every country, alongside with millions of stockholders. BSE_e are not contained by the actual combined cost of the world's physical and intellectual assets. The combined value of stocks (financial instruments) traded on all the major stock exchanges (in New York, London, Tokyo, Berlin, Shanghai, Moscow, etc.) is several times larger than the combined value of the physical assets that they represent. The BSE_e defy prediction explicitly because despite availability and accessibility of all the enormous amount of financial and economic data, generated every second, statistically it is not homogeneous. It consists of "real" data (reflects the physical assets, and "virtual" data (reflects the speculative component and human perception of what the stock may be priced in the future), without any means to parse it into real and virtual. That parsing the data into homogeneous sets is impossible is reflected in the famous mantra: «*All costs are already accounted for in the current price of stock*». There are, obviously, many other reasons for the financial market to be a classical black box, but their consideration is beyond the scope of this paper.

Modern economic and technical systems as they become more complex and interdependent are becoming more vulnerable in respect to accidents and catastrophes. The consequences of industrial catastrophes continuously grow with time due to the negative synergy of damaging effects (leak/rupture + fire + blast + destruction of assets + pollution of air/soil/water + loss of life-days, limbs and life + loss of revenue, trust, reputation and prestige). Many BSE_{ph} became transboundary, international. Industrial and natural catastrophes are becoming a serious destabilizing factor that prevents sustainable development of cities, regions and countries.

The rate of occurrence of industrial accidents is still high despite all the efforts to pull it down. According to [14] in 90% of automobile accidents the culprit is the driver(s); in aviation and river transportation up to 83% of catastrophes have human factor HF as the root cause; the same holds true in the energy field – 80% of industrial accidents are due to mistakes of the field personnel; in sea catastrophes 75% of them are related to human mistakes and mishaps; in railway transportation the corresponding number is 50%. According to [15], the HF in technogenic catastrophes frequently comes down to such human traits as stupidity, neglect or greed. All in all, it can be stated that the weakest link in the industrial safety problem is the human factor. Similar numbers apply to accidents in interdependent urban critical infrastructures. Hence, in order to drive down the probability of BSE_{ph} we should turn our attention to the role of HF in their occurrence.

What can be put against this tendency? Foremost, teaching safety culture to all employees of all ranks. And creating an atmosphere that stimulates active search of vulnerable points of the

IOP Conf. Series: Materials Science and Engineering **972** (2020) 012001 doi:10.1088/1757-899X/972/1/012001

infrastructure being maintained. In this context, the popular motto in the 1980s of a giant American corporation: *To err is human. To forgive is not company policy*, spread worldwide (author came across it while lecturing at Milan Politec) as a cartoon in which a designer standing in front of a drawing board is beaten with a stick from behind, today would not be well received by employees or by the broad public. Instead of punishing, a tolerant climate is needed, where everybody freely communicates, and upper echelon of listens to the lower. Forming safety culture is the universal tool for not allowing wrong actions of personnel that lead to mishaps and accidents.

Industrial catastrophes are developing in a continuously deteriorating man-made environment. The health of deteriorating structures and machines is periodically or continuously diagnosed and monitored, providing operators and owners of the assets with robust data using which sound assessment of their residual life can be obtained. This permits controlling the operated system in time, using risk based predictive maintenance methods.

3. The fragility-antifragility concept

N.N. Taleb justly and correctly describes the fragility-antifragility concept as a specific *current* property of any object, tacitly indicating that antifragility is an explicit function of time. Fragility is a well-known property in structural mechanics, that defines low ability of an object to resist external and/or internal harmful factors. In English language there exists a near-synonym of this word: *brittleness*, that has a more negative connotation than fragility (e.g., *brittle* diabetic, but a mentally *fragile* old relative). Its antonym is *ductility* or *viscosity*. Some strong viscous metals like steel alloys and structures made from them become brittle when exposed to very low temperatures (below minus 50 degrees Celsius). This phenomenon brings to life the urgent design problem of creating cold-climate-resistant pipelines, transportation and ground moving machinery for the Far North, Arctic and Antarctica regions.

In Russian language the word resilience – *живучесть* – refers both to live and dead (non-animated) objects and means that a system is able not only to rebound after being released from a stressor, but also continue to exist (stay alive/operate) when being injured/damaged and perform its functions under stressor or duress as designed by humans or God. According to the Ozhegov Russian dictionary *живучесть* describes the ability of a system (say, a battleship), to stay afloat and continue to fire, using its *remaining* internal strengths, while having its scull damaged by adversary's artillery. In this context the word *непотопляемый* (unsinkable) is also used in relation to boats and humans alike. A perfect visualization of this notion can be seen in a Russian doll called *ванька-встанька* which means *the stand-up vanya,* or *неваляшка* which can be translated as *roly-poly* (Fig. 1). Being forced into a horizontal position and then released, it immediately pops up back into vertical position. Hence the name of the doll.

Figure 1. The roly-poly doll that perfectly illustrates the classical resilience concept.

Antifragility, on the contrary, did not exist as a notion before Nicholas Taleb introduced it in 2012 as an antonym to fragility and demonstrates itself as the ability of a system to become *stronger* (up to an unknown point) *after* experiencing a stressor. If the stressor exceeds this unknown level (which itself is a random function of time), the person instead of accruing antifragility may become seriously harmed and loses his initial capabilities. Hence, antifragility is principally more than resilience or robustness. Taleb uses this connotation (antifragility) only in relation to live or complex objects – human beings and the Mother Nature (biological and ecological systems). He also admits that complex systems can be antifragile, obviously meaning, by default, that their complexity must include the human dimension. Antifragility is a trait that *exclusively* exists in *living* objects, because mind-/spirit-less (*dead*) objects have no ability to proactively react to different stressors of economic, financial or emotional, as well as of physical nature so as to become *more resilient/strong* than *before* the stressors were applied. Creative human action is essentially needed to achieve that part of antifragility that is *beyond* resilience of non-animated objects as we know it.

All objects created by human beings cannot become stronger after being treated by a vector of physical or chemical stressors (loads and/or influences),as they are intrinsically fragile, but this statement may soon not be fully true. There is already one case related to a nano-tube carbon composite material that became *stronger after* being physically stressed [16]. An ever-growing amount of research is conducted addressing self-healing and resilience of critical infrastructures, specifically power and information networks. In the literature that emerged after 2012, the antifragility idea as a trendy topic started to be misused, being mechanistically applied to such inert (dead) man-made systems as architectural design and buildings [3]. In the latter case authors seem to confuse *moral longevity* of buildings and cities (their capability to absorb technical innovations – growing automobile and streetcar traffic, elevators, influx of population, etc. *without change* or becoming extinct) – with antifragility. If I buy a pair of Levi's jeans a couple sizes larger than my current waist and then wear them until I feel them tight on me, I cannot claim that these jeans all the time were *antifragile*. In order to be antifragile they must adjust by *themselves*, *without human intervention*, to the size of my waist and self-repair, if need be.

Antifragility comes from and is revealed by mostly yet unknown inner forces of every human or biological being (i.e., bear, turtle, sequoia tree, etc.) or their population (i.e., rats, bees, ants or human communities) as reaction to different types of stressors. These forces are triggered by thirst for life, freedom, pursuit of happiness, all kinds of extreme challenges, lust for power, sex, wealth, fear, hunger, ultimate anger/fury, obsessive love, vanity, extreme circumstances (e.g., low/high ambient temperature or pressure), words of encouragement, patriotism, faith and sense of duty, drugs, and so on. In general, *it is a manifestation of biological vitality which is, by itself, the force of life*. Probably, because of this, *antifragility* was always nameless and word-(expression)less, understood by *default* by humans for millenniums as an intrinsic indelible component of vitality and, hence, never (before Taleb) became a separate issue for a philosophical discourse.

It is not fully known what happens in the body, mind and spirit of an individual when she/he is exposed to some specific stressor/ set of stressors. The outcome (positive or negative) of such event may come down to *overreaction/overshooting*. This phenomenon is responsible for preparing a biological system to become more powerful in the future in relation to the applied specific stressor (load, poison, cold/hot ambient temperature, etc.), when the stressor will be stronger than the experienced last time. It may be not a universal type of response of the system to stressors. In some of the cases, hidden abilities of humans are revealed immediately, in the face of extreme danger, as in a documented case when an Arctic pilot, wearing heavy arctic gear and being attacked by a polar bear, saved his life jumping directly from the snow onto the plane's wing (approximately two meters from the snow surface) and getting into his cabin [17]. Schizophrenics during their bouts of wild insanity accrue the ability to lift weights (throw heavy oak tables on nurses or lift the back of a taxi car to prevent it from moving away), impossible to lift for them when in normal condition. A champion (of the Warsaw Zoo) chimpanzee (average weight—60 kg), that does not have human conscience and does not fully understand the environment and capabilities of his own body jerks from chest a 650 kg bell bar (nearly 11 times his own weight); on average, a Warsaw Zoo chimpanzee jerks 450 kg [17]. Currently the men's world record in this heavy weight discipline (category up to 56 kg) is only 198 kg (3.54 times the record holder's weight). (This huge difference can be partially explained by that chimpanzee's arm muscles are wired differently from human arms muscles, although they are also longer than ours. It seems that this is the price humans pay for having consciousness). Rats, when being decimated by humans, using poison, traps and other lethal means, demonstrate an unusual antifragility as a cohort. They start intensive breeding, and new-born rats reach puberty at the tender age of two weeks, still being blind [17].

It should be noted that if the stressor is exceeding some ultimate level, the person exposed to such a stressor produces such an overshooting that makes him not stronger, but weaker/sicker, and she/he falls into the subset of fragile components of the system, becoming not a system's asset but its burden/liability. The type of PDF that governs the antifragility skills is a mystery (known unknown), but one could speculate that physical antifragility is closer to the Gaussian distribution rather than to the Pareto type distribution: genius is always a very rare commodity. The distribution of cognitive antifragility skills, though, probably could be better described by the Pareto type PDFs. In any case, when creating a smart city, some empirical data on antifragility has to be used as a guidance for selecting the city resilience team.

With the advent of global digitization, AI and robotization, some forecasters predict that by 2060 AI will match human intelligence [18]. The current level of computer science and AI already permits adding some elements of (quasi-) antifragility to *cloud computing* [19] and creating *emotional* robots, that are able, at least partially, to have a semblance of human emotions (quasi-emotions) and, hence, to some extent understand human psyche. It seems that with the advent of time some 21-st century robotic systems may be able to *maintain and repair themselves* to continue to operate «as new» or «as before the stressor was applied» (this is resilience in its classical meaning [20]), and if need be, to reconstruct themselves to become more smart and durable (for instance, during an unmanned expedition on the Moon or Mars), by adding to themselves new modules with new functionalities and higher resistance to unforeseen stressors. This would be an act of what could be called *resilience plus, ultra-resilience, uber resilience* or, best, *supraresilience (*in Taleb's parlance, *antifragility),* realized by a man-made system *without* direct human intervention.

According to Collins English Dictionary the word *supra* comes from Latin and when used as a prefix, means *above, over, on top of, greater than, beyond the limits of, outside of.* Obviously, it perfectly matches the word resilience to describe the result of overshooting/ overcompensating. Hence, *supraresilienc*e is a positive thinking absolute synonym of *antifragility;* therefore, both terms are interchangeable and from now on will be used as such. In addition, the term *supraresilience,* in general seems to be more fit for the physical and especially, engineering world, where the term *resilience* is very well established and currently is a hot topic of research.

In order to achieve the described above feat, the *supraresilient* robot must be equipped with a smart control subsystem that includes diagnostics, monitoring, maintenance, repair modules and a module that is able to upgrade the capability of the robot (a *self-upgrading* module). This subsystem, driven by IT and an AI «brain» brings the needed (quasi-) human dimension into the AI robot to achieve the needed supraresilience/antifragility, as it can, using a two- or three level maintenance/control policy [21], mitigate the damage due to a black/gray swan event and raise the robot performance to a level higher than that of the initial design.

In many branches of modern industry (military, nuclear power, automotive, oil and gas, aviation and the like) there are already unmanned robotized and flexible production lines equipped with these types of modules. They have in place subsystems that autonomously and automatically carry out diagnostics, monitoring and, to some extent, maintenance. For instance, according to data that can be found in the Internet, the performance of the entire fleet of F35 airplanes is tracked by a software system that provides holistic life support of each detail and aggregate of every airplane of the fleet.

The same source forms that the European Airbus company is in the process of equipping all airplanes of its fleet with an array of sensors on every aggregate, overhead baggage bin and each

IOP Conf. Series: Materials Science and Engineering **972** (2020) 012001 doi:10.1088/1757-899X/972/1/012001

passenger chair that generate in real time gigabytes of data related to the airplane performance and the feedback of its passengers and flight attendants. This Big Data is streamed to the cloud and the on-ground servers and immediately analyzed, using the Airbus digital analytical platform *Skywise,* which will also be fed by data from repair shops, spare parts suppliers, airports and airlines.

Using *Skywise*, by 2023 each Airbus plane will have its own *digital twin*. While the carbon/titanium airplane will fly, its quasi-identical digital twin will do the same, but in virtual space. The discrepancies between the current state of the real airplane and the state of its digital ideal copy will be analyzed for detecting early predictors of possible future malfunctions or accidents. This platform opens new perspectives in flight safety, flight traffic control and optimization. This philosophy is also used by Airbus to create a passenger multicopter (flying taxi) for the city mobility project Vahana.

One doesn't need to be a prophet to predict that all the above achievements will be sooner than later emulated and adopted by modern smart cities of all sizes.

According to N. Taleb, antifragility is implemented by using its main tools: *choice* and *option*. Freedom of choice permits choosing something that leads to more benefits than losses, while option = asymmetry (specific non-linearity) + rationality, could be a substitute of knowledge (in everyday life). N. Taleb doesn't describe how to diagnose, measure, and control the antifragility property in humans during their lifetime. It is obvious though that in a community or society antifragility is spread according to some Weibull type PDF that is a function of time. It also should be specific to each type of antifragility (physical, cognitive social). Each person, no matter of what antifragility level, must learn how to recognize the options and the asymmetries around her/him and at the workplace and how to make sound decisions in the chaotic environment of the 21st century, keeping in mind that there are *no free lunches*. In the areas of physical and engineering environment though, specialists have many tools to describe in quantitative terms the options and the asymmetries as well as the tools that help making right and sound decisions. At this point we stop further considering this topic as it is beyond the scope of this paper.

4. Resilience of urban critical infrastructures: current concept

Some 10–15 years ago city managers around the world realized that industrial diagnostics, monitoring and maintaining systems could be relatively easily adapted and used to optimally run and maintain all city production, transportation and life support facilities. At the beginning, they were slow in implementing these ideas but now, with the advent of digitization and AI, have accelerated their efforts to transform their cities – large and small – into smart cities.

In the above context the concept of resilience emerged currently as a central theme of industrial and urban development (there are more than 120 definitions of resilience, most of them are qualitative) Sustainable cities of the 21-st century must first be safe, resilient, and, consequently, smart in order to then become sustainable. For cities to be safe and smart they have to be resilient in the first place [22]. In general, factors that influence city resilience include: the range and severity of hazards; the risk to life, limb, health and property; the vulnerability and exposure of human, social, and environmental systems to different types of hazards, and the degree of (strategic) preparedness of the physical, social and the governance systems to any natural, urban or industrial shocks and stressors and their consequences during an incident, accident, malicious act, or catastrophe.

The architecture of the urban resilience system (URS) [23] mimics the long time existing different monitoring and maintenance optimization systems designed to enhance performance of critical *industrial* infrastructures. The difference is in that the *urban* infrastructure, in its entirety, is a very specific *complex system of interdependent systems* (SoIS), being widely spread over the whole territory of a municipality, being 1) intensely used by the city community and 2) elements of its transportation infrastructures (cars, trams, buses, metro carriages) continuously move. The URS is designed to provide, in the first place, raw and processed data about how this SoIS functions and degrades in time.

System resilience consists of following components: 1) *Structural resilience* (reliability, probability of failure); 2) *Structural safety* (risk level, damage zones, all types of losses); 3) *Physical restoration subsystem* for renewing a damaged system. This subsystem can vary by means of recovery (e.g., democratic or authoritarian), time and cost of restoration, amount of materiel, financial and human resources necessary for restoration; 4) *Psychological restoration subsystem* for the stressed-out society. This subsystem also varies by the amount of psychological assistance and the duration of its provision, as well as by the required human, materiel and financial resources.

Structural resilience is the ability of a system to maintain its functionality after being stressed by loads greater than their design values or after a specific damage. Critical infrastructure has many functions; therefore, CI has both complex (integral) and partial resiliencies. Both full and partial resiliencies are always conditional. Since the properties of CI and the stressors that it experiences are random, the resilience is also random and can be measured as the probability that the CI will retain its function(s) under any specific random effect that is *beyond the design parameters*. The term *robustness* is strongly related to the term *resilience*, as it means that the system is stable – it gives a weak response to a strong disturbance.

The *strategic readiness* of a CI is such a state of the system when it remains reliable, relatively safe and efficient to perform its main design functions, even with *sudden application of extraordinary, way out-of-design* values of external influences. This can be achieved only by including into the CI a deeply echeloned protection subsystem. A case of a one-dimensional stochastic model of failure and recovery of infrastructure as a system is shown in Fig. 2.

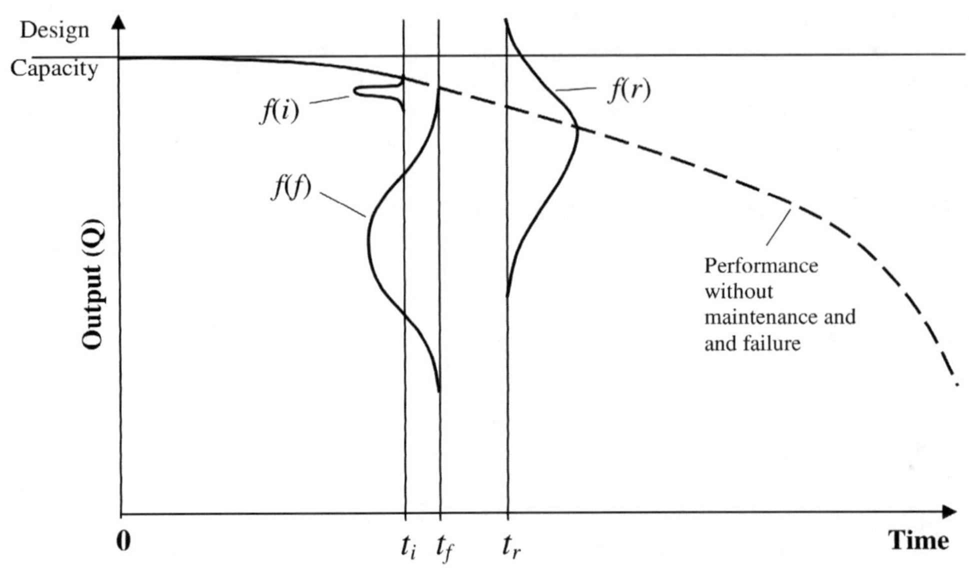

Figure 2. One-dimensional stochastic model of failure and recovery of infrastructure as a system: t_i – time to incident; t_f – time to system failure; t_r – time to recovery; $f(i)$ – PDF of the initial triggering event duration; $f(d)$ – PDF of damage volume due to system failure; $f(r)$ – PDF of recovery quality.

According to the Rockefeller Foundation: «Urban Resilience is the capacity of individuals, communities, institutions, businesses, and systems within a city to survive, adapt, and grow, no matter what kinds of chronic stresses and acute shocks they experience, and even transform when conditions require it.»

Resilient systems refer to those institutions, cities or states that have the ability to reconstruct and to recover, using the right tools, assets and human social and cognitive skills to deal with impacts and *resist, absorb* and *adapt*. The challenge here is how to manage these risks effectively, and how to transform a generic resilience concept into a tailored resilience of a specific infrastructure or a system.The European Commission defines urban resilience (in general) as *«the ability of … a community, a country or a region to withstand, adapt, and quickly recover from stresses and shocks such as drought, violence, conflict or natural disaster»*.

The author's resilience concept adopts a multiple-hazards approach, considering resilience against all types of plausible hazards, and refers not only to reducing risks and damages from disasters [i.e. loss of lives, limbs, health, elements of Mother Nature (flora & fauna) and assets; i.e., components of the second, man-made Nature], but also to the quantified ability to quickly recover to the *pre-event physical* and *psychological* state after a disaster or catastrophe at as minimal as possible cost.

To assess the conditional regional resilience (and strategic preparedness) in relation to a hypothetical catastrophe or accident that occurred within the ICI system under consideration, it is first necessary to *identify* the region whose infrastructure has been damaged. For instance, to estimate the regional resilience in the case of an impact/stressor (hurricane, deliberate actions of intruders, etc.) on a large municipality (which includes industrial and civil buildings and structures that produce commercial products and life support services for the population living in it), following initial data is needed:

- GIS with a database (DB) on the topography, population density (number, gender, age, profession), locations of damaged critical assets and installations, roads, waterways, etc., and how they are interconnected, including information about their owners;
- Type of threats / stressors under consideration (earthquake, flood, technological disaster, resulting in deaths, loss of limb, injuries, pollution of air, water bodies, landscapes, etc.);
- A stochastic description of the considered regional threat (impact);
- The design of the infrastructure in question, including types and number of various structures, buildings and installations that make up the infrastructure and the type of their interdependence (physical, informational) that determine the robustness and innate resilience of each structure and the infrastructure as a whole;
- Statistical characteristics of the physical components of these ICI;
- The volume and form of available resources for mitigating and eliminating the consequences of the disaster (finances, labor, materiel, transport, road-building machines, food, etc.) of the region itself and additionally attracted from outside;
- The socio-demographic description of the society living in the territory occupied by the ICI.

Having the listed above initial data, it is necessary to:

- give a description of the primary failure of a system element that initiated the catastrophe per se;
- define the areas affected by the incident (destruction zones) and assess the social risk and materiel and human losses from the accident, with accounting for the possible cascade development of the catastrophe;
- define the vulnerability / fragility curves (VC/FC) of each CI that got into destruction zone. These FC establish the dependence of the conditional probability of a specified damage of a structure on the magnitude of the received impact. Generalized FC are obtained by integrating specific FC over the entire areas of possible external influences, and the physical properties of the structures and infrastructures' materials. FC, in combination with the results of the analysis of the volume and nature of real damage to structures from one or another type of impact (i.e., earthquakes, hurricanes, technological explosions, etc.) allow (via appropriate calibration) accurate enough assessment of the number of damaged structures and the level of destruction of each individual asset. These problems are solved using statistical dynamics methods.
- Determine the likelihood that the described above region, subjected to a stressor, will recover (from the damage level determined at the previous stage) to the level «as it was before the accident» within the given directive period of time, using a specific amount of allocated

IOP Conf. Series: Materials Science and Engineering **972** (2020) 012001 doi:10.1088/1757-899X/972/1/012001

recovery funds. This stage problems are solved using the statistical operations research methodology.

- Evaluate the time and volume of necessary material and non-material means for bringing the society back to the previous, pre-emergency level of the psychological state of security and well-being, knowing the volume and composition of the destruction of critical infrastructures and human losses.

To solve the above problems following tools are used:

- The «bottom-up» model of regional economic recovery [22,23];
- A full group of events, including all conceivable scenarios with the cascading development of a natural or man-made disaster (involving people whose cognition skills include rich imagination to construct such scenarios);
- A rational (quasi-optimal) plan for responding to an accident and rebuilding a damaged facility, based on considering an as full as possible set of scenarios, including accidents with minimal and maximum consequences.

The above approach allows minimizing the consequences of an accident by optimizing: the distribution of food and water among victims of a disaster; the sequence of repair of electrical networks, gas, water, sewage, etc.; localization, retention and collection of oil spills; road cleaning; maintenance of shelters; distribution of sandbags for the construction of dams, etc. Direct and inverse problems of restoring damaged critical infrastructures of the region also become solvable. *The direct task* gives an assessment of the funds needed to restore the regional infrastructure given the amount of damage and the preset recovery time; *the inverse problem* determines the necessary time for the restoration of regional infrastructure with known volumes of losses and the allocated financial and material resources (see [22,23]). These problems are solved iteratively using the Gantt and similar type networks.

All the above allows for the effective management of regional ICI and of the society related to these ICI, at normal times and times of various crises. Decision makers (DMs) at the level of municipalities and territories receive an instrument of support and substantiation of their decisions. It is no secret that due to lack of time and the lack of flexible tools, many decisions are often taken on a hunch, especially when it is necessary to consider the long-term consequences of decisions. Using the above approach, DMs will be able to track how taken decisions affect the quality of life and the level of satisfaction of their wards / voters.

A typical development cycle of a single accident of a complex infrastructure in time consists of three stages: (1) normal functioning and systematic preparation during its maintenance for the prevention of potential disasters $(0 \le t \le t_0)$; (2) failure of any component that initiates damage to the infrastructure and the development (most often, cascade) of a system accident $(t_0 \le t \le t_1)$, and (3) its recovery to the pre-emergency state or better than (manifestation of *supraresilience /anti-fragility*) $(t_1 \le t \le t_e)$.

Each of the stages reflects, respectively, the ability of the system: (1) to resist the force impact (manifestation of the congenital structural resilience), (2) to absorb the initial damage (by controlling the stressed and deformed state of the system by: duplicating some of its functions, redundancy of multi-component systems, changing the design scheme, dampers, vibration dampers, arrestors of crack growth, etc.), and (3) recovering to the states «as before the accident», «as new», or «much better than new», depending on the plan for using the system in the future and the degree of anti-fragility of the creative DMs who own and / or manage the infrastructure in question. Thus, the system resilience is determined by its ability: (1) to resist mechanical forces, kinematic and degradation effects (the local failure initiating accident is characterized by an insignificant level of damage to the system); (2) to absorb the impact of the initial damage and minimize the cascading/chain failure length and its consequences. This quality is called robustness; and (3) to recover (quickly and efficiently regenerate all its design functional properties).

Strictly speaking, the ability to recover *should not be considered* an attribute of the infrastructure per se (although it is accepted in practically all publications on systems resilience), since restoration of the damaged or destroyed infrastructure is carried out by *another* (construction) infrastructure, while a *third* (financial) infrastructure serves as a source of funds for restoration. The ability to recover should be obviously attributed to the construction and financial branches of an enterprise that is the owner of the damaged infrastructure, or mainly to the construction entity that performed the task of recovery, if it was outsourced. It is also a function of the asset management ingenuity and quality, and of regional governance specifics. Hence, system resilience Res_{Σ} consists of two parts. One of them (Res_{str}) is the resilience of the physical system per se, the other, Res_{rec}, is the resilience of *another* infrastructure, namely the one that conducts or governs the recovery of the damaged system:

$$Res_{\Sigma} = Res_{str} + Res_{rec}. \quad (1)$$

Likewise, the psychological supra-resilience also does not belong to the damaged system, because this restoration is conducted by a fourth, medical infrastructure, although in many cases it is embedded as a matryoshka (nested doll) in the system being recovered.

5. Supra-resilience of urban critical infrastructures

As an intrinsic property of any socio-technological system that explicitly includes human society which consists of humans that are capable to adapt and make decisions based on to some extent free choice as its integral part, supraresilience can serve as the basis and tool for solving the most urgent issues of modern civilization. In order to describe *antifragility* from a mathematical point of view as a logical extension of something that we already know how to describe (this is the most often used mathematical technique in such cases) we can write that antifragility, conditioned on stressor Q

$$\text{Antifragility}(Q) = \text{Resilience}(Q) + \text{Overcompensation}(Q), \quad (2)$$

or, fully equivalent

$$\text{Supraresilience}(Q) = \text{Resilience}(Q) + \text{Overshooting}(Q), \quad (3)$$

Where overcompensation (overshooting) is the ability of the system to *improve its initial quality parameters* (whatever they may be) after being exposed to specific stressors Q (on the condition that they do not exceed some usually unknown ultimate limits. If they do, it injures/breaks the system).

It is also possible to define antifragility and supraresilience as only overcompensation / overshooting.

If we agree on this definition, then the definition of *urban supra-resilience antifragility* using the above European Commission description of resilience formulates as

Urban supraresilience (antifragility) is «the ability of an urban/regional system of systems of critical infrastructures to withstand, adapt, quickly recover from stresses and shocks such as drought, violence, conflict, natural, technological or man-made disaster, and at the same time improve/increase its initial vitality parameters.

It is interesting to note that in China, the only country on the world stage that survived for more than 5000 years without interruption, since ancient times exists a famous adage, which is used as a *curse*, not as a benign wish to one's antagonist: 愿你生活在变革的时代 (Yuàn nǐ shēnghuó zài biàngé de shídài). The English translation sounds as: *May you live in the times of change*. It seems that the Chinese society never cared much about the hidden antifragility of change and aimed at Confucius style *Ordnung*.

The antifragility phenomenon is unique for each person, and for that matter, community, or society, as it largely depends on one's genes or a pool of genes. Moreover, it is a random function/field of the person's age, level of physical and mental health, and the type, level and duration of the stressor imposed on her/him. Hence, for each type of stressor each person reacts by one's own antifragility feedback (overshoot). In order to effectively utilize one's antifragility it is necessary to know her/his boundary between the benign and the unhealthy levels of the stressor at any given time, which is a daunting, impossible task. There are no recipes for estimating these boundaries. Mankind does not know how to reconcile human ambitions and human possibilities, and there are no methods other than *debilitating* to test the strength of people. This statement seems to be true when testing the physical and mental abilities and the creative power of people, the latter being even more difficult, as it also depends on the level of one's education and cultural upbringing. This leaves us with pure empirics when trying to assess the combined (integral) antifragility of an urban society working as a team to create an supraresilient smart city while operating, using and maintaining a system of systems of critical interdependent infrastructures.

The supraresilience is achieved only if there is some extra resource or a genuinely new design for the damaged system that permits *overshooting,* thereby creating a system more resilient than before the event. This is a manifest of the HF *supra-resilience* (creative persons) involved in the project. In general, everything of the man-made Nature that survived time could be considered as antifragility of the past generations accumulated and preserved in built structures and passed on to present and future generations.

6. The source of urban supraresilience

N. Taleb recommends [2] using the antifragility bell bar method when trading stock and conducting other types of stock exchange business, that makes the person smart.This is *advice that should be followed* by all urban DMs when they make macro economic decisions involving stock exchange. In cases related to engineering practices the regional DMs should follow recommended practices from multiple engineering sciences and convergent technologies as related to the areas of their expertise, activity and responsibility. For instance, using structural reliability and resilience methods, they should: (1) Simulate different scenarios of catastrophes. Each scenario produces its own results (loss of production, decrease of RDP, number of injuries, fatalities etc.) of the catastrophe and the resources and time needed to bring the system back to the state «as new», or «as before the catastrophe»; (2) Optimize predictive maintenance using the two-or three-level policy method created for optimizing mechanical systems [21].

The ultimate difference between the problems solved by the considered methods is in that in the sphere of engineering the losses due to a catastrophe never will be as large as in the stock exchange practice. Engineering practice has a sophisticated system of checks and balances in the form of safety design codes and regulations developed over the last 50+ years. The stock exchange business, for obvious reasons, still lacks such tools or methods of their effective implementation.

In order to fully understand the concept of urban antifragility demonstrated as supraresilience, it is necessary to consider the urban system of systems (SoS) of physical critical interdependent infrastructures that includes the human community as a separate social infrastructure in itself, that runs and uses the production/services of the SoS, and, simultaneously, *lives* in the space of these infrastructures (the *urbs* and the *civitas* [3]). When solving this problem, one must consider the interconnection of the Mother-Nature with the second, man-made Nature in an urban environment, looking through an supra-resilient lens that is in the hands of the smart community that runs the show in a city.

In the context of the main theme of the paper (BSC and antifragility) and taking into account its limited size, the remaining space of the paper is spent to describing in highly condensed form following aspects of urban/regional supraresilience: (1) resilience metrics; (2) methods for assessing the volume of damage caused by a specific set of stresses; (3) analysis of interdependence of critical infrastructures; (4) convergent technologies/science needed for conducting research on regional resilience.

7. Resilience metrics

To quantify resilience it is necessary to introduce some transparent, meaningful and simple metrics, easy to calculate and use for all of its components. These metrics can be given in absolute and relative values.

In structural mechanics and reliability design there are just a few quantitative metrics for resilience. Most of them were developed for using in design and are simply ratios of some crucial informative parameters of the design vs same parameters of a reference baseline. They are hard to use, because some of the parameters have to be known at a very early stage of design, before there is any possibility to assess them.

In [22] *structural resilience* was described as *reliability* of a performing structure when loaded by stressors beyond their design values. At the very moment the load on the system exceeds its design value, the structural resilience of the system is equal to its structural reliability and the reliability curve seamlessly transforms into resilience curve. Actually, resilience is described as reliability of a structure, when it is exposed to stresses caused by loads that are above their design values.

Resilience analysis proposed in [23,24] for systems allows assessing physical and relative consequences of the collapse of the quality of functioning of the ICI due to an accident or catastrophe It should be scaled on at least such parameters as: (1) decrease of the volume of goods and services produced by ICI, (2) losses of real regional domestic product RDP, (3) number of lost life-days, injuries, limbs and fatalities (irreversible losses). It demonstrates changes in partial resiliencies and the extent of damage of the infrastructure over time for different probabilities of an incident / accident / catastrophe.

The specificity of calculus of the partial resiliencies is in that while reliability, structural resilience and safety belong to one class of problems, the physical restoration resilience is an *entirely new problem*, which belongs mainly to governance, operation research and logistics. The latter problem is characterized by the time to recovery and its cost. The author proposes using as a resilience parameter for the recovery phase the ratio of the products

$$(\$recTrec) / (\$conTcon), \quad (4)$$

where (\$recTrec) is the product of the actual amount of money and time spent on the recovery of the damaged system; (\$conTcon) is the same, for the same object, but during its construction.

The connection of these two groups of usually independent problems is in that solutions of the first group of problems provide the input to the second group of problems, which is a perfect example of how the infranetics' methodology works (more on this below).

In this paper, following [23,24], resilience is conveniently represented as a dimensionless random variable or function of time, as the ratio of the current resilience to the resilience embedded in the system during its design and realized during its construction and commissioning. Such a dimensionless relative resilience function (variable), by definition, cannot be greater than unity.

When repeating the accident-recovery cycle, the base resilience may already be different from that in the first cycle. Note that for each type of impact or combination thereof, it is necessary to perform a separate assessment of the system resilience with respect to the specific stressor. Only if the system receives (from its supra-resilience/antifragility component) the supraresilience property during its renovation/repair/recovery can it exceed unity. It can turn out to be an emergent property of the system only if the human factor (HF) is explicitly introduced into its composition, namely, those DMs who decide which recovery strategy to choose, to what level of productivity / efficiency and by what (calendar) time to bring back the damaged infrastructure, based on the regional needs and specifics of its governance.

The calculation of resilience as a regenerative ability of a system requires stochastic modeling of emergency response and the process of restoring damaged infrastructure. Here, the baseline is resilience, calculated using «standard» rules and techniques, allowing to put the system into operation at the pre-set time.

All the above reasoning was carried out regarding the generalized productivity of the system. Exactly the same reasoning can be carried out for other types of partial resilience, such as: throughput, volume and speed of providing various services, percentage of population covered by a specific service, number of critical facilities available for use (e.g., housing, grocery stores, medical, educational, cultural, etc. institutions), providing opportunities for private and individual economic activity, etc. A set of partial resiliencies that covers all vital aspects of life of a regional society is considered as full.

For calibrating the algorithm that models the object recovery process, statistics on past accidents and recoveries should be used, whenever and wherever possible. This results in more accurate partial resilience curves of real damaged objects.

8. Methods for assessing the volume of damage caused to an infrastructure by a specific set of stressors

A quantitative assessment of system resilience due to its ability to adapt consists in assessing the degree and nature of the destruction of the whole system by a fixed value of overload. This is done by constructing the so-called fragility/sensibility curves FCs [25], which link the magnitude of the impact on the system with the magnitude of its consequential damage in a probabilistic sense the generalized probabilistic risk assessment (GPRA) methodology can be effectively used for the evaluation of critical infrastructure performance risk. This approach started in the late 1970s as related to nuclear power plants and earthquakes, and are now used worldwide in order to assess safety of different existing or projected civil structures and power plants subjected to strong ground motions. It should be emphasized that there are no limitations to use this approach for other type of stressors and other types of infrastructures having in mind the possibility of a black swan event.

Fragility curves express the conditional probability of failure (CPoF) of a specific infrastructure or its component for a given specific stressor size (wind, wave, snow load, earthquake motion, etc). The bearing capacity of the structure is presumed to be known, via gathering statistical data or conducting computer simulation. Currently it usually is modeled by a Log-Normal distribution [18], but it could be also modeled by a Weibull PDF.

The key components of a GPRA study are (1) stressor hazard analysis, (1) fragility evaluation for each component and substructure and, (1) system analysis and construction of logical fault tree model. In their entirety, these three elements allow for the proper risk quantification of the installation (evaluation of the PoF due to the full set of possible accident scenarios). A hazard analysis yields a family of curves, each corresponding to a confidence level and thus accounting for uncertainty in the estimation of the specific hazard. This PoF can be expressed by convoluting the hazard curve with the fragility curve. The PoF conditioned on stressor level parameter b is given by the cumulative distribution function (CDF) of capacity B, yielding:

$$P_{f|b}(b) = \int_0^s \frac{1}{x\beta\sqrt{2\pi}} e^{-\frac{1}{2}\left(\frac{\log(x/B_m)^2}{\beta^2}\right)} dx = \Phi\left(\frac{\ln(b/B_m)}{\beta}\right) \tag{5}$$

where $\Phi(\cdot)$ is the standard Gaussian CDF. Hence, the FC is entirely defined by median capacity B_m and logarithmic standard deviation β.

The upgraded version of this formula accounts for two types of uncertainties by incorporating two random variables ε_U and ε_R with median equal to one and respective log-standard deviations β_U and β_R (β_U characterizes uncertainty in the knowledge of the median value whereas β_R refers to inherent (epistemic) randomness about the median). In this context, random bearing capacity is expressed as [25]:

$$B = (B_m \varepsilon_U)\varepsilon_R, \tag{6}$$

and the CPoF as:

$$P'_{f|b}(b) = \Phi\left(\frac{\log(b / B_m) + \beta_u \Phi^{-1}(Q)}{\beta_R}\right), \tag{7}$$

where Q = the level of confidence that the CPoF $P'_{f|b} < P_{f|b}$. Equation (6) defines a family of curves, each corresponding to a confidence level Q.

Considering the safety of an infrastructure composed of substructures and equipments (e.g., a petrochemical or power plant), the PoF of the entire system is determined by means of fault trees using simple logical structures of AND and OR in order to combine different events likely to result in global failure.

9. Analysis of interdependence of critical infrastructures

Interdependency is the main hot-spot issue when considering performance of urban critical infrastructures. Currently, there are following possible practical approaches to quantify this interdependence: 1) Analysis of the interaction of CIs in the form of transport networks (see Fig. 3); 2) Study of the interaction and interdependence of CIs using multiple regression theory; 3) Computer simulation of interdependence involving stakeholders (including building a *digital twin/ close relative* of the objects under study) based on the HAZUR-БИЖУР software [26].

The analytical models fit for this kind of research should allow finding in a comparatively simple way and without big errors numerical values of all types of damage and how they change in time. One of the most versatile models here would be the advanced transportation network model [27].

This practical synthetic model describes functioning of interdependent systems of critical infrastructures (ISCI), each of which is represented in the form of a transportation/services network comprised of nodes and directed links. Nodes typically represent physical infrastructure components (the so-called assets, such as electrical power producing plants, oil, gas and water processing units and installations, hospitals, industrial and office buildings, research and education centers, living quarters and the like), which are directly involved in supplying the population and local industries with different products or commodities, and services. The links of the network are modelling the flow (product transportation) between the nodes and may present electrical power grids, main and distribution gas and oil pipelines; water supply and sewage systems, as well as railways and highways, etc. The model considers the conditions and specifics of supply, delivery, overhaul, and demand of resources, such as electricity, hot and cold water, and removal of the industrial and human activity waste [27].

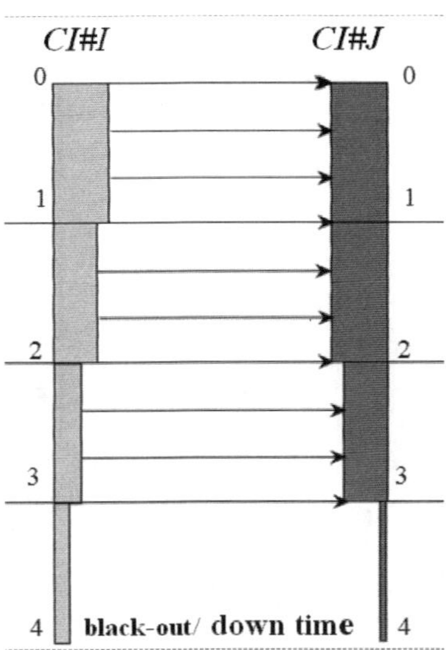

Figure 3. Visualization of interdependence of two CIs.

The specificity of solving such problems is in finding the critical points or thresholds when there appear changes in the normal (design) performance of interacting CIs (see Fig. 3). State (0–1) represents normal operation/performance of both CIs. State (1–2) is a state when the performance of both CIs is still normal, but the reserve of CI #/I capacity is severely depleted. State (2–3) means that any remaining reserve of the CI# I is gone and, hence, the consumption of CI #J is compromised. State (3–4) represents the case when the energy output of CI#I is zero, or close to zero, and the operation of the whole CI# I is brought down. The generic equation that describes the above situations has the form:

$$\sum \mathrm{E}_i(\tau)-\sum Q_i(\tau)\geq \Delta,\ \left[\Delta=0,\ \Delta_1,\Delta_2,\ldots\Delta_N;\ i=1,2,\ldots M,\ 0\leq\tau\leq T\right] \qquad (8)$$

where $\sum \mathrm{E}_i$ is the energy output of the CI #I, $\sum Q_i$ is the energy consumption of the CI# I output, N is the number of performance change triggers, M is the number of nodes in the CI#I network, T is the total duration of joint performance of considered ICIs.

The second approach (multiple regression) is theoretically attractive but very hard to implement, since it needs a lot of reliable data gathered from real life cases. This data can be provided by the third approach, but once it is realized, there is no need for the second approach per se; it could be used for visualizing the results obtained by simulation. The main problem of implementing the third approach is in how to quantify the interdependence of the multitude of critical infrastructures that perform in or out of sync.

10. Designing the regional risk management /governance strategy using infranetics concepts

In order to create an optimal strategy of regional risk management a special tool is needed that would allow using multi- and inter disciplinary approach and various methods from different fundamental, applied and engineering sciences under one umbrella [29]. For this matter in this paper the new convergent science- Infranetics- based on the MAICS-convergent technology (Digital Stochastic **M**echanics, **A**rtificial Intelligence, **I**nformation Theory, **C**ognitive and **S**ocial Sciences) is used. The name Infranetics comes from I*nfra*structures+Cyber*netics*. Infranetics was constructed for solving the central problem of safe innovative development of a region/territory/municipality by creating a methodology of harmonized regulation of regional risk, based on optimal management of systems of interdependent critical infrastructures (ICI). A full description of the method can be found in [28,29]. In Fig. 4 the flow chart of constructing such strategy using the infranetics philosophy and tools is presented.

It can be seen that creation of such strategy involves a heterogeneous set of initial data and knowledge from multiple fundamental, applied, and engineering sciences. When solving this problem for some regions of the Russian Federation, the author came across several surprise info-gaps (most of them known unknowns) in macro-economics [30], demography [31], reliability theory of large systems [32], to name a few, which needed immediate attention, as they were roadblocks preventing reaching the assigned goal. It seems that in this case infranetics concept threw new insight into the research problem; advanced research in several fundamental and applied disciplines; greatly enhanced and expedited the research.

Accumulating results of this kind of interdisciplinary research permits understanding what methods from what disciplines belong to infranetics. The results, obtained by the above scheme, made it possible to effectively manage the SICI in different types of municipalities and territories (see [33–37]).

Infranetics manifests itself as a cross-pollinating complex-sciences discipline. It permits accurate-enough prognosis of the behavior of complex socio-technological and economic systems being designed or already performing, in usual and/or catastrophic situations. Infranetics uses goal-oriented approach according which a unified concept is formed that describes ways and means for solving the problem in consideration. If properly developed, infranetics could become the crucial convergent technology and discipline for governing smart supraresilient municipalities and regions in the 21st century [38].

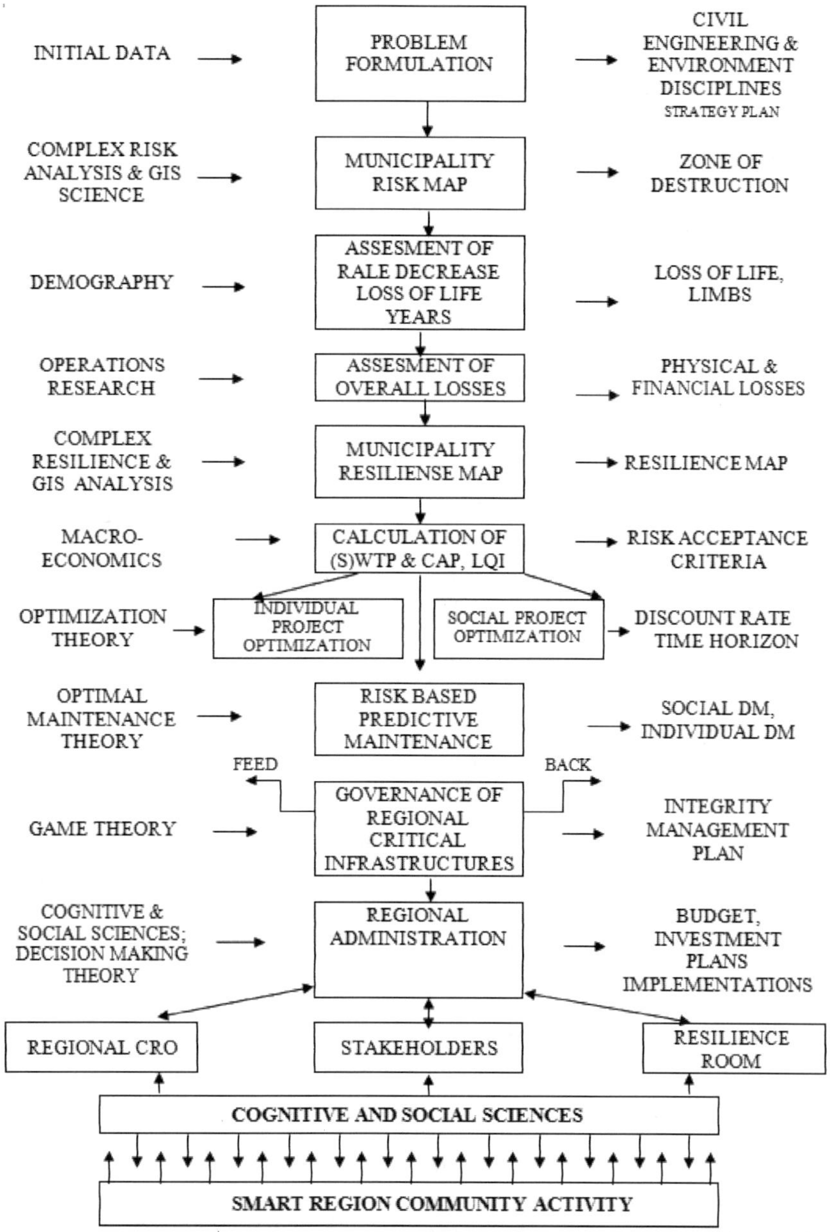

Figure 4. Regional governance strategy via regional resilience and risk management using infranetics principles and tools.

11. Conclusion

The resilience property is the ability of an infrastructure system to continue to perform its design functions in conditions when it is partially damaged and/or deprived of supplies, and with employees and population psychologically stressed. This circumstance is the unconditional and the sole basis for creating a reliable and safe smart territory. The concepts of acceptable risks developed for managing the technological, industrial and natural risks are quite compatible with BSE_{ph}.

Without resilience, it is *impossible* to create a smart region/city capable of sustainable development. Resilience is also necessary for achieving the long-term goals of strategic development of any region, which include developing human capital, improving the quality of life and increasing the competitiveness of the economy through the systematic introduction of services and innovative solutions.

Supra-resilience/antifragility should be a property of every socio-technical system, but it is not a given. All physical infrastructures are fragile, by definition. But if coupled with a supra-resilient component (i.e., high quality human factor) the socio-technical system can obtain needed antifragility traits.

For a socio-technical systems to be antifragile, its supra-resilient components should take care of all the system's fragile components. The supraresilient components in their term have to care of each other, using for that all methods that mankind developed to this day, including the «*skin in the game*» concept [38]. A structure can be designed to last (i.e., possess physical and moral longevity), be adaptable and accommodate new functions through intelligent design, construction, optimal maintenance and effective governance by the cognitive and creative components of the region/city that can face the unpredictable events.

The concept of supraresilience opens the door to meaningful exploration of quantitative dependencies/correlations between physical resilience of a system of systems of ICIs and the psychological/societal resilience of people who live inside and extensively use this system of ICIs in the context of different types of communities – from megacities, to big, medium size and small towns, to villages/settlements and tribal areas.

The effectiveness of the interaction of ICIs cannot be adequately described without direct consideration of the human factor (HF). Which profile specialists are best to pull off and solve this task? This question is open for discussion, but it seems that civil and environmental engineers, structural mechanics specialists and designers are best fit to take lead when tackling this problem.

References

[1] Taleb NN 2007 The black swan: the impact of the highly improbable (London: Penguin)
[2] Taleb NN 2012 Anti-fragile (London: Penguin)
[3] Ivan Blečić and Arnaldo Cecchini 2017 On the antifragility of cities and of their buildings *City Territory and Architecture* 4:3
[4] Aven T 2013 On the meaning of the black swan concept in a risk context *Safe Sci* **57** pp 44–51
[5] Mandelbrot B B 1997 Fractal and Scaling in Finance: Discontinuity, Concentration, Risk (Springer-Verlag)
[6] Aven T 2013 On how to deal with deep uncertainties in a risk assessment and management context *Risk Anal* **33**(**12**) 2082–91
[7] Aven T 2014 On the meaning of the special-cause variation concept used in the quality discourse-and its link to unforeseen and surprising events in risk management *ReliabEng Syst Safe* **126** pp 81–6
[8] Paté-Cornell ME 2012 On black swans and perfect storms: risk analysis and management when statistics are not enough *Risk Anal* **32**(**11**):1823–33
[9] Aven T and Krohn BS 2014 A new perspective on how to understand, assess and manage risk and the unforeseen *ReliabEng Syst Safe* **121** pp 1–10
[10] Hopkins A 2014 Issues in safety science *Safe Sci* **67** pp 6–14
[11] Masys AJ 2012 Black swans to grey swans: revealing the uncertainty *Disaster Prev Manage* 21(3):320–35
[12] Financial Post Deepwater rig worker weeps as he admits he overlooked warning of blast that set off America's worst environmental disaster 2013 http:// business.financialpost.com/2013/03/14/halliburton-worker-weeps-as-he-ad mits-he-overlooked-warning-of-blast-that-set-off-americas-biggest-oil-spil l-in-gulf/?__lsa=42e0-28bb [accessed 15.01.14]

[13] Kaplan S, Visnepolschi S, Zlotin B, Zusman A 1999 New tools for failure and risk analysis: anticipatory failure determination (AFD) and the theory of scenario structuring *Southfield* (MI: Ideation Inter Inc)
[14] Aronov J and Papic L 2018 Reliability and Safety Management of Engineering Systems Through the Prism of Black Swan Theory *In: System Reliability Management, Solutions and Technologies* (Boca Raton: CRC Press, Taylor and Francis Group) pp 103–112
[15] Lee Davis 2002 Man-Made Catastrophes: From the Burning of Rome to the Lockerbie Crash p 402
[16] Carey B, Patra P K et al 2011 Observation of dynamic strain hardening in polymer nanocomposites *ACS Nano* **5**(4): 2715–22
[17] Shwarts S 1969 Public Lecture of the Director (Institute of the Ecology of Plants and Animals, Ural Science Center USSR Academy of Sciences)
[18] Kurzweil Ray 2015 Evolution of the Mind (Moscow: Э)
[19] Brinnolofson Eric and Andrew McAfee 2017 Second era of machines (Moscow: ACT)
[20] Rene Gustavsson and Bjorn Stahl 2009 Self-Healing and Resilient Critical Infrastructures *In: Critical Information Infrastructure Security*
[21] Timashev S A 1982 Reliability of Large Mechanical Systems (Moscow: Nauka)
[22] Timashev S A 2011 Resilience and Strategic Preparedness of Critical Infrastructures *Proc. of the First Int. Conf. on Vulnerability and Risk Analysis and Management ICVRAM and the Fifth Inter. Symp. on Uncertainty Modeling and Analysis ISUMA* pp 764–771
[23] Timashev S A 2016 Infrastructure Resilience: Definition, Calculation, Application *WEEF, IGIP/ICL Conf.* (Pisa Italy)
[24] Timashev S A and Bushinskaya AV 2018 Structural resilience *Proc. Safety and monitoring of technogenic and natural systems: VI All-Russian Conf.* (Krasnoyarsk) pp 36–43
[25] Zentner I, Nadjarian A, Humbert N, Viallet E 2008 Numerical calculation of fragility curves for probabilistic seismic risk assessment *Proc. of the 14th World Conf. on Earthquake Engineering* (China Beijing)
[26] Timashev S A, Alekhin V N, Poluyan LV, Fontanals I, Gheorghe A 2018 Transforming Yekaterinburg into a Safe, Resilient-Smart and Sustainable City *IOP Conf. Series: Earth and Environmental Science* **177**(1) 012001
[27] Timashev S A and Bushinskaya A V 2019 Analysis of Interdependence of Arctic Critical Infrastructures as Transportation Networks *IOP Conference Series: Materials Science and Engineering* **481**(1) 012008
[28] Timashev S A 2018 Infranetics: The New Convergent Science for Risk Based Management of Systems of Interdependent Critical Infrastructures *Proc. of 27th Annual Conf. of the Society for Risk Analysis Europe: Risk & Uncertainty – From Critical Thinking to Practical Impac*t (Östersund Sweden)
[29] Gheorghe A and Masera M 2010 Infranomics *Int. J. of Critical Infrastructures* **6**(4):421–427
[30] Timashev S A, Tyrsin A N, Kozlova O A, Makarova M N, Voronina L N 2014 Methodical Approaches to Assessing the Influence of Budget Expenditures on Average Life Expectancy *Scientific Review* 7–1
[31] Bushinskaya A V and Timashev S A 2019 The Toll of Incidents, Accidents and Disasters on the Average Life Expectancy in Good Health *Proc. of the 29th European Safety and Reliability Conf* doi: 10.3850/978-981-11-2724-3_ 0614-cd
[32] Burukhina O, Bushinskaya A, Maltceva I, Timashev S 2018 Mechanical system reliability analysis using reliability matrix method *IOP Conference Series: Materials Science and Engineering* **365** 042067
[33] Timashev S A 2016 *Infrastructures. Part 1. Reliability and Longevity* (Yekaterinburg: Ural Branch of Russ. Ac. of Sci)
[34] Timashev S A, Bushinskaya AV, Malyukova MG, Poluyan LV 2016 Reliability and Integrity of Pipeline Systems (Yekaterinburg: Ural Branch of Russ.Ac. of Sci)

[35] Timashev S A 2013 Unified quantitative criteria for managing regional risk *Proc. of the 11th Inter. Conf. on Structural Safety & Reliability ICOSSAR* (NY: Columbia University)
[36] Guryev E S, Poluyan L V and Timashev S A 2014 Construction of dynamic risk maps for large metropolitan areas *J. of Risk Analysis and Crisis Response* **4**(2) pp 72–76 (Paris, France: Atlantis Press)
[37] Timashev S A 2014 Average life expectancy as a criterion for regional risk management *J. of Risk Analysis and Crisis Response* **4**(1) pp 10–19 (Paris, France: Atlantis Press)
[38] Taleb NN 2018 Skin in the game (London: Penguin)

IOP Conf. Series: Materials Science and Engineering **972** (2020) 012002 doi:10.1088/1757-899X/972/1/012002

Macroeconomic aspects of maintenance optimization of critical infrastructures

S A Timashev [1,2,3], A V Bushinskaya [1,2]

[1]Sci.& Engng Center "Reliability and Safety of Large Systems and Machines", Ural Branch, Russian Academy of Sciences, Yekaterinburg 620016, Russia

[2]Ural Federal University, 620002, 19 Mira, Yekaterinburg, Russia

[3]Old Dominion University, Norfolk, VA, United States

E-mail: timashevs@gmail.com

Abstract. The main goal of maintenance is prevention, timely detection and elimination of failures and damage. From the point of view of critical infrastructures (CIs), the main purpose of their maintenance is to increase the safety of CIs and / or to ensure life safety. CIs should be optimal in terms of their purpose, cost, as a source of income and profit at all stages of their life cycle, and also acceptable in terms of possible loss of human lives or injuries. The paper considers the assessment of necessary optimal investments in the maintenance (time interval between subsequent maintenance), to increase the safety of life, the so-called life saving costs. The problem of limiting the threat to human life is as follows: "how much money is society ready and able to spend to reduce the likelihood of premature death". A quantitative criterion of risk acceptance is used, in which the marginal cost of life saving is compared with social willingness to pay. This value is determined using the life quality index of population.

1. Introduction

Optimization can be carried out either by a private or public decision maker (DM). An example of a private DM is a company that builds and/or manages an object (and possibly creates risks for society).

Social DMs are government agencies or organizations that manage risks on behalf of society.

The costs and benefits of making a decision, as well as preferences, which form the basis for comparing different alternatives, may vary depending on the interests of the DM.

The paper uses a quantitative criterion of risk acceptance [1], in which the marginal life saving cost is compared with social willingness to pay (SWTP). This value is determined using the life quality index (LQI) of population, which was obtained on the basis of socio-economic assumptions using the GDP per capita and the average life expectancy (ALE)

The LQI criterion defines a threshold that separates effective investments from inefficient investments in terms of saving lives.

Life saving costs can be increased only up to the effectiveness threshold, in our case, SWTP.

Investing in safety measures (in our case, during the time interval between subsequent maintenances) leads to both monetary expenditures and the benefits of a safe life. Cost reductions achieved through investments in security, such as reducing the cost of eliminating the consequences of a failure, are considered to be the benefit of a safety measure. The benefits of improved life safety are determined by a marginal change in the expected number of deaths per year

When making a decision, the following points are distinguished: 1) the criterion for accepting a risk to life and monetary optimization; 2) the interests of private and public decision-makers.

Costs and benefits can accumulate at different points in time, so they have to be discounted.

2. Fundamentals of the theory of discounting and social utility of consumption

The discounted or present value (PV) is an estimate of the value, that is, the current monetary equivalent of the future flow of payments based on the different value of money received at different points in time, the so-called concept of the time value of money.

The monetary amount received today usually has a higher value than the same amount received in the future. This is due to the fact that money received today can bring future income if invested.

The inequality of different amounts of money is numerically expressed in the discount rate.

The discount procedure is the inverse of the accrual of compound interest and has an exponential law. The discount coefficient (factor) is equal to

$$\lim_{N\to\infty}\frac{1}{\left(1+\frac{\gamma}{N}\right)^{Nt}}=e^{-\gamma t}, \tag{1}$$

where γ is called the discount rate (interest rate).

As a tool to compare the benefits received by the society and the costs incurred, the market discount rate is not applicable. In world economic practice, such an instrument is the social (public) discount rate.

The social discount rate is an alternative opportunity for society to use resources between two time periods, a kind of "price" at which society is ready to abandon today's consumption for tomorrow, which reflects the social norm of time preference.

In this paper, we use the method of assessing the social rate of time preferences (SRTP), which is based on identifying society's preferences in terms of consumption.

The social rate of time preferences shows the readiness of the society to abandon consumption at the present time in order to implement the project and receive benefits from its results in the future.

Analytically, this problem is solved by maximizing the function of social utility obtained from consumption at different time periods. In general, when the discount rate is variable over time, this problem has the following form:

$$U(t)=\int_0^t u(c_\tau)\exp\left[-\int_0^\tau \gamma(\theta)d\theta\right]d\tau \to \max_{c_t}, \tag{2}$$

where $c_t > 0$ is per capita consumption at time t; $u(c_t)$ is the social utility function of this consumption; $\gamma(t)>0$ is the interest discount rate of utility consumption at a time t, reflecting the time preferences of the individual; $U(t)$ is, in fact, a function of *total utility.*

Utility (utility) - this is the degree of satisfaction of the needs of individuals, which they receive when consuming goods or services or from conducting any activity.

There are two forms of utility: total and marginal. *Total utility* is the utility obtained as a result of consumption of all units of the good. Total utility increases as consumption increases, but not in proportion to consumption, and the gradually fades away until it reaches zero. *Marginal utility* is the utility that is obtained from the use of another additional unit of the good. In other words, this is an increase in total utility when one additional unit of the good is consumed (that is, *a derivative of the utility function by the amount of good consumed*).

3. The utility function of (whole) life

An utilitarian approach is adopted, according to which social welfare (social utility) is the sum of welfare (sum of utilities) of individual members of society, and marginal utility decreases with increasing consumption (the first Gossen's law). This is the law of diminishing marginal utility - with the growth of consumption of one good (with a constant consumption of all other goods), total utility increases, but the growth rate slows down. The marginal utility (derivative) decreases with increasing consumption, turns to zero at the maximum total utility and then becomes negative, and the total utility, having reached the maximum value, begins to decrease. For example, for a hungry person, the marginal utility of the first bowl of soup is higher than the second, the second is higher than the third, etc. And so it goes with many other goods.

The enjoyment of life or its usefulness in the economic sense is due to the continuous influx of resources available for consumption throughout the whole life. *Therefore, the income necessary for consumption, and the time to enjoy it, are two determining factors in the life quality.*

According to [2, 3], the life quality in the economic sense can be measured using the utility function of life expectancy, depending on the level of consumption. The utility function of the residual (upcoming) life expectancy at age x [2, 3]:

$$U(x,D)=\int_x^D u(c_\tau)d\tau, \tag{3}$$

where $c_\tau > 0$ is the rate *(intensity) of consumption* at the age of τ (\$/year); $u(c_\tau)$ is the *utility* of this consumption from age x to the moment of death D.

The function $U(x,D)$ can be interpreted as total consumption over the remaining life time.

People usually appreciate the opportunity of consumption in the future is less than in the present. This can be accounted for by *discounting the utility function*:

$$U(x,D)=\int_x^D u(c_\tau)\exp\left[-\int_x^\tau \gamma(\theta)d\theta\right]d\tau, \tag{4}$$

where $\gamma(t)>0$ is the discount rate of utility consumption at age t, reflecting the time preferences of the individual.

As noted above, the social discount rate is used as an estimate of $\gamma(t)$. In this paper, we use the so-called Ramsey–Cass–Koopmans mathematical model (Ramsay model), a neoclassical model of equilibrium endogenous economic growth in which the path of consumption and savings are determined on the basis of solving the problem of optimizing households and firms in the conditions of perfect competition

Solving the problem of maximizing social utility for consumption

$$U(x,t)=\int_x^t u(c_\tau)\exp\left[-\int_x^\tau \gamma(\theta)d\theta\right]d\tau \rightarrow \max_{c_t}, \tag{5}$$

according to the Ramsey model [4], we obtain the social discount rate:

$$\gamma(t)=\rho(t)+\mu\delta(t), \tag{6}$$

where $\delta(t)=\frac{\frac{dc_t}{dt}}{c_t}$ is the per capita consumption growth rate; the parameter $\rho(t)$ means the subjective discounting rate of future consumption (parameter of impatience, selfishness). The larger this parameter, the higher the individual appreciates his current consumption in relation to the future consumption. The component $\mu\delta$ shows the increase of social utility received from consumption.

Thus, according to expression (6), the social rate of time preferences depends on three parameters: an individual rate of time preferences (ρ), elasticity of marginal social utility by consumption (μ), and per capita consumption growth rate (δ).

Since the time of death is random, the utility of an individual's life is also a random variable. It can be shown that the average expected discounted utility of the residual life expectancy at age *x* (assuming surviving to this age) can be calculated by the formula

$$L(x)=\frac{1}{S(x)}\int_x^{D_m} S(t)u(c_t)\exp\left[-\int_x^t \gamma(\theta)d\theta\right]dt = \\ =\frac{1}{S(x)}\int_x^{D_m} S(t)u(c_t)\exp\left[-\int_x^t \rho(\theta)d\theta+\mu\delta(t-x)\right]dt \quad (7)$$

where D_m is the maximum age to which people live in the world (or the maximum age of the cohort under consideration (in the region, industry, company); $S(t)$ is the survival function:

$$S(t)=\exp\left(-\int_0^t \lambda(u)du\right), \quad (8)$$

where $\lambda(t)$ is the mortality (intensity) rate (risk function). It is determined on the basis of the Gompertz-Makeham mathematical model [5] depending on mortality by age.

In conditions of perfect competition (ideal market), optimal consumption does not depend on time *t* [6]. In this case, the formula (7) takes the form:

$$L(x)=\frac{u(c)}{S(x)}\int_x^{D_m} S(t)\exp\left[-\int_x^t \rho(\theta)d\theta+\mu\delta(t-x)\right]dt=u(c)e_d(x,\rho,\mu,\delta), \quad (9)$$

where $e_d(x,\rho,\mu,\delta)$ *is the discounted ALE at age x*:

$$e_d(x,\rho,\mu,\delta)=\int_x^{D_m}\exp\left[-\int_x^t[\lambda(\tau)+\rho(\tau)]d\tau+\mu\delta(t-x)\right]dt. \quad (10)$$

With constant discounting, the function $L(x)$ is a monotonically decreasing function. This is confirmed by many theoretical and empirical studies. At a time-dependent discount, the function $L(x)$ increases up to an age of about 25 years, and then decreases [7].

4. Life quality index

The question: how much should be sacrificed from the utility of consumption and other aspects of life quality in order to get a certain increase in life expectancy at the cost of risk reduction?

Income is created by work, and the longer the work, the greater the income, but the less time for leisure/rest. Income is proportional to labor productivity (output in monetary terms per unit of time).

An individual can increase leisure time either by increasing his life expectancy by reducing risk, or by reducing the time he spends on economic production, which leads, in the general case, to lower income.

Nathwani et al. [8] used the hypothesis: "*On average, people work just so much that the marginal cost of their wealth/welfare or the income they receive for their work equals the marginal cost of the time they spent on this work*" (optimization principle of "work-leisure/rest") and identified a measure of life quality as

$$L=f(g)h(t), \quad (11)$$

where g is the consumption (annual income), $t=(1-w)e(0)$ is the leisure/rest time; $e(0)$ is the ALE at birth; w ($0 < w < 1$) is the proportion of life expectancy spent on the (paid) work; $f(g)$ и $h(t)$ are two so far unknown functions of these qualities.

The value of L is called the life quality index [8]. Thus, the LQI is a product of the function $f(g)$ measuring the life quality and the function $h(t)$ measuring life duration.

It is assumed that any change of ALE should be compensated by a corresponding change in g; i.e., any investments (reduction in consumption by dg) in improving (preserving) life should be compensated by a gain de in the ALE (and vice versa), so that L remains unchanged. Formally, this means that the total differential of function L is equal to zero. Accepting some assumptions and conducting a set of mathematical calculations, it can be proved that the LQI is calculated by the formula [8,9,10]:

$$L=\frac{c^q}{q}e(0). \quad (12)$$

Thus, in this case the consumption function

$$u(c)=\frac{c^q}{q}, \quad (13)$$

which corresponds to an isoelastic power function at $q=1-\mu$.

It is generally accepted in economics that a low q value implies that a greater value of the usefulness of life is achieved mainly by increasing life expectancy, while a high q value implies that this usefulness of life is achieved mainly due to consumption.

As a value of c, Nathwani et al. [8] consider GDP per capita (taking into account the purchasing power parity, if necessary), which is an indicator of society's productivity. Then formula (12) takes the following form

$$L=u(g)e(0)=\frac{g^q}{q}e(0), \quad (14)$$

where g is the GDP per capita.

Comparing Eq. (14) with Eq. (9), it can be seen that the latter is the *life utility without discounting at a deterministic life expectancy equal to the ALE at birth.*

Thus, the LQI is a comprehensive social indicator as a function of two social indicators: GDP per capita and ALE in good health. It is assumed that both of these indicators are independent, which may not be the case. However, economists support the idea of independence. For example, they showed that the growth of GDP and ALE have historically developed completely independently; at this the latter is mainly affected by advances in medicine (preventive and therapeutic methods), which are only weakly or indirectly related to GDP [11].

If, in formula (14), the value of ALE at birth e(0) is replaced by a discounted ALE at age x:

$$L(x,g)=u(g)e_d(x,\rho,\mu,\delta)=\frac{g^q}{q}e_d(x,\rho,\mu,\delta) \quad (15)$$

Then the obtained LQI can be interpreted as the discounted utility of the residual life expectancy. Then, the according to Arrow-Pratt, consumption function from (15) will belong to the class of the so-called risk aversion functions with coefficient q of relative risk aversion

5. Value of statistical life and willingness-to-pay

Shepard and Zeckhauser [6] determined the "value of statistical life" (VSL) at age x by dividing equation (15) by marginal utility $du(c)/dc = u'(c)$, which translates its dimension strictly into monetary units:

$$\mathrm{VSL}(x) = \frac{u(c)}{u'(c)} \int_x^{D_m} \exp\left[-\int_x^t [\lambda(\tau) + \rho(\tau)] d\tau + \mu\delta(t-x) \right] dt = \frac{g}{q} e_d(x, \rho, \mu, \delta). \quad (16)$$

It is seen that the VLS decreases with time due to a decrease of $e_d(x, \rho, \mu, \delta)$. The monetary expression of life does not exist - "the price of life is infinite and immeasurable" [12], if we talk about an individual. Here, however, the VSL is considered, as a kind of a formal constant, as a monetary value, which is needed in order to reduce the risk of mortality by unit.

According to statistics, collected over the past hundred years [13], the average population size grows in time exponentially e^{nt}, where $t > 0$ and n is the *population growth rate* (taking into account the effects of immigration and emigration). In order to correctly account for the composition of the population exposed to natural threats and technological accidents and disasters in the PDO territory and its surroundings, the considered variables should be averaged over the PDF of the age distribution $h(x, n)$ of the population (employees) considered as *stable*. The PDF of a specific population can also be obtained from mortality tables. For a stable population, the PDF is

$$h(x,n) = \frac{\exp[-nx] S(x)}{\int_0^{D_m} \exp[-nx] S(x) dx}, \quad (17)$$

In a stable population, mortality does not change with time. Population is stable when $n \approx 0$. Then

$$h(x,n) \approx \frac{S(x)}{e(0)} \quad (18)$$

Then the *social value of statistical life* (SVSL) is defined as

$$SVSL = \frac{g}{q} \bar{E}(\rho, \mu, \delta) \quad (19)$$

where $\bar{E}(\rho, \mu, \delta)$ is the discounted ALE averaged over the age distribution $h(x,n)$:

$$\bar{E}(\rho, \mu, \delta) = \int_0^{D_m} e_d(x, \rho, \mu, \delta) h(x,n) dx \quad (20)$$

In addition, it is possible to evaluate the *social LQI (SLQI)* [14]:

$$SLQI = \frac{g^q}{q} \bar{E}(\rho, \mu, \delta). \quad (21)$$

It must be emphasized that the SLQI, like the original LQI, is not a monetary value and has the dimension «($.)^w(years)» and is interpreted as a utility function. If it is divided by the marginal utility $u'(c)$ in order to turn it into a monetary value, then it coincides with equation (19).

The Willingness-to-pay (WTP) measures a person's willingness to sacrifice one desired attribute, wealth or consumption, to get another desired attribute, in this case, increase life expectancy (improve survival). Let de denote the marginal change in life expectancy and dg the marginal change in consumption. Shepard/Zeckhauser [6] introduced WTP as invariance of $L = L(x, g)$ with respect to loss (increase) in consumption with an increase (decrease) in life expectancy:

$$WTP = dg = -\frac{\dfrac{\partial L(x,g)}{\partial e(a)}}{\dfrac{\partial L(x,g)}{\partial g}} de(x) = -\frac{g}{q}\frac{de(x)}{e(x)}. \quad (22)$$

6. The criterion of social acceptability of investments in risk reduction projects

For decisions regarding investments in life safety, LQI can be interpreted as a two-digit utility function for the average member of the society. It is assumed that a decision with a (marginal) effect on g and e (for example, any investment related to saving a life) is beneficial for society if it leads to an increase in LQI. The requirement that the full differential of LQI function be equal to or greater than zero ($dL \geq 0$) gives rise to the net benefit criterion (acceptability of investments in risk reduction projects) [1,9,10]:

$$\frac{dg}{g} + \frac{1}{q}\frac{de(x)}{e(x)} \geq 0. \quad (23)$$

Inequality (23) is the criterion for the effectiveness and accessibility of specific investments in life safety. Equality in (23) shows what measures preserving human lives for society are necessary and affordable; projects with inequality "<" are not acceptable. Such projects will actually be life threatening and be in conflict with the constitutional right to life. Every time when a small increase in ALE due to some activity preserving human lives (positive de) is associated with more than optimal additional costs (negative dg), it is necessary to look for another alternative for investing in projects to save lives (reduce risk). If a given positive de is feasible for less money than formula (23) provides, then of course this possibility should be realized.

Expression (23) remains valid for the discounted utility of residual life expectancy at age x [equation (15)]:

$$\frac{dg}{g} + \frac{1}{q}\frac{de_d(x,\rho,\mu,\delta)}{e_d(x,\rho,\mu,\delta)} \geq 0. \quad (24)$$

and after averaging over the distribution of ages $h(x,n)$, we obtain a criterion for assessing social willingness to pay (SWP) [1,9,10]:

$$\frac{dg}{g} + \frac{1}{q}\frac{d\bar{E}(\rho,\mu,\delta)}{\bar{E}(\rho,\mu,\delta)} \geq 0 \text{ или } dg \geq -\frac{g}{q}\frac{d\bar{E}(\rho,\mu,\delta)}{\bar{E}(\rho,\mu,\delta)}. \quad (25)$$

Converting (23) express the criterion of effectiveness in terms of limit threshold for changes of GDP per capita, dg [1,9,10]

$$-dg \leq \frac{g}{q}\frac{de(x)}{e(x)} \approx -\frac{g}{q}C_x d\lambda. \quad (26)$$

For technical problems, it is usually easier to estimate the marginal reduction in mortality $d\lambda$ than the effect that the decision has on life expectancy. The de/e ratio is calculated by multiplying the change in mortality by the demographic constant C_x [9, 10], estimated from the mortality (survival) tables.

In formula (26) averaging can be used for age and discounting that is realized by replacing $\dfrac{de(x)}{e(x)}$ by $\dfrac{d\bar{E}(\rho,\mu,\delta)}{\bar{E}(\rho,\mu,\delta)}$.

Until now, the LQI threshold has been determined based on the preferences of the average citizen (per capita). To obtain aggregate (integral) total values, equation (25) should be multiplied by the size of population N_P. Mortality rate λ is then replaced by the expected number of deaths per year, $m = \lambda \cdot N_P$. This allows to determine the threshold for the marginal cost of saving lives at the *infrastructure project level*, *dc* [1]:

$$dc = -dg \cdot N_p \geq SWTP = \frac{g}{q}\frac{d\bar{E}(\rho,\mu,\delta)}{\bar{E}(\rho,\mu,\delta)}N_p \approx -\frac{g}{q}C_x dm. \quad (27)$$

where the *de/e* ratio is calculated by multiplying the change in mortality by the demographic constant C_x, estimated from the mortality (survival) tables.

Acceptance criterion (27) is based on marginal cost *dc* for marginal risk reduction - *dm*, not on average values.

It should also be noted, that criterion (27) is based on the annual expenditures and benefits of saving a life. If these expenditures and benefits arise at different points in time, they must be discounted (lead to a single cut of time) for adequate comparison.

7. The relationship between acceptance criterion and monetary optimization

Since costs and benefits are accrued at different points in time, equation (27) can be reformulated by comparing the present (discounted) values (PV) of future costs and benefits [1]:

$$\begin{aligned} PV(dc_s) &= PV\left[-dg \cdot N_p\right] \geq PV(SWTP) = \\ &= PV\left(\frac{g}{q}\frac{d\bar{E}(\rho,\mu,\delta)}{\bar{E}(\rho,\mu,\delta)}N_p\right) \approx -\frac{g}{q}C_x PV(dm). \end{aligned} \quad (28)$$

This criterion applies to both public and private DMs, even when the latter can have financial costs much higher than the social discount rate implies. The basis for this is the Caldor-Hicks compensation principle, according to which welfare is increased if the decision is beneficial to society as a whole, regardless of whether the investments of private DMs in saving the lives of other people are compensated or not.

In practice, monetary optimization and acceptability of social risk should be combined. It is necessary to clearly distinguish between monetary optimization and the acceptance criterion, which should always be evaluated from a social point of view. In this paper, optimization is performed by maximizing the objective function that represents the individual (private or social) preferences of the decision maker [1]:

$$\max\left\{E\left[\int_0^T \left(b_p(t) - c_p(t)\right)\cdot e^{-\gamma_p t}dt\right]\right\}, \quad (29)$$

where the index *p* indicates that costs *c* and benefits *b* are selected from the view point of the private DM According to decision theory, decisions should be based on the expected values of optimized parameters. Future cash flows are discounted using the selected private discount rate γ_P. Possible loss of life can be included in the monetary parameters of the task, if the DM must pay compensation in case of death.

Acceptability in relation to life safety is ensured by the mandatory use of social acceptability criterion (equation (28)) as the boundary condition for optimization. The marginal costs *dc* of saving a life are defined as the direct investments dc_S needed to improve safety, evaluated from a social point of view.

8. Illustrative example. Impact of probability of failure (POF) on a LQI acceptance criterion

Consider the example from [1] - construction of a steel bridge in the United States. The bridge is financed, constructed and will be operated by a private company which in return is allowed to demand a toll fee from the end-users of the bridge. In this example we will explore the problem from the private and the societal perspective. From the private DM's perspective, a typical cost benefit optimization analysis is performed, as described by Eq. (29). The LQI criterion (Eq. (28)) is used as a boundary condition which is evaluated from a societal DMs perspective.

The benefits of operating the bridge for the private DM are:

$$b(t) = k(t)k_1 e^{\beta} = \left(\frac{k_0 \cdot 24 \cdot 365}{1+e^{-\alpha t}}\right) \cdot k_1 e^{\beta}, \tag{30}$$

where k_1 is the toll fee (US \$6.00 car) and β is the annual increment rate increase of the toll (1%); $k(t)$ is the number of crossing cars per year; $k_0 = 600$ is the traffic density constant (cars/h) and parameter $\alpha = 0.05$ models the increase of traffic.

The construction costs c_c = \$200,000,000. When the structures reaches a minimum performance threshold it is maintained and taken to its original condition, i.e., ''as good as new'' at a cost of \$28,000,000. The maintenance costs $c_m(t, p)$ appear at discrete points in time according to the decision variable p, which is the time interval between subsequent maintenances. The time-dependent failure probability depends on p and is modeled using the Weibull distribution:

$$P_f(t, p) = 1 - e^{-\lambda \cdot \mathrm{mod}(t/p)^k} \tag{31}$$

with a shape factor $k = 2.0$ and a scale factor (failure rate) $\lambda = 0.0005$/year.

The expected failure costs $c_f(t, p) = P_f(t, p) \cdot c_c \cdot 1.2$. are established as the costs of constructing a new bridge, c_c, plus an additional 20% representing the cost of removing, disposing and recycling construction material.

The private DMs goal is to maximize the PV of the net benefits:

$$\max_p \left\{ \sum_{i=1}^{T} \left(b(t) - c_m(t,p) - c_f(t,p)\right) \cdot e^{-\gamma_p t} - c_c \right\}. \tag{32}$$

Fig. 1 shows the objective function for a private discount rate γ_p = 5% and a time horizon T of 80 years. It can be seen that time intervals between subsequent maintenance actions from 2.5 to 19.5 years are feasible from the private DM's point of view. The monetary optimum is reached when p^* (interval between subsequent maintenance) is equal to 8 years.

Now analyze how the bridge POF affects the objective function. Consider the case when the parameter λ parameter changes from 0.00005/year to 0.005/year. Fig. 2 shows how the dependence of the optimal time interval p^* between subsequent maintenance (according to the monetary optimum) on parameter λ. In can be seen, that the optimal p^* decreases sharply for parameter values λ from 0.00005 to 0.002 (from 17 years to 5 years).

In order to guarantee the acceptability of the project, the LQI assessment has to be performed from a societal point of view.

The expected number of fatalities due to bridge failure is computed as the product of the car occupancy (2.5 passengers per car), the probability of dying in case of a bridge collapse (98%), the number of cars over the bridge in a time lap of 1 min and the failure probability of the bridge (assuming flowing traffic):

$$m(t,p) = k(t)/60 \cdot 2.5 \cdot 0.98 \cdot P_f(t,p). \tag{33}$$

The Societal Willingness to Pay (SWTP) is estimated as the product of the marginal change in risk to life $d_m(t,p)/dp$ and the demographic constant C_x multiplied by g/q. For this example, it is assumed that g = \$45 900, q = 0.104 and C_x = 15.795.

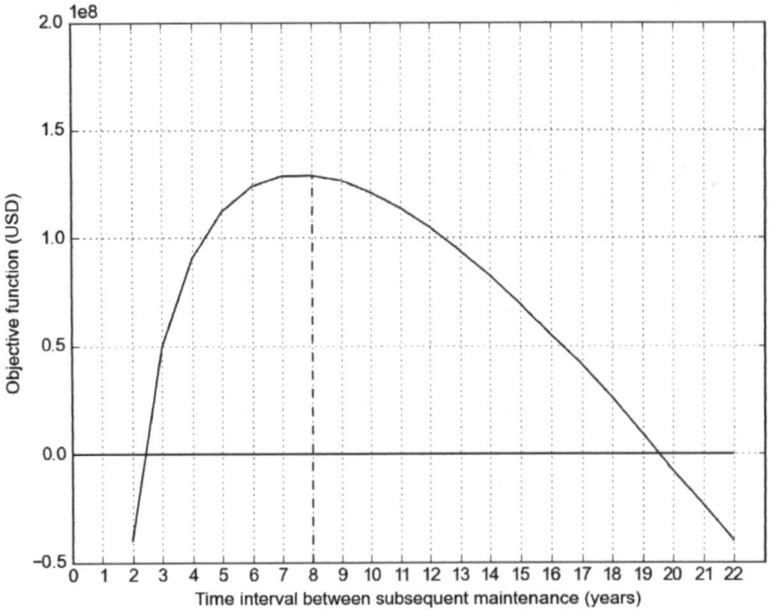

Figure 1. Objective function for a private decision maker over a time horizon of 80 years

Then the marginal change in risk is assessed by formula:

$$\frac{dm(t,p)}{dp} = m(t,p-1) - m(t,p), \; p = 3,4,..,10. \quad (34)$$

The marginal life saving costs are evaluated as the marginal change in maintenance costs for a small change in the decision parameter p, i.e. $dc_m(t,p)/dp$:

$$\frac{dc_m(t,p)}{dp} = c_m(t,p-1) - c_m(t,p), \; p = 3,4,..,10.$$

The acceptable decisions can now be determined by comparing the marginal life savings costs and the SWTP as shown in Figs. 3-5 for scale factor λ = 0.00005/year, 0.0005/year and 0.005/year correspondingly. For comparison, both the SWTP and the marginal life saving costs are evaluated as present values (PV) by discounting to t = 0:

$$PV(SWTP) = -\frac{g}{q} C_x \sum_{t=1}^{80} \frac{m(t,p-1) - m(t,p)}{(1+\gamma_S)^t}, \; p = 3,4,..,10 \quad (35)$$

The acceptance criterion needs to be evaluated from a societal point of view; hence, a societal discount rate γ_S of 2% is used

It can be seen that the present value of marginal life saving costs (black line) is larger than the present value of the SWTP (dashed line) for maintenance intervals smaller than 15 years for scale factor λ = 0.00005; 7.7 years - scale factor λ = 0.0005 and 3.9 years - scale factor λ = 0.005. Therefore, the monetary optimum found from the private decision maker's perspective is not an acceptable decision from a societal point of view, but very close to the acceptance threshold.

In order to get more practical results, this illustrative example should include the compensation costs for human fatalities and the change in construction costs for different design of the bridge – the less /more reliable it is, the less/more the initial costs.

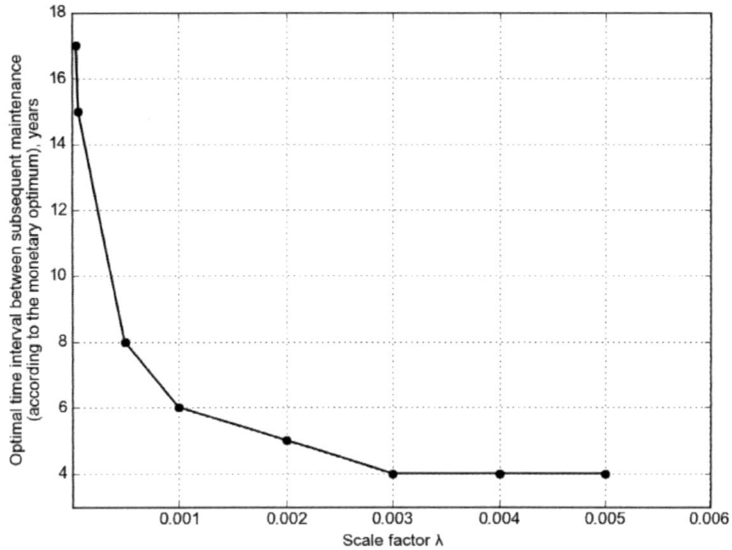

Figure 2. Dependence of the optimal time interval between subsequent maintenance (according to the monetary optimum) on scale factor λ

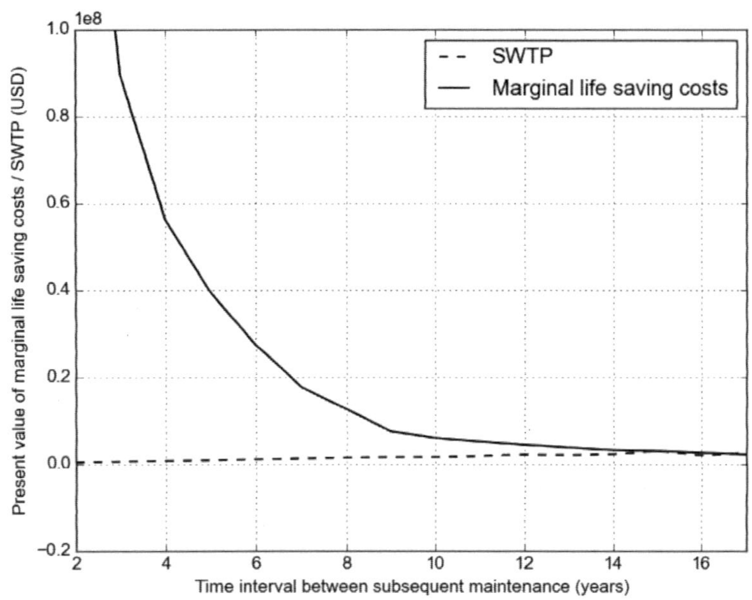

Figure 3. Present value of marginal life saving costs and SWTP for a time horizon of 80 years, societal discount rate of 2% and scale factor $\lambda = 0.00005$/year

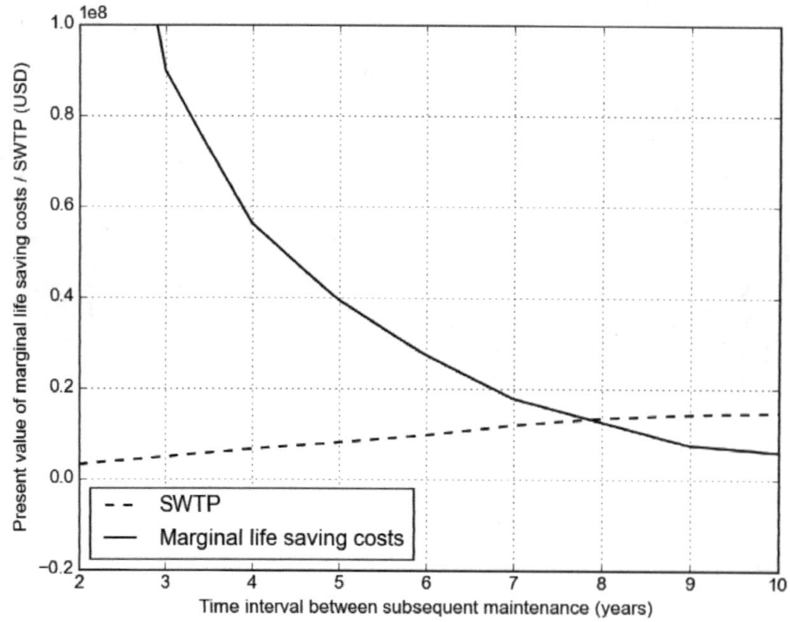

Figure 4. Present value of marginal life saving costs and SWTP for a time horizon of 80 years, societal discount rate of 2% and scale factor λ = 0.0005/year

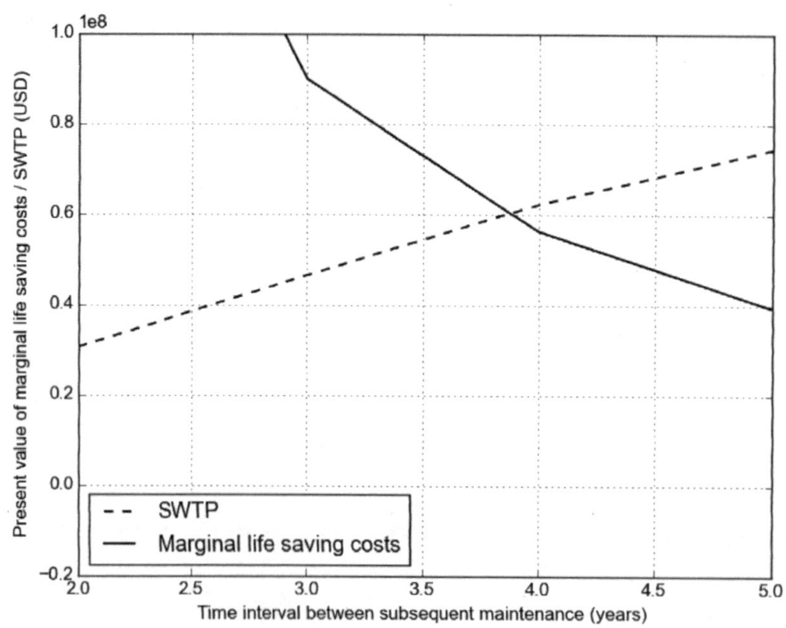

Figure 5. Present value of marginal life saving costs and SWTP for a time horizon of 80 years, societal discount rate of 2% and scale factor λ = 0.005/year

IOP Conf. Series: Materials Science and Engineering **972** (2020) 012002 doi:10.1088/1757-899X/972/1/012002

9. Conclusion

The proposed methodology is based on the differentiation between monetary optimization and the social acceptability criterion, which is included in the decision-making problem as a boundary condition [1].

In a market economy, optimization is usually provided to the individual, at least within an acceptable area defined by the LQI criterion. On the other hand, the acceptance criterion is based on social preferences for saving lives. Hence, marginal cost of saving life should be evaluated from a social point of view. At the project level, they can be defined as direct investments needed to improve safety.

Monetary benefits arising from the same decision (for example, reduced damage from failure) are not included in the acceptance criterion, since this leads to an unnecessarily strong restriction on the optimization performed by individual DM.

From the society view point, the obtained safety level may not always be optimal in monetary terms, since safety of life is the main focus in the criteria for the acceptability of decisions: it ensures the fulfillment of all effective life-saving investments.

10. References

[1] Katharina F and Edgar V and Mauricio S-S and Michael H F 2016 On the assessment of marginal life saving costs for risk acceptance criteria *J. of Structural Safety* **44** pp 37-46.

[2] Yaari M E 1965 Uncertain Lifetime, Life Insurancs and the Theory of the Consumer *J. of Review of Economical Studies* **32 2** pp 137–150.

[3] Usher D 1994 An imputation to the measure of economic growth for changes in life expectancy National Accounting and Economic Theory: The Collected Papers of Dan Usher (UK: Edward Elgar Publishing) **1** pp 105–144.

[4] Ramsey F P 1928 A mathematical theory of saving *J. of Econ* **38** pp 543-559

[5] Makeham W M 1860 On the law of mortality and the construction of annuity *tables J. of Inst. Actuaries* **8** pp 301—310.

[6] Shepard D S and Zeckhauser R J 1984 Survival versus consumption *J. of Manage Sci* **30 4** pp 423-439

[7] Haldar A 2006 Recent developments in reliability-based civil engineering (World Scientific Pub Co Inc) 277 p.

[8] Nathwani J S and Lind N C and Pandey M D 1997 Affordable safety by choice: the life quality method (Waterloo: Institute for Risk Research, University).

[9] Rackwitz R 2008 The Philosophy Behind the Life Quality Index and Empirical Verification *J. of Risk Assessment in Engineering* November.

[10] Rackwitz R 2002 Optimization and risk acceptability based on the life quality index *J. of Struct Safety* **24** pp 297-331.

[11] Easterlin R A 2000 The worldwide standard of living since 1800 *J. of Econ Persp* **14(1)** pp 7–26.

[12] Jakobovits I 1975 Jewish Medical Ethics (New York).

[13] Maddison A 1995 Monitoring the world economy 1820–1992 (Paris: OECD)

[14] Pandey M D and Nathwani J S 2003 Canada wide standard for particulate matter and ozone: cost–benefit analysis using a life-quality index *J. of Risk Anal* **23 (1)** pp 55-68

IOP Conf. Series: Materials Science and Engineering **972** (2020) 012003 doi:10.1088/1757-899X/972/1/012003

Regulation of risks with accounting for the possibility of terrorist attacks on critical infrastructure facilities

E F Dubinin[1] and V I Kuksova[1]

[1]Mechanical Engineering Research Institute, RAS, 4, M. Kharitonyevskiy lane, 101990, Moscow, Russia

E-mail: mibsts@mail.ru

Abstract. An important task of critical infrastructures is ensuring stable functioning of the objects of increased danger which are their part. For increase in safety of critical infrastructures and minimization of consequences from the possible emergency situations initiated by accidents and catastrophes needs to improve methods of assessment and regulation of risks connected with functioning of hazardous facilities, and especially critical facilities at various stages of their life cycle. For practical use the risk-based approach in regulation risks of functioning of critical facilities (CF) in the face of terrorist threats, it is necessary first of all to define the criteria and acceptable levels of risk. The article deals with the state of the system of risk regulation in the Russian Federation in the relevant subject area.

1. Introduction

At the first stage of regulation of risks in the Russian Federation, when assessing the protection of functioning of facilities and territories from emergency situations of various nature the basic concept that determines their safety was acceptable risk, and for regulation of risks used the system of risks, including unacceptable (inadmissible), maximum permissible, admissible, increased, conditionally acceptable, and acceptable risk.

Later, when regulating and managing risks of emergency situations, the range of risks was broken into three areas: zones of unacceptable (excessive), acceptable and negligible risk were allocated. State of protection of objects and territories from emergencies was assessed depending on in what of three areas of the risk level were received results.

At the present stage the range of risks of emergencies (regarding its probabilistic component) breaks into two areas: acceptable and unacceptable risk [1]. So, according to state standard GOST P 51898-2002, assessment of risk – the procedure of check based on results of risk analysis, defining whether the acceptable risk is exceeded.

2. Regulation of acceptable risk of emergency situation

Risk can be assessed for the considered territory or an object, its parts, separate projects, activity, a specific hazardous event. Therefore, in various situations various indicators and methods of risk assessment can be applied. In the regulatory-legal literature differ:

- acceptable risk of emergency situation – risk which is acceptable and reasonable for social and economic development of the considered territory;

IOP Conf. Series: Materials Science and Engineering **972** (2020) 012003 doi:10.1088/1757-899X/972/1/012003

- acceptable risk of an accident – the values of the risk of an accident at a hazardous production facility established or obtained according to a formalized established procedure, which excess characterizes threat of emergence of accident;
- acceptable risk of functioning of technical means and systems at various stages of their life cycle defined as residual risk owing to impossibility to consider and exclude all factors reducing safety because of their accidental and objective nature.

When assessing terrorist risks for CF the concept of acceptable risk of emergency is used. At the same time it is necessary to consider that the immediate goal and the most likely consequence of a terrorist act is an accident at the facility. Therefore, when analyzing terrorist risks, it is advisable to use the currently well-developed regulatory and methodical base for assessment of risk of accidents.

Indicators and methods for calculating the acceptable risk of designing and operating technical means and systems it is advisable to use in assessing the acceptable risk of the functioning of their components, in particular the critical elements of the CF, paying at the same time special attention to justification of safety factors on risks.

The concept of acceptable risk of emergency situation is key for assessment of a state of protection of the population of territorial subjects of the Russian Federation against emergency situations of natural, technogenic and biologo-social character; therefore the main terms, definitions and quantitative values of acceptable risk of emergency are highlighted in a separate state standard – GOST P 22.10.02-2016. Its provisions are intended for use by:

- federal executive authorities and their territorial bodies;
- executive authorities of the constituent entities of the Russian Federation and by local governments;
- research and design organizations and organizations operating *potentially hazardous and critical facilities (italics our)*, defined as such in accordance with the legislation of the Russian Federation.

At establishment of acceptable risk of emergency situations it is necessary to use the following quantitative indices of risk of emergency situation:

- individual risk of emergency;
- social risk of emergency.

Currently, the acceptable individual risk of emergency situations has been established for each subject of the Russian Federation on the basis of statistical data on emergency situations that occurred on the territory of the subjects of the Russian Federation from 1992 to 2014, including technogenic, natural, biologo-social emergencies, technogenic fires and *terrorist acts (italics our)*. For example, the acceptable individual risk for the Sakhalin region is $1.75 \cdot 10^{-4}$ year^{-1}, for the Murmansk region – $9.07 \cdot 10^{-6}$ year^{-1}, for the Sverdlovsk region – $1.23 \cdot 10^{-5}$ year^{-1}, for Moscow – $6.42 \cdot 10^{-6}$ year^{-1}.

The acceptable social risk of emergency situations, established for each subject of the Russian Federation, is 10^{-5} year^{-1}. At the same time, the social risk of emergency situations is considered unacceptable if it is more than 10 times higher than the acceptable social risk of emergency situations.

3. Assessment of acceptable terrorist risk of emergency

The subject of the Russian Federation is a large territorial entity, including construction, industrial and agricultural enterprises, infrastructure facilities and the social sphere, legislative and executive management bodies and other objects. As GOST P 22.10.02-2016 defines acceptable values of risk indicators for the territorial subject of the Russian Federation in general, in relation to separate CF as an elements of critical infrastructure two tasks are relevant:

- determination of acceptable risk of emergency for the concrete CF which is in this given territory;
- selection of the «share» falling on acceptable terrorist risk from all sources of emergency situation.

The solution of the first task depends on at what stage of life cycle there is this given object. If CF is at a stage of operation, reconstruction or technical re-equipment, it must have an approved safety passport of the facility and emergency risk assessment calculations with a detailed justification, including the calculation of individual and social risk indicators. If the calculated risk levels are lower or equal to the established ones, it is concluded that, judging by these indicators, the CF does not require the adoption of special additional measures for strengthening of its safety (the same applies to other regulated types of risk). Otherwise, in accordance with the established procedures, development (correction) of measures to reduce the risk of an emergency is carried out.

At the stage of justifying investments, designing, preparing technical documentation, as well as manufacturing and commissioning in a given territory when determining acceptable individual and social risks of concrete CF it is necessary to proceed from the fact that their actual levels shouldn't be higher than those established by GOST P 22.10.02-2016.

At the solution of the second task it is necessary to consider that sources of the potential increased danger to any CF, along with terrorist threats, are such classes of threats as natural, biologo-social, technogenic, ecological, professional, informational, economic, cyberdangers and other their possible types and combinations. Acceptable individual and social risks for an object are integrated criteria which are determined to a specific CF with its inherent vulnerabilities when exposed to all significant external and internal factors, including various hazards and threats. The approximate scheme of achievement of acceptable risk is shown in Figure 1.

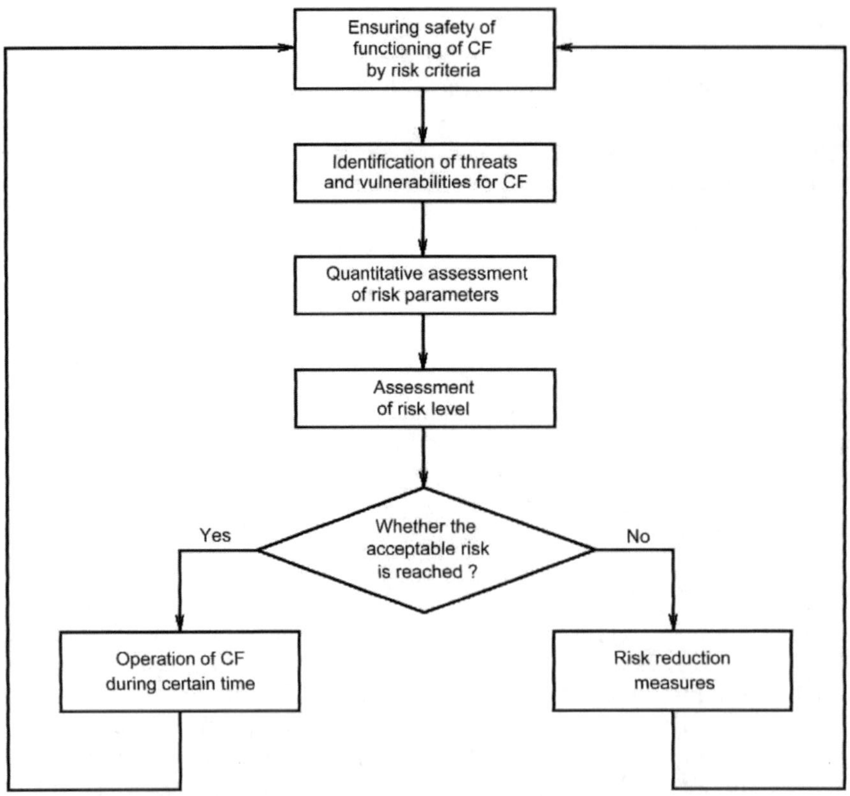

Figure 1. Scheme to achieve acceptable risk.

Values of parameters of acceptable terrorist risk, as well as other types of risks that a concrete object is exposed to, can be obtained by solving the «inverse problem»: the target value of individual and social risk is known; it is necessary on the basis of formulated risk-model of an object to determine the parameters corresponding to the given criteria.

So, the individual risk of emergency can be represented as a function defined on the surface adjacent to hazardous facility [2]:

$$R_{\Sigma}(x,y)=\sum_{i,j}\lambda_i E_{ij}(x,y)P_j \tag{1}$$

where λ_i – the frequency of realization of the i-th scenario; $E_{ij}(x,y)$ – the probability of realization of the j-th impact mechanism at the point (x,y) for the i-th scenario; P_j – the probability of damage during realization of the j-th impact mechanism.

The numerical value of $R_{\Sigma}(x,y)$ for solving the «inverse problem» is determined from GOST P 22.10.02-2016.

Social risk can be, for example, determined by a calculation method according to the data on equipment operating time at hazardous industrial facilities of the study territory and the number of the people getting to zones of possible damage [3]:

$$F=k\frac{T}{N}10^{-4} \tag{2}$$

where F – the frequency of death of 10 and more people (F = 10^{-5} year^{-1}); T – the actual average operating time of hazardous equipment; k – the empirical coefficient of weight of threat (k = 0.05 at threat to society; k = 0.5 at threat to personnel); N – the number of people getting to the affected zone.

The choice of calculated model, allowing, based on the target values of individual and social risks, to determine their acceptable parameters for concrete CF, in each case is determined by specifics of the considered object.

4. Main recommended quantitative indicators of emergencies risk

Besides individual and social risk of emergency situation of the key quantitative indicators of risk of emergency usually are included:

- collective risk of emergency, defined as the mathematical expectation of the death toll as a result of possible impact of all set of damaging factors of sources of emergency situation in the considered territory in a year;
- potential territorial risk of emergency defined as emergence probability in a year in the considered territory of all set of the damaging factors of sources of possible emergency situation with the level which can lead to death of people and to causing material losses;
- economic (variant – material) risk of emergency, defined as the mathematical expectation of the random value of material losses from an emergency situation in the considered territory for a year (variant – for a certain period of time).

When listing the main quantitative indicators of risk of emergency, it is also necessary to single out indicators of technical and environmental risk. Being components of more general characteristics of risk (collective, potential territorial or economic), they have independent significance, extensive literature is devoted to their identification, assessment, and methods of decrease.

According to problems of risk analysis besides the main additional indicators of risk can be applied. So, additional indicators of risk of accident on hazardous production facilities are presented in «Methodical bases on carrying out the analysis of dangers and assessment of risk of accidents on hazardous production facilities» [4].

Practical use of the listed indicators for risk-analysis and the implementation of risk management measures implies the existence of their normative (criterial, acceptable) values.

It is necessary to note the difference in approaches to regulation of a probabilistic and material component of risk (defined as a measure of the danger of emergency situation, combining the probability of emergency and its consequences (losses)).

5. Categorization of objects – rationing of interval values of safety

The extent of losses from an emergency is currently used at categorization (classification) of emergency situations in relation to objects of increased danger.

Example is the classification of emergency of natural and technogenic character by scales accepted by the Resolution of the Government of the Russian Federation of May 21, 2007 No. 304 «About classification of emergency situations of natural and technogenic character».

For determination of category (class) of danger of an object risk matrixes (of threats, of vulnerability) with use of methods of fuzzy logic are formed. Then, on the basis of the calculations made, the category (class) of danger of the object or territory under consideration is determined for different types of emergencies and the necessary measures to manage the identified risks are determined.

Fuzzy methods open up the possibility of using linguistic and scoring variables characterizing the remote consequences of an emergency when compiling matrixes of risk.

Risk matrixes in coordinates «frequency (probability) of refusal – weight of consequences of refusals», «frequency (probability) of an event – material losses» and others are for practical purposes also used [5]. Thus, the indicators characterizing a probabilistic component of risk of emergency are applied to categorization of objects and territories in relation to emergency of various character.

Methodologically matrixes of risk represent set of the interval norms determined by the amount of losses, the concrete region, a time interval (in case of use of probabilistic or frequency characteristics) and other indicators.

Categorization of objects of the increased danger in fact is rationing of interval values of safety, and determination of acceptable values of a probabilistic component of risk is point (limit, boundary, extreme) rationing. Therefore, for example, when developing problems of categorizing objects and territories in applying to emergencies, it is necessary to take into account used in rationing methods for determining the limits of rationing scales of classification signs, choosing the price of dividing for uniform scale, or of multivalued measure of these signs, and for non-uniform scale – for estimating the minimum and maximum price of division.

In a number of cases, special techniques for analysis them are developed to assess risks emergencies. For example, when assessing the key risks of financing terrorism in the Russian Federation the rating system of assessment where risks are considered as function of three factors is applied: threats, vulnerabilities and consequences. For assessment and management of risk three groups of risks are allocated: of high, average and low level. Based on the allocated groups of risks the priority (sequence assessment) of taking measures to management of risk in the relation to time is defined. For the identified high-level risks, it is necessary to take measures to minimize them in the shortest possible time; the group of risks of the average level needs to be controlled and if necessary to take the constraining measures; and in relation to risks of low level it is necessary to begin work on their minimization.

6. Current state of a system of regulation risks emergency. Conclusions

Currently, in the practice of analysis, regulation, and managing risks of emergencies use:

- point (limit, boundary) rationing for indicators characterizing probabilistic component of risk;
- interval rationing for assessment of consequences of emergency situation, applying as one of the main classification signs an indicator of the extent of losses from emergency situation;
- various additional indicators and original techniques of risk assessment, considering specifics of the studied phenomena, objects and territories.

The existing system of regulation risks from emergency is not fully formed and further, from our point of view, will develop in several directions:

- the main will be the interval multiparameter rationing of risks from emergencies. Base for its development is the available reserve in the field of classification and categorization of objects of increased danger. When developing interval norms methods of fuzzy logic, economical and statistical methods, general provisions of the theory and practice of rationing will be used;

- point (limit) rationing of individual quantitative indicators of risks from an emergency will retain its value. Concrete values and range of acceptable risks, apparently, will be periodically updated according to the international standards and rules and also specifics of the concrete temporary period, the considered look or area of emergence of possible risks;
- in some cases for assessment and regulation of risks special techniques will be used. Their application should not contradict the existing regulatory-legal documents.

References

[1] Dubinin E F and Kuksova V I 2018 Assessment of acceptable risks in the view of potential terrorist attacks on critical facilities. *Problems of Safety in Emergency Situations* **6** pp 15-30 (in Russian)

[2] *Techniques of Risk Assessment of Emergency Situations and Standards of Acceptable Risk of Emergency Situations* 2008 http://sra-russia.ru/e_docs/tekhnogennye-chs/metodiki-otsenki-riskov-chrezvychaynykh-situatsiy-i-normativy-priemlemogo-riska-chrezvychaynykh-situ (in Russian)

[3] *Report on the Research Work «Assessment of Technogenic and Complex Risks of Territorial-industrial Formations (on the Example of the Krasnoyarsk Territory)»* 2007 http://scert.ru›f/244/MainPart/Tridvornov.pdf (in Russian)

[4] *Methodical bases on carrying out the analysis of dangers and assessment of risk of accidents on hazardous production facilities* 2016 http://docs.cntd.ru/document/1200133801 (in Russian)

[5] Faleev M I, Malyshev V P, Bykov A A and Kondratiev-Firsov V M 2015 Methodological approaches to the zoning of the territories of the Russian Federation according to the risk levels of emergency situations of natural and technogenic character. *Civil Protection Strategy: Problems and Studies* **5** No **1** (**8**) pp 67-90 (in Russian)

IOP Conf. Series: Materials Science and Engineering **972** (2020) 012004 doi:10.1088/1757-899X/972/1/012004

Route Optimization Methods for Response to Radiological Emergency Situations

O L Tashlykov[1], A N Sesekin[1], S E Shcheklein[1]

[1]Ural Federal University, Mira str. 19, Yekaterinburg, 620002, Russia

E-mail: otashlykov@list.ru

Abstract. The work actualizes the need for regional authorities to be ready to respond to possible radiological emergencies due to the ever-expanding use of radiation sources in various fields of activity (besides nuclear reactors) to ensure the safety of sustainable development of cities, territories and regions taking into account the human factor. Described are the developed routing optimization algorithms using mathematical modelling methods to minimize the radiological impact on emergency formations personnel while eliminating possible radiological emergencies. Presented are the results of computational experiments to assess the reduction of personnel exposure during route optimization of work in radiation-hazardous conditions. The recommendations on the use of the developed algorithms during the operation and decommissioning of nuclear facilities are formulated.

1. Introduction

Any industrial technology, including energy technology, has a certain danger, i.e. the possibility of harmful effects on humans and the environment, poses a threat to the health and life of operating personnel and the public. Nuclear technologies, along with the dangers inherent in other technologies, have a specific hazard (radiation) associated with exposure to ionizing radiation. Thus nuclear technologies determine a special approach to ensuring the safety of its use [1].

The use of radiation sources in various fields of activity, in addition to nuclear reactors, is expanding every day. Despite the precautionary measures envisaged, accidents involving sources of radiation occur more frequently than at nuclear reactors. Such accidents usually affect a small number of people. However, for these people the consequences of the accident can be quite serious.

Currently, international guidelines developed (for example, [2]) contain general response instructions designed to protect the public and emergency workers in various types of radiological accidents, including accidents involving closed and open sources and nuclear power generators, as well as accidents involving transport. The main objectives of emergency response are to reduce the risk or mitigate the consequences of an accident at the place of its occurrence, prevent the development of deterministic medical effects and reduce the risk of stochastic ones by taking protective measures, and also by keeping individual doses received by the public and emergency workers at a level not exceeding the threshold which is set in regulatory documents.

In many radiation accidents that are not related to nuclear reactors, the danger due to radiation is often less than the danger caused by other factors (for example, fire, hazardous chemicals). Therefore, priority should almost always be given to the non-radiological aspect of a radiation accident: saving lives, treating injuries, fighting fire, protecting essential equipment and ensuring personnel safety. As soon as the overall situation is stabilized, immediate actions should be taken to reduce the risk of radiation damage to the population, emergency workers and the environment [2].

Accidents with radioactive sources or material are a broad category, including detected radioactive material, contaminated areas or objects, missing or lost sources of radiation, unprotected sources, accidents in laboratories, in industrial or scientific facilities, transport accident.

Radioactive materials in the form of hermetically sealed sources are used in industry, medicine, scientific research, as well as in the manufacture of many consumer goods intended for sale to the public. They are used for radiography, sterilization, radiation therapy and nuclear medicine; for well logging, in sensors recording the surface level, thickness, density and humidity; in antistatic devices, lightning arresters, smoke detectors. In their activity, all these sources are extremely diverse.

Thousands of transport operations involving the use of radiation and radioactive materials take place all over the world every day. All types of transport are involved in this process to varying degrees – road, rail, air and sea. The main problem in planning a response to a traffic accident is the fact that they can happen anywhere. Therefore, adequate means of emergency response are required, acting mobile within the entire state. A feature of transport accidents is that, in addition to drivers of vehicles, people around are also at great risk. Transportation of radioactive material is governed by strict regulations designed to provide protection in accordance with the danger that the contents of the containers present in the event of an accident. The higher the potential hazard, the more substantial the protection required by the instructions.

In its present form, the fundamental international safety principles, including the provision of radiation safety, are detailed in the "IAEA Requirements and Recommendations. Fundamentals of Safety" №SF-1, 2007 (10 principles in total).

The principles of radiation safety (justification, rationing, optimization, protection of future generations and the environment) are most widely known, since they concern all citizens, not just professionals.

The main provisions of the Recommendations of the International Commission on Radiation Protection (ICRP) of 2007 (Publication 103) include further use of the fundamental principles of justifying and optimizing protection against all controlled exposure situations, which are characterized as planned exposure situations, emergencies and existing exposure situations [3]. Strengthening the principle of optimization of radiation protection, which should equally be applied to all exposure situations, subject to individual dose limitations and radiation risks, is also provided for in national guidelines [4].

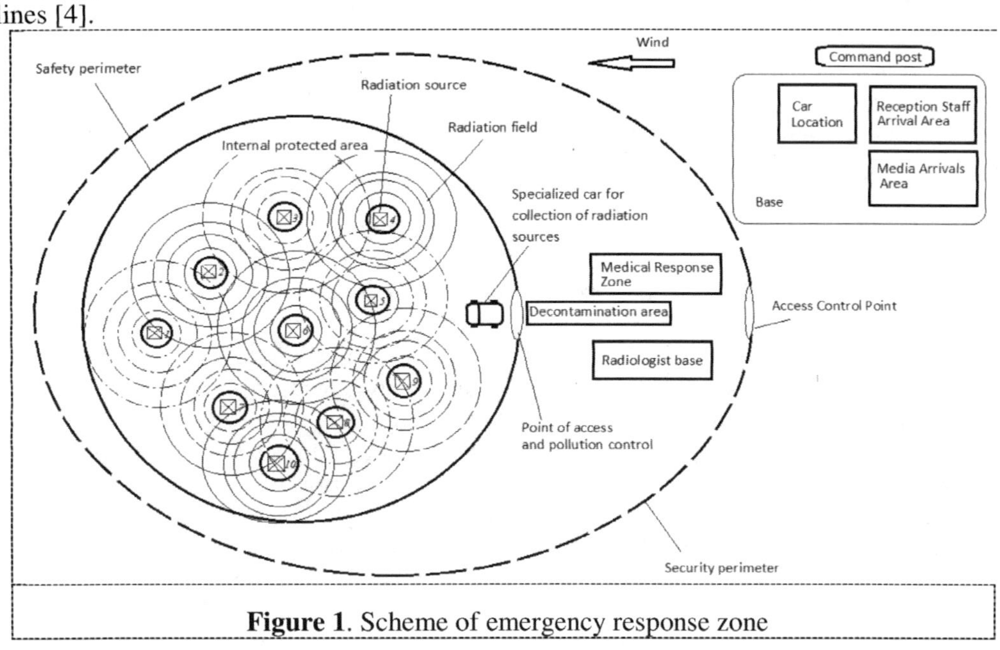

Figure 1. Scheme of emergency response zone

2. Methods and means

The first important stage of emergency response is the establishment of boundaries of safety and security, as well as control over the access of personnel and unauthorized persons to the danger zone. At the same time, it is necessary to constantly monitor the prevalence of pollution, the necessary perimeters of safety and security, protective measures for emergency workers, restrictions on the time they are in the danger zone, protective measures for the public with the help of radiologists.

The best way to control access is to establish physical barriers based on local conditions and the extent to which the risk of exposure can be reduced. Entrance and exit from the protected area should be made through the established control points. These points serve as a gathering place for emergency personnel, as well as radiological monitoring points (Fig. 1).

Dose exposure of personnel during radiation hazardous work is reduced by reducing the residence time in radiation fields, increasing the distance from the radiation source to the person, exposure to the radiation parameter (shielding, deactivation, etc.) [5], [6], [7] . An effective way to reduce personnel dose costs (by 25–40%), which do not require significant material costs, is routing optimization of work in non-stationary radiation fields [8], [9].

The main goal of solving route optimization problems is to find the minimum of the quality functional — the total dose of radiation when performing a complex of works in non-stationary radiation fields:

$$\sum_{i=1}^{n} \dot{H}_i \times t_i \rightarrow min$$

where H_i is the dose rate in the i-th isodose region; t_i is the time when a person is in the i-th isodose region of the radiation-dangerous zone.

The work on the construction of methods and algorithms for solving problems of route optimization, focused on applications in applied problems of nuclear energy, has been conducted since the beginning of the 2000s by specialists from the departments "Nuclear Power Plants and Renewable Energy" and "Applied Mathematics" of the Ural Federal University and the Institute of Mathematics and Mechanics of the Ural Region RAS. During this time, a theoretical substantiation of various variants of route optimization has been carried out, computational algorithms for them have been developed, computational experiments have been carried out, including their implementation on the Uran supercomputer of the IMM Ural Branch of the Russian Academy of Sciences [10], [11], [12]. Possible options for route optimization were discussed (Fig. 2).

As a base, the classic traveling salesman (HQ) problem is used, in which a trader, starting from a certain base, must visit each of the N other cities only once. This one of the most well-known problems of discrete optimization combines simplicity of formulation and computational difficulties, since there is N! possible routes, one of which, or several, gives minimal costs (in the tasks under consideration, this is a collective radiation dose). This task was adapted to determine the optimal trajectory of the worker in radiation fields in order to minimize the "transit" radiation dose.

Further studies switched to the so-called courier task (LC with precedence conditions), as well as to the generalized courier task. In these studies, the dynamic programming (MDP) method was predominant with the preliminary reduction of constraints, which was the original element. Then, variants of the iteration method were constructed, using tunable SQ models and courier tasks and implementing a system of improved bilateral estimates of global extremum. Finally, in connection with the application of the MDP, a method was constructed that does not use the computation of the entire array of values of the Bellman function and, which is very important, is not losing in quality. Further, in connection with the formulation of tasks related to the dismantling of radiation-contaminated equipment of NPP power units decommissioned, theoretical results were obtained concerning accurate and approximate methods for solving route problems with a complicated criterion, including internal work and explicit dependencies on the list of unfulfilled projects. moment of assignments.

IOP Conf. Series: Materials Science and Engineering **972** (2020) 012004 doi:10.1088/1757-899X/972/1/012004

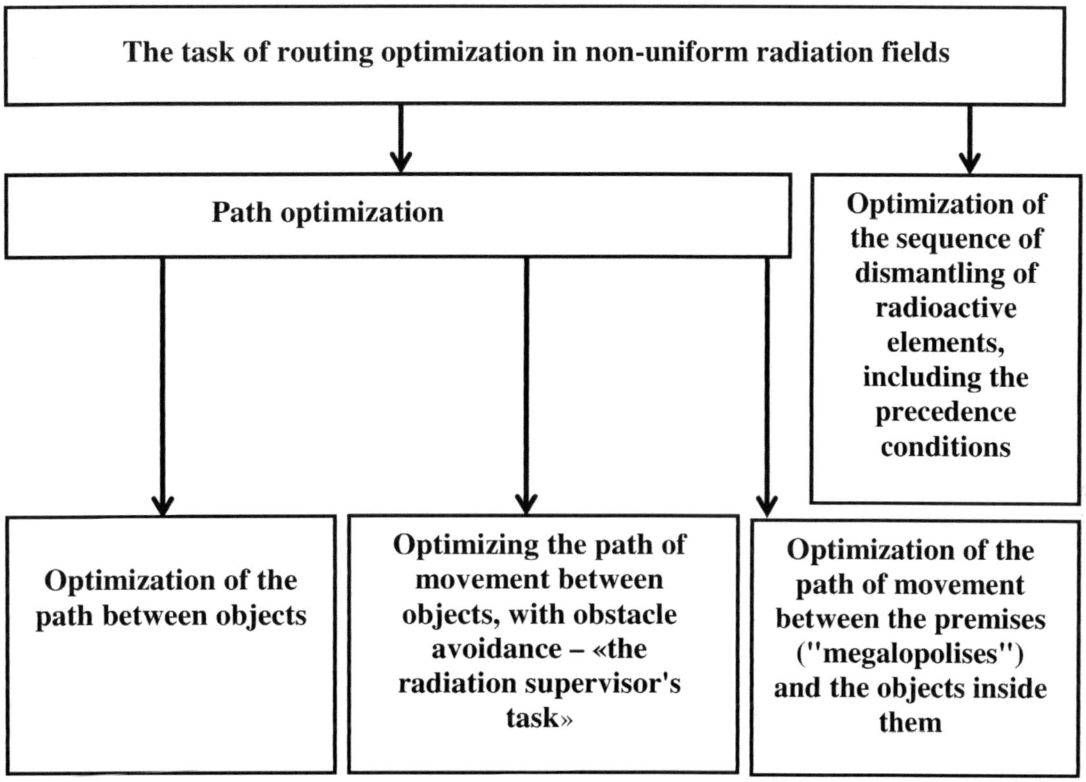

Figure 2. Options for route optimization in non-stationary radiation fields

Recently, an important step was taken: a set of constraints in the form of precedence conditions (traditionally considered difficult for optimization methods) was used "in a positive direction": to overcome computational difficulties, since on this basis the actual reduction of dimension was achieved without loss of quality.

To solve the route problem with precedence conditions and cost functions, taking into account the circumvention of the existing obstacles, the implementation of an independent computation scheme was considered. The studied formulation has as its prototype the well-known intractable task of a traveling salesman, but contains the essential features of the above-mentioned nature. Such a dependence may arise when considering some of the applied problems that are relevant for nuclear energy in the field of minimizing the radiation dose of workers when working in non-stationary radiation fields (for example, eliminating the consequences of radiation accidents, planned work of a radiation supervisor). Also, an important point is the construction of model examples of emergency situations elimination in the accident of the NPP power unit (the initial parameters are taken from the data of radiation reconnaissance and the experience of eliminating the accident at the Chernobyl nuclear power plant). The developed algorithm for solving the problem will allow to automate the construction of the optimal route of movement between given points for emergency rescue specialists (ASF).

Conducting a radiation survey in order to obtain reliable information about the radiation situation both inside the facility and in adjacent territories is the main task at the initial stage of work to eliminate the consequences of an accident at a radiation-hazardous facility or territory. There are various ways to obtain information about radiation fields, the presence of hot spots, etc. Based on these data, it is possible to obtain a cartogram of the radiation situation in the form of isodose lines with the selected resolution.

A promising way to quickly obtain information about the radiation situation is the use of measuring systems for obtaining images of objects emitting photon radiation using mobile measuring systems (gammovisors) [13].

3. Optimization of the "off" sequence (dismantling, shielding) of radiation-contaminated equipment (object, source) in order to minimize personnel exposure.

Radioactive objects are located in separate rooms or on an emergency (contaminated) site, creating different radiation dose rates. At the same time, the labour costs for dismantling each of these objects are different. The total dose of radiation received by workers during dismantling work indoors or on site will depend on the time they are in the radiation fields created by radiation-hazardous objects.

When dismantling works are performed, the radiation fields are formed by separate radioactive elements, therefore when removing (dismantling) any object (element), the parameters of the radiation field change. This feature, which consists in the dependence of the background radiation level in the room when performing the next work on the presence of undemounted (unremoved) objects, makes it much more difficult to solve the problem, unlike the previous one.

To solve the problem of minimizing exposure during the dismantling (removal) of radioactive objects located at the emergency (contaminated) site, it was necessary to develop an original Bellman function construct and program implementation of the MTD procedure [14], [15]. The formulation of this task is the determination of the optimal sequence of dismantling (removal) of radioactive objects in order to minimize personnel exposure.

Consider as an example the finding of the optimal sequence of dismantling for the simplest option - four radioactive objects located on the emergency site. For simplicity, we will not take into account the effect on the dose rate of the distance from the radiation source to the worker.

In the general case, when dismantling N objects in sequence $\alpha(1), \alpha(2), ..., \alpha(N)$, the radiation dose:

$$E = t_{\alpha(1)}(\dot{H}_{\alpha(1)} + \dot{H}_{\alpha(2)} + ... + \dot{H}_{\alpha(N)}) + t_{\alpha(2)}(\dot{H}_{\alpha(2)} + \dot{H}_{\alpha(3)} + ... + \dot{H}_{\alpha(N)}) + ... + t_{\alpha(N)}\dot{H}_{\alpha(N)}$$

where $\dot{H}_{\alpha(i)}$ is a radiation dose rate generated by the m element; $t_{\alpha(i)}$ is the time of the dismantling of the *i*-th element (in this task $t_{\alpha(i)}$ does not depend on the number of the next object to be dismantled, since the time of movement between the objects is neglected). When dismantling object No. 1, its further influence on personnel irradiation in the form of radiation dose rate is excluded, while dismantling object 2, its further influence on radiation in the form of radiation dose rate, etc. is excluded. When dismantling an object, its further influence on irradiation in the form of radiation dose rate is excluded.

As mentioned above, there is *N!* solutions to the problem of route optimization, i.e. For the simplest example (four objects in the room), there are 24 solutions [16]. As a result of calculations of the considered example, the optimal sequence of dismantling (2–1–4–3) was found, corresponding to the minimum dose of radiation (Fig. 3).

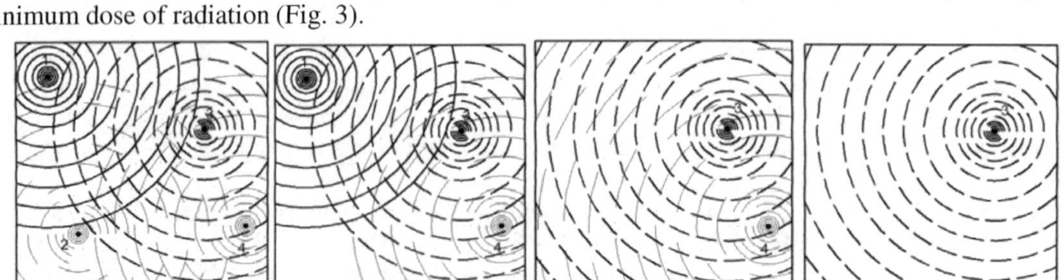

Figure 3. Diagram of the effects of radiation fields created by objects in the room during dismantling

The software implementation of several dozen examples for various radiation-polluted technological systems consisting of 10–20 elements showed the computational efficiency of the developed technique. Optimization of the sequence of dismantling allows reducing personnel radiation doses by 25 ... 40%.

In reality, the removal (shutdown, dismantling) of a radioactive object (source) cannot be carried out in any sequence. There are restrictions associated with the order of dismantling of some objects (for example, located at different elevations, connected structurally). This raises the problem of optimizing the sequence of dismantling with restrictions in the form of precedence conditions. This, to a certain extent, reduces the number of searches in the solution, but required the development of a special approach, which reduces the number of possible “bypass” options.

In the context of eliminating the effects of radiation accidents, in addition to the task of determining the optimal route of movement, it is necessary to solve a combined task that combines both finding the optimal trajectory of movement and determining the optimal “shutdown” (removal or shielding) sequence of the most “significant” sources in terms of generating radiation fields radiation. The main goal of such a combined task in the context of eliminating radiation accidents is to minimize the collective dose of personnel or to observe (not exceed) the established dose budget (limit) for emergency rescue units.

4. Conclusion

- The constantly expanding use of radiation sources in various fields of activity updates the requirement of constant readiness to respond to possible radiological emergencies to ensure the safety of cities, territories and regions, taking into account the human factor.
- The developed routing optimization algorithms using mathematical modeling methods for eliminating possible radiological emergencies are an effective and low-cost way to minimize the radiological impact on the personnel of emergency formations.
- The results of computational experiments to assess the reduction of personnel exposure during work in non-uniform radiation fields using the developed route optimization algorithms show their effectiveness.
- The developed routing optimization algorithms for work in radiation hazardous conditions can be used to minimize radiation loads on personnel during maintenance and repair of radioactive equipment and decommissioning of nuclear facilities.

References

[1] Novikov G A, Tashlykov O L, Scheklein S E 2017 Ensuring safety in the use of atomic energy (Ekaterinburg: Publishing house of Ural University) 552 p.

[2] General instructions for assessing and responding to radiological emergencies 2004 IAEA-TECDOC-1162 (R. - Vienna, IAEA) 204 p.

[3] Publication 103 of the International Commission on Radiation Protection (ICRP). Per from English 2009 Under the general ed. Mf Kiseleva and N.K. Shandaly. (Moskow: LLC PKF "Alana")

[4] MU 2.6.5.054–2017 - Optimization of radiation protection of the personnel of the enterprises of Rosatom State Corporation. Methodical instructions 2017 (Moskow: FMBA) 52 p.

[5] Tashlykov O L Dose costs of personnel in nuclear energy. Analysis. Ways to decline. Optimization 2011 (Saarbrücken, Germany: LAP LAMBERT Academic Publishing GmbH & Co. RG) 232 p.

[6] Tashlykov O L, Scheklein S E, Khomyakov A P, Russkikh I M, Seleznev E N 2015 *Nuclear and radiation safety* **3** (**77**) 17-24

[7] Tashlykov O L, Scheklein S E, Khomyakov A P, Russkikh I M, Seleznev E N 2015 *Global Nuclear Safety* **2** (**15**) 49-55

[8] Sesekin A N, Tashlykov O L, Scheklein S E, Kuklin M Yu, Chentsov A G, Kadnikov A A 2006 *Izvestia vuzov. Nuclear energy.* 2006. **2** 41–48

[9] Tashlykov O L, Sesekin A N, Scheklein S E, Balushkin F A, Chentsov A G, Khomyakov A P 2009 *Radiation safety issues.* **4** 47–57.

[10] Grigoriev A M, Ivanko E E, Chentsov A G, Sesekin A N, Tashlykov O L, Shcheklein S E 2011 *NPP Safety and Training. XII International Conference:* Abstracts (Obninsk: NRNU MEPI) **2** 103-105

[11] Grigoryev A M, Tashlykov O L. 2018 *AIP Conference Proceedings* Vol. 2015. Article Number 020028.

[12] Kropachev Yu A, Tashlykov O L, Scheklein S E 2019 *Izvestiya Vuzov. Nuclear energy.* **1** 119-130

[13 Tashlykov O L, Scheklein S E, Sesekin A N, Kokorin V V, Chentsov A G 2012 *Safety of critical infrastructures and territories: Materials of the V All-Russian Conference and XV School of Young Scientists* (Ekaterinburg: Ural Branch of the Russian Academy of Sciences, Publishing House of the AMB) 181-187

[14] Chentsov A G, Sesekin A N, Shcheklein S Y, Tashlykov O L 2010 *AIP. Conference Proceeding* **1293** 197–202

[15] Tashlykov O L, Sesekin A N, Scheklein S E, Chentsov A G 2009 *Izvestia vuzov. Nuclear energy* **2** 115–120

[16] Korobkin V V, Sesekin A N, Tashlykov O L, Chentsov A G 2012 *Routing methods and their applications in improving the efficiency and safety of operating nuclear power plants* (Moskow: New technologies)

IOP Conf. Series: Materials Science and Engineering **972** (2020) 012005 doi:10.1088/1757-899X/972/1/012005

On the properties of the failure probability function in reliability analysis of constructions and structures with regulation of their stress-strain state

V G Sebeshev, A A Zhbanov and A A Shestakov

Novosibirsk State University of Architecture and Civil Engineering (Sibstrin), 113, Leningradskaya, Novosibirsk, 630008, Russia

E-mail: sebeshev@sibstrin.ru

Abstract. The properties and peculiarities of the failure probability function in dependence on the stress-strain state regulators in various (direct, inverse, optimization) problems of calculating the reliability of constructions and structures have been investigated and described. Predicting the possible changes in the probability of failure due to regulators' varying, on the basis of the formulated signs, allows us to work out the algorithms for finding a rational and optimal regulation of the deformable system state parameters. The qualitative and quantitative illustrations of the identified properties of the failure probability function for multi-span beams with the nonlinear regulation of a bending moments are presented. The possibilities to reveal the areas of permissible values of regulators by the target reliability using a complex reliability indicator that takes into account the structure's volume of material are shown.

1. Introduction

Regulation of the stress-strain state (SSS) of a construction, structure or their separate parts by various technical methods (force or kinematic pre-stressing, variations of the design scheme or loading etc.) is an effective mean mean of improving the technical and economic indicators of a construction systems, which is expressed in reducing their resource consumption (such as a materials volume or a structure cost). Therefore, there has been gained experience in applying the regulation of the state parameters of structures – forces, stresses, displacements and etc. in the practice of modern construction. Modern international [1, 2] and national [3 – 7] normative documents (standards) contain the requirements for quantifying reliability indicators, also they establish a general principles for ensuring the reliability of construction systems, up to the specified (target) reliability indexes implemented in Eurocodes considering the classes of structures. In this aspect, the necessity of developing theoretical and applied issues of reliability analysis of constructions and structures in which the regulation of their mechanical state parameters realizes is obvious. The problems of design deformable systems with regard to the influence of various regulators of their SSS in a deterministic formulation were considered in a sufficiently large number of studies. Their reviews are given particularly in [8, 9]. Theoretical foundations and applied methods for determining the reliability of engineering systems, including construction ones, are also well developed – see [10 – 15]. But studies on the special problem of the reliability of constructions and structures with a regulated mechanical state are few in number.

This work is devoted to studying of the properties of the general reliability characteristics of a system (structure) – the failure probability $P_f = 1 - P_s$ (here P_s is reliability) depending on the regulators, with taking into account the random nature of all design parameters – of the system, loads and regulators.

The purposes of the work are:

- research and description of the character and peculiarities of the failure probability function P_f of a deformable system when varying a finite number of regulating parameters (regulators), considered as arguments of the function P_f, in different definition versions of the reliability analysis problems;
- qualitative and quantitative verification of the formulated properties of the P_f function in a model problems.

2. The main part

2.1. Theoretical bases and methods of the solution of problems

2.1.1. Calculation of reliability P_s or probability of failure $P_f = 1 - P_s$

The main stages of this calculation are:

1) formulation of the serviceability requirements [15] on the basis of non-failure performance general criteria (strength, stability, stiffness etc.):

$$g_j(X) < 0,\ j = \overline{1,m}, \tag{1}$$

where X is the vector of design parameters – random variables characterizing the structure, influences (loads) and regulators; the "$<$" sign means the prevention occurrence of the corresponding limit state (if "$<$" in (1) is changing to "$=$", then (1) sets the boundary of the allowable states area);

2) calculation of the failure probability

$$P_f = 1 - P_s = 1 - \prod_{j=1}^{m}(1 - P_{fj}) \approx \sum_{j=1}^{m} P_{fj}, \tag{2}$$

where P_{fj} is the partial probability of failure by the jth serviceability requirement.

It is generally accepted in the calculations for (2) with the limiting conditions (1) that the concepts of resistance R, load effect (factor) Q and the reserve of serviceability S are used [12 – 15], while requirements (1) are represented as

$$S_j = R_j - Q_j > 0. \tag{3}$$

Load factor Q_j and resistance R_j in (3) generally depend on the design parameters X_s (geometric, stiffness, strength, etc.), impacts q (loads and others) and regulators V, and moreover $X_s \in X$, $q \in X$, $V \in X$. With using of the known stochastic properties of random vectors X_s, q and V the characteristics of the variables Q_j and R_j can be found – such as the bivariate distribution density function $p_{QjRj}(Q_j, R_j)$, means $\overline{Q}_j$, $\overline{R}_j$, coefficients of variation A_{Qj}, A_{Rj}. They determine the probabilistic characteristics of the reserve S – the density $p_{Sj}(S_j)$, mean $\overline{S}_j$ and standard deviation $\hat{S}_j$, after which the probability of failure according to the jth restriction (1) is calculated as

$$P_f = P(S_j < 0) = \int_{-\infty}^{0} p_{Sj}(S_j)\,dS_j \tag{4}$$

or through the reliability index [2] $\beta_j = \overline{S}_j / \hat{S}_j$ (according to [15] – safety characteristic):

$$P_{fj} = 10^{-\varphi_j(\beta_j)}; \tag{5}$$

the expression φ_j (β_j) depends on the type of distribution p_{Sj} (S_j); for example, we use the Laplace function in the case of a normal law: $P_{fj} = 0{,}5 - \Phi_0(\beta_j)$.

The formula for the reliability index is conveniently represented as

$$\beta_j = (1-\xi_j)[A_{Rj}^2 + (\xi_j A_{Qj})^2 - 2\xi_j A_{Rj} A_{Qj} r_{RjQj}]^{-1/2}, \tag{6}$$

where $\xi_j = \overline{Q}_j / \overline{R}_j$; r_{RjQj} – correlation coefficient of R_j and Q_j.

Load factors $Q_1, \ldots, Q_m$ are expressed through effects of q, regulators V, and also the characteristics of the system X_s in the form

$$Q_j = f_{Qj}(q, V, X_s), \quad j = \overline{1, m} \tag{7}$$

using the equations of state (algebraic or differential, linear or nonlinear, in particular, *FEM* equations). For linearly deformable systems with linear regulators [9], dependencies (7) are written as

$$Q_j = Q_{jq} + \sum_{k=1}^{n_V} Q_{j,V_k=1} V_k, \quad j = \overline{1, m} \tag{8}$$

or in matrix form

$$Q = \{Q_1\, Q_2 \ldots Q_m\} = Q_q + \Lambda_Q V \tag{9}$$

where $Q_q = \{Q_{1q}\, Q_{2q} \ldots Q_{mq}\}$ – vector of factors Q from the effects of q;

$V = \{V_1\, V_2 \ldots V_{n_V}\}$ – vector of regulators; n_V – number of quantities V;

$\Lambda_Q = \begin{bmatrix} Q_{1,V_1=1} \cdots Q_{1,Vn_V=1} \\ \cdots\cdots\cdots\cdots \\ Q_{m,V_1=1} \cdots Q_{m,Vn_V=1} \end{bmatrix}$ – influence matrix of regulators V on Q.

Equations (7) describe hypersurfaces, and (8) and (9) describe hyperplanes in n_V+1-dimensional spaces $V - Q_j$. If all Q_j are of the same type, then it is possible to represent (7) – (9) in the general hyperspace $V - Q$. This is convenient for a comparative estimation of changes in Q factors in depending on the varying regulators V. With heterogeneous Q_j, such a combination can be achieved due to their transformation into dimensionless form [16].

The quantities V calculated by (2) – (9) can take any values, but in designs where the regulation of SSS is specifically applied, V are determined by the originally formulated requirements for the results of regulation.

2.1.2. Setting and solving the problem of regulation

The two main types of problems in calculating the regulation of SSS of structures are direct and inverse. The problem of optimal regulation may also be of practical interest and it is usually solved by successive calculations in direct or inverse formulations.

In the direct problem certain values of the regulators V used directly in the calculations from (7) – (9) are specified initially.

The inverse problem is: for a system with known own parameters X_s and impacts q it is necessary to determine the values of the regulators V, ensuring the correspondence to requirements for a selected complex of regulated parameters of the SSS $H = \{H_1\, H_2 \ldots H_{n_H}\}$, initially formulated on the basis of certain criteria set (it is possible that $H \cap Q$).

Equations expressing the conditions (requirements) of regulation

$$\phi_i(H) = 0, \quad i = \overline{1, n_y}, \tag{10}$$

IOP Conf. Series: Materials Science and Engineering **972** (2020) 012005 doi:10.1088/1757-899X/972/1/012005

may be linear or non-linear, homogeneous or nonhomogeneous. The values of H_k ($k=\overline{1,n_H}$) are represented by analogy with (7) as

$$H_k = F_k(q,\ V,\ X_s), \tag{11}$$

in particular, in the linear case $H = H_q + \Lambda_H V$ (here Λ_H is the matrix of the influence of V on H).

Combining (10) and (11), we obtain the regulation equations expressed through regulators V:

$$\psi_i(q, V, X_s) = 0,\ i=\overline{1,n_y}\,. \tag{12}$$

Their solution gives the desired values of V. This is possible for an arbitrary realization of random variables X_s, q, and V, but it is practically to consider their expected values (means). The obtained V are used further in (7) – (9), after which P_f is calculated from (2) – (6).

2.2. Main properties of failure probability function P_f for deformable systems with regulated stress-strain state

In the reliability analysis of structures, there are problems of two main types – direct (verification) and inverse (design). In the direct problem the probability of failure is calculated with initially specified system's parameters and effects (loads). Inverse problems usually do not have a single solution and, as a result, reveal the areas of allowable values of the design parameters for a target (required) reliability. It is also possible to formulate the optimization problem according to the criteria of reliability (or probability of failure).

In the main reliability calculation for the systems with SSS regulators the problem of regulation may appear in both direct and inverse forms, depending on the general setting.

Further only direct reliability calculations are considered, in which the failure probability values, as a rule, are uniquely dependent on certain combinations of regulators (except for cases of strong non-linearity of equations (7), (10), (11)). Variants of problem setting are:

- SSS regulation is carried out with unchanged loads q and system parameters X_S;
- after each changing of regulators the characteristics of X_S are corrected (according to limitations of strength, stability etc.) – such procedure is natural in practical calculations in the design of structures.

In variant 1:

- a) in the case when the functions described by dependencies (7) are smooth, the failure probability function $P_f(V)$ is also smooth; its minimum corresponds to a combination of regulators that provides the most advantageous stress-strain state of the structure, taking into account the restrictions (1) or (3) (modular forms or out-forms [17] can be used to identify this combination);
- b) if the functions (7) have any singularities (discontinuities or jumps of derivatives), in particular, when the parameters of the system X_S change due to the variation of the regulators V (for example, reducing the rigidity during plasticity development with the bilinear dependence $\sigma \sim \varepsilon$), the $P_f(V)$ function also has features (smoothness violations).

In variant 2:

determination of some complex of calculated parameters X_{s0}, in particular, the geometric characteristics of sections G (possibly, $G \cap X_s$ or G are other than X_s) is carried out according to requirements of type (3). If all G are set through one common parameter G_0, then to find it, it is necessary to identify the object (structural element, section, dangerous point) that is the closest to the design limit state. To do this, for some combination of regulators V (arbitrarily specified or corresponding to the required regulation results, according to conditions (10)), the coefficients ξ_j are estimated, and by the largest of them ($\xi_0 = \max \xi_j$) from the corresponding equation of (3) type, in which $\xi_0 = [\xi]$ (here $[\xi]$ is preliminary specified), the value of G_0 is obtained. Note that all $\xi_j \le \xi_0$ in this solution. With a certain combination of SSS regulators the system may be such that two or more design objects may turn out to be "equally dangerous", with the same $\xi_j = \xi_0 = [\xi]$. The equality of the coefficients ξ for a certain pair of restrictions (3) when varying the regulators gives the line of intersection (edge) of two corresponding hypersurfaces, and on different sides of this line the design points of surfaces with the largest from two values are selected for finding G_0. This means that the out-form [17] for characteristics (7) or (3) (in the general case in a dimensionless form) has an edge i.e. violation of the smoothness of its hypersurface. The limit state conditions of two different objects (structural elements, sections, points) use in this case in the areas on opposite sides of an edge. As a result, as noted above, a break (edge) on the hypersurface of the failure probability $P_f\,(V)$ appears too.

In the case of one regulator ($n_V = 1$) the $P_f\,(V)$ diagram is a piecewise line with peak-like segments (figure 1, (a)) – this was shown in [18]. The qualitative view of the indicated peculiarity of the function $P_f\,(V)$ graph in the case with $n_V = 2$ is shown in figure 1, (b) (it can be considered as a particular case when $n_V > 2$ and the fixed values of the other regulators are given).

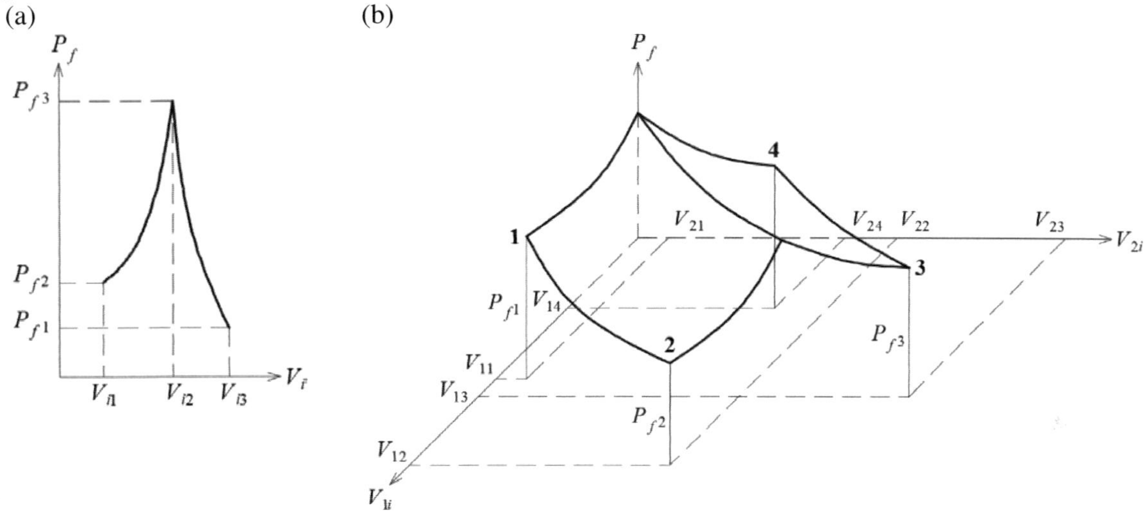

Figure 1. The peculiarity of the failure probability function P_f graph – break (edge) with one regulator (a), with two regulators (b).

If several regulators and serviceability requirements ($n_V \geq 2$, $m > 2$) are used, an equally dangerous state of three or more design objects of the system is possible as a result of improvement of the SSS by using a regulators (leveling the distributions of forces or stresses). This is accompanied by the occurrence of three or more break (edges) on the failure probability $P_f\,(V)$ hypersurface converging at the top (peak), as it is shown in figure 2. If the number of edges increases, the peak sharpens and its height (P_f value) correspondingly increases.

When the regulators change within sufficiently wide limits, as well as when several independent parameters G are determined by the requirements (3), several peaks may appear on the graph of the function $P_f\,(V)$. Comparable peak's values can occur with significantly different combinations of regulators.

The nature of the $P_f\,(V)$ hypersurface in the vicinity of a certain peak depends on the features of the system (more advanced design forms, which are cost-effective in terms of material consumption, can give higher levels of failure probability), as well as on the effectiveness of regulators and stochastic properties of the main design parameters of the system, regulators and loads.

The graphs of the P_f function for large (A_{max}) and small (A_{min}) coefficients of variation of the design variables are qualitatively presented in figure 3, (a). The upper graph not only has significantly larger ordinates, but also has smaller, in comparison with the lower graph, derivations $dP_f\,/dV$ (or gradients in cases of more than one variables V) of the branches on a different sides of the vertex, which corresponds to a wider range D_2 of inadmissible regulator values with a given limiting probability of failure $[P_f]$ – this is disadvantageous in terms of material intensity.

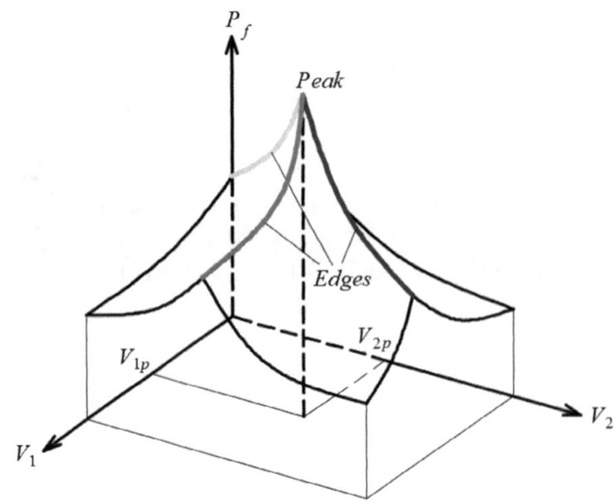

Figure 2. The salient feature of the failure probability function P_f – peak in the case of $n_V = 2$.

In cases where the coefficient $\xi_j > 1$ takes place and the reliability index $\beta_j < -2 \ldots -3$, the appearance of the failure probability values close to 1 is possible, which gives areas of hyperplanes in the form of an almost plateau with $P_f \approx 1$ (figure 3, (b)) – of course, the approach to these areas is unacceptable.

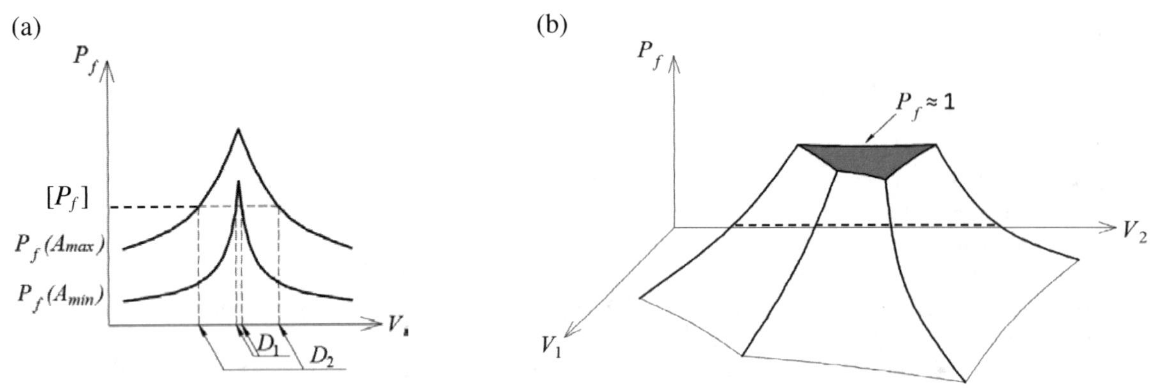

Figure 3. Graphs of the failure probability function for different characteristics of the variability of the design parameters (a), a possible view of $P_f(V)$ surface with the coefficient $\xi_j > 1$ (b).

The same effect would be if all conditions (3) are used without exceptions with the fixed coefficients $\xi_j = [\xi_j]$ for obtaining the parameters G. As a consequence a values of function $P_f(V)$ may be almost flat ($P_f(V) \approx$ const < 1) with the possible deviations due to the different influence of design parameters' coefficients of variation according to various combinations of regulators.

Predicting the nature and features of the probability function of failure based on the peculiarities and properties formulated above allows in the process of structure design to look for areas of regulators' values in which the probability of failure does not exceed the allowable level $[P_f]$. It is essential, especially for optimal regulation of the system's SSS, to take into account the material consumption of the structure, which is achieved by using the integrated dimensionless indicator [18], which takes into account also the volume of material v and the load parameter q, along with the probability of failure P_f:

$$\rho = \rho^* \cdot v^* , \tag{13}$$

where $\rho^* = P_f/P_f^0$; $v^* = v_0^*/q^*$; $v_0^* = \bar{v}/\bar{v}_0^0$; $q^* = \bar{q}/\bar{q}_0^0$;
$\bar{P}_f^0, \bar{v}_0^0, \bar{q}_0^0$ – baseline values.

2.3. Example

The problem of estimating the reliability of structures with regulated parameters was considered in application to statically determinate beams with arbitrary number of spans (figure 4).

Setting of the regulation problem:

- the bending moments in characteristic sections of the beam are taken as the regulated parameters: $H \equiv M = \{M_1\ M_2\ \dots\ M_{n_M}\}$;
- the regulators are the structural and geometric characteristics of the beam's design scheme – the relative coordinates of the hinges and the ratio of the lengths of the spans $V = \{\alpha_1\ \alpha_2\ \dots\ \alpha_{n_\alpha}\ \nu_1\ \nu_2\ \dots\ \nu_{n_\nu}\}$ (here $\bar{\alpha}_i = \bar{a}_i/\bar{l}_i$; $\bar{\nu}_i = \bar{l}_{i+1}/\bar{l}_i$), $n_V = n_\alpha + n_\nu$ (figure 4);
- the influence of the selected regulators on the regulated parameters (moments) is non-linear [9];
- the regulation requirements are set in the form of a ratio between the means of bending moments:

$$\bar{M}_1 = -k_1\bar{M}_2;\ \bar{M}_2 = -k_2\bar{M}_3;\ \dots;\ \bar{M}_j = -k_j\bar{M}_{j+1};\ \dots;\ \bar{M}_{nl} = -k_{nl}\bar{M}_{nl+1} \tag{14}$$

$$\text{or} \begin{cases} |\bar{M}_1| = k_1|\bar{M}_2| \\ |\bar{M}_2| = k_2|\bar{M}_3| \\ \dots \\ |\bar{M}_j| = k_j|\bar{M}_{j+1}| \end{cases}, \tag{15}$$

where $k_1, k_2, \dots, k_j$ – coefficients of regulation that are associated with regulators V, but they are more convenient for use in this example (V values can be expressed in k).

Note that in the case at issue ((15) are homogeneous) the condition $n_M = n_V + 1$ should be satisfied.

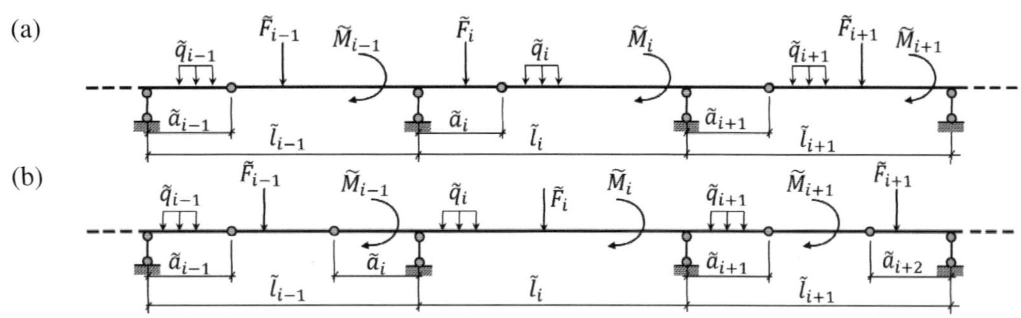

Figure 4. The design schemes of beams with type …s-h-s-h… (a), type …s-s-h-h-s-s-h-h… (b) (s – support, h – hinge).

When the coefficients of regulation k are assigned then the means of regulators obtain from the restrictions (1) or (2) with use of a state equations, nonlinear in the considered problem, while the expected values of the moments contained in a conditions (1), (2) are expressed in dependence of the loads, regulators and beam parameters also from the state equations (nonlinear).

Fundamentally important is that the main design variables were presented in a dimensionless form, which formed the basis of a rational approach to solving the problem.

With known (specified) coefficients of variability of the design parameters it is possible to calculate the system's failure probabilities corresponding to the varying coefficients of regulation.

The two-span beam was considered (figure 5, (a)). Regulators are $\alpha_1 = \bar{a}_1/\bar{l}_2$, $\alpha_2 = \bar{a}_2/(\nu\bar{l}_2)$, $\nu = \bar{l}_1/\bar{l}_2$ ($n_V = 3$), they correspond to the regulation coefficients k_1, k_2, k_3; regulated parameters $H = \{M_1\ M_2 \ldots M_4\}$ ($n_H = 4$); serviceability requirements – by stresses in sections with moments $M_1, M_2, \ldots M_4$ ($m = 4$) when the beam is in an elastic stage of work.

Problem statement is in accordance with the above variant 2 with one geometrical parameter G_0 (the cross section is constant along the beam).

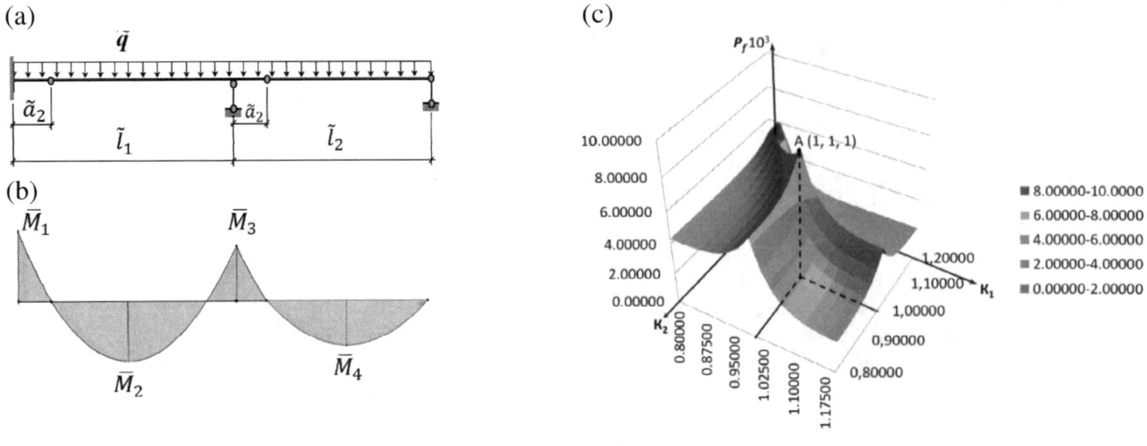

Figure 5. The design scheme of the beam (a), bending moments means diagram (b), the graph of the failure probability function in dependence on the coefficients of regulation (c).

Initial data: $l = 5$ m; $q = 5$ kN/m; $A_l = 0{,}01$; $A_q = 0{,}02$; $A_{\sigma_0} = 0{,}07$; $A_W = 0{,}01$; $A_{\alpha 1} = 0{,}001$; $A_{\alpha 2} = 0{,}001$; $A_{\alpha 2} = 0{,}001$; $A_v = 0{,}01$; $[\xi_0] = 0{,}7$.

The stress variation coefficients at dangerous points of the design cross sections which are needed for calculating the failure probability were determined by the statistical linearization method [15]. The results of the calculations of P_f are presented graphically in figure 5, (c) for a fixed value $k_3 = 1$.

The graphs clearly show the predicted above specific (peak-like) nature of the failure probability dependence on the regulation coefficients. For systems with one and two regulators this feature was discovered earlier, here it is confirmed with three regulators. At point A ($k_1 = 1$, $k_2 = 1$, $k_3 = 1$), which corresponds to the most favorable regulation from the position of minimizing the consumption of materials, the maximum value of the failure probability is obtained.

For the same beam, the complex characteristic (13) of the failure probability was calculated; its dependence on regulators is presented in figure 6, (a). Comparison of this graph with the graph of the probability of failure according to figure 5, (b) shows that taking into account the material's consumption leads to the changes in the mutual arrangement and the ordinates ratio for the surfaces characterizing the probability of failure.

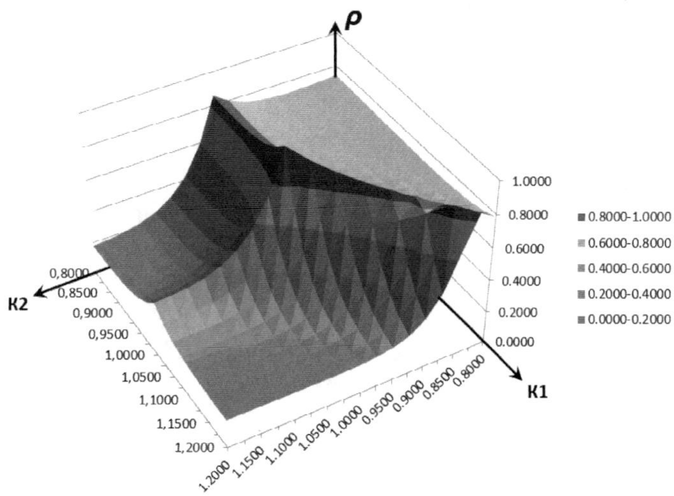

Figure 6. Graph of the complex characteristic of the failure probability.

The influence was studied of design parameters' variability characteristics on the type and features of the function P_f graphs. It is known that the coefficients of variation of loads can vary in fairly wide intervals – much larger than the coefficients of variation of the strength and geometric characteristics.

(a) (b)

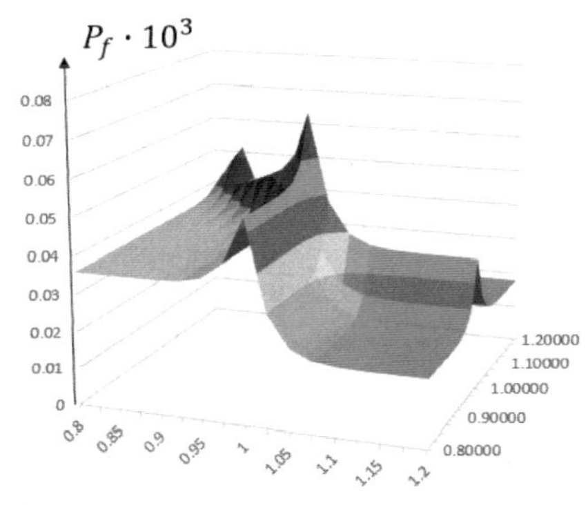

$P_f(1{,}1{,}1, A_{min}) = 0{,}000072$

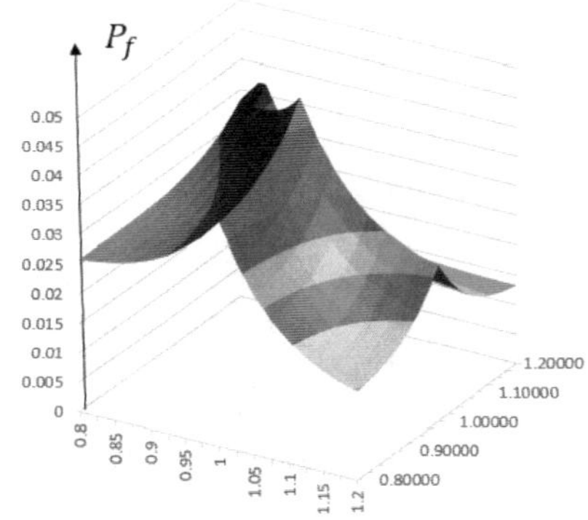

$P_f(1{,}1{,}1, A_{max}) = 0.049131$

Figure 7. Graphs of the failure probability function for different characteristics of the variability at the lowest values of the coefficients of variation (a), at the highest values (b).

For comparison, the following values were taken: $0{,}05 < A_q < 0{,}15$; $0{,}0005 < A_l < 0{,}015$; $0{,}01 < A_w < 0{,}02$; $0{,}001 < A_{\alpha 1/2} < 0{,}02$; $0{,}001 < A_v < 0{,}02$; $0{,}03 < A_{\sigma u} < 0{,}1$. The dependence of P_f on the coefficients of variation is presented in figure 7.

The direct problem of calculating the reliability of the beam was also solved by varying the regulators without changing the geometric characteristics of the sections (for a fixed value of the material strength characteristic) – variant 1 (see above). The graph of the failure probability function P_f in a relatively narrow zone of k_1 and k_2 changes (figure 8) is a smooth surface with a distinct concavity and the minimum point at $k_1 = k_2 = 1$. In the peripheral areas degeneracy is observed in the almost plane form with $P_f \approx$ const. For larger values of k_1 and k_2 the increasing of the P_f is observed (not shown in figure 8).

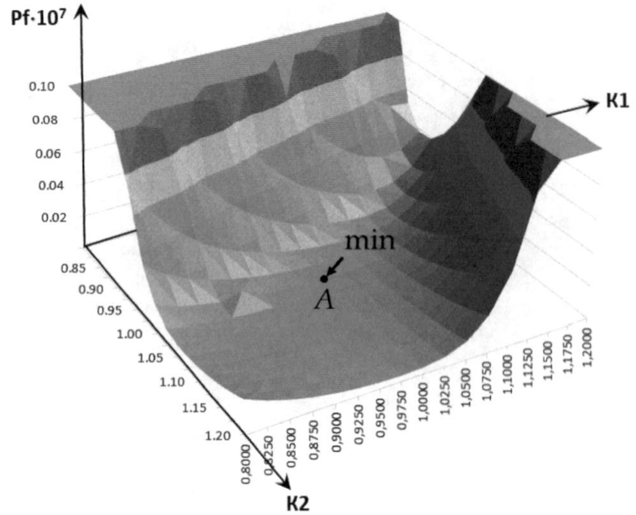

Figure 8. Graph of the failure probability function in direct formulation of problem.

The results of solving the direct problem of reliability analysis of the model system (beam) are in concordance with the predictions on the basis of detected specific features of the failure probability function.

3. Conclusion

1. The analysis has been performed and a description are presented on the main properties and peculiarities of the failure probability function of deformable systems (constructions and structures), in which the stress-strain state is regulated by various methods. The direct and inverse statements of the problems of calculating the reliability and regulation are considered.

2. A method, a design apparatus and algorithms for determining the reliability of beam systems with a different number of non-linear regulators, in various problem statements, including the procedure of using the dimensionless random variables to simplify calculation techniques, are developed. The obtained results of beams' reliability calculations correspond to the predictions of the properties of the failure probability function based on the formulated properties and features.

3. The quantitative estimates of an influence of the design parameters' variability characteristics on the probability of failure are presented. The reliability indicators have been calculated using a complex criterion that takes into account the material consumption of the structure with SSS regulators.

4. The results obtained in this work make it possible to set and solve the problems of optimal regulation of the SSS of constructions and structures with regard to the requirements of reliability.

Acknowledgement
Work is realized in accordance with the Thematic plans of investigation in the Novosibirsk State University of Architecture and Civil Engineering (Sibstrin), Novosibirsk, Russian Federation.

References

[1] BS EN 1990:2002 Eurocode – Basis of structural design (Brussels: Management Centre rue de Stassart 36 B-1050) 87 p
[2] Vrouwenvelder T 2008 *Proc. Inst. Civil Eng. Structures and Buildings* **161** SB4 pp 209 – 214

[3] Offshore Technology Report 2002 HSE Target Level for Reliability-Based Assessment of Offshore Structures during Design and Operation (HSE Books ed HaSE UK)
[4] Ellingwood B R 2001 *Progress in Structural Engineering and Materials* **3** **2** pp 170 – 179
[5] ASCE/SEI Standard 7–05 2005 Minimum Design Loads for Buildings and Other Structures (ASCE Reston VA)
[6] Federal Law of Russian Federation from 12/30/2009 No 383-FL. Technical regulation on the safety of buildings and structures
[7] Standard 27751–2014 Reliability for constructions and foundations. General principles (Moscow: Standardinform) 14 p
[8] Abovskiy N P 1998 *Regulated structures* (Krasnoyarsk: KrasSABA) 420 p
[9] Sebeshev V G 2009 Peculiarities of statically indeterminate systems work and regulation of forces in structures (Novosibirsk: NSUACE Sibstrin) 164 p
[10] Cornell C A 1969 *ACI Journal* **12** **66** pp 974 – 985
[11] Bolotin V V 1971 Use of the probability theory methods and reliability theory for structural analisys (Moscow: Science Publishing) 256 p
[12] Rzhanitsyn A R 1978 The theory of building structures reliability analysis (Moscow: Stroyizdat) 239 p
[13] Ditlevsen O and Madsen S O 1996 Structures Reliability Methods (Chichester: John Wiley & Sons Ltd) 372 p
[14] Nowak A S and Collins K R 2000 Reliability of Structures (New York: McGraw-Hill) 397 p
[15] Raizer V 2009 Reliability of Structures. Analysis and Applications (USA: Backbone Publishing Company) 146 p
[16] Sebeshev V G 2017 *Proc. St. Petersburg Transp. Univ.* **14** **1** pp 165 – 174
[17] Kholopov I S 2008 *Metal Structures* **14** **4** pp 285 – 297
[18] Sebeshev V G 2011 *News Higher Educ. Inst. Construc.* **11** pp 113 – 120

Markov matrixes for random sequences imitation in fatigue testing and design in the problem of safety of steel structures

I Gadolina[1], R Voronkov[2], A Bautin [2], I Serebrjakova [3], A Erpalov [4]

[1] IMASH RAS, Structural materials science Department, Bardina 4, Moscow, Russia

[2] Central Aerohydrodynamic Institute, Structural Integrity Department, Moscow Region, Zhukovsky 1, Russia

[3] Bauman Moscow state technical university, Department Computer Software and Information Technology, Moscow, Russia

[4] South Ural State University, Chelyabinsk, Russia

E-mail: gadolina@mail.ru

Abstract. To guarantee the safety of steel structures it is important to deal properly with the problem of representation of exploitation loading. For metal fatigue testing and design, the proper choice of random (irregular) loading type is very important. The principles of random loading are discussed and some alternative approaches with their pros and cons are shortly reviewed in the paper. As a sound decision for random, but taking into account some specific features of the exploitation loading process, the target Markov method is proposed. According to this method, the important information of the real random processes in the form of the turning point is used for filling the square Markov matrix (analysis phase) and later on, with employing the random number generator, serves as a source for creating of the so-called replicas. The replicas are the random trial for numerical estimation of longevity scatter. Due to the fact, that all these manipulations are performed with the aim of metal fatigue investigation, some important processes' characteristics for fatigue, like irregularity factor, fullness factor and machine part longevities were compared. Some important suggestion for the future development of this method, that is taking into consideration the sequence of the events, is discussed.

Key words. Random processes, metal fatigue, Markov's matrixes

1. Introduction

The process of metal fatigue failure is complex by its nature and possesses huge scatter especially considering the working duration of machines in exploitation under random loading. Even for nominal conditions, the duration before failure might be varied by several times and more. There are two main sources for scatter - the instability of physical material properties and variation of the loading conditions [1]. Scatter of mechanical properties are being investigated during static strength tests (example in [2]) and by regular fatigue testing while building the fatigue curve. To investigate the

IOP Conf. Series: Materials Science and Engineering **972** (2020) 012006 doi:10.1088/1757-899X/972/1/012006

scatter of the results due to the variability of loading condition, it is necessary to perform testing under random loading.

For investigation of the random loads on massive off-shore structures, caused mostly by random waves, the package WAFO (Wave Analysis for Fatigue and Oceanography) was developed. This package includes analysis of the sequence of turning points [3].

Fatigue testing under random loading plays a big role nowadays mostly for two reasons: 1) the test results provide useful data about the fatigue damage accumulation, taking into account the peculiarities of irregular loading; 2) there are excellent technical and control decisions nowadays allowing to realize many options and methods of random loading during the test. The good example of the testing under random loading is given in [4]. Some questions of peculiarities of distributions of random values are discussed in [5]. Extreme loads prediction plays an important role while evaluating the risk in varied fields of industry, considering the different scenarios of the events [6]. In [7] the recordings of random loads in wave energy converter of micro power plants are presented. This fact also indicates the need for developing the methods for simulating the random loading sequence.

In metal fatigue, usually, the researchers consider two fatigue stages with different mechanics: crack initiation and crack growth stages. Effect of sequence in loading history exists at both stages. The degree of degradation during the initial stage depends on the sequence of applied load cycles. On the other hand, from the studies of technological operation influencing on fatigue properties, it is known, that fatigue strength could be more than doubled by a single high overload [8]. Overloads are the cause of residual compressive stresses in the concentration areas and reduce the damage from the rest of the loading cycles, therefore the duration also increases. Even at the greater degree the sequence effect exists at the crack propagation stage.

2. Testing method review

2.1. *Sinus-like fatigue testing*

This type of testing had been performed since the first fatigue testing machine appeared (since ~1850). The result of this testing is a fatigue curve, or additionally, and sometimes alternatively, the estimation of the so-called fatigue limit. Nowadays many scientific sources imply that probably there is no such a thing like a fatigue limit. Further application of these results requires applying the Miner-summation rule [9] to get the reasonable estimation of machine-part longevity

2.2. *Block testing*

Historically the first mode of irregular loading was testing with the blocks. E. Gassner proposed the ideology of block testing for the aviation structures [10]. Block steps are presented in the form of simple sinusoidal loading with the amplitudes vary stepwise. Usually, it is taken 8 steps for each block. The block testing should be planned in such way, that blocks could be repeated several times before failure. The amplitudes and their repetition numbers are estimated on the basis of real loading by applying the cycle counting methods for the transformation of irregular loading process into the set of simple cycles. The example of block is shown in figure 1 (the values of loadings and time are shown non-dimensional).

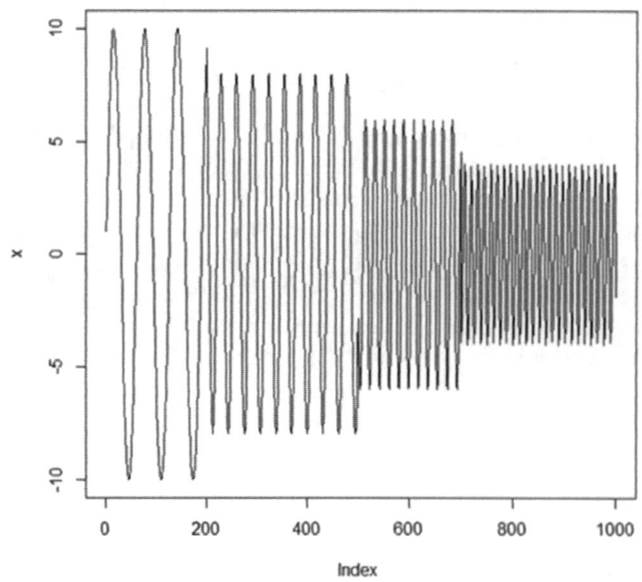

Figure 1. Shortened example of block testing

2.3. One-to one random testing

The idea of testing one-to-one by simply repeating the exploitation loading has many shortcomings. First of all, it is very time-consuming. Machine parts serve years and conclusion of their reliability should be done during few weeks. Even if the researcher excludes time-out and down-states of the machines, the testing time would be extremely long. The second problem is representability of rare events, like overloading. It might come out, that in the particular singe realization the rare event just would not be fixed. In this way, the testing estimation would be overoptimistic. It is also important to have in mind that in reality the process might be different comparing to the single registered realization. For this testing option, the servo-hydraulic machine is required.

2.4. Imitation of random processes using spectral density

One of the modes of early days testing was controlling the power spectral density function of the random exploitation process and later on use this information not only in simulating the random process during laboratory testing [4], but also for fatigue durability estimation [11]. This approach was especially popular in 1970-1980, when in researchers' disposition mostly only the vibration testing machines existed. Those machines were controlled by frequency. Similar methods are still popular nowdays when the parts of the apparatus with minor dimensions are being tested (the solder joints, for example).

It should be noted here, that in situation where the researchers are dealing with the machine parts, where the loading process could be easily registered through the tensor gages, the preferable solution is to perform the cycle counting procedure, namely, rain-flow [12]. While doing this, the information about physical frequencies is being lost. The rain-flow method saves only the information about stress amplitudes and their repetition [13].

2.5. Applying the Markov's matrixes

A good decision for simulating the random sequence of the turning points is the utilization of the Markov's model. Markov's matrixes were proposed as a method for modeling random sequences for academicals research for imitation of random processes with different irregularity factor IF [12, 14]. Irregularity factor features the degree of the complexity of the random process and equals IF=No/N_E. Here N_O is the number of the mean level crossing and N_E is the turning points number. Both values should be taken on the representative enough realization part. In [15] the representative distributions were analyzed for three options: IF = 0.3 (process with complicated structure), IF = 0.7 (medium process) and IF = 0.99 (almost regular process, but the amplitudes vary significantly), figure 2.

In [16] the Markov's modeling is compared to the other random process simulation technique, namely autoregressive moving average modeling. In [16] the loading process is considered to be unceasing, not just the turning points, like in Markov's modeling.

The short parts of standard sequences the examples are shown in figure 2.

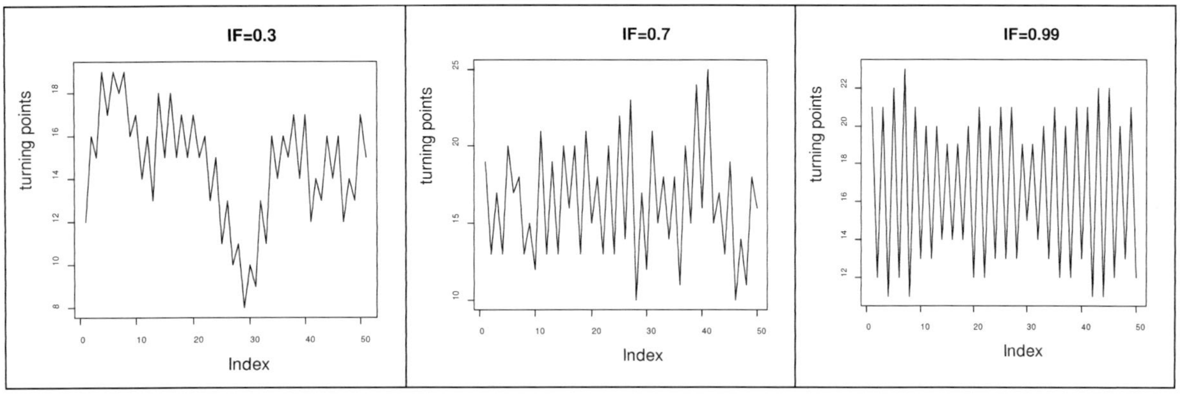

Figure 2. Parts of standard Markov's sequences [14]

3. Method: target Markov's matrixes simulation

The method was developed with the purpose of investigating the scatter produced by the randomness of loading. In this paper target Markov's modelling had been proposed for the better representation of individual peculiarities of the investigated machines loading. Some unique loading features, including randomness and overloadings, deserve to be saved and later on to be employed while simulation of so-called replicas. To create a Markov's matrix the real loading process firstly some treatment should be done.

3.1. Preliminary treating

Before creating the matrixes some steps for process preparation should be done [12]:

- Process discretization. After performing it, instead of the continuous stress process $\sigma(t)$ the researcher gets the set of digital stress reading σ_1, σ_2, …….. σ_z, where z – is the number of readings in the realization;
- Division by classes – that means the finite number of digits, say, N, classes, by which the values σ_i are represented. Value N depends on required representation accuracy and usually is taken as N=2k, where k=5 … 7. While doing this substitution, the researchers implement the preliminary process filtration. The data later on are presented in integer form that speed up the future treatment;
- Turning points definition. After this procedure the initial process is represented by the sequence of turning points (extremums-"the saw"):

$$e_1, e_2 \; ..... \; e_{z1} \quad (1)$$

Note, that z1<z, because of the transformation described above.

3.2. Analysis stage

The analysis implies the creation of Markov's matrix of frequencies on the base of a real process. During the analysis phase, the square matrix [32x32] or [64x64] is filled up on the basis of sequence (1) following the simple rules described below.

For the explanation, a short sequence of events is shown in figure 3, left. On the main diagonal of the matrix figure 3, right, are zeros. The triangle to the right of the main matrix diagonal contains the numbers of ascending ranges (e_i,e_{i+1}), (i∈ 2k - even indexes) from the minimum to the maximum. In the triangle to the left of the main diagonal are the numbers of descending from the maximum to the minimum ranges (i∈ 2k+1 - uneven indexes. In this manner, the analysis – "convolution" of the initial sequence to Markov's matrix is carried out.

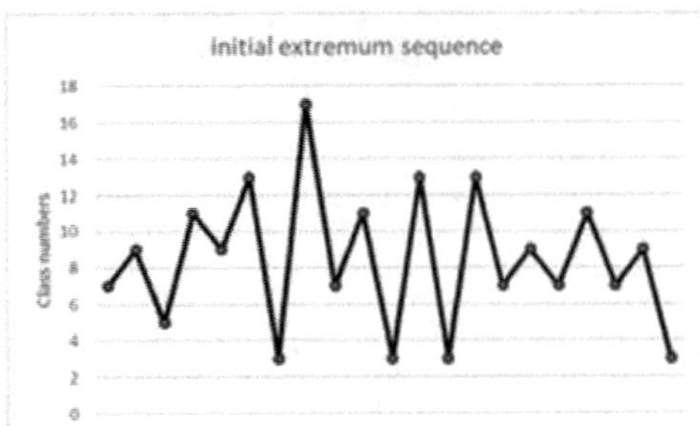

	1	2	3	4	5	6	7	8	9	10	11	12	13	14	15	16
1																
2																
3													2		1	
4																
5											2					
6																
7									1		2		1			
8																
9			1		1		1									
10																
11			1				1		1							
12																
13			2				1									
14																
15							1									
16																

Figure 3. Analysis stage in target Markov's matrixes modeling

It is worth mentioning, that this filled matrix by the form and by its content is very similar to the matrix of ranges, produced by range cycle counting method [12].

3.3. Synthesis stage

After the matrix, like one shown in figure 3, right, has been filled, the next step would be the creation of the series of replicas on the base of the initial sequence. This step might be called "synthesis" because on the base of the target matrix the random sequences are being produced. To investigate the statistical uncertainty, more than one replica had been produced (namely nine, in this example).

The algorithm of this modeling is as follow. The filled matrix and a random number generator are used. The initial sequence minimum (class α) is chosen arbitrarily (figure 4). Next, the ascending range starting from this minimum α is modeled. In the string α contains some frequencies, and their sum equals T_α. The random value RAND * T_α is calculated (RAND is a random number, RAND=0...1). When the accumulated sum of the parts in the line α from the main diagonal to the right exceeds RAND * Tα, it gives a signal that the desired maximum is found and its value is β. This will be the first ascending range: α→β. The next step is to model the next minimum after β. The transition to the main diagonal is performed and the accumulated sum of the parts to the left of the main diagonal T_β is calculated. The newly drawn random number RAND (it will be, of course, the different value), and the RAND * T_β is computed. Similarly, when the sum of frequencies in the line of β starting from the main diagonal to the left will exceeds RAND * T_β, it will determine the following the minimum and its value will be γ. After this step the simulation continues up to the required length of realization z_1, i.e. turning points number.

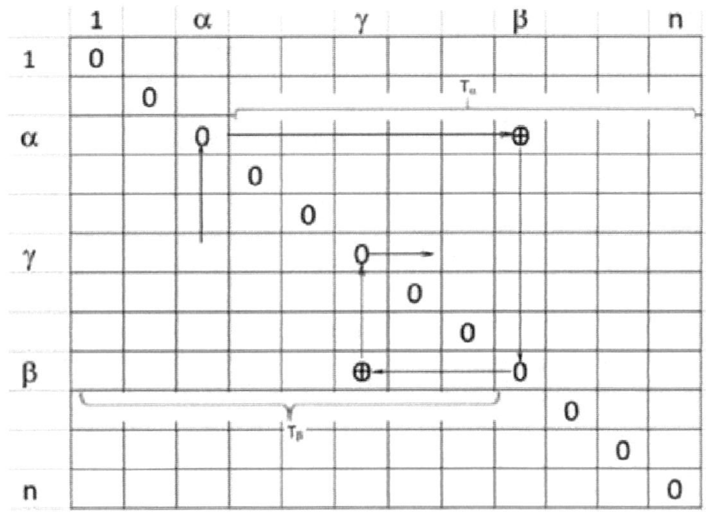

Figure 4. Synthesis stage for modeling the replicas

4. Results and discussion

In accordance to the scope of the paper, that is to create a method for simulation the similar, but varied in some detail's replicas of the initial random process for metal fatigue testing and design, some replicas were modelled. Some important characteristics of realizations were compared and their scatter was analysed. The analysis was performed on the base of stress realization of the part of the transport machine.

The Markov's matrix was filled up accordingly 3.2. Nine replica realizations (1, 2 .... 9) were modelled using the method described in 3.3. The pictures of some replicas processes are shown in figure 5.

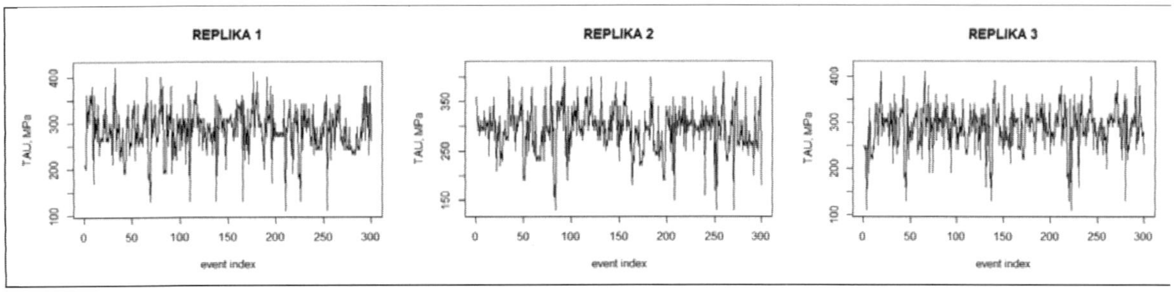

Figure 5. Some target replicas

For the estimation of the loading block, the cycle counting procedure, namely the rain-flow method [8, 12] has been applied. In Table 1 the values V (fullness ratio of the rain-flow distribution) is also shown. Fullness ratio with taking into account the necessity of correction on linear hypothesis of fatigue damage summation [1] (Vcor<1) [18] serves as the equivalent amplitude measure for non-stationary random distribution and is critical for fatigue durability estimation.

$$V_{cor} = \sqrt[m]{\frac{0.25}{h_{sum}} \sum h_i \left(\frac{\sigma_{ai}}{\hat{\sigma}_a}\right)^m} \quad (2)$$

In (2) m- is the fatigue exponent, h_{SUM}– is the total number of cycles in the block; h_i- is the number of cycles at the i –th step; σ_{ai}– is the current value of the stress amplitude at the i –th step; $\hat{\sigma}_a$- is the maximum amplitude in the block.

With known data of fatigue curve of the machine part, known $\hat{\sigma}_a$ and V, it is possible to estimate medium longevity L [cycles], with applying the corrected linear hypothesis of fatigue damage summation, it is possible to estimate the number of blocks until failure λ_b and later on to estimate The longevity L:

$$L=\lambda_b \cdot h_{SUM}$$

The parameters of a convenient fatigue curve were taken arbitrary for the sake of comparison of the results.

Because the longevity of the machine parts is known to be distributed logarithmically normal, the distribution of the random value of decimal logarithm of L, namely, lgL was investigated on the base of the set of replicas. The mean value $lg\ L$ = 0.92; standard deviation $lg\ L$ =0.12 and variation factor $lg\ L$ =0.119. The distribution of lgL is shown in figure 6. This graph shows theoretical and sample quantiles. It can be seen, that the distribution is close to a normal one.

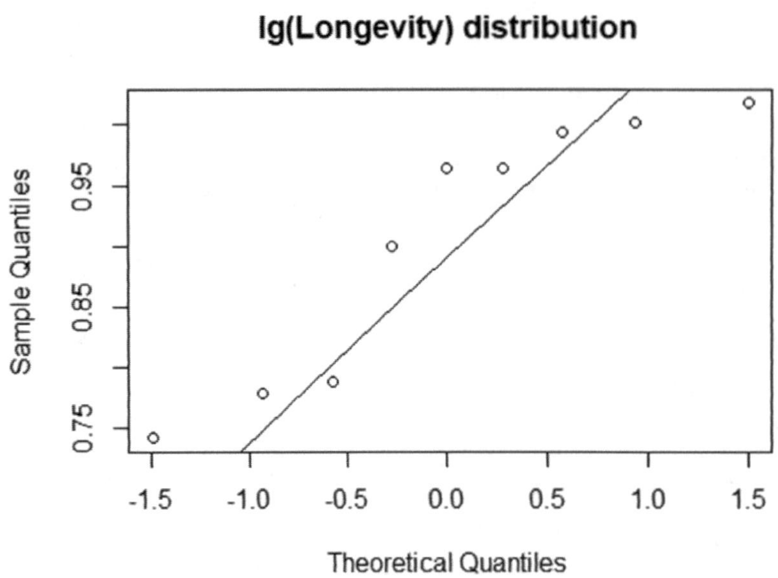

Figure 6. Variability analysis of the longevities estimated by replicas

In Table 1 the summary of initial ("init") process and nine target replicas with the most important characteristics for fatigue is presented.

Table 1. Some characteristics of the replicas

Realization and replicas	$\hat{\sigma}_a$	V	IF	L	Lg L
init	13.5	0.546	0.433	12.44	1.091
1	14.5	0.527	0.497	10.04	1.002
2	15.5	0.537	0.503	6.01	0.779

Realization and replicas	$\hat{\sigma}_a$	V	IF	L	Lg L
3	14.5	0.548	0.493	7.94	0.900
4	15	0.511	0.557	9.86	0.994
5	14.5	0.572	0.53	6.14	0.788
6	14.6	0.535	0.633	9.20	0.964
7	14.5	0.535	0.50	9.22	0.742
8	15.5	0.545	0.57	5.52	1.017
9	14.5	0.524	0.52	10.41	1.017

As can be seen from the Table 1 and Fig.6, although the replica processes demonstrate some scatter in their characteristics, in general, the data are close to the initial ones. Because of the proper choice of physical model, one can expect that this scatter might characterize the scatter of the longevity in exploitation.

5. Further directions for the research

As it was mentioned in Introduction, many investigations have shown, that the consequence of the events might have influenced on the longevity at both stages of fatigue (crack initiation and crack propagation).

It is known that the durability can be increased by residual compressive stresses, which are induced in various ways. Sometimes they are intentionally being produced by special technologies, like shot peening. It worth mentioning, that the impact of the load sequence also needs consideration, especially when the number of cycles to failure is not so big – that is called low cycle fatigue. Some methods, like described in [19] consider the load sequence, because it influences for the estimation of mean stress effect.

The same situation even in the greater degree is true at crack propagation stage. For the crack propagation estimation, the special methods were developed [20-23]. The idea of the models is to calculate effective values of the stress intensity factors (SIF) for the cycles that follow after overloads (i.e. for large-scale cycles, occurring at some segment of the construction). Physically these models based on the formation of a large area of plastic deformation in the top of the crack because of the overload, which slows down the crack growth. In this case, the following after overload cycles form the next zones of plasticity, which are smaller in size. And before crack does not "come out" of the zone of plasticity formed by overload, the growth rate decreases significantly because of the residual compressive stress. These effects are taken into account by the so-called effective values of SIF further are applied in the kinetic diagram of fatigue crack growth.

As it is seen from that short explanation, for both fatigue stages the consequence of the events plays important role. The experiments have shown that under the same distribution of rain-flow cycles the situation with high cycles appearing first will be preferable for slowing down the metal fatigue process. There is a need to create some criteria, which indicate the distribution of the biggest maxima among the realization. In this way, it can be summarized, that there is a need for creating some process parameter, which will show the distribution of overloading during the testing time.

The other fact, which should be considered while modeling by Markov matrix, is the interaction (aftereffect) of the random events of the turning points. It follows from the fact, that the energy is finite, and it is impossible to produce the great leaps from one point to another in no time.

All calculations, as well as simulation programming, were performed in freely distributed programming language and free software environment for statistical computing and graphics program R [24].

References

[1] Kogaev V P 1993 Strength design under non-stationary stresses (Moscow, Mashinostroenie) p 363

[2] Gubenko S I. Ivanov I A and Kononov D P 2018 *Zavod. Lab. Diagn. Mater* 3 The impact of steel quality on the fatigue strength of wrought wheels pp 52-60

[3] Per Andreas Brodtkorb 2000 *WAFO- Paper no.ISOPE -GFC0-02* A Matlab toolbox for an analysis of random waves and loads www.researchgate.net/publication/282371975

[4] Erpalov AV and Shefer L A 2016 *Procedia Engineering 2nd International Conference on Industrial Engineering (ICIE-2016)* Fatigue-based Classification of Loading Processes **150** pp 144-149

[5] Syzrantseva K, Syzrantsev V 2017 *Key Engineering Materials* Determination of parameters of endurance limit distribution law of material by the methods of nonparametric statistics and kinetic theory of high-cycle fatigue **736** 52 DOI: 10.4028/www.scientific.net/KEM

[6] Makhutov N A, Reznikov D O 2019. Assessment of large-scale catastrophes in complex engineering systems *IOP Conference Series: Materials Science and Engineering*, Volume **481** 012002 IOP Publishing. doi:10.1088/1757-899X/481/1/012002 pp 1-8

[7] Hossain I, Velkin V I, Shcheklein S E and Eliseev A V. 2019 Safety 2018. *IOP Conf. Series: Materials Science and Engineering* 481 Structural Design Development of a Float Type Wave MicroPower Plant. IOP Publishing doi:10.1088/1757-899X/481/1/012007 pp 1-6

[8] Fuchs H O, Stephens R I 2002 Metal fatigue in engineering (John Wiley&Sons. New York) p 318

[9] Miner M.A. 1945 *Journal of Applied Mechanics, Trans. ASME* Cumulative Damage in Fatigue **12**

[10] Gassner E. 1939 Dt. Luftwacht, Ausg. Luftwiss **6** pp 61-64

[11] Benasciutti D, Tovo R. 2018 *Mat.-Werkstofftech* Frequency-based analysis of random fatigue loads: Models, hypotheses, reality **49** pp 345-361

[12] GOST 25.101-83 2005 Strength calculation and testing. Representation of random loading of machine elements and structures and statistical evaluation of results (Moscow: Standartinform) p 25

[13] Gadolina I and Zaynetdinov R 2019 *IOP Conference Series: Materials Science and Engineering* Advantages of the rain-flow method at the post-processing stage in comparison with the spectral approach **481** 012005 IOP Publishing. doi:10.1088/1757-899X/481/1/012002 1 pp 1-9

[14] Fisher R, Haibach E. 1983 in book: Behavior of steel under cyclic loads ed Dahl V Modeling loading functions in experiments on the evaluation of materials pp.368-405

[15] Gadolina I V, Monahova A A, Petrova I M. and Serjabrjakova I L 2016 *Proc. of the Conference "Survivability and Structural Materials Science. SSMS* LABORATORY AND SIMULATION EXPERIMENT UNDER RANDOM LOADING AS THE ALTERNATIVE TO TESTING UNDER REGULAR LOADING pp 4-9

[16] Ling Y et al 2011 *Int J Fatigue* Stochastic prediction of fatigue loading using real-time monitoring data doi:10.1016/j.ijfatigue

[17] Dubin D A, Nakaznoi O A, Smirnov I A and Shleev A. N. 2016 *Trudy NAMI* Experimental determination of kinematic and power parameters of loading of system elements suspension of high-speed tracked vehicle **266** pp 45-53

[18] Savkin A N, Sedov A A, Boyko G V and Badikov K A 2017 *AIP Conference Proceedings 1909, 020187* Estimation of contribution of different parties of random loading spectra to damage accumulation https://doi.org/10.1063/1.5013868

[19] Martin J E, Topper T H, Sinclair 1 G M 1971 *Material Research and Standards, MTRSA* Computer Based Simulation of Cyclic Stress-Strain Behavior with Applications to Fatigue **11** pp 23-28
[20] Schijve J, Jacobs P A, Tromp P J 1976 *NLR TR 76065* The effect of load sequence on fatigue crack propagation under random loading and program loading
[21] Svirsky Yu A., Basov V N 1994. *1 st International Conference FUNDAMENTAL RESEARCH IN AEROSPACE SCIENCE* Accounting for non - linearities of damage accumulation in aircraft structures under irregular loading Russia. (Zhukovsky TsAGI) pp 93-96
[22] Willenborg J, Engle R H, Wood H A 1971 *AFFDL-TM-71-1 FBR, WPAFB* A Crack Growth Retardation Based on an Effective Stress Concept
[23] Romanov A N, Nesterenko G I and Filimonova N I 2018 *Journal of Machinery Manufacture and Reliability* Damage Accumulation under Variable Loading of Cyclically Hardening Material at the Stages of Formation and Development of Cracks **47** pp 414-419. DOI: 10.3103/S1052618818050102
[24] R Development Core Team 2018 R: A language and environment for statistical computing. R Foundation for Statistical Computing Vienna Austria. ISBN 3-900051-07-0, URL http://www.R-project.org

IOP Conf. Series: Materials Science and Engineering **972** (2020) 012007 doi:10.1088/1757-899X/972/1/012007

The Influence of Low-Frequency Seismic Phenomena Effects on The Process of Boiling Up the Coolant

I Hossain[*1], P G Sharipov[1], V I Velkin[1], S E Shcheklein[1] and A M Dubinin[1]

[1]Ural Federal University, 19 Mira str, 620002, Yekaterinburg, Russia

*E-mail: hossain.ismail44@yahoo.com

Abstract. This paper studies the low frequency seismic phenomena on the overheating boiling process and the cavitation state of the coolant in a small volume of system. This experiment is basically the imitation of the situation in a nuclear reactor core where distilled water was used as coolant during the experiment. Temperature of the coolant was recorded during the normal state and the time of implementing external low frequency. The stretched state of the liquid due to overheating and the cavitation destruction process were investigated.

Keywords: seismic phenomena, cavitation, overheating, coolant, boiling

1. Introduction

Overheating of the coolant in the absence of circulating cooling process can cause serious damage to the nuclear fuel. When a liquid is heated at a constant temperature or depressurized by static or dynamic methods at a constant temperature, steam bubbles or vapor-filled cavitation bubbles appear and develop over time. This bubble ends up with an explosion known as "steam explosion". After severe earthquake and tsunami Fukushima Dai-Ichi nuclear plant faced a problem of core cooling as the circulating cooling water supply was not available. Even after shutdown right after earthquake reactor cores still continued generating heat and overheating caused serious damage to the fuel rods [1]. Liquid -vapor phase transition is a very common topic in technology and scientific research. Liquid-vapor phase transition leads to the metastable state of the fluid in the presence of continuous increment in heat capacity. This cavitation process of intensive phase transition and steam explosion can cause a great deal of noise, damage to components, vibrations, and a loss of efficiency. In renewable energy sector cavitation also has a detrimental effect on the blade surface of tidal stream turbines [2]. Vibrations have some kind of damaging and negative impacts on machineries and instrumental technology. According to Hou et al. (2013) all of the key performance factors of a fuel cell were negatively affected by vibration, including a 56% increase in ohmic resistance [3]. Thermal performance of heat pipes has been investigated by Prisniakov et al. (2002) They found that while applying 10 Hz–100 Hz frequencies of 3 μm–5 μm displacements the heat transfer coefficients were increased by as much as 5% to 30% [4]. Cavitation can cause extreme damage to mechanical equipment such as impeller [5] as static pressure falls below the vapor pressure of fluid at the prevailing temperature [6]. Any kind of external vibration influences on the overheating liquid section of the vessel which can be under risk and security issues in some extent.

2. Study of Cavitation process and stretched fluid phenomena

Cavitation effects usually happen in a very small amount of gaseous or liquid media with high concentration of energy in a small volume of content. Liquids at negative pressure overheating stretched condition suddenly generate cavitation bubble and ends up with an explosion. Devin, C. (1959), Flynn, H.G. (1964), Neppiras, E.A. et al. (1951) studied on cavitation theory, acoustic cavitation in liquids and bubbles in water [7-9]. Energy accumulates when the bubble expands from the equilibrium radius R_0 to the maximum radius R_{max} under the influence of tensile forces arising in the liquid in the rarefaction phase of the acoustic wave. The stored energy can be estimated as,

$$W = \Delta V \cdot P_0 \quad (1)$$

Where ΔV is the change in volume, when the bubble radius decreases from R_{max} to R_{min}, P_0 is the pressure in the surrounding fluid, which during acoustic cavitation can be assumed to be equal to the static pressure. If we accept the condition, $R_{max} >> R_{min,}$ which usually takes place during cavitation, then for energy we get,

$$W = \frac{4}{3}\pi R^3_{max}.P_o \quad (2)$$

The phenomenon of energy cumulation by a cavitation bubble at the qualitative level can be easily explained. The effect is that the initially stored energy is first transformed into the kinetic energy of the liquid, whose spherically symmetric movement towards the center of a bubble causes an infinite (for a vacuum bubble) increase in the rate of compression, which is expressed in the appearance of singularities in solutions of hydrodynamic equations. Of course, physically infinite speeds Compression techniques are not realized, which is hampered by many factors, the most important of which are the violation of sphericity of motion, heat and gas flow between gas and liquid. For the first time, a mathematical description of the compression process of such a bubble was obtained by Rayleigh. He investigated the dynamics of an empty (vacuum) bubble, Moreover, his model took into account only inertial forces and neglecting viscosity, surface tension of the liquid and pressure of the vapor-gas mixture inside the bubble and had the form,

$$R\ddot{R} + \frac{3}{2}\dot{R}^2 = -P_0/\rho \quad (3)$$

under initial conditions:

$$R(0)=R_0\ ,\ \dot{R}(0)=\dot{R}_0$$

where R (t) is the current radius of the bubble, and the points denote the derivatives with respect to time t, P_0 is the static pressure in a liquid, ρ is its density.

However, the thermodynamic parameters of the medium at the time of bubble collapse reach extremely high values. Apparently, it should be considered experimentally proven that the temperature in a bubble can reach to extremely high range [10-12]. The situation is even more complicated if the processes of heat and mass transfer between gas and liquid are taken into account as well as the compressibility of the liquid which leads to losses energy of a bubble with shock waves. Scripov P.V. mentioned in his book "Metastable state of liquid", film boiling stops when the temperature head to $\Delta t_{kp2} = t_c - t_s$. It turns out to be equal to or usually somewhat lower than the temperature head corresponding to the limiting overheating $\Delta t_n = t_n - t_s$

Therefore,

$$\Delta t_{kp2} = C\ \Delta t_n \quad (4)$$

Where, the coefficient C usually lies in the range of 0.8-1.0

At higher surface temperatures ($t_c > t_n$), liquid cannot contact the heating surface, since when it approaches the surface its spontaneous density of the warm flow occurs when the film boiling mode stops,

$$q_{kp2} = \alpha \, \Delta t_{kp2} \tag{5}$$

where α is the heat transfer coefficient in film boiling mode.

Figure1. shows the dependence $t_n = f(p)$ for water [13]. This figure also shows the Saturation line $t_s = f(p)$ of water. A characteristic feature of the dependence $t_n = f(p)$ is that it is close to a straight line, which ends at the critical point of the state of matter. Limits of temperature values for some liquids are represented in the Table 1 [13].

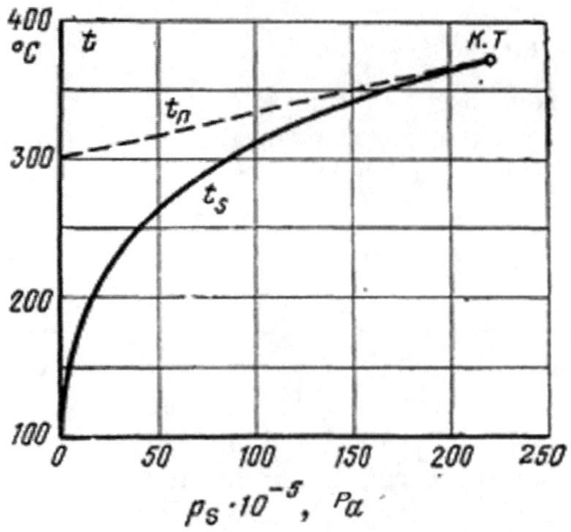

Figure 1. Dependence of the temperature of limiting overheating for water (t_n) vs Pressure (P)

Tabel 1: Limits of temperature values for some liquids are represented

Liquid	t_s (°C)	t_n (°C)	Liquid	t_s (°C)	t_n (°C)
Ethanol	78.3	195	Benzene	80.1	226
Methyl Alcohol	64.5	190	Pentane	36.1	147
Acetone	56.1	181	Hexane	68.7	182
Diethyl ether	34.5	144	Heptane	98.4	215

3. Experimental Methodology

In this experiment we used some beakers as our experimental vessel volume. Pure distilled water was used as working fluid. A heater was set to produce continuous heating. A thermal imager was used to measure and visualize temperature scenario. To record the bubble creation and steam explosion a high-resolution camera was rolling during the whole process of experiment Figure 2.

In order to reduce the error and unusual shortcomings we had to use some chemical mixtures to clean up the experimental vessels. Aqua regia is a mixture of nitric acid and hydrochloric acid, optimally in a molar ratio of 1:3. It is a yellow-orange fuming liquid with high dissolving property. This mixture was boiled in the vessel under 100°C temperature for at least 10 minutes to clean up the vessel thoroughly. Two of the vessels were used to clean up with Aqua regia. Image of the boiling process with chemical mixture is presented in figure 3. Another kind of chemical mixture called Chromic mixture which is a mixture of concentrated sulfuric acid and potassium dichromate was also used to clean up rest of the vessels under temperature 80°C for 10 minutes. Beakers used as experimental vessel are shown in figure 4.

IOP Conf. Series: Materials Science and Engineering **972** (2020) 012007 doi:10.1088/1757-899X/972/1/012007

Figure 2. Image of the experimental section consisting a heater, thermal imager and a high-resolution camera

Figure 3. Cleaning up the vessel with chemical mixture

Figure 4. Beakers used as experimental section

Right after cleaning process with chemical mixtures all the vessels were rinsed and filled with distilled water to set for experiment. The vessel with distilled water was heated homogeneously. Temperature was measured after a certain interval of time and the experimental section was observed attentively. At the same time a high-resolution camera was set to record every detail. External low frequency was applied to implement the seismic phenomena on the experimental section.

4. Results and Discussions

The occurrence of overheating is most likely in the channels of small effective cross section, where there is an additional effect of Laplace forces, which interfere with active vaporization. From the experiment we can see that implemented external low frequency has an impact on the overheated liquid. As we can see, when the coolant came close to boiling temperature due to impurities and non-uniform surface it generates unusual bubbles inside the vessel and the temperature goes down for a moment shown in line figure 5 vessel 2. But incase of vessel 1 the scenario was different. When the coolant came to boil at over temperature 100°C the temperature was rising higher shown in figure 5 vessel 1. The temperature went up to 108°C over the boiling temperature. The zone above the boiling temperature is called Overheating zone.

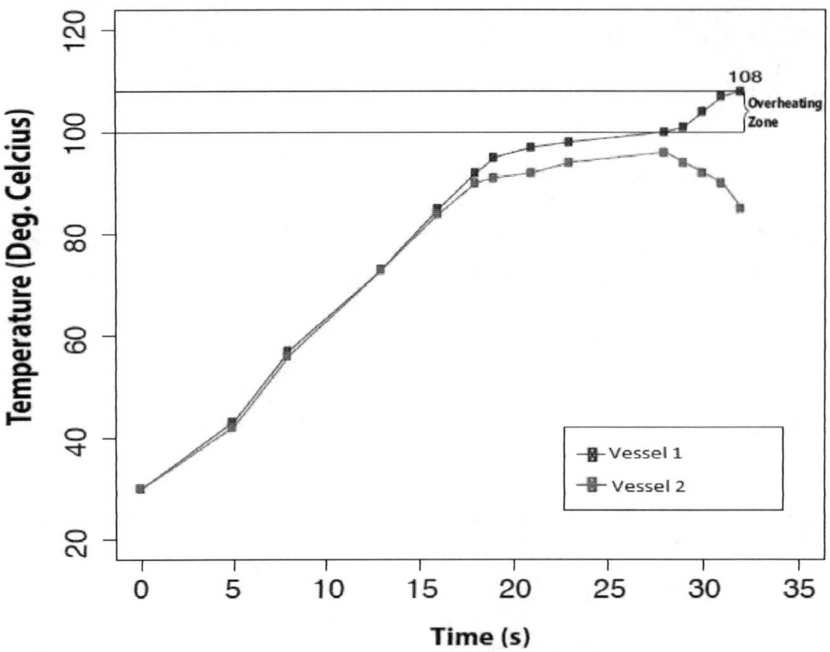

Figure 5. Comparison of temperature and influence of external low frequency in the two vessels of experimental section

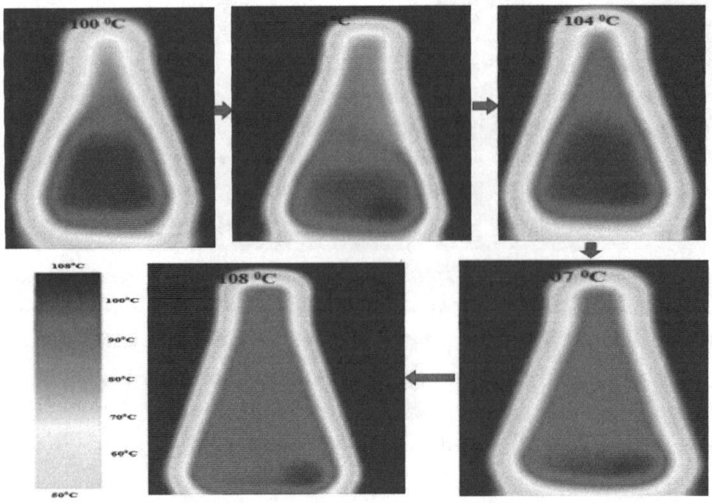

Figure 6. Recorded temperature in the overheated zone

We recorded the temperature with a thermal imager. At this moment we implemented low frequency on the experimental section in case of vessel 1 which lead to a subsequent sharp increase in temperatures and bubble inside ended up with steam explosion. The process was recorded with video camera and the image is presented in figure 6 and 7.

Figure 7. Cavitation process of liquid due to overheating captured with a high-resolution camera

In this experiment we used five pieces of vessels. For the cause of some short comings we could not be successful for all the test sections. Due to the nonuniformity of the side wall, surface and some kind of impurities unexpected bubbles were arising during the experiment and temperature fell down unexpectedly. Chemical material Chromic mixture which is a mixture of concentrated sulfuric acid and potassium dichromate showed better result in case of cleaning up the vessels.

5. Conclusion

Steam explosion due to cavitation in the liquid has a very damaging effect on some mechanical equipment as well as on the reactor core in the absence of circulating cooling process. It can damage fuel rods and lead to core melt down. From the experiment we can conclude that external frequency has a serious impact on cavitation process. It can invoke the process and initiate to explode inside the vessel. Overheating in the coolant generates steam vacuum bubbles and ends up with steam explosion.

References

[1] Mark Holt, Richard J. Campbell, Mary Beth Nikitin 2012 Fukushima Nuclear Disaster (Congressional Research Service)

[2] Buckland H C, Masters I, Orme J A C, Baker T 2013 Cavitation inception and simulation in blade element momentum theory for modelling tidal stream turbines *Journal of Power and Energy* **227** (**4**): 479–485. doi:10.1177/0957650913477093

[3] Hou Y, Hao D, Shen C, Shao Z 2013 Experimental investigation of the steady-state efficiency of fuel cell stack under strengthened road vibrating condition *Int. J. Hydrogen Energy* **38** pp 3767–3772.

[4] Prisniakov K V, Nikolaenko Y E, Prisniakov V F 2002 Vibration actions on heat pipes as cooling element of electronic systems *Proceedings of the THERMES conference* (USA: Santa Fe, New Mexico)

[5] Dular M, Bachert B, Stoffel B, and ˇSirok B 2004 Relationship between cavitation structures and cavitation damage *Wear* vol. **257**, no. **11** pp. 1176–1184

[6] Brennen C E 1994 Cavitation and Bubble Dynamics (New York: Oxford University Press)

IOP Conf. Series: Materials Science and Engineering **972** (2020) 012007 doi:10.1088/1757-899X/972/1/012007

[7] Devin C 1959 Survey of thermal, radiation, and viscous damping of pulsating air bubbles in water *J. Acoust. Soc. Am.* **31** 1654–1667.

[8] Flynn H G 1964 Physics of acoustic cavitation in liquids. Physical Acoustics (Academic Press)

[9] Neppiras E A and Noltingk B E 1994 Cavitation produced by ultrasonics: theoretical conditions for the onset of cavitation. *Proc. Phys. Soc.* (London) 1032–1038

[10] Simon G and Levinsen M T 2003 Parametric dependence of singlebubble sonoluminescence spectra. *Phys. Rev. E.* Vol **68** 046307.

[11] Toegel Rand Lohse D 2003 Phase diagrams for sonoluminescing bubbles: A comparison between experiment and theory *J. Chem. Phys.* Vol **118 № 4** pp 1863-1875

[12] Ruuth S J, Putterman S, and Merriman B 2002 Molecular dynamicssimulation of the response of a gas to a spherical piston: Implications for sonoluminescence *Phys. Rev. E.* Vol **66** 036310.

[13] Scripov V P, Snitsin E N, Pavlov P A, Ermakov G V 1980 Thermophysical properties of liquids in a metastable state (Moskow: Atomizdat) 208 p

Investigation of thermal characteristics of multitubial thermosyphon heat-exchangers

L O Yakovlev[1], V A Klimova[1], S E Shcheklein[1]

[1]Ural Federal University, 19 Mira, 620002, Yekaterinburg, Russia

e-mail: leo00@mail.ru, v.a.klimova@urfu.ru, s.e.shcheklein@urfu.ru

Abstract. Prospects of application and advantages of heat-exchangers based on sealed two-phase thermosyphons are compared with analogies of alternative widespread designs. Theoretical limitations of heat transfer are considered. Experimental facility requirements were formulated. Heat transfer coefficients obtained by calculation, computer simulation and in experiments.

Key words: thermosyphon heat-exchanger, heat transfer coefficient, heat transfer efficiency improvement, experimental research, computational fluid dynamics.

1. Introduction

Although the subject has been well-studied in Russia and abroad, an in-depth investigation of thermosyphon heat exchanges application as a part of heat exchange modules within a temperature range used in utilities systems, heating and ventilation equipment such as domestic hot water (DHW) temperature boosters has, nevertheless, not been conducted. Currently, the plate-type and shell-and-tube heat-exchangers have been used mainly as DHW temperature boosters. The operating theory behind those devices is well-developed and design is close to perfect with the best models achieving a heat transfer coefficient of around 800 $W/(m^2 \cdot K)$. Such high performance is usually achieved by increasing the medium flow velocity in the apparatus channels by decreasing their cross-dimension and that predetermines their main weakness – high sensibility to the water quality. Most often cold water supplied into the apparatus for heating is not treated and is saturated with salts and impurities. That leads to intense scaling, increased hydraulic resistance and lower heat exchange efficiency. Hence the heat-exchangers need purging up to 4 times a year to remove sediments and that is very a labor intensive and costly process which sharply increases running costs. Thermosyphon heat exchange design is free from these limitations and:

- An intermediate heat carrier boiling temperature can be set equal to a calculated temperature of the heated medium (for example 65°C for DHW) and that provides a sharp drop in scaling by bringing the temperature of the wall from the cold medium side down to a significantly lower level than the boiling temperature;
- Double medium separation provides increased reliability;
- High heat-transfer coefficient is maintained by small temperature drop;
- Heat carrier temperature control and adjustment are not required;
- Much more simple purging procedure allows to reduce operating costs.

IOP Conf. Series: Materials Science and Engineering **972** (2020) 012008 doi:10.1088/1757-899X/972/1/012008

2. Thermophysical limitations of heat transfer in the heat-exchanger

The main physical parameters of a vacuum thermosyphon qualitatively affecting its main characteristic (a power transfer) are [1]:

- Diameter to length of evaporating part ratio d/l_e which determines the size of space constraint influencing the boiling process;
- Working fluid (intermediate heat carrier);
- Thermosyphon inner pressure;
- Filling degree of a heating part.

There are also external factors influencing the heat-exchanger efficiency along with the internal features of heat-and-mass transfer inside the thermosyphon:

- Medium temperatures;
- Velocity and direction of flow around the tubes;
- Flow turbulization level.

Research was conducted on stainless steel thermosyphons having outer diameters of 20 mm and wall thicknesses of 1.5 mm. The assembly was vertically orientated and consisted of 4 working thermosyphons and 10 simulators made of heat insulating materials. The thermosyphon length was 1 m, consisting of an evaporating section of 0.7 m and a condensation section of 0.3 m. The absolute pressure inside the thermosyphon was 17 kPa and the calculated boiling temperature was 56°C.

2.1. Limiting heat flux in the thermosyphon' evaporating section

The functional connection between a limiting heat flow and a heating part geometry is well studied and validation is given in [1] based on our own results as well as that of others [2-6] and can be described as

$$q_F^l = 0{,}16 \cdot r \cdot \rho_S^{0{,}5} \cdot \left(\sigma \cdot g \cdot (\rho - \rho_S)\right)^{0.25} \cdot \left(1 - e^{-\frac{d}{l_e} \cdot \left(\frac{\rho}{\rho_S}\right)^{0.13}}\right)$$

where q_F^l – limiting heat flux through the thermosyphon heating surface, r – heat of evaporation, ρ_S – steam density, ρ – liquid density, σ – surface tension coefficient, g – gravity acceleration, d – thermosyphon diameter, l_s – length of thermosiphon evaporating section.

Thus, the limiting heat flux in the thermosyphon evaporating section with given parameters is q_F^l= 51.6741 kW/m^2, with a maximum power output of 1,955 kW for a single thermosyphon and $Q^e_{l\Sigma}$ = 7.822 kW for the thermosyphon assembly.

2.2. Maximum power output of the thermosyphon' condensation section

Following the conclusion [1] made by analysing a number of previous publications it is possible to apply Nusselt's Theory of Film Condensation with provision for the Reynolds number for calculating the heat transfer coefficient during condensation inside a thermosyphon.

An average heat transfer coefficient is α_c = 6053 W/(m^2·K) which agrees with the results obtained by other researchers [7].

Hence, the maximum power output of the thermosyphon condensation section will be 5.9 kW while the maximum power output of the thermosyphon assembly will be $Q^c_{l\Sigma}$ =23.6 kW.

However, according to more recent data, a condensation process in a thermosyphon takes place under different conditions than those occurring under Nusselt's experiments, and that leads to a lowering of the local Nusselt's number by almost an order of magnitude [8]. An acceptable theory to describe that process has yet to be developed so we applied a general conclusion and used Nu and consequently α_c and Q_c as in [8].

Then for the assembly under consideration we have α_c = 605 W/(m^2·K), Q_{c1} = 0.59 kW, $Q^c_{пр\Sigma}$ = 2,36 kW.

2.3. Maximum power output of thermosyphon by external limitations

The internal process parameters for the thermosyphons under consideration do not change, unlike external heat-exchange which is defined by the regime of flow i.e. by direction and velocity of flow at each surface point. Typical medium flow variants are:

- longitudinal medium motion along the thermosyphons (see Figure 1)

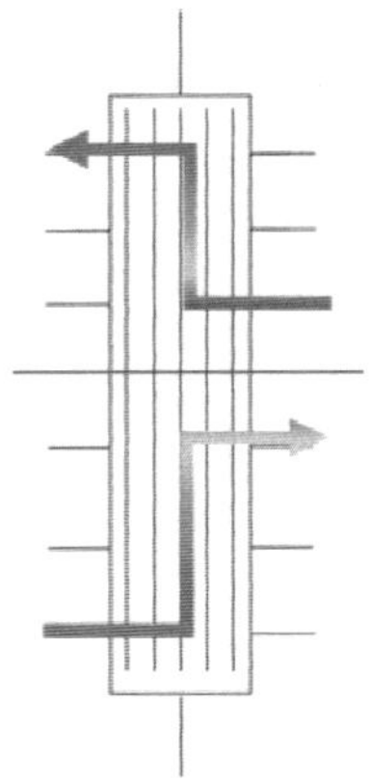

Figure 1. Diagram of longitudinal medium flow in thermosyphons.

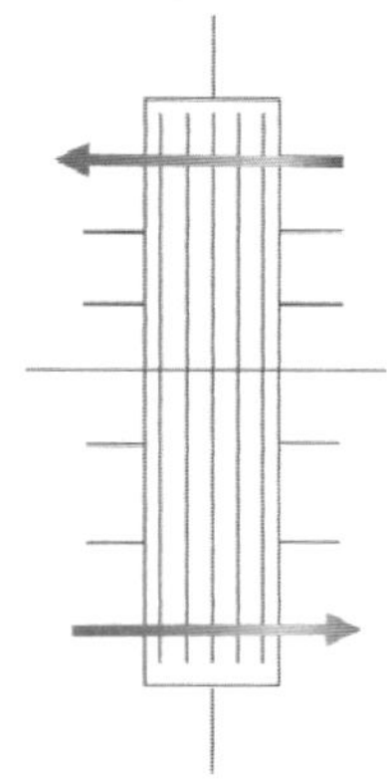

Figure 2. Diagram of cross flow in thermosyphons.

For that regime of flow, we can use the calculation method as for shell-and-tube heat-exchangers [9]:

$$\alpha = 1{,}16 \cdot \left(1210 + 18t_{av}^{int} - 0{,}038\left(t_{av}^{int}\right)^2\right) \cdot \frac{W_{int}^{0.8}}{d_{eq}^{0.2}},$$

where: 'int' is an intertubular space, d_{eq} – equivalent diameter of intertubular space, W_{int} – water velocity in intertubular space, t_{av}^{int} – average water temperature in intertubular space.

Table 1. Power output of a heat-exchanger with longitudinal medium flow.

Volume flow, L/s	Power of evaporating and condenser part			
	Q_e, W	Q_c, W	$Q_{e\Sigma}$, W	$Q_{s\Sigma}$, W
0.1	281.95	122.10	1127.78	488.40
0.2	490.90	212.59	1963.58	850.35

- medium motion across the tube bundle (see Figure 2)

In that case, heat transfer can be described by [7]:

$$\overline{Nu_{ld}} = 1.8 \cdot Re_{ld}^{\frac{1}{3}} \cdot Pr_l^{\frac{1}{3}} \cdot \left(\frac{Pr_l}{Pr_w}\right)^{\frac{1}{4}}$$

where the liquid is «l» and the thermosyphon wall is «w».

Table 2. Power output of a heat-exchanger with cross flow of medium.

Volume flow, L/s	Power of evaporating and condenser part			
	Q_e, W	Q_c, W	$Q_{e\Sigma}$, W	$Q_{s\Sigma}$, W
0.2	819.78	347.47	3279.10	1389.89

2.4. Computer simulation for the thermosyphon power output

FlowSimulation software was used for simulation under the following conditions:

- no temperature losses through the body frame;
- thermosyphons are simulated as rods of infinite heat conductivity.

Figure 3 shows velocities and trajectories of the liquid motion in the body of the heat-exchanger.

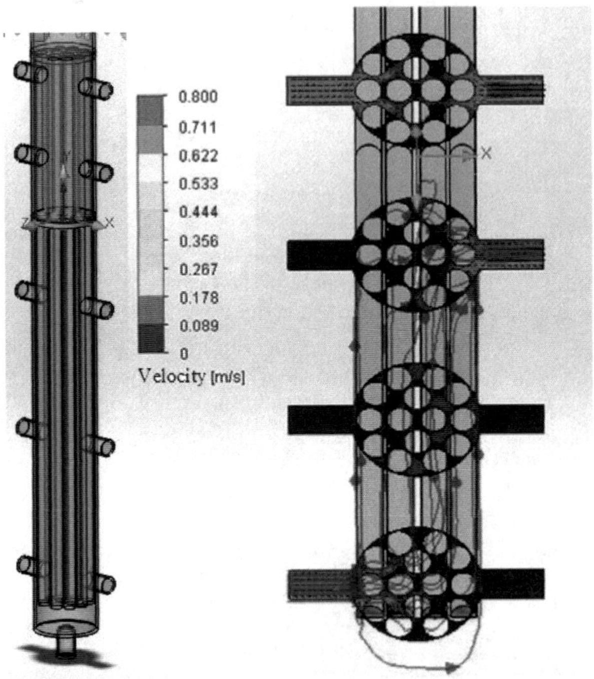

Figure 3. Simulation results at the volume flow rate 0.23 L/s.

Table 3 shows quantitative results of calculation of the heat transfer coefficient for thermosyphons with different orientation inside the heat-exchanger body.

Table 3. Simulation results for heat transfer coefficient, W/m^{2}K.

Tube №	Condenser min	Condenser Surface averaged	Condenser max	Evaporator min	Evaporator Surface averaged	Evaporator max
Flow 0.1 L/s						
1	26.3	409.2	3863	11,5	114.9	998.2
2	0.2	566.2	4015.7	12,3	145.3	4068.5
3	0	593.2	5681.6	7,9	149.3	2964.5
4	0	687.8	9143.7	16,1	153.05	1880.5
Average		**564.1**			**140.6**	
Flow 0.2 L/s						
1	63.9	584.8	3858.7	16.9	142.3	4220.7
2	0.6	983.4	5948.1	12.2	191.6	5210.2
3	21.9	894.4	6370.7	11.5	181.7	3986.7
4	1.86	1142.2	7583.1	15.8	190.8	2935.8
Average		**901.2**			**176.6**	

2.5. Experimental research of the heat-exchanger heat transfer coefficients and power output

Readings were taken under stable conditions; the duration of each measurement was 3 min and reading recording rates were one per second.

Figures 4 and 5 show pictures and diagrams of the thermosyphon assembly and test console.

Temperature was measured using platinum thermal elements of Pt100 type first class and registered by the hardware and software suite. Water flow from both heating and heated circuits side was measured by high accuracy water meters. The pressure in the thermosyphon was calibrated using test pressure gauges–vacuum meters.

a b

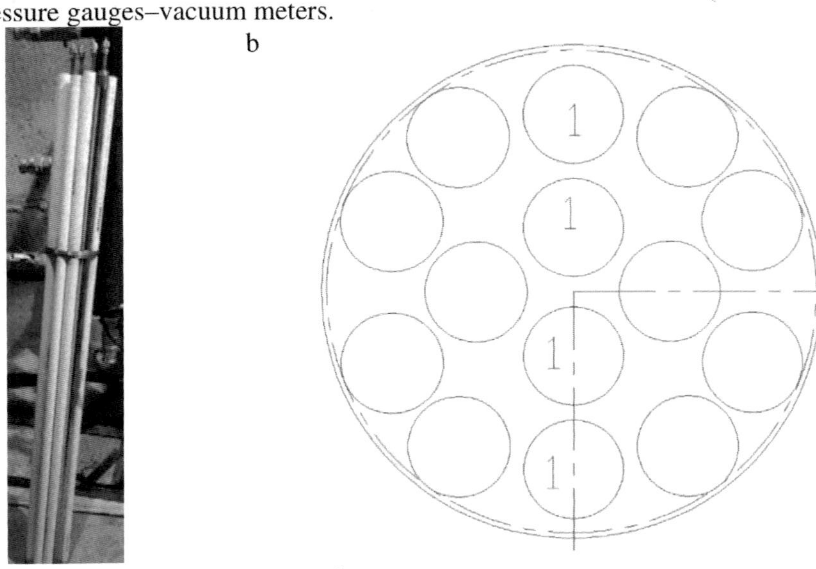

Figure 4. Thermosyphon assembly

a b

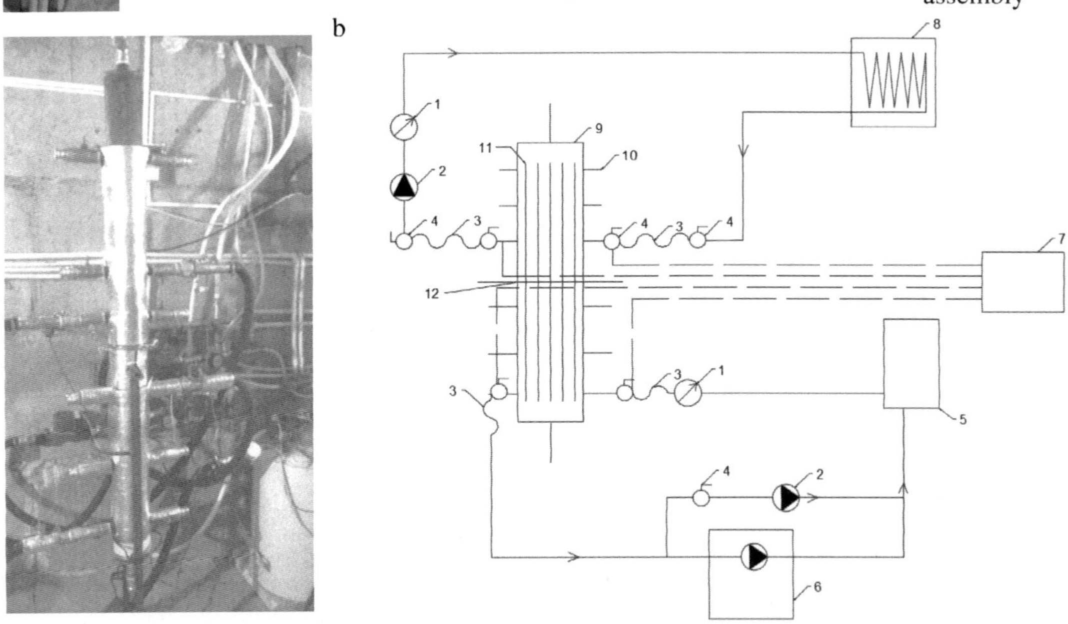

Figure 5. The process flow diagram of test console.
1 – Water flow meter, 2 – Circular water pump, 3 – Flexible hose with quick release connections, 4 – Ball tap, 5 – Insulated heat storage, 6 – Heat source (gas boiler), 7 – Hardware and software suite for registering temperature readings, 8 – Cooling system based on Aluminium radiators, 9 – Heat-exchanger body frame, 10 – Connection port, 11 – Thermosyphon (or simulator), 12 – Quick release body connection.

IOP Conf. Series: Materials Science and Engineering **972** (2020) 012008 doi:10.1088/1757-899X/972/1/012008

Table 4. Experimental values of temperatures, power output and heat transfer coefficients.

Flow, L/s								
	Evaporator	0.1	0.1	0.18	0.2	0.23	0.1	0.2
	Condenser	0.1	0.2	0.18	0.2	0.23	0.59	0.59
Average water temperature of evaporator, °C		67.68	65.65	63.73	62.58	68.02	64.88	63.84
Average temperature drop in evaporator °C		11.68	9.65	7.73	6.58	12.02	8.88	7.84
Average water temperature of condenser, °C		26.13	26.43	26.77	26.94	27.48	27.50	27.33
Average temperature drop in condenser, °C		29.87	29.57	29.23	29.06	28.52	28.50	28.67
Power output Q, kW		1.36	1.79	1.68	1.77	1.89	1.91	2.79
Heat transfer coefficient α, W/(m^2·K)	Evaporator	0.93	1.09	1.45	1.61	1.06	1.54	1.47
	Condenser	0.60	0.80	0.76	0.81	0.88	0.89	1.29

3. Conclusions

Table 5 shows a comparison of the thermosyphon heat-exchanger power output values obtained by calculation, computer simulation and experimental research.

Table 5. Summary table.

Flow, L/s					
	Evaporator	0.1	0.1	0.2	0.2
	Condenser	0.1	0.2	0.2	0.59
Power output, kW	With limitations in evaporator	7.822			
	With limitations in condenser according to [8]	2.36			
	With external limitations of evaporator	1.13	1.13	1.96	1.96
	With external limitations of condenser	1.08	1.39	1.39	2.04
	With computer simulation	1.27		1.77	
	With experimental research	1.36	1.79	1.77	2.79

Clearly, the thermosyphon heat-exchanger demonstrates high performance: heat-exchanger power output shows little dependency on temperature drop and goes over the theoretical limit calculated by using existing equations on external flow conditions. It also goes above the internal limitations described in [8] and matches the computer simulation results.

References

[1] Pioro I, Antonenko V, Pioro L 1991 Efficient heat-exhcangers with two-phased thermosyphons (Ukraine: Naukova Dumka) 245 p.

[2] Bezrodnyi M 1977 Affect of intermediate heat-carrier pressure on critical heat flow in evaporating thermosyphons *USSR. VUZ bulletin. Energetika,* pp 80-84

[3] Stoyanov N Critical heat flux inside closed evaporating thermosyphons 1987 *USSR Engineering and Physics Journal* **53** **№1**. pp. 20-26.

[4] Bezrodnyi M 1979 Investigation of maximum heat flows in tilting evaporating thermosyphons *USSR. VUZ bulletin. Energetika* **№4.** P. 110-112

[5] Savchenkov G, Kunakov V Research of critical heat flux inside low-temperature evaporating thermosyphons *1978 USSR. Engineering and Physics Journal*. **35**, **№3**. pp 397-404.

[6] Bezrodnyi M 1978 About upper limit of the maximum heat-carrier power of evaporating thermosyphons *USSR Teploenergetika Journal* **№8**. pp 63-66.

[7] Isachenko V, Osipova V, Sukomel A 1982 Heat Transfer: coursebook for universitites (Moskow: Egergoizdat) 416 p.

[8] Pashkevitch R, Muratov P 2014 Research on film-type condensation in vertical tubes with upwardy moving steam *Mining informational and analytical journal* (scientific and technical journal).

[9] SP 41-101-95. Boiler house design.

Evolution of the Safety of Fast Neutron Reactors

N N Oshkanov[1], S E Shcheklein[1]

[1]Ural Federal University, Mira 19, Ekaterinburg 620002, Russia

E-mail: nickolai.oshkanov@yandex.ru

Abstract. In its history, fast neutron reactors (BR) have gone through a long evolutionary period. However in general, BRs have gone the way of the evolutionary development. The function of retention radioactivity in case of an accident was performed by BRs safety vessels. For protection against external influences, new BR projects must be constructed containments. In the function of the reactor core cooling, changes have occurred related to the transition of the emergency cooling from the water circuit to the intermediate sodium circuit. Developments in emergency cooling systems are associated with the placement of an emergency heat exchanger inside the reactor, and the use of natural circulation in sodium and air. In the function on reactivity controlling due to the increase in core sizes, difficulties arise with the suppression of positive reactivity caused by sodium void reactivity effect. Therefore, in BR projects, the use of sodium plenum over the fuel part of the core and / or use of heterogeneous cores are needed. With an increase in the core sizes, the risk of jamming the long shifts of emergency rods increases. Therefore, it is advisable to transfer the safety rods to the hydraulic suspension in a sodium stream.

1. Introduction

Fast reactors (BR) from the first days of their existence, especially in the nuclear power plants (NPP) mode, showed sustainability of operation. Their attractiveness is conditioned by the possibility of making new fuel in excess of the consumed fuel, and a great safety potential {1-3].

Since the start of the first fast neutron reactor, the EBR-I (USA), whose energy was used to generate electrical energy, has so far built 17 NPPs with BRs. At present, there are three NPPs in operation (BN-600, BN-800, CEFR) - the rest have either completed their service life or have not been commissioned after construction. 9 BR projects are under study (Russia, France, India, Korea, Japan, China) [4].

Of course, at such a large interval, the safety of the BRs could not but undergo development [2, 5]. Fortunately, these changes were evolutionary in nature, which shows of the initially fundamental basis that was incorporated in the BR projects, starting with the first reactors:

- extremely low probability of destruction of the reactor core, and a small degree of consequences from such destruction
- no need to take measures outside the plant in any accidents
- minimum amount of radioactive waste with reliable treatment technology.

The undoubted world leader in the field of BRs is Russia, whose real results in the field of power reactors are indisputable (BN-350, BN-600, BN-800; projects of BN-1200, MBIR, BREST), which was emphasized at the IAEA conference in 2017 [1].

The most common is the integrated design (pool-type) BR with the location of the core and all the equipment of the primary circuit in one reactor vessel (Figure 1) [5]. All fundamental decisions are saved. In the main vessel of the reactor 10 all equipment of the primary circuits located. To eliminate possible depressurization of the main vessel, it is installed in the safety vessel 11. Core 8 is installed on the diagrid. Sub-assemblies of fuel pins (SA) are installed by bottom nozzles in the diagrid 9, from which sodium enters SA.

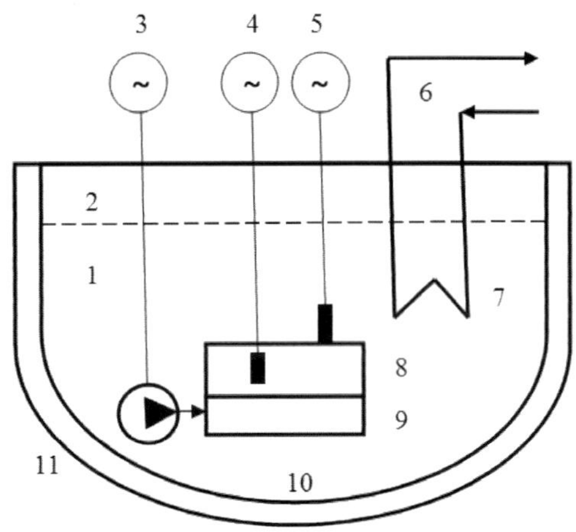

Figure 1. Structure principes of pool-type BR.

1 – sodium, 2 – gas, 3 – pump, 4 – control rod, 5 – shutdown rod, 6 – 2nd circuit, 7 – intermediate heat exchanger, 8 – core, 9 – diagrid, 10 – main vessel, 11 – safety vessel.

The operation of the core is carried out with the help of control rods 4. Emergency protection is carried out using emergency shutdown rods 5. Sodium 1 enters the diagrid with the pump 3. It heated in the core enters the intermediate heat exchanger 7, in which it heats the sodium of the second circuit 6. Sodium has a high boiling point, so it do not need to increase its pressure. To eliminate the entry of air into the reactor during the depressurization of its upper part above sodium, a gas plenum 2 of low pressure is.

The design has undeniable advantages due to factors such as:

- low pressure in the reactor vessel, and a short pipe length of the first circuit (low risk of depressurization);
- the presence of a safety vessel (no risk of loss of coolant);
- large thermal inertia (slow temperature rise in case of accidents);
- simplicity of the decay heat removal
- high boiling point of sodium coolant (no phase transitions during operation).

2. Safety

Consider the evolution of the BR in terms of basic functions:

- retention of radioactivity within the power unit
- the core cooling
- controlling of reactivity.

2.1. Retention of radioactivity within the power unit

This is the most important safety function, which determines the attitude of the whole world to nuclear power. BR has all the barriers to radioactivity established for all types of reactor, in addition they have additional barriers.

2.1.1. Reactor Barriers.

The first barrier is the fuel matrix. It holds solid radionuclides within a tablet with a high retention rate.

The second barrier is a fuel cladding that retains gaseous radionuclides, as well as solid radionuclides released from the tablet for any reason. The integrity of the fuel claddings is constantly determined by the necessary tools.

The radionuclides emerging from the fuel envelope, as well as sodium-activated radionuclides, are retained by the third barrier, the main reactor body.

The fourth barrier is the safety vessel, in which the main reactor vessel is placed In fact, it is the containment of the reactor, because sodium does not boil up due to its high boiling point and low pressure in reactor when sodium leaves the main body, and, as a result, a large volume of containment is not required.

The fifth barrier is the reactor shaft.

The sixth barrier has recently been a core catcher for molten fuel (if that happens).

The seventh barrier is the sodium coolant of the primary circuit. It acts only for the most dangerous of the radionuclides - isotopes of radioactive iodine, which when released in sodium combines with it. The resulting compound has a significantly lower radiation value.

2.1.2. Power unit barriers

The first NPPs with BR were limited to the non-proliferation of radioactivity only by the barriers specified in 2.1.1. It was believed that the safety vessel serves as a containment and no further radioactivity barrier is required. Therefore, the building of the reactor had a general construction purpose and was calculated on general production loads from external influences. We believe that all new BRs should have external containments designed for external influences similarly to other types of modern reactors. For this it is not necessary that they be dense, because during depressurization of BR, there is no pressure increase in the containment due to the absence of a sodium phase transition because the high boiling point of sodium. Recent BRs designs have such containments (Figure 2).

2.2. Core cooling

In normal operation of BR, the heat removal from the core is carried out using a multiloop design. Each loop (Figure 3) consists of a MCP-1 pump for pumping sodium through the core and an intermediate heat exchanger IXH, in which heat is transferred to the secondary coolant and then transferred to the steam generator SG, which produces steam for the turbine

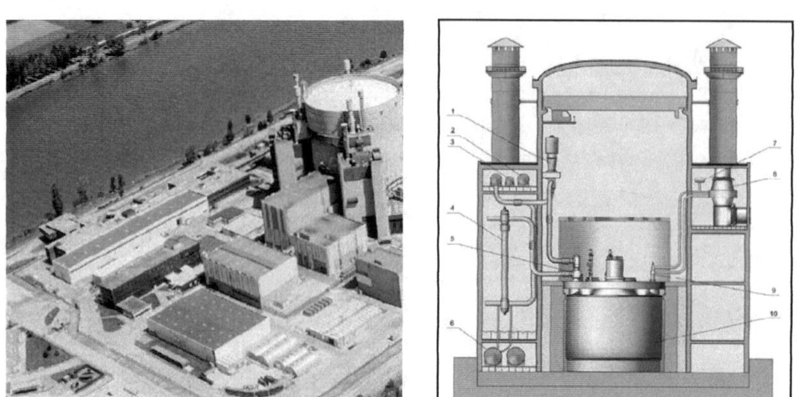

a) SUPERPHENIX (France) [6] b) Design (Russia).

Figure 2. External containment of BRs.

IOP Conf. Series: Materials Science and Engineering **972** (2020) 012009 doi:10.1088/1757-899X/972/1/012009

This design does not change its structure, starting with the first BR. It gradually evolved for emergency heat removal - from first using the water circuit (instead of the feed water pump FWP, an emergency water pump is used) before the heat removal into the air using a heat exchanger connected instead of a steam generator [8].

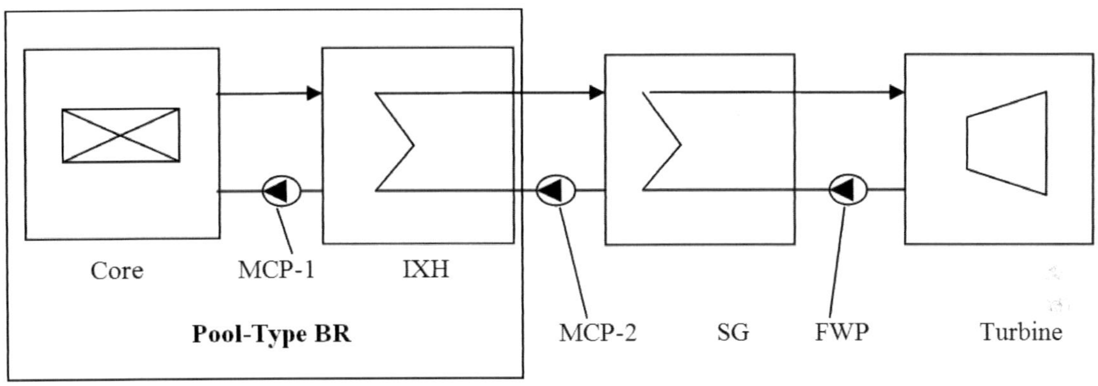

Figure 3. Diagram of a power unit with pool-type BR

However, the fundamental solution is to be applied in new projects related to the placement of emergency heat exchangers in the reactor, and the removal of energy release into the air through natural circulation (Figure 4) [7]. In this variant, there is no need to make switchings under which stress errors or equipment malfunction are possible.

2.3. Controlling reactivity

For controlling reactivity, the effects of reactivity and the reactor emergency shutdown system (reactor emergency protection) are important.

2.3.1. Reactivity effects

Reactivity as the degree of deviation of a reactor from a critical state is directly related to safety. Perturbations affecting reactivity (reactivity effects) excite various transients. The change in reactivity occurs when the change in temperature of the reactor elements, coolant flow, pressure in the gas plenum, the composition, volume and shape and other parameters of the reactor.

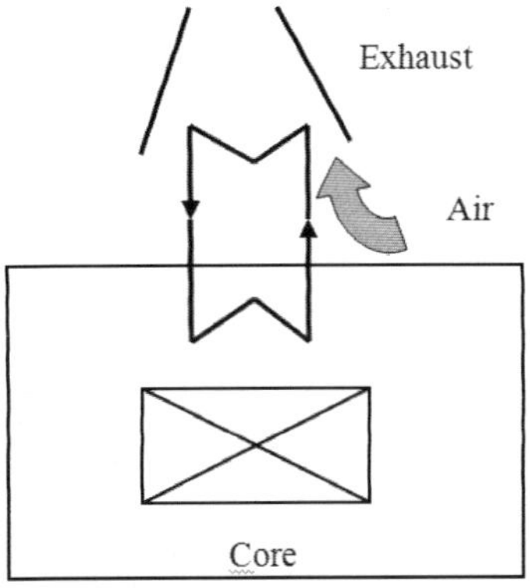

Figure 4. Diagram of the emergency energy removal of the core

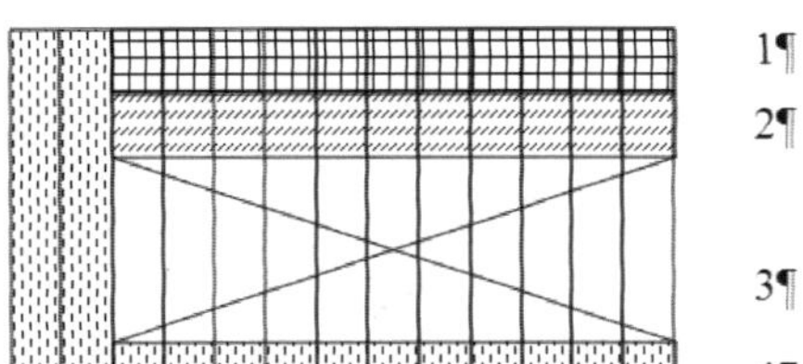

Figure 5. Cut of core with the sodium plenum: *1*– neutron absorber, *2*– sodium, *3* – fuel, *4* – blanket

The temperature effect of reactivity has the greatest effect on reactivity when the temperatures of reactor elements and fuel change. As the temperature of sodium increases, radial expansion of the diagrid, axial expansion of fuel elements, reduction of sodium density, and also the broadening of neutron capture resonance due to the intensification of thermal motion of nuclei (Doppler effect). The expansion of the diagrid, and the axial expansion of the fuel elements increase the volume of the core, causing an increase in neutron leakage from it Due to the Doppler effect, an increasing number of neutrons falls into the expanded resonance region and are captured as the temperature rises. As a result of these phenomena, the reactivity of the reactor decreases.

It should be noted that a decrease in the temperature of the core (for example, with a decrease in power) will cause the opposite effect - release of reactivity. For normal operation, this is compensated by entering the control rods.

Also, the release of positive reactivity occurs with the compaction of the core due to the convergence of SA under external influence (earthquake, shock phenomena, etc.) or the curvature of SA. This phenomenon was observed at the EBR-I reactor [9]. To eliminate this effect, they began to use spacing fins on SA, and subsequent SAs were equipped with contact pads that prevented their convergence [10].

If the above reactivity effects are negative, i.e. as the temperature increases, the reactor power decreases due to the decrease in the density of sodium is not so certain and in some circumstances may cause an increase in power (**the sodium void reactivity effect** (SVER). With a decrease in the density of sodium, sodium boils with the formation of vapor bubbles. This leads to the following parallel processes :

- increase in reactivity due to the growth of fissions due to a hardening of the neutron spectrum and reduction of the nuclei capture on sodium;
- reduced reactivity due to increased neutron leakage because to reduced the nuclei scattering on sodiumi.

These processes compete with each other, and depend on the region of the core. In the center of the core, the effect of hardening of the spectrum prevails over the effect of neutron leakage, which causes an increase in reactivity. At the periphery, neutron leakage prevails, which leads to a decrease in reactivity. These effects also depend on the design of the fuel rods and core. If for small cores, the SVER in the whole core is negative, then for large cores in the whole core, the SVER becomes positive, which creates a safety problem, and can lead to an instantaneous increase in reactor power.

To exclude inserting of positive reactivity - the “Damokl’s Sword” of large nuclear power enginering – it is necessary to apply known technical measures, such as minimizing the sodium volume in the core, flattening it to increase neutron leakage, inserting a sodium plenum over the core to increase axial leakage, the heterogeneous core [11].

One of these solutions is shown in Figure 5 – a change in the design of a fuel element by introducing a sodium plenum, and a neutron absorber above it into the upper part of the fuel element.

2.3.2. Emergency shutdown rods

For emergency shutdown of the reactor, special emergency protection rods (AZ) are provided. The AZ rod is controlled by a sahift, the upper end of which is located outside the reactor, and a AZ rod is attached to its lower end (Figure 6). In normal operation, the shift is held in the upper position by electromagnets that are outside the reactor, while the AZ rod is located above the core. When an alarm is received, the electromagnets de-energize, initiating the movement of the AZ rod in the core forcibly or under its own weight. With the growth of reactor size and, as a result, there is a risk of jamming the shift in a narrow channel in which it moves. Despite the small amount of risk it was considered expedient to implement a hydraulic suspension of some AZ rods (the “float” principle) at the sodium flow (Figure 7). In normal operation, these rods are located above the core due to sodium flow. By reducing the sodium flow below the set level, the rods due to their hydraulic weight are inserted into the core, thereby terminating the chain fission reaction in the core.

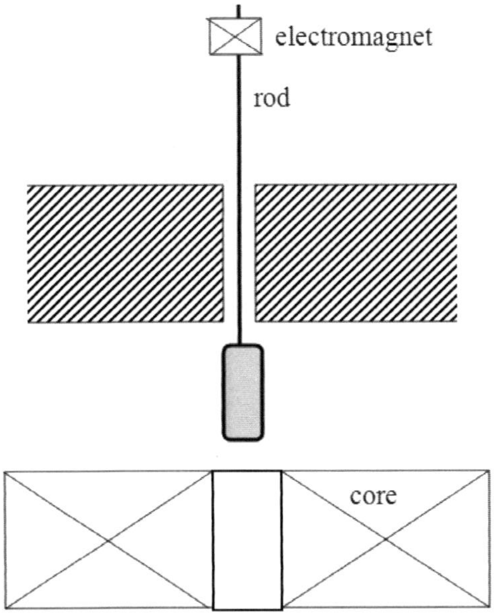

Figure 6. Emergency shutdown rod

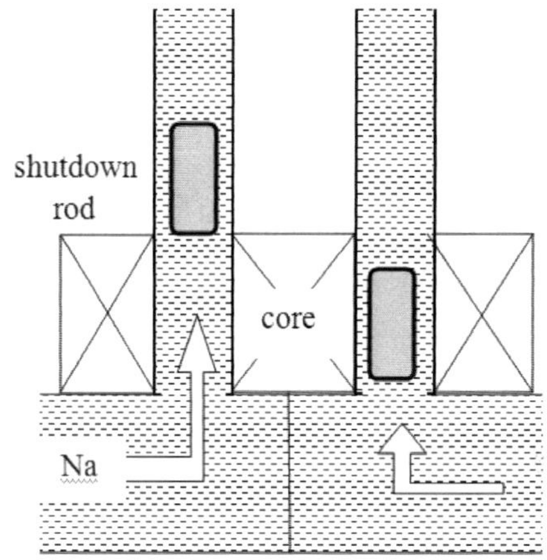

Figure 7. Hydraulic suspension of rod AZ

3. Discussion

We studied the evolution of the BR safety in terms of the basic functions: We note that the basic principles of the design have not changed during the more than 65-year history of BRs. Changes that occurred in the design of equipment were often innovative and increased the quality of DRs, but in general BRs have gone the way of the evolutionary development.

In the radioactivity retention function within the power unit of the first BRs, the reactor buildings did not have containments, which was explained by the sufficiency of the safety vessel. However, in connection with the development of nuclear power, protection against external influences has become increasingly important, therefore, it is necessary to have external containment in new BRs projects.

In the cooling function of the reactor core, evolutionary changes have occurred related to the transition of the reactor emergency cooling down from the water circuit to the intermediate sodium circuit, including using air as a final heat absorber. Recent developments in emergency cooling systems are associated with the placement of an emergency heat exchanger inside the reactor, and the use of natural circulation in the emergency circuit and in the air heat removal path.

In the function on reactivity controlling due to the increase in reactor size, difficulties arise with the suppression of positive reactivity caused by SVER. Therefore, in new BR projects, the use of sodium plenum over the fuel part of the core and use of heterogeneous cores or a combination of these solutions will be required.

4. Conclusions

In their history the BRs have gone through a long evolutionary period. In terms of equipment, of course, there were innovative changes, but in general the reactor did not change the structure laid down in the first reactors.

When developing new projects, it is necessary draw special attention to the use of external containment, the use of an emergency cooling system on the principle of complete natural circulation, and special attention should be drowed to the development of technical measures to achieve a negative sodium void effect.

IOP Conf. Series: Materials Science and Engineering **972** (2020) 012009 doi:10.1088/1757-899X/972/1/012009

Acknowledgments
This work was supported in accordance with Act 211 of the Government of the Russian Federation, contract No. 02.A03.21.0006.

References

[1] International Conference on Fast Reactors and Related Fuel Cycles : Next Generation Nuclear Systems for Sustainable Development 2017 (Yekaterinburg) Introduction FR 17 Conf. Book of Abstracts (Austria, Vienna: IAEA)
[2] Saraev O M, Oshkanov N N, Zrodnikkov A V i et al 2010 Operational experience and prospects for the development of fast sodium nuclear reactors *Nuclear energy* **108** 4 s 191.
[3] Gratchyov A F, Skiba O V, Tsykanov V A et al 2007 Demonstration experiment of 3 BN-600 MOX vibropac irradiation for the excess weapons plutonium disposal *J. Nucl. Sc. and Technol.* pp 504-10
[4] Sodium-Cooled Nuclear Reactors 2016 (Paric : Den Monograph)] p 187
[5] The Experimental Breeder Reactor EBR–II Inherent Safety Demonstration (987. (Amsterdam : North-Holland – Elsevier) p 5
[6] See [4] p 51
[7] Oshkanov N N 2016 Physical and technological features of fast neutron nuclear reactors (Yekaterinburg : Izd-vo Ural'skogo Universiteta) p 11
[8] See [4] p 79
[9] Kollier J, Heuitt J 1989 Introduction to nuclear power (Moskow: Energoizdat)
[10] See [4] p 81
[11] See [4] p 41.

IOP Conf. Series: Materials Science and Engineering **972** (2020) 012010 doi:10.1088/1757-899X/972/1/012010

Safety provision during development of the underground mine in water-bearing karsted rock mass

V V Melnik, A L Zamyatin

Institute of Mining, the Ural Branch of the Russian Academy of Sciences; Mamina-Sibiryaka str. 58, 620075, Ekaterinburg, Russia

Email: a.zamyatin@mail.ru

Abstract. The present article describes issues of up-to-date methods and approaches to ensure safety during development of the underground mine in complex engineering-geological and hydrogeological environments. As a priority to ensure safety while constructing mine workings on the karsted territory is the decrease of water content of the rock massif due to construction of spot water intakes above ground. The efficiency of dewatering depends on quantity and capacity of dewatering wells, The same parameters influence on prime cost, so that in the suggested approach the maximum productivity is reached by selecting drilling sites by means of ground geophysics together with study of geological and hydrogeological conditions. In the presented paper, there is an example of successful drainage of karsted layer by constructing water intake consisting of two dewatering wells, which enabled to start developing the ore shoot, which was on suspension due to the complex environment.

1. Introduction

Occupational safety of development of mineral resources by an underground method directly depends on field hydrogeological environment and possibility to control water entry into mine workings. This issue is particularly critical during mine development related to karsted limestone massif with high water transmissibility, which increases probability of sudden entry of large amount of water into the mine. There are fairly large number of examples of mining in such conditions in the Ural region and they occupy the whole Urals from the northern to southern borders (Severouralsk bauxite mine, Vysokogorsky ore mining and processing works, Uchalinsky ore mining and processing works, etc.). Mainly they are connected with ore mining of natural resources [1-5].

2. Methods of study

One of the effective methods to reduce water content of mine workings, especially in case of mining at a depth of not exceeding 300 meters is deep well dewatering of the aquifer level confined to the roof of the ore body. However, establishing deep well dewatering on the perimeter of the deposit without taking into account structural features of the rock massif often does not solve a problem of water discharge and the majority of wells turned out to be unable to operate due to their low abundance of water.

The best results of drainage of the ore-bearing massif can be reached by a complex of analytic forecasting methods of water inflow into mine workings using current information of mineral exploration and practical application of geophysical methods for determining structural and tectonic profiles of the rock massif and site selection of dewatering well staking [6-7]. It will allow reducing quantity of wells, which will result in significant reduction of expenses for dewatering and, consequently, in cost saving of ore mining.

One of the mines in the Southern Urals, where the described drainage approach was successfully tested with participation of authoring team, may serve as an example.

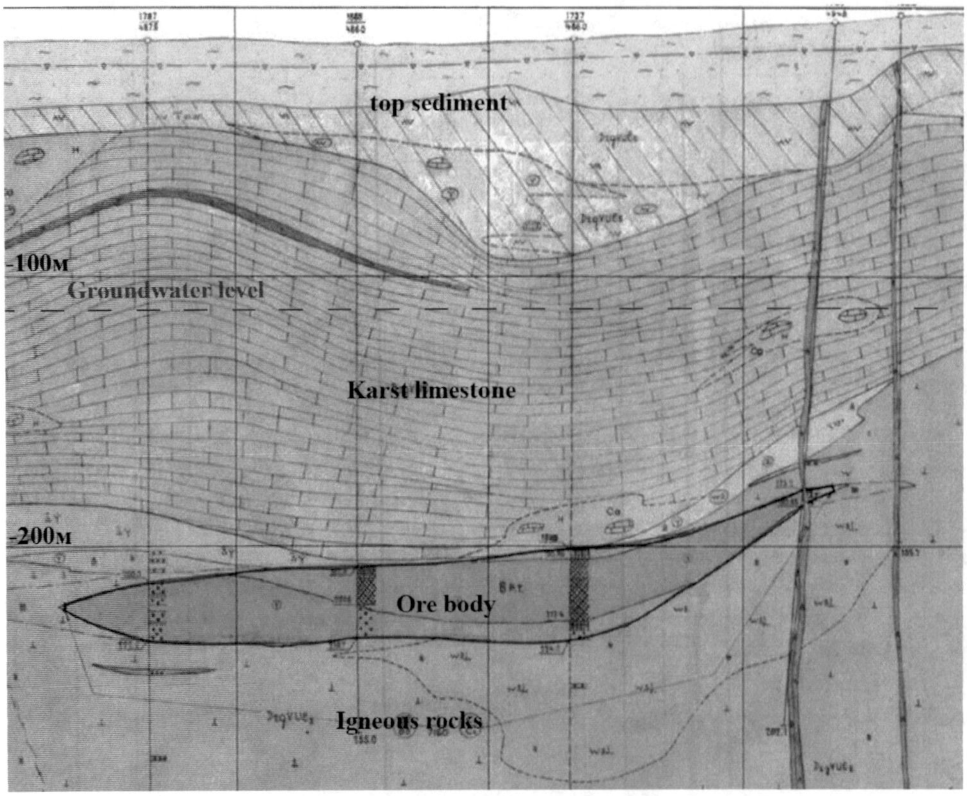

Figure 1. Geologic cross section

The essence of work was that one of the ore bodies of the successfully developed deposit, hosted at depth of a little over 200 m was found within underground water deposit in the karsted limestone massif (Fig. 1).

It was complicated by the fact that the karsted limestones overlaid directly in the roof of the ore body, which could lead to uncontrolled flooding of the mine goaf in case of its mining.

The goal of the research was to provide safe development of the ore body. The main purpose of the work was to reduce to minimum possible water inflows into the mine workings by arranging deep well dewatering with minimum expenses for its installation.

The volume of possible water inflow was difficult to predict, as while running deep well dewatering consisting of two wells (6 and 8), operating for the needs of the enterprise for drinking water supply, the decrease of the water level did not exceed 5 meter per a year though the total output was at the level of 160-180 m^3/h. In addition, water wells were located within the area of possible deformation due to underground mining, which put in question their working efficiency at the beginning of underground mining works.

The degree of karst hazard with regard to emergency outburst in the mine workings is determined by static volume of underground water and limestone filtration properties, which define the response speed of static reserve and dynamic inflow [8-10]. At the current level of underground water in limestones, which is about 100-120 m, the initial water flow from caverns can be within first thousand m^3/h, which presents serous risk for underground work.

Operation of static reserves with high filtration properties of rock and water inflow from the surrounding rock can last for a long time. This time period can be estimated approximately by operation of underground water intake from the wells no.6 and 8. At the average water flow rate from them, about 100 m^3/h, the annual decline of underground water table in its depression pit did not exceed 4-5 m.

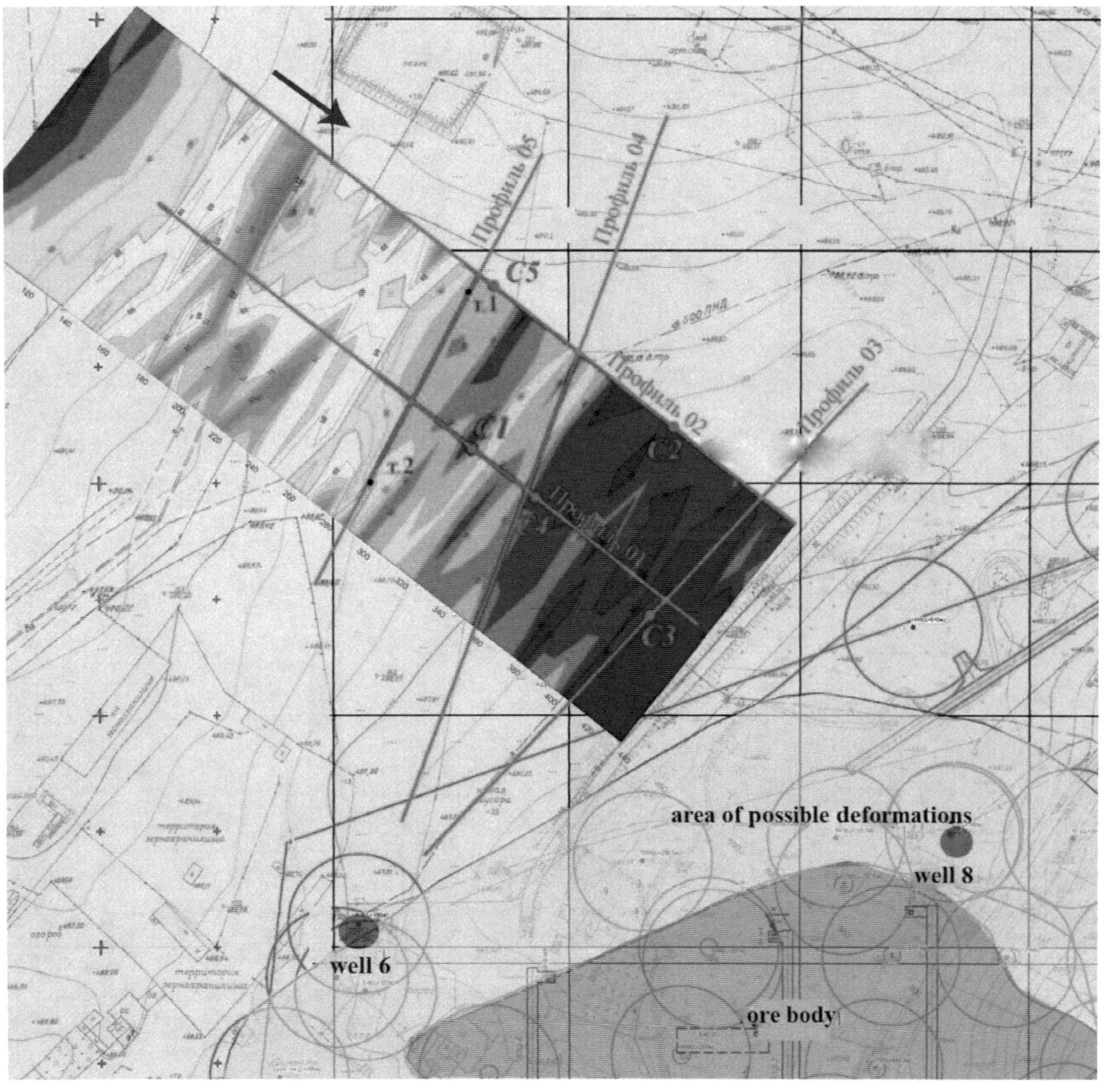

map symbols

C1 - well 15-B
C2 - well 16-B
C3,C4,C5 - alternative well points (not required)
- spectral seismic profiling profiles
- medium gradient tablet
- groundwater direction

Figure 2. Selection of drilling sites of dewatering wells by means of geophysics

In this regard, uncovering of karst structure can create a long emergency situation, which can take several years. A positive factor, which neutralizes this risk, can be the presence of the interbedded layer with thickness up to 70-80 m in the roof of the ore body, which can serve as a water-blocking layer excluding direct exit of the preparatory mine working in the karst hazard area.

3. Research results

The analysis of all survey information, monitoring wells, etc. enabled to define the direction of the main water inflow (north-west) and a complex of geophysical exploration enabled to choose the optimum drilling sites of dewatering wells (Fig. 2).

The first drilled exploration wells (points C1 and C2) showed full conformity with predicted open-cut and the output exceeding 100m^3/h. As a result of drilling at the point C1 of the operation wells 15-B and at the beginning of its operation, the level in all wells started to fall (up to 5 m/month) which enabled to approximately determine the total volume of water entry [8,9].

Putting the well no.16-B into operation should have led to complete water catch in the catchman area, which was located further north than the ore body, providing volume control of the collected water and reducing the risk of ingress of the whole volume of water table into the mine. In addition, the wells no.6 and 8 were in the range of influence of mining works and their loss is a matter of time (Fig. 2).

Putting the well no.15-B into operation resulted in monthly depression of level in the wells no.6, 8 and 15-B within 1.5-2.2 meters, which placed in doubts further use of water intake for water supply.

Putting into operation the second well no.16 drilled at the point C2 allowed to drain the well no.8 for 10 days completely; and as bailing tests showed, it became possible to control the water level there by cutting off one or several wells (Fig. 3).

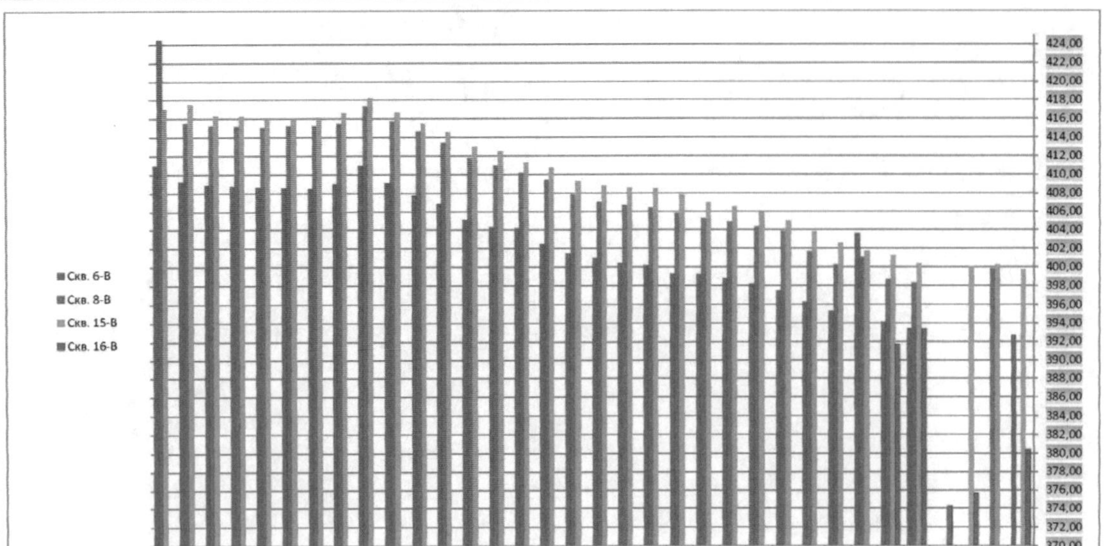

Figure 3. Diagram of level variation in the wells

Commissioning of the well no.16-B that was finished at the end of October 2016 led to drainage of the well no.8 to the depth of possible measurement with the level gauge. The task of ore body drainage considering the obtained results was solved in terms of minimizing the volume of static reserves of underground water. In addition to it, it was decided to perform mining works for development of the ore body no.6 within the planned scope of work meeting safety requirements stated in the project.

At present, only one well no.6-B with the output of not more than 40 m3/h operates for dewatering and water consumption. The remaining wells are deactivated because of fall in level lower than the depth of the pumping equipment.

Based on these observations, the conclusion was as follows: complete development of static reserves up to the depth of about 130 m below the surface, dynamic influx of about 40-50 m^3/h.

As the pumping equipment does not run deeper, karsted pockets where can be quite a large amount of static water remain in the ceiling between the ore body no.6 and the dynamic level in the wells.

One of such pockets (Fig.4) was uncovered during driving of the mine workings along the roof of the ore body with instant outflow of water with clay of about 3000m^3 volume.

Figure 4. Uncovered karsted void

The breakthrough occurred during blasting works; the fact was predetermined that there was no people in close proximity to the object, so that no one was injured. Later, the water inflow from the karsted void reduced to the level of 40 -50 m3/h that corresponded to the dynamic influx observed at the water supply well.

According to the investigation results of the karsted void, its volume, danger calculations of sinkhole collapse at ground surface were done, which showed the absence of such danger. However, after further mining of the ore reserve there appeared high probability of integration of chamber and karst, which together represent danger of collapse. At the technical meeting with participation of authoring team a decision was taken to eliminate the karsted void by filling. At present, the karsted void was completely filled and the water inflow was stopped.

Uncovering of the karsted void proved interrelation of aquifer pumped away by deep well dewatering with the rest of the massif buried above the ore body; and in future this will likely lead to complete drainage of wells and the remaining static volume and dynamic inflow will flow into underground mine workings. It does not present great danger for development of the ore body; repeated water outflow from the re-uncovered karst is possible, however, the main static water reserve was drained by deep well dewatering, and the built water intake facilities in the mine will cope with the task of water removal.

But water supply of the enterprise became an issue. And later an option of water supply from the deep well dewatering was worked out.

One of the option of the alternative source of water supply was accepted the option of draining the ore body no.1 by the method which had proved effective at the sixth ore body where pumped water was used for the needs of water supply of the enterprise. However, close proximity of borehole chosen by the authoring team to refuse heap did not allow to hope for high quality of water and to use it as drinking water.

An exploratory well at the indicated area was drilled and it showed high productivity but a small (about 10-15 meters) remaining layer of the aquifer. Observations for a short period showed gradual decrease of the water level and a complete chemical analysis confirmed bacterial pollution. Considering the absence of infrastructure in the direct proximity from the well (water duct and etc.) and a small water seam, it was decide not to drill an industrial well at this area. Moreover, a tendency for level decrease connected with both development of the first ore body and the influence of drainage of the sixth ore body showed decrease of risk of flooding during mining works at this area.

In this regard, a decision was taken to continue choosing alternative sources of water supply, which at present the authoring team keeps doing being a part of Department of Geomechanics at the Institute of Mining, the Ural Branch of the Russian Academy of Sciences together with geological and hydrogeological service.

Based on the performed investigations confirmed by practical implementation of the results at the Mining Works, it was possible to draw the following conclusions.

4. Conclusion

As a result of the performed work, there was an attempt to solve two tasks: on the one hand, the risk of mine flooding at the beginning of development of the ore body was brought down to a minimum and on the other hand, there was a possibility of deep well dewatering due to moving of water supply wells beyond the area of possible deformation of rock massif away from mining works.

However, as the results of implementation of the chosen approach showed, the task of water supply due to aquifer correlated with karsted limestone massif lying in the roof of the ore body was not simple. As the results of development of the ore body showed, despite moving water supply point beyond the area of possible influence of deformation processes, during mining works and water drainage into mine workings, karsted channels and voids allowed to flush the channels filled with clay, thus decreasing the water level in the wells lower than it is possible to sink the pump.

In addition, water intake from the well no.6 keeps operating but the level decreases gradually. The matter of water supply should be solved with the help of wells located outside the affected area of depression pit from mining works being performed at present, but it will lead to additional expenses for arrangement of water conducts and other infrastructure as well as for additional exploration works.

As far as the first task put before the present investigation is concerned, it was solved entirely. The main aquifer lying above the ore body was processed by four dewatering wells up to the level of dynamic inflows. At present, the full-fledged work is in progress to develop the reserve of the ore body with minimum risks of mine flooding due to which the ore body was on long suspension.

In conclusion, it should be noted that the competent comprehensive approach to the issues of drainage of mine workings, whether it is a mine or an open-cut mine, can significantly reduce capital expenditures for arrangement of dewatering from the mine workings and increase safety of mine development of natural resources and, if possible, use the pumped water for the enterprise needs up to drinking water supply.

Acknowledgements
The research was undertaken within the State Task 075-00581-19-00 subject 0405-2019-0006

References

[1] Dalatkazin T Sh, Kharisov T F 2017 Predictive studies of the effects of flooding of the Turinsky copper mine *Problems of Subsoil Use*. № **3** (**14**) pp 67-75
[2] Melnik V V, Aimbetov M M, Tretyak A V 2008 Ways to reduce the water content of sewage treatment works at the DNA mine *Mining Journal of Kazakhstan* №**5** pp. 20-22.
[3] Zheltysheva O D, Usanov S V, Draskov V P 2016 Measures of protection of buildings and structures from underground mining in the karst massif *Problems of Subsoil Use* №**2** (**9**) pp. 71-76.
[4] Melnik V V 2011 Estimation of the hazard of karst occurrences by geophysical methods *Mining information and analytical bulletin* № **7** pp 151 - 155.
[5] Usanov S V, Usanova A V 2017 Monitoring of surface movement during liquidation and flooding of the mine workings of the Lebyazhinskoye field *Mining Journal* №**1** pp18-22
[6] Melnik V V 2010 Substantiation of geomechanical factors for diagnosing the danger of karst occurrences during subsurface use: Abstract of Ph.D. tech. sciences. (Ekaterinburg) 21 p.
[7] Melnik V V, Zamyatin A L 2008. Study of the structural features of the rock mass in the vicinity of underground structures *News of universities. Mountain Journal* №**8** pp 165-171
[8] Sashurin A D, Melnik V V, Usanov S V, Balek A E 2014 Security issues in the liquidation of mining enterprises *Problems of Subsoil Use* № **3** pp. 60-66.
[9] Zamyatin A L 2014 Experimental studies of the state of the rock mass at the sites of subsoil use. *Problems of Subsoil Use* №**2** pp 29-33

IOP Conf. Series: Materials Science and Engineering **972** (2020) 012011 doi:10.1088/1757-899X/972/1/012011

Heaving Pressure Occurring at Frost Penetration in the Clay Soil in Confined Space

Vyacheslav Kim[1], Marina Kim[2] and Michael Kim[1]

[1]Department of Constructions and Foundation Engineering, Voronezh State Technical University, 20-letiya Oktyabrya str., 84, Voronezh 394006, Russia

E-mail: marskim@yandex.ru

Abstract. The article describes the results of laboratory studies of the frost heaving pressure development at the clay soil freezing in confined space. Intensive construction of industrial and civil buildings in the North and North-East of Russia determines the relevance of research on the topic. The conducted studies allowed us to reveal the peculiarities of frost heaving pressure development, depending on the temperature conditions and physical properties of the clay soil, as well as possible heave deformations. The paper gives dependence of frost heaving pressure development during one-sided and uniform frost penetration in confined space on the degree of its water saturation, temperature conditions and possible deformations. It was revealed that in sandy soils, the formation of frost heaving pressure is caused by the freezing of free (gravitational) water and can be determined by the Bridgman-Tammann formula. In clay soils, the development of frost heaving pressure is associated with the freezing of loosely-bound (film) water and can be determined by the method that we propose both in the absence and in the presence of heaving volumetric deformation. The experimental technique that we offer makes it possible to analyze the development of the frost heaving pressure of the clay soil when it freezes in confined space.

1. Introduction

The development of the rich natural resources of the North and North-East of Russia has led to the relevance of the construction of new territorial-industrial complexes for industrial and civil purposes. During construction in these areas, one inevitably encounters frozen or seasonally freezing soils with the features of frost heaving.

The force impact of freezing heaving soil on underground structures and foundation structures when constructing in these harsh climatic conditions is a problem that requires a scientific and technical applied solution.

In the field of experimental and theoretical studies of the problem of frost heaving, considerable material has been accumulated, which is reflected in [1–7]. The mechanism and dynamics of the development of frost heaving pressure in clay soil when it freezes in confined space is much less studied.

The purpose of this study is to establish patterns of the frost heaving pressure development when the clay soil freezes in confined space and to develop methods for taking this pressure into account in engineering calculations.

2. Materials and Methods

We carried out the studies of the mechanisms of frost heaving pressure development of the soil in confined space under laboratory conditions using water-saturated pulverescent clay of various consistencies (table 1).

Table 1. Physical characteristics of soils in tests 1 ... 4

Test	Soil	Soil density ρ_s (g/cm^3)	Soil humidity W (%)	Porosity index e	Liquidity index I_L	Degree of saturation S_r
1	Firm pulverescent fat clay	1.94	29.6	0.80	0.6	1.0
2	Firm pulverescent fat clay	1.93	30.0	0.82	0.62	1.0
3	Firm-stiff pulverescent fat clay	2.1	24.6	0.61	0.36	1.0
4	Stiff pulverescent fat clay	2.1	20.0	0.54	0.14	1.0

The experimental unit consisted of chambers for placing samples, pressure sensors installed in a removable lid on the top of the sample, temperature sensors located along the entire height of the sample, indicating gage recording the allowed displacement, thermal insulation, heating elements, a set of recording equipment, and the necessary automation equipment for operation of heating elements at the specified modes. The studies were conducted on soil samples of cylindrical shape with a diameter of 0.2 m and a height for the large-size sample of 0.8 m (figure 1, a), and for the small-size sample - 0.3 m (figure 1, b). The chamber for the large-size sample consisted of three links: the upper links with a height of 0.25 m and the lower one with a height of 0.3 m, and for a small-size one - with one link with a height of 0.3 m.

(a) (b)

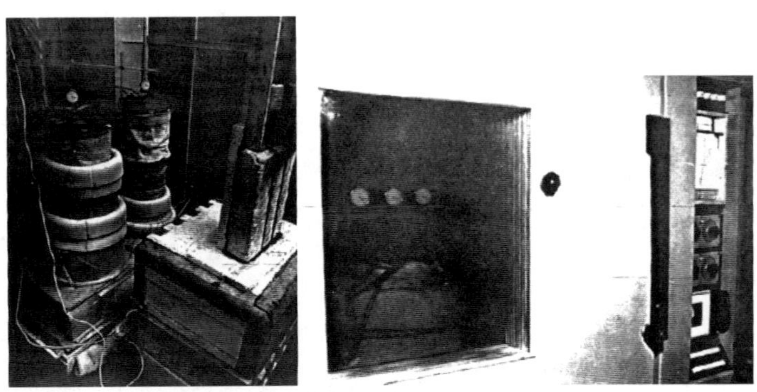

Figure 1. General view of the camera for sample placement:
(a) - large-size unit; (b) - small size unit.

The experimental unit was able to withstand pressures up to 16 MPa and the maximum allowance of relative deformations of the walls and the bottom of the unit did not exceed $1.3 \cdot 10^{-3}$.

Freezing was carried out under conditions of one-sided and uniform frost penetration. The air temperature in the refrigerating chamber varied in the range from + 0°C to −35°C.

The one-sided freezing of the soil took place both in the conditions of the "closed system" - without the influx of moisture into the freezing sample from the outside, and in the conditions of the "open system" - with the inflow of moisture from the bottom into the soil sample. At the same time, we measured the heaving pressure and the allowed displacement. We carried out a number of experiments to determine the free deformation of the heave. When the test was completed, we studied the cryo-texture of the soil.

3. Results

In the course of the laboratory studies, we established the regularities of the heaving pressure development in clay soils freezing in confined space under different temperature and humidity conditions.

To determine the effect of consolidation of thawed soil on the development of heaving pressure, let us consider the results of test 1 and 2, obtained by one-sided freezing of soil samples with a diameter of 0.2 m and a height of 0.8 m. The samples in these tests had almost identical water-physical characteristics. The freezing conditions differed only in the fact that in test 1, the soil froze through without the possibility of the moisture outflow from the sample (closed system), and in test 2, the outflow of moisture from the freezing sample was possible (open system). As can be seen from the comparison of the results obtained (figure 2), the heaving pressure in the case of incomplete soil freezing in test 1 did not exceed 0.4 MPa, and in test 2 it did not exceed 0.2 MPa.

The behavior of the heaving pressure changes significantly when the zero isotherm reaches the bottom of the sample and the soil freezes through completely. Under these conditions, the heaving pressure in test 1 increased from 0.4 to 3 MPa, and in test 2 - from 0.15 to 0.39 MPa. With an increase in temperature at the bottom of the sample and the formation of a thawed soil zone, the pressure in test 2 decreased to 0.1 MPa (figure 2, b).

(a) (b)

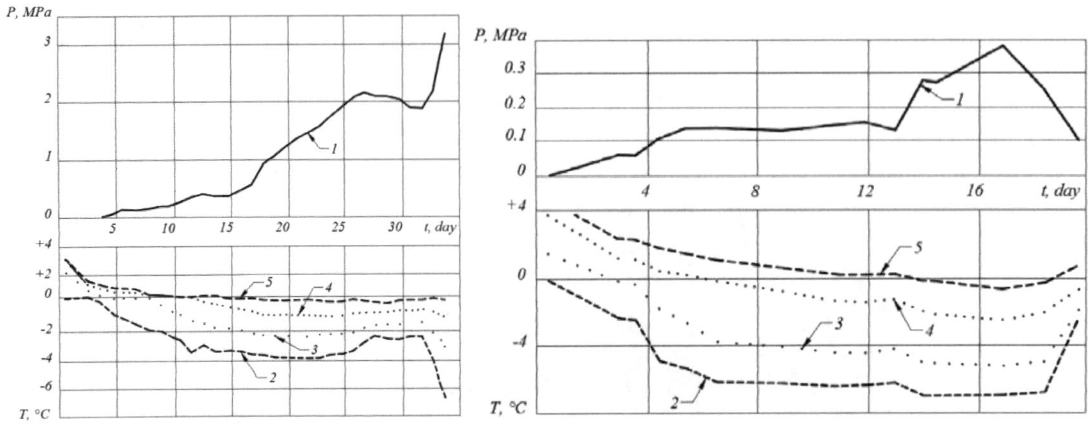

Figure 2. Graphs of heaving pressure and distribution of the soil temperature along the sample depth at one-sided clay freezing in closed and open systems: (a) - test 1; (b) - test 2
1 - heaving pressure; 2 - temperature on the soil surface; 3, 4, 5 - soil temperature at depths of 0.3 m, 0.6 m, and 0.8 m, °C

The dependence of the heaving pressure from the decrease in the temperature of the frozen soil in confined space is clearly seen in tests 3 and 4. In test 3 (figure 3, a), the decrease in the temperature of the soil to −2°C caused the increase in pressure to 2.3 MPa. A further decrease in the temperature of the soil to −32°C was accompanied by the increase in pressure to 15.7 MPa, and then with the increase in the temperature of the soil to −2°C, it decreased to 2.5 MPa. The temperature dependence of pressure is especially distinct when the soil is completely frozen in test 4 (figure 3, b). The temperature regime of soil freezing in this test varied in steps and was maintained at each step for several days to equalize the temperature in the sample and to stabilize the pressure at this step. As can be seen from figure 3, b, the shape of the heaving pressure graph practically coincides with the shape of the graph of the temperature change of the frozen soil. Thus, a stepwise lowering of the soil temperature from −1.2°C to −15°C in the first cycle entailed a similar pressure increase from 0.5 MPa to 11.7 MPa, and the subsequent increase in the soil temperature from −15 to −1.3°C caused a decrease in the heaving pressure from 11.7 to 1.4 MPa. A decrease in the soil temperature in the second cycle from −1.6°C to −16°C was accompanied by an increase in the pressure from 1.4 to 9.5 MPa.

(a) (b)

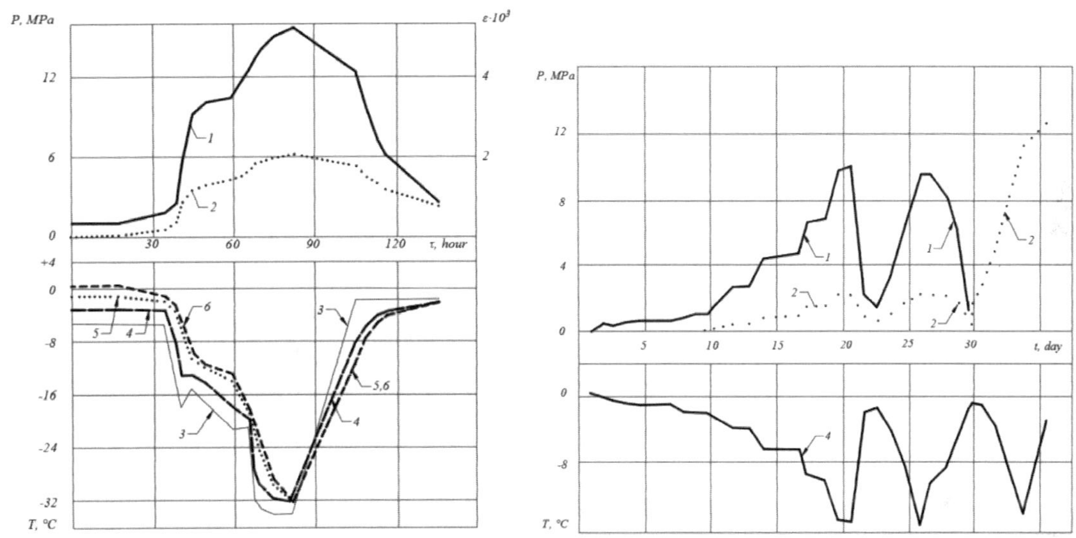

Figure 3. Graphs of the heaving pressure and deformation depending on the negative temperature drop of the frozen soil: (a) - one-sided freezing (test 3); (b) – uniform freezing (test 4)

1 – the heaving pressure; 2 – the allowed relative heaving deformation; 3 - temperature in the refrigerating chamber, °C; 4, 5, 6 – the temperature of the top, middle and bottom of the soil sample °C; 7 - the average temperature of the soil sample, °C

4. Discussion

Using the developed methodology of complex laboratory tests, we established the regularities of the heaving pressure development in clay soils freezing in confined space under different temperature and humidity conditions.

The processes occurring during the freezing of water in soils are described in [8...13] and the processes of freezing of water under pressure, depending on the type of crystallization - in [14, 15].

We examined the processes occurring during soil freezing in confined space and described them in [18, 19].

To quantify the heaving pressure at frost penetration in confined space without taking into account the deformations, we proposed the formula [18]

$$p_i = 1035.76\ln(T) - 6.707T + 0.0031T^2 - 1.66\cdot10^{-11}T^3 - 4210, \quad (1)$$

where p_i - the heaving pressure (MPa), which develops in the freezing soil at the temperature T, K due to the freezing of the unfrozen film water.

To take into account the effect of possible deformations when the soil freezes in confined space, we proposed the formula [18, 19]

$$p_{iu} = k_u p_i \,, \tag{2}$$

where k_u - the reduction factor, taking into account the possibility of deformation, and equal to:

$$k_u = 1 - \frac{\rho_w(\varepsilon_v + 3\alpha_m T)}{0.09\rho_d w_{mmw}} \,, \tag{3}$$

where ρ_d and ρ_w – the densities of the soil skeleton and water respectively, kg/m^3;
ε_v – the relative deformation causing the bound water to freeze;
α_m – the temperature coefficient of frozen soil linear contraction (°C)$^{-1}$;
w_{mmw} – the maximum molecular moisture-holding capacity.

5. Conclusions

The results of the studies to determine the heaving pressure at frost penetration in confined space allow us to draw the following conclusions:

1. In the course of laboratory tests, the features of the development of pressure in confined space were determined depending on the temperature both during the freezing of thawed soil and cooling of frozen soil with incomplete and complete freezing, as well as with the possibility of some allowable deformation.
2. The formation of heaving pressure when at frost penetration in water-saturated clay soils in confined space has a fundamental difference in the case of incomplete and complete freezing. The value of the heaving pressure when the soil is completely frozen is an order of magnitude greater than the pressure values when the soil is not completely frozen in confined space.
3. We found the clear dependence of the heaving pressure when the soil freezes in confined space on the temperature of the frozen soil. Lowering the negative temperature of the soil causes an increase in pressure and an increase in the temperature of the frozen soil causes a corresponding decrease in the heaving pressure.
4. A quantitative estimate of the heaving pressure when the negative temperature of the frozen soil varies in confined space can be given by formulas (1)-(3).

References

[1] Orlov V O 1962 *Cryogenic Heaving of Fine-Dispersed Soils* (Moscow: Academy of Science USSR) p 187

[2] Ershov E D 2002 *General Geocryology. Manual for High School* (Moscow: Moscow University Press) p 682

[3] Orlov V O, Yelgin B B and Zheleznyak I I 1987 *Heaving of Soils in the Calculations of the Foundations of Structures* (Novosibirsk: Nauka) p 133

[4] Orlov V O, Zheleznyak I I, Filippov V D and Fursov V V 1991 *Heaving Soil as the Foundation of Structures* (Novosibirsk: Nauka) p 130

[5] Zheleznyak I I 1990 *Reliability of Frozen Foundations of Structures* (Novosibirsk: Nauka) p 170

[6] Zheleznyak I I 2015 Devices and methods of protecting structures from dangerous cryogenic processes in soils *Proc. of XV Int. Scientific and Practical Conf. "Kulagin readings: Equipment and Technologies of Production Processes"* (Zabaykalsk: Zabaykalskiy State University) pp 272-276

[7] Nevzorov A L 2000 Foundations on Seasonally Frozen Soils. Manual (Moscow: ACB Publ) p 152

[8] Brushkov A V 1984 Formation of the Stress-Strain State of Freezing Soils at Heaving: Cand. of Geolog.-Mineral. Sci. Diss. (Moscow) p 178

[9] Takashi T 1980 Upper limit of heaving pressure derived by pore water pressure measurements of partially frozen soil *Proc. of 2nd Int. Symp. Ground Freezing* (Tronheim) pp 713-724

[10] Radd F J and Ortle D H 1973 Permafrost *2-nd Int. Conf. National Asad* (Washington DC) pp 377–384
[11] Chistotinov L V 1975 Some questions of the thermodynamics of unfrozen water in soils *Geocryological Studies* **87** 24-39
[12] Talamucci F 2003 Freezing processes in porous media: formation of ice lenses, swelling of the soil *Mathematical and Computer Modelling* **37** 595-602.
[13] Starostin E G, Kravtsova O N and Tappyrov N I 2017 Models of equilibrium of water phases in frozen rocks *Science and Education* **2** 41-47
[14] Choukroun M and Grasset O 2007 Thermodynamic model for water and high-pressure ices up to 2.2 GPa and down to the metastable domain *Chem. Phys.* **127(12)** 124506
[15] Dunaeva A N, Antsyshkin D V and Kuskov O L 2010 Phase diagram of H_2O: Thermodynamic functions of the phase transitions of high-pressure ices *Astronomicheskii Vestnik* **44** 222–243
[16] Kim V H 1988 Heaving Pressure Determination at Freezing of Soil in Confined Space: Tech. Sci. Cand. Diss (Moscow: Gersevanov NIIOSP) p 270
[17] Orlov V and Kim V 1988 *Soil Mech. Found. Eng.* **25** 129–135
[18] Kim V and Kim M 2017 Method of evaluation of ground frost heaving pressure on ice-ground barrier of an underground construction and well strengthening walls *Int. Sci. Conf. Energy Management of Municipal Transportation Facilities and Transport EMMFT 2017. Advances in Intelligent Systems and Computing* ed V Murgul and Z Popovic (Springer, Cham) **692** 998-1008
[19] Kim V, Kim M and Kim M 2018 Thermodynamic evaluation of the pressure occurring at frost penetration in confined space of the soil *Proc. of Int. Conf. on Civil, Mechanical and Material Engineering ICCMME 2018, Jeju-do, Korea, 16–18 March 2018* **1973** 02008-1–5

The analysis of computer aided methods applied for pit wall stability rating

P V Koltsov[1], Y S Ivanov[2], N P Pykhteeva[3]

[1,2] Laboratory of pit wall stability and rock movement, JSC Uralmekhanobr, Khokhryakova st., 87, Yekaterinburg, 620063, Russia
[3]Ural Federal University, 19 Mira str., Yekaterinburg, 620002, Russia

E-mail: n.f.pykhteeva@urfu.ru

Abstract. A comparative analysis of the results of the pit slope stability rating carried out under normative technique and Scad Office Slope application is presented. The possibility of combination of normative technique and computerized calculations to gain integrity and computation operability is suggested.

1. Introduction
The problem of pit wall and slope stability is of current importance in mining, civil engineering and other branches of human activity. Theoretical development and improvement of calculation methods for pit slope and dump stability under real mining and geological conditions were developed based on well-established calculation methods (V. Fellenius, V. V. Sokolovskiy, S. S. Golushkevitch, G. T. Shakhunyantz) and research data on slope deformation in field or laboratory conditions [1, 2, 3, 4]. Advanced methods for calculating slope stability are based on the limit equilibrium theory for loose medium together with the limit equilibrium of cohesive soil medium such as rock massif.

2. Methods and materials
Methods of algebraic force addition and force polygon (vector addition) are two known and widely used basic calculation methods currently being used for pit slope and wall stability rating.

The method of algebraic force addition is characterized by sufficient simplicity of graphic construction and subsequent calculations. The method is applied in isotropic (homogeneous) slope stability rating the length (L) of which is 2-3 times more than their height (H) and potential sliding surface has the form of fair curve. The method assumes that the prism of a possible collapse is deformed as a whole and reactions between blocks are not taken into account. This results in the underestimation of design ratio of holding and shear forces (underestimation of determined stability coefficient) by 3-20 % relative to the actual one. However, this method gives acceptable results [5, 6] when the slope height is low (up to 100 m) in loose rocks and internal friction angle values are small ($\varphi < 20°$).

The most reliable and universal method for calculating slope stability in real mining and geological conditions is the method of force polygon (vector addition) advanced by All-Russian Research Institute of Mining Geomechanics and Survey (VNIMI). The method takes into consideration both internal and external forces acting on the limits of possible block mining marked in the prism [7]. The method in question allows for an accurate evaluation of the pit slopes stability or their sites especially in difficult mining and geological conditions, namely the presence of differently oriented contacts of rocks and tectonic deformations in the array; the layered array structure at pit walls; dump (mounds) formation on the sloped or weak base.

To determine the maximum stable parameters of pit walls and quarry faces in homogeneous rock mass, VNIMI developed and constructed graphs of the relationship between the height of the flat slope H and its angle α for different physical and mechanical characteristics of rock mass (density γ, internal friction angle φ and cohesion $C_м$) based on the force polygon method. According to the graphs, it is possible to evaluate the slope stability with the given parameters or to determine the maximum stability parameters at a given safety margin [8]. It is worth noting that this method allows to achieve the results quickly without special training. However, the reverse side of simplicity is the need to average the characteristics of an array, which matters with a complex structure of a pit wall.

Recently, computer-aided methods for pit wall and slope stability calculations have been spreading; the best known are Rocscience (Slide), Geo (Slope), Fine Soft (Geo 5) and others. The apparent advantage of these methods is the possibility of carrying out on-line multivariate calculations (the search for the weakest surface formed in the array at pit walls; the detection of the weakest parts in the slope height). A traditional method, on the other hand, requires much more effort and time of a specialist. However, software application employed for stability rating requires from a specialist theoretical and practical training since the result of computerized analysis heavily depends on the experience of a specialist and may differ from the reality as one cannot describe the natural environment based on typical up-to-date mathematical tools.

The set of forces acting in the slope array (holding and shear) is a complex geomechanical process depending on the set of independent factors the correct accounting of which directly determines the accuracy of stability rating results. Among the main factors affecting the stability rating are the choice of initial physical and mechanical characteristics of mine rock; the shift from the characteristics of rocks in the sample to those in the array; the accuracy of array geomodelling; the provision for the most likely sliding surface; the consideration of the effect of array characteristics (tectonics), hydrogeological conditions, external loads, etc. Therefore, the quality of the initial geomechanical condition modelling, regardless of the applied calculation method, is of great importance for rating the stability of the quarry slopes, pits, dumps or mounds.

Though software packages currently used for slope stability rating fail to meet the requirements of current regulatory documents; they can serve as a powerful auxiliary tool for on-line multivariate calculations thus increasing research data representativeness.

3. Results

To evaluate the possibility of using computer-aided systems for pit slope and slope stability rating, the results of computer calculation in the sofware Scad Office Slope were compared through research with those of manual calculations of the force polygon method (vector addition) recommended in regulatory documentation.

The force polygon method used as a reference in the comparative analysis is specified in the relevant regulatory documentation on issues relating to pit wall stability rating [9, 10, 11, 12, 13].

The computational methods used in the software Scad Office Slope apply the mathematic tools specified both in open sources and in the programme user manual [14, 15, 16, 17]:

- Fedorovskiy – Kurillo;
- Fellenius;
- Bishop simplified;
- Corps of engineers №1;
- Louw and Karafight;
- Janbu simplified;
- Janbu corrected;
- Spencer.

Laboratory of pit wall stability and rock movement has considerable experience in evaluating the stability of pit wall and striping soil bank where the software Scad Office Slope was repeatedly employed to perform calculations as a supplement to traditional methods regulated by reference documents.

Research covered two copper-ore quarries: Yubileiny (LLC Bashkirskaya med) and Novo-Shemurskiy (JSC Svyatogor). To conduct research two specific strip pits (Fig. 1, 2) oriented along the normal to pit wall extension were employed. Pit wall stability was evaluated at maximum quarry depth for each strip pit. Table 1 shows the findings of pit wall stability rating on theYubileiny quarry and Table 2 provides data on the Novo-Shemurskiy quarry.

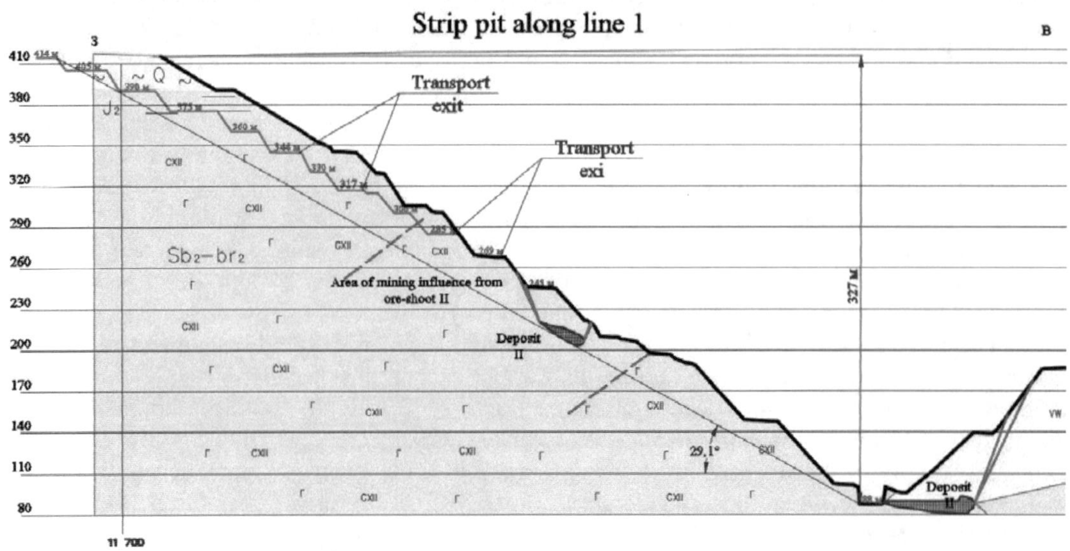

Figure 1. Yubileiny quarry. Calculated strip pit taken along line 1.

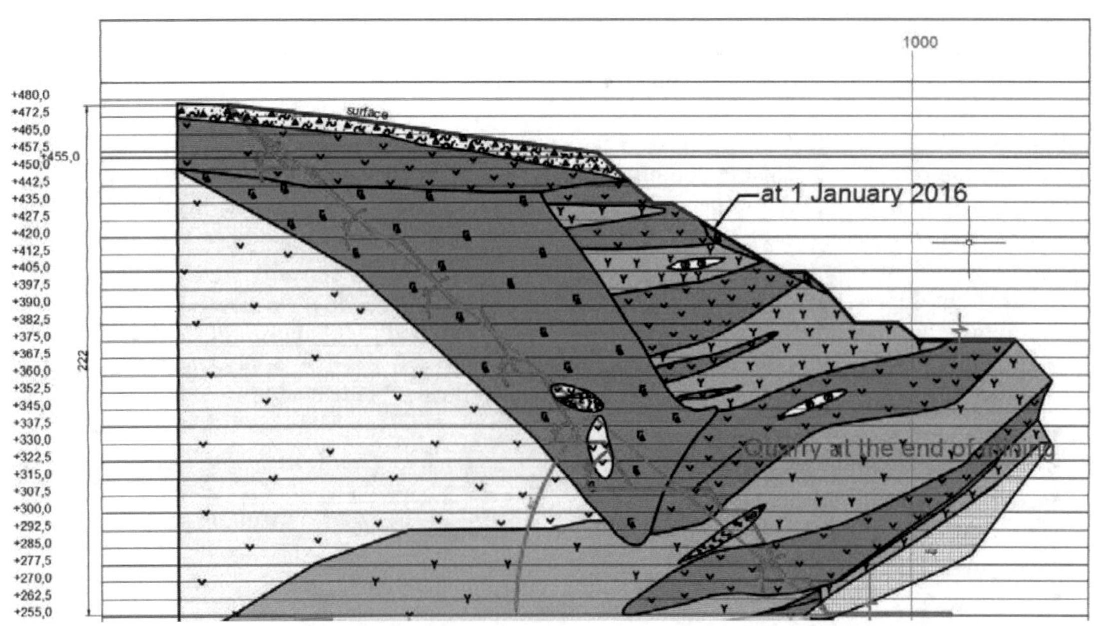

Figure 2. Novo-Shemurskiy quarry. Calculated strip pit taken along line 2.

Table 1. Calculation data on the Yubileiny quarry.

Calculation method	Stability coefficient	
	Strip pit along line 1	Strip pit along line 2
Janbu simplified	1.68	1.65
Spencer	1.81	1.77
Fedorovskiy – Kurillo	1.74	1.70
Polygon of forces (reference)	**2.20**	**2.10**

Table 2. Calculation data on the Novo-Shemurski quarry.

Calculation method	Stability coefficient	
	Strip pit along line 1	Strip pit along line 2
Fedorovskiy – Kurillo	1.95	1.30
Fellenius	1.92	1.30
Bishop simplified	2.02	1.32
Janbu corrected	1.95	1.32
Spencer	2.00	1.32
Polygon of forces (reference)	**1.96**	**1.37**

4. Discussion

Research shows that in general the results of computer calculations are consistent with those of stability rating performed with force polygon method. In this respect, the utmost convergence with classical technologies for quarry mining conditions under study is achieved through the following methods:

- Fedorovskiy – Kurillo;
- Fellenius;
- Bishop simplified;
- Janbu corrected;
- Janbu simplified;
- Spencer.

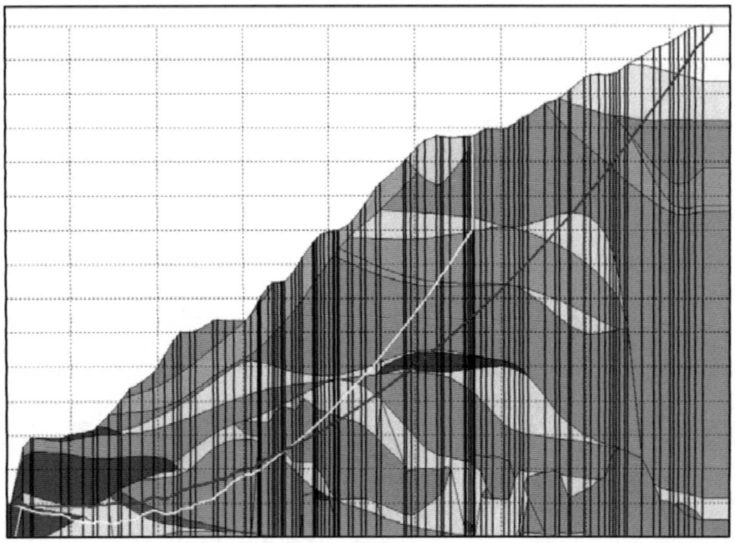

Figure 3. Pit wall modelling and calculation of pit wall stability coefficient in the software Scad Office Slope based on Fedorovskiy – Kurillo method.

Based on computer calculations the stability coefficient values are generally found to be slightly less than those of polygon of forces that is a positive factor as it gives an additional safety margin.

Figures 3 and 4 show the examples of pit wall modelling and computer-aided slide surface modelling on which calculations are made.

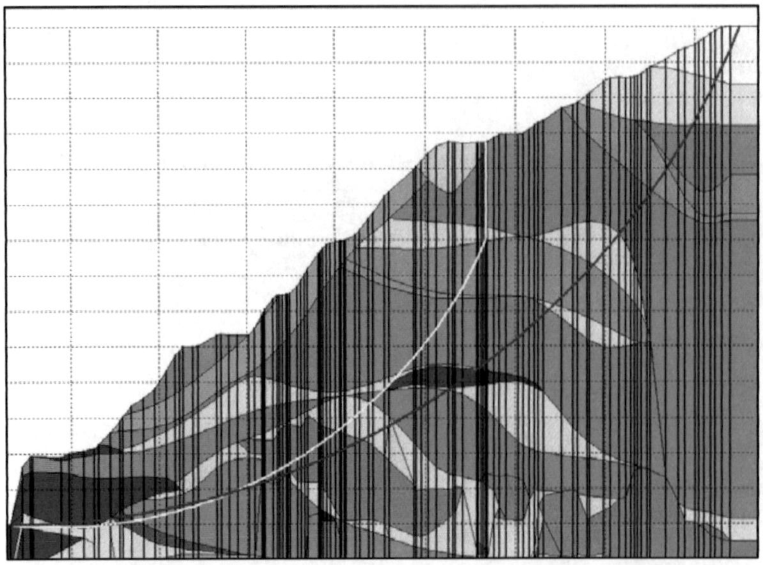

Figure 4. Pit wall modelling and calculation of pit wall stability coefficient in the software Scad Office Slope based on Fellenius method.

In particular, the Novo-Shemurskiy quarry deviations in stability coefficient values did not exceed 0.01-0.07 which is comparable to the accuracy of its determination. Several large deviations reaching 0.33-0.52 were recorded in the Yubileiny quarry. Here the computer calculations and those of polygon of forces performed by different researchers explain the difference in the approach to reference geochemical condition modelling directly affecting the results of calculations. Even when outside experts were involved in checking calculations, the software package introduced an additional coefficient to the stability margin enhancing safety of further operation of a hazardous production facility. This is particularly important when obtaining safety margin coefficients close to the ultimate ones (1.3 – 1.5) where an error as low as one-tenth reclassifies the pit wall from unstable to safe for operation. Moreover, if deformations at the initial stage may be insignificant for non-working pit banks placed in the limit position, the incorrect justification of the pit wall stability with a main exit to the quarry can potentially result in emergencies and the interruption or complete stop in the enterprise operation [18].

5. Conclusions

Close examination reveals that the software package Scad Office Slope having offered acceptable results of the calculations is likely to be applied in the pit wall and slope stability ranking. The application of computer calculations allows:

- enhancing performance effectiveness of slope stability study since the software facilitates completing a great number of calculations with less specialist's effort and time;
- increasing the degree of calculation details including rating the stability of a sufficiently large number of local areas in the height of pit walls or slopes;
- defining accurately the position of the most likely sliding surface that is of great importance for rating the stability of the quarry slopes or mounds where the surface is based on trial method;
- varying quickly enough initial physical and mechanical characteristics observing the dynamics of safety margin changing under initial geomechanical condition changes (in particular, the

IOP Conf. Series: Materials Science and Engineering **972** (2020) 012012 doi:10.1088/1757-899X/972/1/012012

decrease in the rock strength with its watering increasing or underworking of a pit wall by underground mining).

However, the application of software packages for pit wall or slope stability rating within the limits of research works does not preclude the need for the implementation of traditional methods (force polygon method or method for algebraic force addition) regulated by normative documents. Gaining reference values, the methods are employed for checking calculations on the most specific strip pits. Furthermore, the modelling of reference geomechanical conditions – geological and structural modelling of pit wall or slope under study, provision of calculated physical and mechanical characteristics of the rock, allowance for loads of different kinds – is significant to ensure calculation results regardless a calculation method applied.

References

[1] Sokolovskiy V V 1960 Loose medium statics (Moscow: Phizmatgiz)
[2] Golushkevitch S S 1957 Statics of saturated soil limit states (Moscow: Gostechteorizdat)
[3] Shahunyantz G M 1948 The principles of practical computation for free and support slopes
[4] Fellenius V 1993 Solid statics (Moscow: Gosstroyizdat)
[5] Fisenko G L 1965 Pit wall and slope stability (Moscow: Nedra)
[6] Short-term methodology guidelines on variation management of pit walls in nonferrous-metal Quarry: approved by the Ministry of Metallurgy of the USSR 1989 (Moscow: Uni-promed) p. 128
[7] Methodology guidelines on angle detection of pit wall inclination, slope front and dump of the quarries under construction and operation 1972 (Leningrad: VNIMI) p 168
[8] Slope stability control rules for coal strip mine / Mining geomechanics and mine surveying Research Institute: approved by GGTN of the Russian Federation 1988 (Saint Petersburg: VNIMI) p 208
[9] Instructions of safe mining work practice at integrated ore and nonmetal mineral. Referential normative system Norma CS: approved by GGTN of the Russian Federation (database accessed on 03 July 2014)
[10] Instructions for deformation surveillance of pit walls, pit slopes, slope fronts and dumps at mining plants and for planning their stability assurance activities: approved by GGTN of the USSR 1971 (Leningrad: VNIMI) p 187
[11] Methodology guidelines on deformation surveillance of pit walls and dumps, on their result interpretation and on stability forecast 1987 (Leningrad: VNIMI) p 118
[12] Instructions for stability calculations of pit walls during their taking down and for pit wall surrounding custody: approved by the Ministry of Coal Industry of the USSR 1977 (Leningrad: VNIMI) p 55
[13] Methodology guidelines on stability calculation and bearing capacity of dumps 1987 (Leningrad: VNIMI) p. 126
[14] Karpilovskiy V S et al 2012 SCAD Office, SNiP implementation in engineering programmes (Moscow: SCAD SOFT)
[15] Fedorovskiy V G and Kurilo S V 1997 Computational method for slopes and walls *Geoecology* **6** 95-106
[16] Khuan Y Kh 1988 Earth wall stability (Moscow: Stroiizdat)
[17] M.G. Anderson, K.S. Richards, K.S. "Slope Stability: Geotechnical Engineering and Geomorphology", 1987, John Wiley and Sons, N.Y.
[18] Design guidelines for pit walls 2015 ed. John Reed and Peter Stacy (translated from English) (Ekaterinburg: Pravoved) p 544

Regulation of dynamic stress-strain state and reliability of deformable systems with vibration dampers under harmonic loads

Yu A Gerber[1] and V G Sebeshev[1]

[1]Novosibirsk State University of Architecture and Civil Engineering (Sibstrin), 113, Leningradskaya, Novosibirsk, 630008, Russia

jura.gerber@yandex.ru

Abstract. The problem of analysis linearly deformable systems with a finite number of single-mass dynamic vibration dampers (DVD), used as regulators of the stress-strain state (SSS) with different variants of the harmonic load (by location and type) was considered. Equations of state of the system with dampers have been obtained and a method has been proposed to choice a set of parameters (masses and stiffness) for each of the DVD, taking into account the properties of the system, loads, and parameters of other DVD and the specified requirements of regulation. The developed design apparatus has been used for solving a model problem of a dynamic SSS regulation in a rod system with several dampers and for estimation of its reliability with polycriterial serviceability requirements. The quantitative results of calculations are presented. The comparative analysis of the reliability have been performed for unprotected systems and systems with DVD. The possibility of determining the rational location and parameters of DVD under harmonic non-point loading with the target reliability of the structure is shown.

1. Introduction

One of the most dangerous types of impacts on construction systems (structures, building) is dynamic loading, which can cause unacceptable vibrations. An effective way to oppose vibrations is the use of dynamic vibration dampers [1, 2]. In the case when the dynamic load is applied at one point, then when a DVD is installed at the loading point and in load's direction, complete damping of the vibration in the system is possible (ideal DVD). If there are several sources of vibration, then the setting of a one single-mass DVD may not be enough for significant reduction of the displacements level in the system. It becomes possible to improve the dynamic SSS of system noticeably with the help of installation of several dampers. The vibration damper, in its essence, is a very powerful mean of regulating the dynamic SSS of a system. As it was shown by previous studies [3], a system protected by a damper is sensitive even to small random deviations of parameters from their design values. This raises the problem of calculating the characteristics of each damper, taking into account the properties of the system, the load and other DVD. It becomes possible to maintain successful work of the system in a frequencies zone, where large gradients of amplitude-frequency characteristic (AFC) of the unprotected structure take place. By reason of the stochastic nature of all design parameters, an estimation of the reliability of systems protected by a DVD complex becomes necessary.

The purposes of the work are: developing a method for analysis regulation of the dynamic state of linearly deformable systems protected by a finite number of vibration dampers (regulators), taking into account the specific SSS parameters requirements on the system; calculation of the modelling systems reliability with a complex of DVDs; verification of the proposed method.

2. The main part

The solution to the problem of the system SSS regulation
A harmonically loaded arbitrary system with the group of n_d single-mass DVD is considered (figure 1). The number of degrees of freedom for the unprotected system is n, for dampers – n_d.

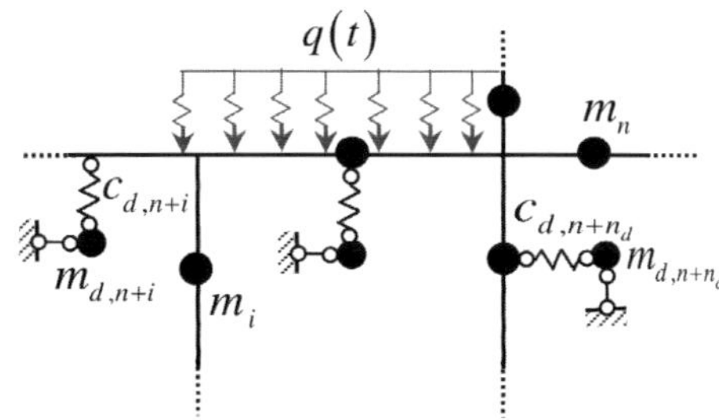

Figure 1. Design scheme of a deformable system with dynamic vibration dampers.

The specific determination of the DVD location depends on the design features of the system and the mode of its vibration under a given load. The qualitative prediction of the dampers' influence leads to the conclusion that it is advisable to install them in places where the dynamic displacement of the unprotected structure are the greatest. The equations of the dynamic state of a system with a DVD complex for forced steady-state vibration [4] are written in the matrix form, in the amplitudes of inertial forces, with a matrix of dynamic pliability, as

$$\left[\delta^*\right]\cdot[J]+[\Delta_F]=0, \tag{1}$$

where $\left[\delta^*\right]$ – system's dynamic pliability matrix of size $(n+n_d)\times(n+n_d)$;
$[J]$ – inertia forces vector; $[\Delta_F]$ – load caused amplitude displacements vector.

The first n equations (1) characterize the amplitude state of the system itself (the first group of equations), and the subsequent n_d equations relate to the dampers (the second group of equations).

Let present the matrix of dynamic pliability of the system in the following form:

$$\left[\delta^*\right]=\left[\begin{array}{ccc|ccc}
\delta^*_{11} & \cdots & \delta_{1n} & \delta_{1,n+1} & \cdots & \delta_{1,n+n_d} \\
\cdots & \cdots & \cdots & \cdots & \cdots & \cdots \\
\delta_{n1} & \cdots & \delta^*_{nn} & \delta_{n,n+1} & \cdots & \delta_{n,n+n_d} \\
\hline
\delta_{n+1,1} & \cdots & \delta_{n+1,n} & \delta^*_{n+1,n+1} & \cdots & \delta_{n+1,n+n_d} \\
\cdots & \cdots & \cdots & \cdots & \cdots & \cdots \\
\delta_{n+n_d,1} & \cdots & \delta_{n+n_d,n} & \delta_{n+n_d,n+1} & \cdots & \delta^*_{n+n_d,n+n_d}
\end{array}\right]$$

or

$$\left[\delta^*\right]=\left[\begin{array}{c}\delta^*_{J_s}\\ \hline \delta^*_{J_d}\end{array}\right]=\left[\begin{array}{c|c}\delta^*_{J_sJ_s} & \delta_{J_sJ_d}\\ \hline \delta_{J_dJ_s} & \delta^*_{J_dJ_d}\end{array}\right], \tag{2}$$

where $\delta^*_{ii}=\delta_{ii}-\left(m_i\omega_F^2\right)^{-1}$ $(i=1,...,n)$; $\delta^*_{n+i,n+i}=\delta_{n+i,n+i}-\left(m_{d,n+i}\omega_F^2\right)^{-1}$ $(i=1,...,n_d)$, $\delta_{n+i,n+i}=\delta_{k,n+i}+\left(c_{d,n+i}\right)^{-1}$ $(k=n+n_d+1,...,n+2n_d)$, k – number of ith damper's attachment point; m_i – the mass of the unprotected system corresponding to the ith degree of freedom; $m_{d,n+i}$ – damper mass; ω_F – forcing frequency. The indexation of parameters is adopted unified in all equations for notational convenience: $i=1,...,n$ – for main points of system; $i=n+1,...,n+n_d$ – for damper masses; $k=n+n_d+1,...,n+2n_d$ – for system points at damper installation places.

In this case the diagonal elements of the block $\left[\delta^*_{J_dJ_d}\right]$ contains unknown parameters of damper – mass (m_d) and stiffness (c_d), which must be selected in such a way to fulfill the requirements for improving the indicators of the system's dynamic SSS. Consequently, the problem of selecting the parameters of the DVD complex is considered as the problem of the system's SSS regulation.

We've used displacement restrictions as the requirements of regulation:

$$\left[\Delta^r_{n+n_d+1}\cdots\Delta^r_{n+2n_d}\right]^{\mathrm{T}}=\left[\left\langle\Delta^r_{n+n_d+1}\right\rangle\cdots\left\langle\Delta^r_{n+2n_d}\right\rangle\right]^{\mathrm{T}}, \tag{3}$$

where Δ^r_k $(k=n+n_d+1,...,n+2n_d)$ – displacements of system points in places of fastening of DVD; $\left\langle\Delta^r_k\right\rangle$ $(k=n+n_d+1,...,n+2n_d)$ – target displacements of points where DVD are placed.

Thus, the regulation equations are written as

$$\left[\Delta^r\right]=\left[\delta^r\right]\cdot\left[J\right]+\left[\Delta^r_F\right]=\left[\left\langle\Delta^r\right\rangle\right]$$

or

$$\begin{bmatrix}\delta_{k1} & \cdots & \delta_{kn} & \delta_{k,n+1} & \cdots & \delta_{k,n+n_d}\\ \cdots & \cdots & \cdots & \cdots & \cdots & \cdots\\ \delta_{k+n_d,1} & \cdots & \delta_{k+n_d,n} & \delta_{k+n,n+1} & \cdots & \delta_{k+n_d,n+n_d}\end{bmatrix}\begin{bmatrix}J_1\\ \cdots\\ J_n\\ \hline J_{n+1}\\ \cdots\\ J_{n+n_d}\end{bmatrix}+\begin{bmatrix}\Delta_{n+n_d+1,F}\\ \cdots\\ \Delta_{n+2n_d,F}\end{bmatrix}=\begin{bmatrix}\left\langle\Delta^r_{n+n_d+1}\right\rangle\\ \cdots\\ \left\langle\Delta^r_{n+2n_d}\right\rangle\end{bmatrix}. \tag{4}$$

The number of regulation equations corresponds to the number of DVDs. In this case only inertia forces are unknown in equations (4) while the remaining parameters may be calculated or specified previously. Since the inertia forces $[J]$ in (1) and (4) are the same, as well as the numbers of equations of the second group in (1) and of the regulation equations (4) are the same, it is possible to change the mentioned groups of equations in places, as a result we have:

$$\begin{bmatrix} \delta^*_{11} & \dots & \delta_{1n} & \delta_{1,n+1} & \dots & \delta_{1,n+n_d} \\ \dots & \dots & \dots & \dots & \dots & \dots \\ \delta_{n1} & \dots & \delta^*_{nn} & \delta_{n,n+1} & \dots & \delta_{n,n+n_d} \\ \hline \delta_{k1} & \dots & \delta_{kn} & \delta_{k,n+1} & \dots & \delta_{k,n+n_d} \\ \dots & \dots & \dots & \dots & \dots & \dots \\ \delta_{k+n_d,1} & \dots & \delta_{k+n_d,n} & \delta_{k+n_d,n+1} & \dots & \delta_{k+n_d,n+n_d} \end{bmatrix} \begin{bmatrix} J_1 \\ \dots \\ J_n \\ \hline J_{n+1} \\ \dots \\ J_{n+n_d} \end{bmatrix} + \begin{bmatrix} \Delta_{1F} \\ \dots \\ \Delta_{nF} \\ \hline \Delta_{n+n_d+1,F} - \left\langle \Delta^r_{n+n_d+1} \right\rangle \\ \dots \\ \Delta_{n+2n_d,F} - \left\langle \Delta^r_{n+2n_d} \right\rangle \end{bmatrix} = 0$$

or

$$\begin{bmatrix} \delta^*_{J_s} \\ \hline \delta^r \end{bmatrix} \cdot \begin{bmatrix} J^n \\ \hline J^{n_d} \end{bmatrix} + \begin{bmatrix} \Delta^n_F \\ \hline \Delta^r_F - \left\langle \Delta^r \right\rangle \end{bmatrix} = 0. \qquad (5)$$

The resulting new system (5) does not include the unknown parameters of dampers. It is important in this case that the inertia forces obtained from (5) correspond to the regulation requirements.

Let's consider the second group of equations, replaced in (1) by the regulation equations (4):

$$\left[\delta^*_{J_d}\right] \cdot [J] + \left[\Delta^d_F\right] = 0. \qquad (6)$$

Each equation (6) contains unknown damper parameters (stiffness and mass), so the solution of the equation leads to the relationship expression between them taking into account the target displacements in the places of dampers fastenings:

$$G_{d,n+i} = \frac{1}{c_{d,n+i}} - \frac{1}{m_{d,n+i}\,\omega_F^2} = \frac{-\sum_{j=1}^{n+i-1} \delta_{n+i,j} J_j - \sum_{j=i+1}^{n+n_d} \delta_{n+i,j} J_j - \Delta_{n+i,F}}{J_{n+1}} - \delta_{k,n+i}\,, \qquad (7)$$

where $i = 1,...,n_d$; $k = n + n_d + 1,...,n + 2n_d$, k – number of ith damper's attachment point.

If one of the damper parameters is preset, the value of the other one may be determined from (7). Thus, the proposed algorithm makes it easy to choose the characteristics for the entire complex of dampers in accordance with the certain initial requirements.

Example

The two degree-of-freedom beam under the uniformly distributed harmonic load (figure 2, (a)) was considered with specified data: $a = 1\,\text{m}$, $EI = 22500\ \text{kN}\cdot\text{m}^2$, $m_1 = 1000$ kg, $m_2 = 2000$ kg, $q = 20\ \text{kN/m}$, $\omega_F = 90\,\text{s}^{-1}$.

The results of realized calculations are presented in Table 1 where the displacements of masses and points *A*, *B* also their dynamic magnification factors (DMF) are given. The bending moment diagrams are shown in figure 2, (b), (c).

Table 1. Displacements and DMFs.

Point	y_{st} (m)	y_{dyn} (m)	DMF
1	0.001833	0.006327	3.45
2	0.001722	0.006633	3.85
A	0.002519	0.008984	3.57
B	-0.00167	-0.007269	4.36

IOP Conf. Series: Materials Science and Engineering **972** (2020) 012013 doi:10.1088/1757-899X/972/1/012013

Note that displacements and forces during dynamic loading increase by more than 3 times in comparison with the corresponding static ones. To improve the system's SSS two DVD are used. They install at the places of maximum dynamic displacements (points A, B) – see figure 3, (a).

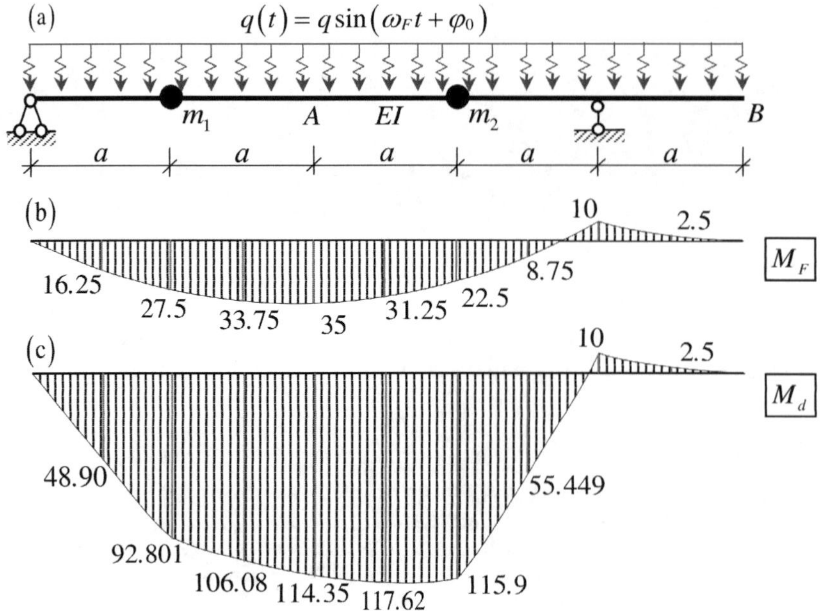

Figure 2. Design model (a); bending moments caused by the conventionally static amplitude harmonic load (b); dynamic bending moments (c).

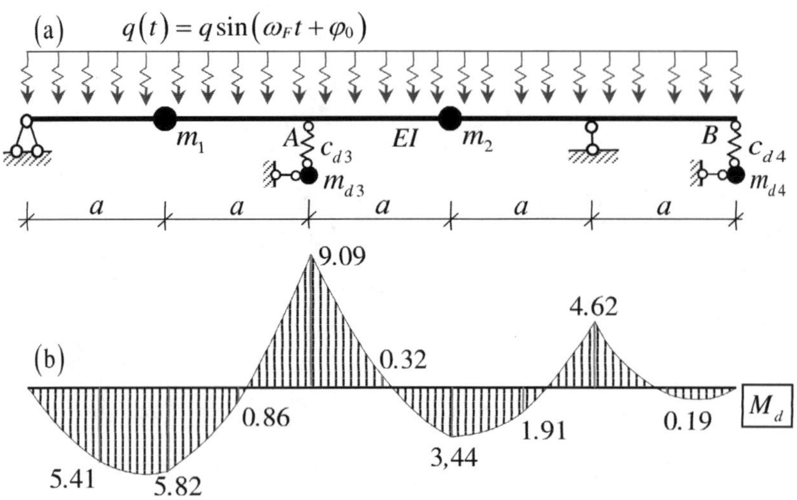

Figure 3. Design model with DVDs (a); dynamic bending moment diagram (b).

The selection of the damper parameters was made according to the above presented algorithm with the requirement that the displacements are equal zero in the points where the dampers are placed (points A and B): $\langle \Delta_5^r \rangle = 0$; $\langle \Delta_6^r \rangle = 0$. From (7) $G_{d,3} = G_{d,4} = 0$ were obtained in this case. The following parameters of dampers were used in the calculation: $m_{d3} = 250\,\text{kg}$; $c_{d3} = 2025\ \text{kN/m}$; $m_{d4} = 250\,\text{kg}$; $c_{d4} = 2025\ \text{kN/m}$. The distribution of the dynamic bending moments is presented in figure 3, (b). Note that their maximum absolute values have decreased significantly – almost 13 times. Varying the parameters of regulation it is possible to choose a combination in which the distribution of forces will be even better. It will positively affect the reliability and durability of the system under multi-cycle loading by reducing the amplitude of the dynamic component of the total stresses [5].

Calculation of reliability
The reliability calculation was performed according to the method described in [6], using the concepts of resistance R, load effect Q and generalized serviceability reserve $S = R - Q$.

The following requirements for non-failure performance of system were used:

- for the bending dynamic moments in the pre-specified design sections of beam – $M_d < [M]$ ($\bar{R}_M = [M] = 140\ \text{kN}\cdot\text{m}$, $\bar{Q}_M = \bar{M}_d(x_i)$);
- for the dynamic displacements of characteristic points – $y_d < [y]$ ($\bar{R}_y = [y] = 0{,}01\,\text{m}$, $\bar{Q}_y = \bar{y}_d(x_i)$).

Hereinafter, the symbol « $\bar{\ }$ » denotes the mean, «ˆ» – the standard deviation; A_X – coefficient of variation of random variable X.

The results of calculations of reliability and failure probability for the unprotected system (figure 2 (a)) for the first criterion are presented in Table 2, and for the second – in Table 3. The calculations were performed with the following probabilistic characteristics: $\bar{m}_1 = 1000\,\text{kg}$; $A_{m_1} = 0.03$; $\bar{m}_2 = 2000\,\text{kg}$; $A_{m_2} = 0.02$; $\overline{EI} = 22500\ \text{kN}\cdot\text{m}^2$; $A_{EI} = 0.04$; $\bar{q} = 20\ \text{kN/m}$; $A_q = 0.06$; $\bar{\omega}_F = 90\,\text{s}^{-1}$; $A_{\omega_F} = 0.01$. All input random parameters are normally distributed.

The probabilistic characteristics of the output parameters were determined by the statistical linearization method (SLM) [7].

Table 2. Results of calculations for the unprotected system (figure 2) (by bending dynamic moments in beam).

x_i (m)	$\bar{M}_d$ (kN·m)	$\bar{S}$ (kN·m)	$\hat{S}$ (kN·m)	β_S	P_S	P_f
0.5	48.900	91.0996	4.3316	21.0314	~1	~0
1	92.801	47.1991	8.4629	5.5772	0.999999	1.222E-08
1.5	106.075	33.9247	9.5051	3.5691	0.999821	0.0001791
2	114.350	25.6502	10.3629	2.4752	0.993342	0.0066582
2.5	117.624	22.3758	11.0387	2.0270	0.978670	0.0213298
3	115.899	24.1014	11.5492	2.0868	0.981548	0.0184515
3.5	55.449	84.5507	5.6856	14.8711	~1	~0
4	-10.000	130.0000	0.6000	216.6666	~1	~0
4.5	-2.500	137.5000	0.1500	916.6666	~1	~0

The total probability of failure by the criterion for limiting bending moments – $P_f = 0.0459546$.

Table 3. Results of calculations for the unprotected system (figure 2) (by dynamic displacements of beam's points).

Point	$\bar{y}_d$ (m)	$\bar{S}$ (m)	$\hat{S}$ (m)	β_S	P_S	P_f
1	0.00633	0.003673	0.000584	6.289054	~1	~0
2	0.00663	0.003367	0.000628	5.359272	~1	~0
A	0.00898	0.001016	0.000748	1.358642	0.912870	0.087130
B	-0.00727	0.002731	0.000649	4.209778	0.999987	0.000013

The total probability of failure by the criterion for target displacements – $P_f = 0.1290917$.

According to the data of tables 2 and 3, the total failure probability under the multi-criteria serviceability requirements is $P_f = 0.16911$, that is unacceptable according to [8].

The results of calculations reliability and probability of failure for the system with DVDs (figure 3, (a)) for the first criterion are shown in table 4, and for the second – in table 5. The calculations were made taking into account the following probabilistic characteristics of the original system (see above) and dampers: $\bar{m}_{d3} = 250\,\text{kg}$; $A_{m_{d3}} = 0.001$; $\bar{c}_{d3} = 2025\,\text{kN/m}$; $A_{c_{d3}} = 0.001$; $\bar{m}_{d4} = 250\,\text{kg}$; $A_{m_{d4}} = 0.001$; $\bar{c}_{d4} = 2025\,\text{kN/m}$; $A_{c_{d4}} = 0.001$.

Table 4. Results of calculations for the system with DVDs (figure 3) (by bending dynamic moments in beam).

x_i (m)	$\bar{M}_d$ (kN·m)	$\bar{S}$ (kN·m)	$\hat{S}$ (kN·m)	β_S	P_S	P_f
0.5	5.408	134.592	3.064	43.922	~1	~0
1	5.815	134.185	6.104	21.982	~1	~0
1.5	0.861	139.139	7.678	18.121	~1	~0
2	-9.094	130.906	9.278	14.109	~1	~0
2.5	-0.325	139.675	5.318	26.265	~1	~0
3	3.445	136.555	1.390	98.211	~1	~0
3.5	1.913	138.087	4.607	29.976	~1	~0
04	-4.620	135.380	10.587	12.788	~1	~0
4.5	0.190	139.810	5.292	26.421	~1	~0

The total probability of failure by the criterion for limiting bending moments – $P_f \approx 0$

Table 5. Results of calculations for the system with DVDs (figure 3) (by dynamic displacements of beam's points).

Point	$\bar{y}_d$ (m)	$\bar{S}$ (m)	$\hat{S}$ (m)	β_S	P_S	P_f
1	0.000090	0.00991	0.0001813	54.665	~1	~0
2	0.000037	0.00996	0.0001344	74.144	~1	~0
A	0.000000	0.01000	0.0001429	69.983	~1	~0
B	0.000000	0.01000	0.0000968	103.288	~1	~0

The total probability of failure by the criterion for target displacements – $P_f \approx 0$.

According to the data of tables 4 and 5, the total failure probability under the multi-criteria serviceability requirements is $P_f \approx 0$, which corresponds to a very high level of reliability.

The comparison of the reliability indices of the unprotected system and the system with DVDs makes it clear that the setting of vibration dampers causes a significant increasing of the reliability in accordance with the design requirements [8, 9]. In addition to the performance criteria considered in the problem solved above, the determination of reliability and durability by the limiting of the material fatigue strength is vital; for this case the technique described in [10] can be used.

3. Conclusion

1. The solution of a problem on determining the characteristics of the DVD complex as a regulators of the dynamic SSS of a deformable multi-degree-of-freedom system under vibration loads is given. The design expressions for calculations the needed masses and stiffnesses of dumpers in accordance with the target limitations of structure points' displacements are obtained in the convenient form for practical use.

2. The proposed methodology and algorithm allow calculating and optimizing the DVD parameters using the specified (required) values of the SSS parameters.

3. The use of a DVD complex significantly improves the dynamic SSS of a multi-degree-of-freedom system and, as a result, leads to the increasing reliability according to the design criteria and requirements of serviceability.

Acknowledgement

Work is realized in accordance with the Thematic plans of investigations in the Novosibirsk State University of Architecture and Civil Engineering (Sibstrin), Novosibirsk, Russia.

References

[1] Korenev B G and Reznikov L M 1988 *Dynamic vibrations dampers: Theory and technical applications* (Moscow: Science Publishing) 304 p

[2] Dukart AV and Oleynik AI *2015 Dynamic vibrations dampers of structures* (Moscow: ASV Publishing) 248 p

[3] Sebeshev V G and Gerber Yu A 2014 *Vestnik Tomsk State Univ. Arch. Civil Eng* **4** pp 93 – 105

[4] Buchholdt H A and Moossavi Nejad S E 2012 *Structural Dynamics for Engineers. Second edition* (London: ICE Publishing) 318 p

[5] Liebowitz H 1971 *Fracture. An advanced treatise Vol. 3 Engineering fundamental and environmental effects* (New York and London: Academic Press. Publishing) 774 p

[6] Vedyakov I I and Raiser V D 2018 *Reliability of Building Structures. Theory and Calculation* (Moscow: ASV Publishing) 414 p

[7] Venttsel E S and Ovcharov L A 2000 *Theory of Probability and its Engineering Applications* (Moscow: Higher School Publishing) 480 p

[8] Eurocodes *Proc. of the Institution of Civil Engineers, Structures and Buildings* 2008 **161 SB4** (August) pp 209 – 214

[9] Standard 27751-2014 *Reliability of Constructional Structures and Foundations. General principles* (Moscow: Standardinform) 14 p

[10] Sebeshev V G and Gerber Yu A 2014 *Problems of optimal structural design: reports of the 3rd All-Russian Conf.* (Novosibirsk: NSUACE (Sibstrin) Publishing) pp 367 – 380

IOP Conf. Series: Materials Science and Engineering **972** (2020) 012014 doi:10.1088/1757-899X/972/1/012014

Lateral-Torsional Buckling Behaviour of Triangularly Tapered Corrugated Web Beam

Andres Camilo Amaya Franco[1], Sergey Kudryavtsev[1]

[1]Ural Federal University, Mira Street, 17a, Yekaterinburg 620002, Russia

Email: acamayaf@gmail.com

Abstract. This paper presents a study of the behaviour of Lateral-Torsional Buckling (LTB) that is one of the common failure modes of large span beams. LTB for non-prismatic I-beams still needs more investigations, due to its sophisticated formulation for cross-section properties. This paper presents the results of numerical (Finite Element Software ABAQUS) and theoretical analysis for LTB of a simply supported I-beam with triangularly tapered corrugated web under concentrated loads. The corrugation densities adopted in this study represent the most commonly used types for such structures in practice of construction. For numerical analysis, the formulas were adopted from the case of a trapezoidal corrugated web, which is successfully verified by different authors. Recommendations of this study are given for simply supported tapered triangularly corrugated web at lateral torsional buckling.

1. Introduction

Tapered or non-prismatic structures have been used principally for bridges and industrial structures. When these kinds of elements are designed with corrugated web that is shown in figure 1, the weight of the structure can be reduced alongside with the improvement in the structure's stability. Although tapered (non-prismatic) beams are not so common and less common is the usage of triangularly corrugated web (TCW) beams for such structures.

The typical types of web used in steel structures are sinusoidal and trapezoidal web, in otherwise TCW is not so common to use in construction despite of all the advantages it has over the others, for example, the web could be thicker than others and have at the same time the better bearing capacity. This is especially important for heavy operating conditions that are characterized by aggressive environment and large values of loads on the structures. Due to its profiled form, corrugated web exhibits enhanced shear stability and therefore eliminate the need for additional transverse stiffeners or thicker web plates [1].

The first detailed studies about LTB of non-prismatic I-girders with flat webs were carried out by Kitipornchai and Trahair (1972) [2], Zhang Lei and Tong Geng Shu [3] have more recent studies about statically determined beams or cantilevers. Elgaaly (1996) was focused on the vertically trapezoidal corrugation. The studies found that the failure of beams under shear loading is due to local buckling of the web for coarse corrugation and global buckling on the web for dense corrugation [4].

There are limited studies about the LTB phenomenon of non-prismatic I-girders even with flat webs have been carried out, as well as the research studies on TCW at suck conditions are very limited too. A numerical analysis is carried out by Sayed-Ahmed (2005) shows that resistance to lateral torsion-flexure buckling of such girders is from 15% to 37% higher than the resistance of plate girders with traditional plane webs [5].

Hasan, Denan and Keong (2015) presented experimental study of the buckling moment resistance ($M_{b,Rd}$) of triangular web profile (TWP), which was determined on the basis of the LTB experimental values of the TWP steel section. The main conclusion was that the section properties such as width, depth, flange thickness, web thickness and corrugation shape of web profile affect the buckling moment resistance, $M_{b,Rd}$, of the steel section. The larger section properties of steel sections increase moment buckling resistance [6].

In recent research, Kudryavtsev (2018) presented an investigation into the behaviour and design of triangularly corrugated steel web beams with web openings. He recommended that it is possible to make openings in corrugated web beams in areas of pure bending or in areas with minimal shear stresses. In this case, it is generally possible to avoid additional reinforcement of the opening [7]. Also, Kudryavtsev (2019) investigated LTB behaviour of prismatic triangularly corrugated web beam and proposed a design recommendation for simple supported case accordance with numeral results [8].

This paper presents a calculation for lateral-torsional buckling on triangularly non-prismatic beam with triangularly corrugated web. The numerical parametric studies analysis is compared with a parametric study of finite element analysis (FEA), which carries out for simply supported beams. This parametric study considered various cases including different type of non-prismatic section, span of the beam and the corrugation density expressed as a ratio between height and length of half-wave of corrugation.

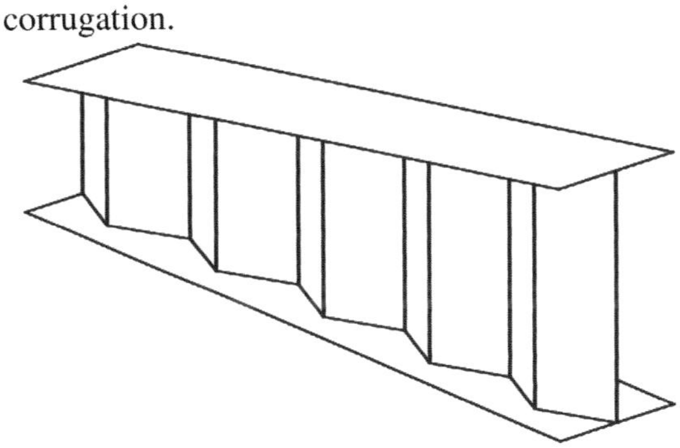

Figure 1. Section of non-prismatic triangularly corrugated web beam.

2. Design of non-prismatic triangularly corrugated web steel beam

The limitations for this study are similar for a steel plate girder with a flat web plate. The destruction of steel beam with a corrugated web may occur for several reasons: due to the exhaustion of the load-bearing capacity of flanges at bending or corrugated web at shear; due to local buckling of compressed flange or at areas of the web; due to overall buckling of the entire structure. In addition, according to the conditions of suitability for normal operation, the maximum vertical deflections of the beam must be limited [8].

Geometrical parameters of the structure and the beam design scheme adopted for this study are shown in figure 2. The AISC Specification [9] covers the design of a doubly symmetric web-tapered I-girder with equal flange. The depth of web varies linearly as follows,

$$h_w = h_{w,\min}\left(1+\frac{l-z}{l}\xi\right), \tag{1}$$

where ξ is the taper ratio calculated by

$$\xi = \frac{h_{w,\max}-h_{w,\min}}{h_{w,\min}} \leq 0.268\frac{l}{h_{\min}} \leq 6.0, \tag{2}$$

in which h_w is the web height that variable over the span of the beam l, $h_{w,\min}$ and $h_{w,\max}$ are the web heights at the each ends of the structure, z is position of cross-section from the lower end of the beam.

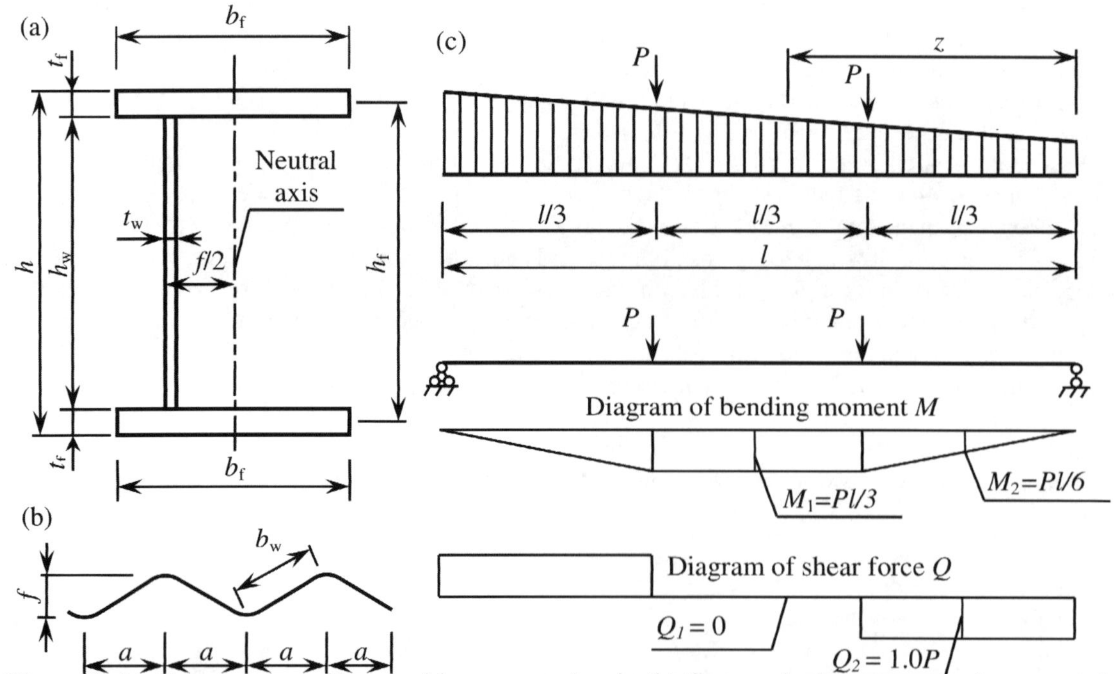

Figure 2. (a) Cross section of beam with corrugated web; (b) Geometrical parameters of corrugated web with triangular shaped corrugations; (c) Considered beam design scheme.

In the study of Samanta and Mukhopadhyay [9] given the formula to calculate the shear modulus G_c for trapezoidal corrugated web that could be adopted for the case of triangularly corrugated web by

$$G_c = \frac{2a}{a+b_w}G, \tag{3}$$

in which $G = 78000MPa$ is shear modulus for steel.

The pure torsional constant I_c of a steel I-beam, according to Lindner [10] could be determined by

$$I_c = \frac{2b_f t_f^3 + h_w t_w^3}{3}. \tag{4}$$

Considering the study of Nguyen [11], the moment of inertia on the weak axis (y axis) $I_{y,c}$ and warping constant of I-beam with trapezoidal web corrugations $C_{w,c}$ are a quadratic function of corrugation height f and varies depending on changing of f from 0 to $f_{\max}$ along the beam span. To simplify the expressions for $I_{y,c}$ and $C_{w,c}$ are given like an average values between case with $f = 0$ and case with $f = f_{\max}$ as follows:

$$I_{y,c} = \frac{t_f b_f\left(2t_f b_f^3 + t_w h_w b_f^2 + 12(f/2)^2 t_w h_w\right)}{6\left(2t_f b_f + t_w h_w\right)}, \tag{5}$$

$$C_{w,c} = \frac{h_w^2 t_f b_f\left(6t_f b_f^3 + t_w h_w b_f^2 + 12(f/2)^2 t_w h_w\right)}{24\left(6t_f b_f + t_w h_w\right)}. \tag{6}$$

Also, Nguyen [11] proposed a formula to evaluate an elastic lateral-torsional buckling strength of an I-beam for a simply supported conditions

$$M_{ocr,c} = \frac{\pi}{l} \cdot \sqrt{EI_{y,c} G_c I_c} \cdot \sqrt{1 + W_c^2}, \tag{7}$$

where $W_c = \frac{\pi}{l}\sqrt{\frac{EC_{wc}}{G_c I_c}}$ represents the effect of warping torsional stiffness.

Nguyen, Nguyen-Van, Han, Choi and Kang [12], adopted formula for I-beam with flat webs according the SSRC Guide [10] to determinate the critical moment of a prismatic I-beam with corrugated webs, warping fixed, lateral bending fixed, and completely fixed end restraint conditions as follow

$$M_{cr,c} = C_{b,c} \frac{\pi^2 EI_{y,c}}{(kl)^2} \cdot \sqrt{\left(\frac{k}{k_w}\right)^2 \frac{C_{w,c}}{I_{y,c}} + \frac{(kl)^2 G_c I_c}{\pi^2 EI_{y,c}}}, \tag{8}$$

where k is the lateral bending coefficient and k_w is the warping coefficient. For free lateral bending $k = 1$. For free warping $k_w = 1$. For lateral bending fixed $k = 0.5$ and for warping fixed $k_w = 0.5$.

Nguyen, Nguyen-Van, Han, Choi and Kang [12] proposed the closed-form expression for moment modification factor $C_{b,c}$ which is applicable for linear and non-linear moment diagrams

$$C_{b,c} = \frac{12.5M_{\max}}{2.5M_{\max} + 3M_A + 4M_B + 3M_C}, \tag{9}$$

where M_{max} is the absolute maximum moment along the beam span l, and M_A, M_B and M_C are the absolute moments at the quarter, the centre, and the three-quarter point, respectively. Consequently, with any end restraint conditions, the values of $C_{b,c}$ are the same. The values of $C_{b,c}$ for the I-beam with corrugations under moment gradient have not yet been proposed [11]. The values of $C_{b,c}$ are could be written by the next equation for a simply supported beam

$$C_{b,c} = -0.078\alpha^6 - 0.19\alpha^5 - 0.274\alpha^4 - 0.182\alpha^3 + 0.364\alpha^2 + 1.219\alpha + 1.829, \tag{10}$$

where α is the end moment ratio i.e. $-1{,}0 \le \alpha \le 1{,}0$.

3. Numerical parametric study

Typical finite element model adopted for the corrugated web beams are shown in figure 3a. Herein, the study was developed in the software ABAQUS. The models are composed of upper and lower flanges and triangularly corrugated webs with different heights.

Table 1. Geometry data of analysis models.

Model	a, mm	f, mm	b_w, mm	t_w, mm	$h_{w,min}$, mm	$h_{w,max}$, mm	b_f, mm	t_f, mm	l, mm
(1)	(2)	(3)	(4)	(5)	(6)	(7)	(8)	(9)	(10)
A1	100	50	111.80	4	220	620	200	10	12000
A2	100	50	111.80	4	320	620	200	10	12000
A3	100	50	111.80	4	420	620	200	10	12000
B1	150	60	161.55	6	220	620	200	10	12000
B2	150	60	161.55	6	320	620	200	10	12000
B3	150	60	161.55	6	420	620	200	10	12000
C1	200	70	211.90	8	220	620	200	10	12000
C2	200	70	211.90	8	320	620	200	10	12000
C3	200	70	211.90	8	420	620	200	10	12000

IOP Conf. Series: Materials Science and Engineering **972** (2020) 012014 doi:10.1088/1757-899X/972/1/012014

For the web were assumed three different corrugation densities and three different heights. For each density just one beam span was considered. All the dimensions of the models are given in Table 1. All models were loaded by two concentrated loads P = 1kN located in the thirds of the upper flange at the centre line of the structure. Boundary conditions were applied on the lower flange and corrugated web simulating simply supported beam. An elastic material was assumed with a yield strength R_y = 235 MPa, modulus of elasticity E = 210000 MPa and Poisson's ratio 0.3.

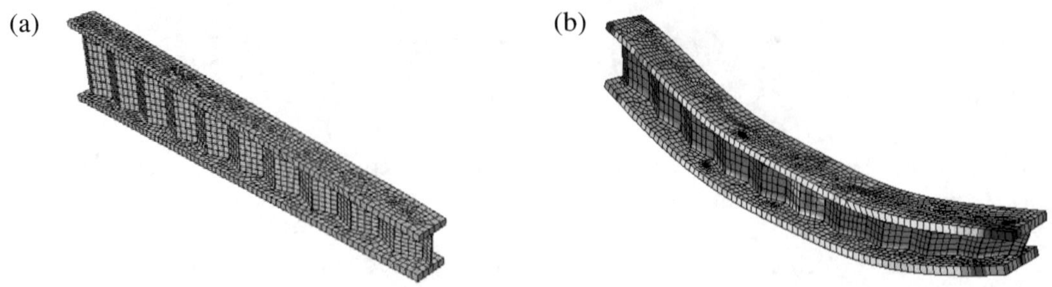

Figure 3. (a) FE model of tapered triangularly corrugated web beam adopted for buckling analysis; (b) Example of the global buckling mode for tapered I-beam with triangularly corrugated web.

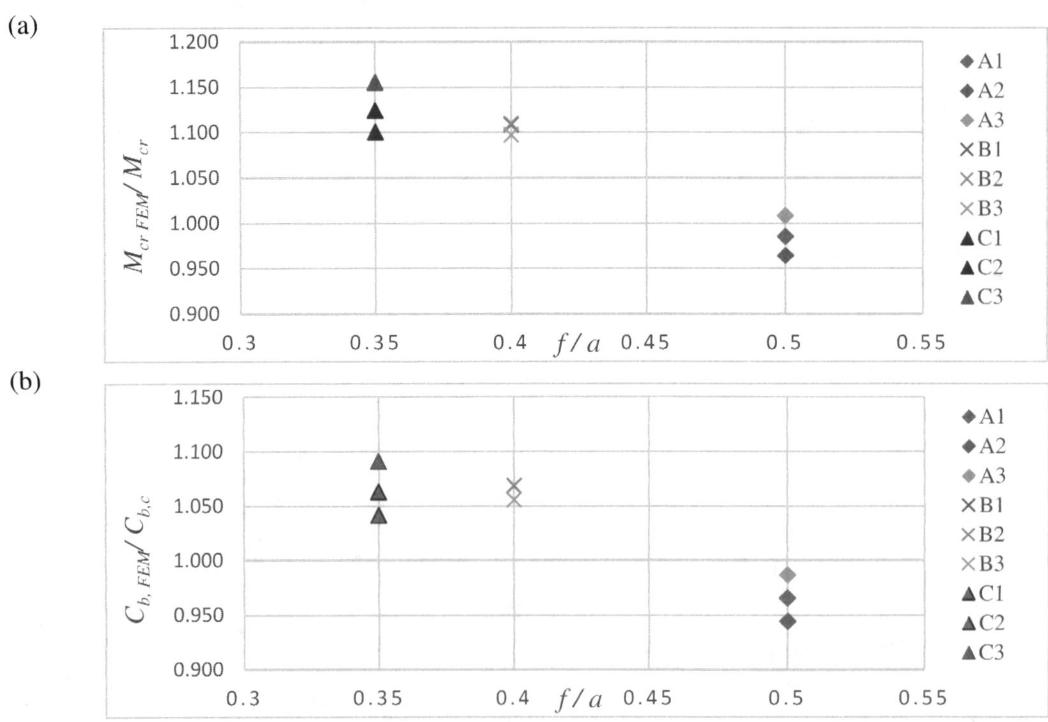

Figure 4. (a) Comparison between $M_{cr,FEM}$ and $M_{,cr}$. (b) Comparison between $C_{b,FEM}$ and $C_{b,c}$.

4. Results of parametric study

In this study, for the tapered I-girder with triangularly corrugated web the critical moment is given by the formula (8). In addition, the values of $C_{b,FEM}$ are calculated as follows

$$C_{b,FEM} = \frac{M_{cr,FEM}}{\dfrac{\pi^2 EI_{y,c}}{(kl)^2}\sqrt{\left(\dfrac{k}{k_w}\right)^2 \dfrac{C_{w,c}}{I_{y,c}} + \dfrac{(kl)^2 G_c I_c}{\pi^2 EI_{y,c}}}}, \tag{11}$$

in which $M_{cr,FEM}$ is the critical moment of I-beam with tapered triangularly corrugated web at lateral torsional buckling. To calculate $M_{cr,FEM}$, the eigenvalues v derived from the FEA were multiplied by the applied load 1kN, so critical force for LTB is $P_{cr} = 1kN \cdot \nu$. For case under consideration

$$M_{cr,FEM} = P_{cr} l/3. \tag{12}$$

Plots with comparisons between $M_{cr,FEM}/M_{cr}$ and $C_{b,FEM}/C_{b,c}$ are given in figure 4 for the different models. The minimum variations are between the models A1, A2 and A3 approximately 3..4%, otherwise the largest discrepancies are found in the models C1, C2 and C3 with approximately 9%, between numerical and analytical analysis.

5. Conclusions

The present study reported the lateral torsional buckling behaviour and design of a triangularly corrugated tapered (non-prismatic) web beams under two concentrated loads on the quarters of the beam. The present work used finite element analysis program to model steel beams with variations in the length and the corrugation web with the same highs for all models.

According to the results, the lateral torsional buckling strength of the non-prismatic I-beam with triangularly corrugated web was determined by using the formulas as the used for a beam with plane web and taking maximum value of M_{cr} with $f = 0$ and $f = f_{max.}$ For this reason, it is important to determine modification coefficient for optimal results to the lateral torsional buckling. Undoubtedly, the study of the buckling resistance for such structures requires further investigation considering initial imperfections, type and location of the applied load, the shape of the bending moment, end support conditions and so on.

References

[1] Kudryavtsev S 2017 Calculation and design of welded I-section with corrugated webs (Yekaterinburg: Publ. of Ural University), 60 p., ISBN 978-5-7996-2056-1 (in Russian).
[2] Kitipornchai S and Trahair N S 1972 Elastic stability of tapered I-beams *J. Struct. Div. ASCE*, **98** (3) pp. 713-728
[3] Zhang L and Tong G S 2008 Lateral buckling of web-tapered I-beams: A new theory *J. Constr. Steel Res.* **64** (12) pp. 1379-1393
[4] Elgaaly M, Hamilton M and Seshadri A 1996 Shear strength of beams with corrugated webs *J. Structural Eng.* **122** (4)
[5] Sayed-Ahmed E Y 2005 Lateral torsion-flexure buckling of corrugated web steel girders *Proc. of the Institution of Civil Engineers – Structures and Buildings* **158** pp. 53–69
[6] Hasan H, De F and Keong C K 2015 Experimental Study on Lateral Torsional Buckling of Triangular Web Profile Steel Section *Applied Mechanics and Materials* **802** pp 178-183
[7] Kudryavtsev S 2018 Influence of web openings on bearing capacity of triangularly corrugated web beam *IOP Conf. Ser. Mater. Sci. Eng.* **365** 042034
[8] Kudryavtsev S 2019 Lateral-torsional buckling behaviour of triangularly corrugated web beam", *IOP Conf. Ser.: Mater. Sci. Eng.* **471** 052016
[9] Samanta A and Mukhopadhyay M 1999 Finite element static and dynamic analyses of folded plates *Eng. Structures* **21** (3) pp. 277–287
[10] Lindner J 1990 Lateral torsional buckling of beams with trapezoidally corrugated webs *Proc.*

Int. Colloquium Stability and ductility of Steel Structures (Budapest, Hungary)

[11] Nguyen N D, Kim S N, Han S R and Kang Y J 2010 Elastic lateral-torsional buckling strength of I-girder with trapezoidal web corrugations using a new warping constant under uniform moment *Eng. Struct.* **32** (8) pp. 2157–2165

[12] Nguyen N D, Nguyen-Van H, Han S Y, Choi J H and Kang Y J 2013 Elastic lateral-torsional buckling of tapered I-girder with corrugated webs", *Int. J. Steel Struct.* **13** (1) pp. 71–79

[13] Galambos T V 1988 Guide to Stability Design Criteria for Metal Structures (New York: Wiley) 4th Edition, Chap. 18.

IOP Conf. Series: Materials Science and Engineering **972** (2020) 012015 doi:10.1088/1757-899X/972/1/012015

About actual contradiction in geotechnical design and optimal way of it resolution

A N Alekhin[1], A A Alekhin[2], V N Alekhin[2]

[1]Ural State University of Railway Transport, 66 Kolmogorova, St., Ekaterinburg, 620034, Russia

[2]Ural Federal University, 19 Mira St., Ekaterinburg, 620002, Russia

E-mail: alekhin.51@gmail.com

Abstract. The paper describes serious and fundamental contradiction in geotechnical design, which manifested itself to the greatest extent when designing the foundations of high-rise buildings in the third quarter of the 20th century has begun to form from the late of the 19th century to the 20th years of the 20th century. Prime cause of this contradiction is in complex physical structure of soils, complex nature of their formation and, therefore, in complex form of their deformation. The article also notes the high technical and economic efficiency of using realistic physically nonlinear soil models in the design of geotechnical parts of buildings and structures. In this case such a designing is most effective when using the parameters of nonlinear models determined from data of in-situ tests.

1. Introduction

Serious and fundamental contradiction in geotechnical design, which manifested itself to the greatest extent when designing the foundations of high-rise buildings in the third quarter of the 20th century has begun to form from the late of the 19th century to the 20th years of the 20th century. Prime cause of this contradiction is in complex physical structure of soils, complex nature of their formation and, therefore, in complex form of their deformation. Once a well-known physicist Lorenz said that if he was asked to make an equation for soil deformation, he would have run away in fright [1]. However, the needs of building design and status of soil base as a supporting element of all structure made the problem of studying the mechanical properties of soils and firstly the problem of studying the character of soil deformation practically and theoretically very relevant. Certainly, as in other technical branches, researches have begun with the simplest assumption of a proportional ratio between a load on the soil (pressure P) and soil settlement S: $S = P/C_z$ (here $C_z = Const$ is proportionality coefficient called "coefficient of soil reaction" «or "bed coefficient"). Graph of the ration between P and S is shown in Figure 1. This ratio proposed by Fuss in 1798 [2], but sometimes not exactly is called as Winkler model [3,4,5], must be more correctly defined as a "contact model of the soil", because it does not reflect the relationship between deformations ε and stresses σ inside soil massif, and provides designers only with a ratio between the surface pressure P on the soil beneath the foundation and settlement S of the part of the surface of soil under foundation. But such a primitive inclusion of the factor of soil massif in designing of buildings (including software) has a number of serious negative consequences: firstly, results of the pressure transfer to the underlying soil layers are not taken into account, but these layers may be weaker than the bearing layer under the foundation;

IOP Conf. Series: Materials Science and Engineering **972** (2020) 012015 doi:10.1088/1757-899X/972/1/012015

secondly, there is not reflected transmission of the pressure to soil bases of neighboring buildings, that not allowing calculation of additional deformations of these buildings, while it is very important in conditions of dense urban development; thirdly, many other important features of loading the soil massif are not reflected, for example, the dependence of the coefficient C_z on the size and shape of the loaded area – phenomenon, which has been well studied experimentally [5, 6, 7, 8, 9, 10, 11]: an approximate diagram of the dependence of the settlement (w) of a rigid square plate at the same pressure under it on the width of the plate (*a'*) is shown in Figure 2; in Figures 3 and 4 are shown similar diagrams for the dependence of settlement (*w*) on the ratio of the area (*A*) of the rigid square plate to its perimeter (*Pr*), i.e. on its shape [8, 9]. This model also does not allow to reflect another very important, well-known and characteristic phenomenon for the soil base, that distinguishes soil from all structural (i.e., artificial) materials, namely, an increase in the rigidity of even a homogeneous soil base with increase in depth [14, 15, 13, 5, 15], as shown in Figure 5 and the consequence of which is a stiffer soil base compared to the case of constant stiffness of soil [14, 5, 15]; later it became clear that all these features of soils are result of significant physical nonlinearity of soils, i.e. dependence of their rigidity on the stress state [15] and relationship of deformation with soil strength [16]. Requirement to take into account in geotechnical designing physically nonlinear type of soil deformation is included in all regulatory documents of the Russian Federation on the design of soil bases of buildings and structures, starting in 1975 [16, 17, 18, 19]. Moreover, in the Russian Federation this requirement, due to its special importance for the safety and reliability of buildings, is included in Article 16 "Mechanical Safety" of Federal Law№ 384-ФЗ "Technical Regulations on the Safety of Buildings and Structures" [20]. Attempts to improve physically linear models, in particular, Fuss-Winkler contact model, by adding in its later version additional coefficient to take into account soil distribution ability [20, 21], did not solve even this problem because essence of the problem was in absence of rations between stresses (σ) and strains (ε) in "contact model" framework and as a result of which new coefficient had the same disadvantages as the coefficient C_z.

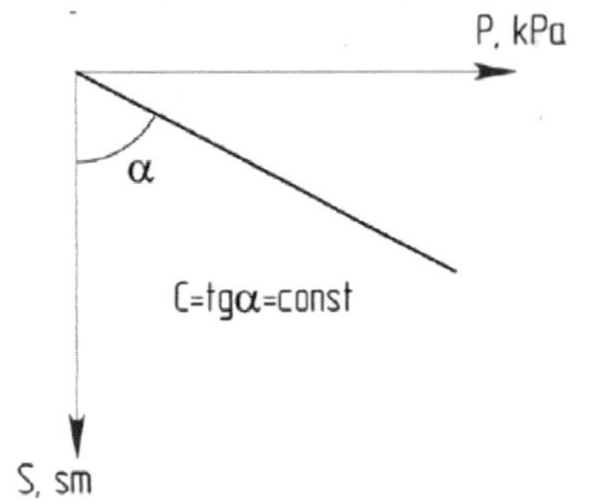

Figure. 1. Graphical interpretation of the Fuss-Winkler model

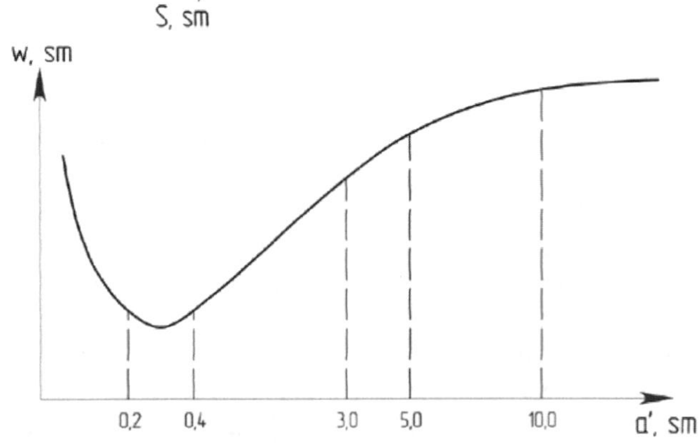

Figure. 2. Diagram of dependence of the settlement of a rigid square plate on its dimensions at the same pressure beneath it [5]

2. Distinction between linear and nonlinear soil models

Thus, further development of geotechnical designing was possible only based on models having relationships between stresses (σ) and strains (ε). But such a model at the turn of the nineteenth and twentieth centuries was only the physically linear Hook's model that he proposed at the end of seventeenth century for calculating deformations of metal, rubber and other similar materials. Nevertheless, by a willed decision, which was not entirely justified, but in absence of other models and with notes about inconsistencies Hook's model to the physical nature of the soil [22, 23] in the 30s…40s of the 20th century for calculating deformations of soil bases exactly Hook's model was adopted [24, 25, 26, 27, 28]. The justification of the volitional decision to apply Hooke's model for soils was not very convincing because, in addition to unconfirmed by more accurate nonlinear calculations [29, 30, 15], of previously obtained with use of Hooke's linear theory of contact stresses beneath foundations and stresses in soil massifs [31, 32, 33], this decision was based mainly on the statement that if in experiments [34, 35, 11, 28] relation between load (*P*) and a settlement (*S*) is directly proportional (linear), then relation between stress (σ) and strain (ε) should be also proportional (linear), as it is defined in Hooke's theory. However, such a logical conclusion is not entirely fair and can be approximately satisfied only under influence of a certain parametric relation linearizing ratio between *S* and *P*. Wherein inverse logical relation ($\varepsilon \sim \sigma \Rightarrow S \sim P$) is certainly true. Semantically linear Hooke's model has following analytical form: $\varepsilon = \sigma/E$ (here E=Const, as in the case of Fuss-Winkler model is a proportionality coefficient which is called "Young's modulus"). Graph of physically linear Hooke's model (*E* = *Const*) in figure 6 is similar to the graphic image of also physically linear (C_z=Const) Fuss-Winkler model in figure 1. However, differential, essentially, Hooke's dependence $\varepsilon = f(\sigma)$ and integral, essentially, Fuss-Winkler relationship $S = f(P)$ differ because strain ε (in general case strain tensor ε_{ij}) can be decomposed into volumetric strain ε де and shape (shear) ε_i deformation in the framework of a continuous medium concept: $\varepsilon_{ij} = \varepsilon + \varepsilon_i$ [36] with corresponding decomposition of the stress tensor: $\sigma_{ij}=\sigma+\sigma_i$.

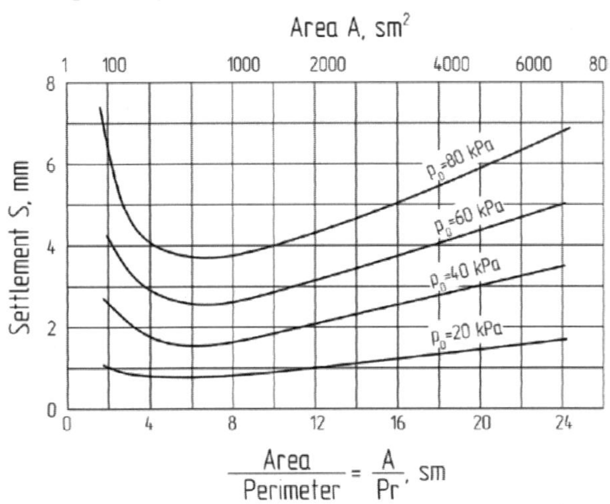

Figure. 3. Diagram of dependence of the settlement of a rigid square plate on its dimensions at the same pressure beneath it (for sand) [10]

Graphically, in coordinates of invariants of stress and strain tensors, such a separation of stresses and strains in the case of a physically linear Hooke model looks like that shown in Figure 7; in this case dependences for both parts of the deformation remain physically linear, independent of the stress-strain state: volumetric stiffness is characterized by a constant bulk modulus $K=\sigma/\varepsilon$ {here σ and ε are first invariants of stress tensor and strain tensor (volume deformation); shaping stiffness is characterized by a constant shear modulus $G=\sigma_I/\varepsilon_i$ (here σ_I and ε_I are second invariants of deviatoric parts of stress tensor and strain tensor). This mechanical model most realistically describes the deformation of elastic-linear (physically linear) materials such as metals, rubber and concrete. Of cause, adoption of such a model for calculating soil deformations was extremely contradictory. By the way, by-effect (philological) consequences of this contradictory was replacement, in the case of soil, of the name "Young's modulus" *E* with "modulus of deformation *E*" [28,37].

IOP Conf. Series: Materials Science and Engineering **972** (2020) 012015 doi:10.1088/1757-899X/972/1/012015

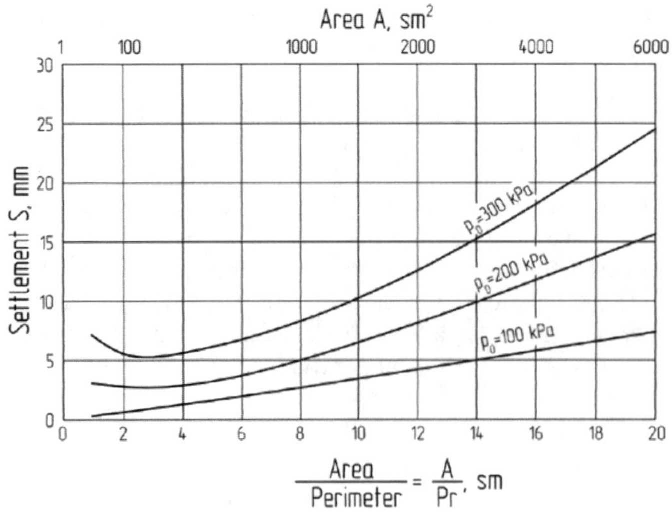

Figure. 4. Diagram of dependence of the settlement of a rigid square plate on its dimensions at the same pressure beneath it (for clay) [10]

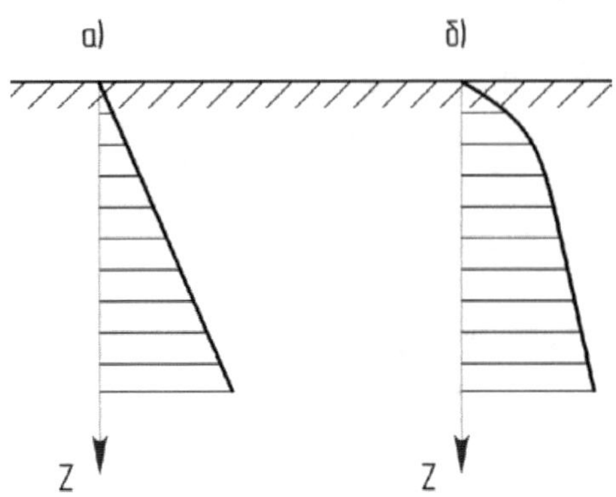

Figure. 5. Increase of the modulus of deformation: a) in a linear form; b) in a parabolic form [14, 5]

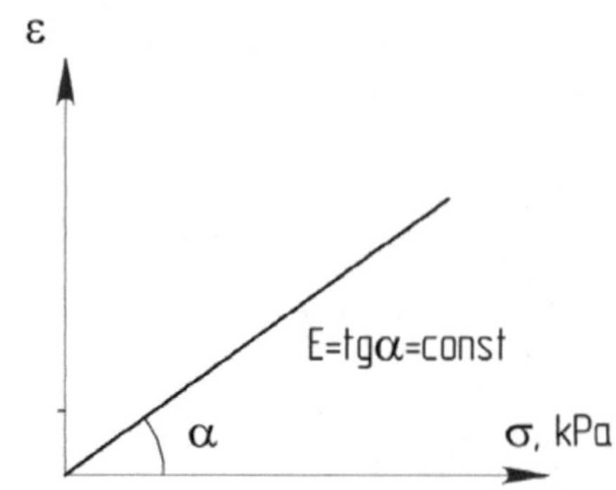

Figure. 6. Graphical interpretation of the physically linear Hooke's model

Below are approximating relations for the bulk modulus K and the shear modulus G:

$$K = \frac{\sigma}{\varepsilon} = \sigma^{1-\alpha}/A_0,$$
$$G = \frac{\sigma_i}{\varepsilon_i} = \sigma_U/(B + \varepsilon_i). \tag{1}$$

Here $\sigma_U = A\sigma + C$ – strength condition of Mises-Botkin for soil; A_0, α, A, B, C – parameters dependent on the soil.

Despite the fruitful technical potential of the Hooke's model, which allowed developing on its basis a large number of geotechnical design methods, both analytical [27, 13, 28,37 38, 39, 40, 41, 42,43,17, 18, 19] and numerical [39, 5, 43], rather quickly serious contradictions appeared from the application of this physically linear model to a substantially physically nonlinear soil. Unfortunately, in a view of the beginning war in 1941, results of investigations of physically nonlinear soil deformation [44, 45], the essence of which as follows from the graphs in Figure 8 and their approximating by formulas (1) consists in dependence of the stiffness characteristics on the stress-strain state, have not been applied in design practice.

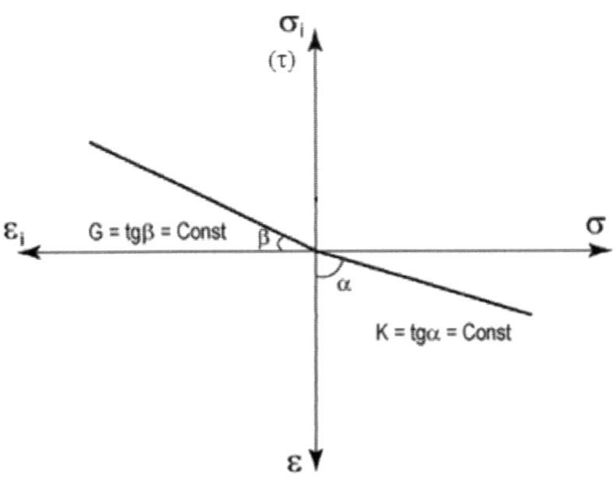

Figure. 7. Diagram of linear deformation of structural materials (metals, as well as rock: : bulk deformation modulus $K = \sigma/\varepsilon$; shear modulus $G = \sigma_i/\varepsilon_I$; σ, ε – first invariants of stress and strain tensors (volumetric deformation); σ_i, ε_i – second invariants of the deviator parts of the tensors of stresses and deformations (shape change – shear).

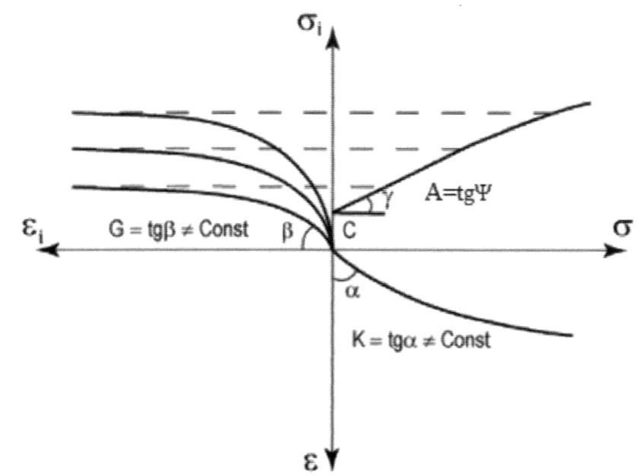

Figure 8. Simplified nonlinear soil deformation diagram; notation here is the same as in Figure 7

At the same time, it was found out that a model with constant stiffness parameter (in particular, $E = const$) cannot, like the Fuss model, reflect increase of the stiffness of a homogeneous soil with increasing depth; the constant value of the modulus E, as in the case of the Fuss model, reduces the actual stiffness of the soil base [15] and overestimates its design deformation. It also turned out that value of Young's modulus (modulus of deformation) E as well as the coefficient C_z depends on the shape and size of the loaded area and hence on the type of test of the soil. The desire to explain and take into account this difference, for example, for the results of soil tests with round rigid plates and pressuremeters, led to the adoption of different module models for describing soil deformations [46, 47], which are simplified versions of a physically nonlinear model, such as the Botkin model described above [44], which, due to a more general structure, which includes not only deformation, but also strength parameters of the soil, reflects the non-invariance of the value of the deformation modulus E with respect to the size and shape of the loaded area [15, 48]. It is no coincidence that, as a result of an extensive analysis, it was established [42], a significant discrepancy between the actual and calculated on the base of Hooke's theory deformations of soil bases of different buildings and structures. In this regard, all soil models that have in their structure a constant Young's modulus E, including non-linear ones (for example, elastic-plastic), retain the drawback of the Hooke model in the form of the non-invariance of the modulus E with respect to various stress-strain state factors. From the other hand, numerous experimental and theoretical investigations of soil deformation within the framework of a physically nonlinear model [30, 31, 15, 48] showed a qualitative and quantitative coincidence of actual and theoretical results even in such details as the dependence of the distribution of contact stresses and stresses in the soil mass on the type and condition of the soil, the concentration of stresses under the edges of the foundations, different in different soils, and the consequent increase in deformation at a certain depth along the axis of the foundation. The prognose of influence of the load onto neighboring buildings is also much more realistic. These successful research results made it possible, in accordance with the Federal Law [20], to use a physically non-linear soil model in real design practice, which ensured both the technical reliability of the structures and significant financial benefits: in the construction of a 70-meter retaining wall, savings amounted to $ 0.55 million, and in the construction of the foundation of a 32-storey residential building, respectively, $ 1.3 million.

IOP Conf. Series: Materials Science and Engineering **972** (2020) 012015 doi:10.1088/1757-899X/972/1/012015

Such an efficiency of using a physically nonlinear model was ensured, inter alia, by determining its parameters from in-situ tests according to the technique described in [49].

3. Conclusion

Nowadays in geotechnical designing, a serious contradiction has arisen between requirements of regulatory documents on the design of soil bases and foundations(including requirements of Federal Law № 384-ФЗ "Technical Regulations on the Safety of Buildings and Structures" about application in geotechnical calculations of a soil model that reflects real, physically nonlinear features of soil deformation, and existing practice of geotechnical design based on the physically linear Hooke's theory, reflecting the deformation of structural materials such as metals, rubber and concrete.

Numerous observations of various structures indicate a significant discrepancy between the actual settlements of their soil bases and calculated ones using Hooke theory.

Taking into account in the designing of physically nonlinear features of soil deformation, the essence of which consists in dependence of soil rigidity on stress-strain state and in the link of soil deformation with its strength allows to build economically and technically optimal foundations and other geotechnical structures, for example, retaining walls; moreover, in this case the cost of geotechnical structures is two or more times cheaper than of those developed using the physically linear theory of Hooke; also, the prognosis of influence on neighboring structures is much more realistic than that predicted by Hooke's theory, with its distribution capacity corresponding to the distribution ability of rubber for a long distance.

To widespread implementation of realistic, economically very profitable physically non-linear soil models in design practice should be promoted by examples of their further successful application in practice, as well as the organization of effective scientific and technical monitoring of buildings under construction, with an analysis of geotechnical results. Undoubtedly, in the process of studying the soil of a construction site before designing, the mechanical parameters of physically nonlinear soil models should be determined, mainly by in-situ methods.

References

[1] Zaretsky Ju 1989 Lectures on modern soil mechanics (Rostov on Don)
[2] Fuss N 1798 Sur la résistance, de provoquer toutes sortes de routes, aux quatre roues et deux roues des chariots (S. Petersburg)
[3] Winkler E 1867 Die Lerne von der Elastizität und Festigkeit (Leipcig)
[4] Winkler E 1872 Neue Theorie des Erddruckes (Leipcig)
[5] Gorbunov-Possadov M, Malikova T, Solomin V 1984 Design of structures on an elastic foundation (Moscow)
[6] Kögler F 1931 Über Baugrund-Probebelastungen. Alte Verfahren neue Erkenntnisse (Berlin)
[7] Kögler F 1933 Baugrundprüfung in Bokrloch (Berlin)
[8] Press H 1929 Druckverteilung in Baugründe (Berlin)
[9] Gerner's experiments with a round pressure area 1932 (Berlin)
[10] Bernatsky L 1935 Applied Geotechnics (Moscow)
[11] Vasiliev B 1937 Soil bases and foundations (Moscow)
[12] Klein G 1954 Design of beams on a solid base, continuously heterogeneous in depth (Moscow)
[13] Klein G 1956 Inhomogeneities, tensile deformations and other mechanical properties of the soil when designing structures on the soil massif (Moscow)
[14] Goldstein M 1937 Mechanical properties of soils (Moscow)
[15] Alekhin A1983 Nonlinear analysis of stress-strain state of soil mass under static loading: Thesis of cand. techn. degree (Sverdlovsk)
[16] Malyshev M 1980 Soil strength and soil base stability (Moscow)
[17] SNiP II-15-74. 1975 Design Standards. Soil bases of buildings and structures (Moscow)
[18] SNiP 2.02.01-83. 1985. Design Standards. Soil bases of buildings and structures (Moscow)
[19] SP 22.13330.2016. 2017. Design Standards. Soil bases of buildings and structures (Moscow)

[20] Federal Law of the Russian Federation №384-ФЗ.2009 ("Regulations on the Safety of Buildings and Structures") (Moscow)
[21] Pasternak P 1926 Die baustatische Theorie biegesteifer Balken und Platten auf elastischen Bettung (Berlin)
[22] Pasternak P 1954 Basics of a new method for designing foundations on an elastic soil base using two bedding coefficients (Moscow)
[23] Terzagi K 1925 Erdbaumechanik auf bodenphysikalischer grundlage (Leipzig und Vien)
[24] Terzagi K 1933 Structural soil mechanics based on its physical properties (Moscow and Leningrad)
[25] Terzagi K 1943 Theoretical Soil Mechanics (New-York)
[26] Terzagi K 1961 Theory of soil mechanics Structural soil mechanics (Moscow)
[27] Gersevanov N 1930 Experience of applying the theory of elasticity for determining permissible loads on the soil (Moscow)
[28] Gersevanov N, Polshin D 1948 Theoretical foundations of soil mechanics and their practical application (Moscow)
[29] Shirokov V, Solomin V, Malyshev M, Zaretsky Ju 1970 Stress state and displacement of a weighty nonlinearly deformable soil half-space under a round hard plate (Moscow)
[30] Solomin V, Kopeikin V, Shmatkov S 1979 Numerical solution for nonlinear problems of relationship between foundations and soil bases (Aachen)
[31] Boussinesq J 1885 Application des potentiels al'etude de l'equilibre et du movement des solides elastiques (Paris)
[32] Minyaev P 1916 On the distribution of stresses in loose solids (Kharkiv)
[33] Schultze E 1961 Distribution of stress beneath a rigid foundation (Paris)
[34] Kurdyumov 1894 Short course of soil bases and foundations (St. Petersburg)
[35] Puzyrevsky N 1929 Theory of stresses in soils (Leningrad)
[36] Mase G 1970 Theory and Problems of Continuum Mechanics (New York)
[37] Gorbunov-Possadov M 1953 Design of structures on an elastic foundation (Moscow)
[38] Tsytovich N 1979 Soil mechanics (Moscow)
[39] Gorbunov-Possadov M 1953 Design of structures on an elastic foundation (Moscow)
[40] Dalmatov B 1969 Designing foundations of buildings and industrial structures (Moscow)
[41] Dalmatov B 1988. Soil mechanics, soil bases and foundations (Leningrad)
[42] Ukhov S, Semenov V, Znamensky V, Ter-Martirosyan Z, Chernyshev S 2002 Soil Mechanics, soil bases and foundations (Moscow).
[43] Gorbunov-Possadov M 1967 Current state of the scientific basis of foundation engineering (Moscow)
[44] Botkin A 1939 Investigation of stress in loose and cohesive soils (Moscow and Leningrad)
[45] Botkin A 1940 On the strength of bulk and brittle materials (Moscow)
[46] Ruppeneit K, Bronstein M, Dolgikh M 1973 Solution of an axisymmetric elastoplastic problem for anisotropic soil (Moscow)
[47] Lushnikov V 1981 Experimental and theoretical research and implementation of the pressuremeter method (Sverdlovsk)
[48] Lushnikov V, Alekhin A 1982 Assessment of the reliability of determining modulus of deformation from data of soil tests (Sverdlovsk)
[49] Alekhin A N 2015 Substantiation of the method of determination of nonlinear soil model parameters from in-situ test data *Proceedings UralNIIproject RAASN* **1** 57-62

Experience in designing the foundations of a multi-storey building on the eluvial soils of the Urals using a model of non-linear soil deformation

A N Alekhin [1], A A Alekhin[2], V N Alekhin[2]

[1]Ural State University of Railway Transport, Department of bridges and tunnels, st. Kolmogorova, 66, Yekaterinburg, 620034, Russia
[2] Ural Federal University named after the first President of Russia B. N. Yeltsin, Institute of Construction and Architecture, Department of Hydraulics, st. Mira, 17, Yekaterinburg, 620002, Russia

E-mail: alekhin.51@gmail.com

Abstract. The article describes an example of the first design of the foundation of a high-rise building in the Middle Urals using a non-linear soil model that reflects the real deformation properties of the soil. The use of an effective model that reflects the real deformation properties of the soil has allowed to reduce the cost of the foundation by more than two times while increasing of its reliability.

1. Introduction

Eluvial soils, as initial weathering product of rocks, have high strength properties and low deformability; in the Urals, their capacity can reach significant sizes.

However mistakenly piles in these solid soils are sometimes used, that often turns out to be economically and technically inexpedient solution. Also important is the fact of negative influence of damage of the eluvial hard layer supporting upper soil layers of the construction site which are basis of water-bearing and other networks of existing and newly constructing buildings.

At the same time now-days requirements for geotechnical design primarily due to the multiple increase in the power of personal computers are aimed at consideration of real mechanical properties of soils in designing foundations. These real mechanical properties, based on the physical nature of soils, are realized in their physically and geometrically non-linear character of deformation.

2. Model of non-linear soil deformation

Geometric nonlinearity associated with large displacements in soil massifs is reflected by geometric relationships [1] in equations of the problem. The essence of physical nonlinearity consists in dependence of soil stiffness on its stress state. Thus, initial stiffness characteristics of soil (bulk deformation modulus *K* and shear modulus *G*) in contrast to Hooke's theory, change with change in the soil stress state. Phenomenologically, in the framework of continuous medium concept about decomposition of common strain tensor ε_{ij} on volumetric strain ε and shape strain ε_i: $\varepsilon_{ij} = \varepsilon + \varepsilon_i$ and on the corresponding decomposition of stress tensor $\sigma_{ij} = \sigma + \sigma_i$, such deformation in coordinates of invariants of stress and strain tensors looks like that shown in Figure 1, opposed to linear deformation studied by Hooke and characterized by the deformation diagram shown in Figure 2. The diagram in Figure 2 is typical for structural materials (metals, concrete, rubber) and for rock.

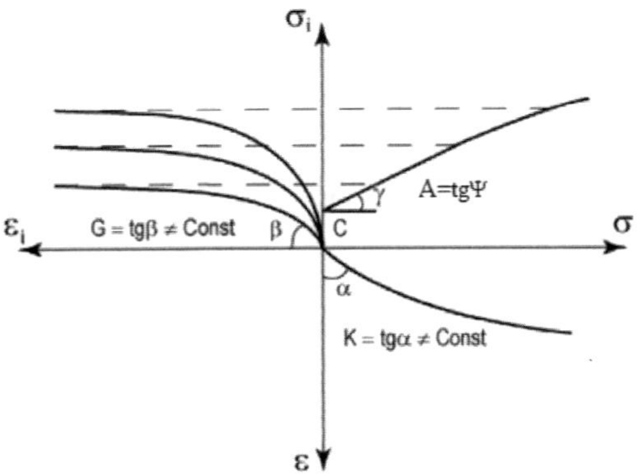

Figure 1. Simplified nonlinear soil deformation diagram

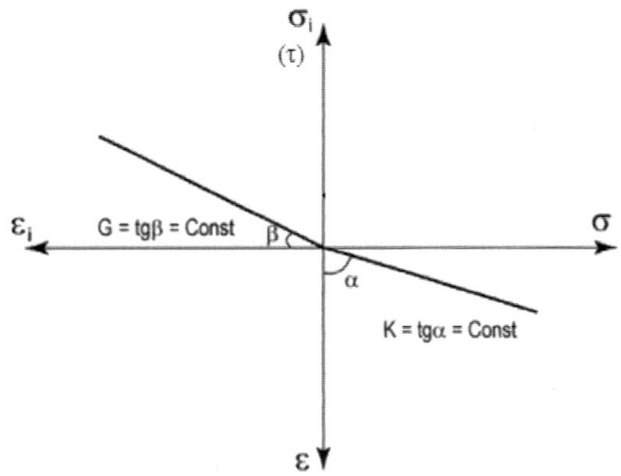

Figure 2. Diagram of linear deformation of structural materials (metals, concrete, plastics, rubber), as well as a rock; notation here is the same as in Figure 1

As follows from the diagram for structural materials and for rock, the basic stiffness characteristics *K* and *G* can be considered as stress-independent constants, while for soils; they are functions of stresses. Graphes in Figure 1 can be expressed using formulas (1) [3,4]:

$$K = \frac{\sigma}{\varepsilon} = \frac{\sigma^{1-\alpha}}{A_0},$$
$$G = \frac{\sigma_i}{\varepsilon_i} = \frac{\sigma_U}{B + \varepsilon_i}. \quad (1)$$

Here $\sigma_U = A\sigma + C$ – strength condition of Mises-Botkin for soil; A_0, α, *A*, *B*, *C* – parameters depending on the type of the soil.

As follows from formulas (1), shear stiffness of the soil *G* is related to its strength, which is absolutely logical, since it is shear stresses in the soil and shear deformations in it that leads to entail soil destruction [5]. According to the theory of continuous media, the main characteristics of soil stiffness (bulk deformation modulus *K*, shear modulus *G*, Young's modulus *E* and Poisson's ratio ν) are related by relations (2) [6]:

$$E = \frac{3KG}{K + G},$$
$$\nu = \frac{K - 2G}{2(K + G)} \quad (2)$$

From relations (1) and (2) it follows, for example, an increase in the deformation modulus *E* with increasing depth [7, 8], which is not provided for by Hooke's theory with constant modulus *E*. Thus, a physically nonlinear theory indicates a greater rigidity of the soil base than follows from the physically linear theory of Hooke.

Dependence of parameters *A*, *B*, *C* of the type of stress state $\mu_\sigma = (2\sigma_2-\sigma_1-\sigma_3)/(\sigma_1-\sigma_3)$ reflects also another important mechanical property of soils, namely, dependence of their rigidity on conditions of loading (type, intensity, loading area); here $\sigma_1>\sigma_2>\sigma_3$ – principal stresses. One of the main consequences of this feature of soils is the difference in the results of their static tests by different kinds of methods. However, the main difference between soils and structural materials is natural origin of soils over millions of years, during which not only their structure was formed, but also their natural stress state.

Such a history of soil formation cannot be reproduced in laboratory conditions in principle, as a result of which the calculated deformations of soil bases according to laboratory data differ from real values by about two times [8]. Therefore, parameters A_0, α, A, B, C of physically nonlinear soil model must be determined from in-situ tests. For this purpose, only pressuremeter and plate loading test devises (preferably automatic similar to those shown in figures 3, 4, 5) must be used since only these devices allow as much as possible to keep natural state of the soil, provided that they are technologically correct installed in the soil massif and tests are methodically correct carried out, taking into account the features of the physically non-linear soil model used to analyze their results.

Figure 3. Automatic pressuremeter device for in-situ tests in bore holes

Figure 4. Automatic self-boring pressuremeter device for in-situ tests in the massif of weak water-saturated soils (including offshore)

Figure 5. Automatic plate loading test device for in-situ tests in bore holes

In the notation of the diagram: bulk deformation modulus $K=\sigma/\varepsilon$; shear modulus $G=\sigma_i/\varepsilon_i$; σ, ε – first invariants of stress and strain tensors (volumetric deformation); σ_i, ε_i – second invariants of the deviator parts of the tensors of stresses and deformations (shape change – shear).

3. Implementation in the design

Based on the above considerations, it was decided to design for the first time in Yekaterinburg spread slab foundation of multi-storey building using, in accordance with requirements of the Federal Law of Russian Federation №384-FZ ("Regulations on the Safety of Buildings and Structures") [9] an effective physically and geometrically nonlinear soil model. Sectional view of the designed foundation is shown in Figure 6.

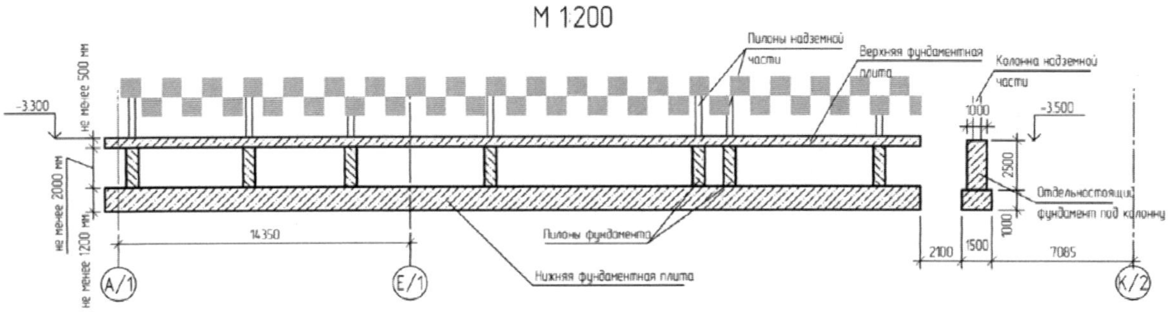

Figure 6. Sectional view of the designed foundation

The box-shaped form of the foundation was adopted off presence of a large thickness of heterogeneous, previously laid for leveling the relief of the site, soils, which had to be constructively overcome in order to transfer the building load to a reliable bearing eluvial soil.

Characteristics of the nonlinear soil model were determined by the method described in [10], main feature of which is special (using minimization methods) processing of the results of in-situ soil tests with solving so-called ill-conditioned problem [11], with regularizing its solution taking into account geotechnical properties of soils, including preliminary approximate values of some parameters of nonlinear soil model according to the results of an auxiliary laboratory mechanical soil tests. To determine the final values of the parameters of nonlinear deformation of soils corresponding to their natural state, data from pressuremeter (see Figure 7) and plate loading test see Figure 8) were used.

Figure 7. Pressuremeter test

Figure 8. Plate loading test

During construction of the building, geodetic observations of its deformations were carried out, which showed good agreement with calculated data based on the physically nonlinear soil model and, on the contrary, serious, similar to those specified in [6], deviations from results of Hooke's linear model. Foundation, designed taking into account the nonlinear properties of soils was almost two times cheaper than the foundation designed on the basis of outdated methods of Hooke's linear theory: cost of the building of the foundation, taking into account of nonlinear properties of the soils was $ 0.7 million (48 million rubles), while on the base of linear theory there was $ 1.3 million (85 million rubles).

During the construction of the building, geodetic observations of its settlements were carried out, which showed good agreement with the calculated data based on the physically nonlinear soil model

IOP Conf. Series: Materials Science and Engineering **972** (2020) 012016 doi:10.1088/1757-899X/972/1/012016

and, on the contrary, serious, similar to those specified in [6], deviations from the results obtained using Hooke model (linear deformation).

Figure 9. Excavation for the box foundation in eluvial soil

Figure 10. Bottom plate of the box-shaped foundation

Figure 11. Construction of first eight floors of the building

IOP Conf. Series: Materials Science and Engineering **972** (2020) 012016 doi:10.1088/1757-899X/972/1/012016

4. Conclusion

Foundations and underground structures should be designed based on non-linear theory of soils. In this case, it is possible to design and build most reliable and cheap their constructions.

Parameters of nonlinear soil model must be determined from in-situ test data.

The use of an effective model that reflects the real deformation properties of soil had allowed to reduce the cost of the foundation by more than two times while increasing its reliability.

Acknowledgments

Authors are grateful to the leadership of "LSR-Yekaterinburg «Group for the technical opportunity and economic support in verifying technical and economic efficiency of the modern geotechnical soil model in real construction conditions.

Authors are also grateful to the management of the Building Institute of the Ural Federal University for the opportunity to publish this message, which is, in our opinion, is important for the modern construction practice of building.

Referenses

[1] Novozhilov V 1958 Theory of elasticity (Leningrad)

[2] Mase G 1970 Theory and Problems of Continuum Mechanics (New York)

[3] Botkin A 1939 Investigation of stress state at cohesionless and cohesive soils *Proceedings of RIIG* (Moscow)

[4] Kriegel H, WiesnerH 1973 Problems of Stress-Strain Conditions in Subsoil *Proceedings of VIII-th ICSMFE* (Moscow)

[5] Malyshev V 1994 Strength of soils and stability of soil bases of the structures (Moscow)

[6] Ukhov S, Semenov V, Ter-Martirosyan Z et al 2002 Soil mechanics, bases and foundations (Moscow)

[7] Klein G 1956 Inhomogeneities, tensile deformations and other mechanical properties of the soil when designing structures on the soil massif *Proceedings of MISI* (Moscow)

[8] Alekhin A 1983 Nonlinear analysis of stress-strain state of soil mass under static loading: Thesis of cand. techn. degree (Sverdlovsk)

[9] Federal Law of the Russian Federation №384-FZ.2009 ("Regulations on the Safety of Buildings and Structures") (Moscow0

[10] Alekhin A N 2015 Substantiation of the method of determination of nonlinear soil model parameters from in-situ test data *Proceedings UralNIIproject RAASN* **1** 57-62

[11] Tikhonov A N, Arsenin V J 1979 Methods for solving ill-conditioned problems (Moscow)

IOP Conf. Series: Materials Science and Engineering **972** (2020) 012017 doi:10.1088/1757-899X/972/1/012017

Forecast of the Development of Sediment Pile Foundations in Water-Saturated Clay Soils

N Troshkova[1], I Maltseva[1], A Panova[1] and G Kochnev[1]

[1]Ural Federal University, 19 Mira street, Yekaterinburg, 620002, Russia

E-mail: n.d.troshkova@urfu.ru, i.n.maltceva@urfu.ru

Abstract. The article presents the methodology and the main results of experimental studies of the laws governing the development of single-pile sediment in time on models under short-term and long-term exposure to loads. The obtained data are compared with the results of observations of precipitation of natural foundations. Indicates the direction of further research on the development of methods for predicting the sediment pile foundations. Established the legitimacy of the design of pile foundations for the maximum allowable deformations in the conditions of water-saturated clay soils.

1. Introduction

Large volumes of capital construction in oil and gas areas of Western Siberia in conditions of weak clay water-saturated soils led to the widespread use of pile foundations. However, using of pile foundations in some cases is constrained by the lack of sufficient data which is characterizing the actual conditions of their work in weak water-saturated clay soils. Especially it is about the questions of the development of sediment pile foundations in time [1-5].

It is known that static tests of piles which are performed on the construction site don't give a complete picture of the pile sediments in the structure under long-term load. To determine the final value of precipitation it is necessary to take into account the time factor. Experimental studies of piles under long-term load and comparison of the final sediment which were obtained experimentally, with theoretical data will clarify the scheme and methods of calculation of pile foundations for deformation [6, 7].

2. Simulation of experiments

The studying of the regularities of the development of pile sediment over time was carried out on large-scale models (scale 1:5) in a soil tray with dimensions of 2,2x4,2m in plan and a depth of 2.5 m. The tray dimensions allowed to create a homogeneous soil mass for simultaneous testing of a series of piles up to 25 pieces. At the same time, under the provided conditions the influence of the tray walls on the stress-strain state of the piles was excluded, as well as their mutual influence in the course of experiments.

The method of filling the tray with soil provided the creation of an array with physical and mechanical properties close to natural (weak clay water-saturated soils). The tray was loaded with loam layers of 10 cm with layer-by-layer compaction and soil moisture. As a result, according to laboratory experiments, the array of soil with the following physical-mechanical characteristics:

specific weight of soil of 16.2 kN/m3, gravimetric moisture content 38.8 per cent; porosity of 51%; the consistency of 0.5; 1.0 degree of saturation; specific traction 15 kN/m^2, angle of internal friction, 38^0.

There were accepted models of piles with cross-section 6x6 cm in length 1.3 m, made of concrete grade 300. Piles are immersed in the ground by pressing with a special installation. The study of the development of pile sediment in time was carried out by testing a series of 8 models with double repeatability of each experiment.

The sequence of experiments is accepted as a following: pile No. 1 is tested according to the method GOST 5686-12 «Soils. Methods of field tests with piles» [8] with increasing load steps to the limit state of the soil (2.8 kN). This test established a General pattern of pile operation and a range of loads on the phases of water-saturated soil in the near-well space was revealed. The obtained data made it possible to specify the number of experiments and the magnitude of loads for testing piles under long-term force action. Short-term and long-term tests are performed on seven models: piles 2-8.

A load on the piles under static tests according to GOST 5686-2012 [8] was carried out with the help of a support frame and a DG-5 hydraulic jack. The force in the jack was measured by a dynamometer DOSM-3. Precipitation during the tests were recorded by deflectometer 6PAO-LISI. The load during static tests was increased after the onset of conditional stabilization of precipitation from the previous loading stage. For conditional stabilization according to GOST 5686-2012, a draft equal to 0.1 mm for two hours of observation was taken. The total duration of such a test with bringing the load to the limit state of the soil was 56 hours. According to the results of experiments conditionally stabilized sediment Δ= 0,1 x S_{ud}=8 mm corresponds to a load of 2.6 kN (here S_{ud} is the ultimate deformation for buildings and structures) [9,10].

For experiments to determine the final value of the single pile shrinkage, a facility was designed that allowed the static load of a certain value to be transferred to the pile model for a long time. The loading was carried out through a special table with reinforced concrete elements with a weight of 0.35 kN. Of the seven tested models, three piles were loaded with a load corresponding to the soil compaction phase (0.7; 1.4; 2.1 kN), two models in the shear phase (2.6; 2.9 kN) and two in the area of the carrying capacity bases with a load of 2.95 and 3.05 kN. During the experiments, the values of conditionally stabilized sediments were measured in accordance with the requirements of GOST 5686-2012 for static tests and full pile sediments from long-acting loads.

3. The results of testing

The results of static load testing of piles, as well as the results of observations of pile shrinkage in time are given in table 1.

Table 1. The results of laboratory observations of the shrinkage of piles in time

Pile number	Load on the piles, kN	Conditional stabilized shrinkage – Δ, mm	The whole shrinkage S, mm	£= Δ/S
2	0.70	0.45	0.95	0.48
3	1.40	1.32	2.04	0.65
4	2.10	1.83	3.00	0.61
5	2.60	9.35	17.30	0.54
6	2.90	15.33	32.80	0.47
7	2.95	18.10	42.53	0.42
8	3.05	25.77	-	-
				$£_{mean}$ = 0.50

According to the test results, series of piles on the long-acting load, the generalized dependence of the magnification of conditionally stabilized precipitation on the load of the close to the static test data of the model No. 1 according to GOST, as well as the dependence on the load of the total precipitation of the piles (figure 1).

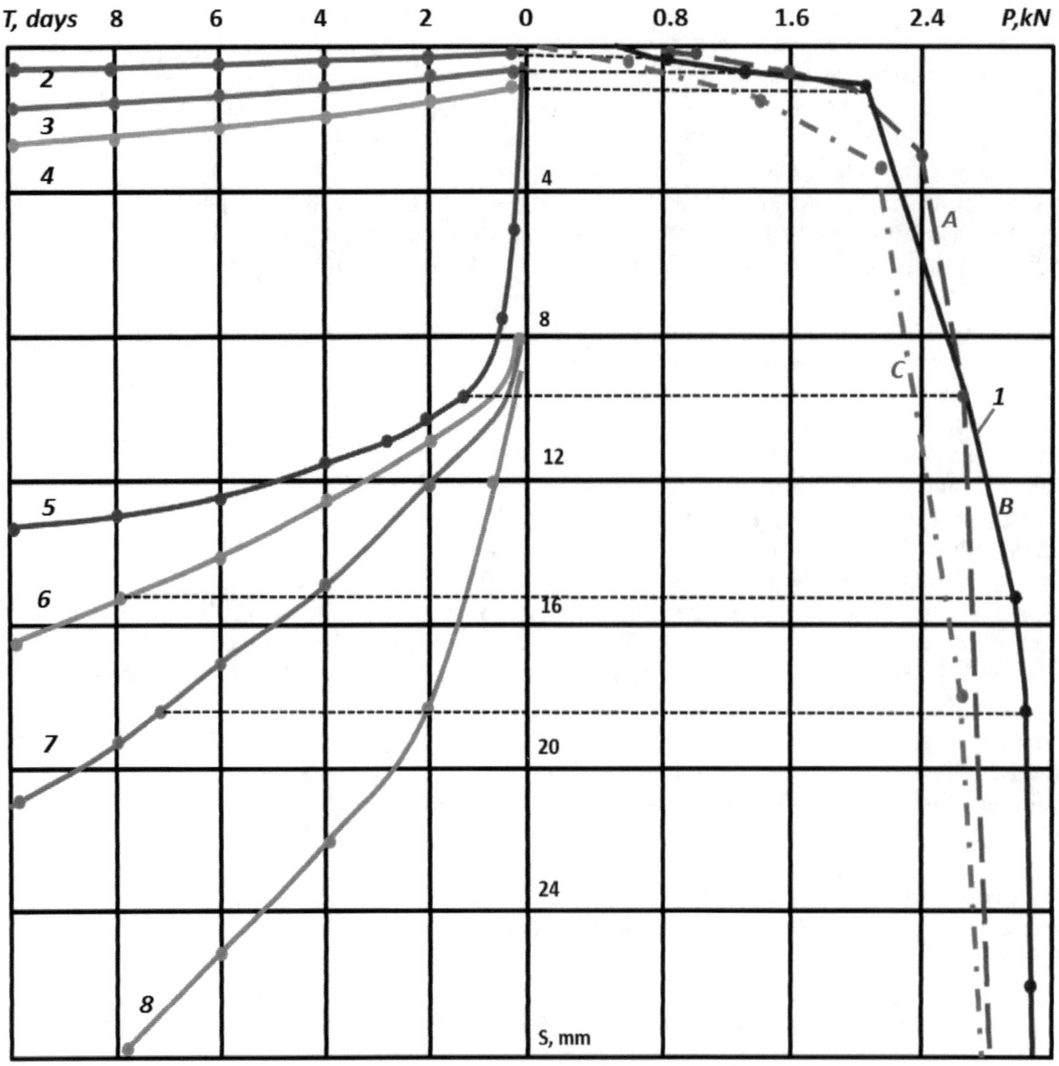

Figure 1. Test results, series of piles.

a, b - the development of deformation of piles in time; b - the dependence of the sediment on the load

(A - test pile No. 1 according to GOST5686-2012; B - generalized graph conditionally stable sediment piles No. 2-8; C - is also full of sediment); the I - phase of consolidation; II - the phase shifts; III - the phase of exhaustion of the bearing capacity of foundation piles (1-8 - numbers of experienced piles).

The obtained results indicate that the total sediment of a single pile is 1.55-2.35 times greater than the conditionally stabilized sediment (stabilization of 0.1 mm in two hours of observation). The average value of the transition coefficient from the conditionally stabilized precipitation to the total precipitation is 2.0.

It should be noted that the stabilization period of precipitation in the shear phase is much longer than in the compaction phase. Thus, the state of complete attenuation of precipitation in the compaction phase occurred after 30 days, and in the second phase of the sediment developed for several months. When the bearing capacity of the base was exhausted, it was not possible to obtain a conditionally stable state of the pile, since significant deformations developed with a small increment of the load. In this case, it is almost impossible to fix the value of the total precipitation and associate it with the concept of conditionally stable (figure 1).

Analysis of the results of laboratory experiments shows that the relationship between the total sediment and conditionally stabilized linear only in the compaction phase (figure 2).

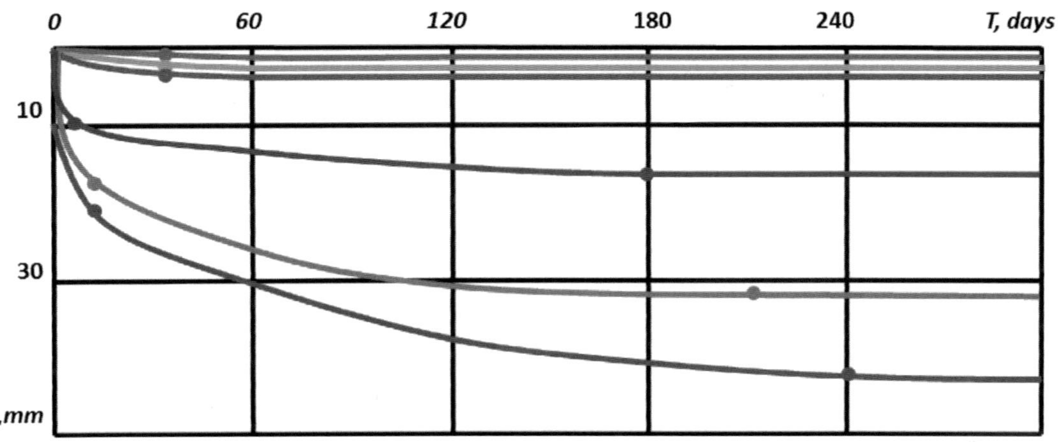

Figure 2. Dependence of the total sediment (S) and conditionally stabilized (Δ).

Transition rates the shrinkage of the pile is obtained in the test for possible sediment in the process of operation can be obtained by observing the development of precipitation of buildings and structures on piles.

Monitoring of the development of shrinkage of two five-storey residential buildings in Tyumen was carried out from the beginning of construction for two years. In engineering - geological respect, the construction site is heterogeneous, which is typical for the regions of the South of Western Siberia. The base is folded layers of sandy loam and loam. Piles in the Foundation composition are hanging/

When the foundations of the hanging piles bearing capacity per pile decreases. Theory and tests show that the total bearing capacity of the Foundation of hanging piles in clay soils is less than the product of the bearing capacity of one pile on their number in the Foundation. This is due to the mutual influence of piles by applying compaction zones around each pile in the foundation.[6-8, 12].

In the construction of foundations used piles with a cross section of 30x30 cm, hammering is made with a rod diesel hammer C-330 with a mass of the shock of 2.5 tons. The bearing capacity of the pile at the construction site is determined by the calculation method and the results of static tests of the piles. Calculation data and results of their comparison with static tests are given in table 2.

Building level is the intermediate point between the flat and a city. It is already not controlling the systems in a flat and still not controlling the overall energy flow of a city. IoT on a building scale is a data operator that gathers, analyses and transmits the information from every flat to suppliers.

Precipitation monitoring was carried out immediately after the completion of the construction and further within 24 months after the construction of the building. The final results of observations of precipitation over time are shown in table 3. Average stabilization of precipitation is accepted in accordance with GOST 5686-2012.

Table 2. The bearing capacity of the experienced piles according to the calculation and the results of static tests on the construction site of residential buildings in Tyumen

Pile number	Mark of pile	Immersion depth, m	Bearing capacity of pile			The percentage of the calculation is relatively static test, %
			On calculation F_d, кN	By results of static tests		
				Draft, mm	$F_{и}$, kN	
1	C70.30	5.6	1712	20	1900	90.1
2	C70.30	4.7	2500	12	2000	125.0
3	C70.30	4.4	2224	13	2400	92.7
4	C70.30	4.1	2268	8.3	2400	94.5
5	C70.30	4.5	2230	8.1	2400	94.1
6	C70.30	5.4	1780	12	2600	68.5
7	C60.30	5.3	2469	30	2560	96/4
8	C60.30	5.0	2439	30	2550	95.6
9	C70.30	4.5	1660	30	2150	77.2

Table 3. Final results of the shrinkage of the pile foundations of residential buildings for 24 months of observations in the city of Tyumen

Object	Mark	Conditionally stabilized shrinkage, mm	The average probation stabilized shrinkage, mm	Full shrinkage of the building, mm (during 24 month), mm
	M-1	8.2		
1	M-2	9.0	9.8	27.0
	M-3	12.2		
	M-4	12.0		
2	M-5	14.2	13.3	32.8
	M-6	13.4		

The precipitation of pile foundations of these houses in time, as shown by observations, 4-5 times higher than the precipitation of single piles. This indicates that in addition to taking into account the time factor when calculating the value of the final precipitation, it is also necessary to assess the influence of the mutual arrangement of piles in the foundation [7].

4. Conclusion

The development of precipitation of natural foundations in time, according to the results of observations of precipitation, confirms the basic laws obtained by laboratory studies. The total draught of buildings for two years of observations in 2-3 times exceeded the size of the deformations recorded at the end of construction of houses. However, the total precipitation of buildings is insignificant in size and 3-4 times less than the maximum permissible.

Experimental studies of many authors [11, 12] show that the bearing capacity of pile foundations from hanging piles can be correctly estimated only on the basis of the maximum permissible sediment of buildings.

Based on the study of the actual pile sediment from the beginning of loading to complete stabilization and comparison of experimental and calculated data, methods for predicting sediment are developed.

In addition, the legitimacy of the design of the foundations of residential buildings on the maximum permissible deformation in conditions of weak clay soils is established.

Reference

[1] Tsytovich N A and Zaretsky Y K 1967 Prediction of the Sediment Rate of the Foundations of Structures (Moscow: Stroyizdat) p 239
[2] Zaretsky Y K 1987 Theory of Soil Consolidation (Moscow: Science) 269 p
[3] Mitchell J K, Campanella R G and Singh A 1968 Soil Creep as a Rate Process. *J. Soil Mech. and Found. Div.* ASCE94, N SM-I
[4] Morgenstern N R 1969 Structural and Physicochemical Effects on Properties of Clays. *Proc. 7-th Mexico* ICSMFE3
[5] Nixon J F and Morgenstern N R 1973 Practical Extensions to a Theory of Consolidation for Thawing Soils. – Proc. 2 nd Int. Conf. on Permafrost (Yakutsk, USSR) (Washington: North American Contribution/National Academy of Sciences) pp 369-377
[6] Dalmatov B I, Lapshin F K and Rossikhin Y V 1975 Design of Pile Foundations in Conditions of Weak Soils (Leningrad: Stroyizdat) 240 p
[7] Randolph M F and Wroth C P 1982 Recent developments in understanding the axial capacity of piles in clay. *J. Ground Eng.*, v 10 (1982) pp 73-87
[8] GOST 5686-2012 Soils. Methods of field tests by piles
[9] SNiP 2.02.03-85, Pile Foundations
[10] SP 24.13330.2011 Pile Foundations. Updated version SNiP 2.02.03-85
[11] Obodovskiy A A 1977 Pile foundation design (Moscow: Stroyizdat) p 111
[12] Bartolomey A A 1982 Calculation of sediment strip pile foundations (Moscow: Stroyizdat) 221 p

IOP Conf. Series: Materials Science and Engineering **972** (2020) 012018 doi:10.1088/1757-899X/972/1/012018

The Role of the Internet of Things (IoT) in Energy Management of a Smart City

I Maltseva[1], K Tkachuk[2]

[1]Ural Federal University, 19 Mira street, Yekaterinburg, 620002, Russia

[2]Institute of Energy Efficient and Sustainable Design and Building, Technical University of Munich, Arcisstraße 21, 80333 Munich, Germany

E-mail: i.n.maltceva@urfu.ru

Abstract. The 2030 climate & energy framework contains three key targets for Europe: to cut greenhouse gas (GHG) emissions and increase the share of renewable energy (RE) sources and promote energy efficiency. These tasks are relevant for Russia as well. As one of the frontrunners, Germany has raised its RE share target to 65% by 2030. At the same time, global energy needs are expected to increase by 30% by 2040. In this context, energy systems should not only be planned for the current supply but also be designed for a substantial increase in energy demand. To achieve these goals and avoid an energy crisis, the general idea is to shift from centralized to decentralized energy systems. Furthermore, energy sectors are currently integrated with Information and Communication Technology (ICT) solutions, especially the Internet of Things (IoT), for efficient and sustainable energy management. The article analyses the role of IoT technology in existing and future energy systems according to three scales: flat, building and city.

1. Introduction

In modern society, one of the main tasks in the field of energy systems is the creation of sustainable energy infrastructure, which is a properly organized eco- and society-friendly technical system that not only transmits energy from producer to consumer, but also fulfils all the environmental and technical requirements. Efficiency and energy saving are particularly important in matters of global climate change and the Agenda 2030 Sustainable Development Goals. In 2014, the European Council set three key targets for the year 2030 [1]: cut at least 40 % of greenhouse gas (GHG) emissions from the year 1990 level, reach at least 27 % share for renewable energies (RE) and gain at least 27 % improvement in energy efficiency. Especially, Germany lifts its RE share target to 65% by 2030 [2].

At the same time, the energy systems should not only be planned for the current supply but also be prepared for a substantial increase in energy demand. Global energy needs are expected to increase by 30 % by 2040 [3]. Further, 70 % of global GHGs are emitted from cities occupying only 2 % of the world's land [4]. Therefore, many efforts have to be made in cities to meet our environmental goals and avoid energy crises. The solution, in this case, cannot be simple. It has to be smart, innovative and fully change the current state of energy infrastructure, which would not be able to meet the future requirements of a zero-emission society.

In this context, the Internet of Things (IoT) plays a significant role in sustainable development [5]. This technology enables sensing, monitoring and collecting real-time energy consumption both on small and large scale by using a ubiquitous computing platform [6,7]. The availability of real-time data offers several opportunities to reduce energy consumption [7]. Further, the energy sector is currently integrated with Information and Communication Technology (ICT) solutions, such as IoT, for efficient and sustainable energy management. This article investigates the role of IoT technology in existing and future energy systems according to three scales: flat, building and city level.

2. Decentralization of energy systems as a step towards smart cities

Major challenges of future energy systems and stricter requirements from policies protecting the environment and fighting global warming will force us to deal with numerous problems we already have right now or will face in the nearest future. In our research have identified four main groups of problems according to tasks that Global Agenda 2030 has formulated:

- Decrease of emissions;
- Maintenance and upgrade of existing grids and stations;
- Increase renewable energy share in the energy mix and support its attractiveness for consumers;
- Support local providers of energy, such as private photovoltaics (PV) panels, geothermal or power-to-grid (P2G) systems.

Analysing existing case studies in this sphere, we came up to the solution, that is already implemented in some countries worldwide [8,9]. The general idea that is proposed by different articles is to shift from Centralized energy systems (CES) to Decentralized energy systems (DES) [10,11]. Main advantages of DES are:

- Low to zero-carbon emissions;
- Larger RE share in the local energy mix;
- Reduced costs for transmission systems;
- Improved efficiency and lower grid loss;
- Increased reliance on distributed generation involving local small-scale providers.

Introducing DES on single house and neighbourhood level will help stabilizing the energy system of a city by providing an opportunity for every consumer to be a supplier for himself. If there is more energy than he needs he can put the energy back into the city grid and receive payment for it. In this case, DES stands head and shoulders above CES as it is basing on RE sources in almost every house connected to the grid. This can be easily done by PV panels and geothermal heat exchangers on a local level and with the help of wind turbines and small hydropower plants on a city level. The reason why we are so interested in renewables when we are speaking about CO2 emissions is the statistic of Life Cycle (LC) GHG emissions obtained by other researches. On the figures 1 and 2, we can clearly see the differences in environmental impact from different energy sources [12].

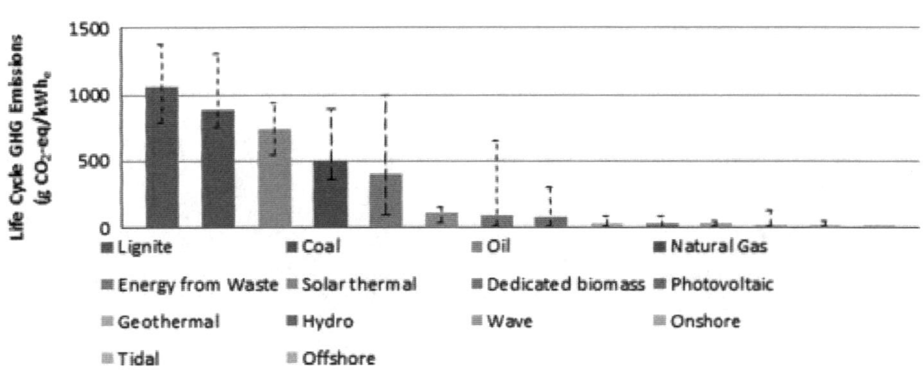

Figure 1. Maximum GHG emission levels of electricity generation methods.

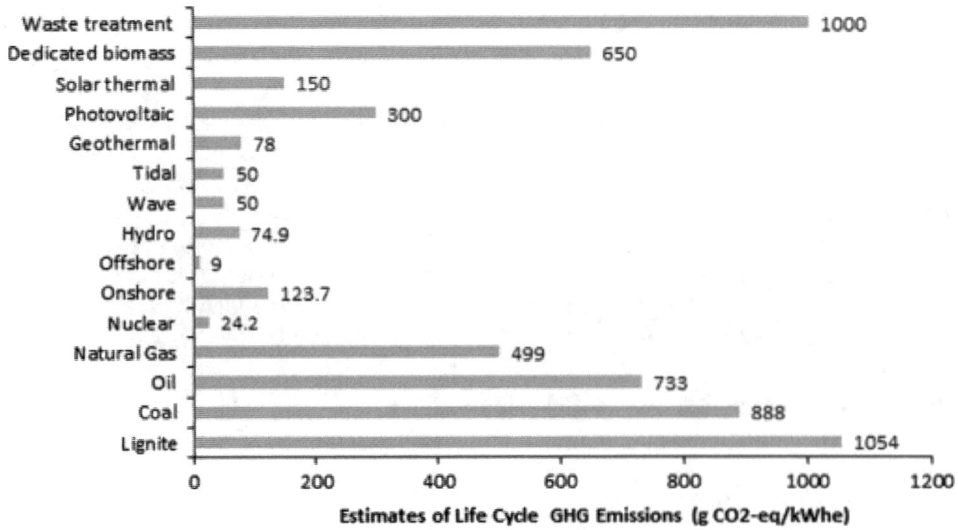

Figure 2. Life cycle GHG emission estimates of electricity generation methods.

Moreover, DES allows a multi-way flow of energy, which allows to redistribute the energy between users according to the current demand and lower the load on main energy roots as DES is built on microgrids. These grids are connecting every house locally, and through those grids it is possible to split the load on some roots and rise on the others. In other words, with DES we are creating the "energy cloud" (figure 3) across the district, neighbourhood or city, whose density depends on the energy demand and there is a lot of different roots for transmiting the needed energy to the consumer. Additionally, operating the energy without a centralized grid and flow direction allows to avoid grid losses (in reference to a centralized plant connected to extra-high voltage grids (EHV)) and save up to 15% of transition losses in energy systems [13].

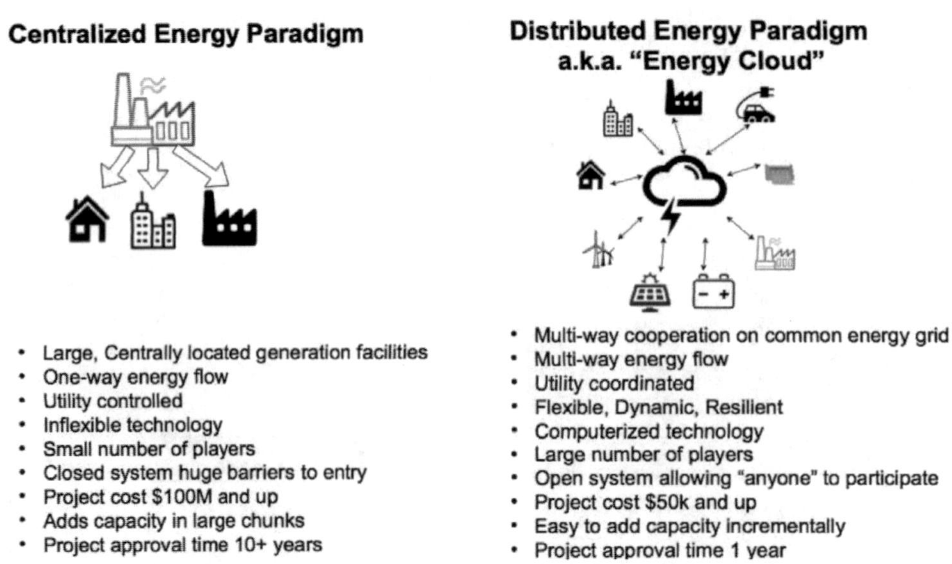

Figure 3. Visual representation of "energy cloud" (https://longtailpipe.com/2016/08/30/ti-enabling-smart-evses-by-adding-wifi-to-charging-station-reference-design/).

In addition, DES will make consumers much less dependent on one supplier of energy and even if something happens with the power plant or with grids, the system will be able to sustain itself by reallocating the energy flow for a period of time that will be needed to solve the problems.

Last but not least, public appreciation plays now a key role in the implementation of emerging technologies on a government level. In order to be accepted, technologies have to be user-friendly, be clear about what they are doing and how they can help. Their design should be appealing, which is another advantage of DES over CES since no one wants to have a big power grid in their backyard.

Considering all the advantages of DES, the authors believe that DES will be the main approach for the transition of the existing energy infrastructure to the new, sustainable and flexible infrastructure of the future.

3. Internet of Things and its part in the future of energy infrastructure

Considering the implementation of DES in the future energy systems we have to consider as well how to operate such a complex system, with big and always changing number of suppliers. The amount of data to deal with will grow exponentially, due to a multi-way flow of energy together with the appearance of new supplies in the grid system, such as private households with PV panels, ground heat exchangers, etc.

Modern technologies already have a solution to deal with Big Data and use it to control complex systems in real-time. The solution, the Internet of Things (IoT), also called the Internet of Everything or the Industrial Internet, is a new technology paradigm envisioned as a global network of machines and devices capable of interacting with each other. The IoT is recognized as one of the most important areas of future technology and is gaining vast attention from a wide range of industries [14]. In our particular case of future of energy infrastructure, IoT approach as well can and should be used as a core operator of all data, which is obtained from all controlling units and sensors from every household, connected to the grid.

3.1. IoT for Energy Management

Taking a closer look at how to implement IoT on a city scale for energy infrastructure, we need first to understand how it is working from inside. Figure 4 shows the IoT architecture in general [15].

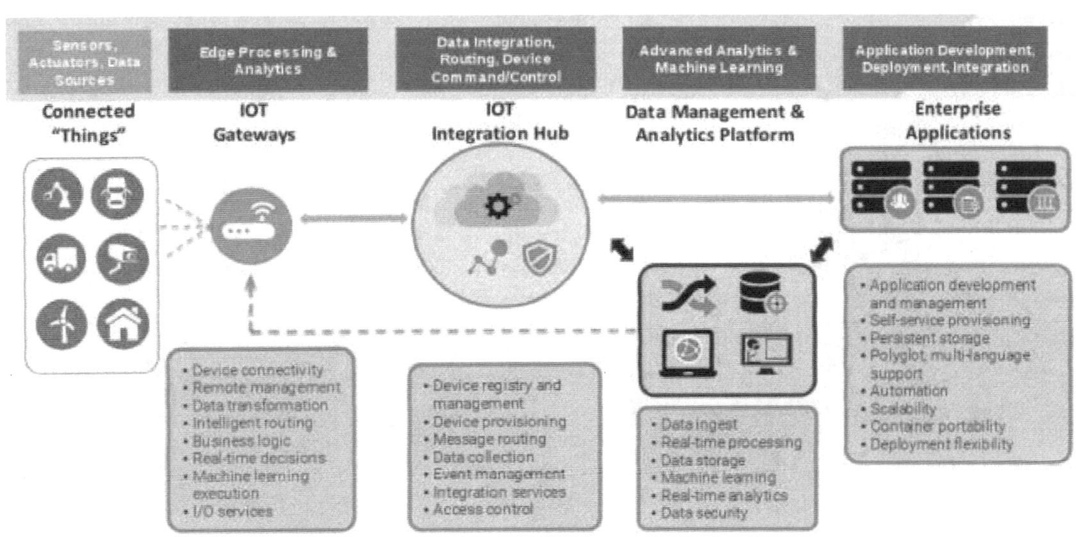

Figure 4. Visual representation of IoT system.

Within the system, there are five main elements:

- Data source – sensors, meters, etc., all that gathers data about the system;
- Control units on data source - real-time reaction on the data obtained from the data source or received response from main IoT hub;
- IoT Hub – main data server, which receives all the data from every data source, collects and provides access to it. As well responsible for registration of new data sources and their connection between each other;
- Data processing and analytics – real-time analytics of data from IoT hub, combined with machine learning. Responsible for the overall performance of the system and telling all control units how to work;
- Enterprise application – front-end part of the IoT system that is responsible application development, deployment and integration.

The same architecture is chosen for our project, only the data source will be specific for our task.

Design process of energy infrastructure always depends on the system we are working with. Detailed communication solutions and connections between the systems cannot be transferred from one implementation to another and always needs some adjustment for proper working. As well there is a considerable need in data to operate the whole system, having to be gathered from real sensors. Immediately upon obtaining the first batch of it, the system control needs to be started and its efficiency has to be improved until it reaches the expected level.

Nevertheless, the authors have done the evaluation of existing IoT technologies and equipment that are generally presented in every system and have divided them into three big groups according to the level they are working with: IoT on a flat level, IoT on a Building level and IoT on a city scale.

3.2. *IoT in a Flat*

IoT systems on a flat scale are mostly working with consumers. Communication between consumer and supplier can be extended under the umbrella of the mobile Internet of Things to include versatile resources, such as smartphones, smart meters and sensors (figure 5) in a flat that can be utilized for obtaining data and controlling energy efficiency. This approach assists in ensuring the awareness about the energy consumption of every person in a flat, specifying how much energy is spent by every system in a flat and how much it costs. In addition, it allows to provide real-time data how much money you have already spent for heating, cooling, etc. and let you know how much will the bills at the end of the month will change, if you change the heating or cooling in your flat. Mainly, this is done with the help of phone apps and smart meters installed in your flat, showing all these parameters on the screen in real-time.

Figure 5. Example of an in-home display and smart meter system (https://www.cse.org.uk/projects/view/1309; https://www.cse.org.uk/advice/advice-and-support/smart-meters).

Smart meters are the key operators of the IoT energy systems. Data obtained through them provides the system with better knowledge of people behaviour in the flat. This knowledge can be used in many different ways and will result in a higher quality of life and at the same time lower energy consumption. For example, predicting the person's location based on an increasing energy consumption in the kitchen as the oven is turned on, the system can automatically turn off the heating in the living rooms. That can result in savings up to 25% of the overall heating costs. Another example can be automated control of cooling systems according to people location and, in addition, automatically turning on and off the ventilation only when needed. This can as well result in substantial savings in energy consumption of a building [16].

Given the above, consumers start using the resources more responsibly and at the same time the data gathered from these smart meters will allow the supplier to produce the exact amount of energy needed by predicting the run-time loads in the system, redistribute the energy flow, produce additionally if needed and in the result cut down the amount of energy waste.

3.3. *IoT in a Building*

Building level is the intermediate point between the flat and a city. It is already not controlling the systems in a flat and still not controlling the overall energy flow of a city. IoT on a building scale is a data operator that gathers, analyses and transmits the information from every flat to suppliers.

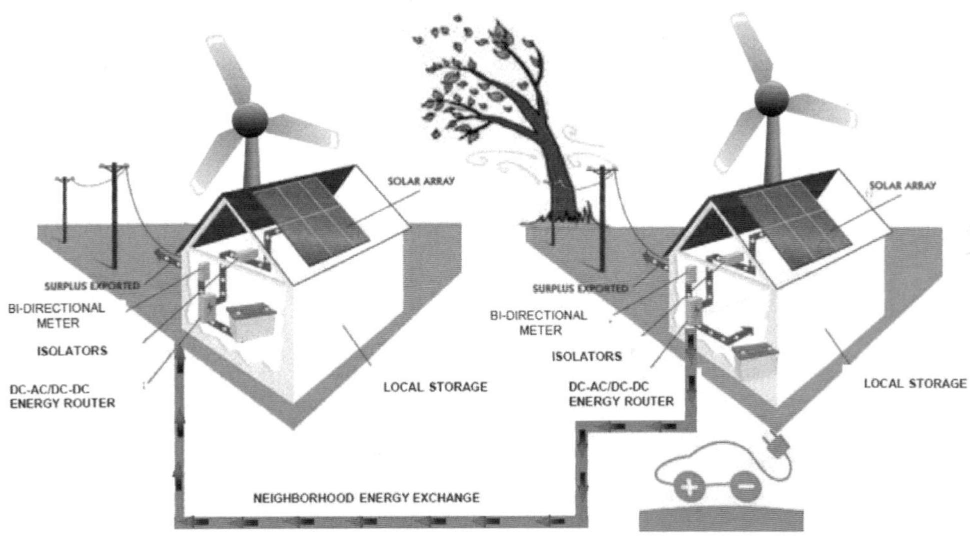

Figure 6. Building scale data operation (https://www.et.aau.dk/research-programmes/microgrids/activities/dc-next-eve-lvdc-microgrids-for-evolved-energy-communities).

The supplier receives from the building already processed data instead of the raw information output from meters. Moreover, some actions towards efficiency were already taken by the IoT processor in the building. This allows the system to perform properly even if the city scale system is not responding or there are some problems in it. Additional benefits from the IoT in the building can be obtained by installing PV panels, ground heat exchangers, air and heat recuperators and local energy storage. These solutions will make the building more self-sustained, i.e. mostly use its own energy produced by renewable sources for the dweller's needs and even put the electricity back in the grid if the production is greater than consumption right now.

3.4. *IoT in a City*

In order to achieve global sustainable development goals, energy systems in a city have to be intelligent. Smart Energy System solution was first introduced by Lund, Andersen, Østergaard,

IOP Conf. Series: Materials Science and Engineering **972** (2020) 012018 doi:10.1088/1757-899X/972/1/012018

Mathiesen and Connolly in 2012 [17]. It is the new paradigm to combine every part of energy grids, such as Smart Electricity Grid and Smart Thermal Grid, for creating a synergy effect among them [18]. All smart systems, such as Smart Grid, Smart Street Lighting, Smart Parking, Smart Water Grid and Smart Thermal Grid, are interconnected on the city level allowing to both communicate and complement each other. Such connection opens new opportunities for creation sustainable and independent infrastructure that is strongly based on renewable sources of energy.

When it comes to operation of all these systems together IoT starts to play a decisive role. The amount of data going through the core control system cannot be operated by human. Numerous parameters, each of which can affect the efficiency of the whole system is controlled by the program based on data coming from every smart meter of every flat in a building of every neighbourhood of a city. In addition, these parameters are constantly changing independently from each other. For example, if clouds appear and PV panels do not produce much energy anymore, this does not mean that the ground heat exchangers and wind turbines will stop as well. The city energy system has to immediately react on this and either open access to energy storages to fulfil the energy demand or request the needed amount of energy directly from nearby power plant.

As a result, many technologies and smart infrastructures, such as Smart Grids, Smart Lighting, Smart Parking and vehicle-to-grid (V2G), can interact with each other to create synergies. As shown in Figure 3-5, PV panels can be used as shade structures over parking lots to charge EVs during day time [19]. EVs in turn can be used as energy storage and send electricity back to the smart grid through V2G if needed. Furthermore, a smart street lighting system based on smart grids contributes to saving energy, and drivers do not have to waste much time on streets to find parking spots, thanks to a smart parking system [20].

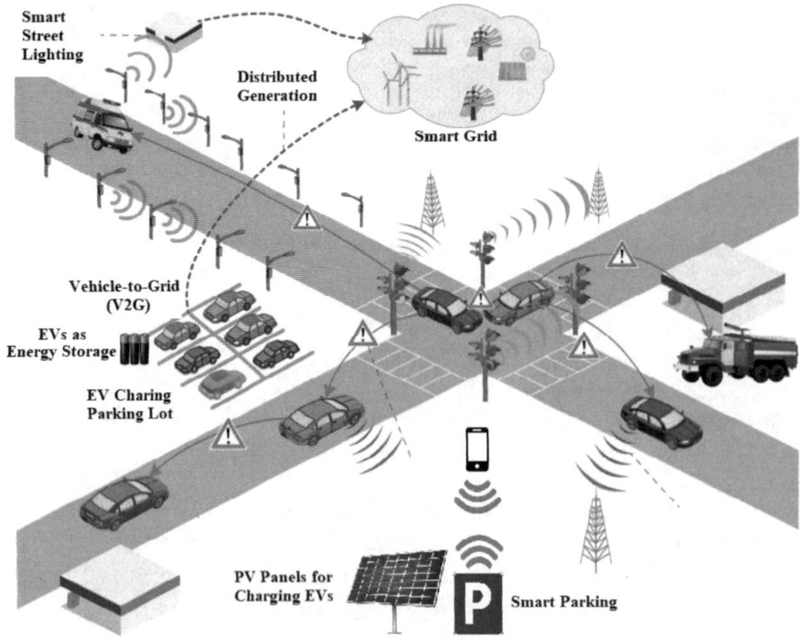

Figure 7. Smart systems and the Smart Grid in a Smart City.

Right now, we cannot fully rely on only renewable sources of energy due to their instability, but making a step towards them will give us an opportunity to tackle the problems connected with them and find the solutions sooner.

4. Conclusion

In order to meet the needs of the growing society of 21st century, energy infrastructure has to be sustainable, efficient and renewable. Reaching these goals is not easy, but there are many small steps we can do right now to reach the target in the nearest future, for example:

- Promote e-mobility;
- Install smart energy-efficient street lighting;
- Refurbish houses to reduce energy consumption, aiming towards 100 % renewable energy;
- Circulate a SmartCity app that creates intelligent links between all services.

These solutions will solve some problems on their own, but without a complex approach that includes step-by-step decentralisation of energy infrastructure, they will only remain as small steps. In order to create a sustainable eco- and society-friendly system, we have to solve both small problems and implement particular solutions. At the same time, we must think on a big scale and change our infrastructure towards fully independent local suppliers and microgrids connection between them. Coordination within the system can be done by a main IoT Hub that collects, reads and controls the system according to obtained information. Automation of these processes will lead to substantial improvement of overall energy performance of system on flat, building, neighbourhood and city scales.

In addition, Global urbanization is one of the main societal trends and all cities must be ready. Major challenges of future energy systems and stricter requirements from policies protecting the environment and fighting global warming will force us to deal with numerous problems we already have right now or will face in the nearest future.

Researches are already tackling these problems and many solutions appearing with technology development. Smart City is one of the perspective solutions, considering urbanization. But the infrastructure has to be ready for it. Such a complex solution requires long preparations before the implementation and the sooner we start them the sooner we will see the result.

In cities, everything is related to energy, thus a number of smart systems have been researched, developed and implemented to efficiently manage energy. This study attempts to investigate some of the smart energy systems, focusing on the role of IoT in these.

Moreover, overall performance of all systems has to correspond to the efficiency goals and the whole control of infrastructure has to be taken on the next level. Authors believe that IoT is the actual future of energy infrastructure. It allows to operate all the data coming from every sensor in every flat and building of the city and adjust the parameters in real-time. In addition, the system greatly benefits from the amount of available data since analysing it the control of the whole system will be very precise and even more efficient than operated by a human

Reference

[1] European Council 23/24 October 2014 Conclusions
[2] CDU, CSU and SPD 2018 Ergebnisse der Sondierungsgespräche von CDU, CSU und SPD: Finale Fassung
[3] IEA 2017 World Energy Outlook 2017
[4] UN 2011 Global Report on Human Settlement 2011
[5] IERC 2012 The Internet of Things 2012: New Horizons ISBN 978-0-9553707-9-3 ed Smith I G
[6] Al-Ali A R 2016 Internet of Things Role in the Renewable Energy Resources *Energy Procedia* **100** 34–38
[7] Shrouf F and Miragliotta G 2015 Energy management based on Internet of Things: practices and framework for adoption in production management *Journal of Cleaner Production* **100** 235–246
[8] Mahapatra S and Dasappa S 2012 Rural electrification: Optimising the choice between decentralised renewable energy sources and grid extension *Energy for Sustainable Development* **16** 146–154

IOP Conf. Series: Materials Science and Engineering **972** (2020) 012018 doi:10.1088/1757-899X/972/1/012018

[9] Williams J 2010 The deployment of decentralised energy systems as part of the housing growth programme in the UK *Energy Policy* **38** 7604–7613

[10] IPCC 2007 Climate Change 2007: Synthesis Report. Contribution of Working Groups I, II and III to the Fourth Assessment Report of the Intergovernmental Panel on Climate Change

[11] Mathews J A 2013 The renewable energies technology surge: A new techno-economic paradigm in the making? *Futures* **46** 10–22

[12] Amponsah, N Y, Troldborg M, Kington B, Aalders I and Hough R L 2014 Greenhouse gas emissions from renewable energy sources: A review of lifecycle considerations *Renewable and Sustainable Energy Reviews* **39** 461–475

[13] Hu J, Harmsen R, and Crijns-Graus W 2017 Developing a Method to Account for Avoided Grid Losses from Decentralized Generation: the EU Case *Energy Procedia* **141** 604–610

[14] Lee I and Lee K 2015 The Internet of Things (IoT): Applications, investments, and challenges for enterprises *Business Horizons* **58** 431–440

[15] Patrick Goemaere and Rajat Ghai 2018 Edge Compute and Software Life-Cycle Management SCTE•ISBE and NCTA page 13

[16] Lachhab F, Bakhouya M, Ouladsine R and Essaaidi M 2018 Towards an Intelligent Approach for Ventilation Systems Control using IoT and Big Data Technologies *Procedia Computer Science* **130** 926–931

[17] Lund Henrik, Andersen A N, Østergaard Poul, Mathiesen Brian and Connolly David 2012 From electricity smart grids to smart energy systems – A market operation based approach and understanding *Energy* **42** 96–102

[18] Mathiesen B V, Lund H, Connolly D, Østergaard P A and Möller B 2013 The design of Smart Energy Systems for 100% renewable energy and transport solutions. Book of Abstracts: 8th Conference on Sustainable Development of Energy, Water and Environment Systems

[19] Birnie D P 2009 Solar-to-vehicle (S2V) systems for powering commuters of the future. *Journal of Power Sources* **186** 539–542

[20] Turner S and Uludag S 2016 Toward Smart Cities via the Smart Grid and Intelligent Transportation Systems 117–166.

IOP Conf. Series: Materials Science and Engineering **972** (2020) 012019 doi:10.1088/1757-899X/972/1/012019

Efficient Waste Management in a Smart City

G B Zakharova[1,2], A V Fedorova[2]

[1] Department of Scientific Research, Ural State University of Architecture and Art, Karl Libknekht str., 23, Yekaterinburg 6200075, Russia

[2] Department of Modeling of Controlled Systems, Ural Federal University, Mira str., 19, Yekaterinburg 620002, Russia

E-mail: zakharova@usaaa.ru

Abstract. National projects for the development of the Russian Federation, including in the field of digitalization, aimed at creating a comfortable ecologically pure urban environment are presented. The evolution of the "smart city" concept is briefly presented: from technological to humanitarian view. Some examples of smart solutions in terms of ecology are given. Indicators of the smart city are analyzed, environmental characteristics are highlighted. The data on Yekaterinburg and actual directions of automation for solving environmental problems in the Sverdlovsk region are shown. The logistical task of garbage picking in multi-agent systems technology is formalized and the model example is presented in the Any Logic simulation system.

1. Introduction

Presidential Decree of May 7, 2018 No. 204 "On the national goals and strategic objectives of the development of the Russian Federation for the period up to 2024" identified 12 areas of the country's strategic development, for which relevant national projects have been developed [1]. National projects affect all major areas of citizens' life, among them such projects as "Demography", "Education", "Healthcare", "Culture", "Science", etc. In this context, we especially note the national projects "Ecology", "Housing and urban environment", "Digital economy of the Russian Federation".

The national project "Ecology" [2] occupies an important place, since the overall environmental situation in Russia is unsatisfactory. The environment in cities and adjacent areas, where 74% of the population lives, is exposed to a significant negative impact from industrial, energy, transport and capital construction facilities. The situation with the accumulation of production and consumption wastes is catastrophic. Annual economic losses due to environmental degradation are 4-6% of GDP. The national project will be implemented in five areas: "Waste", "Water", "Air", "Biodiversity", "Technology".

The key objectives of the national project "Ecology" are the effective handling of production and consumption waste, including the elimination of all unauthorized landfills identified within cities, and the reduction of air pollution in large industrial centers. Attention is also focused on improving the quality of drinking water, cleaning and preserving water bodies, unique water systems, preserving biological diversity and creating new specially protected natural areas, ensuring the balance of deforestation and reproduction of forests. In accordance with these goals, 11 federal projects have been developed, including "Clean Country", "Integrated Solid Municipal Waste Management

System", "Infrastructure for Waste Management 1-2 Hazard Class", "Clean Air", "Clean Water" and others, for which the relevant ministries are responsible.

All these measures are closely related to improving the quality of the urban environment. The Russian government has approved the methodology for calculating the quality index of the urban environment, from which the size of regional subsidies will depend on from 2020. The index was developed by the Ministry of Construction of Russia in the framework of the national project "Housing and Urban Environment" based on 36 indicators. These include the share of emergency housing and the diversity of residential development, the diversity of cultural, leisure and sports infrastructure, the development of the transport system, road congestion and the number of people killed in road accidents, the pedestrian accessibility index and others. An important parameter is the level of landscaping. The Ministry of Construction of the Russian Federation will have to annually ensure the formation of an index of quality of the urban environment.

In March 2019, the Ministry of Construction and Housing and Communal Services of the Russian Federation approved the Smart City standard. The section "Intellectual systems of environmental safety" of this document assumes the development of systems for monitoring the quality of air and water in the online mode. The automation of the management system for solid municipal waste management is also relevant.

Next, we will show how this problem is solved abroad and in Russia, and we will also offer an original approach to automating the management of garbage collection on the basis of a multi-agent modeling approach.

2. Evolution of the concept of "smart city"

The problem of the constant increase in environmental pressure on cities as a result of continuous urbanization and increasing requirements for the quality of the urban environment is called upon to resolve modern technologies of a smart city. In October 2018, the Ministry of Construction and Housing and Communal Services of the Russian Federation approved the passport of the departmental project of digitization of the municipal economy "Smart City", according to which the regions, including the Sverdlovsk region, are actively developing ways to implement this project. In March 2019, the Smart City standard was approved – Basic and additional requirements for smart cities. "Smart City" is developing in line with the national project "Digital Economy of the Russian Federation" [3] and "Strategies for the Development of the Information Society in the Russian Federation for 2017-2030".

The concept of a smart city, formed in the early 2000s, evolved from a predominantly technological approach to a humanitarian one. The new model of the smart city provides not only various ways of applying smart technological solutions, but also the active involvement of residents in their development. A modern smart city is a municipality with a well-developed technological infrastructure, where human life acquires a new quality thanks to smart solutions. This is a systematic approach to the use of information technologies based on data analysis for making management decisions for the purposes of sustainable development and ensuring high living standards.

The British Standards Institute (BSI) describes the smart city as "the effective integration of physical, digital and human systems in an artificially created environment in order to ensure a sustainable, prosperous and comprehensive future for citizens".

At the international industrial exhibition "Innoprom", the Ministry of Construction of the Russian Federation announced an unusual and unique rating of Russian cities: they will be assigned an appropriate IQ. The higher the degree of penetration of digital solutions in the urban environment, including in the sphere of housing and public utilities and landscaping, provides the higher the IQ of the municipality.

3. Examples of smart environmental solutions

Smart solutions for a comfortable urban environment have already been implemented in different countries of the world. Let us show some examples.

The water quality control system in the largest reservoir in Eurasia, Kuibyshevsky, was developed by Airalab Rus, SMART Distributions and Togliatti State University. Using a solar-powered river drone that can move around the river, water quality parameters are measured using the technology of the Spanish company Libelium [4], one of the world's largest manufacturers of intelligent sensors and solutions based on them. In order to provide information on water quality to interested parties (via a web application), a blockchain-based platform and the Internet of Things (IoT) were used. Controlled parameters: water temperature, pH, dissolved oxygen content, electric conductivity of water, ions NH^4 + and NO^{3-}.

Management systems for garbage collection are part of the concept of a smart city [5]. They consist of remote sensors installed in garbage cans and specialized software. Sensors monitor the level of fullness of garbage cans and transmit data to the central server in real time. They are ultrasound and in most cases are powered by solar energy, which allows them to assess the level of fullness of garbage bins offline. The transfer process is carried out over the radio using built-in GPS and GPRS-modules. Further, the specialized software provides detailed detailing of the filling level of each container; on this basis, optimal plans for waste collection routes are built. This allows you to optimize staff working time, reduce the amount of fuel consumed by garbage trucks, reduce the number of purchased garbage collection machines, and reduce equipment and roadbed wear. As a result, according to analysts, the costs of utilities for garbage collection can be reduced by almost 40%.

The main suppliers of automation solutions for garbage collection are companies such as IBM Corporation, SAP SE, Waste Management, Enevo, BigBelly, and Compology. Smart garbage collection systems are implemented in the USA, Spain, UK, Finland. In Russia, the process of automating the collection of street garbage begins to develop.

For the centralized collection of systems for the automation of all sections in accordance with the standard of the smart city, a special resource has been created – the smart city solutions bank [6]. At the end of 2018 – the beginning of 2019, 4 systems appeared in the section "Intellectual systems of environmental safety". Each of them has its own approaches and implementation features.

4. Analysis of Russian cities based on smart environment indicators

In 2017, the Moscow company National Research Institute of Technology and Communications (at the same time consulting center in the field of information and communication technologies) developed smart city indicators (26 parameters distributed in 7 key areas) and the corresponding methodology for analyzing Russian cities based on them [7]. The 16 largest cities in Russia were analyzed, Yekaterinburg ranked 4th in the degree of implementation of smart technologies. In such areas as smart finance, economics, technology, smart people, Yekaterinburg occupies high positions. In areas of smart management, infrastructure and the smart environment is on average and below average.

The direction of the smart environment was estimated on the basis of 2 indicators for which there were open data: the level of activity of residents and the city administration in eliminating illegal dumps and the level of development of monitoring systems and the prevention of threats to environmental safety.

In terms of smart environment indicators, neighboring regions are much more successful in working towards the environment: Novosibirsk, Krasnoyarsk and Chelyabinsk make up the top 3 cities according to these indicators, although they are close to Yekaterinburg in terms of climate and environmental situation.

At present, within the framework of the regional program "Smart Cities of the Sverdlovsk Region", the development of intelligent environmental safety systems is underway. The complex system of digital development of the urban environment of the Sverdlovsk region involves the automation of the management system for solid municipal waste management. The matrix of elements of a smart city will optimize the routes of movement of specialized equipment to ensure efficient use of resources in the collection and disposal of waste, as well as carry out an automatic analysis of the costs of collecting, exporting and disposing of waste and rating taking into account investments in the construction of new infrastructure. Provision is also made to monitor the movement and operation of

IOP Conf. Series: Materials Science and Engineering **972** (2020) 012019 doi:10.1088/1757-899X/972/1/012019

specialized equipment in the online mode. The organization of interaction between regional and municipal authorities, a regional operator for removal of municipal solid waste, waste carriers and landfill sites is needed.

5. Multi-agent approach to system formalization

Next, the proposed formalization of the problem of garbage collection in terms of technology multi-agent systems for subsequent simulation in the system AnyLogic. Figure 1 shows 3 types of intelligent agents that interact in order to optimize resources.

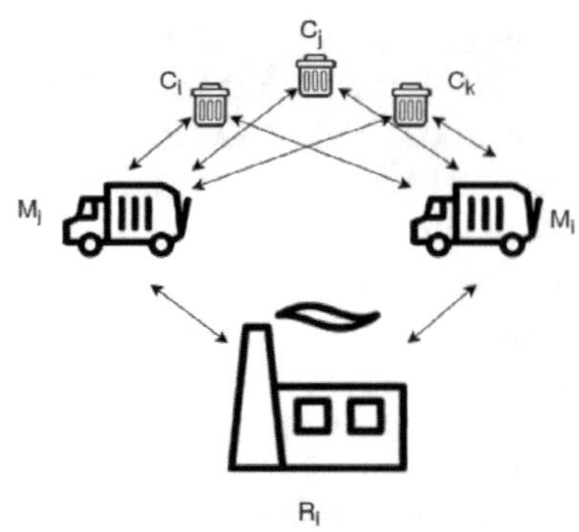

Figure 1. The model in terms of multi-agent systems

The control system for garbage collection is represented as a connected graph, where the vertices are agents, the edges are the connections between them. We distinguish three types of agents: garbage collection points ($C = \{c_i\}_{i=1}^{m}$), garbage collection machines ($M = \{m_i\}_{i=1}^{n}$), recycling sites ($R = \{r_i\}_{i=1}^{k}$). The limitations of the system are directly related to the limitations for each type of agent. The total capacity of the collection points should not exceed the total capacity of the recycling sites. We also introduce restrictions on the fullness of garbage collection points and recycling sites and then consider the sensors of each agent and its communication with other objects of the system.

$$f(x) = \begin{cases} \sum_{i=1}^{m} capacity(c_i) \le \sum_{i=1}^{k} capacity(r_i), \\ \dfrac{\sum_{i=1}^{j} occupancy(c_i)}{\sum_{i=1}^{m} capacity(c_i)} \le 0.95, \\ \dfrac{\sum_{i=1}^{l} occupancy(r_i)}{\sum_{i=1}^{k} capacity(r_i)} \le 0.85. \end{cases}$$

Sensor “garbage collection point” – is a sensor of the current fullness. The agent is able to transfer this data to the agents of the “garbage collection machine” type. The main task of the agent “garbage collection point” is to get rid of the greatest possible amount of garbage during the day, that is, to minimize the time of “useless work”:

$$time\left(\frac{occupancy(c_i)}{capacity(c_i)} \ge 0.7\right) \to min.$$

The sensor of "garbage collection machines" – a sensor of the current fullness. The agent is able to transfer this data to agents of the type “garbage disposal site”, to receive and process data from “garbage collection points”, from “garbage disposal sites”. The tasks of the agents of the “garbage collection machine” include the transportation of the largest possible amount of garbage in the shortest time, that is, minimizing the “idle”, and at the same time reducing the cost of fuel:

$$\begin{cases} time\left(\dfrac{occupancy(m_i)}{capacity(m_i)} \leq 0.7\right) \rightarrow min, \\ fuel(m_i) \rightarrow min. \end{cases}$$

“Garbage disposal sites” are able to receive and process data from “garbage collection machines”, as well as transfer potential load data to agents such as “garbage collection machines”. The task of “waste recycling sites” is to optimize waste recycling:

$$\begin{cases} time\left(\dfrac{occupancy(r_i)}{capacity(r_i)} \geq 0.9\right) \rightarrow min, \\ \dfrac{occupancy(r_i)}{capacity(r_i)} \rightarrow max. \end{cases}$$

This formalization is the basis for the subsequent modeling of autonomous intelligent agents in the Any Logic system.

Thus, the paper presents the results of research in the field of digitalization for the creation of a comfortable environmentally friendly urban environment in the "smart city" system. The actual directions of automation of solving environmental problems in the Sverdlovsk region are shown. The mathematical formalization of the logistic optimization problem when collecting garbage in the technology of multi-agent systems with the goal of modeling in the system Any Logic.

References

[1] National Projects: Key Objectives and Expected Results 2019 Russian Government Website URL: http://government.ru/news/35675/
[2] Passport of the National Project "Ecology" 2019 Russian Government Website URL: http://government.ru/info/35569/
[3] Passport of the national program "Digital Economy of the Russian Federation" 2019 Russian Government Website URL: http://government.ru/info/35568/
[4] Web site of the company Libelium 2019 URL: http://www.libelium.com/ libeliumworld/enviroment/
[5] Clearing Debris Cleverly: IoT News 2019 URL: https://iot.ru/gorodskaya-sreda/uborka-musora-po-umnomu
[6] Smart City Solutions Bank 2019 URL: https://russiasmartcity.ru/projects?category=waste
[7] Indicators of Smart Cities NIITS 2017. Research of the National Research Institute of Technology and Communications 2019 URL: http://niitc.ru/publications/SmartCities.pdf

IOP Conf. Series: Materials Science and Engineering **972** (2020) 012020 doi:10.1088/1757-899X/972/1/012020

Is a sustainable future possible?

Y V Anikin[1], V I Shilkov[2]

[1]Department of water treatment, Ural Federal University, Mira str., 19, Ekaterinburg, 62002, Russia
[2]Graduate School of Economics and Management, Ural Federal University, Mira str., 19, Ekaterinburg, 62002, Russia

E-mail: anikin-urfu@yandex.ru

Abstract. The article presents the issues of transition to a sustainable balanced development of economic systems, taking into account the minimization of environmental losses. A thesis on the internal contradictions of modern economic models is proposed. Problems and negative consequences of the introduction of new technologies are noted. The problems of the interaction of economics and ecology are analyzed. The main obstacles to the transition to sustainable development and a green economy are identified.

1. Introduction

The discussion of the transition to a sustainable environmentally balanced development and a green economy has been taking place over the past few decades.

Environmental scientists, various social movements and the media have long paid attention to the problem of irreversible changes in the environment, which are largely due to negative anthropogenic influences.

Today, it is still possible to argue about whether warming is really coming or, conversely, cooling in the future, but the effects of climate change are often enough to show themselves and thus give rise to a serious discussion of the expected consequences. In the same row are the problems of depletion of energy resources (oil, gas, coal), pollution of surface and ground waters of the land and the World Ocean, air, soil, accumulation of production and consumption waste [1-8]. The modern economy in its most general sense is impossible without taking into account the laws and needs of ecology.

As noted in the report of the World Commission on Environment and Development in 1987, "Our Common Future" [9]: "The environment and the economy are increasingly intertwined with each other - at the local, regional, national and global levels - to form a complex cause and effect ... We do not predict the future; our task is to issue a warning - an urgent warning based on the latest and most reliable scientific data - that the time has come for making appropriate decisions that guarantee the preservation of the resources necessary for current and future generations".

2. Sustainable development's primary dimensions

In recent years, some authors have proposed various models of ecological and economic development. For example, in [10–11] models of a linear (traditional), fractional, cyclic, and green economy are proposed.

These models are characterized by different approaches to the consumption of resources and the processing of industrial and household waste.

An integral part and engine of economic growth is the innovation process. It is innovative technologies that provide access to new materials and types of energy that make it possible for all of humanity to reach new frontiers of economic development.

In connection with the threat of climate change on Earth and the exhaustion of traditional energy resources, it became clear that humanity needs to switch to renewable energy sources (RES).

Table 1 shows the dynamics of investments of countries of the world in renewable energy technologies according to [12-14].

Table 1. Global Trends in Technological Investment (RES), billion dollars

	2004	2006	2008	2010	2011	2012	2013	2014
Solar power	12.0	22.1	60.8	103.3	155.7	144.3	119.8	149.6
Wind power	17.9	39.6	75.2	98.9	84.2	84.1	89.3	99.5
Biomassa and sewage sludge for energy	7.4	12.1	16.9	16.0	17.4	12.4	9.3	8.4
Wood fuel	3.9	28.4	19.2	10.1	10.4	7.0	5.5	5.1
Hydropower <50 MW	2.6	7.6	7.8	5.7	7.2	6.4	5.5	4.5
Geothermal energy	1.2	1.5	1.7	3.0	3.7	1.8	2.2	2.7
Ocean energy	0.0	0.9	0.2	0.3	0.3	0.3	0.2	0.4
Total investment	45.1	112.1	181.8	237.2	278.8	256.4	231.8	270.2
Total installed capacity without hydropower, GW	85	207	200	315	390	480	560	665

As can be seen from Table 1, the trends in global investments in technologies for various types of renewable energy sources are quite heterogeneous. For example, there is almost constant growth in volumes on investments in solar energy, and for investments in biofuels and geothermal energy, fluctuations are observed.

The use of renewable energy is a progressive phenomenon and leads to a decrease in the use of limited natural resources. But on the other hand, quite often new technologies may lead to new dangerous consequences. For example, the use of renewable energy sources (wind, sun) is always presented as an undoubted advantage over the use of oil, gas, coal [15-16]. However, even if we briefly consider the individual pros and cons of renewable energy sources (Table 2), then, for example, for solar energy, negative aspects can be presented in more detail:

- use of toxic elements in the production of solar cells (lead, cadmium, gallium, arsenic, etc.). These elements may be released into the environment;
- battery efficiency is quite low (about 20 %), and 80 % of energy is spent in the form of heat into the surrounding space;
- a device for cooling and energy storage is required, since a large amount of energy is required in the evening and at night when batteries do not work;

In addition, large areas are required to accommodate solar cells. This causes a fairly high cost of 1 kW of energy, and the amount of energy produced by solar energy is less than 1 % of its total worldwide.

In [17] it is noted that “the accelerated development of renewable energy in Russia should be considered as an important factor in the modernization of the economy, including those related to the development of innovative industries, the development of new innovative technologies, the development of small and medium businesses, the creation of new jobs, the improvement of social conditions, improvement of ecology, etc.". Unfortunately, it should be noted that target indicators for the development of renewable energy sources suggest an increase in the share of renewable energy in the country's energy balance by only 2.5% by 2020, which is noticeably less than that of the advanced countries in this area.

Table 2. Selected strengths and weaknesses of renewable energy sources

	POSITIVE EFFECTS	NEGATIVE EFFECTS
SOLAR PANELS	• uses an inexhaustible renewable energy source; • there are no pollutant emissions to the environment.	• high price • large footprint • the need to dispose of toxic waste
GEOTHERMAL SOURCES	• geothermal power plants are more environmentally friendly than thermal power plants	• access to underground sources requires drilling.
HYDROPOWER	• the cost of electricity at hydropower stations is significantly lower than at all other types of power plants	• construction of hydropower plants requires large capital expenditures • efficient hydro stations are far from consumers
SOURCES ON BIOFUEL	• use of agricultural waste, timber processing, food industry and municipal wastewater treatment plants	• reduction of acreage under food crops
Tidal Power Stations	• environmental friendliness and low cost of energy production	• power changes during the day
WIND POWER	• insignificant influence on biocenoses and living conditions of people; • no fuel costs	• weather dependence

In the present economic conditions, the planned terms of return on investments in RES technologies and the level of profitability of such projects are not attractive to investors. Also, no conditions have been created that would motivate Russian industrial enterprises to create new efficient equipment for the renewable energy sector. It is obvious that the influence of state policy in this direction, both as a whole in the country and in the regions, is still not enough.

The report [9] mentioned drought in African countries, which killed 1 million people, as one of the examples of the negative interaction of the economy and the environment. And such examples are not left in the past. Thus, for example, according to [18], for 35 years (from 1965 to 2000) about 4.4 billion people suffered from varying degrees from seven main types of natural disasters. The total economic losses over these 35 years amounted to 895 billion, including over the last decade – 676 billion dollars. It is noted that the greatest social risk (death and injury to people) is characteristic of countries with the lowest level of development. The growth of critical situations is caused not only by the increase in the human population on Earth, but also by the growth of technogenic impacts on the environment.

In subsequent years, mankind experienced a large number of both natural and man-made disasters [19–20]: accidents at nuclear facilities, oil spills from tankers and oil platforms, transportation and industrial disasters that caused significant damage to the global economy and the environment.

The arms race, military conflicts, and now the widespread spread of terrorism in the world also lead to environmental, economic and political crises in a number of countries.

Since the report of the World Commission, environmental pollution has also increased in the air, water environment and in the soil layer. The number of industrial and domestic wastes that are not recycled is growing. New types of pollution have emerged: electromagnetic, noise and light. Due to the imperfection of technologies and the lack of an integrated approach in the development of mineral resources, coal and oil are not fully extracted, a large amount of associated gas is burned. A significant amount of valuable metals is removed to the dumps, the extraction of which is considered uneconomical. Treatment facilities of industrial enterprises and domestic wastewater occupy significant areas for non-utilizable waste.

3. Discussion of the results and conclusion

Why, despite the constant discussion of these problems at various levels, starting with the UN conferences and ending with the teaching of ecology in secondary and higher education, does the ecological situation continue to deteriorate?

In our opinion, the following contradictory trends can be singled out as the main obstacles to the transition to sustainable development and a green economy:

- human society is a "consumer society", and consumption is impossible without constant growth and this is the road to a permanent negative impact on the environment. Most of the energy and material resources of the world economy goes to meet the secondary needs of a minority of the world's population;
- population growth leads to an increase in the negative impact on the environment, and growth restriction is impossible, since in most countries (including Russia) this growth is encouraged as one of the conditions for economic growth;
- inequality between developed and developing countries pushes the latter to economic and material growth, which, as a rule, is connected not with the use of advanced technologies, but with a negative impact on the environment (for example, unreasonable extraction and export of raw materials from developing countries without a forecast for the future);
- government agencies and commercial organizations directly related to economic problems, as a rule, are not directly related to environmental issues and are not interested in a global change in attitudes towards ecology, and current problems require a coherent, integrated approach;
- changes in ecology most seriously affect infrastructure systems (energy, water supply and others), which requires the development of models of their functioning in new conditions, and the modeling should take into account the interaction of infrastructures;
- it is necessary to carry out modeling of the functioning of social systems to determine the behavior of large masses of people in the face of serious environmental changes;
- a healthy environment, preserved for future generations, should be the goal of a green economy at all levels: global, national, regional and local.

References

[1] Kolosova O Yu 2013 Man and society in the face of global environmental problems *Conference SIC Sociosphere collections* **51** 40-43

[2] Kovrizhkina L N 2014 Ways of solving global environmental problems by creating a set of measures for environmental safety *Bulletin of the South Ural Professional Institute* **2 14** 85-91

[3] Kuvshinov Yu A 2012 Noospheric world view and the problem of the global environmental crisis *Successes of modern natural science* **10** 97-100

[4] Kolosova O Yu 2014 The environmental problem: the global nature of the modern era *Science and Modernity* **32(2)** 129-134

[5] Bogdanova A Yu 2015 Absolute overpopulation of land and relative demographic overcrowding in its individual regions as a global social and environmental problem *Conference Sources Sociosphere* **27** 24-25

[6] Pyrsheva M V 2010 Global environmental problems of the world economy *Economics and management: new challenges and prospects* **1** 382-84

[7] Luzgin V P, But E A 2011 Energy and the problem of "sustainable development" J. *Ecology and Industry of Russia* **10** 40-43

[8] Kondratova E S , Shevchenko A A, Rushkova A V 2015 Ecological and energy problems of the development of the industrial complex of the region *J.News of the South-West State University. Series: Economy. Sociology. Management* **3(16)**196-201
[9] Our common future 1987 *Report of the World Commission on Environment and Development* 412 p [Electronic resource] URL: http://www.un.org/ru/ga/ pdf / brundtland.pdf / (appeal date: 10/16/2016).
[10] Chestina T 2016 Economy of Trust *J. Ecology and Law* **2 62** 35-38
[11] Paramonova N 2016Cyclical economy on the threshold of Russia *J. Ecology and law* **2 62** 28-29
[12] Renewable Energy 2016 *Global Report. Main results* [Electronic resource] URL: http://www.ren21.net/wp-content/uploads / 2016/10 / REN21_GSR2016_ KeyFindings_RUSSIAN.pdf / (appeal date: 10/16/2016).
[13] Renewables 2013 Global Status Report KeyFindings_2013_lowres [Electronic resource] URL: http://www.ren21.net/wp-content/uploads/2015/06/KeyFindings_2013_ lowres.pdf / (appeal date: 10/16/2016).
[14] Renewables 2015 Global Status Report [Electronic resource] URL: http://www.ren21.net/wp-content/uploads/2015/07/REN12-GSR2015_Onlinebook_low1.pdf/ (appeal date: 10/16/2016).
[15] Ozhgihin D V 2012 The role of non-traditional renewable energy sources in solving problems of energy and ecology *J. Power and administration in the East of Russia* **4** 165-70
[16] Kosov Yu V 2011 The relationship of energy and environmental security in the current global economic crisis *J. Management consulting* **2 42** 87-95
[17] Fortov V E and Popel O S 2013 Renewable energy sources in the world and in Russia *Proceedings of the first International Forum "Renewable Energy. Ways to improve energy and economic efficiency REENFOR–2013*" [Electronic resource] URL: http://www.reenfor.org/upload/files/ d1d87377ab5ed5c040331420bc9b0a93.pdf / (appeal date: 10/16/2016).
[18] Osipov V I 2001 Natural disasters at the turn of the 21st century *J. Bulletin of the Russian Academy of Sciences* **71 4** 291-302
[19] Korzh V D 2011 New methodology for solving global environmental problems of the *hydrosphere J. Water: chemistry and ecology* **6** 9-19
[20] Davydov B I, Ushakov B N 2011 Is the paradigm agreement possible: ecology-energy-society *J. Medical, biological and socio-psychological problems of safety in emergency situations* **2** 15-22

IOP Conf. Series: Materials Science and Engineering **972** (2020) 012021 doi:10.1088/1757-899X/972/1/012021

The Formation of a Comfortable Urban Environment

L N Kondrat'eva[1], N R Stepanova[2], P V Bochkov[2]

[1]Geotechnical department, Saint Petersburg State University of Architecture and Civil Engineering, 4 2-ya Krasnoarmeyskaya str., Saint Petersburg, 190005, Russia
[2]Ural Federal University, 19 Mira str., Yekaterinburg, 620002, Russia

E-mail: [1]kondratjevaln@yandex.ru [2]n.r.stepanova66@gmail.com
[3]bochkov05@mail.ru

Abstract. The concept of "smart city" and the need to follow this direction in the light of current trends, so as not to miss the chance to improve life in cities. The development and improvement of cities allows humanity to reach a new level in the quality of life and intellectual development. Hence, the population has more prospects to change the surrounding life and their personal. The article reflects the main ideas in the formation of a comfortable urban environment, their relationship and impact on a favorable social climate. The scenario algorithm of practical action for the active citizen on determination of the direction of improvement of comfort of the city environment, establishment of communications of objects of infrastructure of the city environment, Association of the used methods of changes in the concept is offered. The results of the project "Formation of a comfortable urban environment" on the example of the Sverdlovsk region. The stages of organizational procedures are analyzed, the data of analysts are analyzed, conclusions are drawn about the need to involve citizens in the implementation of projects initiated by citizens.

1. Introduction

Cities are created for people, their inhabitants and guests.

The term "urban environment" can be considered as a combination of physical and spiritual space, which includes the city itself. This space has natural and socio-economic features. It has an internal structure, dynamics, evolution [1].

A comfortable urban environment is a space that is maximally adapted to the needs of citizens. The comfort of the city for its residents by several factors:

- transport accessibility,
- availability of necessary services,
- availability of facilities for all categories of the population,
- availability of organized public spaces,
- convenient layout of the street network.

Indicators of satisfaction with the place of residence, quality of life, health and safety of residents are also associated with a comfortable urban environment.

1.1. Problem description

Currently, the formation of a comfortable urban environment is one of the most important problems.

Today, the traditional forms of management of urban and municipal services have almost exhausted themselves. They do not meet modern requirements for logistics, safety and environment. As the percentage of the urban population is constantly increasing, the burden on infrastructure and utilities will continue to increase in the future.

The formation of the concept of "smart city" ("smart city") is an attempt to solve a complex of problems of urban management.

Today there are several definitions of a smart city. However, all formulations say that a smart city is driven by data that allows urban services to improve the quality of life of the population [2].

The data cover such spheres of life of citizens as safety, transport, medical services, municipal services, improvement, etc. data Sources are video cameras, various sensors, information systems, etc.

A "smart" city can be defined as a "city of knowledge", a "digital city", a "cyber city" or an "eco-city" – depending on the purpose of urban planning. "Smart" cities in economic and social aspects are looking to the future [3].

They continuously monitor critical infrastructure for optimal resource allocation and security. They are constantly increasing the number of services provided to the population, providing a sustainable environment that contributes to the well-being and health of citizens. The basis of these services is the infrastructure of information and communication technologies (ICT) [4].

1.2. Relevance practical relevance

The formation of a comfortable urban environment is one of the priority projects in the field of housing and communal services.

The use of foreign experience of cities-colleagues and support and promotion of ideas of their own minds – one of the keys to success in the direction of improvements in the formation of a comfortable urban environment. But before moving on the path of creating a comfortable environment, you need to understand the problem areas and their relationship with all spheres of life.

Many cities in Russia pay great attention to the formation of a comfortable city, for example, such large cities as Moscow, St. Petersburg, Kazan, Krasnodar, Novosibirsk, Yekaterinburg, etc. in Addition to the fact that the largest cities of the country, they are also leaders in the integration of new ideas and technologies in the living conditions of residents.

Social and techno-economic research is required to identify problems. Social studies can be based on surveys, public hearings, etc. Technical and economic studies are based on well-defined indicators and their values showing growth or decline. The data obtained require Desk expert processing to obtain the results. The creation of an expert group can help in this task. It should be a matrix model of management. According to the results, it is necessary to develop a concept of future improvement in the life of the city [5].

1.3. Purpose of research

The purpose of the study is to show that the main driving force in the formation of a comfortable urban environment is the active participation of citizens in the life of the city and its management using intelligent information systems.

The target audience is the residents of the city who are interested in the situation with the territories of the city. Citizens who want to live in a comfortable and safe space. They want to use "smart city" technologies in practice and sustainable development of their territories. Also, representatives of business and city administration, which can be just interested parties, as well as potential customers and investors of projects for the formation and improvement of a comfortable urban environment.

2. Literature review

Mora L., Deakin M., and Reid A. starts filling such a gap by reporting on the findings of a multiple case study analysis, which is conducted into European best practices. In meeting this aim, four European cities considered to be leaders in the field of smart city. The results of this analysis offer a series of critical insights into what strategic principles drive smart city development in Europe and generate scientific knowledge, which helps to overcome the dichotomous nature of smart city research [6].

This study uncovers five main development paths: the Experimental Path; the Ubiquitous Path; the Corporate Path; the European Path; and the Holistic Path. Overall, this analysis offers a comprehensive and systematic view of how a smart city can be understood theoretically and as a scientific object of knowledge able to inform policy-making processes [7].

The article Jonathan Desdemoustier studies how Belgian municipalities understand the concept of Smart Cities in 2016. Based on the groundwork of literature on Smart Cities and the results of a survey of 113 Belgian municipalities, a typology of four understandings of the Smart City (technological, societal, comprehensive and non-existent) is elaborated [8].

A systematic review of the literature on the topic "Can cities become smart without being sustainable? " is presented in the work [9].

The present work aims to determine the existence of commonalities between two modern conceptions of cities, i.e., smart and sustainable. The main contribution of this paper is in considering the smart city as mainly setting the guidelines of a transforming city, while the sustainable city is mostly thought as an approach and a philosophy to modern cities [10].

A systematic review of the literature on the topic "The governance of smart cities" is presented in the work [11].

Albert Meijer and Manuel Pedro Rodríguez Bolívar e stand for a comprehensive perspective: smart city governance is about crafting new forms of human collaboration through the use of ICTs to obtain better outcomes and more open governance processes. Research into smart city governance could benefit from previous studies into success failure factors for e-government and build upon sophisticated theories of socio-technical change [12].

This paper presents an empirical study carried out to assess and analyze the development pattern of 35 smart cities in China using the principal component analysis (PCA) and back propagation (BP) neural network techniques [13].

This study interrogates the smart city as a global discourse network by examining a collection of key texts associated with cities worldwide [14].

The author's article by De Falco Stefano describes the development of smart city strategies in major European cities. Despite its early stage of development, this procedure provides new knowledge, innovative research perspectives, and a conceptual framework for supporting future comparative research and theory-building [15].

Cities worldwide are attempting to transform themselves into smart cities. In the article of Chiehyeon Lim, Kwang-Jae Kim and Paul P. Maglio findings from an analysis of various use cases of big data in cities worldwide and the authors' four projects with government organizations toward developing smart cities. This paper will contribute to urban planning and policy development in the modern data-rich economy [16].

The Internet of Things (IoT) and related big data applications can play a key role in catalyzing and improving the process of environmentally sustainable development. Therefore, the aim of this paper is to review and synthesize the relevant literature with the objective of identifying and discussing the state-of-the-art sensor-based big data applications enabled by the IoT for environmental sustainability and related data processing platforms and computing models in the context of smart sustainable cities of the future. The goal of this study suits a mix of two research approaches: topical literature review and thematic analysis [17].

Silva Bhagya Nathal provides an overview of the architecture components of trends and open problems in smart cities. Consequently, smart city research has focused on identifying core components and building comprehensive frameworks to understand and manage such initiatives. The author shows how involved stakeholders' stake contribute to (or impede) smartness in the planning and outcomes of the process. In smart city literature participation (involvement) of different stakeholders (such as citizens, agencies, and industry) is pointed out as important [18].

3. Research methodology

3.1. Smart cities are the future

Different cities have different priorities and objectives, but all "smart cities" have three things in common [19]:

1. Availability of information and communication technology infrastructure.
2. A well-built and integrated management system should be created in the city.
3. In a "smart" city there should be "smart" users.

The UN estimates that 67% of the world's population will live in cities by 2050. Some megacities of the world are already overpopulated [20].

In order to provide the population with high-quality urban services, administrations are increasingly introducing various information systems.

To summarize the information and a brief description of the smart city, we can distinguish the so-called properties of the smart city, namely: 1) Quality of life, 2) Mobility, 3) Urbanization, 4) Smart technologies, 5) Socialization, 6) Virtualization, 7) Personification, 8) infrastructure Modernization.

Sociologists and economists agree that happy people live in smart cities. The level of confidence of residents in their government, the ability to influence the decision - making process, a sense of security and safety, a low level of offenses and a high degree of responsibility of all participants in the urban environment – these are just a few that distinguishes the daily "smart city" [21].

Modern opportunities allow users to build their own "smart" city in almost any locality – through mobile applications in smartphones, launching their own startups aimed at improving the quality of life and the introduction of digital services within their community.

Conducting various studies and making ratings of the authorities and business of the largest cities should learn from each other, change competencies, share experiences. And then how to apply this knowledge is up to the residents of megacities.

Active participation of citizens in the life of the city and its management with the use of intelligent information systems is the main driving force for the formation of a comfortable urban environment. Now, when designing intelligent systems (IS), we (citizens) are often the end users, which imposes additional requirements to the functional components and user interface [22].

3.2. Modern Internet of things for smart cities

The Internet of Things is a new paradigm of interaction. The main purpose of the Internet of Things is that the objects of everyday life will be equipped with microcontrollers, transceivers for digital communication and appropriate Protocol stacks that will allow them to communicate with each other and customers to solve daily problems. Also, the Internet of Things can interact with a wide range of devices, such as surveillance cameras, control sensors, home appliances, etc. The Internet of Things is used in many areas of life, such as industrial automation, helping the elderly, vehicles, intelligent energy management, traffic management on the road and much more.

The new paradigm is most advantageous when used in the urban environment, as this strategy responds to the challenges of time in the management and automation of public Affairs.

At the moment, the most acute problem is the heterogeneous devices and technologies that are used at the moment. This issue can be solved by the Internet of Things, which can become a single control unit within the urban scale [23].

The situation is aggravated by the economic component. A possible way out of this situation is the design of such services, which combine social utility with economic advantage for the city administration in terms of reducing operating costs [24].

Examples of such services include the following services:

1. Waste management.
2. Air quality control.
3. Noise monitoring.
4. Urban energy consumption.

5. Smart lighting.

The practice of implementing the Internet of Things in the life of the city will create unique services that will allow the city authorities to reduce costs, optimize and automate typical processes of the city.

Smart City services are based on a centralized architecture consisting of heterogeneous devices located in different parts of the city. These devices transmit data to the control center, where data is processed, stored, and managed. A key characteristic of the urban Internet of Things is the ability to work with different technologies, combining them into one vast network.

There are three main parts that make up the urban Internet of Things – a web approach, channel-level technology and devices [26].

The architecture described earlier has found its application in the introduction of the Internet of Things in the city of Padua. This was made possible by the cooperation of municipal authorities and private companies. The main services provided by the smart city of Padua are environmental data collection and street lighting management [27].

These sensors collect environmental data such as carbon dioxide, noise, air humidity, vibration, temperature and the like.

For example, using a mobile application, operators of municipal services can quickly identify a faulty lamp that requires an immediate call to the repair team. These services include a sound map of the city, a map of the city's pollution, a map of the city's energy consumption, readings of temperature sensors throughout the city over the past seven days, etc. This will allow you to see the whole situation, analyzing certain changes in the city after rains, drought, etc. [28].

The technologies discussed are currently close to standardization. The main problem is that the business is interested in the development of the urban Internet of Things and bring investment in these projects. Then you can make people's lives better.

3.3. Project "Smart digital city»

The project "Smart digital city" is an example of a product that provides innovative development of the region and districts of the city, through the introduction of which in all spheres of life of the population of information and telecommunication technologies, it is possible to accelerate the development of the region and improve the quality of life of the population.

The aim of the project is to optimize the conditions for improving the quality of life of people and reduce the use of resources in the region.

The practical significance of the project "Smart digital city" is the development of key areas to create new conditions for the development of the information society.

The conditions of the project lead to the creation of a platform for investing in the digital economy and optimizing the level of electronic services up to 75%, and the level of electronic infrastructure – up to 80%.

The implementation of the project is based on the socio-cultural and territorial characteristics of the region and the zone of economic interests of its residents and production companies [29].

The main objectives of the project "Digitalization of urban infrastructure" are:

- optimization of conditions for improving the quality of life in the region;
- improving the competitiveness of the region;
- improvement of the system of state and municipal management of the region.

Analyzing the main directions presented in the project, it should be clarified that such differentiation of directions is important for the assessment and monitoring of detailed changes in the structure of the city, for the construction of a single information platform, where the spheres of activity will be presented in one interface, which will facilitate the life of the population and simplify the activities of all organizations [30].

The first area is smart healthcare.

The second area is education with digital communications.

The third direction of the project "Digitalization of urban infrastructure" is smart transport.

IOP Conf. Series: Materials Science and Engineering **972** (2020) 012021 doi:10.1088/1757-899X/972/1/012021

Fourth, public security.

The fifth area is digital energy.

Next is the direction, which includes the digital government.

We will present the popular activities in this area:

- popularization of electronic format in the sphere of state and municipal services;
- creation of public-private partnership in the development of digital government;
- transition to the use of a single national population register;
- introduction and expansion of the field for electronic document management for legally significant exchange of electronic documents at the Federal, regional and municipal levels (90 % of the total document management).

Thus, the implementation of the project "Digitalization of urban infrastructure" is a profitable project for the region, aimed at creating favorable conditions for the lives of residents and investment activities. Thanks to this project, it is possible to reduce the level of offenses, improve the transport and environmental situation in the region. It is undoubtedly a potential factor in the socio-economic development of cities.

4. Results and suggestions

4.1. The world's best city experience in the application of innovation

Today, more and more people on the planet pay attention to the environment, pollution of the atmosphere. Cities want to live in a new technological way. Therefore, they strive to get ahead of time, making innovations in their development. So that the residents could adjust to another level of development and feel safe and comfortable [31].

Table 1 shows the cities that innovate in the struggle for safe and comfortable living.

Table 1. Innovation of cities in the struggle for safe and comfortable living.

City	Best practice	Practical example	The Result (Effect)
Oslo (Sweden)	From car Parking to make bike lanes	Increase the number of Bicycle paths, put benches, break mini-parks	More than 700 Parking lots in the city center were eliminated. Installed charging stations for electric vehicles. Made more Parking for disabled drivers. People can't drive into the centre by private car
Buenos Aires (Argentina)	City hall sets flowers on the roads for the safety of pedestrians and cyclists	Boulevard of 20 lanes, today is occupied only by buses. About 100 blocks of the city are exempt from cars. Bright markings are applied to the busy intersections of the city. Vases with flowers are placed along the roads	The time on the way to work of the city residents has decreased dramatically. Free space has become a pedestrian area. The roads have become as safe as possible for pedestrians and cyclists. The authorities are introducing the method of tactical urbanism. It involves inexpensive repairs in a built environment. Repair, designed to improve areas and places of congestion
London (UK)	Reduces the roadway in favor of pedestrians	The government introduced a fee for entering the city centre during peak hours.	A special plan has been created: half of the streets will be freed from cars in favor of pedestrians. The speed of transport will be no more

City	Best practice	Practical example	The Result (Effect)
		Opened the first Bicycle superhighways on busy routes. Transition to zero-emission vehicles	than 15 miles per hour. It is planned to build protected bike paths on most of the main streets. Reducing the number of cars in the city center. Care for the environment and pedestrians
Seoul (South Korea)	Plans to completely get rid of "harmful" cars until 2020	Transformed the highway into a pedestrian road. It is planned to launch 3 thousand electric buses	People choose public transport instead of their own cars. Reduction of harmful emissions in the city center, the transition of citizens to public transport
Madrid (Spain)	Release the Central district from transport	The introduction of restrictions on travel to the Central part of the city transport. Project "low emission zones"	A rapid drop in traffic of 32%. Air purification in the center of the capital
Beijing (China)	"Harmful" cars are limited to travel time and license plates (can drive on certain days of the week)	The administration offers small financial incentives to people who do not travel on their designated day. In new areas only electric cars are allowed	People are transplanted to public transport. Creation of new "clean" areas. Reduction of harmful emissions in the city center. Transition to environmentally friendly cars. Incentive to switch to public transport
Paris (France)	The track near the Seine turned into a footpath and a car park	Restriction on the entry of old cars polluting the environment on weekdays. It is planned to ban diesel and gas cars	Work on reducing pollution by reducing road traffic and eliminating polluting transport. Transition to environmentally friendly cars. Incentive to switch to public transport
India	Full transition to electric vehicles by 2024	Turning a busy street into a pedestrian square. By 2030, all cars must be electric	Rejection of polluting transport. Reduction of harmful emissions in the city center
Copenhagen (Denmark)	Creating a superhighway for cyclists	More than 200 miles of Bicycle paths have already been made. It is planned to make the city completely carbon neutral by 2025	The creation of additional lanes, and superhighway for cyclists to a lower level of car ownership. Orientation of the population to a healthy lifestyle
Berlin (Germany)	The construction of separate villagestyle that are not part of the roads for vehicles	Converted 12 highways in villagestyle	Creating conditions for safer travel. Increasing the number of bicycles on the roads
Georgia	Complete rejection of the production and use of polyethylene	Ban on the production, use and import of plastic bags	The proposal to use biodegradable bags. Environmental protection. Reducing the amount of hazardous waste in the country

4.2. Formation of a comfortable urban environment in the Sverdlovsk region and Yekaterinburg

The project formation of a comfortable urban environment is one of the tasks, the solution of which is an indispensable condition for reaching a new quality of life.

As part of the priority project "Formation of a comfortable urban environment" two areas, the first – the formation of a modern urban environment, the second – the improvement of public places – parks and gardens.

The government of the Russian Federation has been instructed to provide at least 20 billion rubles in the Federal budget for these purposes [32].

Implementation of the project "Formation of a comfortable urban environment" began in 2017. As a result of the project, Sverdlovsk region took the 4th place in the rating of the regions of the Russian Federation and became one of the leading regions in the presence of improvement practices ("5-TOP").

Moreover, one of the projects implemented in the region on complex improvement of yard areas was recommended by the Ministry of construction of Russia for inclusion in the Federal register of best practices (projects) on improvement.

In 2018, in the Sverdlovsk region, the project was carried out improvement of 140 territories – 102 yards and 38 public areas. Taking into account all sources of financing, more than 1.5 billion rubles were spent on their reconstruction.

The results of the project in the region – Sverdlovsk region today is very impressive – a comprehensive improvement of more than 200 yards and more than 70 territories held more than 600 events attended by several thousand residents of the Sverdlovsk region.

The project of complex improvement of the embankment in the city district of Verkhnyaya Tura is recognized at the Federal level as the best from the Sverdlovsk region. The next competition is declared the six completed projects – leisure centre "Color" in Bogdanovich, the embankment of the river Turia in the city, the walk of Fame in Severouralsk, and courtyard areas of Pervouralsk, Krasnouralsk, and Yekaterinburg.

It is planned to allocate 1,733 billion rubles from the Federal budget in 2019 to support the improvement projects of the Sverdlovsk region [33].

The Ministry of energy and housing and communal services of the Sverdlovsk region annually conducts organizational activities aimed at the implementation of the project. At the beginning of the calendar year, meetings of commissions are held to summarize the results of the first and second stages of the selection of applications of municipalities for the right to receive subsidies. Also, at the beginning of the year, an agreement on the provision of subsidies from the Federal budget to the budget of the Sverdlovsk region is signed.

Public areas to be improved as a matter of priority were selected taking into account the opinion of citizens in the framework of the rating vote held on March 18, 2018.

Large-scale work was carried out, as a result of which about 530 thousand proposals were received from residents of the Sverdlovsk region. The total number of citizens who took part in the rating vote was more than 890 thousand people. The winners were 58 public territories. The government approved the distribution of subsidies to support municipal programs for the formation of a modern urban environment.

In order to monitor the implementation of the priority project "Formation of a comfortable urban environment", the Ministry of construction of Russia has created an appropriate module in the state information system of housing and communal services. Work in this module is mandatory for all municipalities. The content of the system is systematic.

The only control point required for filling – is a complete inventory of the yard and public areas of municipalities.

In accordance with the requirements of Federal legislation, an interdepartmental Commission of the Sverdlovsk region has been established. The purpose of this Commission is to monitor the implementation of the priority project. Monthly commissions are considered intermediate results of project implementation in the region in terms of municipal formations according to the results of work of the Commission to municipalities, are given appropriate legal instructions to specify the terms [34].

Local authorities were instructed to create on their official websites a section "Formation of a comfortable urban environment", where residents would have the opportunity to receive detailed information about the program, answers to their questions and leave their proposals on the objects of improvement. A positive trend is the factor of increasing the involvement of the population in the implementation of the project. In order to increase the information content of the population of the Sverdlovsk region published more than 500 materials in the media. There are such events as the Grand opening of landscaped areas, the festival "Go out for a walk", the action "international day of neighbors".

Thus, the main principle of the project "Formation of a comfortable urban environment" is public participation, dialogue, organized with residents and local communities.

The city of Yekaterinburg has great potential in the implementation of the project of formation of a comfortable urban environment. The city has a huge number of areas that need improvement: from the reconstruction of courtyards to the creation of significant public spaces. The infrastructure of the city needs improvements, because they will affect the social and economic climate of the city. For example, in a city with a population of more than 1.46 million people, the level of motorization accounts for more than 40% of the total population and every year it increases. Congestion of streets becomes higher, free space for pedestrians in the yard and house adjoining zones becomes less. This affects the quality of life of residents. For example, unloading of the streets from the transit of cars, the creation of neighborhoods are maximally adapted to the needs of residents, pedestrians and cyclists, as did the municipality of Barcelona, may favorably affect the statistics of the health of the residents, sustainability of the environment [35].

4.3. Ways of manifestation of an active life position of the citizen

We present different types of participation in the life of the city from the standpoint of social responsibility of a concerned member of society.

Table 2 presents the types of manifestations of the active life position of the citizen, taking into account the regulatory legal aspects [36].

Table 2. Different types of participation in the project "I am a resident».

Actions	Algorithm of actions
1. I can include my yard in the state program	1.1. To hold a General meeting of homeowners on inclusion in the state program: 1.1.1 Select list of preferred contractors 1.1.2 to Determine the form of participation of the owners 1.1.3 to Decide whether we include the improved objects in the General property, whether we are ready to contain it further 1.1.4 Select our official representative 1.2. Apply to the municipality 1.3. Take a direct part in the works and their control
2. I can become a member of the public Commission	2.1. Submit an application to the municipality 2.2. Get an answer
3. View the work plan on the website	3.1. See the site plan and list of works 3.2. Check their real performance and quality 3.3. Report to the municipality, region or Ministry of construction of Russia on the result of the audit 3.4. Get reply to your email
4. I can participate in the selection of the object for	4.1. View the work plan on the website 4.2. Vote for the work you like best

Actions	Algorithm of actions
improvement	
5. I can apply for the inclusion of the object in the program	5.1. View the work plan on the website 5.2. To offer the property in the improvement program 5.3. Get an answer
6. Become an activist and support the development of your city	6.1. Take care of the municipality 6.2. Download information materials from the site and study them

We will determine the sequence of methods of development and stages of implementation of actual projects of active citizens taking into account the existing problems. For ease of visualization, we present the material in Table 3.

Table 3. Algorithm of development and stages of project implementation for active citizens.

Used method	The stages of the project	Urgent problem
1. Identifying problems in the formation of a comfortable urban environment; 2. Setting project goals and objectives; 3. Analysis of the present state of the territory; 4. Analysis of public opinion on improvement; 5. Definition of the main ways of improvement; 6. Development of the layout in the program of landscape design; 7. Analysis of resource provision of the project; 8. Search for potential investors; 9. Evaluation of the results of project activities	1. Identifying problems in the formation of a comfortable urban environment; 2. Setting project goals and objectives; 3. Analysis of the present state of the territory; 4. Analysis of public opinion on improvement; 5. Definition of the main ways of improvement; 6. Development of the layout in the program of landscape design; 7. Analysis of resource provision of the project; 8. Search for potential investors; 9. Evaluation of the results of project activities	1. A large amount of garbage in the Park area; 2. The lack of illumination of the city; 3. The lack of a sufficient number of ballot boxes and benches; 4. Painted graffiti objects; 5. No control of walking Pets; 6. The lack of a dedicated and refined spaces; 7. The total neglect of the site (overgrown path, destroyed the main stage, dirty area around the pond, the poor state of the reservoir) 8. The lack of gentrified places

It is also recommended to conduct surveys of citizens on the Internet regarding any territory in the city. Based on the responses of respondents may be formed from the list of facilities that require urgent improvement.

The city can be zoned on the plots, blocks, in which it is possible to introduce gradually developed concepts. The reconstruction of the courtyards plan in the period overlapping with the works on reconstruction and capital repair of engineering networks, structures and buildings. With the help of the concept to link the architectural structures of the present and the past, adding tactical urbanism, architectural and landscape ideas and modern technologies – we get a creative, comfortable urban environment.

5. Discussion of the results

Any project (the relevance of which is proved), initiated by active citizens, can have a huge number of positive effects. The expected results of the project can be seen in Table 4.

Table 4. Expected results from the project initiated by active citizens.

Expected result from project implementation	Description
1. Social	• creating an attractive urban environment • promoting a more active civil society • creation of popular public spaces in the city • improving comfort and quality of life • providing opportunities for more profitable purchase of residential real estate to the population
2. Economic	• increase of investment attractiveness of the city • increase in tax revenues from previously unused territories • identification of problems hindering the development of the city • creation of new jobs • solving the problem of lack of territories • business incentives

The results of the implementation will contribute to the establishment of contact between representatives of the administration, business and residents of the city for effective cooperation in the formation of a comfortable urban environment. This, in turn, will have a huge positive effect on the spatial development of the city, improving the quality of life of citizens and investment attractiveness.

6. Results and suggestions

Today there are several definitions of a smart city. However, all the formulations coincide in the fact that a smart city is managed by data that allows urban services to improve the quality of life of the population.

The formation of a comfortable urban environment in the city is happening today and now. Every resident of the city should take part in it.

Comfortable urban environment can be associated with the success of the city as a whole, as well as many modern cities, defined by three main elements:

- know where your city is going, understand its purpose and its development potential;
- to encourage the development of leadership;
- encourage risky decision-making within a system based on the principles of accountability and transparency.

All of the above gives people the opportunity to creatively participate in the formation of a comfortable urban environment of the city.

References

[1] Medvedkov Y 1978 *Human and urban environment* https://search.rsl.ru/ru/record/01007634253 (in Russian)
[2] Business portal "TAdviser" 2019 http://www.tadviser.ru (in Russian)
[3] Digital magazine "ITUNEWS Magazine" 2019 http://www.itu.int/ru/Pages/default.aspx (in Russian)
[4] Glossary (terms and definitions) to the draft Concept for the development of information and communication infrastructure and technologies in the Russian Federation 2011 http://minsvyaz.ru/ru/documents/3464/ (in Russian)
[5] Landry H 2011 *The Creative city* (Moscow: Publishing house "Classics–XXI") (in Russian)

[6] Mora l 2019 Strategic principles for smart city development: A multiple case study analysis of European best practices *Technol Forecast Soc* **142** P 70 DOI 10.1016/j.techfore.2018.07.035
[7] Mora L 2018 Smart-City Development Paths: Insights from the First Two Decades of Research *Green Energy Technol* P 403 DOI 10.1007/978-3-319-75774-2_28
[8] Desdemoustier J 2019 Municipalities' understanding of the Smart City concept: An exploratory analysis in Belgium *Technol Forecast Soc* **142** P 129 DOI 10.1016/j.techfore.2018.10.029
[9] YIgitcanlar T 2019 Can cities become smart without being sustainable? A systematic review of the literature Sustain Cities Soc **45** P 348 DOI 10.1016/j.scs.2018.11.033
[10] D'Auria A 2018 Modern Conceptions of Cities as Smart and Sustainable and Their Commonalities Sustainability-Basel **10** P 12497 DOI 10.3390/su10082642
[11] Ruhlandt Robert Wilhelm Siegfried 2018 The governance of smart cities: A systematic literature review *Cities* **81** P 1 DOI 10.1016/j.cities.2018.02.014
[12] Meijer A 2015 Governing the smart city: a review of the literature on smart urban governance *International Review Of Administrative Sciences* **82**(**2**) DOI: 10.1177/0020852314564308
[13] Li Xia 2019 Towards sustainable smart cities: An empirical comparative assessment and development pattern optimization in China *J. Clean Prod.* **215** P 730 DOI 10.1016/j.jclepro.2019.01.046
[14] Joss S 2019 The Smart City as Global Discourse: Storylines and Critical Junctures across 27 Cities *J Urban Technol* **26** P 3 DOI 10.1080/10630732.2018.1558387
[15] YIgitcanlar Tan 2019 Can cities become smart without being sustainable? A systematic review of the literature Sustain Cities Soc **45** P 348 DOI 10.1016/j.scs.2018.11.033
[16] Lim C 2018 Smart cities with big data: Reference models challenges and considerations *Cities* **82** P 86 DOI 10.1016/j.cities.2018.04.011
[17] Bibri Simon Elias 2018 The IoT for smart sustainable cities of the future: An analytical framework for sensor-based big data applications for environmental sustainability *Sustain Cities Soc* **38** P 230 DOI 10.1016/j.scs.2017.12.034
[18] Silva Bhagya Nathali 2018 Towards sustainable smart cities: A review of trends architectures components and open challenges in smart cities *Sustain Cities Soc* **38** P 697 DOI 10.1016/j.scs.2018.01.053
[19] The it portal of the company "Infosistemy DZHET" 2019 http://www.jetinfo.ru (in Russian)
[20] Independent publication about technologies and business "Rusbase" 2019] http://rb.ru (in Russian)
[21] Internet resource "Peter-Service" 2019 http://www.billing.ru (in Russian)
[22] Smart cities are the future today 2015 http://www.jetinfo.ru/jetinfo_arhiv/smart-city-nashi (in Russian)
[23] Xu L. D. 2014 Internet of Things in Industries: A Survey *IEEE Transactions on Industrial Informatics* https://ieeexplore.ieee.org/document/6714496
[24] Kortuem G. 2010 Smart objects as building blocks for the Internet of thing *IEEE internet computing* https://ieeexplore.ieee.org/document/5342399
[26] Harisson C. 2010 *Foundations for Smarter Cities IBM Journal of Research and Development* https://ieeexplore. ieee. org/document/5342399
[27] Kumar N 2018 Smart cities in India: Features, policies, current status, and challenges *Technologies for Smart-City Energy Security and Power* (ICSESP) https://ieeexplore. ieee.org/ document/8376669 (access date 25.12.2018).].
[28] Cenedese A 2014 Smart City: An urban Internet of Things experimentation *Proceeding of IEEE International Symposium on a World of Wireless, Mobile and Multimedia Networks* https://ieeexplore.ieee.org/document/6918931
[29] Kozhevina O 2018 Digital factors in the development of smart cities *Management in Russia and Abroad* **3** P 36 (in Russian)
[30] Pronin A 2018 Report on the conference "Digital culture open cities" *News of Higher Educational Institutions Ural Region* **2** P 161 (in Russian)

IOP Conf. Series: Materials Science and Engineering **972** (2020) 012021 doi:10.1088/1757-899X/972/1/012021

[31] 11 cities that have changed dramatically to save the planet https://www.adme.ru/svoboda-puteshestviya/11-gorodov-kotorye-kruto-izmenilis-radi-spaseniya-planety (in Russian)
[32] Official website of the Government of Sverdlovsk region 2019 http://midural.ru/news/list/document137248/ (in Russian)
[33] Ural news 2019 http://ural-news.net/society/2018/12/17/187520.html (in Russian)
[34] Website of the Governor of Sverdlovsk region 2019 http://gubernator96.ru/news/show/id/5431 (in Russian)
[35] Gershman A 2018 Revolution of the streets of Barcelona (in Russian) https://gre4ark.livejournal.com/610126.html (date accessed: 10.03.19)].
[36] Website of the Ministry of construction of Russia 2019 http://www.minstroyrf.ru (in Russian)

IOP Conf. Series: Materials Science and Engineering **972** (2020) 012022 doi:10.1088/1757-899X/972/1/012022

Roadside infrastructure unit as an autonomous small-sized smart grid. "Holonic" approach

D Panyukova[1], A Sladkowski[2] and O Shiryayeva[1]

[1]Department of Automation and Control, Kazakh National Research Technical University named after K.I. Satbayev, Satbayev str. 22a, Almaty, 050013, Kazakhstan
[2]Department of Logistics and Transport Technologies, Silesian University of Technology, Krasińskiego str. 8, Katowice, 40-019, Poland

E-mail: haleth@mail.ru

Abstract. Along now being constructed Western Europe to Western China (WEWC) Corridor big territories are totally uninhabited. It leads to absents of any central power supply. While transport infrastructure and control on the highways should be provided. The article is devoted to autonomous power grid of such road unit as a transport control post "Rubej". Firstly, the scientific analysis of renewable energy sources of the republic of Kazakhstan and justification of exact power plants for the post is provided. After that the research is focused on a principle of power grid control. The "Multi-agent system" (MAS) technique was proven to be promising approach to the smart grid's design. The one of the recently offered MAS techniques is the Holonic architecture. While it was adjusted only for power grids with central grid connection. In the article autonomous smart grid based on Holonic approach is offered and described. The optimization problem for the control of autonomous small-sized smart grid is reckoned. The achieved results can ease the control system's design and scaling process, as it should be provided even for small-sized grids. In the end financial actuality is proved.

1. Introduction

Planned length of WEWC Corridor is 8 445 km. While 2 787 km are passing through Kazakhstan territories with very rarefied population: 17 476 141 people at 2 724 000 km2 (the ninth place worldwide for territory) [1]. Consequently, most of the highway will go through areas without any central power supply. It leads to entanglements in power supply for any remote element of road infrastructure. In particular, modern supporting equipment for transport control post "Rubej" became more intellectual and demand stable electric power supply.

According to Igor Lepeha – chairman of Committee of Administrative Police of Ministry of Internal Affairs of the Republic Kazakhstan – there are still no enough "Rubej" control posts on the territory of republic. For existing roads 44 posts are necessary, while only 35 is active. And even them not always have proper living conditions. Whereas more traffic routes are constructed for WEWC Corridor, and most of them bypassing functioning posts. [2]

Plants based on renewable energy sources (RES) can provide enough amount of power. It includes combination of wind turbines and solar plants depending on geographical location of an exact post. According to the researches at a southern areas of Kazakhstan there are up to 3000 hours effective for solar plants, which leads to nearly 1 500 kW/m2 annually. At the same time half of the territory is suitable for wind resources' utilization as average wind speed is more than 4 m/s. [3]

With RES implementation uncertainty of power grid parameters arises. Power storage by electric accumulators and convenient reserve fuel generator can partly solve the instability of the power sources [4]. With such a complexity of equipment manual control can provide some inappropriate dispatching of the grid elements. While control system design for every exact autonomous power grid's set claims too much human resources. A universal optimization technique for a grid can become a proper solution for the problem.

The second part of the article is devoted to description of such roadside infrastructure unit as a transport control post. On that basis its power consumption is calculated. In the third part information about power supply elements is provided, their choice and technical characteristics are justified. Afterwards the control approach for the autonomous power grid on renewable energy sources and the optimization problem as an implementation on the control post are offered. In following part reckoned power grid structure and elements' interaction is discussed. Next part is devoted to compare possible investments with other solutions. Finally, implementation possibilities and future improvement of the power grid is considered.

2. Power needs of a transport control post

Stationary transport control post "Rubej" is specially designed building with all necessary operational and technical equipment to accomplish functions of: Traffic control, ensuring road security, an offence registration, ensuring officers' welfare.

Traffic control function is provided by automated system "Post-Rubej". It fully automatizes registration and search of passing traffic in data bases about stolen vehicles, tax dodgers. Above that, stopping technique is automatized too.

Main functionality of the system is:

- Fixation and recognition of a state number;
- Record of traffic data into local data base (DB);
- Simple but secure access to the DB;
- Search of recognized numbers in DB about tax dodgers;
- Search of recognized numbers in any other local DBs;
- Automated control system for traffic regulating equipment;
- Integration to existing information systems of the Ministry of Internal Affairs. [5]

Hardware of the system consists of digital camera with high sensibility in infrared diapason, speed radars, infrared lighting, computation modules. [6]

Road security is ensured partly by the "Post-Rubej" system, partly by officers' work. Consequently, no additional hardware is needed. An offence registration requires a computer, a printer and a connection to the Internet. For a light system power saving LED lightings and movement indicators are expected. While to ensure officers' welfare the post can be equipped with refrigerator and microwave cooker.

2.1. Power calculation

The power consumption of previously mentioned equipment can be divided into two main groups as was offered for an autonomous power supply at Tunisian petroleum platform [7]: mandatory and welfare. Mandatory consumption is a sum of demands from exact operating hardware. For the transport control post they are automated system "Post-Rubej" and an offence registration's hardware. While welfare consumption consists of lighting and welfare equipment. Such division generates one more possibility for control in an autonomous power grid. The possibility is to make recommendation on or even switch off the welfare demand. Difference between the platform's and the post's supplies are in the level of demanded power, in availability of fuel and in possibilities to connect with other grids – central as a consumer or neighboring micro-grid as a supplier. For the post level of power supply is lower, but the electricity's quality should be much better. The fuel is available on demand, while for the platform it is limited. The future possibility to become a part of bigger grid for the post's power grid is much higher, as the road infrastructure usually is growing.

The exact electric power demands for both parts are shown in table 1.

Table 1. Daily electric power needs of stationary transport control post "Rubej"

Equipment	Amount	Required power, W	Working hours	Watts per hour
		Mandatory part		
Digital camera	2	27.0	24	1 296.0
Speed radar	2	3.8	24	182.4
Infrared lighting	2	1.2	12	28.8
Computation module	2	30.0	24	1 440.0
Traffic light	2	20.0	24	960.0
Boom gate	2	150.0	4	1 200.0
Officer's computer	1	120.0	24	2 880.0
Printer	1	100.0	1	100.0
Wireless connection module	1	30.0	24	720.0
Total for mandatory part	-	482.0	-	8 807.2
		Welfare part		
LED lighting	4	20.0	10	800.0
Refrigerator	1	150.0	4	600.0
Microwave cooker	1	700.0	4	2 800.0
Total for supplying part	-	870.0	-	4 200.0
Total	-	1 352.0	-	13 007.2

While the calculation of power peak level is done with:

$$P_{peak} = \sum_{i=1}^{n} P_i \tag{1}$$

where P_{peak} is a peak power, W; P_i is a required power for i^{th} equipment unit, W; n is a number of equipment units.

Also for supplying plant's choice daily required electric energy is necessary:

$$E_{total} = \sum_{i=1}^{n} E_i = \sum_{i=1}^{n} P_i \cdot h_i \tag{2}$$

where E_{total} is a total electric energy needed, W·h; E_i, P_i and h_i are an electric energy needed, a required power and working hours of ith equipment unit, W·h, W and h respectively; n is a number of equipment units.

Total peak power demand is up to 1.4 kW and daily power consumption is calculated just higher than 13 kW·h. Mandatory consumption, that is less than 0.5 kW – peak demand and less than 9 kW·h – daily, cannot be ignored. As for welfare needs, that are less than 1 kW in peak and less than 4/5 kW·h – daily, it's supplement can be controlled or even switched off, if both main generation is absent and accumulators' charge is not enough. A user can be consulted by a smart grid when and on which level the power grid can assure welfare power consumption for him to choose which welfare needs can be ignored.

IOP Conf. Series: Materials Science and Engineering **972** (2020) 012022 doi:10.1088/1757-899X/972/1/012022

3. Power needs of a transport control post

Research of the RES potential in Republic of Kazakhstan shows that mainly territories have average wind speed above 4 m/s [3]. So, use of small power wind plants is appropriate there as have been justified by previous research [8]. At the same time, on most geographical areas daily solar radiation is in average 0,3 kW·h for m2 [3]. That proves the possibility to utilize solar plants.

Whereas power demand is less than 1.5 kW in peak, power supply system as a complex of three parts is offered:

- Main supplying plants – wind turbine and several solar plants;
- Buffering equipment – electric accumulator;
- Reserve supplier – fuel generator.

The main criteria of choosing power plants are matching RES potential at exact geographical position and providing all the peak power demand at the nominal operating time. The potential should be calculated by a real data or by RES maps [3]. But, preliminarily, more solar plants and smaller wind plant is recommended in southern parts, while vice versa – in northern and western regions.

Consequently, both wind turbine and solar plants' complex should match minimum nominal power level of 2 kW. While average wind speed is nearly 4 m/s, which limits our wind turbine's choice only for low-powered generators.

For a wind utilization autonomous wind power plant (WPP) "LOW WIND 2.5/3.5 kW" was chosen. Wind speed limits for nominal work is between 2.5 and 25 m/s. It's nominal and maximum power is 2.5 and 3.5 consequently and output is 24 V of DC [9]. The plant can provide more stable work by the small winds and cover the power demand from both the post and the accumulators at nominal speed. Although at average wind speed of 4 m/s its output will be just higher than 0.5 kW [9].

According to the mandatory consumption that is equal to almost 9 kW·h daily 8 solar panels with 260 W power are chosen [10]. It leads to total coverage over the control post's electric power demand in sunny hours, while charging accumulators, if consumption is not in peak.

RES plants are supplying the system with electric power of such properties it should be processed by an inverter. The inverter changing unstandardized currency into proper AC or DC. For the plants' implementation the Tripp Lite APSX3024SW inverter is found. It's nominal and peak input power is 3 and 4.5 kW respectively as the plants' maximum output is just bigger than 3 kW. The inverter provides pure sinusoid with 230 V [11].

At the time when neither sun nor wind is not in charge electric accumulators should be used. 8 pieces of „CHALLENGER A12-200" accumulator are chosen. By 12 V and 200A·h they are providing 19.2 kW·h [12]. Consequently, the accumulators can fully supply the control post for 35 hours without RES plants' input.

The last part of supplying system is a fuel generator. The main criterion is to fully supply the control post, if neither RES plants nor accumulator can provide electricity. As the control post consumption is less than 1.4 W at a peak time the diesel generators are unsuitable. Also the electronic devices demand electric power of a good quality. Within this context SAP·35 inverting petrol generator was chosen. It's nominal power in 3.5 kW with ideal 220 V and 50 Hz of AC. In addition to that, in a condition of small loads the generator reduces fuel consumption up to 40 % [13]. Other advantages of the petrol generator are the wider temperature limits and higher popularity of petrol in automobile sector of Kazakhstan.

All chosen for the supplying system devices can provide energy enough for the control system demand as can be seen in table 2.

The peak supply of WPP covers all consumption and accumulators charge, while solar plants can provide energy just for peak consumptions. If neither of the sources will be unavailable, accumulators can supply full system's demand for 1,4 days and mandatory demand for more than 2 days. If in two days the RES will be still unavailable, the reserve generator can be used. And, while the reserve generator works on petrol, the availability of the resource is high enough.

IOP Conf. Series: Materials Science and Engineering **972** (2020) 012022 doi:10.1088/1757-899X/972/1/012022

Table 2. Electric power supplying plants

Equipment	Amount	Nominal power, W	Peak power supply, W	Other parameters
Wind plant "LOW WIND 2,5/3,5 kW"	1	2500.0	3500.0	Wind speed: 2.5-25 m/s
Solar cell array "LDK-260PA"	8	260.0	2080.0	Nominal voltage: 24 V; Size: 1636×986×35 mm
Inverter "TRIP LITE APSX3024SW"	1	-	-	Nominal input power: 3 kW; Peak input power: 4.5 kW
Accumulator „CHALLENGER A12-200"	8	-	-	Nominal voltage: 12 V; Nominal capacity: 200 Ah
Reserve inverting petrol generator "SAP·35"	1	3500.0	3500.0	Nominal output: 220 V, 50 Hz of AC

4. Smart grid based on "Holonic" approach

First step to design a control system is to choose an approach of elements' interaction within a controlled object. In power systems the "Multi-Agent System" or MAS technique is offered as the advanced power grid's structure [14, 15]. It gives possibility of control and adjusting to lower level of the grid (even till a client or a house) and decreases the data processing loads on the central grid's controlling system. The disadvantage of classical MAS is in necessity of different mathematical description for the grids' elements at every level.

The offered in the article MAS approach is based on "Holarchy" theory. It is similar to "system-subsystem" structure from classic MAS in dividing a whole object into smaller parts. The difference is that smaller parts – Holons – in the Holonic architecture have the same mathematical description as the object itself. It leads to easier scalability of the designed control system. Recently the Holonic architecture was adjusted for optimization of materials' flows in manufacturing systems and their control [16]. The architecture was offered for thermal power systems as the thermal power grids have similarity with manufacturing in materials' flows [17]. As there is straight connection between materials' and energy flows the next step was to research the possibilities of Holonic architecture's implementation in electric power grids. Consequently, the perspectives of such approach for electric power was justified [18]. The concept of smart grid's optimization in multi-agent power grid by Holonic architecture was researched [19, 20]. But neither of them discussed the autonomous power grid optimization and control, while the transport control post requires autonomous electric power supply.

For Holonic architecture the elements of power grid should be categorized into consumers, suppliers and elements with mixed characteristics. The main example of mixed element is accumulator as at some point of time it can be a supplier or a consumer. Holons of the higher level can be a combination of lower level Holons based on Holons' type or their physical connection within a power grid.

For the control post's power grid as a Holarchy 5 lower level Holons, as can be seen from figure 1, are offered.

Lower level's Holons are:

- 3 supplying Holons: WPP, Solar plans and reserve generator;
- 1 consuming Holon: all power consuming equipment;
- 1 mixed Holon: combination of all accumulators.

While solar plans', consuming and mixed Holons have even lower Holons. For solar plants' Holon they are every exact solar cell arrays, for consuming Holon – mandatory and welfare consumptions, for mixed Holon – all 8 accumulators.

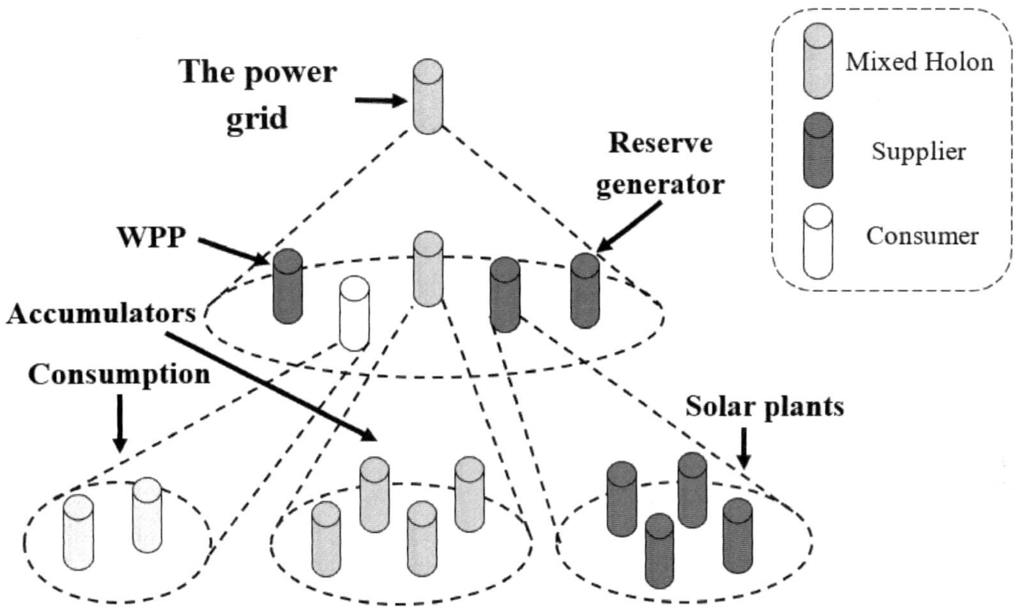

Figure 1. The control post's power grid as a Holarchy.

4.1. Mathematical description of the power grid as a Holon

The real power consumption of the power grid is uncertain value, but with known peak value and total daily amount. In such small power system recommendations about welfare consumption's value in exact time period for person as a consumer can be given. While, with possible future redesign for supplying neighboring entities, the consumption forecasting as uncertain value will be necessary [15].

The supplying plants on RES have uncertain output values too as they depend on weather conditions.

For solar panels' output such dependency will be:

$$P_{ss} = f[G(\beta,\alpha), T_C] \tag{3}$$

where P_{ss} is an output power of a solar array, W; $G(\beta,\alpha)$ is an incident irradiance on the plane of the array, W/m^2; T_C is a cell temperature, °C. [20]

The WPP's output is also a function from weather conditions:

$$P_{sw} = f[\rho, v_w] \tag{4}$$

where P_{sw} is an output power of a plant, W; ρ is an air density, kg/m^3; v_w is a wind speed, m/s. [21]

As can be seen from equations (3) and (4) both power supplies have straight connections with weather conditions and are calculated for next hours with use of weather forecasting from Internet. That gives certain error to the calculation within the control system.

The uncertainty of the system is balanced by power accumulators, which are working as a buffer to collect and store power surplus and use it when power plants are off. Electric accumulators' charging limits have to be taken into consideration. As the capacity of them are decreasing while they are over- or undercharged.

4.2. Optimization problem for the Holon

There are two possible ways to balance the system at the power lack: to use power from batteries and to make recommendations on or even switch off the welfare power consumption. The choice of the system should be based on weather forecast and on accumulators' conditions.

The power supply and consumption in the power grid is considered negative and positive value respectively to calculate the resulting power balance. Then the total power at the exact time in the grid can be calculated as a sum:

$$Pt = \sum_{i=1}^{n} P_{si} \tag{5}$$

where P_t is a total power in the grid, W; P_{si} is a power of ith Holon, W; n is a number of Holons in Holarchy.

Then the optimization problem of a Holarchy without any power storage is formulated as minimization of the sum (5) online:

$$Pt = \sum_{i=1}^{n} P_{si} \to \min \tag{6}$$

The control actions will be switching on/off or demanding exact power value from the Holons.

The optimization problem (6) is the mathematical description for Holons of middle level from figure 1: the solar plants, the consumption and the accumulators.

While for autonomous power grid there is no external output for the surplus. And the optimization problem (6) will be specified as:

$$Pt = \sum_{i=1}^{n} P_{si} \to 0 \tag{7}$$

With implementation of accumulators control action are extended by choosing a state of accumulators between supplying or consuming. That demands not an online but planned control. To calculate the exact period of planning the chosen accumulators' capacity have to be considered. Fully charged they can supply the grid for 28 hours and have enough energy to supply mandatory need for more 24 hours.

Consequently, the final optimization problem is formulated like:

$$\begin{gathered} \sum_{j=1}^{28} P_{tj} \to 0 \\ SoC_{\min} < SoC_j < SoC_{\max} \end{gathered} \tag{8}$$

where P_{tj} is a total power in the grid at j^{th} hour, W; SoC_j is an accumulator's state of charge at j^{th} hour, %.

Then the practical calculation of control system for the smart grid is to find optimal controlling values for next 28 hours on the basis of the formulated optimization problem. There are various techniques to calculate it: classical or advanced. Classical optimization techniques are fully automatized by special software and can be directly implemented to a controller. While advanced techniques are fully described by exact researches and can be implemented into controller's software too. For example, hyper-spherical search was proved to be the best for non-linear optimization [22].

5. Power grid control and interaction between elements

The control system consists of 9 elements as shown in figure 2: mandatory and welfare power consumption, WPP, solar plants, accumulators and fuel generator as power elements; dispatching block, controller and internet connection as controlling elements.

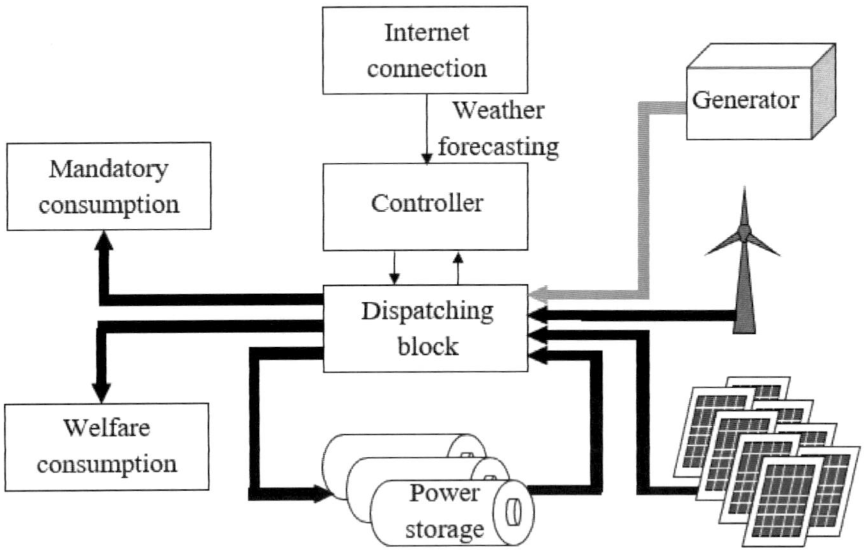

Figure 2. The smart grid structure and interactions between elements.

Within the power grid as a Holarchy there are three Holons on which system cannot affect: WPP, solar plants and mandatory consumption. Three other Holons – welfare consumption, set of accumulators and fuel generator – can be used as a controlling instruments. All of them are connected through dispatching block, where physical switching is happening. While the inverter is also a part of dispatching equipment. The decision for the dispatching block is provided by controller, where current and future states of the grid is processed and exact solution based on previously stated optimization problem is found. To calculate the future power supply of the grid weather forecast through internet connection is used. The connection of fuel generator differs from others as it operates manually, if plants on RES are not giving energy for more than 2 days.

There are more advanced and expensive smart grid structures with load forecasting and data analytics as in CompactRIO complex [23]. It's actual for a grid with several consuming entities or very unstable consumption. While at the control post power is demanded on a regular basis with almost exact amount. And the welfare part of consumption can be regulated by straight interaction with the officers.

6. Financial actuality

The alternative way to supply the control post with power is to provide connection to a central grid. The cost of 1 km of electric lines will be at least 7 921 € [24]. While the small-sized power grid cannot connect straight to the central power line. It leads to increase in the cost by more than 1 802 € for a power substation [25]. After that investment, also monthly payment for electricity should be done. The offered power grid cost is less than 12 500 €, as it needs 10 325 € for power equipment (4 485 € – for wind plant, 1 160 € – for solar complex, 1 340 € – for inverter, 3 163 € – for accumulators, 177 € – for fuel generator) [9-13] and less than 1 000 € – for controlling equipment [26]. While operating expenses will be only for service and petrol's cost in situation of total RES' absents for 2 days.

7. Conclusion

The number of roads in Republic of Kazakhstan is growing fast because of the WEWC Corridor. Most of the roads demand electric power supply to provide proper infrastructure. While with small population of the country road networks are wider presented than electric power networks. It leads to increasing investments of the government or limitations in posts positioning on a road.

On the republic's geographical territories RES, especially solar radiation in southern parts, are highly represented. RES' potential and calculation of power demand of the transport control post was compared. It was stated that the autonomous power supply on RES' plants can fully supply the post. As a result, the complex of power plants on wind and sun energy for the control post's supply is offered. The inverter to ensure proper quality of output electric power is chosen. For periods of time when there is no power production from RES the accumulators as a power buffer is stated. The chosen accumulators can provide full power supply for the system for 35 hours or full supply for 1,4 days with 2 days more for only mandatory supply.

Design of any exact control system is a complex engineering problem. While with proper mathematical description of an object and, if there are, an optimization problem the complexity of the task decreases. The MAS strategy is proved to be a promising designing instrument for power grids of any size. While Holonic architecture gives opportunity to divide system into parts and formulate same mathematic description for both system and its parts. The previous researches justified optimization techniques only for systems with connection to central grid. In the article, after applying this technique on the control post's power grid and receiving structure of the Holarchy, the optimization problem for autonomous power grid with power storage is deduced. While calculation of optimal controlling values can be provided online with special software within an implemented controller. The functional scheme of the smart power grid is reckoned. As a result, offered approach and formulated optimization problem gives the possibility to design control system ones and scale it to other power grids with less research. Also it provides possibility to optimally dispatch the power grid with minimum human participation.

The known alternative for the remote small-sized power grids is to provide electric over-head line from a nearest central grid's connection. Also some simplest transformer sub-station is needed. So, the possible cost of such connection is rising upper than 9 723 € for equipment for 1 km distance. While offered power supply on RES with all power and control equipment will cost nearly 12 500 €. And with almost no operating costs, beside service and some fuel, the autonomous power grid is beneficial in all expenses. In fact, such remote post can have some neighboring entities, that are distant from central grid too. That means that such power grid can provide energy to some neighbors as clients, which leads to additional financial and social effects. It demands only small redesign in power plants and control algorithm, which have been eased by the offered Holonic approach.

References

[1] About_Kazgeology.pdf. Available at: http://zentralasien.ahk.de/uploads/media/About_Kazgeology.pdf

[2] Lepeha *There are no enough "Rubej" control posts on roads of Kazakhstan* Available at: https://tengrinews.kz/kazakhstan_news/dorogah-kazahstana-hvataet-kontrolnyih-postov-rubej-lepeha-288704/ (In Russian)

[3] Karatayev M and Clarke M I 2016 A review of current energy systems and green energy potential in Kazakhstan *Renewable and Sustainable Energy Reviews* **55** 491-504

[4] Nurpeissova G B, Panyukova D V and Nurpeissov A T 2018 Smart Techniques' Implementation for Small Entities and Central Power Grid in Almaty, Kazakhstan: Challenges and Possibilities *Balkan J. Of Electrical & Computer Engineering* **6-4** 242-6

[5] Automated system of traffic control "Post-Rubej" – SMEU Astana Available at: http://www.smeu-astana.kz/ru/rubezh (In Russian)

[6] "VIDEOCONTROL-Rubej" complex Available at: http://www.ollie.com.ua/videocontrol/index.html#3 (In Russian)

[7] Abidi M G, Smida M B, Khalgui M, Li Zh and Wu N 2017 Multi-agent oriented solution for forecasting-based control strategy with load priority of microgrids in an island mode – Case study: Tunisian petroleum platform *Electric Power Systems Research* **152** 411-23

[8] Nurpeissova G B, Nurpeissova T B and Panuykova D V 2014 Estimation of wind energy resources at Kazakhstan territory for autonomous power supply of low power consumers

Proc. 4th Int. Multidisciplinary Scientific GeoConf. on Energy and clean Technologies (Bulgaria, Albena) vol 1 pp 213-20
[9] Official site of MAP Energy - LOW·WIND·24·2·5-3·5 Wind generators
Available at: http://www.invertor.ru/zzz/item/low_wind_24_2_5-3_5 (In Russian)
[10] LDK-260PA (China), 260 W, 24 V. cheap solar panels in Astana by "Internet-store "EcoEnergy.kz" company
Available at: https://eco-energy.kz/p36648239-ldk-260pa-kitaj.html (In Russian)
[11] Inverter and charging device from serial model Tripp Lite APSX3024SW
Available at: http://www.all-generator.ru/invertor/tripp-lite-apsx-3024-sw.shtml (In Russian)
[12] CHALLENGER A12-200 accumulator (AGM). 200 Ah 12 V. cheap general purpose accumulators in Astana by "Internet-store "EcoEnergy.kz" company
Available at: https://eco-energy.kz/p48536577-challenger-a12-200.html (In Russian)
[13] Inverting petrol and diesel generators (modernized for ASS
Available at: http://www.invertor.ru/page3.html (In Russian)
[14] Coelho V N, Cohen M W, Coelho I M, Liu N and Guimarães F G 2017 Multi-agent systems applied for energy systems integration: State-of-the-art applications and trends in microgrids. *Appl. Energy* **187** 820-32
[15] Sanjari M J, Karami H and Gooi H B 2016 Micro-generation dispatch in a smart residential multi-carrier energy system considering demand forecast error *Energy Conversion and Management* **120** 90-9
[16] Cardin O, Derigent W and Trentesaux D 2018 Evolution of holonic control architectures towards Industry 4.0: A short overview *IFAC-PapersOnLine* **51-11** 1243-8
[17] Vasek L and Dolinay V 2016 Holonic Model of DHC for Energy Flow Optimization *IFAC-PapersOnLine* **49-27** 413-8
[18] Ferreira A, Ferreira Â, Cardin O and Leitão P 2015 Extension of holonic paradigm to smart grids *IFAC-PapersOnLine* **48-3** 1099-104
[19] Ansari J, Gholami A and Kazemi A 2016 Multi-agent systems for reactive power control in smart grids *Int. J. of Electrical Power & Energy Systems* **83** 411-25
[20] Howell Sh, Rezgui Y, Hippolyte J.-L., Jayan B and Li H 2017 Towards the next generation of smart grids: Semantic and holonic multiagent management of distributed energy resources *Renewable and Sustainable Energy Reviews* **77** 193-214
[21] Arcos-Aviles D, Pascual J, Guinjoan F, Marroyo L, Sanchis P and Marietta M P 2017 Low complexity energy management strategy for grid profile smoothing of a residential grid-connected microgrid using generation and demand forecasting *Appl. Energy* **205** 69-84
[22] Ahmadi S A, Karami H, Sanjari M J, Tarimoradi H and Gharehpetian G B 2016 Application of hyper-spherical search algorithm for optimal coordination of overcurrent relays considering different relay characteristics *Int. J. of Electrical Power & Energy Systems* **83** 443-9
[23] Thiyagarajana K and Saravana Kumar R 2016 Real Time Energy Management and Load Forecasting in Smart Grid using CompactRIO *Procedia Computer Science* **85** 656-61
[24] Construction of overhead lines (OHL) (In Russian)
Available at: http://www.masterenergoservice.com/stroitelstvo-vozdushnyh-liniy-vl
[25] Buy a dead-end transformer substation KTP-1, cost in the Ukrain
Available at: https://krsl.com.ua/portfolio-archive/tupikovaja-podstancija/ (In Russian)
[26] OVEN PLC110 [M02]. Cost (In Russian)
Available at:
https://www.owen.ru/product/programmiruemij_logicheskij_kontroller_oven_plk110

IOP Conf. Series: Materials Science and Engineering **972** (2020) 012023 doi:10.1088/1757-899X/972/1/012023

Suitability of map data for processing of pedestrian analysis questionnaire in small towns

Z Kramářová[1]

[1]Department of Civil Engineering, Faculty of Technology, Institute of Technology and Business in České Budějovice, Okružní 517/10, 370 01 České Budějovice, Czech Republic

E-mail: kramarova@mail.vstecb.cz

Abstract. From a historical point of view, pedestrian movement is the oldest way one moves from place to place. It is a basic, most natural and environmentally friendly means of transport, which is unattractive for many individuals in terms of today's trends. Walking is absolutely indispensable for the creation of cities, especially their public space, as it alone allows people to interact with their surroundings. Public space in cities is an essential element that ensures the existence of a settlement, its life and the life of its inhabitants. If we consider as a premise the fact that cities are created for people's living, public space fulfills the functions of socializing, communicating, passing, etc. - that is, a function that supports the community of society. The analysis of the permeability of settlements, especially of small towns, is one of the most important documents for the creation and regeneration of the public space. A questionnaire survey is one of the methods for its quality processing, which is most easily done by plotting the map. Choosing a suitable map background that is sufficiently understandable to the general public is the first prerequisite to obtain quality data for this analysis.

1. Introduction

In order to better understand the whole issue, it is necessary to define the basic concepts first.

1.1. Walking versus running

Walking is the basic way people move from point A to point B. Historically, it is the oldest way to move a person. This movement uses the support of the lower limbs, respectively their alternating movement. Whittle [1] defines gait as a locomotive manifestation and as a method of locomotion, which is characterized by alternating movements of the lower limbs. Similarly, other authors, Vaughan, Kolář, etc., define walking similarly [2, 3]. However, the definition of wikipedia must be mentioned like the most comprehensible and comprehensive way for the general public:

„ *Walking is a locomotion movement using the lower limbs and is inherent in bipedal animals like humans. Allows you to move from one location to another. Walking consists of one step at a time, one leg at a time, moving one by one. Walking is usually slower than running, but while running an untrained person can only exercise for a short time, he can walk for many hours without a break.* "
(Source: [4])

Very close to walking is running. It is a similar movement, but it is faster, more demanding and less comfortable. Due to the scope of the text, professional definitions of run will not be given. The main difference between walking and running lies in contact with the ground. During the run, there is a phase where none of the limbs touches the ground (support). However, one foot is always in contact with the ground when walking. Contact with the ground minimizes the vertical movement of the head and thus stabilizes it (Figure 1). Therefore, when walking, it is very easy to observe the environment that a person is going through.

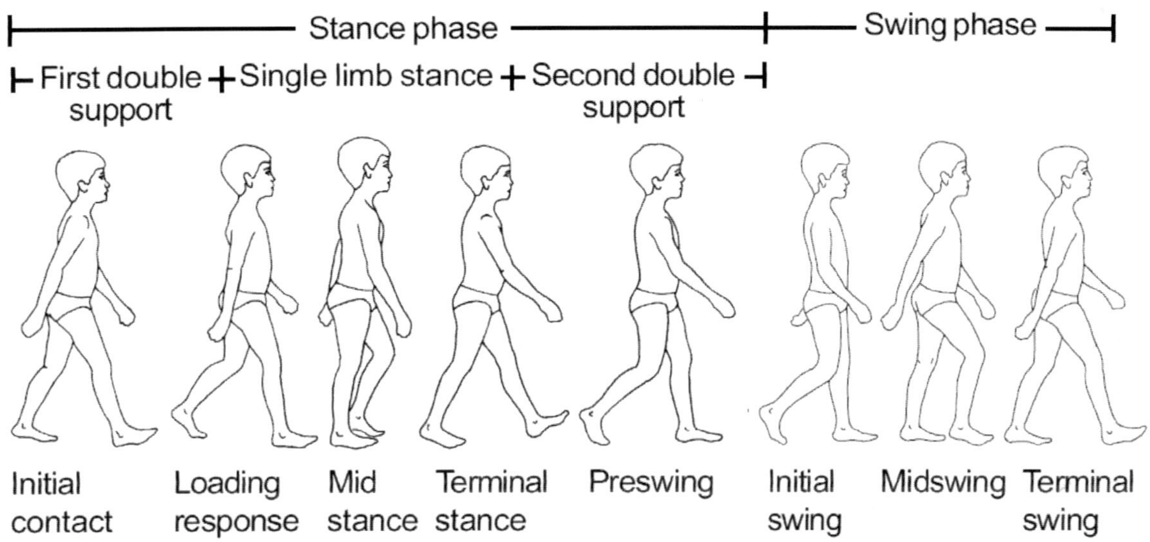

Fifure 1. Step cycle phases. (Source: [3])

At the same time, walking is a significantly slower way of moving than other translational ways of movement (running, cycling, driving,...). Therefore, it imposes considerably higher demands on public space quality of the moving individual. When walking, unlike other types of motion, the reaction distance is minimal (based on the reaction time and the speed of movement), so one can stop almost immediately on the spot and react - to see what intrigued him, greet another person, talk, etc.

For these reasons, walking is indispensable as one of the main elements that help to shape the environment of man, his living space, and thus the city and hence his public space.

1.2. Public space

The structure of the city can be divided into two basic elements - buildings and space between buildings. The space between buildings can be further divided into private, dedicated and public. A private space is a space used by the owner or a person close to him. The owner's consent is a condition for use and access to this area. Examples include a private garden, a courtyard, an atrium.

The space reserved depends on the date of reserved greenery, ie. greenery accessible under certain conditions. Similarly, this space is only accessible to a specific group. Examples include the school garden and courtyard, the castle park without free entry, the greenery in the factory areal, etc.

Public space is by its very name intended for all. Access to such space is not limited by age, gender, religion, or any other factor. These include street space, squares, embankments, pedestrian zones, parks, etc.

At present, public space can be divided into the physical public space and the virtual public space [5]. From the point of view of the contribution, especially the physical public space, which can also be called public place, is significant. Act No. 128/2000 Coll., The Act on Municipalities (municipal establishment) in §34 defines public space in the Czech Republic in a binding manner.

„ Public areas are all squares, streets, marketplaces, sidewalks, public green areas, parks and other areas accessible to everyone without restrictions, that is, serving the general purpose, regardless of ownership of the area.“
(Source: [6])

The second legislative reference can be found in Decree No. 501/2006 Coll., On General Requirements for Land Use. This Decree lays down general requirements for the use of land, delineation of land and land and placement of buildings in the territory. This Decree says in §7:

„Areas of public spaces

(1) Public space areas are usually self-defined in order to ensure conditions for the adequate location, extent and availability of land plots and to ensure conditions for their use in accordance with their importance and purpose.

(2) Public space areas generally include existing and proposed plots of each type of public space and other land of related transport and technical infrastructure and civil facilities compatible with the purpose of public spaces. For each two hectares of built-up area of housing, recreation, civic amenities or mixed dwelling, the associated area of public space of at least 1000 m2 is defined with this built-up area; this area does not include roads.“
(Source: [7])

Another interesting definition is given in the Glossary of Terms, which is maintained on its website by the Department of Urban Design, Town and Regional Planning of the Faculty of Civil Engineering CTU in Prague.

„ Public space is the space of society's life, the space of social communication. It is a place where society "happens", where there is contact and communication between people, to share, inspire and create everything that transcends individuals. Public space is a widely open space where a distinction is made between high and low, beautiful and ugly, important and insignificant. This space is an open space in a very broad sense: it is open to all members of the society, but also to a space where this society opens up to things and to strangers; this space is also open to a wide range of activities and uses; it is a void where something unexpected can happen and is also a space where society opens up for the future. … “
(Source: [5])

Obviously, the world's leading and Czech urban designers also define public space, but due to the legally binding definition and the great similarity of these definitions, other examples will not be mentioned.

The third definition is a definition that understands public space in particular as regards its possible interactions between the environment and the individual and their interaction. This is the second very important aspect of public space. This is to be understood not only in terms of physical spatial significance (place of residence and transport), but also in terms of socialization (place for contact, life and action). In addition, public spaces can of course be perceived, analyzed and created from other perspectives - such as transport, environmental, air quality, aesthetic, functional, etc.

2. Pedestrian movement analysis – map requirements

Due to the fact that the contribution is linked to the implementation of the project TAČR TL02000559 "Safe Cities for Pedestrians and Seniors", which deals with the public space in small towns of the Czech Republic, all the facts will be presented for small towns. A small town in the Czech Republic is understood to be a seat with a town status and a population of between 5,000 and 29,999.

Materials for analyzing pedestrian movement in a territory can be obtained in many ways. The three most commonly used are questionnaire surveys, local surveys at critical points, and traffic diaries.

Keeping a diary is based on recording individual routes and data when and how they were implemented. As a rule, this evidence is recorded verbally in the "start - over - destination" manner. Very little is done using map drawings (only in digital form).

Local survey at critical points is based on profile census - survey of traffic flow intensity. It is a local survey where a worker at a critical point physically calculates a passage for a predetermined time. This method does not require mapping to the map except for the critical point position plot.

Data collection for pedestrian movement analysis using a questionnaire survey is probably the most commonly used method. It can be implemented in two basic ways. Either it is a verbal description of the journey, when the respondent verbally describes the individual routes - their frequency, route ("start - through - destination"), period, etc. Or these data are recorded directly into the map by direct plotting of the route and for example color resolution. This data collection method can be successfully implemented both in paper and digital records, eg via a web interface. Among disadvantages of this method is also the necessary orientation in the presented map, ie. "respondent can read map".

Because it is important for the analysis that the sample is relevant, it must include respondents of all ages, gender, education, etc. in a balanced population mix. In practice this means that the questionnaire must be processed in the simplest way possible so that it is understood by children, adults and seniors, people with lower and higher education, technical and humanitarian orientation population, etc. This is not difficult at all, but the opposite is true .

From the practice of conducting local surveys over the last 20 years, it should be noted that the level of map orientation varies widely among the general public. It is inconceivable for a technician (a builder, an architect, a land-use planner) can not read the situation on the map / plot, but the general public does not share this need. Today's modern applications such as GPS navigation are very "recording", which guides the individual visually and verbally from place to place without having to read the map.

Figure 2. Example of changing the legibility of the map when changing color. (Source: [8], editing: own)

Elements that facilitate orientation include larger green areas (parks, alleys), important buildings (office, school, health centre, shop, church, library, cinema, post office, etc.), large areas and spaces (squares, car parks, roundabout, traffic lights, etc.), sports facilities, areas and playgrounds (winter

IOP Conf. Series: Materials Science and Engineering **972** (2020) 012023 doi:10.1088/1757-899X/972/1/012023

stadium, summer stadium, tennis courts, workout, swimming pool, playground, etc.), high-rise objects (factory chimney, transmitter, lookout, etc.) memorials, statues, ATMs, bus stops. In summary, these objects can be characterized as important objects visually, aesthetically or attending (destination). These are the objects the public notices and according to which they are oriented in the area. If these points are clearly shown in the map, it will greatly facilitate the orientation of the respondents on the map.

Another criterion for usability is the colour and contrast of the map. Map orientation in black/white is very limited by its own display options. If only the black/white contrast can be used, the display and expression capability is very limited (Figure 2). Perception of the colour spectrum is an individual matter. However, it can be generally stated that most of the population prefers light colours rather than rich dark colours. In addition, light colours make it possible to add a textual description to the colour background (Figures 7, 8, 9, 10). Dark text on a lighter background is also more readable than light text on a dark background.

3. Types of maps and their analysis

The basic selection criterion was the representation of the whole built-up area of the town on a sheet of max A3. The second condition was to differentiate between individual blocks of buildings and individual streets. The last criterion was the relative ease of availability of the map - the map must be either freely available or available commercially (purchase). Based on these three criteria, these maps were selected for analysis:

- Cadastral map (Figure 3) - has been chosen because of its frequent use in land-use studies and projects by architects and civil engineers.
- Basic map of the Czech Republic 1:10 000 (Figure 4) - is an official map with the closest smaller scale to the cadastral map;
- Map data of Mapy.cz and Google maps servers (Figures 5, 6) - are most commonly used by the general public.
- Maps from the information centres of Chotěboř and Milevsko (Figures 7, 8, 9 and 10) - are maps specially designed for easier orientation of tourists and other visitors to the city.
- Orthophotomap (Figure 11) - a very popular map by the general public, which is freely available on the Internet in many views, at the same time displays real reality at a given time point.

Unsuitable from the point of view of insufficient details and size, tourist maps and cycling maps were excluded, the map of the Czech Republic 1:25,000 and other smaller scales, road maps (maps of cities in road maps only detail the city centre, not the whole seat) and other similar maps.

For better clarity and comparison, the display of approximately the same area of the city centre Chotěboř was chosen (Figures 3, 4, 5, 6, 7, 9 and 11). This is not the case with Milevsko maps (Figures 8, 10).

The observed criteria were selected on the basis of public interviews in field surveys over the past 20 years. These are, in particular, the contrast and colour of the map, street names and local names, important objects (e.g. schools, post offices, sports facilities, etc.), larger green areas (e.g. parks, alleys, greenery at monuments) and works of art (e.g. statues, monuments, etc.). Other orientation elements in space include objects that evoke a strong emotion - positive or negative. These are places that the respondent perceives as either very positive (such as building a house he likes very much by his appearance) or as a very negative one (e.g. a ruin of an house). Unfortunately, they do not exist in the analyzed maps because they are not officially labeled or categorized.

Figure 3. Example of cadastral map. (Source: [9])

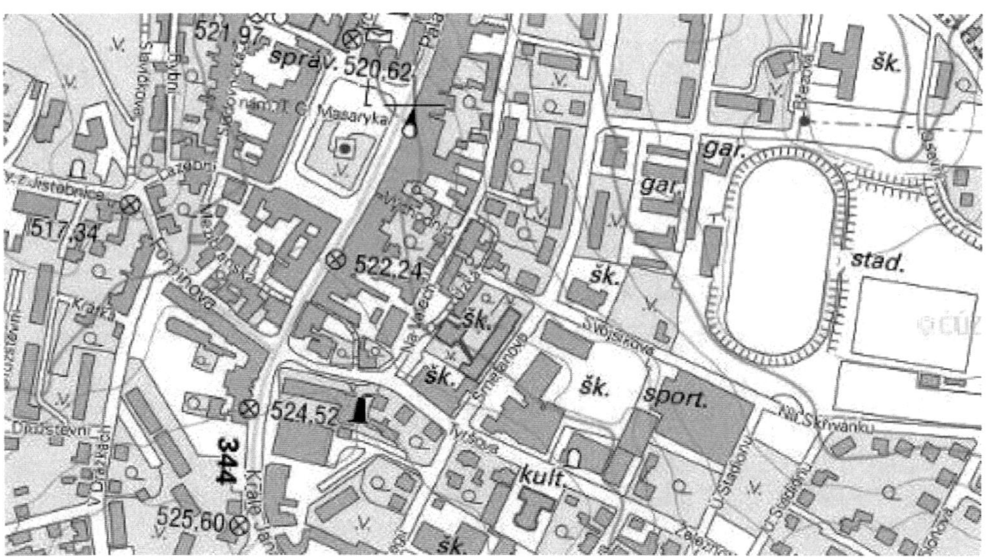

Figure 4. Example of a basic map of the Czech Republic 1:10 000. (Source: [9])

IOP Conf. Series: Materials Science and Engineering **972** (2020) 012023 doi:10.1088/1757-899X/972/1/012023

Figure 5. Example map from Mapy.cz. (Source: [8])

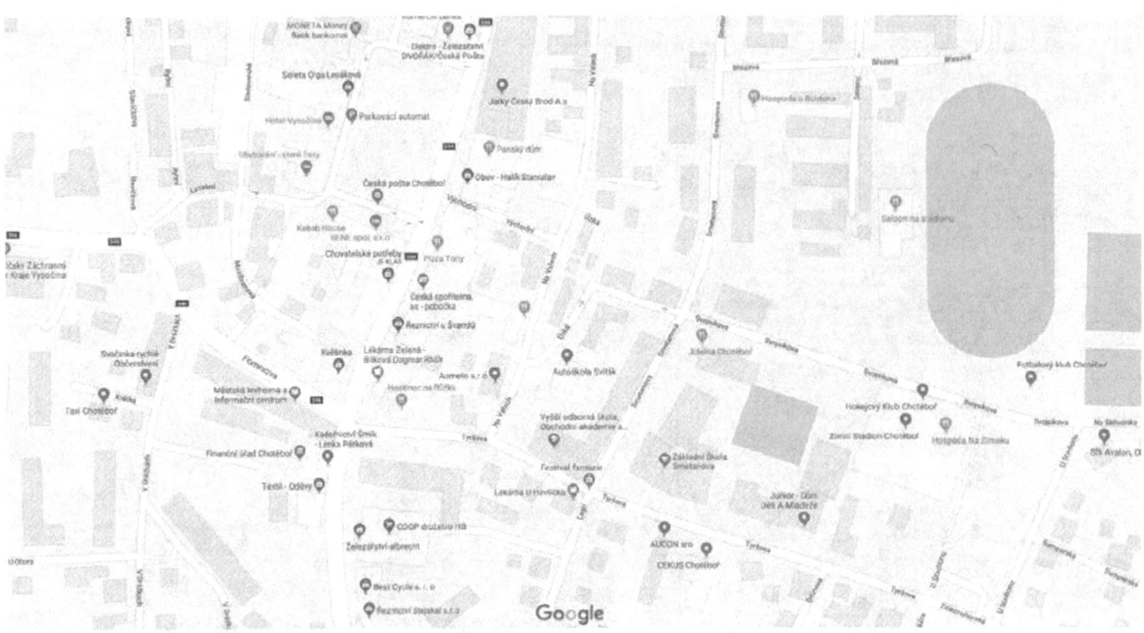

Figure 6. Example map from Google maps. (Source: [10])

IOP Conf. Series: Materials Science and Engineering **972** (2020) 012023 doi:10.1088/1757-899X/972/1/012023

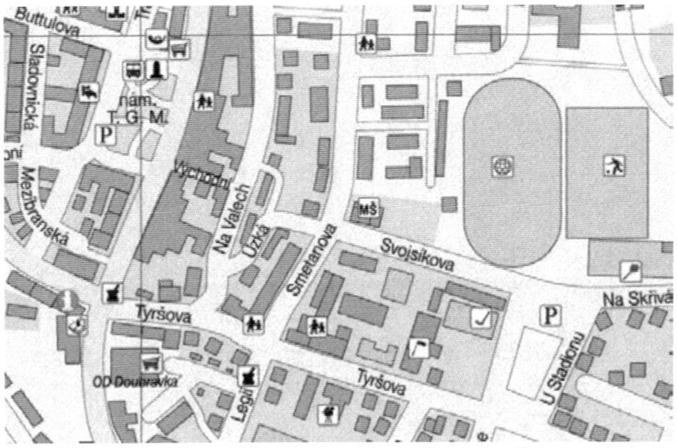

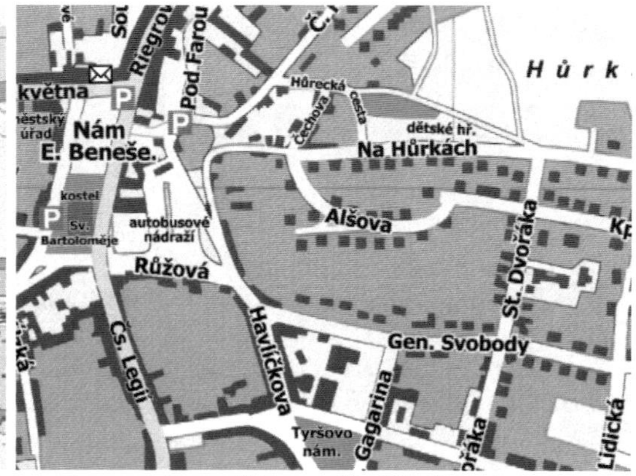

Figures 7 and 8. Examples of maps from the information centers of town Chotěboř and town Milevsko - information maps. (Source: [11, 12])

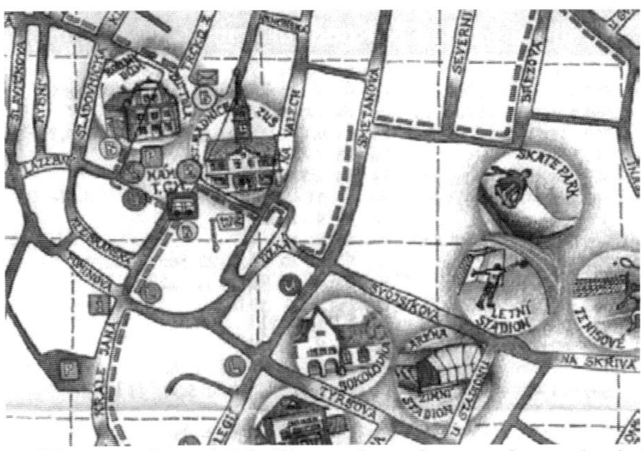

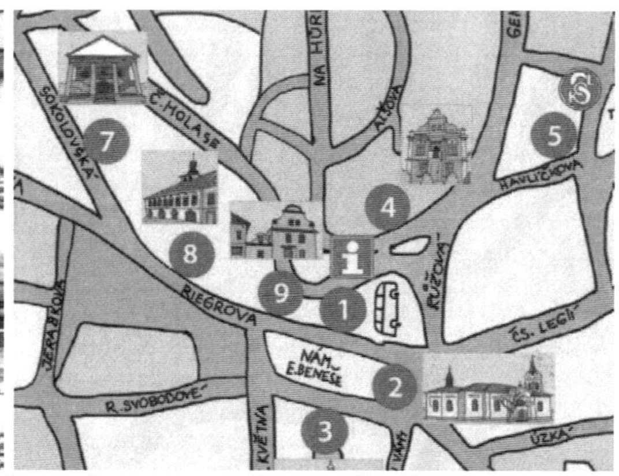

Figures 9 and 10. Examples of maps from the information centers of town Chotěboř and town Milevsko – „draw" maps with pictures. (Source: [11, 12])

Figure 11. Example of ortofotomap. (Source: [9])

kind of map / observed phenomenon	Cadastral map	Basic map of Czech Republic 1:10000	Mapy.cz	Google maps	Information map of town Chotěboř	Information map of town Milevsko	Picture map - town Chotěboř	Picture map - town Milevsko	Ortofotomap
colour		xxx	xxx	xxx	xxx	xxx	xxx	xxx	xxx
light colours		xxx	xxx	xxx	xxx	x	xxx	xxx	x
contrast	xx	xx	xxx	x	xxx	xxx	xxx	xxx	xx
good view of urban structure		xx	xxx	x	xxx	xxx	xxx	xxx	xx
view of street infrastructure	xxx	xxx	xxx	xxx	xxx	xxx	xxx	xxx	xxx
description of street names			xxx	xxx	xxx	xxx	xxx	x	
locally used names					x	x	x	x	
view of inner area (e.g. townsquare)			xxx	xxx	xxx	xxx	x	x	xxx
road resolution		xxx	xxx	xxx	xxx	xxx			
view of individual plats	xxx								
view of individual buildings	xxx	xx	xxx	xxx	xxx	xxx			xxx
view of importaint buildings (school, restaurant, office, etc.)		x	xxx	x	xxx		xx	x	
description of importaint building names		x	x	xxx		xxx			
view of small architecture or sculptures		xx	xxx	xxx	xxx		xx	x	x
view of bus stops			xxx	x	x	x			x

Figure 12. Comparison of observed phenomena on individual maps. (Source: own)

Note: xxx very good view, in full
xx worse view, incomplete range
x very light view, incomplete information
nothing shown at all

The comparison shown in Figure 12 shows that none of the analyzed map data meets all the highest quality criteria. The most successful one is the map of Mapy.cz and also the Information Map from the Chotěboř Information Centre. Both maps are very useful for easy orientation of the general public. If we would like to improve the quality of the map for the questionnaire survey, it seems to be the best way to create a map based on the above mentioned maps. These were to be supplemented by the names of important objects and possibly a reference to a photo of a positive or negative point (see above).

If the survey is to be combined - designed to get pedestrian movement data and, for example, feeling data (localization of sites perceived positively, negatively, pride points, etc.), the use of photos is totally inappropriate. This is due to the inconsistency of the information and thus the overall distortion of the information obtained. Therefore, it is very important to first decide on the data that will be obtained by the questionnaire and only then to find the most suitable map.

4. Conclusion

The analysis of pedestrian mobility is one of the most important, but little processed documents for the revitalization of public space and the solution of conflicts of pedestrians and road transport in small towns in Czech Republic. One of the most widespread methods of processing this analysis is the method of questionnaire survey with drawing of pedestrian movement into the map. One of the prerequisites for obtaining relevant data is the easy and good orientation of the general public in the used map.

The analysis of the selected map types showed that there is no map that meets all the criteria. Maps of Mapy.cz server and locally prepared information maps for tourists and visitors of the city were the most suitable. For even better orientation, it would be necessary to add further description, possibly other points, based on the local survey of the city.

The processed analysis of the maps will serve as a basis for the compilation of the real pedestrian mobility questionnaire and its implementation in the framework of the project "Safe Cities for Pedestrians and Seniors". One of the aims of this project is to create a methodology for the regeneration, revitalization and creation of safe public spaces in small towns in the Czech Republic.

Acknowledgments
The research described in this paper has been financed by TAČR - program ÉTA -project TL02000559 Safe and secure cities for pedestrians and senior citizens, (Bezpečná města pro chodce a seniory).

References

[1] Whittle M W 2007 *Gait Analysis: An introduction.*4.ed. (China: Elsevier) ISBN 9 780 7506 8883 3
[2] Vaughan Ch L Davis B L OĆonnor J C 1999 *Dynamics of human gait.* 2. ed. (Cape Town, South Africa: Kiboho publishers) p 9 ISBN: 0 620 23558 6
[3] Kolář P et al. 2009 *Rehabilitace v klinické praxi.* 1. ed. (Praha, Czech Republic: Galén) . ISBN: 978-80-7262-657-1
[4] Wikipedie open encyklopedy 2019 *Chůze.* In cs.wikipedia.org [online]. Avaible from https://cs.wikipedia.org/wiki/Ch%C5%AFze [cit. 2019-05-03].
[5] Department of Urban Design, Town and Regional Planning 2019 *Urban dictionary.* In www.uzemi.eu [online]. Available from http://www.uzemi.eu/pojmy/verejny-prostor [cit. 2019-05-03].
[6] Czechia 2000 *Act No. 128/2000 Coll., Act on Municipalities (municipal establishment).* [online] (Praha, Czech Republic) Available from https://www.zakonyprolidi.cz/cs/2000-128 [cit. 2019-05-03]
[7] Czechia 2006 *Decree No. 501/2006 Coll., On general requirements for land use.* [online] (Praha, Czech Republic) Available from https://www.zakonyprolidi.cz/cs/2006-501 [cit. 2019-05-03]
[8] Mapy.cz 2019 *Map of Czech Republic.* In www.mapy.cz [online]. Available from https://mapy.cz/zakladni?x=15.6721989&y=49.7189213&z=18&l=0 [cit. 2019-05-03].
[9] State Administration of Land Surveying and Cadastre 2019 *Consultation of the Cadastre of Real Estate – map view* In www.cuzk.cz [online]. Available from https://nahlizenidokn.cuzk.cz/VyberKatastrMapa.aspx [cit. 2019-05-03].
[10] Google maps 2019 *Map of Czech Republic.* In www.google.com [online]. Available from https://www.google.com/maps/@49.7189464,15.6718553,18z [cit. 2019-05-03].
[11] Town Chotěboř 2019 *Chotěboř – official website* [online]. (Chotěboř: Czech Republic) Available from: https://chotebor.cz/ [cit. 2019-05-03].
[12] Town Milevsko 2019 *Milevsko – official website* [online]. (Milevsko: Czech Republic) Available from: http://www.milevsko-mesto.cz/ [cit. 2019-05-03].

IOP Conf. Series: Materials Science and Engineering **972** (2020) 012024 doi:10.1088/1757-899X/972/1/012024

Analysis of the motivation of the residents of Saransk for the transition to the use of "smart bikes"

A V Maryonkin[1], A V Razumov[1]

[1]Faculty of architecture and construction, National Research Mordovia State University; MRSU, 68 Bol'shevistskaya street, 430005 Saransk, Russia

E-mail: alexeymaryonkin@gmail.com

Abstract. This article aims to analyze the main motives of using a bicycle as sustainable public transport. Consider the prerequisites for the introduction of "smart transport" for the territory of Saransk. More than 1000 citizens took part in the survey. The majority of respondents stated low satisfaction with the bicycle infrastructure, but are interested in the implementation of the "smart bike" system in the vicinity of the city. The analysis provides an understanding that residents are concerned about environmental issues, affordable transport and personal health. That speaks about the high potential of development of this system in the city of Saransk.

1. Introduction

Bicycles are another part of a formed urban environment and perceive life and movement differently. They represent a faster form of travel but their users still remain pedestrians. By that cyclists promote development of "smart cities" with a healthy, safe and sustainable urban matrix.

Today cyclists are an important part of city life. Everything is connected with various processes taking place in the world, such as depletion of fuel, air and ocean pollution and the deterioration of the health state of city residents. Therefore, the prevalence of cycling over the car looks reasonable and farsighted. We need cities being comfortable for cyclists and this can be achieved without much effort and expenses in most cities of the world [4].

There are already some cities which are actively developing cycling infrastructure and introducing the concept of "smart" mobility into urban life. One of these cities is Copenhagen where the tradition of cycling has never given way to automobilization progressing in the last century. The main citizens' reason for the use of bicycles is the fuel crisis in Europe in the 1970s [1]. At the moment bicycles take a significant part of urban traffic what highly reduces car traffic which is the lowest in Western Europe. The Copenhagen experience is being used as a platform for discussions about a city for elaboration a sustainable development of the territory [2].

The basic principle of the formation of the bicycle network policy is the search for a place, the development of a route through the streets with high urban activity and safety all along the route. Another important aspect of sustainable development is the green routes, designated paths in parks, urban recreation and along unused railway tracks. They work seasonally and are considered to be backup in case of overloading of the main tracks.

For the formation of a sustainable bicycle network it was decided to narrow the lanes. Most of the city four lane arteries were redesigned into two lane arteries with a bicycle lane and an extended sidewalk. Abundant landscaping has also been added what makes the environment safe for citizens.

The inclusion of cycling in the city transport system solves the problem of sustainable development of urban areas. By that the mobility issue has also been solved and it gave an opportunity to bring bicycles into city buses and subways which added variety of trips and their combining [6].

Safe traffic and coexistence with cars is the main factor for sustainable bicycle development. Trucks and buses have to have special bicycle mirrors and numerous information companies remind drivers of the need to pay attention to cyclists, especially at the crossroads [9].

Also one of the most significant factors of safe cycling routes is the volume of traffic. According to the observations it is revealed that the presence of a big number of cyclists increases the vigilance of drivers, this has a positive impact on road safety.

Cycling gradually led to the expansion of the range. Many new types of bicycles have appeared on the street including tricycles for transporting children and goods, bicycles for disabled people and taxi-bicycles. They require more space and older cyclists, so as parents transporting children by bicycles, need an increased guarantee of safety. Despite these new requests bike continues to be an excellent wheeled vehicle which takes the least place on the streets per person [12].

Till now many cities have switched to renting various urban bicycles by easy interaction with a mobile phone. The idea behind the system of "smart" city bikes is to let inexperienced cyclists to drive around the city in a relatively safe, stable bicycle environment without causing problems with storage and maintenance. The growing number of problems connected with sustainable development, climate change and health will likely lead to the situation when more and more cities will form a new culture of city life and city traffic. The development of cycling is the obvious answer to the many problems faced by cities around the world [3].

2. Sustainable development of "smart cities"

2.1. Forming "smart mobility"

The system of "smart city" in the modern world is becoming a popular topic in the development of cities resulting into both positive reviews and alertness. This article raises the question of the work of urban mobility in a "smart city". "Smart" mobility is one of the main components of what is commonly called "smart city" in the modern world [5].

Both personal and public bicycles, as one of the types of urban transport, become to be more and more popular in the whole world [7]. The reasons for this are quite obvious. This is the democracy and availability of this type of transport, its cheapness, environmental friendliness, promotion of a healthy lifestyle and reduction of the number of cars on the roads and traffic jams in urban arteries. The state systems in the Russian Federation, in turn, do not yet offer the sharing of bicycles in necessary quantities, which can be used by citizens along with the existing transport systems, such as the metro, buses, taxis, and so on.

In the world practice the "smart bike" of public use tends to reduce car dependency and improve the environmental situation in cities. The Netherlands is a particularly clear representative of the transition to a new sustainable urban mobility where about 400,000 cyclists ride the streets every day. [11].

Sustainable mobility of this system requires the acceptance of sustainable development of the required infrastructure and the tendency of reducing the number of private cars in the city [14]. Thereby, not only architects, urbanists, politicians, but also market analysts fixed their eyes on this. Marketers introduce visual advertising for the bicycles they represent, urban infrastructure, and adjust the field of view of road users.

However, this system cannot be used for each city with the same structure, it is necessary to take into account many factors, such as climate, the requests of the citizens, geography, financial situation in the region and competent promotion of sports as a way of life.

For the long-term use of bicycles of new mobility, a number of mandatory things is required: cycling routes that comply with the regulations, timely maintenance of bicycles and infrastructure, demographic and operational conditions, safety of terminals and legal protection of the cyclist as a full participant in road traffic.

Also, the bike sharing system requires constant investment, support and feedback from users. It should be noted that after the crisis in 2008 the use of public bicycles and all other urban modes of transport increased [8]. This proves once again that the bicycle, as a public transport, is developing and has everything what is necessary for the further growth of caring residents, and with them the necessary financial and creative investments.

Urban mobility issues are major for cities due to rapid urbanization and the development of new regulatory options for a sustainable urban transport system. According to that "smart mobility" is one of the factors forming "smart city" and affects the quality of life of citizens.

The goals of "smart mobility" are formed from 6 categories: pollution decrease; decrease of traffic jams; increase of safety; reduction of noise pollution; improvement of the speed of movement; reduction of transport costs of citizens.

It is worth noting that "smart mobility" intends to help smart cities achieve their specific goals by reducing urban environmental impact and improving the quality of life of citizens. It results into impossibility of unstable mobility plan being "smart" and meeting the modern characteristics of a "smart city" [10].

The introduction of "smart" technologies in the system of urban public bicycles dispelled concerns and solved many problems, such as vandalism and theft of bicycles. And alongside this, they have increased the demand among tourists due to fast and easy control via the Internet.

But in order to be confident of success and the expansion of a sustainable "smart bike" zone, it is necessary to analyze and determine the desire and need of citizens to use urban public bicycles as a permanent vehicle.

Sources identified several barriers to the transition to the formation of a "smart bike", they include travel time, lack of safety, heavy traffic, lack of infrastructure, weather conditions which are not preferable for driving, and terrain features. One of the main factors is the well-built system of public use of bicycles in all parts of the city, the presence of a sufficient range of "smart bikes" in areas of continuous demand and comfortable access to them. The combination of a favorable landscape and weather conditions are equally important, as well as the availability of a guarantee for a bicycle and the possibility of renting for a long period of time, what stimulates healthy and environmentally friendly means of transport.

2.2. "Smart Bikes" in Europe

In the past, European studies have already compared the impact and comparison of cyclists with those who prefer other modes of transport (for example, car, bus, skateboard). However, in addition to research, it is important to have more personal contact and interaction with citizens.

It was the modern Copenhagen city bike rental system that was the first and inspired over 600 cities throughout the world. The network of "smart bikes" has become commonplace for Copenhagen residents so much that except of stations for renting a bike, you can "pick up" the ways as well. It does not make any sense to steal "Smart bikes" because they have a peculiar design that immediately catches the eye of passersby. Already today, bicycles are equipped with electric motors, charging them from the sun or a conventional outlet. They quickly started to be in use by tourists and elderly people to travel to a fair [3].

Such European cities as Paris (1,200 stations), London (570 stations only in the central part of the city), Barcelona, Berlin, Amsterdam, Vienna are famous for their comfortable bicycle infrastructure. The introduction of "smart bikes" in Barcelona has reduced the use of cars in the city by 4% over the year. At the moment, the implemented "smart mobility" in various European cities has already become a necessary connecting part of the transport system with a human creating a comfortable environment where cyclists do not compete with cars for a place in the urban infrastructure but work together.

But the creation of comfortable conditions for using bicycles is not able to reduce the motorization of the city. For example, there is an attempt to unite the bicycle and public transport into a single system in Spanish Seville. Passengers of the city bus station can get a rental bike for free during the whole day using their bus ticket, and students at the University of Seville have discounts of up to 100% on renting a bike during the school year.

In some places where there is not enough space on the roads, car drivers have to give way to cyclists. City authorities of Paris allowed cyclists to drive when it is red light at a number of crossroads. The statistics collected during testing over several years has shown that the throughput of crossroads has increased markedly due to this measure.

Architects together with urbanists came up with a fantastic project for the future of London – a network of elevated bicycle infrastructure over the existing railway lines with a total length of 220 kilometers. The new road junction will not have any intersections with highways, and two hundred exits from the elevated cycle highway to ordinary streets will provide convenient access to new roads. The project seems to be completely unrealistic, but it is quite feasible. It has already been supported by the operators of the railway infrastructure of Great Britain and the city transport service [29].

The most striking example of this is Ljubljana, which in a few years has become “The green capital of Europe”, making a bet on bike lanes and pedestrian zones, which serve as a kind of incentive for choosing a bicycle.

Such projects are planned in Russia as well. The Moscow Cycle Sustainable Territory “Green Ring” will connect largest green parks of the capital with public transport and embankments. The project involved the Dutch specialist in “smart” urban planning, Johan Dipens. In the project he proposes to revise urban planning completely and move to the concept of “smart city” through the introduction of sustainable development of the territory. Equipping the public transport park with “smart bikes”, providing benefits for people refusing cars, holding festivals, lectures on the subject of ecology and further development of “smart mobility” are the main points.

3. Methods

3.1. Survey of respondents

The purpose of this article is to analyze the motivation of citizens to use public bicycles in the city of Saransk, Russia. Before analyzing and identifying patterns, it is necessary to collect the initial information. This collected data will serve as the basis for any further analysis on the formation of a sustainable urban area.

In order to analyze the interest and to identify the characteristic factors of the availability of the implementation of "smart bikes" in Saransk, the research method includes a survey in social networks, statistics and questioning in the rental points of bicycles. Following the example of the program “архитекторы.рф”.

In the summer the city residents actively use bikes to enjoy their free time. The situation changes with the beginning of rains and cold snaps. Social request for bikes disappears. The survey took place in the warm season. More than 1400 residents of the city who rented bicycles were surveyed.

The survey questions included the following topics: environmental pollution reduction, joint movement of bicycles with urban and private vehicles, quality of life, economy, environmental friendliness, the desire to use the bicycle as the main mean of transport, the lack of bicycle infrastructure, the distance of possible cycling, climate, landscape, clothes for bike rides. Also there were a number of questions about the reasons that may incline a citizen to use the “smart bike” system in the near future. Each of the questions needed to be rated on a 5-point Likert scale [15].

The data collection took place in July 2018 and January 2019 in social networks and at the bicycle stations. The study was voluntary. Respondents were informed that there were no right or wrong answers. Everyone was also informed about the anonymity of the answers and that the answers would be considered together and used for scientific purposes. The average time to answer the questions was 10 minutes.

Answers from the online and print versions were combined and counted. The number of the respondents is 1 400, 560 of them were questioned online. A "snowball technique" was used to guarantee a large girth of the respondents. Each respondent was asked to send the poll to other people.

According to the survey results, it was clear that residents more often use the following types of transport: public transport, own car, walking, cycling on their own bike, and riding a public bike. The lowest result in popularity among the respondents is a motorcycle, scooter and skateboard.

This research is important for understanding the scale of requests of various groups of the population, relying on their interests and needs it is possible to reduce the discontent and ensure sustainable development of the urban area.

3.2. Public Support

The development of "smart bikes" was supported by representatives of small trade, a number of sports organizations, students and can contribute to the formation of an attractive, comfortable and safe infrastructure. For example, residents may be directly involved in popularizing a bicycle in and around the city.

Entrepreneurs, in turn, can reduce the financial burden, stimulate the roadway and thereby increase the motivation for "smart bike" sustainability.

This also applies to improving the quality of the "green" infrastructure. Examples are public urban landscaping or gardening communities, providing access to local food products, introducing new environmentally friendly methods for building sustainable development of the territory.

In the middle lane you need to take care of winter trips, but designing heated bicycle paths can scare off the idea because of the financial component. Still premature cleaning of the coating from icing can motivate people to choose a bicycle as a means of transportation.

With the support of the public, people on bicycles can form communities, and by joint methods, through communication, continue to shape the environment on their own. Constant communication among themselves will allow us to determine where the cycling network will take place and to develop "smart mobility".

4. Analysis

4.1. Data Received

Based on shopolog.ru statistics, the demand for the purchase of personal bicycles increased by 2 times in 2018. A total of 1,400 people took part in the survey. In the course of which it turned out that about 35% of students ride a bike. Most of these use it to enjoy the trip (60%); as a way to reach their destination - 25%; to improve your fitness - 15%. The peak of cycling activity occurs in the summer: 9% of respondents ride a bicycle very often and 31% periodically, the rest are completely uninterested in moving around the city on a bicycle. In autumn, the number of functioning bicycle rental points decreases noticeably on the streets of Saransk. Only 2% of respondents use a bicycle as a permanent vehicle.

In winter, 0–1% of respondents continue to use the bike. The sharp decrease in the number of active cyclists in the cold season is due to several factors: weather conditions 86%, lack of necessary infrastructure 78%, fear of hypothermia 63%.

Public bicycles are used for small and medium movements around the city. The average rental time is 1 hour 20 minutes, and an average of 143 trips per day in summer. The highest peak of use is from 12:00 to 18:59 pm, 67% of bicycles are used during this period of time.

39.7% of respondents answered that they get to work or study on a personal car, and by public transport 47.5% and only 12.8% get on foot. When asked about the vehicle they would prefer to get to work / study, 44.2% of the citizens spoke for the bicycle, provided that the weather was favorable.

Cyclists prefer to ride in the company of family and friends 70.3%; 11.3% chose a single trip; 18.4% of respondents were indifferent.

About 57% said they never or rarely change the course of their daily cycling routes. Accordingly, the choice of route is highly dependent on the person's habit. The most frequently chosen factors were the length and time of the required route, as well as the presence of green spaces.

Answers to health-related questionnaire questions showed that citizens more and more often think about a favorable and clean movement component. Most cyclists agree that stress can have a negative impact on health 89%, and a minority will even feel a direct effect 11%. Cyclists tend to be heavily exposed to noise 63%. Consequently, a negative impact on the body is not always perceived immediately, but has a cumulative nature. In addition, a larger number of respondents 84% stated that they are interested and will, if possible, work on reducing urban noise. Also 85% of those who completed the survey, we would like to use the application to plan the route, which makes up the route individually for you, taking into account the state of your health.

In this context, an application for smartphones and inexpensive sensors for bikes prove to be reliable and smart solutions for collecting data about your health and physical activity.

The most important advantage of "smart bikes" is the simultaneous collection of objective and subjective data on place and time using geolocation. In addition, the display of effective tips indicates shortcuts, congestion and the status of the cyclist's pulse. [13].

Most citizens 64.2% approved the initiative to create a park of "smart bikes".
The majority 75.7% of the participants intended to continue using the bike in the city.

The most significant factor for the citizens is "Health and the environment". Because every meter of the path leads to individual and social and environmental benefits, which significantly improve people's health, reduces cash spending and increases the social responsibility of citizens. The health benefit of driving is weight control, reducing the risk of cardiovascular disease.

The second factor - "Social Impact" - dissatisfaction with urban public transport, and adopting for themselves as an example developed European cities that have long since switched to "smart bikes". Social influences on citizens will set an example for others, encouraging the use of a bicycle as a factor associated with moving around the city, which leads to the optimization of a favorable social connection. The social influence factor is directly related to the tendency of users to develop sustainable practices. The behavior of people who are in society based on the principle of conformity, that is, citizens repeat what is considered normal and practiced by others.

4.2. Sharing

Sustainable transport planning emphasizes the long-term development of the city and involves the population and other interested parties to develop and find the best tools, products, efficient services and design new successful models of public "smart" urban transport.

The essence of this unifying program is that the participants are based on the exchange of experience, the purpose of which is to encourage the use of ecological, practical, quiet and healthy public transport.

The system complies with the principle of sharing bikes: people use bicycles according to their needs and with low costs of using them.

This attracts tourists and investors, and this is one of the most important conditions for the formation of a sustainable environment. As a result of the analysis of the conditions of public transport and human functioning, the key factors are: growing expectations of residents regarding the quality of life and taking part in their formation, air quality, demography and health.

The strategy of developing "smart mobility" envisages three directions: informing citizens, developing civil and social relations, and supporting cyclists' initiatives. However, in practice, the proposed actions may be selective in nature, and be considered comprehensively.

The idea of integrated planning and implementation of cycling routes, aimed at achieving long-term goals and creating a single, sustainable mobile cluster, is of concern to all who have encountered the problem of developing urban areas. Changes have occurred in the perception of the necessary tools – the strategy places an emphasis on the need to use innovative methods of consulting citizens and introducing new tools in the framework of e-government, which in turn covers the field of mobile development for smartphones.

5. Conclusion

The collected analysis shows that residents adequately perceive the introduction of a model of a public "smart" urban bicycle. The most important factor was "Health and the environment" - the citizens are united by a sense of contribution to the development of the city, clean air and health. In addition, bicycles take up less space, which affected maneuverability and storage. Each kilometer traveled leads to individual and social benefits, because cycling is an "active" mode of transport that provides significant health benefits and, therefore, reduces the social costs of society. Financial and social benefits motivate well the choice of permanent cycling.

The second factor - "Social Impact" - refers to mobility as a sustainable practice, and includes variables associated with the use of the bike in favor of sustainability.

They are examples to others, encouraging the use of a bicycle as a factor associated with self-presentation, which leads to an optimization of favorable social relations. The "Social Impact" factor is directly related to the tendency of users to develop sustainable practices. The behavior is based on the principle of conformity, that is, citizens practice what is considered normal and practiced by others, despite their understanding of instability.

In addition, the use of a bicycle is an independent factor of behavior. That is, a person under different weather conditions has the choice of using transport, which gives an understanding of freedom.

The results also show that the bike sharing system is well received when it has a high potential for development to become part of and at the same time find its identity. However, public "smart bikes" must be constantly monitored, and its development must be targeted.

Thus, the concept of "smart bike" has the opportunity to develop in Saransk. Sustainable "smart city" can constantly develop and create new social environments with the participation of citizens and the administration.

References

[1] Albino V 2015 Smart city: definitions, dimensions, performance and initiatives *J. Urban Technol.* **22** P 3-21
[2] Allwinkle S 2011 Creating smarter cities: an overview *J. Urban Technol.* P 1-16
[3] Brown T 2009 Instrumentation and motivations for organized cycling: the development of the Cyclist Motivation Instrument *J. Sports Sci. Med.* **8** P 211-218
[4] Bührmann C 2007 New seamless mobility services: public bicycles Cologne: Rupprecht Consult Forschung & Beratung GmbH building innovation ecosystems of smart city *J. Knowl. Econ* **4** P 217-231
[5] Gehl J 2010 *Cities for people Devices* (Washington: Island Press)
[6] De Sousa A 2014 Perception of barriers for the use of bicycles *Procedia -Proc. Soc. Behav. Sci.* P 304-313
[7] Kanter R M and Litow S S 2009 Informed and interconnected: A manifesto for smarter cities *Harvard Business School Working Paper* **9** P 1-27
[8] Cowley R 2017 The smart city and its publics: Insights from across six UK cities *Urban Research & Practice* **11** P 1-25
[9] Lee D 2013 An integrated service-device-technology roadmap for smart city development *Technol. Forecast. Soc.* **80** P 286-306
[10] May A 2013 Balancing prescription and guidance for local transportation plans *Proc. Inst. Civ. Eng.* P 36-48
[11] Meschik M 2012 Sustainable cycle tourism along the Danube cycle route in Austria *Tour Plan Dev.* P 41-56

[12] Midgley P 2011 *Bicycle-sharing Schemes: Enhancing Sustainable Mobility in Urban Areas* (New York: Commission on Sustainable Development) P 2-13
[13] Pojani D 2015 Sustainable urban transporting the developing world: beyond megacities *Sustainability* **7** P 784-805
[14] Watson R 2014 *Future 50 ideas to know about* (Moscow: Phantom Press)
[15] Fishman E 2014 Barriers to bike sharing: an analysis from Melbourne and Brisbane *J Transport Geogr* **41** P 325-337

IOP Conf. Series: Materials Science and Engineering **972** (2020) 012025 doi:10.1088/1757-899X/972/1/012025

The Use of Municipal Solid Waste as Secondary Energy Resources on the example of the Housing Complex Novopatrushevo, Tyumen

K V Afonin[1], T S Zhilina[1], A A Zagorskaya[2], M N Pavlova[1], A N Shchekin[2]

[1] Department of Heat and Gas Supply and Ventilation, Tyumen Industrial University, Lunacharskogo st.,2, Tyumen, 625001, Russia

[2] Department of Industrial Safety, Tyumen Industrial University, Lunacharskogo st.,2, Tyumen, 625001, Russia

E-mail: pavmashanic@gmail.com

Abstract. The article considers the feasibility of using biomethane, obtained on the basis of municipal solid waste (MSW) for the production of thermal energy. Research suggests the possibility of using biogas directly at the place of generation. The article also proposes a solution for the processing of numerous sludge of household waste water, which allows to improve the operating conditions of domestic sewage networks. The authors carry out detailed calculations for a residential complex under construction in the city of Tyumen. This paper determines the energy efficiency of a selected waste management scheme.

1. Introduction

Since the last century, the volume of production has grown significantly, which has stimulated the introduction of new technologies in all spheres of human life. This was facilitated by engineering, chemical and biological sciences. On their basis, technologies were developed and put into operation from the production of durable materials to the processing of raw materials. Modern technological progress, thanks to the synthesis of engineering and ecology, allows the use of waste as a useful raw material, transforming them into thermal or electrical energy.

According to statistics, about one ton of waste accounts for every person on the planet Earth every year. If we add up, we get 7.7 billion tons per year. The content of the Tyumen region in the total amount of waste is about 2.5 million tons, of which only 73.5% have been neutralized [2,4]. The main method of disposal of industrial and municipal waste on the territory of the Tyumen region are landfills, of which only 74% meet sanitary and environmental standards. The rest of the waste is located in unauthorized places.

To date, municipal solid waste (MSW) is a serious problem for Tyumen, due to their significant volumes (up to 56% of the total waste) [1, 11]. The development of the waste management system is stagnated by the inertia of the entire HLI system, as well as by the low culture of the population in the field of waste management. Therefore, MSWs are decomposed outdoors at landfills within the city or in the suburbs.

IOP Conf. Series: Materials Science and Engineering **972** (2020) 012025 doi:10.1088/1757-899X/972/1/012025

As a way out of the current environmental problem, it is proposed to process waste that is part of MSW into biogas obtained by methane fermentation of biomass. Biomass digestion occurs under the influence of 3 types of bacteria. In the food chain, subsequent bacteria feed on the waste products of the previous ones. The first type is hydrolytic bacteria, the second is acid-forming, the third is methane-forming. The composition of biogas: 55% -75% methane, 25% -45% CO_2, minor impurities H_2 and H_2S. After purification of biogas from CO_2, biomethane is obtained, its properties close to natural gas [3,19].

This method is universal because it allows you to dispose of almost all types of waste. As a result of waste processing under anaerobic conditions, another currently relevant problem is solved - the application of organic fertilizers to agricultural land. Namely, the reacted substrate after biogas production can be applied to the fields without any preliminary tests [17,19].

Biogas is used as a fuel for the production of: electricity, heat or steam, or as an automobile fuel. Now biogas plants are used mainly in sewage treatment plants, farms, poultry farms, distilleries, sugar factories, meat processing plants. The use of methanogenesis for the digestion of MSW is the next stage in the development of the production of secondary energy resources [3, 16] .

The use of a biogas plant allows reducing the size of sanitary protection zones many times, as well as reducing CO2 emissions to the atmosphere [12, 20]. In addition, the combustion of biogas can produce both electricity and heat, which leads to savings in resources. This aspect can be taken into account in the construction of autonomous energy sources in remote regions of Siberia and the Arctic. Biogas plants can be placed in any regions of our country, they do not require the construction of expensive gas pipelines of complex configuration.

2. Materials and methods

In our case, the biogas plant is used as equipment for generating thermal energy. This facility will be able to provide with heating an entire building or a small area. The residential complex "Novo-Patrushevo" was chosen as the object for calculation and research.

The raw material for biogas production was a mixture of sorted MSW and waste collected from domestic sewage water. The latter are a serious problem for operating organizations, annually causing blockages and serious accidents in sewer networks.

Thus, the extraction and co-processing of household sewage sludge at the exit from buildings can also lead to an increase in the reliability of the sewerage system as a whole.

In the first stage, the recyclable components (plastic, glass, metal, etc.) were recovered from the waste. Sewage sludge underwent decantation, after which the waste mixture was ground. The location of the biogas plant and the boiler house in accordance with the requirements of [8-10, 12, 15, 18] is shown in Figure 1.

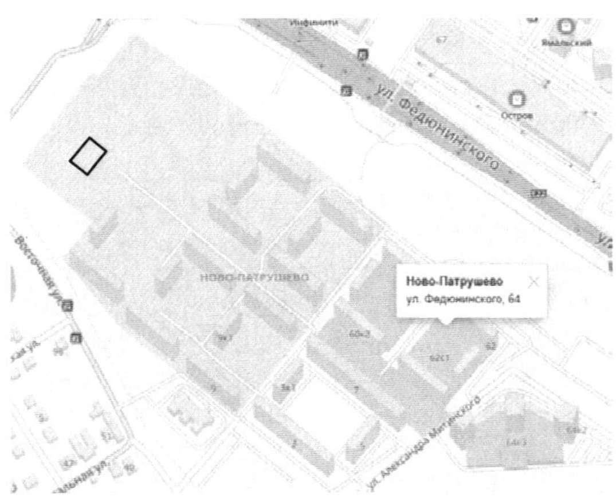

Figure 1. Placement of a biogas plant in the Novo-Patrushevo residential complex

3. Results

Table 1 shows the total cost of thermal energy to ensure the selected complex. The total number was 22896.3 KW (22.89 MW or 19.7 Gcal / h) [5-7,13,14].

Table 2 presents the mass of generated MSW, as well as the volumes of biogas and the approximate amount of thermal energy from its utilization. The total amount of energy produced on the basis of biomethane is 1171.849 kW (1.17 MW or 1.008 Gcal / h).

Table 1 - Calculation of the consumed heat energy

Object name	Total area of building	Average amount of heat loss per m2		Total amount heat	Integrated indicator kWh / m2
		single section	multisection		
st. Fedyuninsky 62 h.1 (GP 31.2)	24995,52	70	58	1449740,16	1971,716
st. Pavla Sharova, 9 h.1 (GP 39.3)	18194,4	70	58	1055275,2	2218,181
st. Pavla Sharova, 9 (GP 39.2)	36262,8	70	58	2103242,4	2218,181
st. A. Mitinsky, 7 (GP 40.1)	87376,32	70	58	5067826,56	1971,716
st. A.Mitinskogo, 5 (GP-40.4)	25484,76	70	58	1478116,08	2218,181
st. A. Mitinsky, 3/1 (GP-40.2)	21871,08	70	58	1268522,64	2218,181
st. A.Mitinskogo, 3 (GP-40.3)	60918,9	70	58	3533296,2	2218,181
st. Fedyuninsky, 60 (GP 32.1)	41772,96	70	58	2422831,68	1971,716
st. Fedyuninsky, 60 h.1 (GP 32.2)	21871,08	70	58	1268522,64	2218,181
st. Fedyuninsky, 60 h.2 (GP 32.3)	56016	70	58	3248928	1971,716

Table 2 - The amount of biomethane produced on selected objects

Object name	Number of inhabitants	Amount of MSW	Amounts of scums	Methane amount tons per day		Produced thermal energy, MJ
		tons per day	tons per day	MSW	Scums	
st. Fedyuninsky 62 h.1 (GP 31.2)	448	0,577	0,023	0,724	0,188	249,9944
st. Pavla Sharova, 9 h.1 (GP 39.3)	392	0,505	0,020	0,633	0,165	218,7451
st. Pavla Sharova, 9 (GP 39.2)	882	1,136	0,045	1,425	0,371	492,1764
st. A. Mitinsky, 7 (GP 40.1)	1344	1,731	0,069	2,171	0,565	749,9831

st. A.Mitinskogo, 5 (GP-40.4)	392	0,505	0,020	0,633	0,165	218,7451
st. A. Mitinsky, 3/1 (GP-40.2)	392	0,505	0,020	0,633	0,165	218,7451
st. A.Mitinskogo, 3 (GP-40.3)	1470	1,893	0,076	2,375	0,618	820,294
st. Fedyuninsky, 60 (GP 32.1)	840	1,082	0,043	1,357	0,353	468,7394
st. Fedyuninsky, 60 h.1 (GP 32.2)	392	0,505	0,020	0,633	0,165	218,7451
st. Fedyuninsky, 60 h.2 (GP 32.3)	1008	1,298	0,052	1,628	0,424	562,4873

Compensation of heat losses of each building in the complex due to the use of biogas is shown in Fig.2.

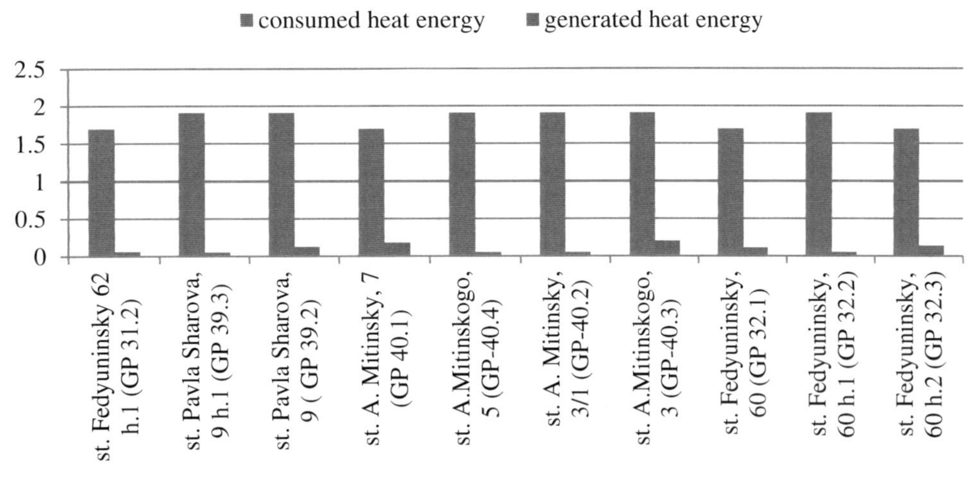

Figure 2. Comparison of the required and produced secondary heat energy, Gcal per day

4. Conclusion

The evaluation of the results of the calculation of heat energy consumption and its production on the basis of biomethane allow us to draw the following conclusions:

- The use of biogas makes it possible to compensate for an average of 5.56% of the thermal costs of supplying a residential complex during the period of stable methane generation (Fig. 2)
- Placement of facilities for methanogenesis in residential areas is possible and meets technical, sanitary and environmental requirements.
- The proposed method allows you to dispose of up to 95% of municipal solid waste and solve the problem of placement and operation of landfills in cities.

Reference

[1] Report "On the environmental situation in the Tyumen region in 2017" 2018 (Department of subsoil use and ecology of the Tyumen region)

[2] Afonin K V, Zhilina T S, Zagorskaya A A 2017 Reduction of environmental impact of solid domestic landfills of residential area due to their recycling. *MATEC Web of Conferences. "International Science Conference SPBWOSCE-2016 "SMART City""* (In Russ) p 3
[3] Shchukina T V 2012 Biogas – prospects and manufacture possibilities. *News of universities. Applied chemistry and biotechnology* (In Russ)
[4] Afonin K V, Zhilina T S, Zagorskaya A A 2014 Calculations of emissions solid waste landfills. *Basic Research* (In Russ) p 987-988
[5] Livchak V I 2005 Justification of the calculation of specific indicators of heat consumption for heating of multi-storey residential buildings *ABOK* **2** (In Russ)
[6] SP 124.3330.2012. Heating network (Moskow: Ministry of Regional Development) (In Russ)
[7] SP 60.3330.2012. Heating, ventilation and air conditioning (Moskow: Ministry of Regional Development) (In Russ)
[8] SP 89.13330.2016. Boiler installations (Moskow: Ministry of Regional Development) (in Russian)
[9] SP 42.13330.2016. Town planning. Planning and development of urban and rural settlements (Moskow: Ministry of Regional Development) (In Russ)
[10] SP 41-104-2000. Designing of autonomous sources of heat supply (Moskow: Ministry of Regional Development) (In Russ)
[11] Resolution of the Tyumen Regional Duma No. 775 of December 20, 2012 (On the information of the government of the Tyumen region on the implementation in the Tyumen region of the federal law "On production waste and consumption"). (In Russ)
[12] GOST 53790-2010 Unconventional technology. Energy biowaste. General technical requirements for biogas plants (Moskow) (In Russ)
[13] Guidelines for calculating the quantitative characteristics of emissions of pollutants into the atmosphere from landfills for solid household and industrial waste 2004 (Moskow) (In Russ)
[14] Guidelines for the study of the properties of solid household and industrial waste 1995 (Moskow)
[15] SP 2.1.7.1038-01. Hygienic requirements for the design and maintenance of landfills for household waste. (Moskow: Ministry of Regional Development) (In Russ)
[16] Eder B, Schulz H 1996 Biogas plants. Practical guide (Switzerland: Zorg Biogas)
[17] Latysheva N I 2009 Scientific substantiation of the ecological and hygienic scheme for assessing the functioning of landfills MSW. Public health and habitat. (In Russ)
[18] SP 131.13330.2012. Building climatology (Moskow: Ministry of Regional Development) (In Russ)
[19] Afonin K V, Zhilina T S, Zagorskaya A A 2017 Recycling of residential areas in order to obtain secondary energy and material resources *Bulletin of FEFU Engineering School* (In Russ)
[20] Renewables 2016. Global status report. URL: http://www.ren21.net/ – 08.08.2017

Alternative Energy Sources in Resolving Environmental Problems and Providing Safety of Single-Industry Towns

M B Permyakov[1], T V Krasnova[1]

[1] Department of construction production, Nosov Magnitogorsk State Technical University, Lenin Prospect 38, Magnitogorsk, 455000, Russia

E-mail: permyakov.1965@mail.ru; toma.krasnova.70@mail.ru

Abstract. Environmental protection and reasonable use of natural resources stipulates for shaping consumer culture correcting the way of life of a modern society in compliance with the environmental requirements of modern life. It is necessary to introduce the measures efficiently regulating economic activity and orienting it towards environmentally correct forms. It is generally known that the key factor of the further civilization development is an efficient use of energy sources. In this connection, a reasonable way of developing national economy is the replacement of traditional "non-renewable" energy sources by alternative ones. The way of the economic development is the search for new ways of providing the national economy with a sufficient amount of energy with a corresponding reduction of the general volume of its use.

1. Introduction

Many states in the world strive for a reasonable and environmentally correct use of energy resources. For example, the energy capacity of the gross domestic product in Europe for 10 years of a new millennium from 252 kg/1000 $ (in 2000) reduced to 179 $ (in 2010) [1]. In Great Britain some companies successfully apply the system of target energy monitoring while in the European Union there are the following market mechanisms in place: system of "green certificates" (encouraging companies to use renewable energy sources) and the system of "white certificates" (encouraging the reduction of energy sources consumption and greenhouse gas emission) [2].

Energy sources are traditionally divided into non-renewable (traditional) and renewable (non-traditional, alternative). Non-renewable sources include: coal, oil, gas and atomic energy. Non-traditional (alternative) and renewable sources are: biomass, wind, ocean, sun, water motion, geothermal energy, secondary energy sources.

According to preliminary estimates, the potential capacity of non-renewable resources in Russia is as follows: mineral coal – 86 %, oil – 2 %, gas – 2.5 %, nuclear fuel – 8.5 %. Renewable resources have the following potential capacity: hydroenergy – 57 %, solar energy – 7.5 %, wind energy – 18 %, biomass – 17.5 % [3].

2. Rationale

As of today traditional energy sources are generally used, such as: oil, coal, natural gas. However, the growth of global economy can be limited because of the lack of energy sources and continuously growing prices for them [4]. "The environmental consequences of traditional energy sources use cause an urgent problem". The planet population and energy consumption keep growing. "The general

IOP Conf. Series: Materials Science and Engineering **972** (2020) 012026 doi:10.1088/1757-899X/972/1/012026

energy consumption for the following 15 years is going to grow more than 1.5 times". In future the use of traditional energy sources (mostly heat-electric generation plants), hazardous substances emissions can reach disaster proportions [5].

The energy capacity of the gross national product in Russia is 2-4 times higher than the one in the USA and Western Europe, the efficiency of use of electric&heat energy is lower, correspondingly, by 2-6 times. Energy saving is the way to natural energy preservation. In the modern world the term "energy efficiency" is being increasingly used to denote a type of cost efficient activity directed at reasonable use of fuel&energy resources with the "upgrade of existing technologies and compliance with the requirements set to environment protection". Today the companies successfully apply such energy sources as solar batteries, biological gas units, mini-hydroelectric plants, wind generators. "People strive to use alternative energy sources to gain the maximum possible amount of energy from renewable resources" [6]. The critical point in renewable energy use was reached in 2015 in the USA. There the cost of electric power energy kW manufacture from solar electric plants and wind electric plants (taking into account subsidy assistance) was equal to the level of electric power production at burning of fossil fuel types. However, even without public subsidies the cost of a kWh of solar electric plants would be equal to 7.2 cents, wind electric plants – 3.7 cents, comparing with 6.1 cents at gas energy and 6.6 cents - coal energy [7]. Today the share of renewable energy sources in the world is equal to 9.3 % while the largest share is taken by hydroelectrical plants – 6.8 % of global energy. The other shares account for solar energy, wind energy, geothermal energy and biological fuel. The most developed countries in the area of renewable power generation: Germany, China, Japan, USA and Italy. More than a half of all global energy generated by solar power plants are divided these 5 countries [7]. The European Union sets a task to increase the electrical energy generation by means of renewable energy sources up to 30 % of general energy consumption. In this view solar energy is considered to be the most profitable. The solar energy peculiarity is the possibility to generate heat and solar energy during a whole year round [8]. Solar electric plants are virtually noiseless and do not emit any substances in the atmosphere [5; 9].

3. Research objective

To consider the possibility of applying alternative energy sources in resolving environmental problems and providing safety of single-industry towns.

4. The theoretical part

A single-industry town is a populated area depending on one or several crucial companies. The exceptions can include: town - transport nodes, museum towns, science cities, resort cities/towns which can also have a status of single-industry towns without large industrial companies located there [10]. It is a common situation when Russian single-industry towns directly depend on crucial companies in these towns. The strategy of economic development, energy provision, development of social infrastructure, territory environmental state - all these factors, as a rule, result from the activity of such crucial company.

Consider a typical example of a Ural single-industry town – Magnitogorsk. The crucial Magnitogorsk company is MMK - Magnitogorsk Iron and Steel Works. The energy complex of MMK includes: a heat electric generation plant, central power plant and air-vapor electric plant. In addition, this complex includes a workshop of electric grids and substations, steam power and gas, water supply and energy workshop, the center of energy-saving technologies and electrotechnical laboratory. Energy from electric power plants provides 70% of in-house electric energy for MMK, and the "net cost of it is virtually two times lower in comparison with a purchased one, these plants fully meet the company's demands in heat energy - heat and vapor, and also cover the city heat demands by 70 %" [11]. The main natural source of energy used by the company is coal. It should be mentioned that in Russia each second kWh is produced with the use of coal while in the European part 90 % of electric power is generated on the basis of gas [12]. The technology of coal burning applied by modern industrial plants, in particular, coal-dust firing and coal burning in the fluidized bed, has a number of

IOP Conf. Series: Materials Science and Engineering **972** (2020) 012026 doi:10.1088/1757-899X/972/1/012026

flaws. One of the main flaws are the presence of gold slag dumps and ash occurrence in the atmosphere. Such gold slag dumps contain: SiO_2, Al_2O_3, CaO and many other substances [13]. Polluted aerial environment conditions a large number of diseases: various types of dust diseases, bronchial asthma, various kinds of dust bronchitis and many others. "In the metallurgical industry pollution of the aerial environment at the work places is quite a critical problem". Metal production is accompanied by "emission of various hazardous gases into the aerial environment" [14].

Modern directions of the environmental policy at industrial companies can be divided into: organizational (providing for a technological procedure, manufacturing discipline, management and control over resource losses); technological (industrial efficiency and re-organization which do not require large investments and made at the expense of current costs); investment (stipulation for a critical reconstruction of technologies requiring large investments). The investment direction of the environmental policy of companies stipulates for "long-term cost demanding measures, for example, introduction of new technologies (biogas, geothermal waters, solar energy, wind energy, heat pump use, etc.)" [15].

The problem of upgrading industrial processes of manufacturing companies are relevant for Russian single-industry towns. The point is not only in increasing the production quality and volumes. Here we speak about the development of the environmental aspect of industrial processes based upon the preservation of natural resources, environmental and industrial safety, human health preservation. Indeed, the application of renewable energy sources on industrial scale in a single-industry town is hardly achievable now. But it is possible to partially introduce the innovations into conventional economic areas. Today Magnitogorsk and its crucial company "MMK" taken as an example are fully provided with energy from non-renewable sources. The environmental pollution indicators are close to the critical limits. The great part of pollution is caused by production wastes. The application of renewable energy sources, such as solar energy, at manufacturing facilities located in the industrial area and adjacent territories will provide for partial "unloading" of the energy complex "MMK". The application of solar batteries in the infrastructure facilities of the metallurgical plant can become the first step towards the plant upgrade.

5. The practical part

Solar energy can be obtained and applied by means of using solar batteries. As of today there are various kinds of solar batteries: mono-crystalline silicone-based batteries; poly-crystalline silicone-based batteries; belt silicone-based batteries; amorphic silicone-based batteries; film batteries; cadmium telluride batteries. Silicone-based poly-crystalline batteries MSK-200 with the dimensions of 805x1575x40 mm at much lower prices than the mono-crystalline silicone ones due to the content of many admixtures. The useful life of such batteries is approximately 30 years. The relevance of solar batteries use is largely owing to the environment situation and safety as the application of such type of energy obtaining allows one partially stop using non-renewable energy sources. Such transition has a positive impact not only on the environment situation but also on the scientific progress development, for example, a wide introduction of electromotive cars in daily use. Therefore, after some refining of this technology, it can become a fully functional energy source. Solar batteries have an advantage over common electric plants - the batteries can supply energy to adjacent facilities which eliminates the necessity of transporting energy from the power plant to its destination, consequently, the energy losses are minimum. Also, in case of unexpected emergencies the aftermath will not be so critical as in case of electric power plants because the electric power plant failure leaves the whole districts and even cities without energy. The risk of emergency situations at industrial facilities is high, the reliability of their element structures depends on multiple factors. If we consider metallurgical plants, there is always emergency risk [16; 17; 18; 19]. When we use solar batteries a comparatively small number of facilities will be left without energy in case of emergency. Therefore, at optimizing the technology of solar battery use one can obtain quite large energy volumes without significant economic expenditures.

Consider solar batteries in detail. A solar battery consists of several solar converters representing semi-conductors. They directly convert solar energy into electric current. Some of such elements combined in one device and represent a solar battery. The temperature rise has a bad impact on the operation of solar batteries decreasing their capacity. That is why at the installation of batteries on flat surfaces it is necessary to leave a free space at every side f the battery to provide air circulation and heat removal. In addition, for the purpose of decreasing the sensitivity towards the temperature drop at plate manufacture, various admixtures are used such as boron, copper, arsenic, cadmium and others. Snow has positive and negative impact on the operation of solar batteries. The negative impact is explained by the fact that it can become cluttered the space between the battery and surface to which it is fastened and thus prevents ventilation. The positive impact is that the battery absorbs light from all sides, consequently, it can absorb the light reflected by snow with its rear side which leads to the growth in performance. The angle of inclination is important for the efficient operation of solar batteries. The optimal angle depends on the land and varies from 15 to 90 degrees. As a rule, this angle is approximated to the land latitude. According to the solar atlas the southern angle of the sun in the Chelyabinsk region, where MMK is located, is 45° in summer and 9° in winter. That is why the installation of plates should be made under the angle of 30°, therefore, it can cover most of luminous flux for all seasons.

The application of solar batteries at the facilities located in the MMK territory is the idea requiring its development. As of today there is already a project of a multilevel parking in the area of the 5th control post of Magnitogorsk Iron and Steel Works developed by Nosov Magnitogorsk State Technical University.

Consider advantages and disadvantages of solar batteries installation. The advantages include that it is necessary only to install batteries properly, they need no fuel. Automatic regulation and noiseless operation can be also referred to the advantages. Solar batteries do not require turning on and off. Electricity is reserved in special accumulators and thus, energy is available day and night. The battery can occupy a large area of the roof surface making it practical and useful. The advantage of such battery is also a long useful life (not less than 20 years). After 20 years its capacity and amount of processed energy decrease by 70-80 %. As the batteries do not have complex mechanisms and moving parts, they are not subject to break and wear. In addition, the advantages include its continuity (stability) and universal access as sun is virtually everything and its light is free in contrast to fuel. Sunlight is a reliable energy source. The battery will generate electric power every day until the dusk even at gloomy weather (although the amount of energy will be less).

Apparently, this system is not ideal and also has its disadvantages:

- need of quite intense light. The batteries will generate energy in any case, but it is more efficient to locate them at the sites with lots of sunshine (non-shaded areas and on the southern side, if possible);
- in winter the efficiency of batteries decreases;
- quite a large area required for their installation;
- quite a high cost as of today.

One of the most critical disadvantages of solar batteries is that the largest amount of energy is generated at daytime while the highest demand for its energy falls on the evening. This disadvantage is offset by the use of accumulators allowing for preserving energy stores and its use any time.

Using an alternative energy source as the main one for providing the infrastructure object of a metallurgical plant with energy is a new and prospective idea. The students demonstrated a creative approach to the project suggesting the installation of a range of solar batteries on the roof of a future multilevel parking in the MMK territory. These batteries will be able to provide power for the whole facility.

The project of the car parking complex stipulates for the installation of solar panels MSK-200, 805*1575*40. They are installed in a row at the angle of 30 degrees with the working surface oriented towards the south. The panels are located in special fastening stands from stainless steel at the height of 500 mm from the roof surface. It is stipulated to prevent panel covering with snow and for a good operation in winter. The rows of solar batteries are located at the distance of 3,000 mm from each other, it is necessary for panel ventilation, preventing the shadowing of some rows because of the close positioning of others and for the convenience of snow cleaning from the roof.

It is assumed that the generated energy will be used for:

- car parking lightning inside and outside;
- ramp heating in winter;
- electrical device power supply in the territory.

Under the project the modules of solar batteries are fully functional;

- within a temperature range -50°C to +75°C;
- within the range of atmospheric pressure up to 85-107 kPa;
- within the range of relative humidity 0-100 %;
- at maximum rain intensity up to 5mm/min;
- at maximum load snow+wind up to 2,000 Pa.

6. Conclusions

The application of solar batteries at infrastructure facilities of an industrial plant allows gradual upgrade of industrial processes and the economic organization orienting the vector of its development towards a modern and relevant direction. The use of alternative energy sources facilitates shaping of consumer culture correcting the way of life of modern society in compliance with the environmental requirements of the time.

The application of new technologies and materials in construction and reconstruction of the infrastructure facilities of industrial companies will probably set the problem of searching for architectural&designer solutions which, in their turn, can encourage young architects to generate creative ideas. Using alternative sources as design elements of the architectural environment will be supported for the application of additive construction technologies. In any case all the aforementioned will contribute to the development of the architectural environment of single-industry towns modifying them and making them modern and more attractive for the youth [20; 21; 22; 23]. Searching for the options of applying alternative energy sources will facilitate the process of solving environmental problems and safety issues of industrial companies as well as the development of single-industry towns in compliance with modern requirements.

References

[1] The concept of long-term socio-economic development of the Russian Federation until 2020 [Electronic resource] materials of the VIII International student electronic scientific conference "Student scientific forum" URL-address: http://base.garant.ru/194365/ (date of application: 24.10.2014)

[2] Bashmakov I 2009 *Economic Issues,* Russian resource of energy efficiency: scale, costs and benefits [Electronic resource] URL-address: https://doi.org/10.32609/0042-8736-2009-2-71-89 No. 2. P. 71-89.

[3] Aglukon I H et al 2015 *Energy and resource saving in construction* Kubelkova G V, Voronin K M (Magnitogorsk: Magnitogorsk state technical University.UN-TA im. G. I. Nosova, S.)

[4] Strebkov D S 2009 *Matrix solar cells In 3 volumes. Volume 1* (Moscow: GNU All-Russian research Institute of agricultural electrification)

[5] Toshkova D V 2017 Renewable energy sources *Collected articles of International scientific-practical conference: Scientific mechanisms for the resolution of problems of innovative development* (Ufa: AETERNA) P 156-158

[6] Kharitonova O D 2015 Non-traditional and renewable energy sources *Efficient use of energy resources as a factor in the development of the Russian economy* (Ekaterinburg : UrFU) **Vol 1** P 240-243

[7] World Bank open data [Electronic resource] URL-address: https://data.worldbank.org/ (date of application: 24.10.2014)

IOP Conf. Series: Materials Science and Engineering **972** (2020) 012026 doi:10.1088/1757-899X/972/1/012026

[8] Batrakov A V and Rtischeva A S 2013 The Use of solar collectors for heating private homes *Energy and resource saving. Energy supply. Non-traditional and renewable energy sources* (Ekaterinburg: UrFU) P 340-342
[9] Shilov I A 2016 *Ecology* (Yurayt)
[10] Permyakov M B and Krasnova T V 2018 Conceptual design of Russian modern monotowns' architectural space *IOP Conf. Ser.: Mater. Sci. Eng.* **451** 012153
[11] Teplova M 2016 *Energy of creation. The energy management Department of the Magnitogorsk metallurgical combine was 85 years old* [Electronic resource] URL-address: http://magmetall.ru/contribution/19797.htm (date accessed: 17.04.2019)
[12] Cherepanova O A, Khasanov R R, Semenov NA and Bogatova T F 2015 Global trends of coal energy development *Energy and resource saving. Energy supply. Non- traditional and renewable energy source* (Ekaterinburg: UrFU) **Vol 1** P 259-260
[13] Zakharov R V, Kalmykova E A and Kartavtsev S V 2015 Clean coal thermal power plant with the use of technology of electroslag melt *Energy and resource saving. Energy supply. Non-traditional and renewable energy sources* (Ekaterinburg: UrFU) **Vol 1** P 116.
[14] Gulov D Yu, Kosimov B I and Sultanov O O 2017 Calculation of prevented economic damage from diseases associated with air pollution in the workplace in industry and agriculture *Priority areas of energy development in agriculture* (Kurgan: publishing House of Kurgan state agricultural Academy) P 106-113
[15] Kordyukova L S and Baranova T S 2015 Energy Saving as a basis of effective economy of the country *Energy and resource saving. Energy supply. Non-traditional and renewable energy sources* (Ekaterinburg, December 16-19, 2014) (Ekaterinburg: UrFU) **Vol 1** P 145-148
[16] Permyakov M B, Ilin A N, Andreev V M, Voronin K M and Krasnova T V 2018 Assessment of reliability and accident risk for industrial buildings *MATEC Web of Conferences* **251** 02007 DOI https://doi.org/10.1051/matecconf/201825102007
[17] Permyakov M B 2014 Methods of building residual life calculation *Advances in Environmental Biology* **8(7)** P 1983-1987
[18] Permyakov M B 2014 Building residual life calculation at hazardous production facilitie *Advances in Environmental Biology* **8(7)** P 1969-1973.
[19] Yeriomin K I, Nischeta S A, Permiakov M B., Naschiokin M V, Alfiorov D L and Ilyina O Yu 2000 Study of the state of stress and strain of cyclically loaded metal structures by the tensometric method *Nondestructive Testing and Evaluation* **16(1)** P 1-14
[20] Krasnova T V and Permyakov M B 2019 Creative approaches in the formation of the image of the urban environment by means of architecture and design *Modern science-intensive technologies* **2** P 89-93.
[21] Permyakov M B, Krasnova T V and Dorofeev A V 2018 Additive technologies in construction and design of architectural environment: present and future *Actual problems of modern science, technology and education* **Vol 9(2)** P 2-5
[22] Krasnova T V and Permyakov M B 2018 Technology of Developing an Artistic Character in Design and Architecture Using Graphic Design Concepts *Modern problems of science and education* **1** P 11 (in Russian)
[23] Permyakov M B and Chernyshova E P 2013 Architectural town-planning factor and color environment *World applied sciences journal* **Vol 27(4)** P 437-443

Energy Efficiency Increasing of Premises through Type of Illuminating Lamps

Y A Ivanov[1], A Y Morozov[1], E A Malyar[1] and E V Ovchinnikova[1]

[1]Ural Federal University, 19 Mira Street, Ekaterinburg 620002, Russia

E-mail: e.a.maliar@urfu.ru

Abstract. The study is considered issues of energy efficiency improving of exhibit and other halls with a large number of people. In such cases maintenance of required parameters of intensity of illumination is accompanied by heat emissions from the lighting, constituting significant part of the total heat gain in the room that increases load on the system of microclimate maintenance. By the example of an exhibit hall in a multifunctional complex it is shown that replacement of incandescent lamps for energy-saving lamps allows to reduce energy consumptions for lighting and also capital and operating expenses on the microclimate control systems. At the same time, temperature and humidity conditions and air velocity are supported, optimal conditions for staff, equipment and machinery are created.

Key words: Energy efficiency of premises; lighting; microclimate; air changes; economic comparison; energy-saving.

1. Introduction

The premises of the exhibition, concert and conference halls are connected with the staying of a large number of people. Therefore, it needs to maintain certain level of lighting and microclimate parameters (temperature and humidity conditions and air velocity).

For such premises, heat emissions from the lighting constitute significant part of the total heat gain and negatively affect the indoor microclimate state.

In such cases, the replacement of incandescent lamps to energy-saving is one of the measures aimed to the room energy efficiency improving, maintenance of comfortable microclimate parameters and creating of optimal conditions for the staff, equipment and machinery [1]. At the same time heat emissions into the room are reduced, that allows to decrease energy consumptions as for lighting as for preparing and distributing of supply air.

2. Object of research

An exhibit hall in the multifunctional complex with the total area is 1675 m2 (with administrative premises) with 168 people staying. It was chosen as the research object. Indoor air temperature for cold period of year (CP), transient conditions (TC) and warm period (WP) equals twz=+20°C.

This article is considered two options of hall lighting: the first is energy-saving lamps, the second is incandescent lamps. And air changes for these options are compared.

Option 1: Energy-saving lamps (EL).

1. Heat gain from artificial lighting is defined as:

$$Q_l = E \cdot F_r \cdot q_l \cdot \eta_l \quad (1)$$

where E - normalized intensity of lighting, accepted according [2]; for exhibit halls $E = 300\ lx$;
F_r- room square, m^2, $F_r = 1675\ m^2$;
q_l- specific heat emission per 1 Lux, $W/(lx \cdot m^2)$, accepted $q_l = 0.038\ W/(lx \cdot m^2)$, according [2];
η_l- part of heat entering to the room, accepted $\eta_l = 1$ as heat source is indoors, [2].

$$Q_l = 300 \cdot 1675 \cdot 0.038 = 19100\ W$$

2. Heat gains from people, W:

Sensible heat emission:

$$Q_s = q_s \cdot n \cdot k \quad (2)$$

where q_s – sensible heat emission from one person, W, accepted according [3].
n – number of people in the room;
k – corrective factor k =1, the calculation is carried out at maximum heat emission.

$$Q_s = 105 \cdot 168 \cdot 1 = 17640\ W$$

Total heat emission is determined by analogy: Q_t=34440 W
Moisture emission: W = 23520 g/h
Carbon dioxide emissions: G = 12600 g/h
3. Heat gains from solar radiation are calculated according [3] and equal $Q_{s.r.}$=16230 W
Received heat gains are shown in the table 1.1.

Table 1.1. Calculation of heat emissions using energy-saving lamps

Period of the year	Heat emissions, W						Pollution emissions, g/h	
			From people		Calculated		W	CO_2
	$Q_{s.r.}$	Q_l	Q_s	Q_t	Q_s	Q_t		
1	2	3	4	5	6	7	8	9
CP	-	19100	17640	34440	36740	53540	23520	12600
TC	-	19100	17640	34440	36740	53540	23520	12600
WP	16230	19100	17640	34440	52970	69770	23520	12600

Option 2: Incandescent lamps (IL).

1. In such case $q_l = 0.067\ W/(lx \cdot m^2)$ and value of heat emission from incandescent lamps is:

$$Q_l = 300 \cdot 1675 \cdot 0.067 = 33670\ W$$

The values of other heat gains and pollution emissions will remain unchanged. The calculation results are shown in the table 1.2.

Table 1.2. Calculation of heat emissions using incandescent lamps

Period of the year	Heat emissions, W						Pollution emissions, g/h	
			From people		Calculated		W	CO_2
	$Q_{s.r}$	Q_l	Q_s	Q_t	Q_s	Q_t		
1	2	3	4	5	6	7	8	9
CP	-	33670	17640	34440	51310	68110	23520	12600
TC	-	33670	17640	34440	51310	68110	23520	12600
WP	16230	33670	17640	34440	67540	84340	23520	12600

IOP Conf. Series: Materials Science and Engineering **972** (2020) 012027 doi:10.1088/1757-899X/972/1/012027

In this way incandescent lamps replacing on energy-saving lamps decreases heat emission from lighting for 14.5 kW.

Air change calculation for considered variants of lighting

Option 1: Energy-saving lamps (EL)

1. Air change rate for assimilation of sensible heat excess, L_h m^3/h, is determined as:

$$L_h = \frac{3.6 \cdot Q_h}{1.2 \cdot (t_l - t_{in})} \quad (3)$$

where Q_h – excess of sensible heat emission, W;
t_l, t_{in}– temperatures of supply and exhaust air, °C.
CP:

$$L_h = \frac{3.6 \cdot 36740}{1.2 \cdot (23 - 18)} = 22040\ m^3/h$$

TC: $L_h^{TC} = 13780$ m^3/h
WP: $L_h^{WP} = 31780$ m^3/h

2. Air change rate for assimilation of total heat excess, L_{hf} м^3/ч is determined as:

$$L_h = \frac{3.6 \cdot Q_{hf}}{1.2 \cdot (I_l - I_{in})} \quad (4)$$

where Q_{hf} – excess of total heat emission,W;
I_{in}, I_l - enthalpy of supplied and exhaust air, J/kg.

Cold period

t_{ext}=-32°C
I_{ext}=-31.6 kJ/kg
$I_{вв}$=t+(1.96t+2500t)d
$-31.6 = -32 + (1.96 \cdot 32 + 2500) \cdot d_{ext}$
$d_{ext} = 1.64 \cdot 10^{-4} kg/kg = 0.17 g/kg$
t_{in}=18°C
t_l=23°C
$Q_{hf} = 53540$ Вт $= 53{,}54\ kW$
$W = 23520\ g/h = 23{,}52\ g/s$

Determination of heat and humidity ratio:

$$E = \frac{Q_{hf} \cdot 3600}{W} \quad (5)$$

$$E = \frac{53.54 \cdot 3600}{23.52} = 7648\ kJ/kg$$

Determination of the point characterizing the supply air parameters:
d_{in}=0.17 g/kg
I_{in}=19 kJ/kg

Determination of the point characterizing the exhaust air parameters:
d_l=1.2 g/kg
I_l=26 kJ/kg

$$L_{hf} = \frac{3.6 \cdot 53540}{1.2 \cdot (26 - 19)} = 22910\ m^3/h \quad (6)$$

The process of air parameters change in CP is shown on figure 1.

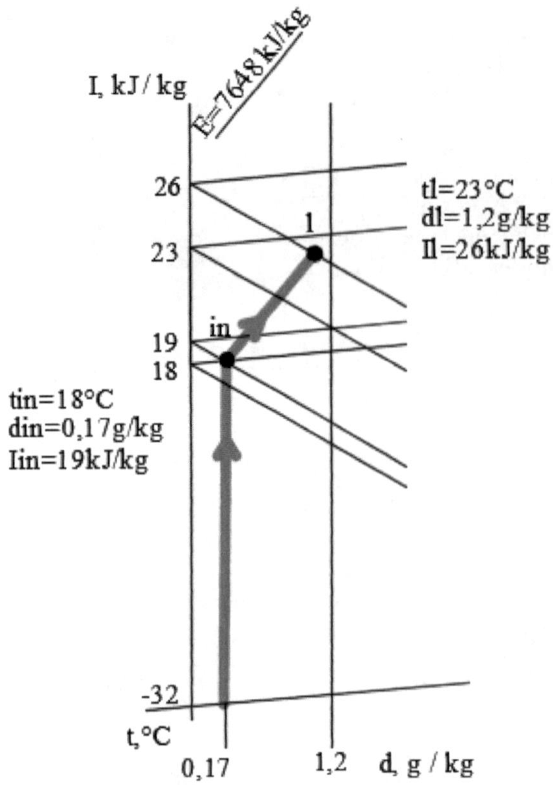

Figure 1. The process of air parameters change in CP.

Air changes rate for TC and WP are determined by analogy.

TC: $L_{hf}^{TC} = 12360$ m^3/h

WP: $L_{hf}^{WP} = 36590$ m^3/h

3. Air change rate for assimilation of moisture excesses, L_w m^3/h, is determined as:

$$L_w = \frac{W}{1.2 \cdot (d_l - d_{in})} \tag{7}$$

where W – moisture excesses, g/h;

d_{in}, d_l – moisture content of supply and exhaust air, g/kg. Values of air moisture content d_{in}, d_l are determined with I-d- diagram of humid air.

$$L_w = \frac{23520}{1.2 \cdot (1.2 - 0.17)} = 19030\ m^3/h$$

TC: $L_w^{TC} = 11530$ m^3/h

WP: $L_w^{WP} = 19600$ m^3/h

4. Air change rate for assimilation of pollutions, L_G m^3/h, is determined according [4]:

$$L_G = \frac{m_{po}}{q_l - q_{in}} \tag{8}$$

where m_{po} – mass of harmful substances entering the room air, g/h;

$q_l = 2g/m^3$ – concentration of harmful substances in exhaust air (for institutions);
$q_{in} = 0.8g/m^3$ – concentration of harmful substances in supply air (concentration of harmful substances is accepted for large cities).

CP, TC and WP:

$$L_G = \frac{12600}{2 - 0.8} = 10500m^3/h$$

The results of air changes rates calculation using energy-saving lamps are shown in table 1.3.

Table 1.3. The results of air changes rate calculation using energy-saving lamps

Period of the year	Air supply volume, m^3/h			
	on excess of sensible heat emission L_h	on excess of total heat emission L_{hf}	on moisture excesses L_w	on the mass of pollutions L_{CO2}
CP	22040	22910	19030	10500
TC	13780	12360	11530	10500
WP	31780	29900	19600	10500

Option 2: Incandescent lamps (IL).

Air change calculation for this option is performed similarly to option 1. The results of air changes calculations using incandescent lamps are shown in table 1.4.

Table 1.4. The results air changes rate calculation using incandescent lamps

Period of the year	Air supply volume, m^3/h			
	on excess of sensible heat emission L_h	on excess of total heat emission L_{hf}	on moisture excesses L_w	on the mass of pollutions L_{CO2}
CP	30790	29190	19600	10500
TC	19240	18580	11530	10500
WP	40520	35720	19600	10500

The maximum value is accepted as calculating.

Option 1: 31780 m^3/h.

Option 2: 40520 m^3/h.

In this way replacing of incandescent lamps to energy-saving decreases air supply rate for 8740 m^3/h. Consequently, it needs smaller air handling unit, also capital and operating expenses are reduced.

3. The research results

Performing economic comparison of variants is based on recommendations [5]. As initial information it is accepted:

- time of electricity consumption per year is τ =2920 hours;
- the number of lighting lamps in the exhibition hall: N=300 pieces;
- incandescent lamp capacity: P_{IL}=60 W;
- energy-saving lamp capacity: P_{EL}=11 W;

- electricity tariff in Yekaterinburg: T_e=3.96 rub/kW·h;
- cost of energy-saving lamp: 95 rubles

Electricity consumption per year is determines as:

$$W = \tau \cdot \mathrm{P} \cdot N \tag{9}$$

$$W_{IL} = 2920 \cdot 60 \cdot 300 = 52560\; kW \cdot h$$

$$W_{EL} = 2920 \cdot 11 \cdot 300 = 9636\; kW \cdot h$$

Energy costs per year:

$$Z = W \cdot \mathrm{T}_e \tag{10}$$

$$Z_{IL} = 52560 \cdot 3.96 = 208140\; rub/year$$

$$Z_{EL} = 9636 \cdot 3.96 = 38160\; rub/year$$

Saving electricity in natural terms:

$$\Delta W = W_{IL} - W_{EL} \tag{11}$$

$$\Delta W = 52560 - 9636 = 42924\; kW \cdot h$$

Energy cost savings:

$$\Delta Z = Z_{IL} - Z_{EL} \tag{12}$$

$$\Delta Z = 280140 - 38160 = 241980\; rub./year$$

Cost of energy-saving lamps:

$$EL = N \cdot 95 \tag{13}$$

$$EL = 300 \cdot 95 = 28500\; rubles$$

The payback period with energy-saving lamps is:

$$\tau_p = \frac{EL}{\Delta Z} \tag{14}$$

$$\tau_p = \frac{28500}{241980} = 0.112\; years = 43\; days$$

4. Conclusion

In this way replacement of incandescent lamps to energy-saving lamps in an exhibit hall leads to energy cost savings and pays for itself in 43 days. Herewith heat emissions from lighting are reduced for 14,5 kW, that allows air change rate to decrease for 8740 m3/h. Consequently, power of microclimate control system, capital and operating expenses are reduced.

This variant was considered by customer and accepted for implementation on the facility.

References

[1] Matiyashchuk S V 2010 Commentary to the Federal Law No. 261 "Energy Saving and Energy Efficiency Improvement and Amendments to Certain Legislative Acts of the Russian Federation" (Moskow: Yusticinform)

[2] SP 52.13330.2011 Natural and artificial lighting (Moskow: Ministry of Regional Development)

[3] Pavlov N N, Shiller Yu I 1992 Reference Designer. Internal sanitary facilities. Part 3 Book 1. Ventilation and air conditioning, - 4th edition (Moskow: Stroiizdat)

[4] Pavlov N N, Shiller Yu I 1992 Reference Designer. Internal sanitary facilities. Part 3 Book 2. Ventilation and air conditioning, - 4th edition, (Moskow: Stroiizdat)

[5] The complex of methodical positions and recommendations on the calculation of economic benefits from the implementation of energy saving in housing and industrial power 2010 (Moscow)

IOP Conf. Series: Materials Science and Engineering **972** (2020) 012028 doi:10.1088/1757-899X/972/1/012028

Features of the calculation of heat losses through the filling of window openings

A D Sukhov[1], M G Ushakov[1]

[1]Ural Federal University, 19 Mira Street, Ekaterinburg 620002, Russia

E-mail: suhov-ad@bk.ru, paritet.m@mail.ru

Abstract. The article is devoted to the calculation of heat losses through the filling of a horse aperture. The scheme of formation of the "cold bridge" through the filling of the window opening was presented. Two window designs were calculated and analysed using numerical simulation. It was calculated to reduce the cost of the project by changing the constructive solution. The article examined various ways to take into account the position of the window opening when calculating heat losses. A formula was proposed for the correct calculation of heat losses through the filling of a window opening. Separately highlighted issues that need to be considered in further research.

1. Introduction

In connection with the depletion of natural energy reserves, the urgency of the problem of saving fuel and energy resources and improving the energy efficiency of buildings increases. To solve this problem, it is necessary to conduct various energy-saving measures: the use of effective and environmentally friendly thermal insulation materials; the use of energy efficient, reliable and durable enclosing structures, reducing heat loss; improving the quality of thermal design of buildings.

Full calculations of the values of reduced resistance to heat transfer of structures, coefficients of heat engineering uniformity, temperatures on the inner surface of the building envelope when designing buildings should be performed based on the calculation of temperature fields. The need of this kind is primarily related to ensuring the safety of the building, preventing freezing, moisture, mold on the inner surface of the enclosing structure in the area of heat-conducting inclusions (exterior corners, joints, window slopes), as well as reduce heat loss through these places. Calculations of temperature fields help to avoid thermal defects in advance (at the design stage).

The computational methods developed decades ago need adjustments that will make them more adapted to variant design solutions. Modern calculation programs that are continually developed, allow to do it.

The development of programs for engineering and scientific calculations allows for more detailed calculations using existing engineering calculation methods. In this material, the ANSYS Steady State Thermal design complex will be used to determine heat losses through the wall with a window opening in the living room.

Determining the heat losses of the premises of any building is the basis for the selection of heating equipment and the calculation of the required power of the heating system, which affects the design solutions for outdoor heating networks. Consequently, the set of subsequent decisions that affect the total cost of the project depends on the accuracy of calculating the heat loss of the building.

This article analyzes the results of the thermal analysis of two design options for installing a window opening.

2. Features of the calculation of heat loss

The given resistance to heat transfer of the window opening in the method of determining heat losses through external enclosing structures is taken according to the thermal calculation [2, app. E]or based on the results of field tests in an accredited laboratory[2, app. K]. On the basis of calculations and field tests conducted by manufacturers of window units, tables of reduced resistances to heat transfer of fillings of window openings were compiled, which, in most cases, designers use. However, these tables do not take into account the various ways of installing a window opening in the wall structure. In fact, this factor will affect the actual amount of heat loss of the room.

If the filling of the window opening begins to shift towards the inner surface of the wall so that between the window and thermal insulation from the outside air a part of the supporting wall of the wall protrudes (Fig. 1), the heat loss of the room to be calculated will increase due to the appearance of a "cold bridge". Moreover, if the window cover will be partially shifted towards the thermal insulation of the outer wall, on the contrary, the heat loss will be less.

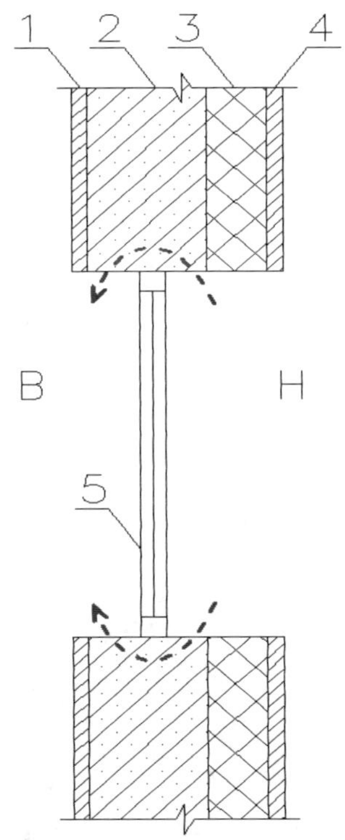

Figure 1. Scheme formation of "cold bridge"
Notation: 1 - internal finishing; 2 - wall bearing layer; 3 - thermal insulation; 4 - outer finishing layer; 5 - window cover; B - internal air; H - outside air.

In order to assess the influence of the installation method of filling the window opening on the total heat loss through the outer wall of the calculated room, we will perform a numerical calculation in ANSYS Steady State Thermal for two options for the location of the window cover (Fig. 2and 3). In the first version, the outer border of the window sash is overlapped with thermal insulation, while in the second it is offset by 50 mm towards the outside air. Both installation options are allowed. The difference in the cost of mounting these options is practically absent.

The calculation was performed for the climatological data of the city of Yekaterinburg for a living room with an outer wall of 6.0 x 3.3 m and one window opening. Outdoor temperature is -32°C [1]. The calculated grid of models was built in ANSYS Meshing. A stationary process was considered.

The temperature fields obtained as a result of numerical calculation of the models are presented in Fig. 4 and 5.

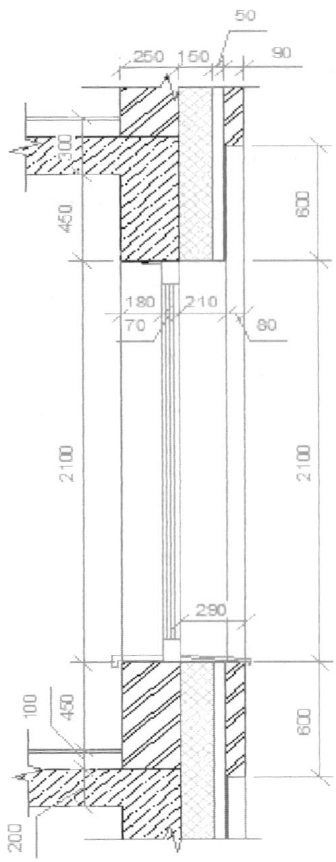

Figure 2. Option №1 install window sash.

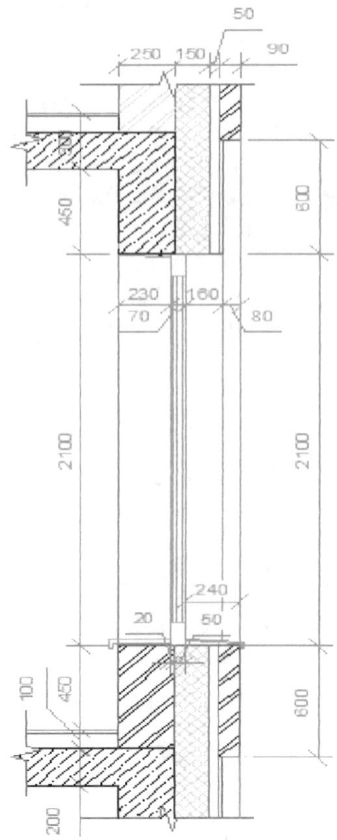

Figure 3. Option №2 install window sash.

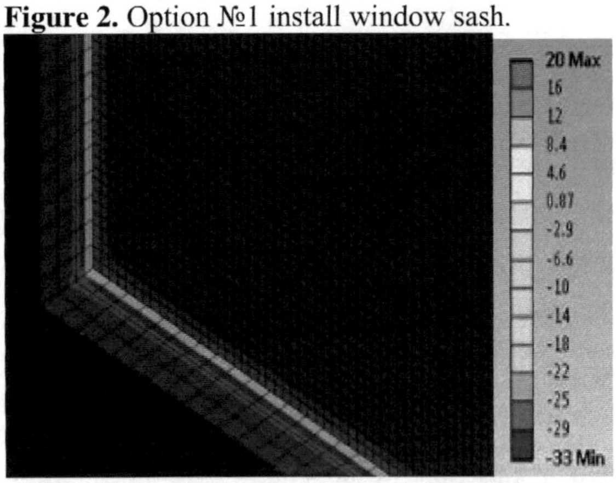

Figure 4. Temperature fields of the window from the outside (option № 1).

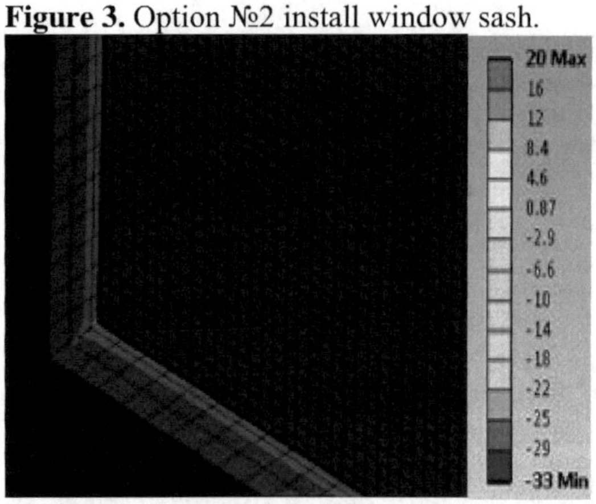

Figure 5. Temperature fields of the window from the outside (option №2).

The calculation in the program showed that the heat loss through the outer wall for the first option is 870 W, and for the second - 730 W. When determining the heat loss using the standard “manual” calculation method, the result is 850 W. The value of heat loss for the first option, obtained in the design program, is close to the value of 850 W. However, when you change the position of the window sash in the opening in the second version, heat loss is reduced by as much as 120 watts.

Let us analyze what this calculation result means. This outer wall construction is taken from the project of a 25-storey residential house under construction. On one floor there are ten windows of this design. Under each window is installed one sectional steel radiator. Reducing the heat output of a sectional heater by 120 W (by about 8-9%) will lead to a decrease in the number of sections by at least one. That is, the total number of sections of all radiators will be reduced by 250 pieces. This will mean significant savings for the developer. In addition, such a reduction in the calculated power of the heating devices will lead to a certain decrease in the diameters of the heating system pipelines. In addition, the selection of equipment for individual heat point will be carried out for a load of less than 30 kW (25 floors × 10 devices × 120 W = 30 kW) because of this unaccounted factor.

Therefore, the location of the filling of the window opening in the construction of the opening must be taken into account in the calculation of heat losses. The current rules provide for the calculation, which is based on the calculation of temperature fields[2, app. H]. But this method is rather complicated for an ordinary designer who is not familiar with the calculation in complex engineering software systems, such as ANSYS. In addition, it seems irrational every time to make calculations of this kind, when most of the designs of the exterior walls of residential buildings are similar.

In addition, a set of rules [3] allows solving this problem.But this solution will be incorrect, since this linear heterogeneity will be taken into account in the construction of the wall, and not the window. That is, heat losses will be determined incorrectly in rooms with a large number of windows and in rooms without them.Also, this method is quite time-consuming, because the designer for the calculation will have to determine the total length of all the window slopes on the facade. Because of these shortcomings, it is not rational to use this set of rules in this case.

The actual heat transfer resistance of filling the window opening will be most easily determined by the following formula:

$$R^{w}_{fact} = R^{w}_{red} \cdot m \tag{1}$$

R^{w}_{fact} –actual resistance to heat transfer filling the window opening,$\frac{m^2 \cdot ^\circ C}{W}$;

R^{w}_{red}– reduced resistance to heat transfer filling the window opening,$\frac{m^2 \cdot ^\circ C}{W}$;

m – correction factor that takes into account the position of filling the window opening in the thickness of the wall.

The correction factor in the formula (1) can be determined by calculations in engineering programs, or by field tests in accredited laboratories. It can be set for each specific position of the window sash, or it can be a function of its actual displacement from a certain standard position.

This method allows the designer to take into account each window in the calculation of heat loss separately and correctly determine the load on the heating for each room.

3. Current issues

This paper leaves a number of questions, the solution of which will lead to an improvement in the quality of design:

1) How to take into account the influence of the position of filling the window opening in the wall on the value of its resistance to heat transfer?

2) What will be the economic effect of a more detailed calculation of the required capacity of the heating system of the building?

3) How does the filling position of the window opening affect the amount of infiltration of outside air into the room?

4) How can I optimize the filling position of the window opening in the wall structure?

IOP Conf. Series: Materials Science and Engineering **972** (2020) 012028 doi:10.1088/1757-899X/972/1/012028

4. Conclusion

This article analyzes the results of the heat engineering calculation of two design options for installing a window opening in a wall. The calculation methods used do not take into account this factor. According to the results of numerical calculations revealed a significant difference in heat loss of the room for these options. This nuance significantly affects the cost of the project. The position of filling the window opening in the wall construction must be considered when calculating. The resulted algorithm of calculation of an ejection stream will allow making exact selection of the ventilating equipment for the accepted technical decision.

There are several open problems associated with the correction factor in the proposed formula. It is important to choose in what form its presentation will be: in the form of a function, in the form of a table or in another form. In addition, in the future it is necessary to conduct laboratory tests for various types of window constructions in order to determine the values of this coefficient.

Improving the quality of building design allows us to save significant amounts of energy resources. That is why it is impossible to neglect the possibility of computer modeling in building construction processes.

Reference

[1] Ministry of Regional Development of Russia 2012 SP 50.13330.2012 Thermal protection of buildings (Moscow: FCS) p 139 [In Russ]

[2] Ministry of Regional Development of Russia 2012 SP 60.13330.2012 Heating, ventilation and air conditioning (Moscow: FCS) p 76 [In Russ]

[3] Ministry of Regional Development of Russia 2012 SP 230.1325800.2015Structures enclosing buildings. Characteristics of heat engineering heterogeneities (Moscow: FCS) p 72 [In Russ]

[4] GOST 17079-88 Residential and public buildings Indoor microclimate parameters (Moscow: Gosstroy of Russia, GUP TsPP) p 14 [In Russ]

[5] Ministry of Regional Development of Russia 2001 SANPiN 2.1.2.1002-00 Sanitary and epidemiological requirements for residential buildings and premises (Moscow: Gosstroy of Russia) p 15 In Russ [In Russ]

[6] Chigarev A V 2004 ANSYS for engineers: a reference manual (Moscow: Mechanical Engineering-1) p 512 [In Russ]

[7] Fedorov N N 2017 Basics in ANSYS 17 (Moscow: DMK Press) p 210 [In Russ]

[8] Malyavina E G 2007 Building heat loss (Moscow: AVOK-PRESS) p 146 [In Russ]

[9] Pilipenko N V 2016 Heat loss and energy efficiency of buildings and structures (St. Petersburg: ITMO University) p 54[In Russ]

[10] Eryomkin A I 2000 Thermal regime of buildings (Moscow: ASV) p 368 [In Russ]

[11] Lykov A V 1967 Heat exchange theory (Moscow: High school) p 600 [In Russ]

[12] Pekhovich A I 1968 Calculation of the thermal regime of solids (Leningrad: Energy) p 304 [In Russ]

[13] Kudinov V A 2005 Analytical solutions of heat and mass transfer and thermoelasticity problems for multilayer structures (Moscow: High school) p 340 [In Russ]

[14] Fokin K F 2006 Building heat engineering of the enclosing parts of the building (Moscow: AVOK-PRESS) p 256 [In Russ]

[15] Bogoslovskiy V N 1979 Building thermal comfort (Moscow: Stroyizdat) p 248 [In Russ]

[16] Samarskiy A A 2009 Computational heat transfer (Moscow: Librokom) p 784 [In Russ]

[17] Tabunshchikov Yu A 2002 Mathematical models of thermal conditions in buildings (Moscow: AVOK-PRESS) p 194 [In Russ]

[18] Patankar S V 2003 Numerical solution of problems of heat conduction and convective heat exchange during flow in channels (Moscow: Izdatelstvo MEI) p 312[In Russ]

Assessment of pollutants discharges and air pollution in the Tyumen region (without considering autonomous districts) and its urban districts

V A Sapega[1]

[1]Department of Technosphere Safety, Industrial University of Tyumen, Lunacharski str., 2, Tyumen, 625001, Russia

E-mail: sapegavalerii@rambler.ru

Abstract. The article provides an analysis of pollutants discharges in the Tyumen region, in the Tyumen region without considering autonomous districts, and in its urban districts for the period of 2013-2017. The researcher compares the number of stationary emission sources throughout the Tyumen region in 2017 and in 2013. The increasing amount of pollutants discharges has also been ranged from the Tyumen region (129083 tons) to its urban districts, namely Tyumen (16890 tons), Tobolsk (13723 tons). The decrease in emissions was noted in Ishim and Yalutorovsk in 2017 compared to 2013. It has been determined that the total amount of discharges in urban areas contained the average portion of solid discharge up to 5.0%, and of gaseous and liquid discharges up to 95.0%. Carbon monoxide and hydrocarbons (without VOCs) prevailed among gaseous emissions in the Tyumen region without considering autonomous districts (29.1 and 26.1% of all gaseous emissions respectively). Testing revealed the increase in the number of trapped and neutralized pollutants in the Tyumen region without considering autonomous districts and including urban districts in 2017 compared to 2013. In the analyzed period, the atmosphere in Tyumen and Tobolsk was characterized as slightly polluted.

1. Introduction

Nowadays one of the vital problems of ecology is the problem of the relationship between man and the environment [1-3]. The main environmental problem in most urbanized areas of this country is the increased level of air pollution [4-6]. Natural ecological processes and habitats are the most distorted in urban areas. The growth of cities and lack of investment in environmental protection worsen the situation [7, 8].

Industries and other types of economic activity of people cause indoor and outdoor air pollution [9-14].

Atmospheric pollutants (gaseous, aerosols, particulate matter) get into the air from both natural and anthropogenic sources [15-19]. Annually the total average amount of pollutant emissions throughout Russia makes up 20.8 million tons from stationary sources, and 13.2 million tons from mobile sources. The main air pollutants are nitrogen oxide and sulfur dioxide. On average, sulfur and nitrogen emissions per unit area amount to 0.38 and 0.12 t / km2 per year from stationary and mobile sources respectively [20].

The risk of damage to human health is the highest of environmental risks. Air pollution is the most serious environmental problem for human health in the near or middle term, as its part, according to various sources, makes up 80 to 90% of all polluting factors [21].

In view of the mentioned above, it is necessary to study the changes in the basic parameters of air pollution in various regions and, in particular, in the Tyumen region and its urban districts. Data from the Report on Ecological Situation in the Tyumen Region in 2017 [22] and the Statistical Bulletin on Environmental Protection in the Tyumen Region for 2013-2017 were used for the research [23].

The area of the Tyumen region (without autonomous districts) is 160.1 thousand km2. The population of the region was 1499.8 thousand people at the beginning of 2018. The average population density is 9.4 persons / km2. The largest cities of the Tyumen region are Tyumen, Tobolsk and Ishim.

2. Results and Discussion

Industry is the most important factor in the development of the Tyumen region economy. But remarkable production growth trends in the key industries do not have a significant negative impact on the environment of the region.

It should be noted that the number of stationary sources of pollutant emissions throughout the Tyumen region was 128413 in the period of 2013-2017, and 21327 units in the Tyumen region without considering autonomous districts, which is 16.6% of the total number throughout the region (table 1). Over the analyzed period, there was observed an increase in the number of stationary sources of pollutant emissions. In particular, 4.700 stationary sources of pollutant emissions appeared in the Tyumen region without considering autonomous districts in 2017.

Air pollutant emissions from stationary sources over the Tyumen region made up 2341123 tons on average over the period of 2013-2017. And 129083 tons were discharged throughout the Tyumen region without considering autonomous districts (5.5% of total emissions in the region) (table 1). In general, there was a decrease in emissions by 414477 tons in the Tyumen region in 2017 compared to 2013, and an increase of 3.842 tons was observed in the Tyumen region without considering autonomous districts.

Table 1. Characteristics of the main indicators of air pollution in the Tyumen region (stationary sources).

Indicator	The Tyumen region			The Tyumen region without autonomous districts		
	2013-2017	%	2017 compared to 2013 ±	2013-2017	% (of the indicators in the Tyumen region)	2017 compared to 2013 ±
The number of stationary sources of pollutants discharge – total, units	128413	100.0	9917	21327	16.6	4700
Air pollutants discharged from stationary sources - total, t	2341123	100.0	-414477	129083	5.5	3842
including:						
solid	88449	3.8	-24621	7221	8.2	36
gaseous and liquid	2252674	96.2	-389856	121862	5.4	3851
of them: sulfur dioxide	11783	0.5*	-	2250	19.1	-

Indicator	The Tyumen region			The Tyumen region without autonomous districts		
	2013-2017	%	2017 compared to 2013 ±	2013-2017	% (of the indicators in the Tyumen region)	2017 compared to 2013 ±
carbon monoxide	899039	39.9*	-	35522	4.0	-
nitrogen oxide	229005	10.2*	-	24293	10.6	-
hydrocarbons (without VOCs)	783796	34.8*		31851	4.1	-
volatile organic compounds (VOCs)	324698	14.4*	-	26573	8.2	
of them (of all emissions) discharged without purification, t	2340211	99.9**	-414604	128482	5.5	3791
trapped and neutralized pollutants, t	129997	5.6**	123774	125464	96.5	116094

* - of gaseous and liquid - total
** - to discharged - total

Judging by the data, gaseous and liquid discharges prevail in total discharges throughout the whole region - 96.2%.

The part of solid discharges in the Tyumen region without considering autonomous districts was 8.2 % of the discharges throughout the whole Tyumen region. Gaseous and liquid ones made up 5.4%. There was registered a decrease by 36 tons in solid discharges as well as in both gaseous and liquid ones in the whole region in 2017 compared to 2013. However, the same period was marked by a slight increase, by 3851 tons, in all of the discharges in the Tyumen region without considering autonomous districts.

Over the given period among the discharged ingredients carbon monoxide prevailed in the Tyumen region - 899,039 tons (39.9% of gaseous and liquid discharges). Following were hydrocarbons (without VOCs) - 783796 tons (34.8% of gaseous and liquid discharges). The same ingredients were most often observed in the discharged of the Tyumen region without considering autonomous districts (35,522 and 31,851 tons, respectively).

The amount of air pollutant emission without purification was 99.9% of all emissions in the Tyumen region. Over the period of 2013-2017 unpurified discharges amounted to 128.482 tons in the Tyumen region without considering autonomous districts that is 5.5% of emissions without treatment in the Tyumen region. Thus, proceeding from the ratio in the Tyumen region without considering autonomous districts, the bulk of all emissions and emissions without purification was observed in autonomous districts Khanty-Mansiysk and Yamalo-Nenets. Untreated emissions both in the Tyumen region as a whole and in the Tyumen region without considering autonomous districts increased by 128.482 and 3.791 tons in 2017 compared to 2013.

The portion of trapped and neutralized pollutants from all emissions is very low and amounts to 5.6% in the Tyumen region as a whole. The emission rate in the Tyumen region excluding autonomous districts amounted to 125464 tons for the period of 2013-2017. In this period the indicator increased both in the whole region (by 123774 tons) and in the Tyumen region excluding autonomous districts (by 116094 tons).

The analysis of the number of stationary sources of pollutant emissions in 2013-2017 showed that the portion of organized sources was 64.1% of their total number in the Tyumen region excluding autonomous districts (table 2).

Table 2. The number of stationary sources of pollutants discharges in the Tyumen region excluding autonomous districts and urban districts in 2013-2017.

Region, urban district	Number of stationary emission sources - total			Of them organized	
	units	%*	2017 compared to 2013 г. ± units	units	of stationary sources - total, %
Tyumen region without considering autonomous regions	21327	100.0	4700	13669	64.1
including					
Tyumen	7212	33.8	1141	4589	63.6
Zavodoukovsk	890	4.2	338	547	61.5
Ishim	796	3.7	-60	464	58.3
Tobolsk	826	3.9	-36	464	56.2
Yalutorovsk	404	1.2	-75	272	67.3
In urban districts on average	2026	9.5	262	1267	62.5

* - of the indicator for the Tyumen region without considering autonomous regions

Tyumen and Zavodoukovsk were registered as urban districts having the majority of stationary emission sources, namely 7.212 units and 890 units respectively. More than half of the stationary emission sources in all urban districts are organized. Their average share in urban districts is 62.5%. In 2017, compared to 2013, the number of stationary emission sources increased in Tyumen and Zavodoukovsk (by 1141 and 338 units, respectively) and decreased insignificantly in other cities.

Tyumen and Tobolsk were top cities among urban districts by the total amount of pollutant emissions in 2013-2017: 16890 tons and 13723 tons respectively (table 3). In terms of proportions, gaseous and liquid emissions prevail - from 84.1% (Ishim) to 97.3% (Tobolsk), while in the Tyumen region excluding autonomous districts their part was 94.4%.

Most of both solid and gaseous and liquid pollutants discharges were also noted in Tyumen (916 and 15,974 tons respectively) and Tobolsk (367 and 13,356 tons respectively).

On average, emissions in urban districts amounted to 6,650 tons. 95.0% of them were gaseous and liquid discharges.

The increase in total amount of emissions including solid, gaseous and liquid discharges in 2017 compared to 2013 was noted in Tyumen, Zavodoukovsk and Tobolsk. In particular, total amount of emissions in the above-mentioned cities increased by 913, 285 and 1650 tons, respectively, and in urban districts - on average by 496 tons.

The largest reduction in pollutant emissions over the given period was noted in Ishim - by 228, 62, and 165 tons (respectively, emissions are total, solid, gaseous and liquid). Analysis of emissions by individual ingredients of pollutants showed that carbon monoxide and nitrogen oxide predominated in gaseous emissions over urban districts. Thus, nitrogen oxide emissions in urban districts were measured from 76 tons (14.0%) in Yalutorovsk to 6849 tons (42.9%) in Tyumen, and on average in urban districts their content in gaseous emissions was only 45.3%.

Sulfur dioxide is the least emitted – 2250 tons (1.8%) in the Tyumen region without considering autonomous districts. Its part in urban districts gaseous emissions is from 0.2% (Yalutorovsk) to 3.1% (Tyumen) only.

IOP Conf. Series: Materials Science and Engineering **972** (2020) 012029 doi:10.1088/1757-899X/972/1/012029

Table 3. Ingredients characteristics of pollutants discharge into the atmosphere from stationary sources in the Tyumen region without considering autonomous districts and urban districts over the period of 2013-2017.

Region, urban district	Discharged - total			Including																Trapped and neutralized pollutants			
				solid			gaseous and liquid			of them										total		of them utilized	
										sulfur dioxide		carbon monoxide		nitrogen oxide		hydrocarbons (without VOCs)		volatile organic compounds					
	t	%*	2017 to 2013 ± *t*	*t*	%*	2017 to 2013 ± *t*	*t*	%*	2017 to 2013 ± *t*	*t*	%**	*t*	%**	*t*	%**	*t*	%**	*t*	%**	*t*	2017 to 2013 ± *t*	*t*	%***
The Tyumen region without autonomous regions	129083	100,0	3842	7221	5.6	36	121862	94.4	3851	2250	1,8	35522	29,1	24293	19,9	31851	26,1	26573	21,8	125464	116094	122096	97,3
including Tyumen	16890	13,1	913	916	5.4	291	15974	94.6	622	499	3.1	4535	28.4	6849	42.9	523	3.3	3354	21.0	64189	71324	62122	87.1
Zavodoukovsk	1367	1,0	285	207	1.1	5	1160	84.9	250	30	2.6	454	39.1	174	15.0	197	17.0	259	22.3	118	27	112	94.9
Ishim	673	0,5	-228	107	15.9	-62	566	84.1	-165	7	1.2	299	52.8	119	21.0	18	3.2	116	20.5	1316	-150	1315	99.9
Tobolsk	13723	10,6	1650	367	2.7	184	13356	97.3	1466	202	1.5	3332	24.9	7086	53.0	242	1.8	1930	14.4	8129	12633	7709	94.7
Yalutorovsk	596	0,5	-137	55	9.2	-1	541	90.8	-136	0.9	0.2	303	56.0	76	14.0	5	0.9	152	28.1	1765	1826	1765	100.0
over urban districts on average	6650	5,2	496	330	5.0	83	6319	95.0	407	148	2.3	1785	28.2	2861	45.3	197	3.1	1162	18.4	15103	17132	14605	96.7

The amount of trapped and neutralized pollutants discharges increased over the analyzed period by 11.6094 tons in 2017 compared to 2013 in the Tyumen region without considering autonomous districts, and by 17.132 tons in urban districts. The most part of captured and neutralized emissions is utilized - 97.3% in the Tyumen region without considering autonomous districts and 96.7% in urban districts.

The value of the atmospheric pollution index (API) for Tyumen and Tobolsk, the two largest cities of the Tyumen region without considering autonomous districts, is presented for the period from 2013 to 2017 (table 4).

Table 4. Atmospheric pollution index and characteristics.

Year	Tyumen		Tobolsk	
	Atmospheric pollution index(API)	Atmospheric pollution characteristics	Atmospheric pollution index(API)	Atmospheric pollution characteristics
2013	7	slightly polluted	2	clean
2014	8	polluted	2	clean
2015	6	slightly polluted	3	slightly polluted
2016	4	slightly polluted	3	slightly polluted
2017	4	slightly polluted	3	slightly polluted

The data reveal that Tyumen atmosphere was mostly characterized as slightly polluted in the analyzed period. At the same time, the pollution index decreased (from API = 7.2013 to API = 4.2017). In Tobolsk a slight reduction in the air pollution indicator was observed over the given period of time. In particular, in 2013-2014. the atmosphere was characterized as clean (API = 2), and in 2015-2017 it was slightly polluted (API = 3).

3. Conclusion

The average number of stationary emission sources in the Tyumen region without considering autonomous districts amounted to 21.327 units (16.6% of total emission sources in the region) in 2013-2017, and among urban districts the majority of them 7212 located in Tyumen. The number of stationary emission sources there increased in 2017 compared to 2013. Ishim, Tobolsk and Yalutorovsk are an exception.

The largest amount of pollutants discharges from stationary sources in 2013–2017 was recorded in the Tyumen region without considering autonomous districts. The rate rose up to 129083 tons (5.5% of total emissions in the region). Tyumen was recorded among the urban districts with the largest amount of emissions, namely 16890 tons, i.e. 13.1% of emissions in the Tyumen region without considering autonomous districts. The amount of emissions increased in 2017 compared to 2013. Ishim and Yalutorovsk are an exception.

Gaseous and liquid discharges prevailed in urban districts. Their range was from 84.1% (Ishim) to 97.3% (Tobolsk). Carbon monoxide and nitrous oxide were predominant elements in gaseous emissions.

In 2017 compared to 2013 the amount of trapped and neutralized emissions increased by 1.16094 tons in the Tyumen region without considering autonomous districts, and in urban districts - by 14.605 tons. Most of the captured and neutralized pollutants were utilized. According to the air pollution index in the analyzed period the atmosphere in Tyumen and Tobolsk was characterized as slightly polluted.

References

[1] Molchanova E V 2009 The impact of environmental and socio - economic factors on the health of the population of the region (the example of the Republic of Karelia) *Ecology of urbanized territories* **4** pp 31-6

[2] Bondarenko V L 2013 Criteria of environmental safety in natural-technical systems "natural environment - object of activity - population" *Ecology of urbanized territories* **1** pp. 39-45
[3] Lopatnikov D L 2006 Ecological Perspectives of the Post-Industrial World (Moscow: AVF) 312p
[4] Bityukova V R 2007 Environmental problems of Russian towns *The Bulletin of Moscow State University Ser. 5 Geogr.* **1** pp 13-21
[5] Kolomyts E G 2000 *The state of forest ecosystems in a large city* Proc. of the Samara Scientific Center of the Russian Academy of Sciences (Samara vol 2) **2** pp 311-8.
[6] Bezuglaya E U 1980 *Meteorological Potential and Climatic Features of Urban Air Pollution* (Leningrad: Gidrometeoizdat) p 183
[7] Voronich S S 2012 Assessment of the impact of mobile sources of emissions on the atmosphere (the example of Moscow) *Ecology of urbanized territories* **3** pp 103-7
[8] Ivchenko T V 2014 Urban greening as compensation for air pollution by motor transport *Ecology of urbanized territories* **1** pp 30-3
[9] Vetoshkin A G 2008 *Processes and Devices for Environmental Protection* (Moscow: Higher school) 639 p
[10] Schonfeld B E 2012 On the issue of dust components in emissions into the air from the mining industry *Ecology of urbanized territories* **3** pp 88-92
[11] Krasnova T V 2013 Petrol stations as sources of air pollution in Orenburg *Ecology of urbanized territories* **2** pp 49-55
[12] Panov V P 2008 *Theoretical Foundations of Environmental Protection* (Moscow: Academy) 320p
[13] Bogdanov V N 2006 *Geoinformation Mappping of the Urban Environment orman* (Novosibirsk: Publishing House of the SB RAS) 80 p
[14] Grushko L M 1987 *Harmful Inorganic Compounds in Industrial Discharges into the Air* (Leningrad: Khimiya) 327 p
[15] Khaustov A P 2014 *Environmental Monitoring* (Moscow: Yurait) 637 p
[16] Petrova I F 2009 Mapping categories of pollution (the example of air pollution) *Ecology of urbanized territories* **5** pp. 21-5
[17] Ryaboshapko A G 2002 Ammonia discharges into the air from anthropogenic sources in the USSR: *Problems of environmental monitoring and ecosystem modeling* **18** pp 59-80
[18] Polonskiy V M 1979 *Current researches of mercury vapor extraction from gases and ways of normalization of mercury concentration in gas emissions* Coll. works of the republican scientific and technical conf.: Current state, prospects and trends in cleaning air and water basins of nonferrous metal enterprises in Uzbekistan (Tashkent) pp 75-7
[19] Polonskiy V M 1996 *Main trends in creating an information system for monitoring environmental condition* Scientific and technical conf.: Automated information systems in construction and maintenace of buildings, structures and life support objects (Samara) p 48
[20] Polonskiy V M 2006 Air Protection (Moscow: Publishing House of the Association of Construction Universities) 152 p
[21] Ganyukov S P 2012 Justification of a new approach to assessing the risk to public health due to the impact of technogenic discharges and heir transformation products *Ecology of urbanized territories* **4** pp. 62-9
[22] Goverment of the Tyumen region 2018 *Report on the Ecological Situation in the Tyumen Region in 2017* (Tyumen) 240 p
[23] Department of the Federal State Statistics Service of the Tyumen Region, Khanty-Mansiysk Autonomous Okrug – Ugra and the Yamalo-Nenets Autonomous Okrug 2018 *Environmental Protection in the Tyumen Region* (Tyumen) 152 p

IOP Conf. Series: Materials Science and Engineering **972** (2020) 012030 doi:10.1088/1757-899X/972/1/012030

Increasing Ecological Safety of Lead Production

Ju I Tolstova[1], L G Pastukhova[1], K V Kozyrin[1], R V Osipov[1]

[1]Ural Federal University, 19 Mira Street, Ekaterinburg 620002, Russia

E-mail: l.g.pastuhova@urfu.ru

Abstract. Air pollution as a result of lead evaporation from the surface of melts is considered. Conditions for aerosol diffusion by convective and circulation flows are given. The method of analogy between processes of heat and mass transfer was used for the calculation of evaporation. A program set was developed and new data were obtained on the amount of lead transferring into the air from the melts open surfaces and during transportation in ladles with considering of the cooling

1. Introduction

Lead production, including smelting of rough lead and its further refining, is accompanied by environmental pollution with various substances. Among these substances, highly toxic lead compounds predominate [1]. Lead plants emissions are of serious hazard to the environment. They cause air pollution and consequently soil and water pollution.

The source of lead in the working zone air is the open surfaces of the melts. In melting shops technological processes are carried out at the temperature of 800-1350 degrees Celsius, and the lead goes into the air mainly in the form of aerosols due to evaporation [2]. Combined ventilation is now used, which consist of local and general ventilation. Local ventilation removes harmful substances from the melt overflow nodes, and general ventilation provides aeration. A significant amount of aerosols comes from ladles with melts during their transportation between the melting units.

Lead refining is performed at temperatures of 330 – 900 degrees Celsius. Lead transfers in air as a result of dust formation when slug crusts overloading, melt stirring and ladles transporting. Using of local ventilation is difficult due to the conditions of refining. Ventilation of refining workshops is performed with general ventilation. This method of ventilation is not effective enough in spite of ventilation rate.

The air from local ventilation systems is cleaned in bag filters and electrical precipitators, and then released into the atmosphere. The most dangerous are the low sources of lead emissions, that is, aeration femerells. Femerell emissions are poorly dispersed and cause pollution of the surface layer of the atmosphere. It complicates the organization of supply systems air intake and restrict the aeration use.

Reduction of environmental pollution can be achieved on the basis of complex technical solutions including technical measures and improvement of the ventilation for emissions reducing and the replacement of low emission sources.

2. Experimental

The dynamics of the melt transport in the ladle was estimated on the base of technological process timing in the melting shop, Ust-Kamenogorsk lead-zinc plant [1]. Evaporation and cooling of melts evaluated using methods of mathematical and computer modeling [5, 6].

Data on the amount of dust in air and the lead content in this dust were obtained from in-place tests at different stages of refining [2]. Sampling for dust and lead in convective flows over refining tanks was carried out by aspiration-weight method. The content of dust and lead in the samples was determined by the polarographic method (sensitivity of the method - 0.5 mg / l, error 10%).

Possible pollution of atmospheric air, water and soil in the emission dispersion zone estimated on the basis of the emission dispersion calculation using the Methodology for calculating the concentrations of harmful substances contained in the enterprises emissions.

3. Results and discussion

Using the method of analogy of the processes of heat and mass transfer and the basis of the molecular-kinetic theory [7, 8], the authors developed a method for calculating the evaporation of metals from the melts surface [9]. The obtained dependences allow determining the intensity of lead vapor release taking into account the temperature and percentage of lead in the melt. Figure 1 shows the results of calculations for the main melts of lead production, namely, matte, slag and rough lead with lead content of 12%, 3% and 98%, respectively. As shown in the figure, the intensity of evaporation depends substantially on the temperature and the melt type.

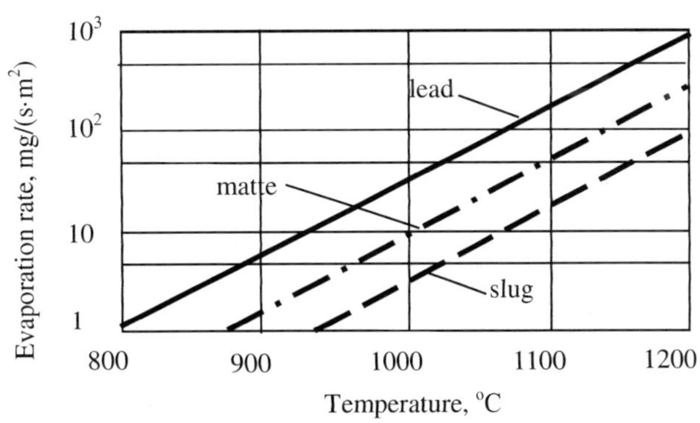

Figure 1. Evaporation rate of lead.

During transportation of melt or its settling, the melt surface cools and the evaporation intensity decreases. To assess the dynamics of these processes, regularities of nonstationary heat transfer for a sphere [10, 11], equivalent to the ladle in volume, were used in the range of the numbers Bi = 30 ... 300 and Fo = 0 ... 0.2, with considering the dependence of the thermophysical characteristics of melts from the temperature. The temperature change of the melts surface in the ladle with cooling is shown in Fig. 2. We note here that the cooling of the surface occurs rather quickly. In a minute the temperature of slag became 0.35 (fraction of the original value), of matte - 0.62; of lead - 0.75. Thus the nature of the surface temperature decrease depends on the thermophysical properties of the melts.

IOP Conf. Series: Materials Science and Engineering **972** (2020) 012030 doi:10.1088/1757-899X/972/1/012030

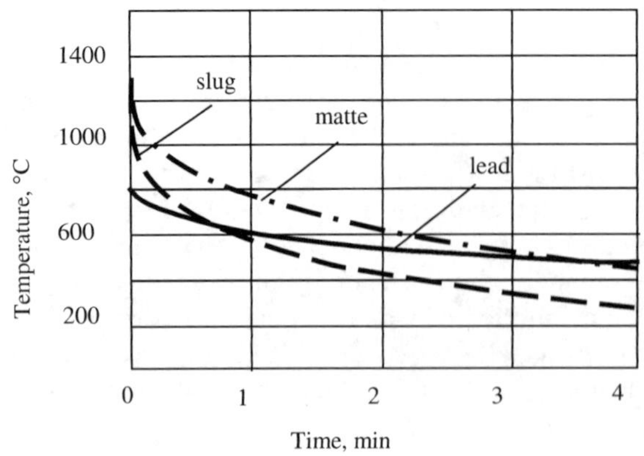

Figure 2. Cooling of the melt surface in ladles.

The data presented (Figures 1 and 2) make it possible to estimate the dynamics of the lead evaporation processes during ladles transportation and settling of melts (Fig. 3). The maximum intensity of evaporation corresponds to the moment when the ladle is filled with matte because of a high initial temperature. For slag the initial intensity of evaporation is less due to the smaller content of lead. The lowest intensity of evaporation is at the moment of casting the ladle with rough lead, but the decrease in the intensity of evaporation occurs slowly. For all melts, formation of lead aerosols due to evaporation depends on the residence time of ladle out of the shelter.

In light of the foregoing, an addition was made to the technological regulations on the need for the sludge to be placed under the shelters of local exhaust ventilation for at least one minute. It allows reducing the transferring of lead to the air from 36 to 1 mg/s.

Estimation of air pollution during refining was obtained on the basis of full-scale studies performed jointly with the Central Research Institute for the Prevention of Pneumoconiosis (Ekaterinburg) in the refining shop of the Ust-Kamenogorsk lead-zinc plant (Kazakhstan). Table 1 shows the amount of incoming dust and the lead content in different stages of the technological process.

Table 1. Lead emission by stages of the refining process

Technological operation	Max. melt temperature, °C	Lead emission from one pot, mg/s	Lead emission from 1m^2 surface, mg/s
Ladling out of lead	800	23.3	2.23
Dressing:			
- coarse dressing of lead	500	1.3	0.124
- fine dressing of lead	600	10.0	0.952
I alkaline refining	600	4.8	0.457
Desulverization:			
- mixing	540	12.5	1.19
- settling	330	1.2	0.124
- foam removal	330	5.0	0.476
Tellurium removal:			
- sodium blending	380	5.0	0.476
- settling	380	0.5	0.048
Bismuth removal			
- mixing	380	6.5	0.619
- dross removal	380	0.7	0.067
II alkaline refining			
- nitrate loading	550	16.0	1.52
Fusion removal	380	0.4	0.038

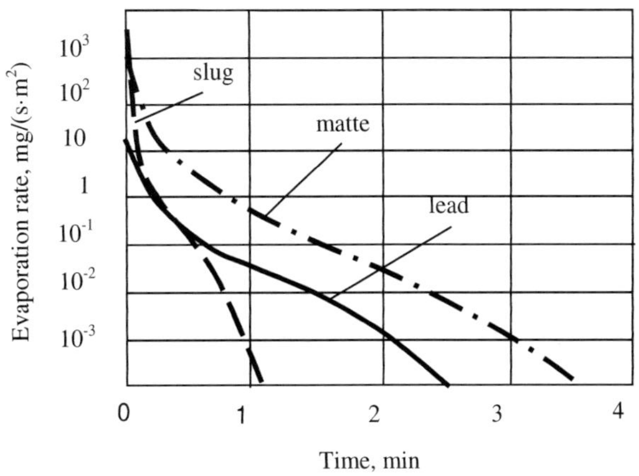

Figure 3. Dynamics of lead evaporation from ladles

Taking into account the technological arrangements in the smelting shop, it is possible to reduce the required air exchange to the air volume removed by local suction. This excludes a necessity for additional suction from the upper zone [9]. In order to reduce the contamination of the working area by recirculation flows, the supply air must be supplied directly to the working zone by panel-type air diffusers. At that, lead concentrations in the working area tends to the maximum permissible values (MPCs). It also eliminates the need to remove contaminated air through the femerell.

In order to achieve the maximum permissible concentration in the working area, air exchange must be ensured in the flow volume in the convective stream above the melting pot at the level of ceiling, that is, in the absence of recirculation. Minimization of air exchange in the refining shop is achieved on the basis of an analysis of the operating conditions of the melting pots. Draught can only be realized over melting pots operating in an intensive mode [12]. Removal of air is provided from the upper zone by means of an exhaust manifold with receiving openings equipped with valves with remote control. All the exhaust air is sent for cleaning to the gas cleaning department. Thus the polluted air emission through the femerell is excluded.

The environmental effect of the proposed activities is as follows: holding the ladles with melts under covering before transportation, exclusion of the release of unpurified air through the femerells, reducing the amount of air being removed.

At present, there are no MPC standards concerning atmospheric surface layer air of industrial sites. The recommendations of the Code of Regulations SP 60.13330.2012 "Heating, ventilation, air conditioning" is used as a criterion of air quality. According to the rules, use of unpurified air for supply air systems and aeration of industrial buildings can be possible, if the concentration of harmful substances does not exceed 0.3 MPC (working area). Since the value of MPC (working area) for lead is 0.01 mg / m3, the concentration of lead in the surface layer should not exceed 0.003 mg / m3.
Calculation of emission dispersion revealed that the increase in lead concentration during femerell emissions is 0.0053 mg / m3 in the smelting shop and 0.027 mg / m3 in the refining shop, which exceeds the standard by 1.8 times in the first case, and 9 times in the second case.

It is possible to achieve a corresponding decrease in lead concentrations in the atmospheric surface layer on account of preventing the emission through the femerell, which gives the lead release of 1.14 tons per year. Replacement of the aeration femerell in the refining shop makes it possible to eliminate lead emissions in the amount of 3.3 tons / year.

As a result of air pollution in the lead production, pollution of soils and water occurs. This is due to the precipitation of lead aerosols into the soil, and then in the water with rain effluent. According to the data on the fertilizers removal from agricultural, the removal of hazards varies widely and depends on climatic conditions, the type and texture of the soil, and ranges from 0.1 to 30%. Thus, some of the lead aerosols are found in nearby reservoir. For example, up to 4 tons of lead dust is annually transferred in the reservoirs located near the Ust-Kamenogorsk lead-zinc plant through the femerell of the smelting and refining shops.

The introduction of the proposed measures almost completely excludes the influence of workshop ventilation emissions on reservoirs and soil, as the release of polluted air without cleaning is prevented.

4. Conclusion

1. It has been established that rooms in the production of rough lead air contamination of production rooms occurs mainly as a result of evaporation from the melts surface during transportation in ladles. As to the refining process, contamination occurs as a result of dust formation under intensive operating conditions of process equipment.

2. The results of calculating the melts surface cooling show that the nature of the surface temperature decrease depends on the thermophysical properties of the melts and the lead content. Cooling of the melt surface leads to a decrease in the intensity of evaporation, therefore, when estimating air pollution, the residence time of ladle out of the shelter should be taken into account. The time increase allows reducing the transferring of lead aerosols by a factor of ten and higher.

3. In the smelting shop, the exposure of the ladles under the shelter allows to reduce the air volume to the value of removed air by local exhausts. It will enable to eliminate the emission through the femerell. Removal of contaminated air should be carried out by local exhaust ventilation systems with subsequent purification.

4. Minimization of air exchange in the refining shop is achieved using an analysis of the operation mode of melting pots. Draught can only be realized over melting pots operating in an intensive mode. Removal of polluted air should be carried out by a system of general ventilation with the following cleaning.

5. Elimination of emission from femerells allows obtaining an ecological effect by reducing the pollution of the industrial site air and eliminating the influence of workshops on the water and soil of the surrounding area.

References

[1] Zajcev V J and Margulis E V 1985 Metallurgija svinca i cinka (Moscow: Metallurgija)

[2] Ivanovskij M, and Sorokin V and Subbotin V 1976 Isparenie i kondensacija metallov (Moscow: Atomizdat)

[3] Smirnov M R 1977 Rafinirovanie svinca i pererabotka poluproduktov (Moscow: Metallurgija)

[4] Shivrin G N 1982 Metallurgija svinca i cinka (Moscow: Metallurgija)

[5] Kafarov V V 1979 Osnovy massoperedachi (Moscow: Vysshaja shkola)

[6] Kikoin A K and Kikoin I K 1976 Molekuljarnaja fizika (Moscow: Nauka)

[7] Lykov A V 1972 Teplomassoobmen: Spravochnik (Moscow: Jenergija)

[8] Miheev M A and Miheeva I M 1973 Osnovy teploperedachi (Moscow: Jenergija)

[9] Shumilov R, Tolstova Ju, Pastuhova L 2002 Ventiljacija nachinaetsja s tehnologii. Sovershenstvovanie ventiljacii plavil'nyh cehov svincovyh zavodov *Inzh. Sist.* **8** 27-29

[10] Frank-Kameneckij D A 1987 Diffuzija i teploperedacha v himicheskoj kinetike (Moscow: Nauka)

[11] Chepmen S Kauling T 1960 Matematicheskaja teorija neodnorodnyh gazov (Moscow: Izdatel'stvo Inostrannoj Literatury)

[12] Pastuhova L ,Tolstova Ju and Shumilov R 2003 Ventiljacija nachinaetsja s tehnologii. Osobennosti ventiljacii cehov rafinirovanija svinca *Inzh. sist.* **10** 54-55

[13] Taylor J and Zuncel A 1988 Environmental Chelenges for the Lead-Zinc Industry *J. of Metals* **40** 189-97

[14] Shuzo M, Shinsoke K and Yoshihiro I 1988 Numtrical Simulation of Room Air Flow wiht Generalized Curvilinear Coordinates *Mon. J. Inst. Ind. Sci. Univ.* **40** 293-6

[15] Nielsen P, Restivo A and Whitelaw J 1978 The velocity characteristics of ventilated rooms *J. Fluid Engng.* **100** 291-8
[16] Pozin G 1992 Mathematical Modelling of Heat and Air Processes in Michanically Ventilated Spases *RoomVent* 361-9
[17] Batonneau Y, Laureyns J, Merlin J and Bremard C 2001 Self-modeling mixture analysis of Raman microspectrometric investigations of dust emitted by lead and zinc smelters *Anal. Chim. Acta* **446** 23-37
[18] Manceau A, Boisset C, Sarret G, Hazemann J and Mench M 1996 Direct determination of lead speciation in contaminated soils by EXAFS spectroscopy *Environ. Sci. Tech.***30** 1540-52
[19] Batonneau Y, Bremard C, Gengembre L, Laureyns J and LeMaguer A 2004 Speciation of PM10 sources of airborne nonferrous metals within the 3-km zone of lead/zinc smelters *Environ. Sci. Tech.* **38** 5281-9
[20] Roberts T, Hutchinson T, Paciga J, Chattopadhyay A, Jervis R, Vanloon J and Parkinson D 1974 , Lead contamination around secondary smelters: estimation of dispersal and accumulation by humans *Science* **186** 1120-3
[21] Lilis R, Fischbein A and Eisinger J 1997 Prevalence of lead disease among secondary lead smelter workers and biological indicators of lead exposure *Environ. Res.***14** 255-85

Development of the energy efficient residential low-rise house concept for the Ural region

D A Vasileva[1] and N P Nikitina[1]

[1]Ural Federal University, 19 Mira str., Ekaterinburg, 620000, Russia

E-mail: darya.vasilyeva95@gmail.com

Abstract. In the article the problem of energy-efficient construction in Russia is considered. The energy-saving technologies and renewable energy sources acceptable in construction in the Ural region analysis is carried out. Main reasons of the irrational consumption of thermal energy are highlighted and the strategy of the energy-efficient construction development in the Urals is described. As a result of the research, a conceptual project of an individual residential house was developed.

1. Introduction

The problem of ecological and energy-efficient construction development in the Russian Federation from year to year is becoming increasingly popular and widely discussed. There are different ways to develop energy efficiency depending on the climatic conditions of the country. Primarily it depends on the characteristics of energy consumption. While in the southern countries the main proportion of energy is spent on air conditioning, in Russia up to 40% of countrywide energy resources goes to heat buildings in the winter and large amounts of carbon dioxide are emitted into the atmosphere, which leads to a "greenhouse effect". Therefore, the environmental situation, the economic growth of the country and the living standard of citizens depend on the results of solving this problem. It is obvious that technologies aimed at reducing heat loss and automating the processes of thermal energy consumption will provide the greatest economic effect.

Despite numerous discussions of this problem, there are only a few dozen implemented projects in Russia using energy efficient technologies and materials. In most projects, “energy efficiency” is limited only by improved glazing and an automated lighting system.

2. Problems of energy-efficient construction growth lack in Russia

Why the problem is still not solved? There are five key reasons hindering the development of energy-efficient technologies in the Russian construction industry.

The first one is the lack of a competent regulatory framework. In 1996 the State Duma adopted federal law No. 28 of April 3, 1996 “Energy-saving”, in which the issue of energy efficiency was mentioned for the first time. The main objects of regulation were: efficient use of energy resources, starting from their extraction and ending with consumption; control over the efficient use of energy resources through the state; creation and development of alternative options for expensive and rare energy resources; the creation and development of energy-efficient technologies, metering and control of energy resources, the introduction of new modern energy-efficient materials, products, structures; ensuring the uniformity of measurements in the field of energy resources.

Due to the above requirements, their main solution was the widespread use of multilayer wall structures. Thus, in accordance with modern building codes, the required resistance to heat transfer, for example, of walls was increased by 3-3.5 times in comparison with the old standards.

The second reason that hampers the development of energy-efficient construction is that it is expensive at relatively low utility rates for Europe. The cost price per square meter in an energy-efficient eco-house is higher than usual on average by 20%, while the payback comes at least after 20 years.

The third reason is the lack of cheap domestic materials that meet the requirements of energy efficiency and environmental friendliness. Over the past 20 years, investment in the research sector in the field of construction has been minimal. Therefore, most materials are either imported or produced in Russia using foreign technologies.

The fourth is the complexity of the process of introducing energy efficient technologies. The development of energy efficiency requires a serious technical, scientific approach. It is necessary to consider the existing experience and project it on the climatic features of the country. The task is complicated by the lack of qualified personnel: they are mainly represented by employees of foreign companies or Russian specialists trained abroad.

The last problem is social. The majority of Russian citizens are not aware or not enough aware of this problem. The other part is not ready to sacrifice personal amenities and well-established household habits for the sake of improving the energy efficiency of their homes and using environmentally friendly materials. Not every citizen of the Russian Federation is concerned about the problem of ecology in the country and the world, the results of which we can observe every day in almost every city in the country.

Obviously, government support is necessary for the effective implementation of an energy saving strategy. The government has to stimulate domestic producers of building materials so cheap products can appear on the market. It is also extremely important to develop a mechanism for interaction with the scientific sector, since at present relatively few specialists deal with this problem.

3. Development of energy efficient construction in the Urals

The greatest impetus to the development of energy-efficient construction in Russia will come from the development of an energy-saving and energy efficiency strategy for each region based on its climatic features. Let us analyze energy efficient technologies acceptable for the Ural region.

3.1. *Architectural and technological strategies for energy saving in the Urals*

Currently, according to expert estimates, the specific heat losses in buildings are distributed as follows: up to 40% due to organized and unorganized infiltration of heated air, up to 30% due to insufficient heat transfer resistance of enclosing structures, up to 30% due to wasteful use of hot water and unregulated mode of operation of heating systems [1]. The complex of energy saving measures corresponding to the energy saving strategy in the Ural region can be divided into five main groups:

- town planning solutions;
- architectural and planning solutions;
- constructive systems;
- engineering systems;
- renewable energy sources.

The combination of energy-saving urban planning activities includes: the organization of the planning structure of neighborhoods from semi-enclosed residential groups that open to the south side of the horizon; compactness of the complex development of micro-districts due to an increase of the construction density; application of wind-shelter in the formation of residential groups to reduce infiltration heat losses from wind exposure; the use of blocking buildings, allowing to significantly reduce their heat loss; optimization of the location of the network of service institutions in the form of compact public, commercial, sports and recreational, cultural and entertainment, and other centers of various levels of service; integrated development of underground space [2].

IOP Conf. Series: Materials Science and Engineering **972** (2020) 012031 doi:10.1088/1757-899X/972/1/012031

Significant impact on the specific heat loss in residential and public buildings has their space-planning decisions and, in particular, the ratio of the enclosing structures area to the total area of buildings, placing them on the relief and relative to the cardinal points. The recommended complex of architectural and planning solutions includes: the choice of the optimal shape of buildings, characterized by a high compactness factor and ensuring minimal heat loss in the winter period and minimal heat gain in the summer period; the choice of the optimal orientation of buildings to the cardinal points, taking into account the prevailing wind direction in the winter period; reduction of the area of external enclosing structures by reducing the perimeter of external walls due to the rejection of the facades roughness and of the "architectural openings"; communication of premises without excessive corridors, halls and dark rooms [3].

The most rational types of energy-efficient exterior walling are multi-layer composite structures of walls and coatings using mineral efficient materials. Warming of external walls - the most expensive and labor-intensive process - reduces the heat loss by about 12–15%. Heat loss through the windows reaches 50% of the total heat loss through the building envelope, so the first step is to improve the heat-shielding qualities of the windows. For example, the use of windows with heat-reflecting glass in residential construction allows reducing heat loss through them up to 40% of energy.

Approximate calculations show that with the cumulative implementation of measures to modernize engineering systems, it is possible to reduce the heat consumption in residential and public buildings for heating and heating of inflowing or infiltrating air by 30–40%. At the same time, one-time capital costs will be significantly (from 2 to 10 times) lower than the costs of increasing the thermal resistance of the walls.

A significant energy saving effect can be realized through the use of renewable energy sources [4]. These types include: the energy of the sun, wind, heat of the seas and oceans, land (geothermal energy), and hydropower.

Thus, the main strategies for the development of energy conservation in the construction and operation of buildings and structures are as follows:

1) A systematic approach and an economically sound sequence of implementation of a interrelated and interdependent energy-saving measures of urban planning, architectural planning, constructive, engineering and operational nature set.

2) The program-targeted method of developing and implementing a system of energy-saving measures aimed at obtaining the final result – maximum savings of non-renewable fuel resources with minimal expenditure of funds and time to achieve this goal.

3) Orientation of scientific, design and practical activities on energy saving to the most energy-intensive area of fixed assets operation, the implementation of energy-saving technologies in which more than 90% of the potential effect on energy-saving provides through the modernization and reconstruction of buildings, structures, utilities, communications and energy facilities.

3.2. Renewable energy sources for the Urals

The economic costs of heating and hot water in the Ural region reach 40% of the total energy consumption. Two options for heating and hot water installations will be considered further: on the basis of a solar collector and on the basis of a heat pump.

For heat engineering calculation and calculation of heat losses, we take as a basis a three-storey individual residential house with an area of 280 m^2 (figure 1). The composition of the wall consists of: wooden frame, mineral wool insulation Tizol EURO-Vent, facade system with a ventilated gap. Heater Tizol EURO-Vent [5]: 70 mm – in the composition of the outer wall (residential premises), 80 mm – for combined bathroom, 50 mm – garage, 200 mm – in the composition of the coating (residential floor).

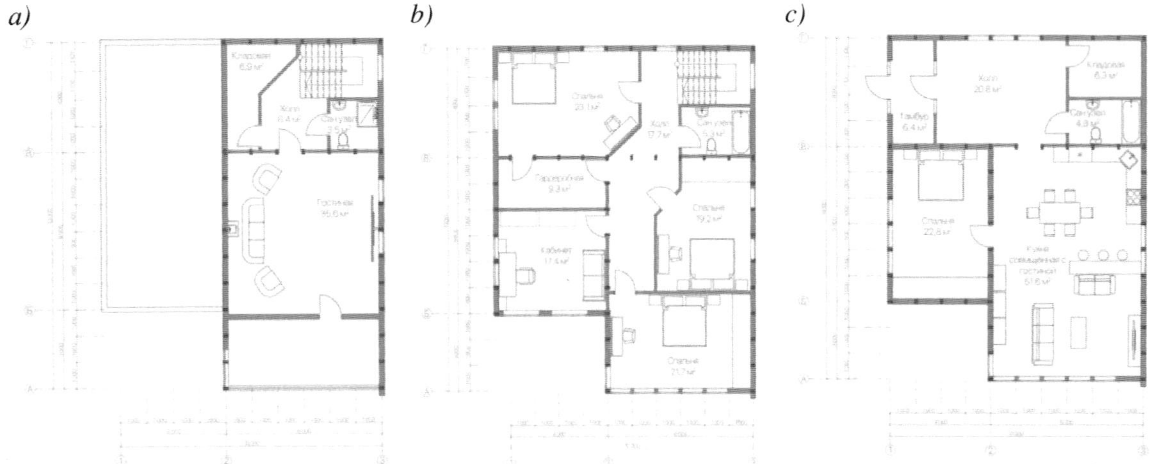

Figure 1. Plans for an individual residential building: *a* – first floor, *b* – second floor, *c* – third floor.

With such building parameters, the heat loss through the fence and for infiltration amounted to 9.7 kW, which must be compensated by heating system. The consumption of hot water per person is an average of 4 m^3 per month. For example, in a house adopted for the calculation four people live permanently, then such a family will spend 16 m^3 of hot water per month or 192 m^3 per year [6].

Next, we will consider two options with using renewable energy sources installations: a solar collector and a heat pump, and also without the use of renewable energy sources - an electric boiler (figure 2).

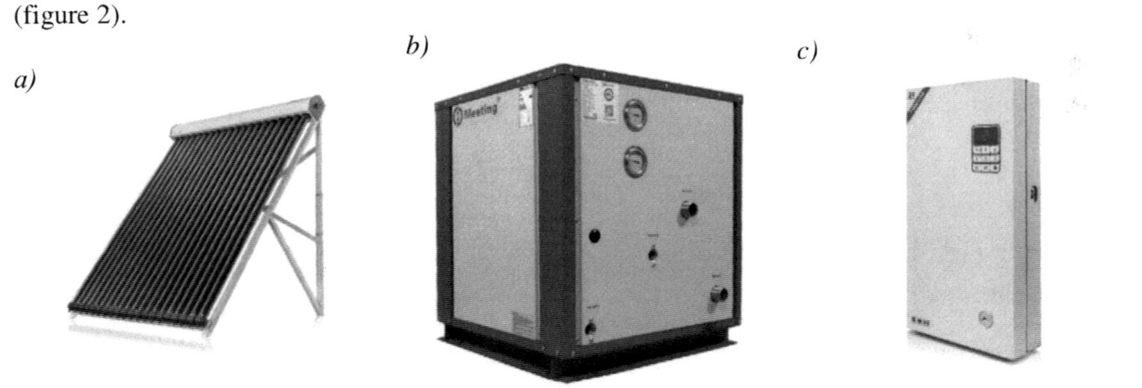

Figure 2. *a* – Solar vacuum collector SCH-24-15 for year-round use; *b* – Meeting-MDS60D water-to-water heat pump; *c* – Electric boiler class Comfort Plus.

Solar collectors are a source of free and clean energy. They convert direct and diffused sunlight into heat. Taking into account the seasonal changes in solar radiation in the Urals region 14% of the total generated heat will be produced in the winter, 29% in the spring, 36% in the summer and 21% in the autumn. In December and January, heat generation is minimal, and will amount to 3.3% per month of total heat production per year [7]. That is why the solar collectors are rarely used for heating in this climate.

We obtain that the system with the solar collector for heating and hot water will require 6 704 kWh per month in the heated period (from October to April) and 288 kWh per month in the unheated period (from May to September) [8]. The cost of electricity for heating and hot water are shown in Table 2.

IOP Conf. Series: Materials Science and Engineering **972** (2020) 012031 doi:10.1088/1757-899X/972/1/012031

The principle of operation of a water-to-water heat pump is the selection of heat energy from a low-grade heat source and its direction to heat the building. The coefficient of efficiency of heat pump is greater than one by 3-4 times.

When using heat pumps for heating and hot water, 1 852 kWh per month during the heated period and 223 kWh per month during the unheated period [9] are required, costs are shown in Table 2.

Let us compare the costs with the option of heating and hot water without the use of renewable energy (Table 1).

Table 1. Energy consumption for heating water per month, kWh.

Period of time	Type of heater		
	Solar collector	Heat pump	Electric boiler
October-April	6704	1852	7910
May-September	288	223	953
TOTAL, kWh (per year)	**48 368**	**14 079**	**60 135**

Required installation costs and costs during operation, (Table 2) payback period [10], (Figure 3).

Table 2. Costs and service life of various heating systems and hot water supply.

Costs	Solar collector	Heat pump	Electric boiler
Installation cost, rub.	332 800,00	179 200,00	39 800,00
Service, rub.	30 000,00	50 000,00	10 000,00
Electricity costs per year, rub.	188 151,52	54 767,31	233 925,15
Service life, year	15	15	5

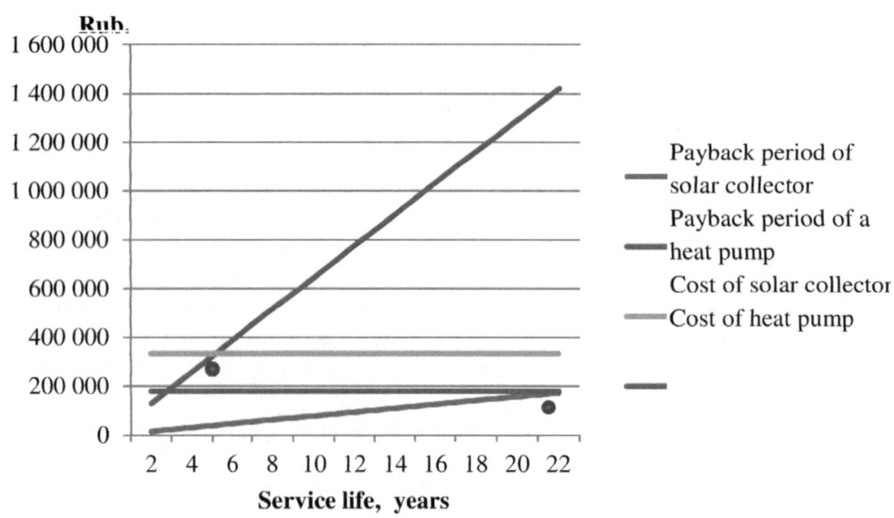

Figure 3. Payback periods.

From this chart it is clear that a system with a solar collector will pay off in 21 years and a system with a heat pump in 1.5 years. So the payback period of the heat pump will exceed 14 times the payback period of the solar collector. Such a large gap appears due to the fact that the solar collector power is not enough for heating in winter in the conditions of the Ural region [11]. Therefore, in these climatic conditions, it is more reasonably to use a heat pump or a combination of a heat pump and a solar collector [12].

4. The concept of an energy efficient low-rise house for the Ural region

The complex of energy-saving measures, consisting of five main groups: urban planning solutions; architectural planning solutions; structural systems; engineering systems; as well, the analysis of renewable energy sources made it possible to form the concept of an energy efficient low-rise house for the Urals region (Figures 4-6).

Figure 4. Conceptual project of energy-efficient residential building.

The concept of an energy efficient house for the Ural region includes:

- an effective thermal envelope of the building, with an optimal heat transfer coefficient of external walls, and a compact layout, which create a closed loop of the building that provides an increase in energy efficiency;
- pitched roof with a warmed attic - the most effective way to cover an individual house in these climatic conditions, allowing to place solar collectors at a certain angle;
- the presence of a warmed portal and a heat-reflecting triple glazed window, which allows to reduce the largest part of heat losses attributable to windows and doors;
- the location and size of the windows according to the cardinal points, which contributes to the most efficient use of solar radiation;
- installation of a vacuum solar collector system and a heat pump that make the heating and hot water supply of a residential building autonomous.

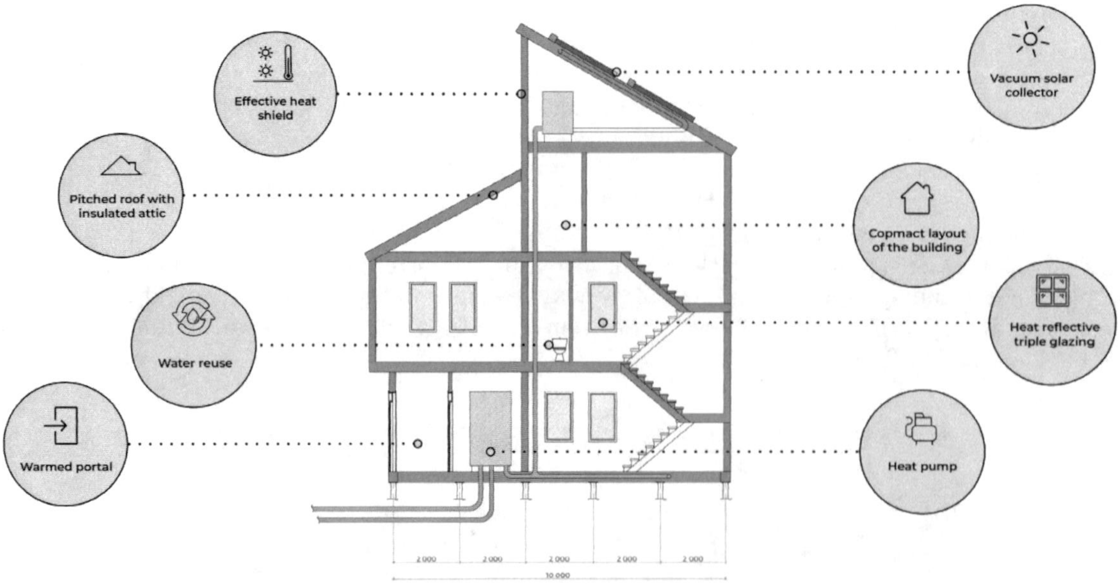

Figure 5. The scheme of energy-efficient low-rise houses for the Ural region.

Figure 6. Facade solutions for an energy-efficient low-rise house for the Ural region.

Energy-efficient technologies and renewable energy sources acceptable to the Urals was analyzed. The presented concept of an energy efficient house corresponds to the energy saving and energy efficiency strategy for low-rise construction in the Ural region, based on its climatic features. That, in turn, is a definite impetus to the development of energy-efficient construction in Russia as a whole.

References

[1] Bulgakov S N 1999 Energy-efficient building systems and technologies *AVOK* **2** (in Russian)

[2] Vatin N.I., Gorshkov A.S., Nemova D.V. Energy Efficiency Formula // Construction of Unique Buildings and Structures. 2013. № 7. (in Russian)

[3] Gorshkov A S, Derunov D V and Zavgorodniy V V 2013 Technology and organization of construction of a building with zero energy consumption *Construction of unique buildings and structures* **3** (in Russian)

[4] Velkin V I, Tyagunov G V, Scheklein S E and Ukhov A L *Energy Efficient House. The patent of the Russian Federation for utility model RUS 61760 09.03.2005* (in Russian)

[5] SP 50.13330.2012 "Thermal protection of buildings" (in Russian)
[6] SanPiN 2.1.4.2496-09 “Hygienic requirements for ensuring the security of hot water systems. Change to SanPiN 2.1.4.1074-01 ".(in Russian)
[7] Sheryazov S K, Velkin V I, Semenov A Yu and Chernov N A 2012 Bases of research of the system of energy supply with the use of renewable sources *International scientific journal Alternative energy and ecology* **4**(**108**) P 147-149
[8] Ovchinnikov Yu.V. The method of calculating the solar collector in an environmentally friendly and energy-saving system of individual heating. (in Russian) https://docplayer.ru/39470555-Metodika-rascheta-solnechnogo-kollektora-vekologichnoy-i-energosberegayushchey-sisteme-individualnogo-otopleniya.html
[9] Patlakh V V Calculation and design of heat pumps http://patlah.ru/etm/etm24/a_energia/teplovoi_nasos/teplovoi_nasos.htm (in Russian)
[10] Sultanov I A Ways to account the payback of the project (in Russian) http://projectimo.ru/upravlenie-investiciyami/srok-okupaemosti.html
[11] Kalinin I M 2003 Energy-saving heat pump technologies *Ecological systems* **6** P 12–18 (in Russian)
[12] Velkin V I and Danilov V Yu 2012 Experimental studies of vacuum solar collector under conditions of negative temperatures *International Journal of Alternative Energy and Ecology* **11**(**115**) P 28-31 (in Russian)

Development of innovative proposals for improving the environmental situation in the Novovoronezh NPP region using the comparative analysis of waste management at Kudankulam NPP in India and Bruce NPP in Canada

M G Dobrosotskikh[1], **D I Emelyanov, V R Kuznetsov, E V Rodina**

Department of technology, construction organization and realty's expertise, Voronezh State Technical University, 20 let Oktjabrja, 84, Voronezh, 394006, Russia

E-mail: dobrmax@yandex.ru

Abstract. Paper provides a comparative analysis of three nuclear power plants: Novovoronezh NPP in Russia, Kudankulam NPP in India, and Bruce NPP in Canada. The basic parameters (site plan, biosphere diversity, types and capacity of reactors, amount of electricity generated and its cost) are compared. Types of waste generated at all three nuclear power plants and methods of their treatment are analyzed. Brief overview of some of the security systems used at these power plants is conducted. On the basis of comparative analysis, proposals for improving the environmental situation in the Novovoronezh NPP region (e.g. improvement of the wastes management) are made, with the proposal of nuclear fuel cycle scheme enhancement.

1. Introduction

In the today's world the use of atomic energy is an integral part of any developed country's lifestyle, but generation of electric power this way has a damaging impact on the environment and human health. In this regard, each country has developed its own policy to minimize the adverse effects of atomic radiation on the biosphere. In this context one of the important measures is assigned to waste management.

The comparative analysis of waste management presented in this paper is based on the data from three nuclear power plants, namely: Novovoronezh NPP in Russia (Concern *Rosenergoatom),* Kudankulam NPP in India and Bruce NPP in Canada. The main sources of information for the comparative analysis are [6,5,23]. The choice of these particular NPPs is justified by the fact that the Novovoronezh NPP (here and after NV NPP) is one of the oldest nuclear power enterprises in the Russian Federation (the largest producer of electrical energy in the Voronezh Region) became the first power plant in contemporary Russia with the rector of the new generation VVER-1200 [32]. The Indian Kudankulam NPP was chosen because at the moment it is named to be the largest Russian-Indian economic cooperation facility. Since October 2002, the leading Russian design organizations and research centers are taking part in construction of Kudankulam NPP (e.g. the chief designer is Moscow Institute *Atomenergoproekt*, the general designer is *Gidropress* office, the scientific head is *Kurchatov Institute* [19]). In total, more than 100 Russian enterprises are involved in the Kudankulam project (including Voronezh architectural and construction offices). The Canadian Bruce NPP was chosen due to the fact that numerous Internet resources name it to be the one of the most powerful power plants in the world. Moreover, Canadian legislation has strict regulations and guidelines for ensuring the safety of its nuclear power plants, enforcing continuous environmental monitoring and ecological improvement in NPPs regions.

The main objective of the comparative analysis of the three above-mentioned NPPs is the methods of their waste management with the aim of identification of the positive and negative components of the technological processes applied, which affects the safety of strategic atomic infrastructure and the surrounding biosphere respectively.

The site of NV NPP (Fig. 1) is located on the left bank of the River Don, 40 km south of Voronezh and 5 km south of Novovoronezh – the city with the estimated population of 31.5 thousand people (in 2018), most of which works and lives in close proximity to the nuclear power plant.

Figure 1. Topographical view of the Novovoronezh NPP [11]

NV NPP (one of the oldest power plants in Russia) has been built in three phases, namely NPP-1, power units: 1 (VVER-210 in 1964), 2 (VVER-365 in 1969); at the NPP-2, power units 3 and 4 (VVER-440 in 1971 and 1972), at the NPP-3, power unit 5 (VVER-1000 - in 1980). In 1984, after 20 years of operation, the power unit 1 was decommissioned, following by the decommission of power unit 2 in 1990 and power unit 3 in 2016. Since 1995, NV NPP has been gradually upgrading its power units to bring them in line with modern security standards. From 2007 to March 22, 2019, two power units (6 and 7) of the new generation "3+", with the most powerful reactor facility in today's world (VVER-1200) were built at the site of the NV NPP [32]. This epoch-making event allowed Russia to take the leading role in the atomic high-tech industry.

At the moment, the NV NPP has three power units (4, 5 and 6 (in commercial operation since 2017), with the reactor of the 7th unit being brought to the minimum controlled power level on March 22, 2019 (Fig. 2). The VVER-1200 reactor is an evolutionary project that emerged from the VVER-1000 options, which were built for foreign customers in the 1990s and 2000s (e.g. Bushehr NPP, Kundankulam NPP, Tianwan NPP).

To construct the VVER-1200, each reactor parameter of VVER-1000 has been slightly improved with many modern technological safety systems introduced to reduce the likelihood of radiation leak beyond the hermetic dome reactor compartment (containment) in any possible accidents. The reactor uses a passive heat removal system from the reactor, a passive system for the catalytic removal of hydrogen and a core melt trap. The double containment shell, in which the inner shell prevents the leakage of radioactive substances in case of accidents and the outer shell resists natural and man-made influences (such as tornadoes or the fall of an aircraft) became another innovative feature of the new reactor. VVER-1200 is distinguished by increased power (by 20%) with approximately the same equipment size, a service life of 60 years, the ability to maneuver power in the interests of the power system, high capacity factor (installed capacity utilization factor), the ability to work for 18 months without fuel overload and other improved specific indicators.

The Kudankulam NPP (located in the south of the Indian state of Tamil Nadu, whose capital (and the fourth largest city in India) is Chennai (formerly Madras) is being built as part of the Russian-Indian agreement from November 20, 1988 and its addendum dated June 21, 1998 year [19]. The customer is the Indian Atomic Energy Corporation. In the period from 2002 to 2013, 1 and 2 power units with VVER-1000 reactors with a total capacity of 2000 MW has been built (Fig. 3). The 3rd and 4th power units of the Kudankulam nuclear power plant are scheduled for launch in 2020 and 2021, respectively [28].

Figure 2. Actual view of Novovoronezh NPP [20]

Figure 3. Actual view of Kudankulam NPP [17]

For construction of the Kudankulam NPP dry and barren land has been acquired, however, its location on the very shore of the oceanic waters (Fig.4) possess a certain threat to the marine life. The size of the nearby resident population is 12 thousand people. In 2004, in the immediate vicinity of the power plant (1500 m), a small seaport has been built with the purpose to eliminate the risk of damage during transportation of components coming from Russia and nuclear fuel [8].

IOP Conf. Series: Materials Science and Engineering **972** (2020) 012032 doi:10.1088/1757-899X/972/1/012032

Figure 4. Topographical view of Kudankulam NPP from above [15].

The Bruce nuclear power plant is located on the eastern shore of Lake Huron, about 18 km north of Kincardine and 17 km. southwest of Port Elgin in Bruce County, Ontario, and covers an area of 932 hectares (2,300 acres) (Fig. 5).

Figure 5. Topographical view of NPP "Bruce" from above [12].

In close proximity to the station the agricultural, recreational and rural residential areas are located. The population size is 65 thousand people. Recreational land use includes Inverhuron Park with residential cottages in the village of Inverhuron (south of Bruce Power) and in the Baie du Doré / Scott Point area (north of Bruce Power). It should be noted that over the past 50 years the nuclear energy was safely generated at the plant site, first through the Douglas Point NPP (1968-1982), and then through the Bruce A and B NPPs, which were introduced into operation from 1977 to 1979 and from 1984 to 1987, respectively.

At the Bruce power plant, there are 8 rectors of the CANDU type (Canada Deuterium Uranium) (Fig. 6) - a heavy-water nuclear water reactor produced in Canada. CANDU uses heavy water as a moderator, which makes it possible to use ordinary natural uranium as a fuel. Unlike most water-cooled reactors (for example, VVER), the CANDU is a channel-type reactor, which allows replacing the used fuel with fresh fuel without stopping the reactor. The primary coolant can be both heavy and ordinary water.

IOP Conf. Series: Materials Science and Engineering **972** (2020) 012032 doi:10.1088/1757-899X/972/1/012032

Figure 6. Actual view of the Bruce NPP [27]

Table 1 presents the site plan summery of three power selected plants (e.g. comparison by terrain type, number of inhabitants and diversity of flora and fauna). From the comparison presented it is clear that all the power plants are located close enough to the residential areas. A wide range of biological diversity both on land and in water presented around the NV NPP and "Bruce" power plants. "Kudankulam" in India is located on a comparatively deserted and biologically poor land area, but it has a direct impact on the biosphere of the ocean, located directly in its water basin. In 2011, Tamil Nadu residents have massively protested against the construction of NPP, among them were the local fishermen, who expressed the concern that the water temperature would rise due to the power plant's operation and the fish would leave these places [27]. This once again confirms the need for the careful monitoring and control and waste management.

Table 2 presents the comparative power capacity characteristics of the three NPPs.

Table 1. Site plan summery of the three selected NPPs.

	NV NPP	Kudankulam NPP	Bruce NPP
Terrain type	Fields/Forest/City	Wasteland	Forest/City
Waters around	River	Ocean	Lake
Distance to the settlement, km	5	8	0
Population size, thousand pers.	31	12	65
Variety of flora	average	low	high
Variety of fauna	average	low	high
Variety in waters	average	low	high

Table 2. Comparative power capacity characteristics of the three NPPs.

	NV NPP	Kudankulam NPP	Bruce NPP
Number of active power units	3	2	8
Reactor type	VVER-1200	VVER-1000	CANDU 750, 791
Most powerful reactor (Net power), MWt	1114	917	817
Total capacity, MWt	2449	1834	6258

IOP Conf. Series: Materials Science and Engineering **972** (2020) 012032 doi:10.1088/1757-899X/972/1/012032

2. Methods of waste management and emissions.

2.1. Waste management methods at Novovoronezh NPP

Waste at nuclear power plants are divided into radioactive (RW) and non-radioactive waste (NRW). RW is divided into solid (SRW) and gaseous (GRW). During the operation of the plant, as well as the decommissioning of old power units, the generated solid RAW is collected, reprocessed and stored in temporary storage facilities before being transferred to specialized organizations for disposal. For the primary collection of SRW, tripods and supports are used in which plastic or kraft bags are inserted, along with plastic or kraft bags being used in the in the form of self-packing, which after filling are loaded into collection containers. The drainage collection system of organized and unorganized leaks from technological systems and the reactor compartment equipment, and special water treatment, is designed to receive and pre-clean the collected water from mechanical impurities for subsequent processing in evaporation plants.

For dealing with GRW, the existing ventilation system of the NV NPP units provides:

- means to ensure fire and explosion safety, including means for prevention of flame spread and combustion products through air ducts and ducts;
- means for regulating the performance of the blower devices (gas blowers, ventilation units, fans, compressors, etc.);
- redundancy of filtering and absorbing elements of equipment with a chain to ensure the replacement or regeneration of filters without stopping cleaning;
- local air cleaning, removed from the places of possible formation and accumulation of GRW (e.g. from repair or dismantling jobs).

Technical characteristics of fans in the supply and exhaust systems are chosen in a way so that during the normal operation of ventilation systems, the total flow of exhaust air exceeds the air flow. Air intake by air supply ventilation systems is carried out from the roof of the machine room through air supply shafts. The air is cleaned of dust by fabric filters. The air that is removed from controlled access zone premises by the exhaust ventilation systems is emitted into the atmosphere through a ventilation pipe with a height of 120 m and a mouth diameter of 3 m.

NRW at the NV NPP are divided into 5 hazard categories, which are:

Hazard Category 1 includes: used mercury-containing fluorescent lamps, metallic mercury (drained from mercury-containing devices), mercury thermometers (which are not reused, but transferred to enterprises have the appropriate disposal license).

Hazard Category 2 includes: waste batteries with non-drained electrolyte, spent battery fluid (which is transferred to specialized waste treatment plants or other organizations have the appropriate license for this type of activity).

Hazard Category 3 includes: waste oil, wiping oily rags and greasy gloves, oily soil, ferrous metal (copper); Waste oil products are sent for processing and use in organizations that have a license to carry out activities in the field of hazardous waste management. Used oil filters are placed in special metal containers in open areas and transferred to enterprises that have the appropriate license for the right to handle this type of waste. Non-ferrous and black scrap metal is collected in containers or piles in designated areas (ferrous scrap is stored separately from non-ferrous scrap metal) and subjected to radiation monitoring.

Hazard Category 4 includes: used car tires, slag wool, and household garbage. Spent car tires are accumulated at a special site, then transported to a company that has a license to collect and process car tires.

Hazard Category 5 includes: small fractions of ionexchange resins and iron oxides (sludge after neutralization unit and chemical water treatment plants), black scrap metal, non-ferrous scrap metal (aluminum), sediment from vehicle washing, cost estimation paper documentation from industrial areas, rubber hoses, abrasive dust from grinding and grinding tools and equipment, stubs of electrodes, construction waste (broken brick, plaster, concrete pieces, glass, etc.), glass wool, coastal pumping sludge, hay, wood waste from the cutting of tree branches, food waste of canteens. All this waste is accumulated at the designated locations in the NPP units, and then disposed to municipal dump (within its stated disposal limits) on enterprise that have the appropriate disposal license.

During waste loading and transporting, conditions are provided to exclude their loss and dusting (the filled body is covered with tarp).

All RW and NRW exported from the territory of the NV NPP are monitored for radiation control. Material passes must be made for each waste removal carry out from the territory of the NV NPP, with indication of waste type and its total quantity in cubic meters.

2.2. Waste management methods at Kudankulam NPP

RW management at Kudankulam NPP focuses on the effective reduction of radioactivity (decontamination) and minimization of secondary waste, where the promising new technologies introduced include the synthesis and use of specific sorbents (inorganic, magnetic and biological technologies), the use of electro-oxidative methods for organic waste, cryogenic distillation to separate radionuclides from waste gases, liquid-based membranes, etc. [5]. Facilites for the treatment of solid radioactive waste (SRW) are equipped with equipment for separation, repackaging, compacting, incineration and embedding of radiation sources.

To eliminate GAO (volatile radionuclides such as iodine, ruthenium, etc.) adsorbers are used. All waste gases pass through highly efficient particulate filters, which are designed to be more than 99.9% efficient for submicron-sized particles. Leakage of radiation waste at this Indian NPP is minimized, due to the fact that the main cooling water circuit at the NPP is a closed loop, any radioactivity resulting from the fuel enters the filters and ion exchangers, thus does not pose any danger to the facility or its personnel (without any chance of entering into surrounding biosphere). Other liquid waste from the technological processes evaporates with a mildly condensate remaining, which is recyclable at the plant and safe for disposal. If any SRW is contaminated with radioactivity, it is collected and incinerated or compressed to reduce its size; then the waste is conditioned by fixing it in cement concrete. In a few years they will be considered for burial in a near-surface repository (if enough time passes for the decay of short-term radioactivity) [5].

2.3. Waste management methods at Bruce NPP

The following types of RW and NRW are identified at the Bruce NPP [6]:

- Hazardous RW (oils, chemical lighting lamps and ballasts (some of them are recycled)
- Recycled NRW (glass, plastic, metal, cardboard, paper, wood, batteries and electronics)
- Organic NRW (compost)
- Radiological RW
- Waste for NRW landfills (for those items that are neither hazardous, nor recyclable, neither organic nor radiological).

To minimize the amount of waste sent to landfills every day, the Bruce NPP has implemented a number of waste management methods, such as:

1) Sending to recycling and then recycling of hazardous waste (chemicals, oils, batteries and fluorescent lamps). Pick up is conducted by a number of contractors with whom the NPP cooperates and who are responsible for the environmentally safe transportation, recycling, waste disposal, etc.

2) Minimization of recyclable waste by applying the principles of reduction, reuse, recovery and recycling.

3) Composting of organic waste, including food, paper towels and biodegradable containers.

4) Managing and minimizing the amount of radiological waste through effective materials management, decontamination and separation, with the use of the reduction, reuse, recycling and recovery principles where possible. RW is handled in a safe, environmentally responsible manner that takes into account the ALARA principle (As Low As Reasonably Achievable), which provides for maintaining as low as possible and attainable levels of both individual (below the limits set by current standards) and collective doses of radiation, taking into account social and economic factors.

Table 3 provides comparative data on waste disposal and treatment at three NPPs.

Table 3. Comparative data on waste disposal and treatment at three NPPs.

	NV NPP	Kudankulam NPP	Bruce NPP
Total amount of waste treated %	5,4%	Data is inaccessible	62,2%
Independent waste recycling	Low level	Low level	High level
Reuse/recycling of waste by other organizations	Yes	Yes	Yes
Reuse of recycled materials	Low level	Low level	High level
Total amount of waste, tones	6830	Data is inaccessible	1967

Table 4 provides comparative data on the radiation situation and the quantity / cost of electricity in the regions of NPPs.

Table 4. Data on the radiation situation and quantity / cost of electricity in the regions of power plants

	NV NPP	Kudankulam NPP	Bruce NPP
Maximum radiation background, mSv/year	0,78	0,042	0,00167
Presence of reactor safety systems that do not require operator intervention in an accident	Yes	No	No
Presence of modern monitoring systems	Yes	Yes	Yes
Population size of the region receiving electricity from NPP, mil. persons	2,34	72,1	14,3
Total capacity of reactor MVt (see Tab..2)	2449	1834	6258
The amount of electricity per person, KW/hour per person	1,05	0,25	0,44
Average cost of KW/hour, USA cents	5	5	13

3. Conclusions

From the results of the comparative analysis, it is clear that in all countries, security issues are taken seriously, however, despite the many similarities in the waste treatment methods, there are also some differences.

Unfortunately, all measures for the protection of the environment and safety at NVNPP and Kudankulam NPP are only mandatory, indicated by the regulatory documentation and legislation of the countries, but not proactive.

The particular difference between the Canadian Bruce NPP and the other two is the highest priority in ensuring safety in the first place (this applies to both employees and the environment). This NPP constantly seeks to reduce its emissions by upgrading the equipment for radiological monitoring of wastewater. The transition has been made to the use of NuclearIQ, a laboratory information management system that will provide maximum quality assurance and quality control when recording data obtained as a result of sampling and analysis. Furthermore, annual testing of monitoring capabilities is being carried out, along with the Canadian Nuclear Safety Commission (CNSB) conducting its independent environmental monitoring that analyzes samples near the Bruce power plant to confirm the safety of the power plant. Also, due to the increased attention to the issues of reducing the amount of materials used and efforts to separate materials, Bruce NPP managed to significantly increase the amount of material sent for recycling, rather than throwing it away as waste, which is another cardinal advantage over the other NV NPP and Kudankulam NPP.

Table 3 clearly shows that the total waste at the Bruce NPP is more than 3 times less (3.47 times) than the total waste of the NV NPP, while the capacity of the Canadian nuclear power plant is almost 2.5 times more. Most importantly - more than 62% of the total amount of waste at NPP Brus is recycled, which is 11.5 times more than at NV NPP. The reduction of radioactive background data at three NPP to one unit of measurement (the readings of the background radiation at the Russian NPP presented in Micro X-rays / hour, at the Indian NPP in mSv/year, and at Canadian NPP in Milievert / year (which is 1000 times less than 1 MicroSiever) allowed to find out that the radiation background at NV NPP is 18.5 times higher than at Kudankulam NPP and 467 times higher than at Bruce NPP. However, Russian legislation states that these figures do not exceed the allowable values and natural background.

At present, the NV NPP uses an open nuclear fuel cycle scheme and waste disposal methods (Scheme 1).

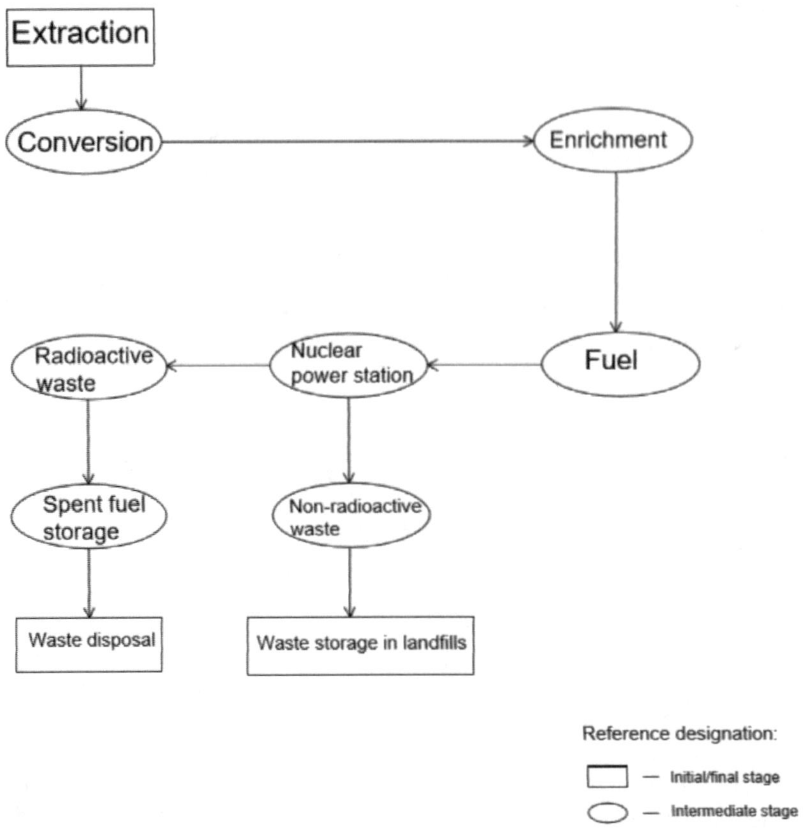

Figure 7. The existing fuel cycle and waste management at the NV NPP

Considering the comparative analysis conducted the more efficient scheme is proposed for improving waste management at NVNPP (Scheme 2).

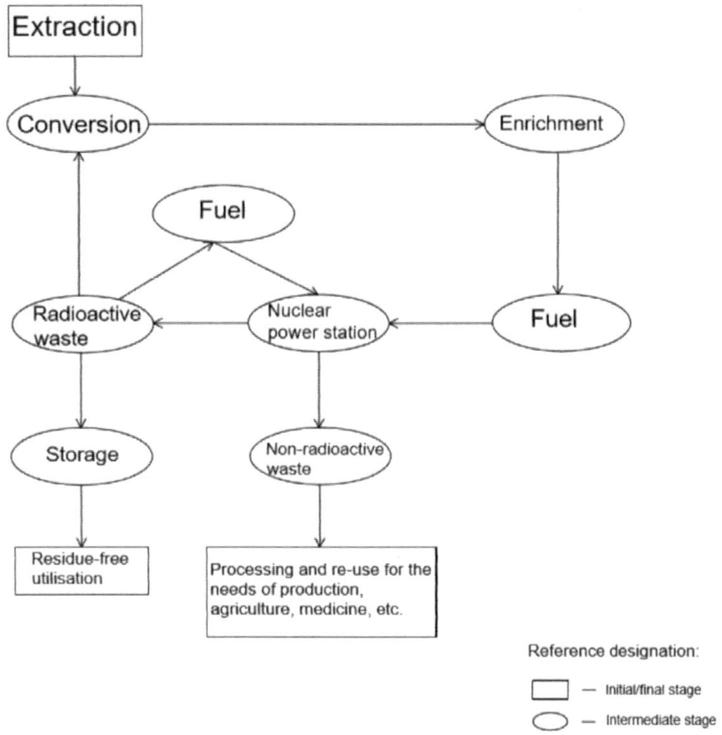

Figure 8. Proposal for the improvement of the functional fuel cycle and waste management at the NV NPP

Effective proposals for improving waste management at NV NPPs can be:

1) Reduction to waste-free production, i.e. sending all non-radioactive waste for processing for possible subsequent use in various fields (agriculture, medicine, manufacturing, etc.).

2) Reprocess of RW for possible subsequent re-use (stages of closed NFC include exposure of spent nuclear fuel on the NPP territory for 3–10 years; temporary controlled storage of SNF in autonomous storage facilities at the radiochemical plant, processing of SNF with separate (or total) fissionable nuclides from it and fission products of commercial interest, solidification and disposal of waste).

Reprocessing of the nuclear fuel spent provides certain economic benefits, restoring unused uranium and drawing into the power industry the accumulated plutonium. This reduces the amount of highly radioactive and hazardous waste that must be properly stored, which also has a certain economic and environmental feasibility. The spent nuclear fuel, containing about 1% plutonium, considered to be a very good nuclear fuel, which does not need any enrichment process, it can be mixed with depleted uranium and supplied as fresh fuel assemblies for loading into the reactors. It can be used for loading and in breeder reactors (covertors and breeders).

Other NRW generated at NPP should be send to storage facilities, using decontamination measures, and then recycled (instead of many years of burial). Moreover, to minimize the amount of stored waste at NV NPP it is recommended to exclude the disposal of radioactive waste, increase the types and volumes of recyclable materials, and apply the waste treatment, using third parties services only for the disposal of residual waste (to minimize waste transportation), respectively, thereby improving the ecological situation in the region.

References

[1] AERB Clearance for Hot-Run of Kudankulam Nuclear Power Plant Unit-1: June 30, 2011.
[2] AES Bryus. https://www.seogan.ru/aes-bryus.html (access date: 01.05.2019).
[3] Atomic energy. Novovoronezh NPP-2. http://www.atomic-energy.ru/organizations/novovoronezhskaya-aes-2 (access date: 02.05.2019) [In Russ]
[4] Nuclear construction sites. http://publicatom.ru/blog/stroyka/1453.html (access date: 3.05.2019) [In Russ]
[5] Bhabha Atomic Research Centre. http://www.barc.gov.in/egreport.pdf (access date: 04.04.2019).
[6] Bruce Power. 2018 Environmental Protection Report. https://www.brucepower.com/2018-environmental-protection-report/ (access date: 29.03.2019).
[7] Bruce Power. Radiation and the Environment. https://www.brucepower.com/radiation-and-the-environment/ (access date: 28.03.2019).
[8] Business lines. Financial Daily from THE HINDU group of publications. https://www.thehindubusinessline.com/2004/01/17/stories/2004011700940600.htm (access date: 10.05.2019).
[9] Appendix to the technical specifications for the technical project of blocks 1 and 2 of the Kudankulam NPP technical specification for the development of the Kudankulam NPP project, blocks 3,4, Atomstroyexport», 2013. [In Russ]
[10] Electronic library of the history of Rosatom. http://elib.biblioatom.ru/text/byulleten-atomnoy-energii_2006_v3/go,7/ (access date: 37.04.2019) [In Russ]
[11] Google maps. https://www.google.ru/maps/place/%D0%9D%D0%BE%D0%B2%D0%BE%D0%B2%D0%BE%D1%80%D0%BE%D0%BD%D0%B5%D0%B6%D1%81%D0%BA%D0%B0%D1%8F+%D0%90%D1%8D%D1%81/@51.2793861,39.1792277,10297m/data=!3m1!1e3!4m5!3m4!1s0x4124b97e12897057:0x2763da55192e9318!8m2!3d51.2657075!4d39.2116551 (access date: 27.05.2019).
[12] Google maps. https://www.google.ru/maps/place/%D0%91%D1%80%D1%8E%D1%81+%D0%9F%D0%B0%D1%83%D1%8D%D1%80/@44.3272628,-81.5886129,479m/data=!3m2!1e3!4b1!4m5!3m4!1s0x88284bf009bf52d1:0xcb1870fd26b33c16!8m2!3d44.327259!4d-81.5864242 (access date: 27.05.2019).
[13] Instruction No. 18-CD on supply, transportation and disposal of industrial waste at nuclear power plants. 2003. Novovoronezh NPP. [In Russ]
[14] Unit Converter. https://www.translatorscafe.com/unit-converter/RU/radiation/28-24 (access date: 10.05.2019) [In Russ]
[15] Kudankulam Nuclear Power Plant, India. https://ejatlas.org/conflict/kudankulam-nuclear-power-plant-india (access date: 03.04.2019).
[16] Marine Impact Assessment Study for KKNPP 3 to 6 August 2011. https://www.npcil.nic.in/WriteReadData/userfiles/file/Marine_Impact_Assessment_Study_for_KKNPP_3_to_6_August_2011.pdf (04.05.2019).
[17] National industrial portal. http://uprom.info/wp-content/uploads/2017/02/DSC_6840.jpg (access date: 27.04.2019) [In Ukrain]
[18] Nuclear Power Corporation of India Limited. https://www.npcil.nic.in/index.aspx (access date: 02.04.2019).
[19] Atomtechenergo. http://www.atech.ru/projects/kudankulam/ (access date: 10.05.2019) [In Russ]
[20] OOO "MNU-1 Korporacii AK "ESKM". http://mnu1.ru/sites/default/files/foto/nvaes1-2.jpg (access date: 17.04.2019) [In Russ]
[21] Environmental safety report for 2010. Novovoronezh NPP. 2011 [In Russ]
[22] Environmental safety report for 2016 https://www.rosatom.ru/upload/iblock/14a/14adadf1ee2b27110a19616bdee1cb34.pdf (access date: 28.03.2019) [In Russ]

[23] Environmental safety report for 2017. http://www.rosenergoatom.ru/upload/iblock/1b0/1b04fad014abe6ec24d67fb9a4a00753.pdf (access date: 29.03.2019) [In Russ]
[24] The radiation situation at the enterprises of Rosatom. http://www.russianatom.ru/ (access date: 10.05.2019) [In Russ]
[25] Regulation of external control of polluting substances entering the environment No. 16-EPA. Novovoronezh NPP. 2005 [In Russ]
[26] Regulations for internal environmental control of Novovoronezh NPP No. 17-OOS. 2008 [In Russ]
[27] Ratings VivaReit. https://vivareit.ru/wp-content/uploads/2015/07/bruce_aes.jpg (access date: 27.04.2019) [In Russ]
[28] RIA Novosti. https://ria.ru/20111013/458046025.html (access date: 10.05.2019) [In Russ]
[29] RIA Novosti. Nuclear technology. https://ria.ru/20171130/1509900391.html (access date: 10.04.2019) [In Russ]
[30] RIA Novosti. Nuclear technology. https://ria.ru/20150316/1052852303.html (access date: 6.05.2019) [In Russ]
[31] "Concern Rosenergoatom". Annual report. http://www.rosenergoatom.ru/shareholders/raskrytie-informatsii/godovye-otchety/ (access date: 02.05.2019) [In Russ]
[32] Novovoronezh nuclear power plant. http://www.rosenergoatom.ru/power plants_projects/sayt-novovoronezhskoy-aes/ (access date: 27.04.2019) [In Russ]
[33] Russia's Nuclear Export Programme. https://www.sciencedirect.com/science/article/pii/S0301421518304245 (access date: 12.04.2019).
[34] The VVERs at KudanKulam. https://www.sciencedirect.com/science/article/pii/S0029549306000781 (access date: 37.04.2019).

Implementation of Zero Liquid Discharge Systems in Russian Industry

V Aksenov[1], N Tsarev[1] and E Maksimova[1]
[1]Ural Federal University, 19 Mira street, Yekaterinburg, 620002, Russia

E-mail: v.i.aksenov@urfu.ru

Abstract. The article focuses on the implementation of zero liquid discharge (ZLD) systems precluding the pollutant release to the environment at metallurgical, machine-building and metalworking enterprises. The need to change over to such systems is due to two main factors: 1. Significant savings in fresh water used in production; 2. A decrease in the number of all kinds of pollutants reaching the environment with various discharges (purified, insufficiently purified and completely untreated). The main feature of ZLD systems at industrial enterprises is the presence of tailing management facilities, e. g. concentrated wastewater treatment facilities; industrial wastewater sludge treatment facilities; non-recyclable oily wastewater and sludge incineration facilities; reclaimed (recycled) industrial wastewater treatment facilities etc. The development stages of ZLD systems in Russia are presented. In 1973, for the first time in the world's practice, in the Verkh-Isetsky Metallurgical Plant in the city of Sverdlovsk (now Yekaterinburg), the ZLD developed by Prof. V. I. Aksenov and his colleagues was launched and has successfully operated in the cold rolled transformer steel workshop. As an example of wastewater treatment facility as part of a ZLD system, some technical installation solutions are described in order to treat chemically contaminated acid ferrous-containing wastewater generated in pickling carbon steel. The focus is on the feasibility of implementing local ZLD systems instead of central ZLD systems. The advantages of such local systems are: possible regenerating exhausted process solutions at the place of their formation, obtaining demineralized water, the local treatment of chemically polluted wastewater, lower investment and operating costs.

Keywords: Steel pickling; Industrial wastewater treatment; Sludge dewatering; Zero liquid discharge system

1. Introduction

The 1990s of the last century were marked by the widespread transition of industrial enterprises of the world's developed countries to the zero liquid discharge (ZLD) systems to prevent the discharge of pollutants into the environment. One of the first formal zero waste policies was created in 1995 when Canberra, Australia endorsed a goal of «No Waste by 2010» [1].

The need to switch to ZLD systems is due to two main factors [2]:

- significant savings in fresh water used in production;
- a decrease in the number of all kinds of pollutants entering the environment with various discharges (purified, insufficiently purified and completely untreated).

In developed countries, the constraints preventing an increase in water consumption always include the high cost of fresh (natural) water, huge penalties and other consequences of a discharge into waters (discharges in excess of permitted pollutant limits) and instant adverse publicity. Therefore, the ZLD systems are being built in the USA, EC etc., after which efforts are taken to use the treated water to the fullest extent [3].

The transition to ZLD systems involves the introduction in industrial enterprises of:

- the process equipment for the operation of which water is not required (for example, dry gas cleaning installations to replace wet gas cleaning installations) or less water is needed. In addition to the reduced water consumption, such equipment makes it possible to eliminate the formation of industrial wastewaters or significantly reduces industrial wastewater generation;
- the local open, closed and combined water recirculating cooling systems in individual production units and workshops (our experience shows that investing in such cycles pays off within a few months);
- local industrial wastewater treatment facilities for certain categories of industrial wastewaters containing heavy metal compounds, organic constituents etc., and end-of-pipe industrial wastewater treatment facilities for release from an enterprise to wastewaters not contaminated by any specific constituents (for example, surface stormwater runoff);
- treatment facilities for surface stormwater runoff and (or) treatment facilities for the advanced treatment of municipal wastewater designed to use purified water after those facilities to replace fresh (natural) water;
- units designed to eliminate the accumulated environmental damage making it possible to obtain some useful industrial products from industrial wastewater sludges stored for a long time;
- units designed to treat the industrial wastewater sludges that eliminating the need to store the industrial wastewater sludges in hazardous waste landfills.

At the same time, the ZLD systems implementation in the Russian industrial enterprises should be preceded by fairly costly preliminary overhaul or replacement of the main production facilities of water industry systems (pumps and pumping stations, water cooling towers, wastewater clarifiers etc.), elimination of water leaks and unauthorised pollutant discharges into water bodies and on the terrain (for example, regular industrial wastewater overflow from clarifiers) [4]. In addition, it is also necessary to carry out administrative and technical measures to bring the environmental documentation available at the enterprises in accordance with the existing provisions of the federal law of Russian Federation «On Environmental Protection» and other related regulatory documents on the best available techniques (BAT) implementation.

2. Features of the Zero Liquid Discharge Systems of Industrial Enterprises

To make a ZLD system of an industrial enterprise, tailing management facilities are used:

- units designed to treat the concentrated wastewater (for example, spent solutions from the pickling processes and the plating processes);
- units for the treatment of industrial wastewater sludges, the operation of which results in not waste being sent to a hazardous waste landfill but recoverable industrial products;
- units for the incineration of unrecyclable oily wastewaters and sludges with waste gas treatment systems that exclude any environmental pollution by combustion products;
- reclaimed (recycled) industrial wastewater treatment facilities (purification and demineralization units (for example, brine from regenerating water softeners or blowdown from the operation of recirculating wet cooling towers);
- units for surface stormwater runoff treatment.

The main tailing management facilities are those for the concentrated spent process solutions treatment (for example, spent pickling solutions) and industrial wastewater sludges, since more than

IOP Conf. Series: Materials Science and Engineering **972** (2020) 012033 doi:10.1088/1757-899X/972/1/012033

80 % of soluble compounds are concentrated in the spent process solutions, and about the same amount in the industrial wastewater sludges. The treatment of this kind of waste alone largely solves the environmental pollution problem.

Tailing management facilities make the water industry system of an enterprise more complicated; the system turns into a chemical and engineering facility designed to produce water for the industrial water supply system and treat the resulting concentrates of demineralization plants, industrial wastewater sludges and other tailing.

Tailing management facilities at industrial enterprises require:

- some significant investment and operating costs;
- high-quality and expensive basic process and pumping equipment, instruments and meters, automation systems;
- highly qualified engineering and operations personnel.

3. Stages of Development of the Zero Liquid Discharge Systems in Russia

The first ZLD systems in the USSR were built to treat the liquid waste with a low radioactivity level, abroad — a reclaimed industrial wastewater treatment facility for municipal wastewater across Lake Tahoe (USA) [5] and a unit for the treatment of saline mine waters in Upper Silesia (Poland) [6, 7]. They were used as the prototype of the modern ZLD systems.

The entire experience of the construction and operation of ZLD systems of our country's industrial enterprises can be divided into three stages, from 1973 to 1990, from 1991 to 2000, from 2001 to the present.

The first stage, especially in the 1980s, was typical of the rapid construction of ZLD systems. The systems created are well described in the papers of that time [8, 9]. We should recall that in 1973, for the first time in the world's practice, in the Verkh-Isetsky Metallurgical Plant in the city of Sverdlovsk (now NLMK Group's VIZ-Steel in the city of Yekaterinburg), a ZLD system developed by V. I. Aksenov and colleagues in the cold rolled transformer steel workshop was launched and successfully operated [10, 11].

The estimated cost of the system at that time amounted to 29 million roubles. Almost at full capacity, it was possible to launch the entire group of treatment facilities in six months due to such gifted water engineers as Yu. V. Bobylev and his colleagues. At that time, the plant manager V. S. Ozhiganov provided invaluable assistance. It is impossible to overestimate the obtained experience of using ZLD systems in domestic and foreign practices.

During the second stage after the collapse of the USSR in 1991, the ZLD systems activities were practically ceased.

By the beginning of the third stage, the main production facilities of the Russian industrial enterprises, including almost all of the ZLD systems of industrial enterprises created during the Soviet Union, were physically and morally obsolete. The decline in the water industry led to the fact that at present in Russia, according to official data, only 11 % of wastewater discharged into water bodies meets all the regulatory requirements. In our opinion, the current status is even worse.

4. An Example of Acid Wastewater Treatment Technology as Part of a Zero Liquid Discharge System

As an example, let us consider a facility for the treatment of chemically contaminated acid ferrous-containing wastewater generated during the carbon steel pickling operation (Figure 1).

The chemically polluted wastewater purification technology at the initial stage of treatment involves wastewater blending followed by a uniform supply to the mixer for reagent treatment. Depending on the composition of the initial flow and process problems to be solved, some alkaline reagents (calcium hydroxide, sodium hydroxide, sodium carbonate), ferrous-containing coagulant, flocculant and other reagents can be applied to the flow treatment.

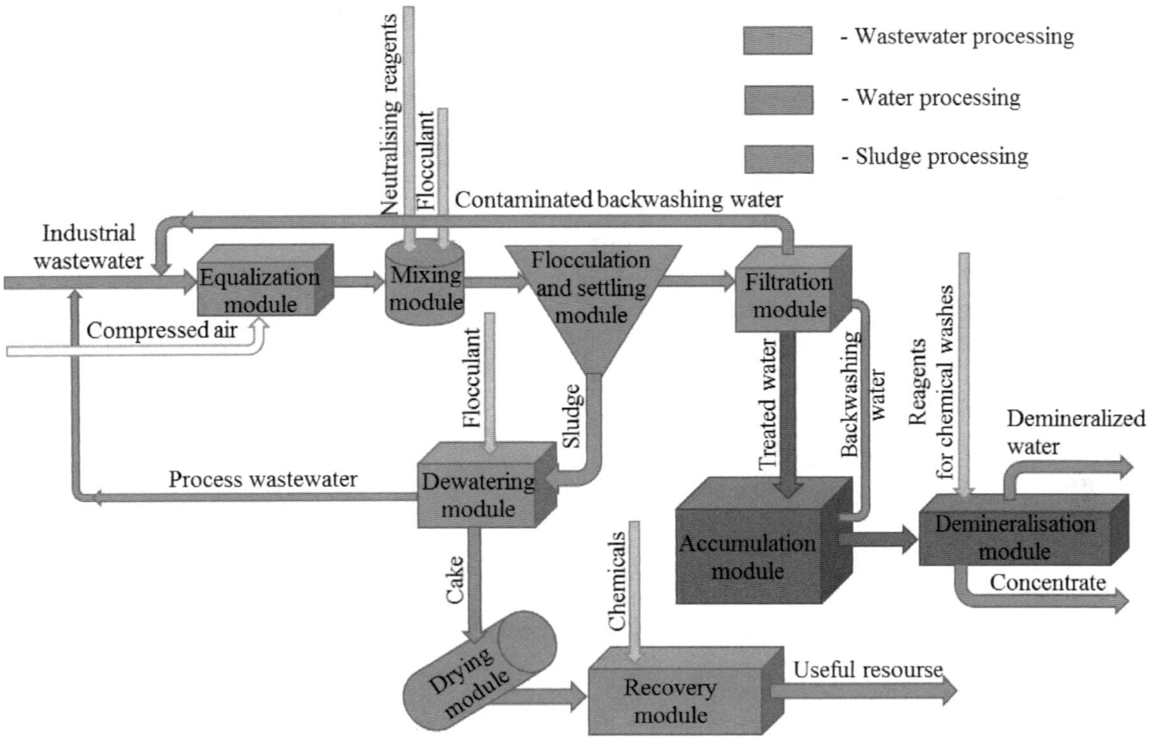

Figure 1. Process flow diagram of a facility for the treatment of the acid ferrous-containing wastewater generated during the carbon steel pickling.

The water treated with reagents enters the flocculation and settling module, where the main amount of toxic pollutants is removed from the discharge. At the same time, the wastewater flocculation and sedimentation facilities can be very diverse, e. g. free-standing and combined with settling tanks of the flocculation chamber, clarifiers of various designs, clarifiers with a suspended sediment layer etc.

For the purification of clarified water, deep bed sand filters with various designs or ultrafiltration units are arranged (filtration module).

Usually, water cannot be reused due to the water quality after a physical and chemical treatment with by the salt component content; therefore a water demineralization module is needed.

To obtain the demineralized water, the filtered water is directed to a demineralization module (a reverse osmosis system or an evaporator). Then, the demineralized water is fed for reuse (steam generation in the installation of steam boilers, use in pickling plants, or for other purposes). The resulting salt concentrate shall be recovered [12–14].

The sludge from the flocculation and settling module is sent to the dewatering module. After that, the cake (dewatered sludge) is dried in drying module, if necessary; the cake (sludge) composition is adjusted at the recovery module, and then the cake (sludge) sent for recycling.

5. Local Zero Liquid Discharge Systems

The capacity of the first ZLD systems of industrial enterprises were up to several hundred cubic meters per hour, since at that time it was possible to construct relatively economical multi-stage evaporators with only high productivity.

The mass implementation in the last 15–20 years in various industries of falling film evaporators, without and with vapor compression, heat pumps, membrane systems and other equipment gave an impetus to the development of energy-efficient evaporation facilities [15] and membrane facilities

[16–18] with a capacity from 10 to 4,000 l/h. Such fully automated units with low productivity are developed and manufactured by companies in Germany (H2O GmbH [19] etc.), France (Degremont Technologies [20] etc.) Israel (IDE Technologies [21] etc.), United states (Aquatech [22], Oasys Water [23] etc.), and other countries. As a result, it is now possible to create local ZLD systems for an individual plant or a industrial site of an enterprise instead of central ZLD systems for a whole enterprise.

Local ZLD systems have several significant advantages:

- in introducing local ZLD systems with the installation of demineralization at any separate production site during the purification of the resulting industrial wastewater, a concentrate will be formed, which can be recycled or returned to the process. For example, the concentrate of reverse osmosis unit, with which the rinsing water of the chrome-plating process are treated, and can be returned to a plating coating bath;
- the demineralized water produced at the local desalination unit as part of the local ZLD system can be immediately returned to the process. At the same time, the replacement of demineralized process water can significantly enhance the quality of products (for example, galvanic and paint-and-lacquer coatings);
- gradual elimination of the chemically polluted and (or) mineralized wastewaters from the industrial enterprise discharge including the treatment of those wastewaters at local treatment facilities increases the efficiency and reliability of the end-of-pipe wastewater treatment plant of industrial enterprise and reduces the man-induced impact on the environment;
- the creation of local ZLD systems requires smaller investment and subsequent operating costs, and also makes it possible to reduce the implementation of a ZLD system investment project.

6. Conclusion

In conclusion, it should be noted that currently Russian industrial enterprises lose the leading position in the creation of ZLD systems. While, virtually, no such systems have been created in Russia in the last five years, in the EU and countries and the USA there are thousands of such systems were put into operation. The reason for lagging behind is due to the lack of economic incentives for industrial enterprises. Let us hope that the situation will somehow change after the entry into force of all the new provisions of the federal law of Russian Federation «On Environmental Protection» and other legislation governing the introduction of the best available techniques (BAT) with a gradual increase in environmental charges and penalties for environmental law violations.

7. References

[1] Platt B and Lakhani M 2004 *Report from Institute for Local Self-Reliance for Global Anti-Incinerator Alliance (GAIA)* (Quezon City: Institute for Local Self-Reliance for Global Anti-Incinerator Alliance (GAIA)) p 85 Available from: http://www.no-burn.org/wp-content/uploads/Resources-up-in-Flames.pdf.

[2] Aksenov V I, Arkhipova O A, Sidorova I A, Nichkova I I 2005 Creation of Closed Water Supply Circuits for Metallurgical Enterprises *Stal* (9) 83–5.

[3] Tong T and Elimelech M 2016 The Global Rise of Zero Liquid Discharge for Wastewater Management: Drivers, Technologies, and Future directions *Environ. Sci. Technol.* **50**(13) 6846–55.

[4] Aksenov V I, Arkhipova O A, Sidorova I A and Nichkova I I 2005 Organizational Problems of Water Handing Facilities at Metallurgical Enterprises *Stal* (8) 96–8.

[5] Culp R L and Roderick R E 1966 The Lake Tahoe water reclamation plant *J. Water. Pollut. Control. Fed.* **38**(2) 147–155.

[6] Gzyl G, Janson E and Łabaj P 2017 Mine Water Discharges in Upper Silesian Coal Basin (Poland) *Assessment, Restoration and Reclamation of Mining Influenced Soils* ed J Bech, C Bini and M A Pashkevich (Amsterdam: Academic Press) chapter 17 pp 463–486.

[7] Ciesielczuk J 2014 Coal Mining and Combustion in the Coal Waste Dumps of Poland *Coal and Peat Fires: A Global Perspective* ed G B Stracher, A Prakash and E V Sokol (Amsterdam: Elsevier Science) vol 3 chapter 16 pp 464–473.
[8] Voznjuk G G and Gorshkov V A 1983 Mine Water Utilization in the USSR National Economy *Mine Water Environ.* **2**(1) 23–30.
[9] Hrudey S E Water Reclamation and Reuse 1982 *J. Water Pollut. Control. Fed.* **54**(6) 654–673.
[10] Aksenov V, Denisov A, Nichkova I and Denisov Y 2018 Zero Disharge Systems: Innovation Trends of Industrial Wastewater Re-use for Metallurgical and Steelmaking Plants *Water & Wastewater Asia* May/Jun 39–42 Available from: http://www.waterwastewaterasia.com/ebook/WWA_MayJun2018/2/index.html.
[11] Zero Discharges Into Water: NLMK Group Companies Introduce a Closed Loop Water System [Internet] [cited 2019 may 5] Available from: https://lipetsk.nlmk.com/en/responsibility/ecology/key-cases/zero-discharges-into-water-nlmk-group-companies-introduce-a-closed-loop-water-system/
[12] Gilron J 2016 Brine Treatment and High Recovery Desalination *Emerging Membrane Technology for Sustainable Water Treatment* ed N P Hankins and R Singh (Amsterdam: Elsevier) chapter 12 pp 297–324.
[13] Kwon K, Han J, Park B H Shin Y and Kim D 2015 Brine Recovery Using Reverse Electrodialysis in Membrane-Based Desalination Processes *Desalination* **362** 1–10.
[14] Vane L M 2017 Water Recovery From Brines and Salt-Saturated Solutions: Operability and Thermodynamic Efficiency Considerations for Desalination Technologies *J. Chem. Technol. Biotechnol.* **92**(10) 2506–18.
[15] Thome J R 2017 A Review on Falling Film Evaporation *J. Enhanced Heat Transf.* **24**(1–6) 483–498.
[16] Hajbi F, Hammi H and M'Nif A 2010 Reuse of RO Desalination Plant Reject Brine *J. Phase. Equilib. Diffus.* **31**(4) 341–7.
[17] Le N L, Nunes S P 2016 Materials and Membrane Technologies for Water and Energy Sustainability *Sustain. Mater. Technol.* **7** 1–28 Available from: https://www.sciencedirect.com/science/article/pii/S2214993715300105
[18] Razali M, Kim J F, Attfield M, Budd P M, Drioli E, Lee Y M and Szekely G 2015 Sustainable Wastewater Treatment and Recycling in Membrane Manufacturing *Green Chem.* **17**(12) 5196–205 Available from: https://pubs.rsc.org/en/content/articlehtml/2015/gc/c5gc01937k
[19] H2O [Internet] [cited 2019 may 5] Available from: https://www.h2o-de.com/
[20] Degremont [Internet] [cited 2019 may 5] Available from: http://www.degremont.com/en/homepage/.
[21] IDE Technologies [Internet] [cited 2019 may 5] Available from: https://www.ide-tech.com/en/
[22] Aquatech [Internet] [cited 2019 may 5] Available from: https://www.aquatech.com/
[23] Oasys Water [Internet] [cited 2019 may 5] Available from: http://oasyswater.com/

Solving problem of airflow in the open air space in double pitched roofs of passive objects with concrete cover

J Plachy[1], J Šál[1]

[1] Department of Civil Engineering, the Institute of Technology and Business in České Budějovice, Okružní 517/10, 370 01 Czech Budweis, Czech Republic

E-mail: plachy@mail.vstecb.cz

Abstract. The paper deals with air flow in open air gap in pitched roofs with folded roofing. The sloping roof is thermally insulated to the passive standard. Double-skinned roofs work with the drainage of moisture from the air gap based on the difference in water vapor pressure that is generated by the warming of the air gap. Moisture removal is suppressed due to the large thickness of the thermal insulation. The paper compares experimentally measured values with values obtained by calculation using the appropriate computer software. For experimental verification of the air flow, a model of the roof cladding was constructed, on which air flow in the air gap was verified. There was no air flow on the roof model. In the theoretical calculation the flow occurred.

1. Introduction

Currently, the trend of building construction towards passive buildings is heading. The recommended heat transfer coefficient values for passive buildings are Upas, 20 = 0.15-0.1 W / (m2.K). [1]. These buildings are covered with flat roofs with cladding or pitched roofs with folded roofing. This covering can be heavy (concrete, ceramics) or light (metal).

Roofs can be made as single-shell or double-shell. For double-skinned roofs, a ventilated air gap is designed between two roof claddings.

Double-skinned ventilated roof structures work on the principle of water vapor moisture removal depending on the air flow. The air flow is caused by the difference in the specific air masses in the gap due to the warming of the air from the interior, due to the radiation of the night sky, the influence of the outside air flow and the different amount of water vapor in the air. Previously, warming from the interior was critical. Warming is greatly reduced with increased thermal insulation thickness. In order to know how the roof works, it is necessary to answer the question of how the air flow in the ventilated roof structure occurs when the insulation thickness is higher.

According to our information nobody has dealt with this issue in detail yet and our article. [2]. This article deals with air gap flow in double-skin roofs with lightweight sheet metal roofing. This contribution follows the previous research. The standard ČSN 731901: 2011 [3] solves the problem of air flow in the air gap by the lack of air gap thickness without further explanation.

2. Material and methodology

The airflow verification methodology in the openair gap is based on two points. The theoretical verification will be done by calculation in the relevant software and experimental verification on the assembled roof section model.

2.1. *Theoretical verification*

For the calculation of the design of double-skinned roofs, the Czech Republic has traditionally used the methodology ČSN 73 0540-4. [1] The methodology of this standard does not allow for the spatial complexity of roofs, the influence of sunlight or the effect of radiation exchange with the sky.

The theoretical verification was carried out in the program MEZERA, which works according to the above standard and is accessible at VŠTE. The program has a Windows user interface. The program works on the principle of one-dimensional stationary conduction of heat and water vapor. More information about the program is available on the website of K-CAD. [4].

Program output:

- airflow velocity in the air gap
- air gap temperature

all depending on:

- thermal insulation thickness
- air gap thickness
- roof slope.

2.2. *Experimental verification*

In the VSTE laboratories, a real roof cladding model was constructed to simulate boundary measurement conditions. The model size is 2 x 3 m. Shown in Figure 1. The model was installed indoors to eliminate the effect of wind. The model can be imagined as a sample roof section representing a field between 2 rafters with a variable slope of 0 ° to 35 °. The thermal insulation is placed above the rafters

Figure 1. Roof model for experimental verification of air gap airflow. Source: own.

The roof structure consists of:

- concrete tile covering,
- battens 40x60 mm,
- contra battens 40x60 mm,
- air cavity (of adjustable height with levelling screws), thickness 40 - 150 mm,

- additional waterproofing layer (AWL) - breather underlay, three-layer high diffusion under-roof membrane (diffusion film and two layers of polypropylene film). Breather waterproof roofing underlay manufactured by thermally bonding outer spunbonded polypropylene layers and inner layer of microporous polypropylene film for pitched supported.
- thermal insulation of hard polyurethane foam boards, λD = 0.021 W / (mK), thickness = 220mm
- vapor barrier, four-layer film (reinforcement grid and two layers of polyethylene laminate film and reflective aluminium foil). Vapor barrier manufactured as multilayer polyethylene sheet reinforced with tear-resistant mesh with heat reflective aluminized surface to be used against vapor and moisture penetration in pitched construction.
- rafters

The air gap in which the air temperature and the air flow rate were measured is made adjustable. Its thickness can be adjusted from 60 to 140 mm with 4 adjustment screws. The model was clad to achieve the necessary conditions for the roof. The cladding was made with an oriented strand board (OSB) on which 180 mm thick expanded polystyrene (EPS) boards were attached. This cladding was made from all four sides and covered from the top. The cladding was not made under the structure. Indoor boundary conditions were provided by the heated laboratory interior. The exterior boundary conditions were ensured by a closed stripped space in which cooling is located.

An aggregate and liquid vaporizer were used to cool the insulated space. In the model the evaporator was placed above the roof covering so that it could be moved as required. The need to be able to change the pitch of the roof meant the connections to the evaporator needed to be flexible.

The measurement will be continuously recorded using the Ahlborn ALMEMO 5690-2M universal measuring center. Eight probes were used for the measurements. These probes were calibrated before the experiment. Six probes measured only the temperature (probes 1, 2, 3, 5, 6 and 7) and the remaining two probes measured the air flow velocity (probes 4 and 8). The probes were positioned in the air gap (inlet, middle and outlet port) and in the cooled compartment (above the covering) and inside (under the roof structure). The positions of the probes are shown in Figure 2.

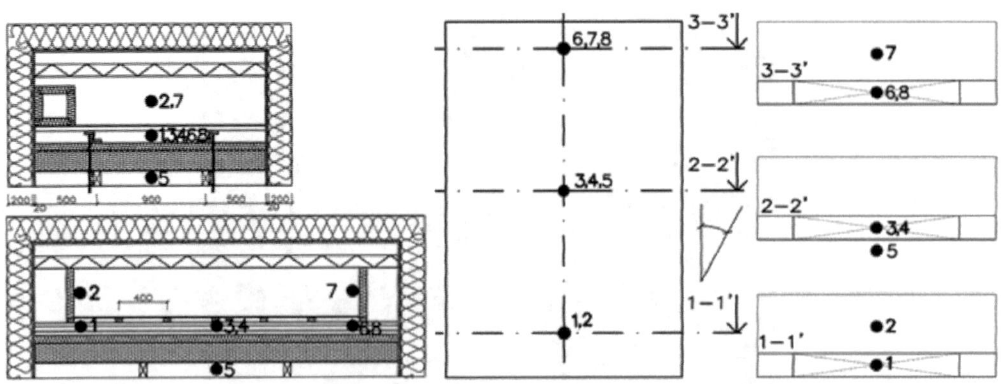

Figure 2. Probe position for temperature and air velocity measurement. Source: [3].

Probe description:

Probe 1 ALMEMO ZA 9020-FS Thermo E4, - temperature °C (accurate to 0.1°C)
Location: air gap, at the inlet port;

Probe 2 ALMEMO ZA 9020-FS Thermo E4, - temperature °C (accurate to 0.1°C)
Location: inside the model in the same place as the evaporator (exterior) above the inlet port;

Probe 3 ALMEMO ZA 9020-FS Thermo E4, - temperature °C (accurate to 0.1°C)
Location: air gap, in the middle of the length of the model;

Probe 4 ALMEMO FVAD 35 TH5, - air velocity m/s, pressure and temperature of the air (this sensor has a control function for the air temperature measured at probe 3) (accurate to 0.01 m/s)
Location: air gap, in the middle of the length of the model;
Probe 5 ALMEMO ZA 9020-FS Thermo E4, - temperature °C (accurate to 0.1°C)
Location: under the model, between the rafters (interior);
Probe 6 ALMEMO ZA 9020-FS Thermo E4, - temperature °C (accurate to 0.1°C)
Location: air gap, at the outlet port;
Probe 7 ALMEMO ZA 9020-FS Thermo E4, - temperature °C (accurate to 0.1°C)
Location: inside the model in the same place as the evaporator (exterior) above the outlet port;
Probe 8 ALMEMO FVAD 35 TH5, air velocity, air pressure and temperature (this sensor has a control function for the air temperature measured at probe 6) (accurate to 0.01 m/s)
Location: air gap, at the outlet port.

3. Results of tests

3.1. Theoretical verification.

The theoretical calculation is based on the above-mentioned publication [2]. The calculation has been extended by further gradients. Three slopes of 10, 22 and 30 ° were selected. The slopes were chosen with regard to the real application of the concrete covering. The thickness of the air gap of 100 mm was chosen with respect to ČSN 731901: 2011 [3].

The theoretical calculation in the MEZERA program was made for boundary conditions on the roof model:

Rafter length: 2415 mm, Inlet opening: 0.092 m2, Outlet opening: 0.092 m^2, Hole rise: 0.42 m for 10 inclination, Hole rise: 0.90 m for 22 ° inclination, Hole rise: 1.21 m for inclination 30° θe = -17°C (-50°C), rel. humidity 85% (95%), θ_{ai} = + 21°C, rel. humidity 50%, wind speed v = 0 m/s, Aerodynam. factors C1 / C2 = 0,6 / -0,3.

The results are shown in Table 1, 2.

Table 1. Air flow temperature in air gap depending on air gap length. Source: Own.

Rafter length (m)	**0.00**	**0.51**	**1.02**	**1.53**	**2.04**	**2.55**
Air temperature (°C) for 10 ° inclination.	-17.00	-16.86	-16.76	-16.68	-16.62	-16.57
Air temperature (°C) for 22 ° inclination.	-17.00	´-16.90	-16.81	-16.74	-16.69	-16.64
Air temperature (°C) for 30 ° inclination.	-17.00	´-16.91	-16.83	-16.77	-16.72	-16.67

Table 2. Airflow velocity (m / s) in the air gap depending on the air gap height and roof slope. Source: Own.

Roof slope (°)	**10**	**22**	**30**
Air velocity (m / s)	0.0804	0.1117	0.1282

3.2. Experimental measurements

Three measurements were taken for about 12 hours. Measurements were made for th. air gaps of 100 mm and slopes of 10°, 22° and 30°. See results. Table 3,4. The table contains maximum temperatures.

Table 3. Air flow temperature in air gap depending on air gap length. Source: Own.

Rafter length (m)	**0.00**	**1.27**	**2.55**
Air temperature (° C) for 10 ° inclination.	-17.00	-15,90	-13.60
Air temperature (° C) for 22 ° inclination.	-17.50	-15.70	-15.00
Air temperature (° C) for 30 ° inclination.	-17.30	-16.40	-14.50

Table 4. Airflow velocity (m / s) in the air gap depending on the air gap height and roof slope. Source: Own.

Roof slope (°)	**10**	**22**	**30**
Air velocity (m / s).	0.0	0.0	0.0

4. Discusions

The theoretical calculation showed air flow in the gap with selected heat transfer coefficient U = 0.11 W / (m^2.K). The calculated speed was 0.0804-0.1282 m/s. Experimental airflow was not measured. There was no flow in the gap of the roof structure model, although different temperatures at the beginning and end of the gap were measured. The temperature gradient was from 2.5 to 3.4°C. Output temperatures must be re-verified. The device in the gap was able to measure the flow rate from 0.01 m/s). Heating the cavity from the interior was not sufficient to move the air based on the difference in specific air weights in the gap. The different result can be justified by the software used, which works with one-dimensional stationary heat conduction. This simplification is likely to be reflected in the result.

5. Conclusions

The experiment showed that if the roofs of passive houses are not exposed to the radiation of the night sky and the flow of outside air, there is no air movement in the gap.

If further research confirms that there is very low air flow in these structures, it will be necessary to consider this issue in more detail. Increased relative humidity in the ventilated roof gap leads to additional problems. These include the formation of condensate, the increased moisture content of wooden structures, thermal insulation and thus the life of these structures.

References

[1] ČSN 730540-2:2011. Thermal protection of buildings – Part 2: Requirements. Prague: Institute for Standards, Metrology and Testing. 2011-11-01. Sorting character 730540-2.
[2] ŠÁL J, Plachý J and Bednářová P 2017, Problems of airflow in the open air space between double skin roofs of passive objects. 17th International Multidisciplinary Scientific GeoConference Sofia: Stef92 technology LTD vol 17 pp 367-374 ISBN 978-619-7408-13-3.
[3] ČSN 731901:2011. Designing of roofs – Basic provisions. Prague: Institute for Standards, Metrology and Testing. 2011-02-01. Sorting character 731901.
[4] K-CAD spol. s r.o. Available from: http://kcad.cz/cz/uvod/. [Quoted 2018-11-27]

Defining the Principles of Organic Shaping while Classifying Organic Architecture

T Y Bystrova[1]

[1]Ural NII project RAASN, 50a Lenina str., Ekaterinburg, 620075, Russia

E-mail: taby27@yandex.ru

Abstract. The creation of organic forms brings architectural objects closer to the organization and needs of a human being. With all the variety of design approaches, it is possible to identify the basic principles of organic shaping, helping to avoid superficial "mimetic" solutions. Classification of approaches of organic shaping in the architecture of the 19th–21st centuries provides a complete analysis of the material necessary to derive such principles, as well as the conceptual one-sidedness, in which one possible solution is prioritized. Developed by the author tables help to master a wide range of areas and approaches of organic architecture.

Keywords: organic shaping, organic architecture, co-proportion with a human being, ecological architecture, classification of organic architecture

1. Introduction

The loss of the integrity of architectural objects and complexes, increase of visual and physical "pollution" of space, the impossibility of creating urban ensembles for economic reasons – all of these processes raise the question of the creation of architectural objects and systems that are commensurate with the mental and psychological organization of a human being, bringing a sense of harmony, nurturing and enhancing it [see: 26]. As a paradigm shift for urban development (from industrial guidelines to neo-industrial) these trends have an opportunity for implementation.

At the same time, environmental trends in culture are gaining strength, a healthy lifestyle is becoming popular when a number of companies, places, cities need to communicate their "advancement" in this area. The easiest way to do this is by means of organic architecture, receiving, if not widespread, but confident spread in the world.

The concept of "organic architectural shape" comes from the idea of the necessary integrity, internal structure and deep connectivity of parts of the object of any scale, designed by an architect. We are not limited to an interpretation of the organic approach declared by L. G. Sullivan and F. L. Wright, finding sources in the architecture of much earlier periods [6].

The methodology of the paper is based on the analysis of both the objects of architecture and the creative attitudes of their authors. The research is hampered by the lack of recognizable features of the external form, which could be clearly attributed to the object to the organic architecture: buildings can have very different looks but can be made using organic shaping. A small number of objects corresponding to the parameters of organic architecture in modern Russian cities complicates professional reflection.

IOP Conf. Series: Materials Science and Engineering **972** (2020) 012035 doi:10.1088/1757-899X/972/1/012035

The first attempt to review all organic architectural approaches belongs, most likely, to B. Zevi (1918–2000), after moving to the United States under the influence of the ideas of F. L. Wright and L. Sullivan [34].

In the Russian-language theoretical literature of the beginning of the 21st century, we can find a small number of references to organic architecture, mainly in relation to the objects and programs of F. L. Wright, who owns the introduction of the term "organic architecture" [30], as well as other proponents of the organic approach in modern architecture – A. Gaudi [29], El. Saarinen [16], R. Steiner [24] M. Budzynsky [3]. The principles of organic architecture are stated in [28], but they are not confirmed by the analysis of objects. The most informative works are by [19], [30].

The following works published abroad should be acknowledged: W. Auer, L. Bachman, G. de Bruyn, M. Zbasnik-Senegacnik, L. Kroll, H. V. Kruft, R. Motro, A. Njoo and others, whose individual approaches disclose their ideas, creative principles, including about regional versions [21]. We have not found a systematic presentation of the development or classification of organic architecture in these sources. Sometimes useful information is contained in interviews with architects [15]. An extensive database of photos of organic architectural solutions is collected in [14]. The most structured information on the topic is in the Volkania website.

In this paper the following sources were also used:

- references on the history of aesthetics, enabling the development of key concepts of the study [5; 9; 10; 32];
- the authoritative writings of theorists and practitioners of architecture, allowing to understand its essence, features independent of style or technologies [7], [8], [23], [25].
- the critics of modernist architecture, R. Venturi and his co-authors [31], R. Koolhaas [17] et al., which has parameters being far from the organic architecture. The latter makes it possible to draw a fundamental distinction that helps in defining the organic architecture of the 20th–21st centuries: organic architecture is a reaction to all versions of modernist architecture and is opposite to it, first of all, as a creative program and a way of project thinking. At the same time, the study of the evolution of architectural thought shows the deep connection of organic architecture with the ancient tradition.

Conceptual and methodological framework influenced by the ideas of Ch. Alexander [1] and N. A. Salingaros [26]. It lies in the fact that organic shaping makes architectural objects closer to the physical, mental and psychological structure of a person, and therefore satisfy his needs and contribute to his comfort.

The comparative-historical approach ensures the balanced vision of the subject of the research on the basis of the idea of invariant characteristics of objects of organic architecture. The morphological analysis allows us to assess the sequence of implementation of organicism as a design and methodological prescription by the architect.

The study of the works of architects who share an organic approach or are inclined to it, allows us to revise the traditional classification of architects in different directions. Generalizing the characteristics of organic architecture by different researchers [2; 11; 13; 18] shows, the organic form and its design do not require following the rigid style features, and therefore may be generated by authors who share approaches to different styles [4; 12; 20; 27]. In other words, **the organic form in architecture is not so much the result of following a rigid set of rules, as the implementation of conceptual and philosophical, scientific and theoretical set-ups.** The movement in this direction is of **meta-project** nature, i.e. it is of primary importance in relation to the definition of formal stylistic features of an architectural object [22; 28; 33].

Hence, the strive for organic shaping can be found (to various degrees) in the works of architects belonging to different approaches. We think that the main requirement, which must be observed, is awareness, a professional reflection of this desire. This is not always the case. Designers may have spontaneous aspirations for organicism being influenced by the best examples or authoritative opinions, as well as by the aesthetic and compositional grounds.

Results. The comparative historical morphological analysis allows classifying representatives of organic architecture by historical periods and regions, highlighting the invariant principles of the approach and specific historical features.

So, for the antique architecture and the subsequent classical traditions: morphological and visual primacy of the whole to its parts, which is determined by the comprehension of the whole by an architect; the individuality of the object when it is executed on the basis of the theory of proportions and the standard algorithms of construction; the isomorphism of an architectural object to the organization of the natural world.

The Western European Medieval organic form is a structural clarity and subordination of dimensions; integrity, clarity, proportionality (F. Aquinas); the presence of a symbolic aspect of the form, which is created by an architect just as the world was created by God.

In the organic architecture of L. G. Sullivan and F. L. Wright, the primary focus is on the visual and spatial relationship of the object with the environment with insufficient elaboration of internal structural relations.

R. Steiner and his followers (until present) are characterized by a reference to the first phenomenon; shaping the building according to its natural environment; the use of the principle of self-similarity; refusal to imitate the external shape of natural objects.

A. Gaudi and his followers are known for working with structures similar to the structures of the living followed by the decoration with local natural materials; transformation of natural forms into abstract, geometric curvilinear shapes, introduced into architectural solution and providing its connection with other forms of a particular culture; individualization of both the whole and its elements of different size; the principle of "sacred ugliness" of the form [29].

The architecture of S. Kalatrava is based on the understanding that organic architecture does not have stable formal features. It is focused on the person not only as a carrier of needs but also – in the unity of his physical and mental features – as the starting point of any architecture. In this regard, it is not a method of design, but rather a certain way of creativity [Ivy]. It is characterized by real, not imitative, the expressive pronunciation of forms, just like nature, it does not stand still. Dynamics is about the organization, not chaos. It is not accidental that the researcher P. van der Ree characterizes deconstructivism and organicism as two types of architects' works with dynamic forms that are opposite in terms of idea and result [30].

K. Alexander and N. Salingaros put physical and mathematical correspondence of architectural forms to the laws of nature at the forefront. Along with the traditional indicators, they operate with the vocabulary of the information approach, which allows evaluating, for example, modernist forms, as a variant of the degradation of architecture, due to the loss of its necessary multilevelness and connectivity.

In general, the organic architecture of the second half of the 20th century becomes more diverse; it overcomes the only interpretation as an imitation of natural forms which is replaced by an understanding of organic architecture in the unity of structural, aesthetic, social and environmental aspects, taking into account the external relations of the designed object. At the beginning of the 21st century, there is an active development of regional versions of organic architecture; regional trends develop alongside international; the desire of architects to create complexes - microcosms, arranged according to the laws of the organization of the living, is increasing; attention to the users of architecture is increasing. At the same time, due to the general growth of environmental trends, organic architecture becomes image-building, especially for public buildings and structures; the form at this stage involves the inclusion of technical components, including those contributing to the "sustainability" of the building.

2. Discussion

The classification of various organic architectural approaches is made according to the authors' understanding of the concept of "organicism". The chronological division is advisable only in the early stages of the collection of material because it allows you to identify specific historical and regional

IOP Conf. Series: Materials Science and Engineering **972** (2020) 012035 doi:10.1088/1757-899X/972/1/012035

identity through the comparative analysis. In order to generalize the material and highlight the invariant principles of organic formation, it is necessary to take steps of a higher level, namely, to isolate the General points present in the architects of different eras and regions. In its turn, the identification of the degree of consistency of organic shaping in the projects of different architects allows us to further develop recommendations for the use of an organic approach in various projects, different types of buildings, etc.

We emphasize that, following the majority of architects and theorists, we generally withdraw from consideration mimetic, imitative forms, only externally resembling natural and, in fact, are examples of styling – works with the outer shell, capable of enhancing the commercial attractiveness of the object.

Classification is made in three directions of organic shaping, further presented in tables. These are:

- a form as *a rationally (consciously) organized whole conditioned by structural goals* (Table 1). This aspect requires an additional explanation of the term "rational", in the 20th century traditionally identified with the industrial nature of the creation or machine organization of architectural form. In fact, rationality since ancient times implies the realization of the goal, the purpose of the object in all its diversity of semantics and functioning for the human being, consistent expression of this goal by architectural means;
- a form as *a whole close to nature*, including *sensual-emotional* and even irrational and decorative elements (Table 2), what K. Alexander calls "roughness" or "inaccuracy" of form. These qualities are associated with a deviation from regularity or perfect symmetry, can be caused by taking into account the natural or urban environment. Surprising as it may seem, they strengthen the internal connectivity, and hence the stability of architecture, bringing necessary for life asymmetry and hierarchy in its forms. This option in Western European architecture, which is under the significant influence of classical tradition, appears quite late and is reflected in the examples. In addition, it echoes the Taoist-Zen interpretation of the form, gradually penetrating into the world architecture since the end of the 19th century;
- a form as a whole, which is dominated by spontaneous, irrational, even mystical moments, especially when it comes to cult architecture (Table 3). Paradoxically, almost all representatives of the organic approach come to this interpretation (for example, here: [26]). Images of their architecture, however, do not recreate the appearance of living organisms but have their own logic and expressiveness.

Table 1. Rationalistic shaping in the organic architecture of different periods and regions

Time and region	An object of organic architecture	Information about the object and comments
Antiquity, proportionality, which defines not only functional but also the semantic subordination of parts to the whole, including nature ("cosmos")		Ancient temple. Garni, Armenia. I century. Photo: A. Manukov. The architect makes the base and steps higher, enhancing the effect of the already huge height of steep mountains

IOP Conf. Series: Materials Science and Engineering **972** (2020) 012035 doi:10.1088/1757-899X/972/1/012035

Time and region	An object of organic architecture	Information about the object and comments
Late Middle Ages. Western Europe. The whole, not reduced to the sum of its parts, and therefore has a deep symbolic meaning within the theocentric worldview		Amiens Cathedral. France. Detail. Architects: Robert of Luzarches, Thomas de Cormont, Renaud de Cormont. Beginning: 1220. The impossibility of reducing the whole to parts is clearly visible, and so, like the Universe created by God, the whole structure is organized.
XX century. The USA. The organicity of the human environment, the connection of all elements of the architectural form, including the material		Dana-Thomas House. Architect F. L. Wright. Springfield, the United States. 1902. Source: website architime.ru
Twenty-first century. Western Europe. Design that creates and organizes the necessary space.		Mediopadana railway platform by architect S. Calatrava. Italy. 2018. Photo: MEGLIO

IOP Conf. Series: Materials Science and Engineering **972** (2020) 012035 doi:10.1088/1757-899X/972/1/012035

Table 2. Form shaping based on the understanding of form as a whole, including sensory-emotional components

Time and region	An object of organic architecture	Information about object and comments
The end of the nineteenth century, Europe and Russia		Home of Talberg. Architect E. Saarinen. Finland. 1897-98. Slight asymmetry and rhythmic disturbances enhance emotional expressiveness [Jetsonen J.,].
The second half of the 20th century, Europe		The main building of the faculty of Humanities, Catholic University named after Peter Pazman. Architect I. Makovec. Hungary. 1995.
The 2000s, the architecture of symbiosis		The building of the cancer center. Swansea, UK. 2006. Architect K. Kurokawa. Source: http://www.archplatforma.ru/
2010s., China		Gallery in Shanghai, China. 2009-2011. Architect: Atelier Deshaus (LiuYichun / Chen Yiming) URL: http://www.deshaus.com/ en /works15.HTML

IOP Conf. Series: Materials Science and Engineering **972** (2020) 012035 doi:10.1088/1757-899X/972/1/012035

Table 3. Interpretation of the architectural shape, dominated by spontaneity

Time and region	An object of organic architecture	Information about object and comments
The 1920s, Europe		Residential building. The Complex of The Goetheanum. Architect R. Steiner. Germany. 1924-28. Photo: H. Kuhle.
The 1980s, the USA		Lencioni Residence. Sanger, California, USA. Architect Arthur Dyson. 1986.
The 1990s, Europe		Residential building "in the Meadows". Bad-Soden, Austria. 1990–93. Architect F. Hundertwasser. Source: http://housing.totalarch.com/
The 2010s, China		Exhibition center of OTOG. Erdos, China. Architect K. Wong. 2019. Source https://www.arch2o.com/

Along with the above, it is possible to classify objects of organic architecture by the method of creating the basic elements of the shape. Summarizing and systematizing historical and modern material, we identify eight main options. We show them on a scale from the most "speculative" to more and more empirically-based, leaving the chronology behind:

- Movement from the idea of the architect of living objects, *imitation of the structure* and the appearance of the living, often with the loss of its structural and organisational features (R. Steiner, Z. Hadid, representatives of parametrism, etc.)

- *Geometric* differentiation of the integral architectural form (antiquity, versions of the classical approach, E. Saarinen et al.) in accordance with the laws of the world order, as it is speculatively and mathematically understood and expressed by the architect.
- Approximation to organic shaping through the achievement of the *structural integrity* of the object (F. Candela, K. C. Yoon, Avanto Architects, etc.)
- Creation of a shape as a *structured proportional whole*, which determines its parts, just as the forms of the living are organized (Gothic and later Gothic versions; Russian modern).
- Geometric abstraction of architectural shape from the *original natural forms* (A. Gaudi, Akihisa Hirato, etc.).
- *Modeling* of the structure of architectural objects according to the laws of the living (B. Hetger, S. Calatrava, G. Lynn, etc.)
- The materialization of regularities of the surrounding landscape in the architectural shape (F. L. Wright, F. Hundertwasser, A. Blok, T. Alberts, etc.)
- Development and calculation of the architectural shape as a part of the landscape, becoming a whole in relation to it (environmental approach in Russian architecture, B. Prince, G. Keller, Annti Lovag, Christian Müller Architects etc.).

On the basis of these and a number of other classifications, we have derived such invariant principles of creating an organic shape in architecture as the primacy of the whole in relation to its parts; the rationality of the form; the aesthetics of the whole ("proportionality", "clarity"); isomorphism in relation to a human being, proportionality at different levels; connection with the natural and/or urban environment and interaction with it; the lack of strict style features.

3. Conclusion

The work allows us to identify the principles of creating an organic shape in architecture, which can later be used in determining the design and artistic and aesthetic solutions in buildings of various types, including taking into account the regional component.

The methodology of the article, which caused the movement from the classifications of understanding the organic shape to the definition of invariant principles, showed its productivity. At the next stage of the work, the author plans to focus on the trends in the development of the organic architecture of the present time, as well as to raise the question of the possibility of incorporating structures and technologies into buildings that have an organic shape.

Of course, all options for organic solutions are favorable for people and have a great future.

Acknowledgements

The paper is written by supporting of Russian Academy of Architecture and Construction Sciences (RAASN), research 1.6.5. "The mastering of new concepts of shaping in the architecture of Russia of the XXI century" of the Basic Research Plan of the RAASN and the Ministry of Construction of Russia.

References

[1] Alexander Ch 1977 *Pattern Language. Towns, Buildings, Construction* (New York: Oxford University Press)
[2] Batty M and Longley P 1994 *Fractal Cities. A Geometry of Form and Function* (San Diego: Academic Press)
[3] Bembel I O 2015 Ideas of organic architecture and the work of Polish architect Marek Budzinsky *Forma magazine* **1** P 21–26 http://www.forma.spb.ru/magazine/articles/d_0010/main.shtml Circulation (in Russian)
[4] Bultman C *Ist Santiago Calatrava ein organischer Architekt* https://www.academia.edu/Download
[5] Bychkov V V 2003 *Aesthetics. Short course* (Moskow: Project) (in Russian)

[6] Bystrova T Yu 2018 *From Modernism to Neo-Rationalism: Architectural Concepts of the 20th–21st Centuries* (Moskow-Ekaterinburg: Cabinet Scientist) (in Russian)

[7] Bystrova T Yu 2013 The Evolution of Organic Architecture: Interpretations and Modifications *Academic Bulletin UralNIIproekt RAACS* Part 1 #1 P 37–44; Part 2 #2 P 45–51 (in Russian)

[8] Dobritsyna I A 2007 *From Postmodernism to Nonlinear Architecture. Architecture in the context of modern philosophy and science* (Moskow: Progress-Tradition) (in Russian)

[9] Dolgov K M 2004 *Reconstruction of the aesthetic in Western European and Russian culture* (Moskow: Progress-Tradition) (in Russian)

[10] Eco U 2004 *Evolution of medieval aesthetics* (Moskow: Azbuka) (in Russian)

[11] Brittain-Catlin T 2017 *Feelings Are a Better Way to Discuss Architecture Than Conceptsy* https://thewire.in/127360/feelings-better-way-discuss-architecture-concepts/

[12] Friedensreich Hundertwassers Welt der Farben und Rundungen *Volkania. Organische Architektur. Eine Sammlung* (in German) http://volcania.wordpress.com/2011/03/03/friedensreich-hundertwassers-welt-der-farben-und-rundungen/

[13] Muthesius A 1996 Für ein natur- und menschengerechtes Bauen – Hundertwasser Architektur. Fragmente (Köln: Benedikt Taschen Verlag) (in German)

[14] Hess A 2006 *Organic Architecture: The Other Modernism* (Salt Lake City: Gibbs Smith)

[15] Ivy R An Interview with Santiago Calatrava *Architectural Record* https://haiart.wordpress.com/2009/03/08/an-interview-with-santiago-calatrava/

[16] Jetsonen J and Jetsonen S 2014 *Saarinen Houses* (New York: Princeton Architectural Press)

[17] Koolhaas R 2013 *New York outside of itself: Manhattan Retrospective* Manifesto (Moskow: Strelka Press) (in Russian)

[18] Kroll L 1986 *The Architecture of Complexity* (London: BT Batsford Ltd)

[19] Kurbatov Y Time of Organic Architecture *Form. Architecture and design for those who understand* http://www.forma.spb.ru/magazine/articles/d_0010/main.shtml (in Russian)

[20] Bruyn G de, Betz O, Nebelsick J H, Drack M and Limpinsel M 2016 *Making Life "Visible": Organism Concepts in Biology and Architecture as the Basis for an Interdisciplinary Synopsis of Constructional Biomimetics* (Springer)

[21] Njoo A H G 2008 *Organic Architecture: its origin, development and impact on mid 20th century Melbourne architecture. A thesis submitted in fulfillment of the requirements for the degree of Master of Architecture* (RMIT University)

[22] 2008 *On Growth and Form. Organic Architecture and Beyond* Philip Beesley and Sarah Bonnemaison (Halifax: Tuns Press and Riverside Architectural Press)

[23] Pearson D 2001 *New organic architecture: the breaking wave* (University of California Press)

[24] Rudolf Steiner als Architekt *Anthromedia. Internetportal Anthroposophie* http://www.anthromedia.net/de/themen/anthroposophie/goetheanum/der-goetheanum-bau/architektur-des-ersten-und-zweiten-goetheanum/ (in German)

[25] Ryabushin A V 2005 *Architects of the turn of the millennia* (Moskow: Art. XXI century) (in Russian)

[26] Salingaros N A 2014 *Algorithmic Sustainable Design. Twelve Lectures on Architecture* (Portland: Sustasis Foundation)

[27] *Santiago Calatrava's anatomical structures, biomorphims* http://personal.strath.ac.uk/j.wood/Biomimetics/Inspirational%20people/Santiago%20Calatrava_files/Santiago%20Calatrava.htm

[28] Shuvalov V M, Komarova Yu A, Golovataya O V and Kenich O 2016 Organic Architecture in Harmony with Man and Nature *RUDN Bulletin. Ser.: Engineering research* **4** P 98–104 (in Russian)

[29] Sirlot H-E, Vivas P and Pla R 2013 *Gaudi. Introduction to its architecture* (Barcelona: Transgle Postals) (in Russian)

[30] Van der Ree P *Mensch und Natur als Inspirationsqelle. Ursprung, Entwicklung und Aktualitaet der organischen Architektur* (in German) http://www.organische-architektur.org/pags/docs/77.1336550571.pdf
[31] Venturi R, Scott Brown D and Izenour S 1977 *Learning from Las Vegas* (Cambridge, London: The MIT Press)
[32] Wölflin G 1912 *Classical Art: Introduction to the Study of the Italian Renaissance* (Moskow: Brockhaus-Euphron) (in Russian)
[33] Zbasnik-Senegacnic M and Kitek Kuzman M 2014 Interpretations of organic architecture *Prostor* **22**(**2**) P 290–301
[34] Zevi B 1950 *Towards an organic architecture* (London: Faber&Faber Limited)

IOP Conf. Series: Materials Science and Engineering **972** (2020) 012036 doi:10.1088/1757-899X/972/1/012036

Ordination Architecture Analysis: Calculation of Proportionality of Order Form Elements

A V Dolgov[1]

[1]Ural NII project RAASN, 50a Lenina str., Ekaterinburg, 620075, Russia

Email: mail@uniip.ru

Abstract. Architectural ordination is part of the theory of architecture dating back to antiquity but in many ways distorted today. The generalization of approaches to it gives two directions of research, namely, abstract (the search for universal laws of proportionality) and empirically-specific (the analysis of the proportions of an individual object). Adhering to the second approach, the author develops his own version of ordination analysis based on major-minor ratios. The calculations are confirmed by the results of measurements of buildings of different epochs and tested in the design practice presented in the article. Objects created on the basis of the theory of ordination can belong to different styles, but, undoubtedly, harmonize the human condition and the urban environment.

Key words: architecture, harmonization, ordination, proportioning, architectural order, major-minor order, proportionality, system relations.

1. Introduction

The need for a harmonious subject-spatial environment refers to basic human needs. As urbanization progresses, this environment becomes increasingly discrete and chaotic, disfiguring the environment, leading to mental and visual stresses. One of the possible forms of harmonization is the classical architecture, the harmonizing effect of which is generally recognized. However, the existing pluralism of architectural formation has put an in-depth study of the theoretical foundations of classical architecture, in particular, the concept of ordination, to the periphery of professional knowledge.

At the same time, the use of forms of style architecture based on order principles still stays at the forefront of modern architecture. Both at the beginning of the XIX, XX centuries, and at the beginning of the XXI century, architectural classics were again updated in the works of modern architects responding to social and aesthetic inquiries. The requirements of modern developers and customers to the quality of theories and pre-design concepts found out that most outstanding authors of treatises on the rules of orders essentially took separate chapters from Vitruvius as a theoretical basis and considered the correct graphic representation of samples in the system of scales devised by themselves to be the basis of the rules [11; 28]. We believe it is necessary and possible to overcome such narrowness and develop the theory and practice of proportionality of order forms. Therefore, along with a theoretical search, the article presents the results of its design testing by the author of the article.

IOP Conf. Series: Materials Science and Engineering **972** (2020) 012036 doi:10.1088/1757-899X/972/1/012036

2. Methodology

Despite the fact that the concept of the whole as a system of forms with diverse connections, has long been rooted in aesthetics and related branches of knowledge, architectural theory has not developed this issue within its disciplinary boundaries.

Starting with Vitruvius, then L. B. Alberti, J. da Vignola [11], A. Palladio [22] and their followers [3], the proportionality of the main parts of the order is declared, but not disclosed. The founders do not go further than stating the most stable ratios of column height to entablature height, column height to architrave height and other paired comparisons. Moreover, they are different for different authors. O. Choisy [8], V. P. Zubov [32], G. N. Emelyanov, A. G. Gabrichevsky, and others paid attention to this. The diagram of the dependence of the height of the architrave on the height of the column, illustrating the linear relationship between them [8] is well-know. In the comments to the treatise of G. S. Lebedeva [19] provides various statistics of dimensional relationships of cornices, architraves, friezes, columns, intercolumnia, for different buildings and authors, comparing them with each other but not trying to find common roots of proportionality parts of any particular order.

Professional literature, one way or another covering the problems of linear relations of order forms, is colossal in its scope (for a bibliography, see [29]). It is possible to distinguish two main wings of interest to us. Proponents of the first position continue the Renaissance-Classicist tradition and approach the study of the subject of proportioning more abstractly trying to deduce universal laws. These include R. Wittkower [30], A. Zeising [31], S. Colman [10], J. Hambridge [16], V. P. Zubov [32], V. Radzyukevich [23], L. I. Taruashvili [25], I. S. Shevelyev [24]. Other authors come from empirical data, studying specific objects of architecture: O. Balakshin [2], M. Ghyka [14], Ch. B. McGrew [21], K. N. Afanasyev, writing about Old Russian architecture [1], N. I. Brunov [5] and [6]; [20]. Our work was carried out based on the second, concrete empirical approach.

Let us note the connection of the scientific interests of specialists in proportioning with the stylistic processes of the second half of the XIX (Europe) and the 1930s (USSR), while we, above all, are interested in the possibility of using knowledge of proportions to create a stable, harmonious living environment in modern cities. The relevance of research on the problem is proved by the materials connecting the searches of the XVIII, XIX centuries with modern architectural processes [7]. The direction connecting the process of architecture and music is promising [12; 13; 27]. The proposals by L. Krier on the optimization of the urban environment based on the idea of proportions are interesting [18].

The difficulty lies in the fact that, in the presence of the work of the proportionalists abroad and in Russia, the traditions of the support of a particular object have been lost. It is not by chance that M. Cohen wonders where proportional studies disappeared, the peak of which falls in the middle of the XX century [9]. It was the absence of specifics that made possible the discussions of the type that the supporters of Renaissance views R. Wittkower and Le Corbusier, who relied on the Fibonacci series, led on the theme of divine proportion [17].

Moreover, attempts to identify the consistency of the linear relations of individual elements of the order were made based on minor-major relations. For example, G. D. Grimm (1865–1942) constantly used minor-major relations when analyzing the proportions of orders and their parts, including when analyzing the modern architecture of the 1930s [15, *123–135*]. However, these were minors and majors derived from the "golden section" relationship, and not from the minor-major order peculiar to a specific order or construction. Naturally, the proportion alien to a particular subject could not clarify anything.

In general, there are no works in domestic and foreign architectural science explaining the system's linear relations of order form parameters and their parts, taking the main order values inherent in a particular order, the order height and column height, or their derivatives. Meanwhile, their correlations are easily determined through measurements, absolutely visual and individual for each architectural object. It is in this direction that we go in deciphering proportionality and building the whole and its parts using the method of architectural ordinates and the theory of initial linear relations developed and tested by the author in the projects of 2016–2018: correction of the facades of the Tour Theater in

V. Pyshma, Sverdlovsk region; the project of the church in V. Pyshma; the development of facades of Sandunovskie baths in Yekaterinburg; Cathedral in Honor of the Resurrection of Christ in the city of Kokshetau, Kazakhstan, etc.

3. Results

During the study, we obtained two types of results, namely, measurement-computational and design-architectural.

Ordination analysis was carried out to study the peculiarities of the major-minor order based on the images of orders contained in the sources and their full-scale survey. It allowed us to collect and systematize many previously separated data. These are graphic representations of orders in the treatises of A. Palladio [22], J. Da Vignola [11], H. Blum [4], A. Desgodetz, F. de Chambray and others.

From the author's measurements, the dimensions of the building and porticos of the Corinthian order of the Alexander Nevsky Cathedral of the Novotikhvinsky Female Monastery (Yekaterinburg, mid-XIX century, arch.: D. Visconti, I. Charlemagne), as well as measurements of the bases and pedestals of the Vienna classicist buildings (Figure 1): Austrian Parliament (arch. T. von Hansen); University of Vienna (arch. H. von Ferstel); The Vienna Opera (architect S. von Sicardsburg); buildings of the imperial museums (architect G. Semper and C. Hasenauer); Karlskirche church (arch. J. B. Fischer von Erlach) and others have been analyzed.

Figure 1. Examples of measurement bases and pedestals of historical buildings of Vienna (photos by A. V. Dolgov).

Author's developments were tested in the design of a number of objects, both classical and modern (Figures 2, 3). Patterns found in the course of the study are capable of practical implementation, and this is the main result of the study.

4. Discussion

Studies by the author's method of ordination analysis allowed proving the proportionality of the linear parameters of the basic elements of orders and the forms derived from them, linking them with the simplest mathematical formulas following up the author's theory of initial linear relations. As a result,

IOP Conf. Series: Materials Science and Engineering **972** (2020) 012036 doi:10.1088/1757-899X/972/1/012036

it was determined that the formulas of proportionality of the linear parameters of the order and its parts are similar to the system of relations arising from the arbitrary division of a segment into two parts: larger and smaller (major and minor). That is, the linear parameters of order forms are subject to the laws of the minor-major order, forming coherence according to dimension – the harmony of unity of the plurality of forms.

Figure 2. Cathedral in Honor of the Resurrection of Christ. Kokshetau, Kazakhstan. 2016–18. Arch. A. V. Dolgov. Author's photograph.

Figure 3. The main proportional relations on the southern facade of the Cathedral in Honor of the Resurrection of Christ. Kokshetau, Kazakhstan. 2016–18.
Based on the proportions of the "golden section". Drawing of A. V. Dolgov.

IOP Conf. Series: Materials Science and Engineering **972** (2020) 012036 doi:10.1088/1757-899X/972/1/012036

All classical orders possess coherence according to dimension, showing this quality in one, two and three-dimensional spaces. At the same time, the linear parameters of the forms that constitute architectural orders are in correlation relations, that is, a change in the value of a parameter entails systematic changes in the values of other parameters.

The degree of such a correlation may be different, especially in architectural practice; however, in theory it is possible to create architectural objects (bodies) with a very high degree of correlation. Such, for example, are architectural orders. In addition to these, high correlation properties have been established in a number of antique buildings, in the first place, the Parthenon [12].

4.1. Notions «major» and «minor» as basis of ordination analysis

These facts point to the inherent independent linear system of order forms, which goes back to the ratio of the multiplicity and difference of the main major (height of the order) and the main minor (height of the column of the same order). The system is not subject to external imperatives, for example, the "golden section" (unless it is the main order ratio). The dimensions of a major and a minor originate from the arbitrary division of the segment A into two parts: M – greater (major) and m – smaller (minor), i.e. $A = M + m$; also we allocated a value $\Delta = M - m$.

In the course of ordination analysis of architectural forms, these quantities serve as indicators showing the presence of linear dimensions in the analyzed form, related to each other similarly to the ratios of the initial linear quantities. This refers to the pair ratios of the multiplicity and difference of A, M, m, Δ between themselves.

If there are quite a few such ratios and differences, this indicates the dominance in the horizontal and vertical dissection of the architectural form of a certain general structure, expressed in repeatability at different scale levels of the same ratios of the linear dimensions of the elements of architectural forms.

It has been established that the comparison of the height of the entire order (M) with the height of the columns (m) appears most clearly in the order composition. We called it the main order ratio $M : m$, where $M > m$ always.

Calling M : m the main order relation, we recognized that it is the whole order structure that depends on it, therefore, it can be defined as a major minor order.

On the basis of the concepts highlighted, an ordination analysis of the classical architecture can be carried out, in order to then use the knowledge gained in modern architectural practice. We present below one of the options in order to give an idea of the main steps and results.

4.2. Ordination analysis of the Theater of Marcellus in Rome using the major minor approach

The building was built in 12 BC and gives an example of a Doric order (Figure 4). Characteristic divisions according to type 1:

- along the top of the architrave: a) architrave with a capital (m), frieze with a capital (M), b) architrave (m), triglyph (M);
- height of the capital (M) and architrave (m);
- cornice (m) and frieze (M);
- echinus with annulets (M), neck of a capital (m);
- width of the taennius architrave (m) and the trigriff globule (M);
- capital abacus plinth (M) and abacus cornice (m);
- and other divisions.

Characteristic divisions according to type 2:

- cornice height (Δ);
- height of the supporting elements of the cornice (Δ);
- height of abacus cornice of the capital (Δ);
- astragal fascia height with fillet (Δ);
- other horizontal divisions.

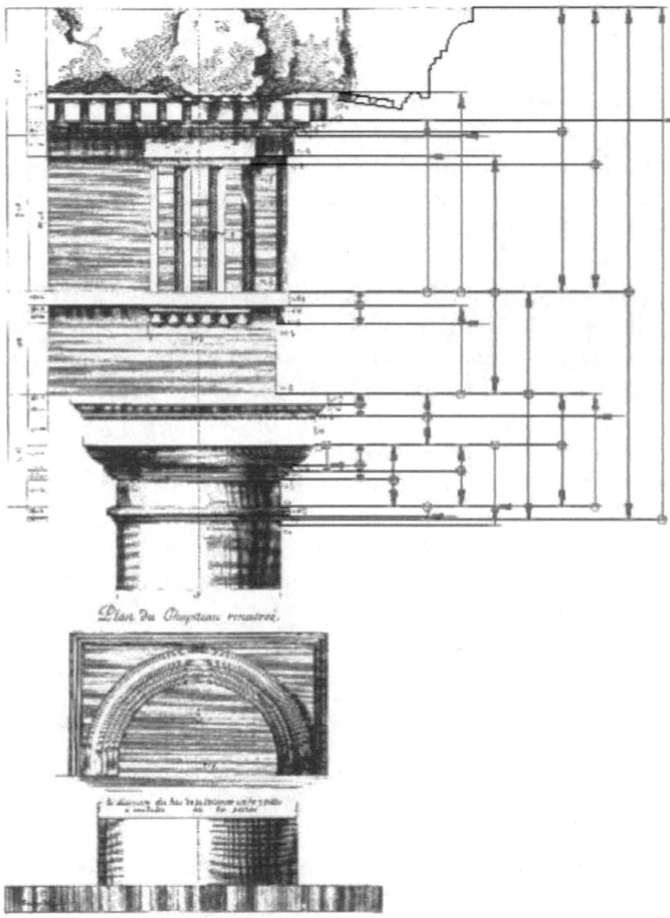

Figure 4. Ordination analysis of the Doric order of the Theater of Marcellus. Initial drawing [26].

Similarly, we analyzed the architecture of different orders, gradually accumulating empirical data on ordinal relations.

4.3. Universality of major minor design

Measurements and design confirm the presence of minor-major pair relations that arise when the order forms are divided into parts, in the overwhelming majority of cases. At the same time, within a particular object (order), subordination to the constant relation $M : m$ is maintained.

Dimensional relationships exist because in each order and its fragments we constantly find various combinations of minor-major variables: A, M, m, Δ.

It is characteristic of such combinations that the same dimensions, fixing the dissection of order forms, can appear in different qualities. For example, Δ, being the height of the entablature obtained by cutting off the height of the column m from the height of the order M when the height of the entablature is divided into parts, also acts as A (thus, we can determine the characteristic lines of the average separation of the entablature), and as M (thus, we can determine the cutoff position from the entablature plate), and as m (thus, we can search for the pair measure to «m» outside the entablature; as a rule, it turns out to be some important part of the capital of the column, or the entire height of the column).

Consequently, in the major-minor structure, characteristic of order forms, we find that the same value can be divided many times, using ratios similar to those of the initial linear quantities. At the same time, all such divisions of the form will consist of a set of segments included in the major-minor order. There may be an infinite number of their combinations, but these will be combinations of the values established by the major minor order.

5. Conclusions

As a result of the study, both partial and general patterns of organization of order forms were identified, some of which are stable and repetitive in nature; it has been found out that these patterns serve to harmonize order forms among themselves in a single order composition; it is confirmed that, knowing the essence of sustainable forms, their correlations and combinations, an architect acts more meaningfully and is able to create new compositions of not only order but also of abstract nature.

The analytically determined properties of linearly independent systematics of the parameters of order forms were not previously known. Their further revelation requires expanding the scope of research on the study of linear plane and volumetric parameters of real architectural objects, which will deepen the understanding of the principles and mechanisms of harmonization of sets of various forms subordinate to the laws of linear systems of architectural orders and can radically change ideas about the role of formative order patterns in architecture.

The completed studies confirmed the prospects of this area of work related to the return to the theory and practice of modern architecture of knowledge and skills on the fundamental principles of the organization of architectural forms, known in antiquity and currently lost, which is designed to significantly improve the aesthetic qualities of design and deepen the understanding of architectural advantages of historic buildings and structures.

Acknowledgements
The paper is written by supporting of Russian Academy of Architecture and Construction Sciences (RAASN), research 1.4.6. "Investigation of the formative principles of the linear systems of architectural orders" of the Basic Research Plan of the RAASN and the Ministry of Construction of Russia.

References

[1] Afanasyev K N 2002 *Construction of Architectural Forms by Ancient Russian Architects* (Moskow: Ladomir) (in Russian)
[2] Balakshin O B 2008 *The Da Vinci Codes – a New Role in Natural Science? The Unexpected About the Golden Section: Harmony of Asymmetric Similarities in Nature* (Moskow: LKI Publ) (in Russian)
[3] Barbaro D A1938 *Commentary on Ten Books on the Architecture of Vitruvius* (Moskow: All-Union Academy of Architecture) (in Russian)
[4] Blum G V 1936 *Columnae or Description and Application of Five Orders* Gabrichevsky A G (Moskow: All-Union Academy of Architecture) (in Russian)
[5] Brunov N I 1973 *Monuments of the Athenian Acropolis. Parthenon and Erechtheum* (Moskow: Iskusstvo) (in Russian)
[6] *Building Drawings of the Temple of Apollo in Didyma* http://ancientrome.ru/ publik/article.htm?a=1415875893 (in Russian)
[7] Carvais R 2014 A digital edition of the didactic knowledge of construction: a critical edition of the courses of Antoin Desgodets, professor at the Academie Royale d`Architecture (Paris, 1719-1728) *Proceedings of the First Conference of the Construction History Society* Editors: Campbell J, Bill N, Draper K, Fleming P, Pan Y, Andrews W (Cambridge: Queens College) P 59-69
[8] Choisy O *Architecture of Ancient Greece. Exterior Architecture of the Greek Temples* http://antique.totalarch.com/choisy/grece/16 (in Russian)
[9] Cohen M A 2014 Introduction: Two Kinds of Proportion *Architectural Histories* **2**(**1**) P Art. 21

IOP Conf. Series: Materials Science and Engineering **972** (2020) 012036 doi:10.1088/1757-899X/972/1/012036

[10] Colman S 2013 *Harmonic Proportion and Form in Nature, Art and Architecture* (Dover Publications)

[11] Da Vignola Giacomo Barozzi 2005 *Canon of the Five Orders of Architecture* Gabrichevsky A G (Moskow: Publishing House «Architectura-C») (in Russian)

[12] Dolgov A V 2015 Example Analysis Ordination of Architectural Forms Panthenon in Rome *Academic Bulletin of UralSRIproject of RAASC* **2** P 59–61 (in Russian)

[13] Dolgov A V 2016Theory of Initail Linear Relations *Academic Bulletin of UralSRIproject of RAASC* **4** P 33–37 (in Russian)

[14] Ghyka M S 1936 *Aesthetics of Proportions in Nature and Art* (Moskow: All-Union Acad. of Architecture) (in Russian)

[15] Grimm H D 1935 *Proportionality in Architecture* (Moskow: ONTI. Main editorship of building architecture) (in Russian)

[16] Hambridge J 1936 *Practical Applications of Dynamic Symmetry* / Brunov N (Moskow: All-Union Acad. of Architecture) (in Russian)

[17] Kirk Irwin J 2015 Ratio and the Divine Proportions: Le Corbusier and Rudolf Wittkower. *Proceeding of International Congress "Le Corbusier, 50 years later"* http://ocs.editorial.upv.es/index.php/LC2015/LC2015/paper/viewFile/743/1293

[18] Krier L 2008 *The Architectural Tuning of Settlements* (London: The Prince's Foundation for the Built Environment)

[19] Lebedeva G S 2003 *The latest commentary on the treatise of Vitruvius «Ten books on architecture»* (Moskow: Editorial URSS) (in Russian)

[20] *Mathematics and Architecture of the Parthenon* http://www. artmatlab.ru/articles.php?id=l 18&srn=2 (in Russian)

[21] McGrew Ch B 1949 *Doors and Portals in Italian Architecture: Measurements and Photographs* (Moskow: Publ. USSR Academy of Architecture) (in Russian)

[22] Palladio A 1936 *Four Books on Architecture in Two Volumes* (Moskow: Publ. All-Union Academy of Architecture) (in Russian)

[23] Radzyukevich A V 2012 Proportional-Metrological and Geometric Features of the Shaping of the Roman Pantheon *Architectural Shaping and Geometry* (Moskow: URSS) P 183–194 (in Russian)

[24] Shevelyev I Sh 1995 *Shaping. Number. The form. Art. Life* (Kostroma: Publ. «DiAr») (in Russian)

[25] Tarusshvili L I 1995 Aesthetics of an Architectural Order (Moskow: Architectura) (in Russian)

[26] Desgodetz A 2010 *The ancient buildings of Rome* Vol 2 (Gale ECCO)

[27] Vera M A C 2009 Proportion, Symmetry and Mathematics in the Renaissance Theory of Construction: Vignola's Treaty of Architecture and its Musical Mirror *Proceedings of the Third International Congress on Construction History, Cottbus* http://www.bma.arch.unige.it/PDF/CONSTRUCTION_HISTORY_2009/VOL1/CAJIGAL-VERA_Migue-Angel_VW_2_Paper-revised-2_layouted.pdf

[28] Vitruvius Marcus Pollio 1936 *Ten books On Architecture* Translation by Mishulin AV (Leningrag: OGIZ, SOTSEKGIZ) (in Russian)

[29] Weber R and Larner S 1993 The Concept of Proportion in Architecture: An Introductory Bibliographic Essay *Journal of the Art Libraries Society of North America* Vol **12**(**4**) P 147–154

[30] Wittkower R 1978 *Art and Architecture in Italy: 1600 to 1750* (London: Penguin books)

[31] Zeising A 2015 Neue Lehre von den Proportionen des menschlichen Körpers (Norderstedt: Vero Verlag)

[32] Zubov V P 2001 Architectural Theory of Alberti (Sankt Petersburg: Aleteya) (in Russian)

Zero Liquid Discharge (ZLD) Industrial Wastewater Treatment Systems as Sustainable Development Basic Ecological Components

V Aksenov[1], N Tsarev[1], I Nichkova[1] and E Tatyannikova[1]

[1]Ural Federal University, 19 Mira street, Yekaterinburg, 620002, Russia

E-mail: v.i.aksenov@urfu.ru

Abstract. The problems in processing the acid ferrous-containing rinsing water and spent pickling solutions are considered. The first-ever in metallurgy zero liquid discharge (ZLD) industrial wastewater treatment systems of a transformer cold rolling plant at the Verkh-Isetsky steelworks in Sverdlovsk (now — LLC VIZ Steel is a part of Novolipetsk Steel Group of Companies) is described. The experience of the VIZ Steel, particularly in relation to industrial wastewater treatment and sludge processing, is useful for designing ZLD industrial wastewater treatment systems of metallurgical, machine-building and metalworking enterprises. The recommendations for organizing the ZLD industrial wastewater treatment systems in the pickling plants with the repeated use of the treated rinsing water and processing the spent pickling solutions and the sludges are presented. For practical calculations of clarifiers the empirical formulas are offered. Process flow diagram of acid ferrous-containing rinsing water treatment is suggested. Acid ferrous-containing rinsing water treatment includes the following processes: rinsing water neutralization to pH 9–10,5 by 5 % calcium hydroxide and partial ferrous oxidation by air; addition of anionic flocculant and active gypsum priming to the neutralized rinsing water; settling of rinsing water conditioned by reagents in clarifiers; treatment of clarified water on deep bed sand filters; reclaimed water treatment.

Keywords: Zero liquid discharge (ZLD) systems; Acid ferrous-containing rinsing water treatment; Rinse pickling solution; Industrial wastewater sludge processing

1. Introduction

The creation trend of industrial plants zero liquid discharge (ZLD) systems is clearly traced in majority of advanced countries of the world during the last 15–20 years [1–4]. The reasons, causing beginning of widespread ZLD systems development, are, on the one hand — catastrophic natural water bodies pollution with industrial, domestic and agricultural wastewater [5–8] and, on the other hand — appearance of various technological equipment (multi-stage evaporating devices [9], reverse osmosis installations [10–12] etc.), which application provides acceptable economic indicators of ZLD systems. Nowadays an implementation of ZLD industrial wastewater systems is the only rational solution of the industrial water reclamation and reuse issue.

Russia has experience of ZLD systems development and exploitation The first-ever in metallurgy ZLD systems of a transformer cold rolling plant, which is still in exploitation has been put into operation at the Verkh-Isetsky steelworks in Sverdlovsk (1973) (now — LLC VIZ Steel is a part of Novolipetsk Steel Group of Companies) [13].

In the early 1980s a small group of recycling experts started talking about the idea of «Total Recycling». Zero waste concepts followed. By 1990, activists in the Philippines were already using the term zero waste. One of the first formal zero waste policies was created in 1995 when Canberra, Australia endorsed a goal of «No Waste by 2010». Since 1995, zero waste has been endorsed as a goal by governments in New Zealand; Denmark; Seattle, Washington; Del Norte County, California; San Francisco, California; Santa Cruz County, California; Edmonton, Alberta; Ottawa, Ontario; and Nova Scotia. Furthermore, a number of national and international businesses have adopted some zero waste principles [14].

2. Zero Liquid Discharge System of VIZ Steel (Novolipetsk Steel Group of Companies)

Verkh-Isetsky steelworks ZLD system was created during the solving process of industrial wastewater treatment and reuse issue. Industrial wastewater of Verkh-Isetsky steelworks can be divided into nine categories: four-high rolling mill wastewater containing palm oil or its substitutes; electric insulation coating units and decarburization annealing units wastewater, which contain fine magnesium oxide; wastewater from rolling mill carters washing; containing oils and suspensions; acid ferrous-containing rinsing water from pickling plant; spent pickiling solutions from pickling plant; spent oil emulsions; spent degreasing solutions; degreasing installation washing water; oil emulsion systems washing water. During the ZLD system development the industrial wastewaters were integrated by types of pollution in four groups:

- spent pickling solutions;
- rinsing water;
- wastewater containing oils;
- wastewater polluted by mainly mechanical impurities.

For each wastewater group the appropriate treatment facilities have been designed and constructed, placed in several buildings. After treatment water is reused in production. Liquid and solids wastewater sludges are exposed to mechanical dewatering, drying and utilization. Oil-containing waste goes for burning, and salted bleed water — for evaporation with receiving desalinated water and concentrate of salts.

The issue of rinsing water and spent pickling solution treatment and reuse is not solved at the most metallurgical plants of Russia. Contrary to the known and approbated recommendations [15, 16] rinsing water and spent pickling solution are mixed with each other and neutralized with calcium hydroxide. Then, in best case, neutralized wastewater is settled with subsequent dumping of poor quality clarified water into water body and sludge pumping to the storages. In worst case — pumping to the storages without sludge removal. As the way of chemical metal-stripping procedure by pickling in acids (usually in sulfuric and hydrochloric acid) remains the basic, consumption of acids increases by these purposes every year. So the volumes of acid ferrous-containing rinsing water and its sludge placed in storages are also increase. Because of this, the technical solutions realized during development of the Verkh-Isetsky steelworks ZLD system are relevant at present time.

3. Results and Discussion

3.1. Rinsing Water Treatment

It's important to pay attention to that low-concentrated rinsing waters (0.6–0.8 g/L of H_2SO_4, 1.8–2.5 g/L of Fe^{2+}) should be processed separately from concentrated spent pickling solutions (40–150 g/L of H_2SO_4, 80–150 g/L of Fe^{2+}) with receiving purified water which is reused and dewatering sludge which is directed to the utilization.

The main stage of acid ferrous-containing rinsing water treatment is their neutralization. With calcium hydroxide using significant solid phase quantity is formed. And it is necessary to separate it in the clarifiers. At the same time the structure of the formed flocs is of great importance: large and dense particles quickly settle and the sludge occupies smaller volume and in further easily gives to mechanical dewatering.

IOP Conf. Series: Materials Science and Engineering **972** (2020) 012037 doi:10.1088/1757-899X/972/1/012037

In most cases about measure of different factors influence on the flocs structure judge by the particles settling velocity. In case with considered acid ferrous-containing rinsing water the settling velocity depends on quantity and quality of ferrous hydroxide flocs as they settle much more slowly than other impurity (figure 1).

Sludge formed during neutralized rinsing water settling undergoes changes in time. Flocs with very small density are formed initially, then they are gradually crushing during 2–3 h. After 20–25 hours flocs begin to be integrated, further the sizes of flakes change slightly, their dispersion decreases and density increases.

Flocs sedimentation and further formed sludge mechanical dewatering processes can be significantly intensified due to the part ferrous oxidation by air with receiving more dense and heavy magnetite sludge Fe_3O_4.

Another way to increase efficiency of neutralized rinsing water clarification is its flocculation treatment [17–19].

Due to the anionic polyacrylamide with low molecular weight adding to neutralized with calcium hydroxide rinsing water it is possible to intensify ferrous hydroxide flocs sedimentation.

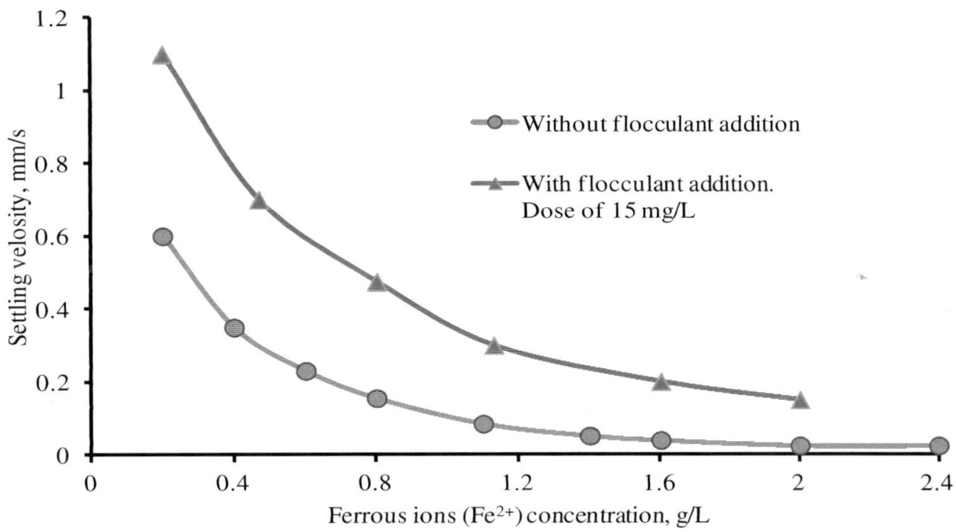

Figure 1. Effect of ferrous ions concentration on settling velocity.

For practical calculations of clarifiers the empirical formulas are offered with (1) and without (2) flocculant addition:

$$\upsilon_s = (0.47 + 2.3C_{Fe})^{-1} \quad (1)$$

$$\upsilon_s = (1.1 + 8.8C_{Fe})^{-1} \quad (2)$$

where υ_s = settling velocity, mm/s

C_{Fe} = ferrous ions (Fe^{2+}) concentration in rinsing water, g/L.

Influence of dosed anionic flocculant quantity on the settling velocity is presented in the figure 2.

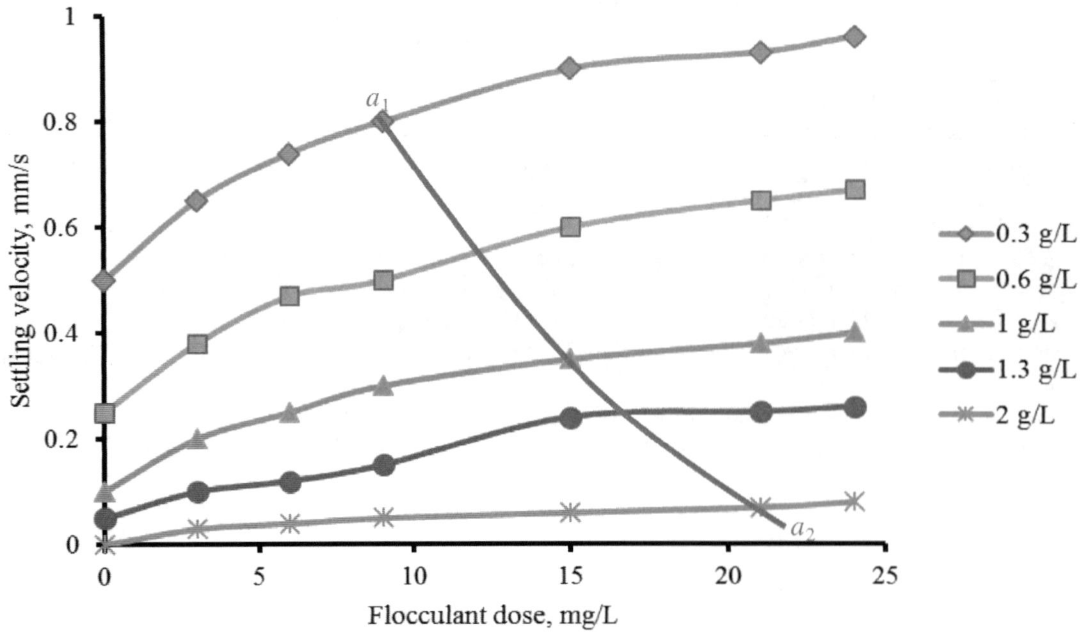

Figure 2. Effect of flocculant dose and ferrous ions (Fe^{2+}) concentration on settling velocity.

The curve a_1–a_2 demonstrates optimum quantities of the added anionic flocculant for rinsing water with various ferrous ions (Fe^{2+}) content.

Proceeding from data presented on the figures 1 and 2, it is possible to note that use of anionic flocculant increases the flocs sedimentation speed by 2–4 times. At the same time sludge volume practically doesn't change.

Taking into account the received researches results and also forty-year successful exploitation experience of Verkh-Isetsky steelworks ZLD system [14–16] it is recommended to organize acid ferrous-containing rinsing water treatment at pickling plant as illustrated in figure 3. Acid ferrous-containing rinsing water treatment includes the following processes:

- rinsing water neutralization to pH 9–10,5 by 5 % calcium hydroxide and partial ferrous oxidation by air;
- addition of anionic flocculant and active gypsum priming to the neutralized rinsing water;
- settling of rinsing water conditioned by reagents in clarifiers;
- treatment of clarified water on deep bed sand filters.

After such treatment water will have pH=10,5–11, total hardness — 30–40 mEq/L and total alkalinity — 10–16 mEq/L. For reuse of such treated water it is necessary to adjust its salt composition. The water amount directed to further demineralization will be defined by ZLD system water salt balance.

Ferrous-containing sludge from clarifiers after oxidation by air oxygen for receiving magnetite sludge is dewatering by means of a plate-and-frame filter-presses, then dried and directed to utilization.

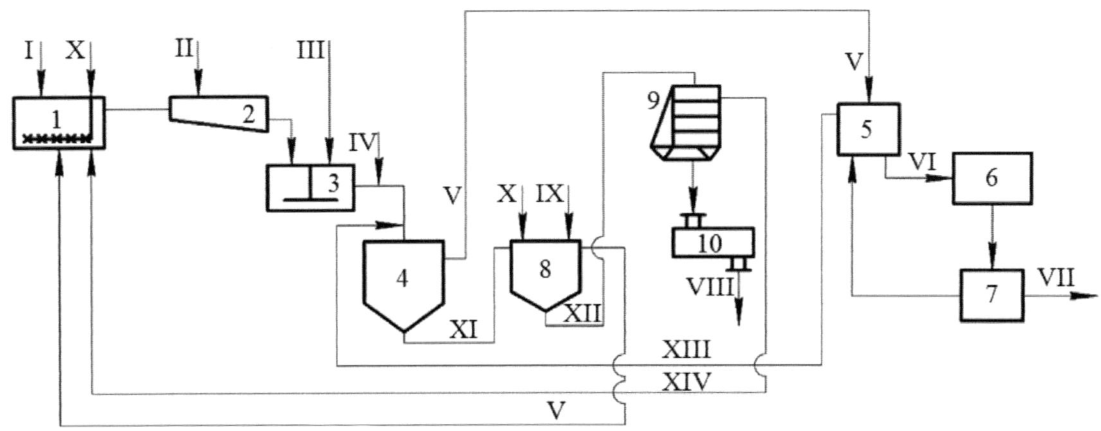

1 — equalization basin; 2 — mechanical mixer; 3- mechanical flocculator; 4 — clarifier; 5 — deep bed sand filter; 6 — purified water tank; 7 — pumping station; 8 — magnetite sludge receiving chamber; 9 — plate-and-chamber filter-press; 10 — cake dryer; I — rinsing water; II — calcium hydroxide; III — recirculated sludge; IV — flocculant; V — clarified water; VI — filtered water; VII — purified water directed to the demineralization; VIII — cake directed to utilization; IX – steam; X — air; XI — sludge; XII — magnetite sludge; XIII — backwash water; XIV — filter-press filtrate.

Figure 3. Process flow diagram of acid ferrous-containing rinsing water treatment.

3.2. Spent Pickling Solutions Treatment

In case of hydrochloric acid pickling spent solutions treatment can give either ferric chloride concentrate (liquid crystal) [20], which is known as recognized coagulant, or with its thermal method processing — 20–25 % hydrochloric acid solution [21, 22], which is reused at pickling plants, and as solid waste — fine iron oxide, which is easily utilized. Such technology is introduced on PJSC Severstal [23] and JSC Magnitogorsk Iron and Steel Works [24].

In case of sulfuric acid spent pickling solutions processing is implemented with receiving Ferrous sulfate ($FeSO_4 \cdot 7H_2O$), which is utilized [25, 26].

4. Conclusion

It should be noted that industrial wastewater and sludge processing technologies during the ZLD industrial wastewater systems creation are very various and depend on many factors: industrial wastewater characteristics, industrial enterprises possibilities in use of purified water, possibilities of concentrates (brines etc.) and sludges utilization at the industrial enterprises or in other organizations.

5. References

[1] Mansour F, Alnouri S Y, Al-Hindi M, Azizi F and Linke P 2018 Screening and Cost Assessment Strategies for End-of-Pipe Zero Liquid Discharge Systems J. *Clean. Prod.* **179** 460–477.

[2] Popuri A K and Guttikonda P 2016 Zero Liquid Discharge (ZLD) Industrial Wastewater Treatment System *Int. J. ChemTech. Res.* **9**(11) 80–86.

[3] Rajamani S 2016 Novel Industrial Wastewater Treatment Integrated with Recovery of Water and Salt Under a Zero Liquid Discharge Concept *Rev. Environ. Health* **31**(1) 63–66.

[4] Sankar D, Balachandar M, Anbuvanan T, Rajagopal S, Thankarathi T and Deepa N 2017 Condenser Cooling System & Effluent Disposal System for Steam-Electric Power Plants: Improved Techniques *Membr. Water Treat.* **8**(4) 355–367.

[5] González-Fernández B, Rodríguez-Valdés E, Boente C, Menéndez-Casares E, Fernández-Braña A and Gallego J R 2018 Long-Term Ongoing Impact of Arsenic Contamination on the Environmental Compartments of a Former Mining-Metallurgy Area *Sci. Total Environ.* **610–611** 820–830.

[6] Tiwari A K, De Maio M, Singh P K and Mahato M K 2015 Evaluation of Surface Water Quality by Using GIS and a Heavy Metal Pollution Index (HPI) Model in a Coal Mining Area, India, *B. Environ. Contam. Tox.* **95**(3) 304–310.

[7] Lezier V, Gusarova M and Kopytova A 2017 Water Supply of the Population as a Problem of Energy Efficiency on the Example of the Tyumen Region of Russia *IOP Conf. Ser.: Earth Environ. Sci.* **90**(1) 012069 Available from: https://iopscience.iop.org/article/10.1088/1755-1315/90/1/012069/pdf.

[8] Smith L et al. 2017 Mitigation of Diffuse Water Pollution from Agriculture in England and China, and the Scope for Policy Transfer *Land. Use Policy* **61** 208–219.

[9] Thome J R 2017 A Review on Falling Film Evaporation *J. Enhanced Heat Transf.* **24**(1–6) 483–498.

[10] Hajbi F, Hammi H and M'Nif A 2010 Reuse of RO Desalination Plant Reject Brine *J. Phase Equilib. Diffus.* **31**(4) 341–7.

[11] Le N L and Nunes S P 2016 Materials and Membrane Technologies for Water and Energy Sustainability *Sustain. Mater. Technol.* **7** 1–28 Available from: https://www.sciencedirect.com/science/article/pii/S2214993715300105.

[12] Razali M, Kim J F, Attfield M, Budd P M, Drioli E, Lee Y M and Szekely G 2015 Sustainable Wastewater Treatment and Recycling in Membrane Manufacturing *Green Chem.* **17**(12) 5196–205 Available from: https://pubs.rsc.org/en/content/articlehtml/2015/gc/c5gc01937k.

[13] Zero Discharges into Water: NLMK Group Companies Introduce a Closed Loop Water System [Internet] [cited 2019 may 5] Available from: https://lipetsk.nlmk.com/en/responsibility/ecology/key-cases/zero-discharges-into-water-nlmk-group-companies-introduce-a-closed-loop-water-system/.

[14] Platt B and Lakhani M 2004 *Report from Institute for Local Self-Reliance for Global Anti-Incinerator Alliance (GAIA)* (Quezon City: Institute for Local Self-Reliance for Global Anti-Incinerator Alliance (GAIA)) p 85 Available from: http://www.no-burn.org/wp-content/uploads/Resources-up-in-Flames.pdf.

[15] Aksenov V I, Arkhipova O A, Sidorova I A and Nichkova I I 2005 Organizational Problems of Water Handing Facilities at Metallurgical Enterprises *Stal* (8) 96–8.

[16] Aksenov V I, Arkhipova O A, Sidorova I A, Nichkova I I 2005 Creation of Closed Water Supply Circuits for Metallurgical Enterprises *Stal* (9) 83–5.

[17] Bratby J 2016 *Coagulation and Flocculation in Water and Wastewater Treatment* (London: IWA Publishing) p 538.

[18] Teh C Y, Budiman P M, Shak K P Y and Wu T Y 2016 Recent Advancement of Coagulation–Flocculation and its Application in Wastewater Treatment *Ind. Eng. Chem. Res.* **55**(16) 4363–4389.

[19] Wei H, Gao B, Ren J, Li A and Yang H 2018. Coagulation/Flocculation in Dewatering of Sludge: A Review *Water res.* **143** 608–631.

[20] Pietrelli L, Ferro S and Vocciante M 2018 Raw Materials Recovery from spent Hydrochloric Acid-Based Galvanizing Wastewater *Chem. Eng. J.* **341** 539–546.

[21] McKinley C and Ghahreman A 2018 Hydrochloric Acid Regeneration in Hydrometallurgical Processes: A Review *Miner. Process. Extr. M.* **127**(3) 157–168.

[22] Machado R M, Gameiro M L F, Rodrigues J M, Ismael M R C, Reis M T A and Carvalho J M 2017 Recovery of Hydrochloric Acid from Galvanizing Industrial Effluents *Sep. Sci. Technol.* **52**(8) 1333–1340.

[23] Severstal [Internet] [cited 2019 may 5] Available from: https://www.severstal.com/eng/about/

[24] Magnitogorsk Iron and Steel Works [Internet] [cited 2019 may 5] Available from:

http://eng.mmk.ru/about/about_the_company/.
[25] Bartholomew F J 1952 Sulfuric Acid Recovery from Waste Liquors *Ind. Eng. Chem. Res.* **44**(3), 541–545.
[26] Kumar G S, Basu D, Hung Y-T and Wang L K 2010 *Waste Treatment in the Iron and Steel Manufacturing Industry Handbook of Advanced Industrial and Hazardous Wastes Treatment* ed L K Wang, Y-T Hung and N K Shammas (Boca Raton: CRC Press) chapter 2 pp 37–70.

Problems of the implementation of BIM technologies in Russia

E A Chesnokova[1], V V Khokhlova[1], N A Ponyavina[1] and A S Chesnokov[2]

[1]Department of Technology, Construction Management, Expertise and Property Management, Voronezh State Technical University, 20 let Oktyabrya st., 84, Voronezh, 394006, Russian Federation

[2]Department of Applied Mathematics and Mechanics, Voronezh State Technical University, 20 let Oktyabrya st., 84, Voronezh, 394006, Russian Federation

E-mail: zhdamirova@vgasu.vrn.ru, kivikviki1@mail.ru, ponyavochka@vgasu.vrn.ru , selches@inbox.ru

Abstract: This article is devoted to the latest technology of computer-aided design - Building Information Modeling (Building Information Modeling or Building Information Model). It discusses the prerequisites for the emergence of information modeling, describes its advantages and capabilities compared with the traditional two-dimensional design method and are presented in the figures. Currently in Russia there is a problem of introducing BIM. In this paper, we identified the main problem of introducing BIM technologies in Russia, which is associated with a decrease in the productivity of designers when switching to a new design technology, as well as possible solutions to this problem. In addition, the difficulties arising from attempts to move to BIM-technologies are described. The main conclusion is the indisputable advantage of BIM-technologies over 2D design.

1. Introduction

In the modern world, the development of technology brings a number of changes in various areas of activity [1-3]. The emergence of technology information modeling occurred as a result of the necessity processing for man the huge and steadily growing flows of information used at the design stage and subsequently used throughout the life cycle of buildings and structures. The flow of incoming information is becoming more and more and a person is forced to look for new ways of processing and compiling the received data. Information technology progress improves computer technologists, which in turn is instantly reflected in the design and construction activities [4-6]. At the moment, it is not enough to have only a paper project of the building being created, since information continues to flow even when the building has already been built and is being actively exploited, interacting with the environment.

2. Information model of the use of BIM technologies

In the current realities, when modeling objects, it is necessary to take into account not only the architectural-design features, but also all the data of different contents associated with it as a single information complex. This entails the need to structure of data large volume with different typologies.

IOP Conf. Series: Materials Science and Engineering **972** (2020) 012038 doi:10.1088/1757-899X/972/1/012038

More in demand will be information model, which will remain relevant at any point in time and allow you to make changes throughout the life cycle of the property (figure 1).

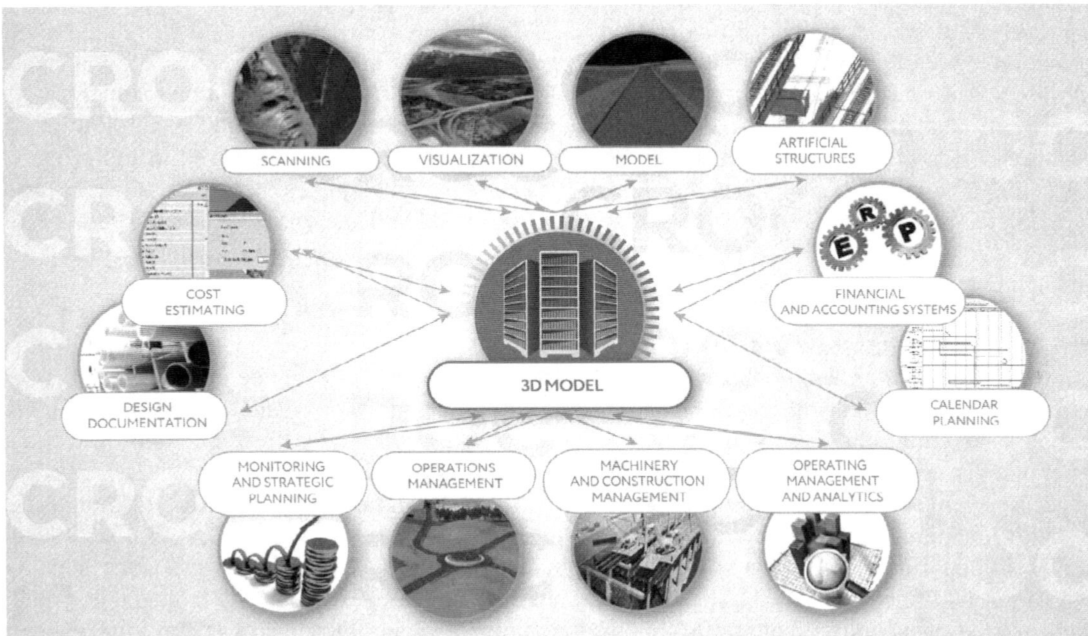

Figure 1. Information model structure [10]

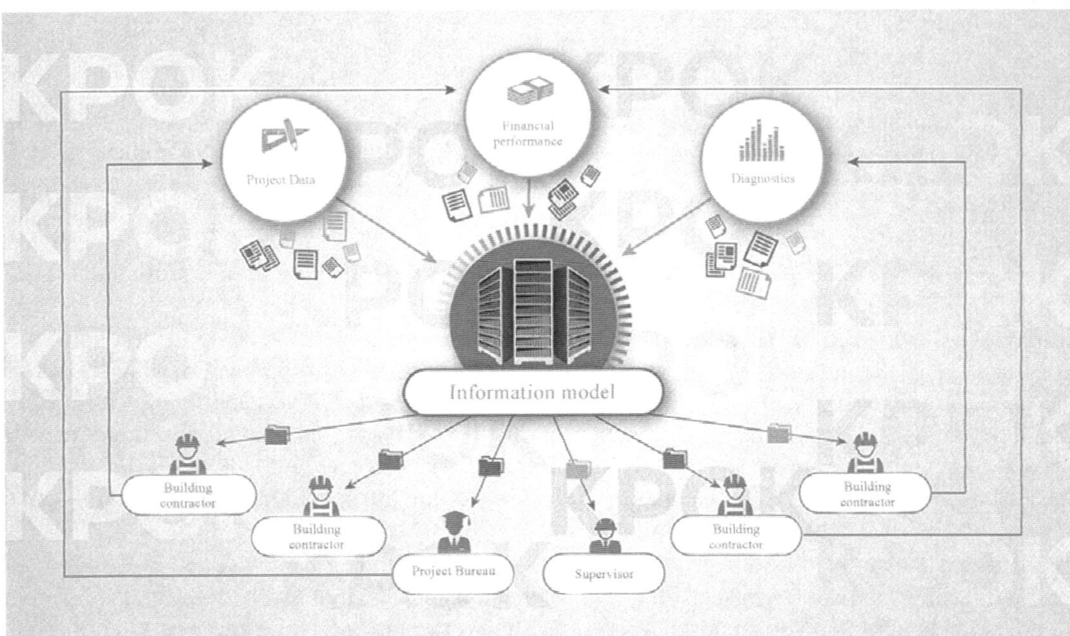

Figure 2. Integration of information in the framework of BIM [10]

IOP Conf. Series: Materials Science and Engineering **972** (2020) 012038 doi:10.1088/1757-899X/972/1/012038

However, it should not be just a copy of a cardboard layout showing only the architectural appearance and spatial solutions of the object, but should contain all information about the building, its structural elements, materials and equipment, quantitative and qualitative characteristics. Accordingly, it should be a single living model, able to respond to the changes and automatically complement and adjust the entire model (figure 2) [7-10]. In connection with these requirements, information modeling of buildings and structures appeared.

Awareness of the need to use information modeling in the Russian construction industry is only beginning to gain momentum. The overwhelming majority of design organizations already use this concept of design, which cannot be said about companies that are directly involved in the process of constructing a building. Today, when developing a project, information design in design organizations is reduced to modeling a 3D layout of engineering networks and the architectural appearance of a building, less often – to all main sections within the framework of creating complete design and estimate documentation.

3. Problems of the implementation of BIM technologies

The introduction and mass application of BIM technologies requires huge initiative efforts and associated risks caused by the lack of mandatory expertise of the 3D model and the regulatory framework governing all aspects of working with the information model, from its creation to its use throughout the entire life cycle of the property [11-14]. But even if the regulatory framework is created, rapid implementation and widespread use will not occur. Organizations that do not work in BIM, unfortunately, do not receive an adequate level of motivation. Here the psychological factor plays a special role. An ordinary employee interested in the development and success of his organization will find it easy to master a new design environment. However, some managers want to switch to a new technology without loss of productivity of their subordinates at the initial stage of implementation, which is unlikely (the graph of labor productivity is presented in Figure 3) [15].

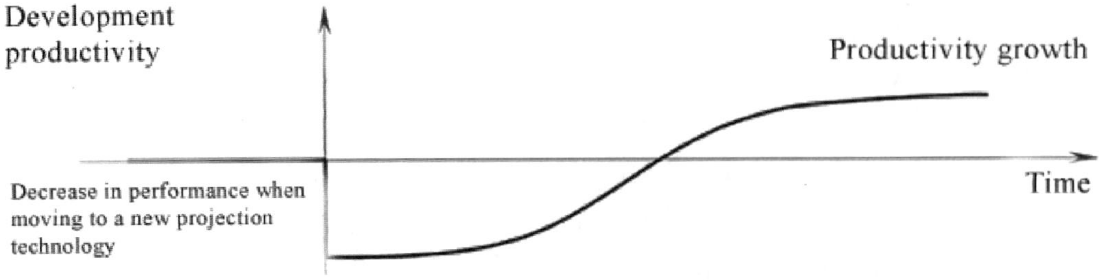

Figure 3. Graph of the productivity of designers in the transition to new software

In practice, it turns out that when switching to BIM, as a result of a decrease in productivity, we have a decrease in the volume of work with a simultaneous decrease in the level of wages. In such a situation, employees will try to maintain their income levels, and perceive BIM as the cause of their financial problems and oppose their implementation in every way. In this case, the introduction of any new technology will not happen, despite the best efforts of management [11-13].

According to Vladimir Talapov's book, BIM Basics: An Introduction to Building Information Modeling [16], in Russia there are three ways to implement BIM:

- Private company;
- With the company's own efforts, but a specialist is invited, who coordinates the entire implementation process and trains employees to work in the software package;
- Attracting third-party organizations.

Indeed, in order for the introduction of new software, the curve in figure 3 should be more even and with minimal difference, it is recommended to make the transition not by the whole organization at once, but by departments that ensure the full output of the product of the project activity. You can invite a specialist who has at least the slightest understanding of the technology of information modeling, even without practical experience in the field of design. Such a specialist will support and accelerate the transition to BIM.

To make the transition to BIM, you need a proper understanding of the laws of this process, the optimal combination of ways to optimize and mitigate the process of implementing the information modeling programs described above.

The psychological factor is the main problem arising in the transition to BIM design, and, unfortunately, the most formidable. Unreasonable distrust of the new software, misunderstanding of the emerging opportunities for working in the BIM complex, its potential for development and use aggravates attempts to switch to an innovative type of modeling [17].

There are a number of other difficulties that are listed below.

The introduction of BIM technology always incurs costs. The main costs are in purchasing software licenses and various training courses. And this is an inevitable phenomenon, as well as a decrease in the productivity and efficiency of workers in the new software package, and accordingly, the costs associated with a decrease in productivity. However, the costs of information modeling will quickly pay for themselves, and the subsequent economic effect of using these programs will increase the profit of the organization.

Also, one of the factors that impede the transition to BIM is the judgment that information modeling consists in unintelligent execution of mechanical work. This judgment is dangerous because highly qualified specialists who do not have skills in this software package will be uncompetitive compared with young designers who have experience in information modeling programs, but who make a number of design errors due to the lack of necessary knowledge and experience. This will lead to a decline in the construction sector and design culture.

The resulting factor from the above is the problem of a fuzzy level of responsibility for the work performed. Undoubtedly, the concept of information modeling implies the correction of inaccuracies in the project due to the simultaneous work of a large number of employees and the interrelation of all designed elements, but because of these advantages, unlike the traditional design method, where each project participant individually works with the project, BIM blurs the personal level responsibility of the employee in case of revealing a critical error. In addition, there is a blurring of the control points of the project stages, it is quite difficult to track the result at a certain point, the work is carried out in parallel by different employees and, collectively, this leads to an increase in cost risks, which also causes some concern and mistrust on the part of users.

4. Conclusion

Of course, the information modeling technology has changed and optimized the process of designing, erecting and operating the structure. The use of BIM programs has led to the achievement of an enormous economic effect associated with a reduction in risks and costs, a decrease in the design and construction terms, an increase in the project's profitability, income and quality. The number of computer programs based on the BIM concept is currently growing in direct proportion to the number of users of these software. But the massive introduction is just beginning to gain momentum. To achieve this goal, firstly, it is necessary to overcome a number of difficulties associated with the designers' distrust of this complex, personnel training in BIM programs and the development of a design and construction culture in general. Despite the slow introduction of BIM technologies in Russia, there is a high probability that they will soon completely replace traditional 2D design methods.

References

[1] Astafieva N S Kibireva Yu A and Vasilyeva I L 2017. Advantages of use and difficulties in implementing information modeling of buildings. *Construction of unique buildings and*

structures. vol **8** pp 41-62 (Electronic materials URL: http://unistroy.spbstu.ru/index_2017_59/3_Kibireva_59.pdf).
[2] Mishchenko V Ya, Barkalov S A and Kurochka P N 1997 Modeling and automation of organizational and technological design of construction production (Voronezh: VGASU) p 120.
[3] Matreninsky S I , Mishchenko V Ya and Chernyshov E M 2015 The systemic approach to modeling of compact build-up development areas and planning of their renovation. *International Journal of Energy and Environmental Engineering*. vol **6(9)**. pp 32-43.
[4] Mishchenko V Ya, Emelyanov D I, Tikhonenko A A and Startsev R V. 2012. Stochastic algorithms in solving multicriteria tasks of optimization of resource distribution when planning construction and installation works. *Scientific Herald of the Voronezh State University of Architecture and Civil Engineering. Construction and Architecture.* Vol **1 (25)**. pp 92-97.
[5] Matreninsky S I and Mishchenko V Ya 2014 Feasibility study of the configuration of built-up urban areas with the aim of their renovation. *Recent Advances in Urban Planning, Sustainable Development and Green Energy . "Energy, Environmental and Structural Engineering Series"* vol **30**. pp 58-63.
[6] Emelyanov D I, Ponyavina N A and Chesnokova E A 2017 Solution of the task of planning complicated production processes at the enterprise based on methods of network planning. *Proceedings of higher educational institutions. Technology of Textile industry.* Vol **5(371).** pp 28-32.
[7] Surovtsev I S, Mishchenko V Ya, Ponyavina N A and Drapalyuk D A 2011 Development of types of formation of organizational and technological system of real estate reconstruction. *Scientific Herald of the Voronezh State University of Architecture and Civil Engineering. Construction and Architecture.* vol **1(9)**. pp. 86-95.
[8] Mishchenko V Ya, Ponyavina N A and Nazarov A N 2010 Simulation of sustainable organizational and technological system reproduction of urban infrastructure. *Scientific Herald of the Voronezh State University of Architecture and Civil Engineering. High technology in ecology.* **Vol 1**. Pp 74-80.
[9] Ponyavina N A, Zubareva Yu V and Zuev E A 2018 Applying IOT on the stage of operation of real estate objects. *Construction and Real Estate.* (Voronezh) vol **2-1**. pp 29-33.
[10] CROC - System integrator. Information modeling of a building object (BIM) (Electronic materials URL:https: //www.croc.ru/solution/business-solutions/bim/? Tab = 3).
[11] Khokhlova V V, Chuntonova E V and Chesnokova E A 2018 Methodical bases for the creation and functioning of an inter-regional BIM-cluster *In the collection: Scientific support of the Voronezh region Collection of works of winners of the competition of research works of students and postgraduates of VSTU in priority areas of science and technology*. (Voronezh) pp 90-99.
[12] Chesnokova E A, Khokhlova V V, Chuntonova E V and Beret E ZH 2018 Application of BIM-technologies in modern construction. *Construction and real estate*. vol **1-1.** pp. 47-54.
[13] Chesnokova E A, Khokhlova V V and Mishchenko AV. 2018. Economic efficiency of using BIM-technologies. *Construction and Real Estate*(Voronezh) vol **2-1** pp 71-75.
[14] Chesnokov A S and Tarasov A A 2018 Application of bim technologies in the design of high-rise buildings. *Construction and Real Estate.* vol **2-1**. pp 125-128.
[16] Talapov V V 2011. BIM Basics: An Introduction to Building Information Modeling. (Moscow: DMK Press) p 392.
[15] Talapov V V 2010 What influences the implementation of BIM in Russia. *"CAD and graphics."* vol **11** pp 12-16 (Electronic materials URL: https://sapr.ru/article/21805)
[17] Korovin M D and Shavva A A 2016 The complexity of the transition to BIM design *Actual problems of the humanities and natural sciences.* vol **12-3** pp 124-7 (Electronic materials URL: https://publikacia.net/archive/2016/12/3/33).

IOP Conf. Series: Materials Science and Engineering **972** (2020) 012039 doi:10.1088/1757-899X/972/1/012039

Improving the calculation of the cost of the construction project using BIM technology

V A Belyakov[1,2], V B Salnikov[1,2] and R T Galiakhmetov[3]

[1]Department of materials sciences in construction of the Institute of new materials and technologies, Ural Federal University, 19 Mira street, Yekaterinburg 620002, Russia

[2]LLC "Institute for planning, architecture and design», 9 Sheynkmana street, 2nd floor, Yekaterinburg 620014, Russia

[3]Engineering Department, Ural State Agrarian University, 42 Karl Liebknecht street, Yekaterinburg 620075, Russia

E-mail: belyakow@rambler.ru

Abstract. In the scientific article the authors summarize the results of theoretical research and practical implementation in the design of residential and public buildings BIM – information modeling technology. The questions of application of digital models for control of financial expenses at all stages of project implementation in construction are considered. An attempt is made to harmonize BIM technology with the Comsol Multiphysics software package. A review of existing studies on this issue in Europe, Asia, the United States and Russia. With the participation of the authors of the article developed a set of specialized families for the system Autodesk Revit, which allow you to perform a comprehensive preliminary study of the construction site of residential buildings. The approach providing increase of efficiency of determination of estimated cost of works on construction of object of construction based on several generalized principles is offered. This innovative approach has been tested in the design of several large construction projects in the Ural region of Russia, implemented in 2017-2019.

1. Introduction

BIM technology (Building information Modeling – building information modeling) involves the construction of one or more virtual building models in digital form. The use of BIM allows with a high degree of reliability to recreate the object itself with all the structures, materials, engineering equipment and processes occurring in it and to debug the main design solutions on the virtual model [1].

The use of models facilitates the design process at all stages, providing more thorough analysis and control. The information model exists throughout the life cycle of the building, and even longer. The information contained in it can be changed, supplemented, replaced, reflecting the current state of the building [2-4]. For the correct operation of the BIM-approach need relevant information sources. Besides static documents (like imagery) need a point of integration between CAD (design), control systems, accounting (e.g. automated complex software 1C), trouble ticket system, email, or regulations of contractors and so on.

In Russia, such scientists and designers as V. V. Talapov, M. S. Panteleeva, V. V. Alenkov, V. Shakhramanyan, V. Sharmanov and others were engaged in the study of BIM technology in relation to the construction and calculation of the cost of construction materials in the project [2-5]. At the state level, adopted a plan for the phased implementation of BIM in the field of industrial and civil construction, approved by the order of the Ministry of construction of the Russian Federation № 926/PR of 29 December 2014.

The world's greatest development in the study of BIM information modeling is currently being celebrated at Brisbane University of technology in Australia, MARA University of technology in Malaysia, as well as at a number of higher education institutions In new Zealand and South Korea [1,6-9]. USA, United Kingdom, Finland, Denmark, Norway, Netherlands, UK, Hong Kong & Singapore are among the countries heading towards significant BIM uptake. In these countries, such an approach as design, construction and operation in one integrated development environment has already been adopted at the state level.

2. Results and discussion

The specialists of "Ural Federal University named after First President of Russia B. N. Yeltsin" together "Institute for planning, architecture and design" (LLC "Inpad") has developed a digital model of the object "Multifunctional building of the Congress centre of the international exhibition centre "Ekaterinburg-EXPO", with the use of BIM. The Congress center, which is being built by Sinara-development, will create the most comfortable conditions both for the organizers of the events and for the audience. In July 2019 The Congress center will become a venue for the global summit of production and industrialization GMIS-2019, which will bring together more than 10 heads of state.

Earlier, based on the results of calculations using BIM tools, an innovative model of strengthening the ground base for five nine-storey residential buildings in the right Bank of the city of Berezniki was developed, which was highly appreciated by the specialists of the Federal Science and High Technology Center and Department of state expertise of Perm region [3]. At the construction site, the creation of a soil mass with the help of soil-cement elements was performed, which was not previously used in the design and construction of buildings in this region.

Technical specifications for the design were set strict requirements for the economic and functional efficiency of the object:

- total area of the building – 41.6 thousand square meter;
- the number of floors to four floors and one basement floor;
- the main multifunctional hall in the basic configuration-4000 seats;
- Concert version of the hall with the organization of the fan zone for 5000 people;
- 15 conference rooms and 26 meeting rooms for public and exhibition forums;
- Catering complexes and restaurant yard;
- short terms of construction and installation works.

According to the design results, the following results were achieved:

- the technical equipment of the multifunctional hall allows to organize not only business events, but also entertainment, theater, Philharmonic concerts at the highest international level, meeting all the specific requirements for the organization of a spectacular space for architectural acoustics, visibility, comfort and safety.
- clarification of the consumption of materials for construction and installation works in the direction of a reasonable reduction;
- elimination of errors in design solutions due to the possibility of checking conflicts at the stage of development of the building model;
- reducing the cost of the object in terms of square meter.

The use of a digital model of the building made it possible to make a refined calculation of the impact of adverse engineering-geological conditions of the construction site.

In the process, the authors investigated the issue of harmonization of BIM technology with the COMSOL Multiphysics software package, which is actively used in the design of energy-saving structures [10].

A sample of stationary heat engineering calculation of 3D-model of the building using this software package is shown below (Figure 1).

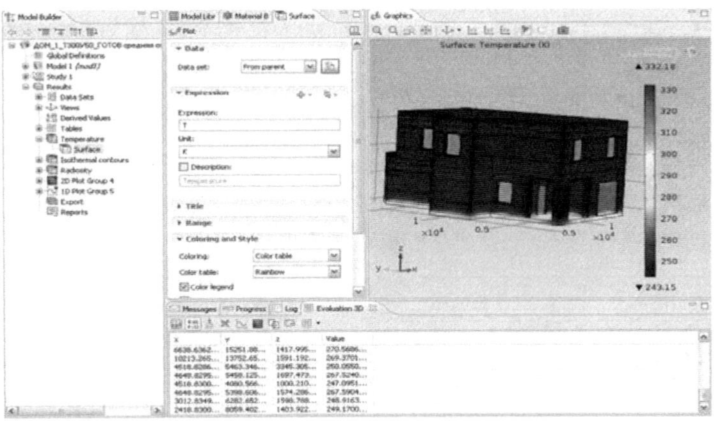

Figure 1. Window with the results of thermal stationary 3D calculation

To solve the PDE, COMSOL Multiphysics uses the finite element method (FEM).

Design and detailed documentation for all design objects is carried out in the licensed Autodesk Revit software package. Elements of the digital model of the building is shown below (Figure 2).

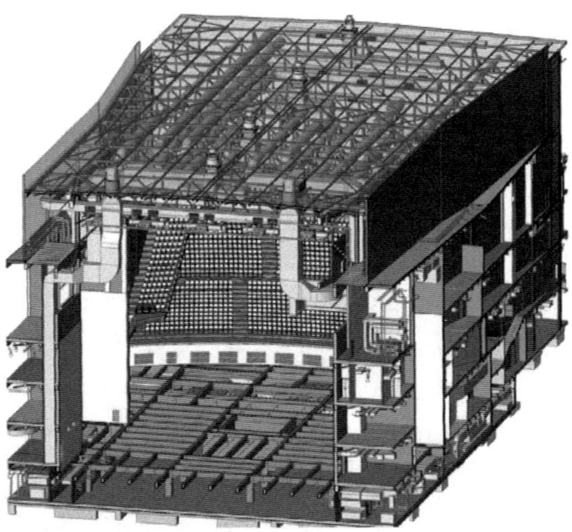

Figure 2. General model of the international exhibition centre building. View in section with internal engineering networks of heating and ventilation.

The task of determining the economic efficiency of the project at the earliest stage of work was a priority in the planning. No less important task is the planning and control of costs at the stage of project implementation. The use of BIM tools, namely, families, including data classified in accordance with the requirements of the customer, allows you to control costs online, at the same time, this work – budgeting and control of financial costs – carried out by specialists of the customer or specially involved organizations. As you know, one of the main advantages of designing using BIM is the ability to calculate the amount of building materials used with high accuracy: with the correct construction of the model, it is possible to extract any set of data originally incorporated into it.

IOP Conf. Series: Materials Science and Engineering **972** (2020) 012039 doi:10.1088/1757-899X/972/1/012039

One of the options for the implementation of this possibility for the future calculation of estimates is the use of classifiers.

The classifier is an approach adopted by agreement of the customer and the designer to systematize information about materials and products used for the construction of the facility.

The formulation of the classifier, in the form as it is now offered by the customer, is reduced to coding information for a specific method of calculation (estimate method, resource, per unit volume, etc.). Further processing - creation of sheets with codes in drawings and assignment of information to a specific element, for example (conditionally):

- Wall;
- Brick;
- Design thickness of 250 mm;
- Wire mesh reinforcement;
- On cement-sand mortar M100;
- Fire resistance EI60;
- So forth.

The customer has the possibility of further use of this information. It should be taken into account that the designer (architect, designer, engineer-specialist in engineering networks) there is a significant additional amount of work associated with the correct distribution and sorting of certain building materials within the classifier.

The designer often does not have, and, perhaps, should not have the qualification of an engineer-estimator. The order of definition of these or those parameters of materials, works and elements for determination of cost of works according to the existing estimated standards is often absolutely not obvious. It is also worth considering that at different stages of the project, as a rule, each has its own method of calculating the cost, based on its previous experience.

As a result, designers fill in the parameters of customers (similar, but each has its own requirements), trying to combine them with the internal standards of the organization. Also, they need to take into account the standards of designations, design and use the resources of the program for partial automation. As a result, there is an excessive overflow of information, often duplicating each other in one form or another. Well, the most negative point is the increase in the amount of additional, unpaid work of the employee. As a result, either the quality of work suffers or the term of performance of work is violated.

In order to quickly and effectively check construction estimates, we need a national program to create new regulatory frameworks and their quarterly update. In the United States, for example, thus updated regulatory framework RSMeans, which can be found for the money anyone [6].

In Russia, the reform of the pricing system and estimated regulation in the construction sector, carried out in 2015-2018, despite a long time period and a significant amount of money spent, has not achieved its goal — the transition to a new pricing model.

In order to facilitate cost estimation and verification of cost estimates using Building information Modeling, a "fair price" principle should be incorporated into the regulatory framework for each type of construction and installation work with reference to a specific region. This will allow, in turn, to conduct tenders for the selection of contractors based on a comparison of the consolidated unit prices, as well as to check the claimed estimated cost [4, 11].

The authors of the article proposed and tested in practice a new approach to calculating the cost of construction of the object using BIM design. At the moment BIM models of objects "Multifunctional building of the Congress center as a part of the international exhibition center "Yekaterinburg-EXPO", Apart-hotel "Belorechensky" in Yekaterinburg, Techno-shopping center JCB, showroom "Mitsubishi Motors Corporation" are prepared.

3. Conclusion

For the design of residential houses, the authors with the participation of specialists of the Institute of design, architecture and design (Inpad LLC) [11] developed a set of specialized families for Autodesk Revit system, which allow to perform a comprehensive preliminary study of the site:

1. Determine the insolation deduction for the building site-the volume, the placement of objects within which will not violate the insolation of the existing building.

2. Place within the boundaries of the land residential sections of different configurations, the main parameters of which (number of storeys, floor area, type of commercial property located on the ground floor) can be dynamically changed.

3. Get a full set of basic technical and economic indicators of the planned development, including the area of the required sites, Parking lots, the calculation of the number of places in preschool institutions and schools.

On the basis of the implemented research results, the authors found that the following approach is the most effective for the correct definition of the economy of the construction object at all stages of design using Building information Modelling project:

1. Calculation of the cost of objects at the stage of the concept and conceptual design is carried out on the basis of integrated performance indicators on the basis of objects of analogues.

2. Exclusion of changes in the type of materials used after the design stage "Preliminary design". At the same time, already at the Project stage, with the proper use of BIM construction technologies, it is possible to obtain the volumes of the main materials used, on the basis of which it is possible to prepare an enlarged estimate.

3. Preparation of estimates based on the volumes obtained from the BIM model, by specialized organizations.

4. Control of expenditure of funds during construction works is possible by agreement of the executed volumes, which are entered in the BIM model by means of automated control, or by employees of the design organization in manual mode.

Thus, Building information Modelling technology has shown the possibility of achieving high speed, volume and quality of design and construction, as well as significant budget savings.

References

[1] Ismail N A A, Drogemuller R, Beazley S and Owen R 2016 A Review of BIM Capabilities for Quantity Surveying Practice *MATEC Web of Conferences* 1-7

[2] Panteleeva M S, Unosheva A V 2016 BIM-technology and peculiarities of strategic management construction enterprise *Materials of the XI International research and practice conference «Science, Technology, and Higher Education. Strategic Studies Institute»* 52-56

[3] Salnikov V B, Belyakov V A and Berngardt K V 2016 Empowering optimization of design solutions with the introduction of BIM *II International Conference «Safety Problems of Civil Engineering Critical Infrastructures* 210-213

[4] Talapov V V 2016 Technology BIM: a single model and related misconceptions *Construction technology* **1–2 (111–112)** 74–80

[5] Alenkov V V, Chuprynousky V P, Shaklein A G, Ovsyannikov M L, Chebotarev, E M, Yartsev D I and Kolesnikov A N 2018 On using structured BIM information to protect the health and safety of workers in construction (International Journal of Open Information Technologies vol 6) **5** 39-50

[6] Bryde D, Broquetas M and Volm J M 2013 The project benefits of Building Information Modelling (BIM) (Int. J. Proj. Manag. vol 31) pp 971–980

[7] Kalinichuk S and Tomek A 2013 Construction Industry Products Diversification by Implementation of BIM (Int. J. Eng. Technol. Innov. vol. 3) **4** pp 251–258

[8] Abbasnejad B and Moud H I 2013 BIM and Basic Challenges Associated with its Definitions, Interpretations and Expectations (Int. J. Eng. Res. Appl. vol. 3) **2** pp 287–294

IOP Conf. Series: Materials Science and Engineering **972** (2020) 012039 doi:10.1088/1757-899X/972/1/012039

[9] Thurairajah N and Goucher D 2013 Advantages and Challenges of Using BIM: a Cost Consultant's Perspective *in 49th ASC Annual International Conference Proceedings.*
[10] Salnikov V B and Belyakov V A 2016 The thermal calculations of building constructions with application of the program complex COMSOL Multiphysics (Yekaterinburg: Publishing house of Ural University) p 48
[11] Salnikov V, Romanova E and A Fedorov 2016 The experience of implementing BIM technology *Stroykompleks Srednego Urala* **10 (202)** 76

IOP Conf. Series: Materials Science and Engineering **972** (2020) 012040 doi:10.1088/1757-899X/972/1/012040

BIM within current building facilities and infrastructure

V Nyvlt[1]

[1]Department of Civil Engineering, Institute of Technology and Business in Ceske Budejovice, Okružní 517/10, České Budějovice 370 01, Czech republic

E-mail: vlad.nyvlt@gmail.com

Abstract. One of the main reasons for introducing BIM into existing buildings is cost savings in the construction management and maintenance phase (operating phase). Operational phases have the biggest impact on the total life cycle costs. The introduction of BIM and the creation of a building information model, albeit at a higher price, will be counterbalanced several times by a more efficient way of managing the building during its life cycle, the possibility of creating a healthier indoor environment for building users and better access to necessary information, for example in the case of changes to completed buildings. The BIM model is a digital equivalent of a real construction – real digital twin. It contains geometric data in the form of a 3D model and non-geometry data, including construction management and support documents. The BIM model allows access to building usage information. The information obtained is more accurate. It keeps building element information up to date, enabling more informed decisions. It supports comparison of different solutions and their energy needs. Renovation can be done more efficiently. Construction information is up-to-date, so we can use the time needed to process different solutions instead of finding the necessary information about the construction.

1. Introduction

The building information model is not just a 3D model, but a database of information that is used throughout the construction lifecycle. All process participants contribute to this database during the construction and use of the object with up-to-date information. The advantage of this shared data is collaboration without losing data and maintaining access to their current version. The 3D model is only one way of presenting these shared data. By providing all the elements of the 3D model with structural, material and utility features, construction schedule positions, unit price, inspection and replacement schedule, investment and operating costs, etc., it is possible to create a model that also serves within operation and building maintenance phases.

The common "center" of the BIM method is CDE (Common Data Environment). This environment contains all the information like a 3D model technical data and other documents, but also communication between co-workers and individual processes in given phases of construction.

BIM's task is to bring together all the participants involved in the construction, implementation and operation of the building. Throughout the life cycle, the continuity of all individual phases can be maintained. The benefit of the BIM method to the builder (investor) is primarily to control the project and finance at each stage, or, for example, to process requests and changes to the building faster. For other workers, such as chief planners (architects, engineers, engineers), HVAC designers, structural engineers, budget workers, contractors, etc., the greatest advantage is the ease of communication with other participants, easier incorporation of changes, and better recording and editing requirements.

The BIM model data is destined for further use within building operations. In contrast, the current processing of data ends with the handover of the printed version of the actual construction. For example, when using BIM consistently and correctly, the facility manager may have all requirements for changes, defects, or backlogs. Data is available to all participants throughout the construction lifecycle. Facility management can thus use them not only to optimize operations, but also for timely planning of maintenance work, inspections and changes to completed structures (reconstructions).

2. Use of BIM for existing buildings

Models for existing buildings are made for rebuilding, reconstruction and maintenance. The detail and focus depends on the selection of the targeting method. The most widely used method and the roughest measurement is the measurement using a laser rangefinder. Furthermore, it is possible to use photogrammetric measurement or measurement using total station. The most accurate measurement is the combination of 3D laser scanner and photogrammetric measurement. This focus can be used both from the interior and from the exterior of the building. For photogrammetry of higher floors or a roof plane, it is possible to use a drone. Another aid to creating a model can be the original project documentation, where there is a risk that the project was not fully complied with during construction.

One of the first companies in the Czech republic that publicly showcased its experience with BIM's production of measured data was di5 architects and engineers. An example was creation of the Prague State Opera model (see Fig.1) The model was created in Revit 2011. To create the BIM model, data from existing documentation, historical documentation, photo documentation and geodetic surveying of the perimeter of the building were used. Furthermore, laser scanning and visual inspection of the building. First of all, source data was added to the project, the model itself was made up of walls, columns, and components. In addition to geometric information, the model also includes photographs and documents about the building attached to it, or information about the age of the individual structures (using the phase tool) [1].

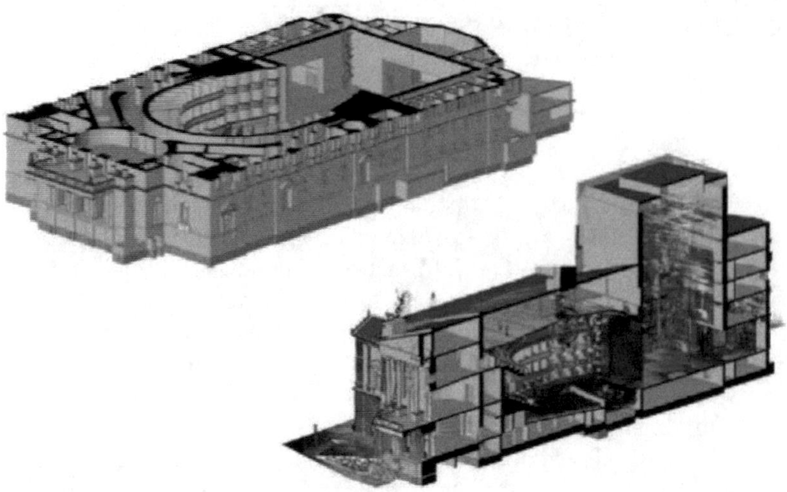

Figure 1. Model of Prague State Opera, source [1].

3. Public sector and BIM implementation

Public procurement began to require BIM support in managing database of public buildings and infrastructure. This initiative started in northern European countries, USA, Australia, and according to

[3] it is a clear schedule till 2027 in Czech republic, too. This is also related to the development of modeling from measured data [4].

Taking data in this way is fast, accurate and detailed. Data is collected using a ground laser scanner or unmanned aerial vehicle. When using the Revit program, we have to pay attention to the fact that it is not a program directly designed for creating an existing object model and it is necessary to set boundaries when the point cloud deviation cannot be neglected. Other ways of focusing from scientific articles and manuals are known, such as manual rangefinder, photogrammetry, or tachymetry.

The BIM model is useful for maintaining and managing object data in one file. Unlike conventional documentation, where data is transferred to 3D and may be lost. If a change occurs in the object, the change is entered into the model. When you enter a model, the change appears in all parts at once and does not need to be corrected separately in each drawing. In a BIM model, there cannot be a situation where floor plans, cuts, and views are contrary to one model generation. Thus, the drafting documentation, tables and lists are updated when the change is made.

The existing building model is the best means of refurbishment or rebuilding and a designer tool to design the best solution. When redesigning and completing the original model, the model is transformed into a model for building construction, which once again has been completed as a model for building maintenance. Sometimes it can also be used to compare the original project with the actual design.

The model is not only used for generating drawings, but also for solving visual aspects (color variations of components, interior, exterior and other visual designs) and for virtual tours. Compared to photo inspections, the model is specific in that it can also show hidden installations, such as air conditioning, wiring and other building equipment.

It is best to use a laser scanner for historic buildings. With a fairly detailed model, conservationists can preserve all historical features. Thanks to the BIM model, the elements are thoroughly described. When performing a sufficiently detailed model, it is easier to carry out maintenance and reconstruction of the building.

4. Scanners for BIM model creation

3D scanners are devices for capturing the shape and texture of physical objects and then converting them into digital form for further processing on a computer. The principle is based on scanning individual points of the object surface. A point cloud is created after all points are scanned. The model is also reconstructed using points and simultaneously using a suitable polygon mesh. One of many camera technologies, X-rays, magnetic microtomographs, lasers and touch sensors is used to earn points. Depending on the technology used, methods such as X-ray, ultrasound, laser, optical or mechanical 3D scanners are called. The most common method is to use non-destructive methods. These methods do not destroy individual scanned elements. All methods have advantages and disadvantages in terms of limitation and price. Outputs from 3D scanners are not only used for construction. Difference between point cloud and completed captured picture is on Fig. 2 [5].

Laser Flight Scanners or ToF Technology Principles are very similar to sonar. The scanner points at the scanned object and the laser beam is directed through the object to the nearest sensor. We measure the flight time of the beam to the object and back to the scanner that is the track time. The point has a fully defined point position in three-dimensional space. This method helps us to use the laser beam scanner to record the entire body using rotating parts. The rotating portion may be the entire scanner head in the case of another scanner, or the rotating coupled two mirrors, or the more advanced rotating optical reflecting prism, typically has the shape of a regular n-angle. The density at which the laser beam covers the surface of the real body determines the quality of the digitized body. A color camera can also be included to capture solid body information. The measurement result is a cloud of points. The cloud is composed of thousands of points in three-dimensional space. It is a dimensionally accurate geometric imitation. The acquired data is further transferred to an information model that can be further processed in CAD software [6].

Today, 3D scanning occupies an important position in civil engineering. He also finds application in other fields such as archaeology, medicine, film industry and engineering. This technology evolves very quickly and provides users with many new possibilities. The scanner analyses and collects data about the object, gets information about its shape and possibly colour. For example, such data can serve as a very accurate and detailed basis for the creation of replicas of sculptures or other major cultural monuments. With digitized objects, you can work with different SW, including those commonly used by architects and designers. They have the opportunity to transfer any item to their project and visualization. At the same time, there is a possibility to print the scanned object with a 3D printer in a reduced form.

Figure 2. Laser scanning, left part is completed picture, right part just scanned point cloud [5].

5. BIM Implementation Process for Existing Buildings and Infrastructure

Owners / Facility Managers / Administrators perform site maintenance activities according to legal requirements and provide routine activities. Such activities became so complex that without the appropriate systems that support the facility managers' work, they are virtually impossible. If a properly trained facility manager takes care of the BIM introduced into an existing building, it is likely that using this system will make it easier to obtain and archive managed object data.

Introducing BIM into existing buildings is a very time-consuming process. The continuity of implementation of BIM is influenced mainly by the readiness and timeliness of the documents, staffing of key and trained persons, and last but not least, by information technology equipment. In most cases, the documents are not up-to-date, they are incomplete or even missing completely. In such cases, the documentation must be updated or even created.

Another factor affecting the length of the process is the occupancy of key people. Anyone involved in implementing BIM into an existing building must be able to operate one of the CAD systems that supports 3D and BIM drawing, followed by a CAFM (Computer Aided Facility Management) system that is compatible with the CAD system and can be interconnected. Interconnection BIM and CAFM systems can convert graphical and alphanumeric data.

Building information model is just a database of information, including data from the initial design, through construction, building management and eventual changes to completed buildings (reconstruction) to demolition, including the ecological disposal of the building and restoration of the space. It covers therefore the whole life cycle of the building. All participants in the building process enter information into this database. For this database to be fully functional, all participants from the entire building cycle should upload the up-to-date information. The big advantage of collaboration and access to building information is the confidence that data is not lost and access to the current version is maintained. Therefore, information that is important to other participants in the building process is shared [7].

Steps to Implement BIM for Existing Buildings:

1. Evaluation of the current state of passport data and drawing documentation.
   The existing drawing documentation is evaluated. The timeliness of the data is determined and further action is determined accordingly.
2. Defining relevant passport data for FM / BIM.
   It is necessary to set the goals that facility management / BIM should fulfill. According to these goals, the exact structure and detail of the data is then determined.
3. Passportization of the building
   Focusing and documenting the actual state of all real estate, ie all buildings, outdoor areas, etc., in a predetermined range of data and structures. A well-arranged database of the current state of the building will be created. The database provides all the information on the building and technical condition, from individual building structures to internal installations to utility lines. It is one of the possibilities of property registration, which allows to monitor not only the house as a whole, but also its individual parts in the form of separate housing units and non-residential premises. For example, it may reveal imperfections in areas that are not commonly used, are not visible from the outside, or are not visible at first sight.
4. Completion / acquisition of drawing documentation.
   This point is based on the current state of the drawing documentation. It is absolutely necessary to visit the building together with its complete measurement. Based on the information obtained, the necessary documentation will be created by the designer. It contains not only the necessary drawings, which clearly characterize the building, but also includes technical and accompanying reports with substantial information.
5. Digitization of 2D drawing documentation for 3D model.
   Digitization of the documentation on the 3D model is possible both from the old documentation (drawings) and from the 2D electronic format (dwg).
6. Filling the CAFM system with passport data, connecting the CAFM system with the passport data obtained. It is also necessary to fill the system with manuals, guarantees for built-in products, technical specifications, personnel agenda, overview of workplaces, etc.
7. Interconnection with drawing documentation.
   All acquired data, manuals, technical product specifications, etc. must be linked to the drawing documentation. By linking BIM and CAFM, the system can convert both graphical and alphanumeric data.
8. Setting of facility management processes in CAFM system.
   These processes include, but are not limited to, cleaning, maintenance, revision, staffing, duplicity detection, property management, and more. A necessary prerequisite for correct process management and facility management activities in the CAFM application is, in addition to the selected methodologies and process settings, the definition of specific objects for which processes are implemented (buildings, facilities, workers, departments and organizations).
9. Own management and operation.
   This is followed by controlling all necessary processes in the object using the BIM and CAFM systems.
10. Create BIM Model Position.
    You need to create a new BIM job position. The BIM model must always be up-to-date, so the administrator needs to take care of the model and keep it up to date. All data collected must remain usable throughout the construction lifecycle.
11. Continuous updating of data and processes.
    Data and processes must be continuously updated.

12. Analysis and simulation, evaluation.

The final step is to decide on the other processes necessary for the construction and investment to function.

Fig 3 shows complex environment of sharing and linking data within whole building lifecycle.

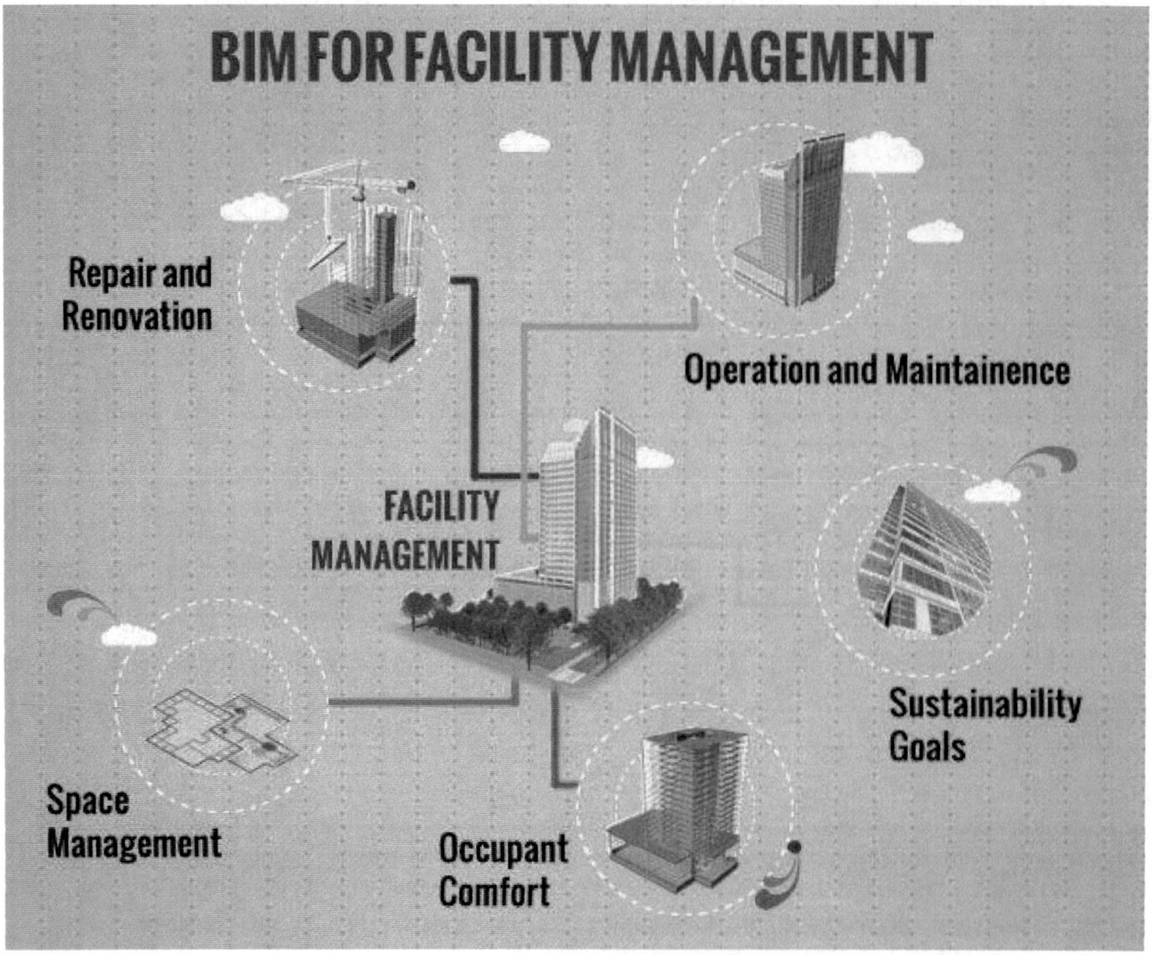

Figure 3. Sharing and linking data within whole building life-cycle [8]

6. Conclusions

The current practice of using paper records for all documentation of buildings is being gradually replaced by digital models with constantly evolving technology. These models allow for faster and cheaper management of all properties. Therefore, the process of digitization of already-built buildings has also become part of modern trends that go beyond maintenance and property management. This digitization is greatly aided by the current rapid development of laser scanning and the development of SW, which allows point clouds to convert, at least in part, to 3D models, BIM models. This process, of course, goes beyond the digitization of the entire area, where both the buildings themselves, the surrounding area, the terrain, other infrastructure, etc. will be advantageously linked to data. This paper outlines the beginning of this approach that has just begun.

Acknowledgments
The research described in this paper has been financed by TAČR - program ÉTA -project TL02000559 Safe and secure cities for pedestrians and senior citizens, (Bezpečná města pro chodce a seniory).

References

[1] Pivec D 2014 Modern methods off 3D model creation based on laser scanning [presentation on conference Revit in practice, Prague, 2014
[2] bimtech.cz 2019 available at www.bimtech.cz/bim [online] cit 15.5.2019
[3] ODBOR71100 2017 Concept of BIM implementation in the Czech Republic. Ministry of Industry and Trade, September 2017. Available from: https: //www.mpo.cz/assets/cz/stavebnictvi-a-suroviny/bim/2017/10/ Concept-introduction-methods-BIM-v-CR.pdf.
[4] Bartuska L, Hanzl J and Lizbetinova L 2016 Possibilities of Using the Data for Planning the Cycling Infrastructure. World Multidisciplinary Civil Engineering-Architecture-Urban Planning Symposium (WMCAUS), Prague, CZECH REPUBLIC, Procedia Engineering Volume: 161 Pages: 282-289
[5] cept.cz 2019 available at http://www.cept.cz/cinnosti/laserove-skenovani/ [online], cit. 15.5.2019
[6] CAD.cz 2019 CAFM Systems - IT Facility Management Support [online] cit 3.5. 2019 Available from: https://www.cad.cz/pdmplm/7-2007/1311-cafm-systemy-it-podpora-facility-managementu.html
[7] SVJ: Passportization in Practice [online] cit 4.5. 2019 Available at: https://www.svjaktualne.cz/33/pasportizace-v-praxi
[8] BIM for Existing Buildings [online] cit 3.5. 2019 Available from: https://www.tzb-info.cz/bim/18480-zavadeni-bim-u-jiz-existujicich-staveb

IOP Conf. Series: Materials Science and Engineering **972** (2020) 012041 doi:10.1088/1757-899X/972/1/012041

Optimization of constructions with the grids of lenticular pipes

A S Marutyan[1], E V Galdin[2]

[1]Department of Project Grant Organization, Institute of Service, Tourism and Design (branch) of North Caucasus Federal University in Pyatigorsk, 357500, Pyatigorsk, St. 40 let Octyabrya, 56, Russia.

[2]Department of Design, Institute of Service, Tourism and Design (branch) of North Caucasus Federal University in Pyatigorsk, 357500, Pyatigorsk, St. 40 let Octyabrya, 56, Russia.

E-mail: al_marut@mail.ru, galdin@list.ru

Abstract. Presented a new technical solution, which is a highly efficient technical and economic indicators and massively used in industrial civil engineering. Development work and optimization design calculations of lattice structures using video surveillance from core pipes. Their new technical solution has been developed, the originality of which is confirmed by patent examination. Worked gussetless node connections of belts and grids with direct connection of core elements to each other without misalignments, as well as constructive eccentricities limited by 0.25 height of belt elements, which allows not to take them into account in the calculations, provides an increase in the degree of unification of the nodes of the upper and lower carrier zones designs. The whole diagram of changes in the design parameters of the lenticular pipes is given during the transformation of their cross sections from vertical to horizontal configurations, including a transition through a circular shape. The prospects of using a new technical solution in light metal structures of buildings and structures are shown. The area of its rational use is outlined, where, using alternative design of known and new solutions, a quantitative assessment of the resources of their carrying capacity is given.

Key words: calculation of optimal parameters, shaped pipes, core systems, trusses, lattice girders, light metal structures.

1. Introduction

Due to its high performance and aesthetic properties, rectangular and square pipes (welded profiles) have found mass distribution in the domestic and foreign construction practice. Due to the constructive form, they very effectively combine increased thin-walledness with not less increased rigidity. This can explain the fact that already in 1964, 80 thousand tons of square and rectangular profiles were used at construction sites in England [1]. Since then, the total number of round, square and rectangular pipes have been supplemented with new profiles. In particular, GOST R 54157-2010 "Profile steel pipes for metal structures. Specifications" regulates such profiles as oval, flat-oval, semi-oval, semi-plano and some others [2]. During the same time, light metal structures were developed and introduced for the complete delivery of spatially pivotal (structural) systems of the Moscow Architectural Institute, "Kislovodsk" of round pipes, as well as "Molodechno" truss and subrafter trusses of rectangular profiles [3].

The most widely used in tubular structures were gussetless node connections with direct connection of the lattice rods to the belts. Here, in order to avoid pushing (pulling out) the diameter of the grate pipe, at least

1/3 of the diameter of the belt pipe should be taken [4, p. 38]. In the construction of rectangular (square) pipes, this restriction is tightened twice, that is, the width of the lattice rod should be taken at least 0.6 of the transverse size of the belt [5, p. 26]. At the same time, the cutting of the end edges of rectangular and square pipes for their direct junction in the unit-free units, as a rule, is ensured by flat cuts at the angles specified by the design. The manufacture of gussetless nodes of round pipes requires additional expenses for figured cutting of the end edges due to the complex mating outlines of cylindrical shells. Therefore, in practice, quite often to simplify the faceless units, the ends of round pipes are subjected to partial or complete flattening. The formation of transitional and flattened sections of the rod elements of the tubular section of the gratings is recommended to ensure the slope of the transition section 1/6 ... 1/4 [3, p. 152; 4, s. 47-49; 6, s. 102]. Victor Ivanovich Trofimov has a priority in the creation of faceless assemblies with the flattening of all core elements in relation to structural structures. To obtain a knot, the ends of the pipes are flattened and cut in braces at an angle equal to the inclination of the brace. The rods are closed on the end faces and fixed in the conductor with the formation of free space with a width of about 3 cm, which is filled with melting metal. The time of welding of a single node is about 20 ... 25 minutes, and its weight does not exceed 2 ... 2.5% of the weight of the whole structure [7].

Similar design of connecting nodes without gussets is also characteristic of steel bar structures of coatings that have a longer history in construction science and technology, and the priority in their creation belongs to Vladimir Grigorievich Shukhov [8]. The lattice of such structures is made of a single round steel rod in the form of individual V-shaped, W-shaped (zigzag) elements or a continuous “snake” and at the points of inflection it has areas quite sufficient in length for welding the grid to the belts [9]. Therefore, in search of new technical solutions, it is very effective to take into account the practice and experience of steel bar structures of coatings, replacing zigzag lattices from rods with lattices of shaped tubes similar in shape. For the direct connection to the belts and the formation of gussetless nodes, the tubular profile in the places specified in the design is flattened and double-zigzag give a zigzag appearance. The length of the strip (tape) blank tubular profile (strip) can be chosen so that it is enough for the entire length of the structure or the entire length of the shipping mark. Compared with bar, such grids have a higher bearing capacity (especially in compression), which allows increasing the load on the structure or at a fixed load to reduce its material consumption.

At the level of optimization and variant design, several versions of roof trusses of run-through and rootless coverings with belts made of rectangular and square pipes, as well as with triangular and diagonal zigzag outlines of profile pipes (rhombic, round, oval, flat oval, lenticular) were worked out [10]. The solutions found were convincing enough to pass a patent examination with a positive outcome. All of them have their own distinctive features, but a number of common features allows us to conditionally consider them in a single cycle. Lines of bends in the power plane of the bearing structure form sheet hinges, close to ideal hinges (the calculated length of the compressed element is equal to its geometric length), and from the plane - frame mounts, close to rigid fittings (the calculated length of the squeezed element is half the length of its geometric length). Then, for stability (flexibility or rigidity) of the core elements of the gratings, which are the same from the plane and in the plane of the structure, it is possible to select cross sections, the inertia radii of which differ in terms of their values by two times. In practice, this is quite feasible with a noticeable positive effect due to the conversion of square pipes into rhombic, and round pipes into oval, flat-oval, lenticular. Additionally, for fabricated profiles, the technology of their manufacture using compressed air is applicable [11]. After the publication of the positive decision of the patent examination, the supporting structures with lattices of lenticular pipes are presented here as the proposed (new) solution, and the structures with lattices of oval and flat-oval pipes are the prototype of the new solution. On the other hand, such a representation of a technical solution, as well as its analogues and prototype, can be viewed as summing up some results of a small cycle started from rhombic pipe grids. Their knots are rather universal, unified, due to the sheet hinges, they eliminate the need to take into account the rigidity of the nodal, non-faceted joints of welded structures made of tubular profiles and help reduce the consumption of structural material [12].

Thus, the main idea of this publication is to identify the relevance, prospects and effectiveness of the systematic use of zigzag grids of core pipes for the development and research, design and implementation of new forms of supporting structures of buildings and structures.

2. Literature review

The proposed solution relates to the construction and can be used in the manufacture of trusses, girders, columns, arches, frames and other lengthy supporting grid structures of pipes. Currently, the most economical and common tubular structures include trusses of coverings (floors) of rectangular and square pipes (closed arc-welded profiles), whose effectiveness especially increases in the presence of off-site loads [6; 13, p. 461-463; 14, p. 294-296; 15, s. 29-35; 16-29]. In these truss structures, type assemblies are faceless joints with direct connection of core lattice elements to one of the four faces of the waist pipe, which is a source of stress concentration. Such junction is more often one-sided, where compliance with the ratio of transverse dimensions given in the introduction contributes to a more uniform distribution of stresses. A further decrease in the stress concentration ensures a reversal of the square pipe of the belt by 45 ° around the longitudinal axis when it occupies a rhombic position relative to the plane of the structure (Fig. 1, a) [30]. Even earlier, the rhombic position was adopted for the core elements of the lattice (Fig. 1, b) [31–34].

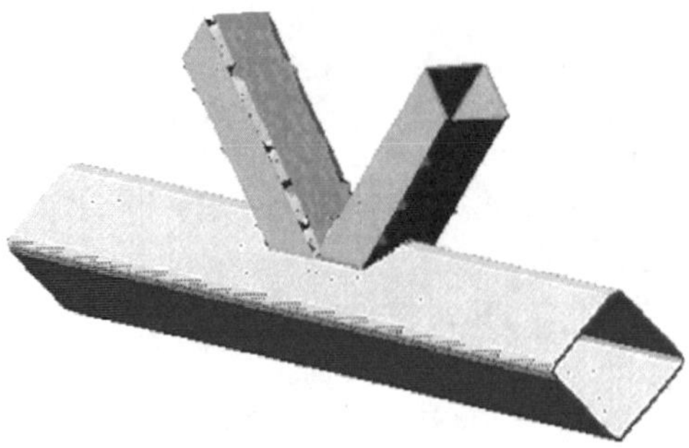

a

b

Figure 1. Axonometry of the node (a) and a snapshot of a fragment of a truss from square pipes (b)

The direct connection of the gratings to the belts to the tubular trusses was initially successfully tested in rod girders of coatings (Fig. 2) [9; 14, p. 370; 15, s. 25; 35; 36, p. 56-59].

To increase the carrying capacity (carrying capacity) of the run, its modifications were developed, including several zigzag lattices in the form of flat continuous "snakes" made of round rods or a pair of lattices in the form of one spatial continuous "snake" made of ribbon (strip) rods reinforced with stamped ribs (Fig. 3) [37, 38]. In addition, continuous "snakes" of lattice structures can be profiled with an open or closed cross section (Fig. 4) [39, 40].

In relation to bare surfaces, a truss (lattice) design has been developed, the belts of which are made of triangular pipes, and the grille has two layouts: from W-shaped (V-shaped) round rods or from round tubular rods with ends flattened in the plane of the structure and cut off at an angle of inclination of the braces (Fig. 5) [41].

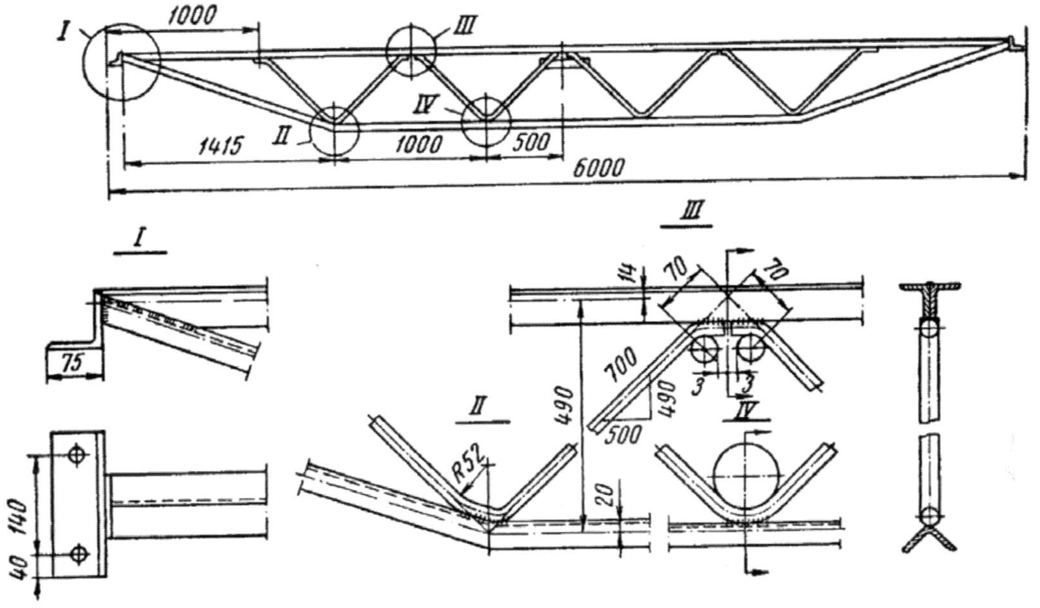

Figure 2. Steel truss construction (lattice girder) using rods

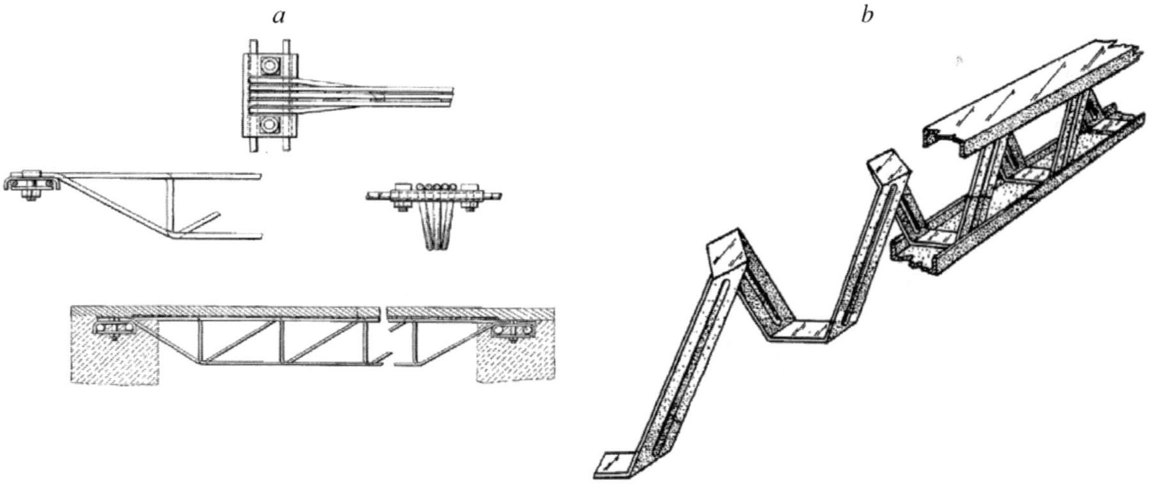

Figure 3. The design of runs with grids of three flat "snakes" (a) and one spatial snake (b)

a

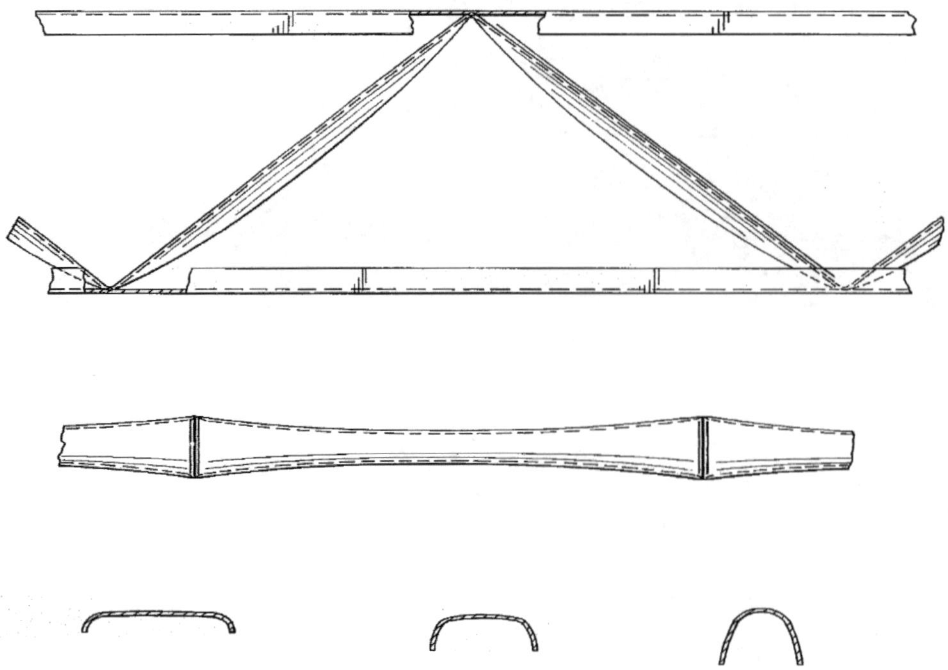

b

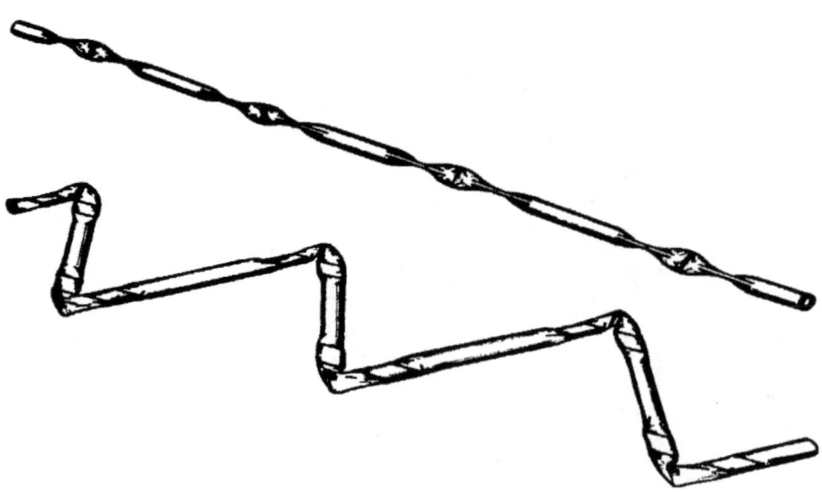

Figure 4. The design of the girders with grids of flat "snakes" open (a) and closed (b) cross-section

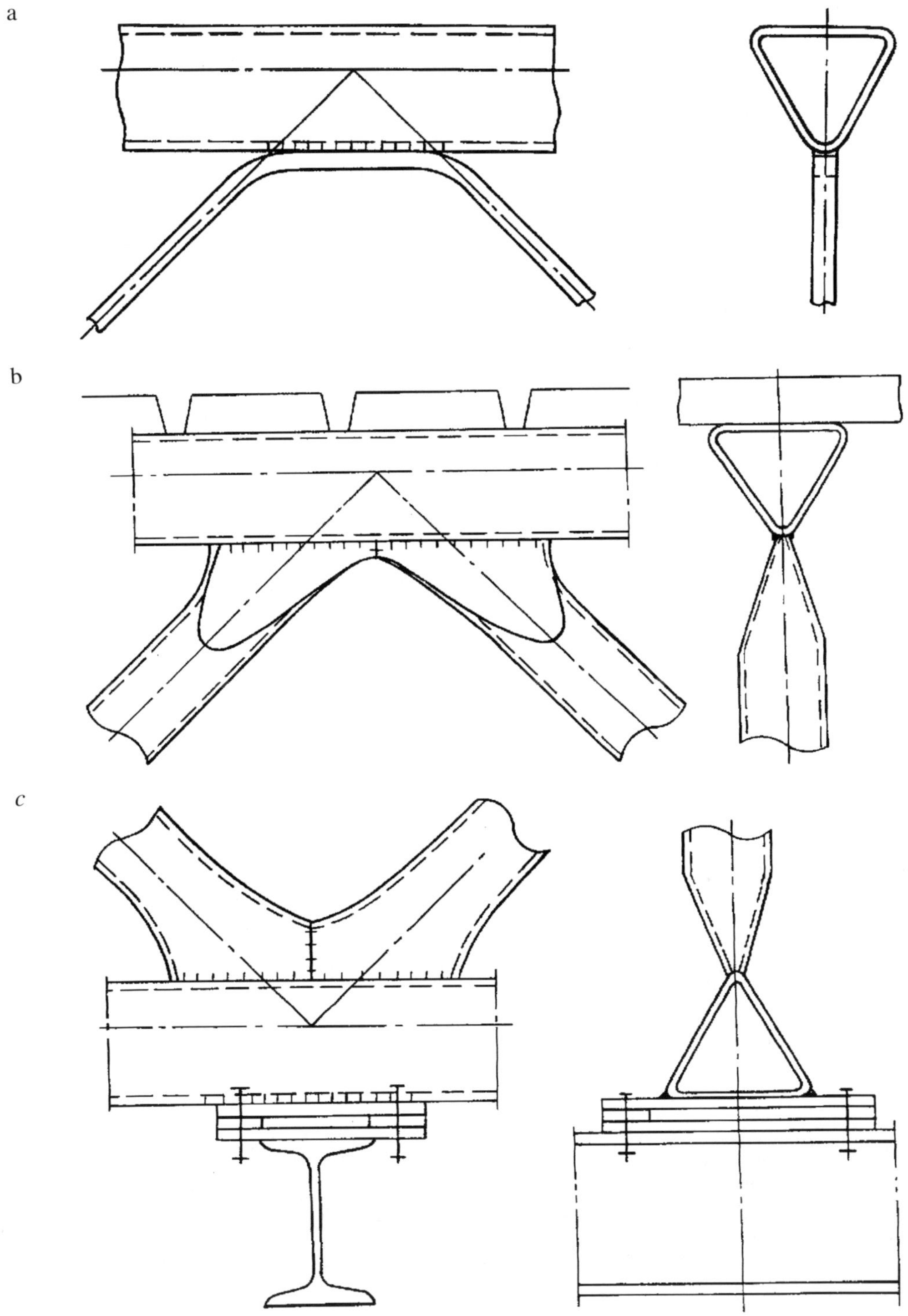

Figure 5. Scheme nodes truss (lattice) design of purlinless coverage: *a* - the top belt with a bar lattice; *b* - upper belt with a tubular lattice; *c* - lower belt with a tubular grid and suspension crane equipment

IOP Conf. Series: Materials Science and Engineering **972** (2020) 012041 doi:10.1088/1757-899X/972/1/012041

Belt elements of rectangular tubes in purlinless coverage work more efficiently than similar elements of triangular profiles. Therefore, a supporting structure with a lattice in the form of a continuous "snake" of rhombic pipes, whose flattened sections provide the local (local) stability of rectangular and square belt elements in gussetless nodes, has been developed. If we take into account the fixing of the lattice rod elements (designs hinged in the plane and rigid from the plane), the greatest positive effect will be in a thin-walled tubular section of a rhombic shape with a diagonal ratio of 1/2, where the larger diagonal is located in the design plane and the smaller one is from the plane (Fig. 6). For the running coverings, the supporting structure with a lattice of rhombic profiles was modified with the replacement of belt elements from rectangular (square) pipes with rhombic (Fig. 7) [42].

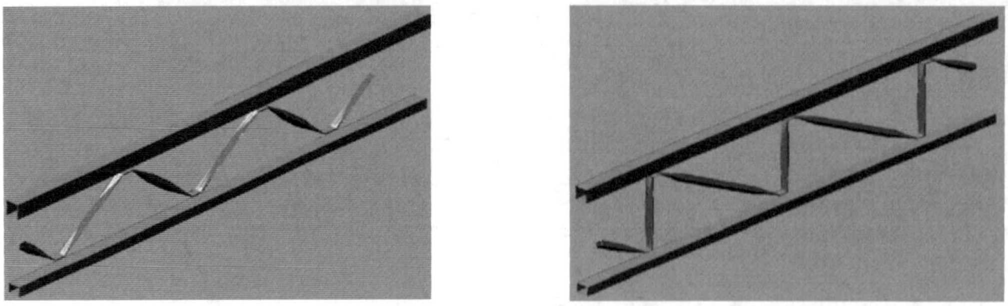

a *b*

Figure 6. Axonometry of the supporting structure with a triangular (a) and diagonal (b) grid of a rhombic pipe

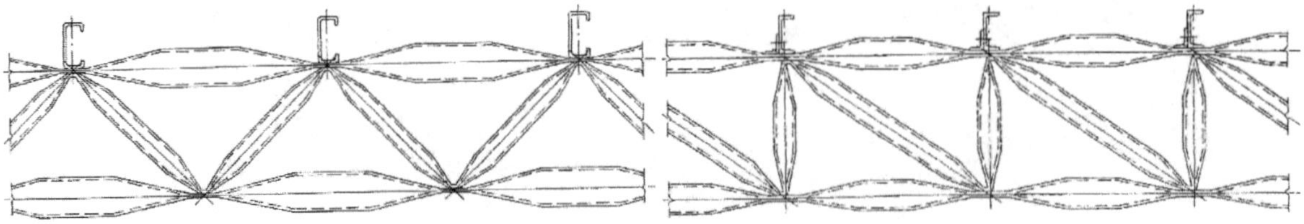

a) *b)*

Figure 7. Diagrams of the supporting structure of rhombic pipes for the purlinless coverage of a triangular (a) and diagonal (b) lattice

If zigzag lattices from rhombic profiles in supporting structures are replaced with lattice made of round, oval or flat-oval tubes of similar in shape (Fig. 8), due to the absence of angular zones of cross sections, the stress concentration and metal hardening will decrease, and the margin of safety will increase. The maximum positive effect will be in a thin-walled tubular section of an oval shape with a ratio of dimensions of 1 / 2.5 and a flat-oval shape with a ratio of dimensions of 1 / 2.63 (where the larger envelope is located in the plane of the structure and the smaller one is from the plane). In such lattices, all other things being equal, the flat-oval profile is more compact than the oval profile [43, 44]. Flattening and double bends of tubular lattice elements provide the layout of gussetless nodes joints without constructive eccentricities (Fig. 8, a), characteristic of rafter truss and subrafter trusses from rectangular (square) bent-welded closed profiles, which eliminates the appearance of bending moments and positively affects the consumption of structural material. However, at the same time, to increase the degree of unification of the nodes of the upper and lower belts, it is fully justified to use constructive eccentricities (Fig. 8, b), limited to 0.25 height of the belt elements, which allows not to take them into account in the calculations [5, p. 24].

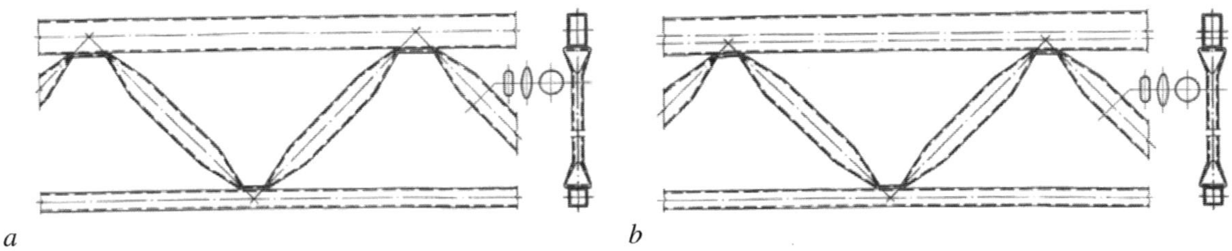

Figure 8. Schemes of fragments of supporting structures with grids of round, oval and flat-oval pipes: a - without eccentricities in nodal joints; b - with constructive eccentricities for unification of the nodal joints of the upper and lower belts

For the manufacture of oval profiles, you can use round longitudinal pipes, which, after molding and welding, orient at a certain angle relative to the minor axis of the oval (approximately 39°32 '). After orientation of the weld, the round tube enters the two-roll stand, where it is ovalized with dimensions of 440 × 650 mm (Fig. 9, a, b) [45]. Such a method of manufacturing tubular profiles should be refined to optimize oval and flat-oval pipes for truss (or beam) structures (Fig. 9, c, d) [46, 47].

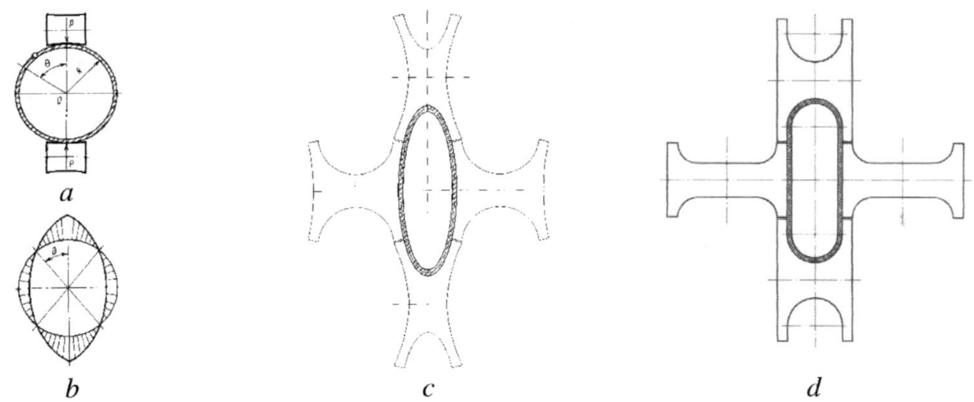

Figure 9. Scheme of ovalization of a round pipe (a) and a diagram of bending moments in cross section during its unraveling deformation (b), as well as a reduction scheme for oval (c) and flat oval (d) pipes

As already noted, the supporting structures with grids of oval and flat-oval pipes served as a prototype for a new technical solution, which are similar designs with grids of lenticular profiles [48]. Such profiles can be manufactured according to the technologies adopted in the practice of pipe production, including the above. However, the practical interest is noted in the introduction technology of their manufacture using compressed air.

Thus, oval tubes are 76 ... 85% more expensive than "cylindrical shells of lenticular cross-section" (Fig. 10) [49, 50]:

- according to tabular data from a journal publication for November 2012, a ton of steel mini-shells cost 20,000 rubles [51];
- according to the price list of 04/01/2008 OJSC Moscow Pipe Plant (Moscow, Barclay St., 6) a ton of an oval pipe cost 37090 rubles (37090/20000 = 1.85);
- according to the price list dated July 27, 2012 LLC "StalIntech" (Moscow, Federativny Prospect, 5, building 1) a ton of an oval pipe cost 35,200 rubles (35,220,200 = 1.76).

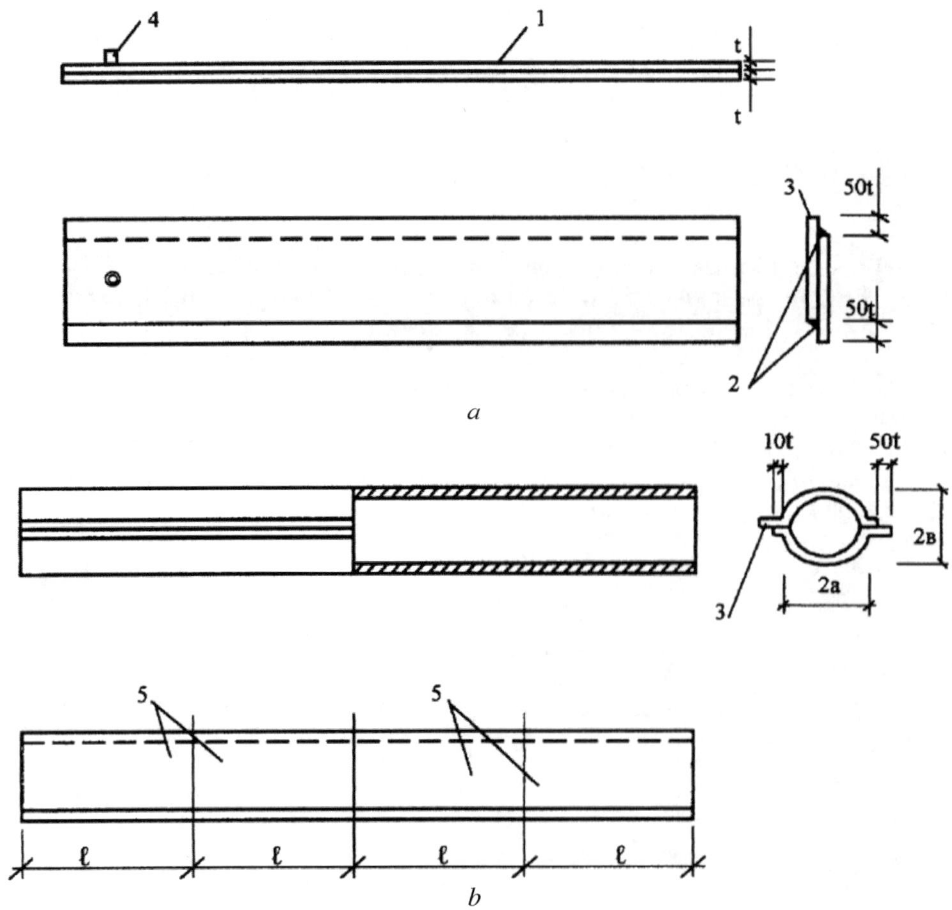

Figure 10. Schemes of formation of cylindrical shells of a lenticular cross-section of rolled materials: a - billet to blow up; b - preform after blowing up; 1 - half-foldable blanks; 2 - welds; 3 - consoles; 4 - fitting; 5 - shell sections

Completing the review part and comparing the new technical solution with its prototype, it can be concluded that the curved profile tubes are quite promising for optimization in relation to the gratings of supporting structures of buildings and structures.

3. Results of research

The technical result of the proposed solution is to reduce the complexity of the manufacture of supporting structures and reduce their costs. It is achieved by the fact that in supporting structures, including belts and lattices rigidly attached to them, made with flattened ends of curved zigzag outline elements, the core elements of lattices have ocular sections with an aspect ratio of 1 / 2.414, where large dimensions are located in the planes of the structures, and the smaller ones are perpendicular to these planes (Fig. 11) [48].

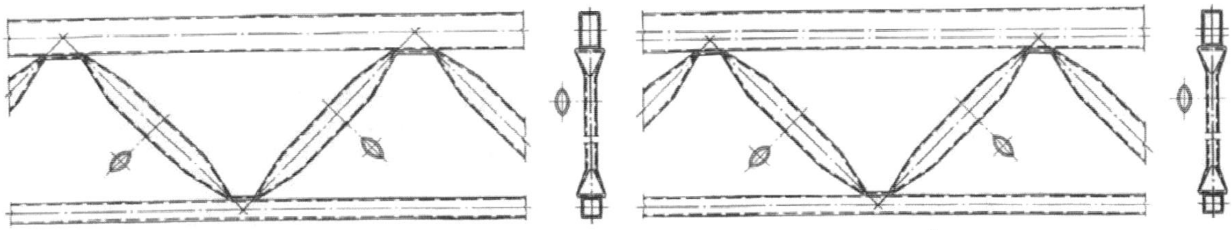

Figure 11. Schemes of fragments of supporting structures with gratings of lenticular tubes: a - without eccentricities in nodal joints; b - with constructive eccentricities for unification of the nodal joints of the upper and lower belts

To derive the above ratio and quantify the resources of the bearing capacity, it is advisable to calculate the axial inertia moments *Ix* and *Iy,* as well as the sectional area *A* of the curved tubular profile. The cross section of this profile can be considered as hollow in the form of lentils (Fig. 12) [52]:

$$I_x = R^3 t(2\alpha - \sin 2\alpha)\,; \tag{1}$$

$$I_y = R^3 t(2\alpha(2 + \cos 2\alpha) - 3\sin 2\alpha)\,; \tag{2}$$

$$A = 4\alpha R t\,, \tag{3}$$

from where

$$R = \frac{A}{4\alpha t}\,, \tag{4}$$

where R – is the arc radius of the cross section of the lenticular tube along the centerline; t – wall thickness of the lenticular tube; α – half of the central angle that tightens the arc of the cross section of the lenticular tube along the middle line.

To the reference formulas we need to add one more:

$$n = \frac{U}{V}\,, \tag{5}$$

here n – is the ratio of the smaller to the larger; U – smaller envelope of the cross section of the lenticular pipe along the midline; V – larger envelope of the cross section of the lenticular pipe in the midline.

To obtain a reduced ratio, to simplify the calculations, it is advisable to consider a special case when $\alpha = \pi/4$ or $\alpha = 45°$, when

$$I_x = R^3 t(2\alpha - \sin 2\alpha) = R^3 t(\pi/2 - \sin(\pi/2)) = R^3 t(3.14/2 - 1) = 0.57 R^3 t \tag{6}$$

$$I_y = R^3 t(2\alpha(2+\cos 2\alpha) - 3\sin 2\alpha) = R^3 t((\pi/2)(2+\cos(\pi/2)) - 3\sin(\pi/2)) = \\ = R^3 t((3.14/2)(2+0) - 3\times 1) = 0.14 R^3 t \tag{7}$$

$$\frac{I_x}{I_y} = \frac{0.57}{0.4} = 4.0714285 \approx 4 \tag{8}$$

with errors $100(4.0714285 - 4)/(4.0714285...4) = 1.75...1.79\%$;

$$\frac{i_x}{i_y} = \sqrt{\frac{0.57}{0.14}} = 2.0177781 \approx 2 \tag{9}$$

with errors $100(2.0177781-2)/(2.0177781...2)=0.881...0.889\%$;

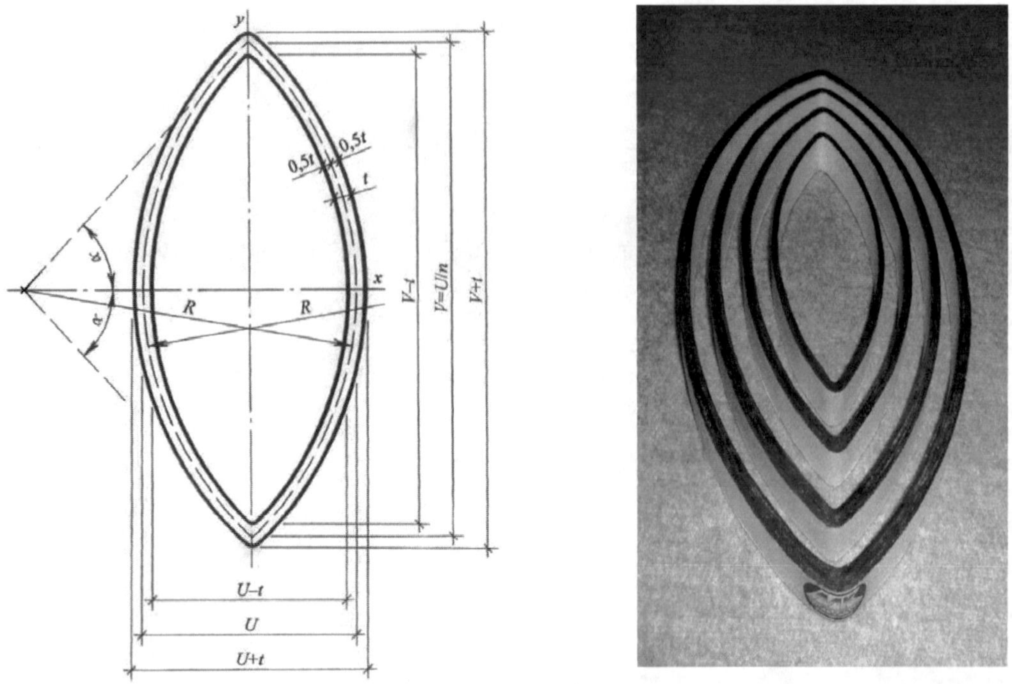

a *b*

Figure 12. The design scheme of cross sections of the lenticular pipes (a) and a snapshot of the slice of different-sized profiles (b)

$$R=\frac{A}{4\alpha t}=\frac{A}{4(\pi/4)t}=\frac{A}{3.14t}=0.3184713\frac{A}{t}; \quad (10)$$

$$I_x=0.57R^3t=0.57\left(0.3184713\frac{A}{t}\right)^3 t=0.0184113\frac{A^3}{t^2}; \quad (11)$$

$$I_y=0.14R^3t=0.14\left(0.3184713\frac{A}{t}\right)^3 t=0.004522\frac{A^3}{t^2}; \quad (12)$$

$$i_x=\sqrt{\frac{0.0184113A^3}{t^2A}}=0.135688\frac{A}{t}; \quad (13)$$

$$i_y=\sqrt{\frac{0.004522A^3}{t^2A}}=0.0672458\frac{A}{t}. \quad (14)$$

As can be seen, there is a two-fold difference in the inertia radii of the considered cross section with an error that is quite acceptable in practical calculations. Therefore, it remains to determine the dimensions of this section and their relationship. The larger dimension of the cross section along the middle line V on the design diagram of a hollow profile in the form of lentils is a diagonal of a square with a side equal to the radius of the arc R, and half of the smaller envelope U is the difference of the same radius and half of the same diagonal (or half of the larger dimension), that is

$$V = \frac{2R}{\sqrt{2}} = \frac{2\times 0.3184713A}{t\sqrt{2}} = 0.4503864\frac{A}{t}; \qquad (15)$$

$$U = 2(R - \sqrt{R^2 - 0,25V^2}) = 2(0,3184713 - \sqrt{0,3184713^2 - 0,25\times 0,4503864^2}) = = 0.1865562\frac{A}{t} \qquad (16)$$

$$n = \frac{U}{V} = \frac{0.1865562}{0.4503864} = 0.4142135 \approx 0.4142 = \frac{1}{2.414}. \qquad (17)$$

4. Testing of the design circuit

The results can be tested using the Huygens formula for arc length [53]:

$$p \approx 2l + \frac{1}{3}(2l - L) = \frac{8l - L}{3} = \frac{8\sqrt{0.25U^2 + 0.25V^2} - V}{3} = \frac{8\sqrt{0.25n^2V^2 + 0.25V^2} - V}{3} = = \frac{8\times 0.5V\sqrt{n^2+1} - V}{3} = \frac{V\left(4\sqrt{n^2+1} - 1\right)}{3} = 0.4503864\frac{A}{t}\times\frac{4\sqrt{0.4142^2+1}}{3} = 0.4998609\frac{A}{t} \approx \approx 0.5(A/t) \qquad (18)$$

with errors $100(0.5 - 0.4998609)/(0.5...0.4998609) = 0.028\%$,

where p – is the length of one of the two arches of the lenticular profile along its midline, $p = 0.5(A/t)$; l – a half arc chord, $l = \sqrt{0.25U^2 + 0.25V^2}$; L - chord of the whole arc, $L = V$.

The verification calculation is rather correct, since its relative error slightly differs from the error of the Huygens formula, which is approximately equal to $\alpha = 45°$ the angular parameter which is 0.02%.

5. Comparison of the calculated parameters of the new solution and its prototype

After testing, the obtained results for the new technical solution are interesting to compare with the similar results for its prototype, which is an oval tube with attitude $n = 0.4 = 1/2.5$. Then, taking with, in relation to the prototype can be written down [43]:

$$U = 0.182\frac{A}{t},\ V = 0.455\frac{A}{t},\ n = \frac{U}{V} = \frac{0.182}{0.455} = \frac{1}{2.5};$$

$$I_x = 0.0203346\frac{A^3}{t^2},\ I_y = 0.0050281\frac{A^3}{t^2},\ \frac{I_x}{I_y} = 4.044192 \approx 4$$

with errors $100(4.044192 - 4)/(4.044192...4) = 1.093...1.105\%$;

$$i_x = 0.1425994\frac{A}{t},\ i_y = 0.0709090\frac{A}{t},\ \frac{i_x}{i_y} = 2.0110197 \approx 2$$

with errors $100(2.0110197 - 2)/(2.0110197...2) = 0.548...0.551\%$.

Comparison of the results allows to conclude that the estimated parameters of the prototype in absolute value exceed the calculated parameters of the proposed solution:

$$\Delta U = \frac{0.182}{0.1865562} = 0.9755773,$$

IOP Conf. Series: Materials Science and Engineering **972** (2020) 012041 doi:10.1088/1757-899X/972/1/012041

$$\Delta V = \frac{0.455}{0.4503864} = 1.0102436;$$

$$\Delta I_x = \frac{0.0203346}{0.0184113} = 1.1240671,$$

$$\Delta I_y = \frac{0.0050281}{0.004522} = 1.1117228;$$

$$\Delta i_x = \frac{0.1425994}{0.1356882} = 1.062201,$$

$$\Delta i_y = \frac{0.070909}{0.0672458} = 1.0543822,$$

that is, under the given conditions, when $A = const$ and $t = const$, and $i_x = 2i_y$ ($I_x = 4I_y$), for the lenticular pipes, the cross sections in height are 1.024% more compact than those of the oval, while for oval pipes, the geometrical characteristics of the sections are $5{,}4...12{,}4\%$ larger than those of the lenticular.

6. Example of optional design

As an object of alternative design for comparing the proposed (new) technical solution with its counterparts and prototype, a steel truss from shaped pipes (bent-welded profiles) with a span of 18 m was adopted to cover an industrial building:

- truss with a grid of square pipes [36, p. 157-172];
- truss with a rhombic pipe grille [42];
- truss with a round pipe grille [43];
- a truss with an oval pipe grille [43];
- a truss with a grating of a flat-oval pipe [44];
- a farm with a grill of a lentil tube,

where the transitional and flattened sections of the core elements of the lattices of tubular profiles, it is advisable to form with a slope of 1/4 [6, p. 102].

The results of the calculation of the consumption of structural material are summed up in a tabular form (table 1), from which it is obvious that modifications with lattices of oval, flat-oval and curved pipes are more economical. The cost of making a tube is cheaper, so all other things being equal, a new technical solution is more preferable.

Table 1. Consumption of structural material (steel S245) on rod elements

Cross section	length,	Qty,	Weight, kg				Note.
mm	mm*	PC.	1 м	1 pc.	all	total	
□ 160×120×5	9000	2	20,7	186,3	372,6	880,4 (100%)	Belts
□ 120×4	7500	2	14,3	107,3	214,6		
□ 100×4	2390	8	11,8	28,2	225,6		Grate
□ 80×3	2390	4	7,07	16,9	67,6		
□ 160×120×5	9000	2	20,7	186,3	372,6	851,7 (96,7%)	Belts
□ 120×4	7500	2	14,3	107,3	214,6		
◊ 80×4	2390	12	9,22	22,04	264,5		Grate
□ 160×120×5	9000	2	20,7	186,3	372,6	831,0 (94,4%)	Belts
□ 120×4	7500	2	14,3	107,3	214,6		
○ 102×3,5	2390	12	8,50	20,32	243,8		Grate
□ 160×120×5	9000	2	20,7	186,3	372,6	799,3 (90,8%)	Belts
□ 120×4	7500	2	14,3	107,3	214,6		

Cross section mm	length, mm*	Qty, PC.	Weight, kg 1 м	1 pc.	all	total	Note.
Oval	2390	12	7,38	17,64	212,1		
125,6×52,4×3,5**							Grate
(○ 89×3,5)							
□ 160×120×5	9000	2	20,7	186,3	372,6		Belts
□ 120×4	7500	2	14,3	107,3	214,6	799,3 (90,8%)	
Flat-oval	2390	12	7,38	17,64	212,1		
113,5×45,5×3,5**							Grate
(○ 89×3,5)							
□ 160×120×5	9000	2	20,7	186,3	372,6		Belts
□ 120×4	7500	2	14,3	107,3	214,6	799,3 (90,8%)	
Lenticular	2390	12	7,38	17,64	212,1		
124,4×53,6×3,5**							Grate
(○ 89×3,5)							

* Dimensions are in axes without detailing.

** Dimensions of oval, flat-oval and lenticular pipes are indicated by the external dimensions of their profiles.

The obtained results of the optimization and variant design show that the proposed technical solution allows depending on the design decisions by appropriate selection of the ratio of the dimensions of the section of the tubular pipes, as well as the arrangement of these dimensions in the planes of the supporting structures to adjust their stress-strain states. Such regulation is not limited to the given values $1/2,414$ ($n = 0,4142$) and $1/1$ ($n = 1$), but has a very wide range (Fig. 13) and, in general, can ensure the effectiveness of further optimization of the supporting structures of buildings and structures.

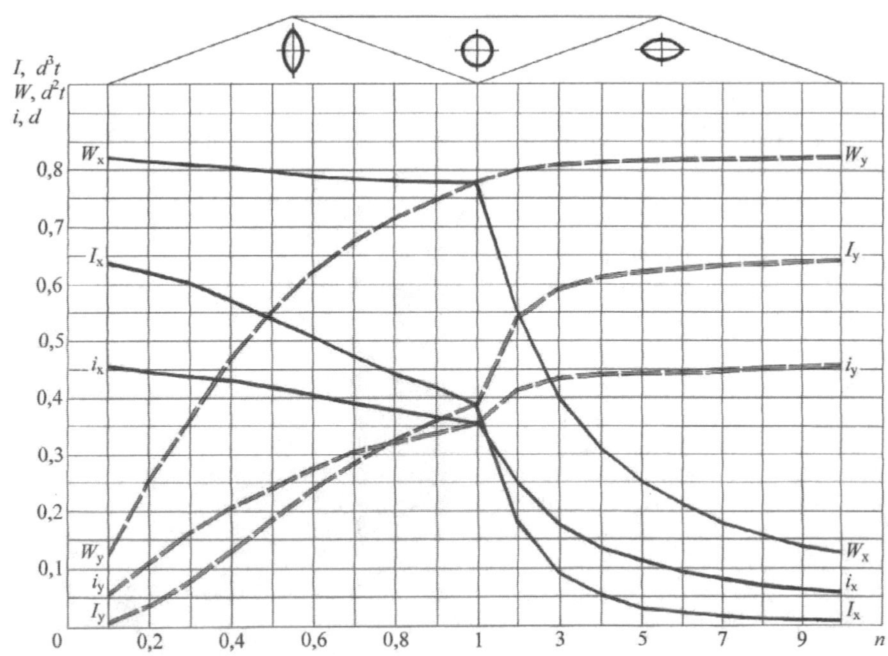

Figure 13. Graphs of changes in the static characteristics of sections of lenticular pipes depending on the

ratio of their dimensions, where d – is the diameter of a round pipe along its midline when $A = const$ and $t = const$

7. Conclusion

Thus, summing up some results, we can formulate a number of key conclusions.

1. Review and analysis of the research, development and design work, as well as optimization and variant design show that the systemic use of the proposed zigzag outlines of shaped pipes seems to be quite relevant, promising and effective for developing and researching, designing and introducing new forms load-bearing structures of buildings and structures.
2. The sheet hinges in the gussetless nodal joints provide the approach of the lattice structures to their design schemes (models) in the form of hinged-rod systems, which increases the degree of reliability and structural safety of buildings and structures.
3. An approximate calculation of the occipital sections confirmed its correctness and simplicity for the practical solution of problems of optimizing load-bearing structures.
4. Selection of the optimal parameters of shaped pipes and their use in supporting structures instead of square, rhombic and round profiles provides an increase in rigidity (or decrease in flexibility) of the core elements of the grids.
5. Comparison of the proposed and well-known solutions on the same basic object revealed a decrease in the consumption of structural material (steel), which ensures the maintenance and increase of the competitiveness of the supporting structures during their modernization.

References

[1] Mel'nikov N P, Vinkler O N, Levitanskii I V 1973 Kholodnognutye zamknutye svarnye profili – vysokoeffektivnyi material dlya legkikh konstruktsii *Promyshlennoe stroitel'stvo* **6** pp 24-27

[2] GOST R 54157-2010. 2011 Truby stal'nye profil'nye dlya metallokonstruktsii Tekhnicheskie usloviya (Moscow Standartinform) p 74

[3] Trofimov V I, Kaminskii A M 2002 Legkie metallicheskie konstruktsii zdanii i sooruzhenii (razrabotka konstruktsii, issledovaniya, raschet, izgotovlenie, montazh) (Moscow: Izd-vo ASV) 574 p

[4] Rekomendatsii po proektirovaniyu stal'nykh konstruktsii s primeneniem kruglykh trub 1973 (Moscow: TsNIISK im. V.A. Kucherenko) 94 p

[5] Rukovodstvo po proektirovaniyu stal'nykh konstruktsii iz gnutosvarnykh zamknutykh profilei 1978 (Moscow: TsNIIproektstal'konstruktsiya) 44 p

[6] Packer J A, Wardenier J, Zhao X-L, van der Vegte G J and Kurobane Y 2009 Construction with hollow steel sections. Design Guide for rectangular hollow section (RHS) joints under predominantly static loading. (CIDECT) 156 p

[7] Trofimov V I, Begun G B 1972 Strukturnye konstruktsii (issledovanie, raschet i proektirovanie (Moscow: Stroiizdat) pp 47-48

[8] Shukhov V G 1977 Stroitel'naya mekhanika. Izbrannye trudy (Moscow: Nauka) pp 71-82

[9] Podlipskii A A 1954 Stal'nye prutkovye konstruktsii pokrytii (Moscow: Gos izd-vo lit po str-vu i arkhitekture) 144 p

[10] Marutyan A S 2016 Profil'nye truby novykh modifikatsii dlya stroitel'nykh metallokonstruktsii, vklyuchaya moduli «Pyatigorsk» i «Novokislovodsk» (Pyatigorsk: Izd-vo PF SKFU) 168 p

[11] Rashchepkina S A 2009 *Vestnik MGSU Spetsvypusk* **3** pp 147-150

[12] Pokrovskii A A 2011 Ob uchete zhestkostei uzlov v raschetakh ferm s elementami maloi gibkosti *Stroitel'naya mekhanika i raschet sooruzhenii* **3** pp 31-32.

[13] Gorev V V 2004 Metallicheskie konstruktsii (Moscow: Vysshaya shkola) vol 1 pp 461-463

[14] Kudishin Yu I 2007 Metallicheskie konstruktsii (Moscow: Izdatel'skii tsentr «Akademiya»), 688 p

[15] Trishevskii I S, Klepanda V V 1978 Metallicheskie oblegchennye konstruktsii (Kiev: Budivel'nik), 112 p

[16] Kuz'menko S M 1984 Ratsional'nye printsipy razrabotki tipovykh seriinykh stroitel'nykh konstruktsii na primere proektirovaniya besprogonnogo pokrytiya «Molodechno» *Promyshlennoe stroitel'stvo* **12** pp 34-37
[17] Vedyakov I I 2007 Sovremennye tendentsii razvitiya rossiiskoi industrii stroitel'nykh metallicheskikh konstruktsii *Promyshlennoe i grazhdanskoe stroitel'stvo* **3** pp 8-9
[18] Fermennye konstruktsii sistemy TrasskonTM 2009 (Obninsk: OOO «Ruukki Rus») p 88
[19] Specification for Structural Steel Buildings 2010 *American Institute of Steel Construction* June 22 2010 pp 425-436
[20] Markku Heinisuo, Teemu Tiainen and Timo Jokinen 2013 Tubular truss design using steel grades S355 and S420 35 p
[21] Corona E, Vaze S P 1996 Buckling of elastic-plastic square tubes under bending *International Journal of Mechanical Sciences* vol **38** (**7**) pp 753-775
[22] Reyes A, Langseth M, Hopperstad O S 2003 Square aluminum tubes subjected to oblique loading *International Journal of Impact Engineering* vol **28** (**10**) pp 1077-1106
[23] Karagiozova D, Jones N 2004 Dynamic buckling of elastic-plastic square tubes under axial impact-ii: structural response *International Journal of Impact Engineering* vol **30** (**2**) pp 167-192
[24] Zhao H, Abdennadher S 2004 On the strength enhancement under impact loading of square tubes made from rate insensitive metals *International Journal of Solids and Structures* **41** (**24-25**) pp 6677-6697
[25] Utsumi N, Sakaki S 2002 Countermeasures against undesirable phenomena in the draw-bending process for extruded square tubes *Journal of Materials Processing Technology* **123** (**2**) pp 264-269
[26] El-Hage H, Mallick P K, Zamani N 2006 A numerical study on the quasi-static axial crush characteristics of square aluminum-composite hybrid tubes *Composite Structures* **73** (**4**) pp 505-14
[27] Liu Y-B, Zheng N-N, Liu J-B, Han Q 2010 A simplified approach to define the yield surface of concrete-filled square steel tubes subjected to compression-bending *Chongqing Daxue Xuebao* (**33**) **10** pp 70-75
[28] Kol'zeev A A 2005 Otsenka vliyaniya tolshchiny polok na ustoichivost' vnetsentrenno-szhatykh sterzhnei iz pryamougol'nykh trub *Izvestiya vuzov. Stroitel'stvo* **6** 85-88
[29] Psyuk V V, Nikishina I A 2014 Raspredelenie ostatochnykh napryazhenii v stal'nykh trubakh kvadratnogo secheniya *Sbornik nauchnykh trudov Ukrainskogo nauchno-issledovatel'skogo i proektnogo instituta stal'nykh konstruktsii im. V M Shimanovskogo* **14** pp 129-135
[30] Sokolov A A, Logachev K I, Zin'kova V A 2007 Chislennye issledovaniya napryazhenno-deformirovannogo sostoyaniya uzlovykh besfasonochnykh soedinenii trubchatykh elementov ferm *Promyshlennoe i grazhdanskoe stroitel'stvo* **8** pp 40-41
[31] Ono T, Iwata M, Ishida K 1994 Local failure of joints of new truss system using rectangular hollow sections subjected to out-of-plane bending moment. *Proceedings VI International Symposium on Tubular Structures* (Melbourne, Australia, Rotterdam, the Netherlands) pp 441-448
[32] Davies G, Owen J S, Kelly R B 2001 The effect of purlin loads on the capacity of overlapped bird-beak K-joints *Proceedings IX International Symposium on Tubular Structures* (Dusseldorf, Germany), pp 229-238
[33] Kuznetsov A F, Kuznetsov V A 2011 Pat. RF №116877, MPK E04S3/08 (2006.01). Ferma iz kvadratnykh trub; patentoobl. Yuzhno-Ural'skii gosudarstvennyi universitet №2011150569/03; zayavl. 12.12.2011; opubl. 10.06.2012. Byul. №16.
[34] Baikov D A, Lapshin A A, Kolesov A I 2014 Osobennosti raboty ploskikh ferm iz zamknutykh gnutykh profilei, soedinennykh v uzlakh na rebro *16-i Mezhdunarodnyi nauchno-promyshlennyi forum «Velikie reki' 2014»* (N Novgorod: NNGASU), pp 197-201
[35] Streletskii N S, Geniev A N 1935 Osnovy metallicheskikh konstruktsii (Moscow-Leningrad: Glavnaya redaktsiya stroitel'noi literatury) pp 584-590
[36] Kuzin N Ya 1998 Proektirovanie i raschet stal'nykh ferm pokrytii promyshlennykh zdanii (Moscow: Izd-vo ASV), p 184
[37] Roberts A L 1949 United States Patent №2476386. Floor unit. 19.07.1949.

IOP Conf. Series: Materials Science and Engineering **972** (2020) 012041 doi:10.1088/1757-899X/972/1/012041

[38] Claude G Harding 1981 United States Patent №4291515. Int. Cl E04C2/42. Structural elements. 29.09.1981.

[39] Melvin L Ollman 1975 United States Patent №3882653. Int. Cl E04C3/02, E04CB1/343. Truss construction. 13.05.1975.

[40] Grubb L W 1961 United States Patent №3129493. Methods for manufacture of lightweight structural memders. 20.06.1961.

[41] Orlik V M 1992 Avtorskoe svidetel'stvo SSSR №1760041, MPK E04V1/58, E04S3/08. Stroitel'naya metallicheskaya tonkostennaya reshetchataya konstruktsiya. №4776531/33. 07.09.1992. Byul. №33.

[42] Marutyan A S, Chernov P S, Orobinskaya V N, Galdin E V 2017 Improvement of truss bearing capacity by means of rhombic pipes *Key Engineering Materials* vol **736** pp 171-176

[43] Marutyan A S, Orobinskaya V N 2016 Optimizatsiya konstruktsii s reshetkami iz kruglykh i oval'nykh trub *Vestnik MGSU* **10** pp 45-57

[44] Marutyan A, Kobaliya T, Galdin E 2018 Optimization of flat-oval pipes and perspectives of their application in core structures *MATEC Web of Conferences* **251**, 03015 (2018)

[45] Zarankin V N 1989 Avtorskoe svidetel'stvo SSSR №1466827, MPK V21S37/08. Sposob izgotovleniya oval'nykh svarnykh pryamolineinykh trub; patentoobl. Ivanovskii zavod avtokranov. №4265826/29-27. 23.03.1989. Byul. №11.

[46] Nezhdanov K K, Tumanov V A, Nezhdanov A K, Rublikov S G 2007 Pat. RF №2304479, MPK V21D9/00 (2006.01). Sposob povysheniya nesushchei sposobnosti tsilindricheskoi truby na izgib. №2005115787/02. 20.08.2007. Byul. №23.

[47] Marutyan A, Kobaliya T, Galdin E 2018 Optimization of flat-oval pipes and perspectives of their application in beam structures *MATEC Web of Conferences* **251**, 03016 (2018)

[48] Marutyan A S Pat. RF №2618771, MPK E04C3/08, E04B1/58 (2006.01). Nesushchaya konstruktsiya s reshetkoi iz chechevidnykh trub. №2016105868/ 11.05.2017. Byul. №14.

[49] Denisova A P, Rashchepkina S A 2009 K voprosu uprugoplasticheskogo deformirovaniya stal'nykh polos *Promyshlennoe i grazhdanskoe stroitel'stvo* **9** pp 31-32

[50] Zemlyanskii A A, Denisova A P, Rashchepkina S A, Rashchepkin S V 2001 Pat. RF №2175372, MPK E04N7/06 (2006.01). Sposob sooruzheniya emkosti iz rulonirovannykh materialov. №99117202/03. 27.10.2001. Byul. №29.

[51] Rashchepkina S A 2012 Formoobrazovanie innovatsionnykh metallicheskikh konstruktsii razlichnogo naznacheniya *Promyshlennoe i grazhdanskoe stroitel'stvo* **11** pp 74-76

[52] Pisarenko G S, Yakovlev A P, Matveev V V 1988 Spravochnik po soprotivleniyu materialov (Kiev: Naukova dumka) pp 80-81

[53] Vygodskii M Ya 2006 Spravochnik po elementarnoi matematike (Moscow: Astrel'-AST) pp 342-343

Effect of Limestone Filler Dosage and Granulometry on the 3D printable Mixture Rheology

G Slavcheva[1], O Artamonova[1], D Babenko[1], A Ibryaeva[1]

[1]Colleges of Department of Building Materials Technology and Structures, Voronezh State Technical University, Bldg 84, 20-letiya Oktyabrya St., Voronezh 394006, Russia

E-mail: gslavcheva@yandex.ru

Abstract. The article presents the experimental results concerning the influence of the concentration and a particle size distribution of limestone filler on the rheological behaviour of cement pastes as matrixes of 3D printable concrete. The squeezing test is used in this paper as a rheological behaviour identification tool of cement-based materials in order to evaluate the extrudability and buildability. It has been established that introduction of filler as a factor changing the properties of solid phase in the system "cement + water" is an effective method for plasticity regulation and increasing the 3D printable cement-based materials resistance to load during the printing. Structural stability of cement paste can be improved by 2 times by introduction of filler with efficient ranging of a particle size in dosage of 20 – 40 % by weight of cement. Cement paste with limestone filler having a particle size distribution ranging from 1 to 75 μm is characterized by the optimal plasticity, structural strength and deformability under the load. Using limestone filler with equal particle size impairs these characteristics of cement paste.

1. Introduction

3D build printing is one of the additive manufacturing processes based on a viscoplastic materials deposition via a printing nozzle. Potential advantages of this process include ability to get multicomponent materials of different functionality directly on the construction site, to print the freeform constructions without formwork, to reduce materials consumption and building work content [1-3].

One of the top priority problems defining the ability to introduce the innovative 3D-printing into building practice is the problem of creating nomenclature of mixtures ensuring the implementation of this process. Efficiency and buildability of 3D-build printing depend on the complex parameters of the mixtures at all stages of the process. To ensure the required quality and construction time, management of rheological behaviour of mixtures must be done according to the conditions securing fluidity for their pumping, plasticity for extrusion, integrity of surface during multi-layer casting, structural stability for load accommodation of the upper layers. For this reason, researchers focus on the issues of optimization extrudability, firm stability and structural build up mixture compositions, for example [4-14]. As a result, collecting an array of experimental information, numerous mixtures has been received and tested. Relevant materials commonly comprise a combination of cement paste and bulk materials such as natural, manufactured and recycled aggregates and fillers, natural and recycled fibres. The obtained mixtures are multi-component, their compositions include superplasticizers, viscosity modifying additives, structural build-up regulators, bulk materials with various chemical-mineral compound and dispersion.

However, there is no systemized information and common approach to explanation of the role of each of the components for regulation the buildability of 3D-printable mixtures. Therefore, it is necessary to conduct system research identifying the influence on the set of rheological parameters of the mixtures of each single component used today in order to obtain them.

Author's theoretical approach to identifying the main factors of stability of the 3D printable materials is based on the analysis of the 3D printable disperse systems behaviour in dynamic (during pumping and extrusion) and static (during multi-layer casting) conditions of 3D-printing according to point of view of classic structural rheology of disperse systems [15]. In work [16] we have justified basic means of the rheological behavioural management of 3D-printable mixtures considered as a system "solid phase – liquid phase".

In connection with solid phase, they include its concentration, size of particles and their morphology, chemical-mineralogical composition, physicochemical characteristics of a particle surface. In connection with liquid phase, they include its ionic composition, viscosity, density. According to these means of management, a set of real technological tools has been suggested for management of rheological characteristics of 3D printable concrete. They include the type of binding agent, type and granulometry of filling agents, types and dosages of electrolytes, plastisizers, viscosity modifying additives, etc. For this reason, the main technological means of regulating rheological behaviour of mixtures include three grades of factors. The first one is water/cement ratio as a factor defining the concentration of solid phase in the system. The second one is a concentration and kind of addictives as a factor defining the properties of liquid phase. The third one is a concentration, type and granulometry of fillers and aggregates.

This article is dedicated to the study the influence of the concentration and a particle size distribution of limestone filler on the rheological behaviour of cement pastes as matrixes of 3D printable concrete.

2. Materials and methods

Five types of 3D-printable cement pastes were studied (Table 1). Portland cement CEM I 42.5 (EN 197 – 1 : 2011), plasticizer of Sika trademark based on polycarboxylate ethers, limestone filler ($CaCO_3$ content ~ 95%) were used as initial components of the system. Limestone filler had a particle size distribution ranging from 1 to 75 μm.

Table 1. Tested cement paste composition and specimen identification.

Specimen ID	Components			W/C
	plasticiser / mass cement (%)	limestone filler / mass cement (%)		
		different particle size (DLF)	equal particle size (ELF)	
C – W –P (control)	0.2	-	-	0.24
C – W –P – DLF20	0.2	20	-	0.23
C – W –P – DLF40	0.4	40	-	0.22
C – W –P – ELF20	0.2	-	20	0.23
C – W –P – ELF40	0.4	-	40	0.22

Concentration and granulometry (Table 2) of limestone filler was assumed as a main factor for the change of rheological behaviour of the 3D-printable cement pastes. Consistence of cement paste was assumed as a fixed factor for the experiments.

To evaluate the extrudability and firm stability of cement pastes, the different kind of squeezing tests were used in accordance with the methodology developed in the works [17-20].

Cylindrical samples of fresh cement paste with radius R equal to their height h_0 = 25 mm were used for the implementation of the experiment. For squeezing test, the sample was put between two smooth plates diameter of which corresponded to the size of the sample and was loaded into testing system "INSTRON Sates 1500 HDS". The test was conducted on a fresh sample for all compositions of cement paste directly after their molding.

Table 2. The particle size distribution of limestone filler.

Limestone filler	Granulometry	
	volume of particles, %	size (d), μm
Different particle size (DLF)	10	55
	10	27
	19.8	10
	14.2	2
	7.8	1
Equal particle size (ELF)	90	74.6
	10	≤ 70

High compression speed test using constant plate speed v = 5 mm/s was implemented as the behaviour of the system in the process of extrusion is most adequately modeled with this speed. The curves "compression force N – displacement Δ" obtained during the experiments were interpreted as influence curves of reduced compression load F^* from relative change of height of the sample h_i/R:

$$F_i^* = \frac{Ph_i}{\pi R^2}, \tag{1}$$

where $h_i = (h_0 - \Delta)$, h_0 is initial height of the sample, Δ is transfer in the i point of time, value R was taken as constant and equal to the radius of the sample at the beginning of the experiment.

According to the results of the analysis of the received experimental curves for the studied systems values K_i, called plastic yield value by N. Roussel and C. Lanos [17,18], were calculated at the inflection point of the $F^*(h_i/R)$ curves between "placing phase" and "perfect plastic response phase" zones (Figure 1):

$$K_i\left(\frac{h_i}{R}\right) = \frac{\sqrt{3}F^*}{2}. \tag{2}$$

The squeezing test was conducted with constant strain rate v = 0.5 N/s was implemented as the behavior of the system in the process of multi-layer casting. The strain rate conforms to the average speed of load increase during multi-layer casting of building sites by industrial printers. Methodology of its implementation corresponds to the approaches of A. Perrot [19,20] to evaluation of buildability of the 3D printable mixtures. Squeezing was conducted until the rupture of the samples. The curves "compression force N - displacement Δ" were recorded as a result of the experiments. Based on the obtained experimental curves, structural strength (σ_0) of cement pastes was calculated for the moments corresponding to the start of deformation (Δ = 0.1 mm), and plastic strength (σ_{pl}) was calculated at the beginning of cracking. The value of plastic deformations Δ_{pl} at the start of cracking was determined as a ratio of the displacement Δ of the plates in the squeezing test to the initial height of the sample h_0.

3. Experimental results and discussion

As a result of interpretation of the squeezing test with constant plate speed, we received experimental curves $F^* = f(h_i/R)$ (Figure 1) which correspond to the similar curves of N. Roussel [18]. Analysis of experimental curves $F^* = f(h_i/R)$ for description of rheological behaviour of cement paste during squeezing was conducted on the basis of approaches of fundamental structural rheology of disperse systems [15].

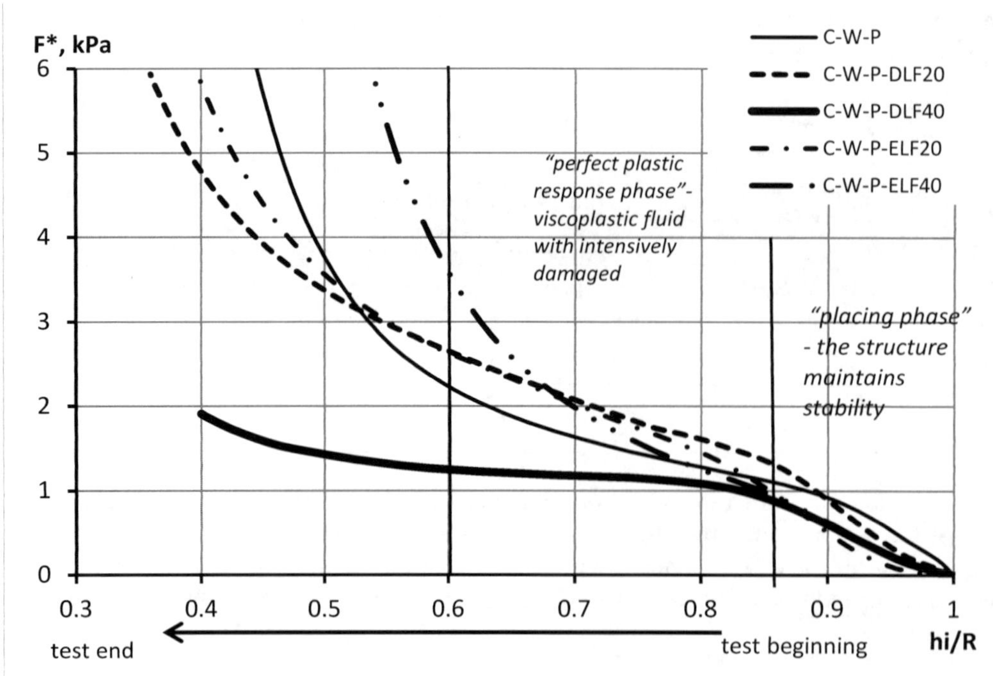

Figure 1. Tested cement pastes $F^*(h_i/R)$ curves.

Under the action of low compression stress on the first section of the curve within deformation range ~ $0.8 < h_i/R < 1$ the structure maintains stability ("placing phase" according to N. Roussel's terminology). According to point of view structural rheology of disperse systems, the "placing phase" can be characterized as a viscoplastic fluid with undisturbed structure. When the stress on the second section increases with $0.5 < h_i/R < 0.8$, the system is plastically deformed while its structure loses its stability ("perfect plastic response phase" according to N. Roussel). This section can be correlated with a viscoplastic fluid with intensively damaged structure in conformity with the structural rheology approache.

The experimental results show the similar kinds of $F^*(h/R)$ curves for all studied cement systems. They have expressed horizontal section of plastic deformation between the two points of inflection. For the systems transfer from stable condition to plastic flow is estimated by load F^* for ~ 1.0 ÷ 1.5 kPa, transfer into the condition of the flow with damaged structure happens with F^* = ~ 1.5 ÷ 3.5 kPa. At the same time, rational values plastic yield value K_i, which is determined at the beginning of viscoplastic fluid of disperse system with undisturbed structure, are within the range of 0.85 ÷ 1.20 kPa (Table 3). These values K_i ensure the best extrability due to their sufficient plasticity and capacity for viscoplastic flow without the damage of the structure under the influence of compression stress. At the same time the mixtures can be extruded through an extruder nozzle with low squeezing force. With values $K_i < 0.85$ kPa cement-based materials would not maintain stability during an extrusion process.

At the same time, the structural strength σ_0 is calculated on the basis of the quantity load N at the beginning of deformation of viscoplastic mixtures can be considered as the main criterion of their buildability (Table 3). The structural strength σ_0 characterizes the ability of the system to maintain stability and resist to a deformation during loading.

Table 3. Rheological properties of 3D-printable cement pastes.

Specimen ID	Plastic yield value K_i (kPa)	Strength(kPa)		Value of plastic deformations Δ_{pl} (mm/mm)
		structural σ_0	plastic σ_{pl}	
C – W –P (control)	0.86	1.10	45.01	0.18
C – W –P – DLF20	1.21	2.41	73.18	0.05
C – W –P – DLF40	0.93	2.34	69.59	0.10
C – W –P – ELF20	0.95	0.15	37.48	0.11
C – W – P – ELF40	0.65	cracking at the test beginning		

Analysis of the received experimental data of the curves "compression force N - displacement Δ" (Figure 2) shows two typical zone of rheological behaviour. The first one is the zone of plastic deformation. The system's ability to deform without destruction is evaluated by the quantity plastic strength σ_{pl} calculated on the basis of the quantity load N at the beginning of cracking. To characterize buildability, it seems reasonable to evaluate the quantity of plastic deformations on this section $\Delta_{pl,}$ which have to be minimized for 3D printable materials. The second one is the zone of cracking and intensively destruction of the structure. The cracking process is characterized by the multi- peak load fluctuations in the second section on "N - Δ" curves.

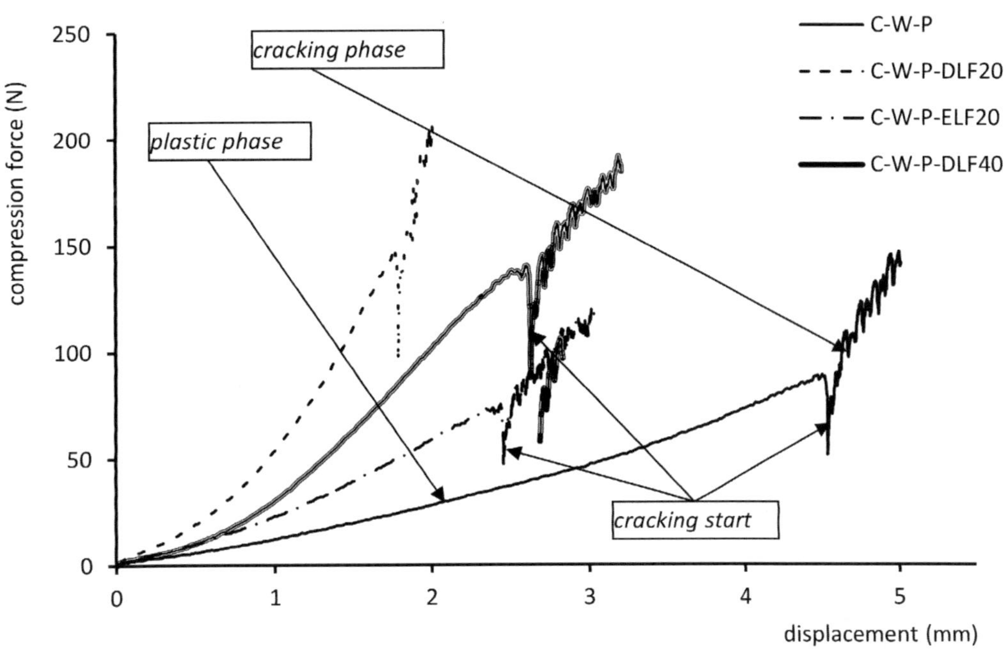

Figure 2. Tested cement pastes "compression force N - displacement Δ" experimental results.

The transition between plastic yield value K_i, value of structural σ_0 and plastic σ_{pl} strength, plastic deformations Δ_{pl} is linked to the concentration and a particle size distribution of filler.

Analysis of the curves "N - Δ" shows that system C – W – P displays the longest plastic phase zone (Figure 2). Interval of plastic phase zone for system with filler decreases by ~2 times comparing with the control systems C – W– P. This effect of decreasing plasticity of cement paste is logically related to the influence of filler. Introduction of filler into disperse system "cement + water" as a regulating factor of solid phase properties changes packing density and molecular interactions

IOP Conf. Series: Materials Science and Engineering **972** (2020) 012042 doi:10.1088/1757-899X/972/1/012042

between solid particles. As a result, flow under stresses becomes difficult for coarser systems containing filler. For systems with DFL (limestone filler with different particle size) plastic yield value K_i, corresponding to the beginning of plastic flow, is in the range from 0.9 to 1.2 kPa, which is more than in the control system without fillers (C – W – P). For systems with EFL (limestone filler with equal particle size) plastic yield value K_i is in the range from 0.65 to 0.95 kPa.

As a result, the system C – W –P – DLF20, C – W –P – DLF40 have the highest values of structural strength σ_0 and the minimum values of plastic deformations Δ_{pl} (Table 3). Introduction of limestone filler with different particle size (DLF) to the cement paste in dosages 20-40% allows to increase the structural strength by 2 times, to reduce plastic deformation by 2 - 5 times in comparison with control systems C – W – P. The system C – W –P – DLF20 has the best indicators such as the highest values of the structural strength $\sigma_0 = 2.41$ kPa, the minimum of plastic deformations $\Delta_{pl} = 0.05$ mm/mm. The structural strength σ_0 of the system C – W –P – DLF40 is the similar with the system C – W –P – DLF20 but plastic deformations Δ_{pl} of C – W –P – DLF40 system is higher by 2 times that plastic deformations Δ_{pl} of C – W –P – DLF20. Introduction of limestone filler with equal particle size (ELF) would not possible to ensure rational characteristics of plasticity and structural stability.

Change of rheological characteristics of cement pastes with the introduction of the fillers depends on kind of their granulometry. If fillers have different particle size, their particles locate between cement particles in a rational way. As a result, dense spatial packing of solid particles into cement paste structure is ensured. Such structuring of the cement pastes solid phase ensures the increase of their firm stability and consequently buildability. In contrast, fillers having equal particle size loosen cement paste structure. According to the experimental data, system C – W –P – ELF20 and C – W –P – ELF40 do not have structural stability ($\sigma_0 < 0.05$ kPa).

4. Conclusion

Analysis of the study results shows that the use of the squeezing test with constant plate speed is effective tool for evaluation of extrudability. Interpretation of the results of this test from the positions of structural rheology of disperse systems allows categorizing the parameters of plastic yield value $K_i \cong 1.0 \div 1.5$ kPa as criterion defining the ability of 3D printable materials to plastically deform without structure destruction and maintain stability during extrusion.

The squeezing test with constant strain rate is effective for the evaluation of buildability. The values of structural and plastic strength, plastic deformations defined by the results of this test characterize the system's ability to hold its form, resist the deformation under compressions load during multi-layer casting.

It has been established that introduction of filler as a factor changing the properties of solid phase in the system "cement + water" is an effective method for plasticity regulation and increasing the 3D printable cement-based materials resistance to load during the printing. Structural stability of cement paste can be improved by 2 times by introduction of filler with efficient ranging of a particle size in dosage of 20–40% by weight of cement. Cement paste with limestone filler having a particle size distribution ranging from 1 to 75 μm is characterized by the optimal plasticity, structural strength and deformability under the load. Using limestone filler with equal particle size impairs these characteristics of cement paste.

Acknowledgements

This research work has been carried out in the frame of the 7.10781.2018/11.12 Project, financed by the Russian government in order to obtain initial scientific results broadening the participation of subordinate educational organisations in the implementation of the National Technological Initiative.

References

[1] Labonnote N, Ronnquist A, Manum B and Rüther P 2016 Additive construction: State-of-the-art, challenges and opportunities *Autom. Constr.* **72** 347–366

[2] Wu P, Wang J and Wang X 2016 A critical review of the use of 3-D printing in the construction industry. *Autom. Constr.* **68** 301-316

[3] Bos F, Wolfs R, Ahmed Zand Salet T 2016 Additive manufacturing of concrete in construction: potentials and challenges of 3D concrete printing *Virt. Phys. Prot.* **11(3)** 209–225

[4] Kazemiana A, Yuan X, Cochran E and Khoshnevis B 2017 Cementitious materials for construction-scale 3D printing: Laboratory testing of fresh printing mixture *Constr. Build. Mater.* **145** 639–647

[5] Khalil N, Aouad G, Cheikh K. El, Rémond S 201Use of calcium sulfoaluminate cements for setting control of 3D-printing mortars *Constr. Build. Mater.* **157** 382–391

[6] Ma G, Li Z and Wang L 2018 Printable properties of cementitious material containing copper tailings for extrusion based 3D printing. *Constr. Build. Mater.* **162** 613–627

[7] Wenga Y, Li M, Tan M and Qian S 2018 Design 3D printing cementitious materials via Fuller Thompson theory and Marson-Percy model *Constr. Build. Mater.* **163** 600–610

[8] Paul S, Tay Y, Panda B and Tan M 2018 Fresh and hardened properties of 3D printable cementitious materials for building and construction *Arch. Civ. Mech. Eng.* **18** 311–319

[9] Le T, Austin S, Lim S, Buswell R, Gibb A and Thorpe T 2012 Mix design and fresh properties for high-performance printing concrete *Mater. Struct.* **45(8)** 1221–32.

[10] Le T, Austin S, Lim S, Buswell R, Law R, Gibb A and Thorpe T 2012 Hardened properties of high-performance printing concrete *Cem. Concr. Res.* **42** 558–566

[11] Feng P, Menga X, Chenb J, Ye L 2015 Mechanical properties of structures 3D printed with cementitious powders *Constr. Build. Mater.***93** 486–497

[12] Malaeb Z, Hachem H, Tourbah A, Maalouf T, Zarwi N. El, Hamzeh F 2015 3D Concrete Printing: Machine and Mix Design *Int. J. Civ. Eng. Technol.* **6** 14–22

[13] Panda B, Chandra Paul S, Jen Tan M. 2017 Anisotropic mechanical performance of 3D printed fiber reinforced sustainable construction material *Mater. Lett.* **209** 146–149

[14] Hambach M, Volkmer D 2017 Properties of 3D-printed fiber-reinforced Portland cement paste *Cement and Concrete Composites* **79** 62-70

[15] Uriev N 2010 Physico-chemical dynamics of structured nanodispersed systems and nanodispersed composite materials. Part 1 *Fizikohimiya Pov. Zash. Mat* **46** 3-23 (rus).

[16] Slavcheva G, Artamonova O 2018 The rheological behavior of disperse systems for 3d printing in construction: the problem of control and possibility of «nano» tools application *I-J.Nanotechnol. in Const.* **10** 107-122 (rus).

[17] Roussel N and Lanos C 2003 Plastic Fluid Flow Parameters Identification Using a Simple Squeezing Test *Appl. Rheol.* **13** 132–141

[18] Zahia T, Roussel N and Lanos C 2005 The squeezing test: a tool to identify firm cement-based material's rheological behaviour and evaluate their extrusion ability *Cem. Concr. Res.* **35** 1891–99

[19] Perrot A, Rangeard D and Pierre A 2016 Structural built-up of cement-based materials used for 3D- printing extrusion techniques *Mater. Struct.* **49** 1213–20

[20] Perrot A, M´elingeY, RangeardD, Micaelli F, Estell´eP, et al. 2012 Use of ram extruder as a combined rheo-tribometer to study the behaviour of high yield stress fluids at low strain rate *Rheologica Acta* **51(8)** 743-754.

Improving the technology installation of slabs of jointless ossature without girders and joints of KUB system civic buildings

N I Fomin[1], K V Berngardt [1], A V Vorobiyov[1], E E Zoteeva [1]
[1]Ural Federal University, 19 Mira str., Ekaterinburg, 620002, Russia

E-mail: ni.fomin@urfu.ru, kv.berngardt@urfu.ru, zoteeva@outlook.com

Abstract. According to the research results of precast and cast-in-situ civil buildings construction, made in KUB constructive system (cross-functional jointless ossature without girders and joints) in the Middle Urals conditions, the need to develop a technologically advanced method for the installation of the column floor slab, as well as erection device for its implementation is identified. While developing a new design and technological solution, methodological means were used (identifying contradictions, methods of eliminating them) of the theory of solving inventive problems (TRIZ). As a result of inventive activity, a mounting method of the column slab of girderless floor construction in KUB system and a mounting device for its implementation has been proposed.

1. Introduction

During the technical survey of unfinished residential building structures of KUB system (Pervouralsk, Sverdlovsk region), as well as erecting process studies of girderless floor construction, contained in this structural system, it was found that the technological reliability of installation process of the column slab is insufficient. The results obtained are consistent with the data of a number of studies [1-3], which emphasize the importance of a quality device for connecting columns and floor without beams. This node point is the most vulnerable place of the frame by the mounting structures criterion. The existing low level of technological reliability is due to the fact, that the known method of floor slab mounting, provided by the series on the jointless ossature without girders and joints KUB-2,5, as well as its modifications (KUB-3V, etc.), includes the sequential installation of two mounting devices: conductor and cleat. These devices provide separately the adjustment of the slab in the plan (conductor) and height (cleat). Adjusting the slab position vertically is carried out by cleat support rods, which have no connection with each other. The main disadvantages of the existing method of installation should be considered: the need to use two mounting devices for slab aligning in the plan and in height; the high labor intensity of the adjustment of the slab over height (small areas of base platform of the adjusting elements (rods) do not allow smooth and precise adjustment). Thus, the need to develop a new method of installation of a column slab and erection device for its implementation has been revealed. The development of new design and technology solutions in the construction field can be carried out by solving the formulated inventive problem, using the methodological tools of the theory of inventive problem solving (TRIZ).

2. Raising and resolution of the inventive issue by the TRIZ method

In 1950-1980 Russian researcher and inventor G.S. Altshuller created the methodical basis of the classical theory of solving inventive problems. The backbone of TRIZ can be described quite generally as follows: in a problem situation, technical contradictions are consistently identified, which are eliminated with the help of a number of special techniques [4-6]. Using terminology ModernTRIZ (MTRIZ) – the modern direction of development of the theory created by the Russian scientist M.A. Orloff [7, 8], the process of solving an inventive problem can be divided into three main stages.

Stage 1. Trend. Identify shortcomings in the existing solution and determine the course of changes. The main disadvantages of the existing solution: the need to use two mounting devices for slab alignment in the plan and in height; low technological reliability, which leads to a high labor intensity of the height adjustment of the floor slab (small areas of base platforms of the adjusting elements do not allow smooth and precise adjustment). The statement of a question: How to make the process of floor slab mounting technologically reliable? Goal setting: to improve the process of floor slab mounting.

Stage 2. Reduction. Establish technical contradictions that are the causes of previously identified deficiencies. Contradictions elimination in the future will allow achieving the goal set above. At this stage, the table of choice of methods for eliminating technical contradictions (A-matrix) and a table of basic techniques for eliminating technical inconsistencies (Ac- catalog) are applied. Two variants of contradictions and methods for their elimination are defined.

Option 1.Question 1.1: What needs to be changed under the conditions of the problem? Answer 1.1: Reliability (system parameter No 27). Question1.2: What gets worse while changing? Answer 1.2: Serviceability (system parameter No 33). Recommended numbers of methods for eliminating technical contradictions according to the selection table: 27, 17, 40.

Technique No 17. The principle of transition to another dimension: the difficulties associated with the movement (or object placement) are eliminated on the line if the object **gets the ability to move in two dimensions**, i.e. on the surface). – This technique can be applied to solve the formulated problem.

Technique No 27. Cheap fragility instead of inexpensive durability: replace an expensive object with a set of cheap objects, sacrificing some qualities (durability, for example). This technique is not applicable to the solution of the formulated problem.

Technique No 40 Use of composite materials: proceed from homogeneous to composite materials. This technique is not applicable to the solution of the formulated problem.

The resulting solution can be summarized as follows: **27**. Reliability (+) **VS 33**. Serviceability (-) = **17**. The principle of another dimension transition.

Option 2. Question 2.1: «What needs to be changed under the conditions of the problem? ». Answer 2.1: Ease of manufacturing (system parameter No 32). Question 2.2: What gets worse while changing? Answer: Device complexity (system parameter No 36). Recommended numbers of methods for eliminating technical contradictions according to the selection table: 27, 26, 1.

Technique No 27. Cheap fragility instead of inexpensive durability: replace an expensive object with a set of cheap objects, sacrificing some qualities (durability, for example). This technique is not applicable to the solution of the formulated problem.

Technique No 26. Copying principle: instead of an inaccessible, complex, expensive, inconvenient, etc. object, use its simplified copies; replace an object or system of objects with their optical copies; if visible optical copies are used, switch to infrared or ultraviolet copies. This technique is not applicable to the solution of the formulated problem.

Technique No 1. Fragmentation principle: divide the object into independent parts; **make object collapsible**; increase the fragmentation degree of the object. This technique can be applied to solve the formulated problem.

The resulting solution can be summarized as follows: **32**. Ease of manufacturing (+) **VS 36**. Device complexity (-) = **1**. Fragmentation principle.

Stage 3. Invention. Key ideas: to provide the possibility of simultaneous adjustment of the position of the floor slab in the plan and in height with a single mounting device consisting of two symmetrical blocks, by combining two existing devices: a cleat and a conductor; to increase the contact area of the adjusting elements of the mounting fixture (use adjusting plates instead of adjusting rods).

Based on the obtained solution of the inventive problem, a method of installation and an assembly device for the installation of the column slab has been developed.

3. Method of installation and erection device for the installation of the column slab

The method of the column slab mounting includes mounting the slab on the erection device attached to the column with stretching screw. While the edges of the hole in the column slab are aligned with the outer edges of the centering plates of the erection device and adjust the position of the slab in height along its four sections, using adjustment screws on the beams of the erection device, with the subsequent implementation of the design connection of the column slab and the column.

The proposed erection device is made of two symmetric blocks, with the possibility of bolting, and temporary fixation by stretching screw. Each block contains a beam with two adjusting screws, as well as at least three centering plates rigidly attached to the top of the beam.

Two sets of erection device are temporarily fixed to the column with stretching screw. When blocks are fixed, the centering plates form a closed boundary around the perimeter of the column, with the same gap between the outer faces of the centering plates and the faces of the column.

At the end of these works and the connection design of the column slab and the column, including the set of necessary concrete strength, the erection device is dismantled. In order to do this, it is necessary to unscrew the stretching screw and separate the blocks of the device.

While performing technical operations according to the proposed technique, as well as using the developed erection device, it is possible to increase the technological reliability of the installation process of the column floor slab as follows:

- significant reduction in the complexity of the floor slab alignment in height, including the uneven surface of the slab, due to the presence of beams with adjusting screws;
- significant reduction in the complexity of the floor slab alignment plan (horizontally), due to the presence of centering plates, rigidly fixed to the beams, which provide forced alignment of the slab horizontally (centering of the slab towards the column);
- ensuring the possibility of technological installation and fixation of erection device due to the presence of symmetric blocks, with the possibility of their bolting to each other, as well as the presence of the stretching screw.

In accordance with the recommendations [9, 10], a graphical explanation of the developed method and erection device is presented in the isometric view and characteristic cross-section (Figure).

The patent search, carried out in accordance with the recommendations [11-13], devices solutions for mounting column slab of girderless floor construction in the KUB system did not reveal any analogs with high relevance. Therefore, an application was prepared for obtaining a patent of the Russian Federation for the invention No 2018146398 based on the developed solution.

IOP Conf. Series: Materials Science and Engineering **972** (2020) 012043 doi:10.1088/1757-899X/972/1/012043

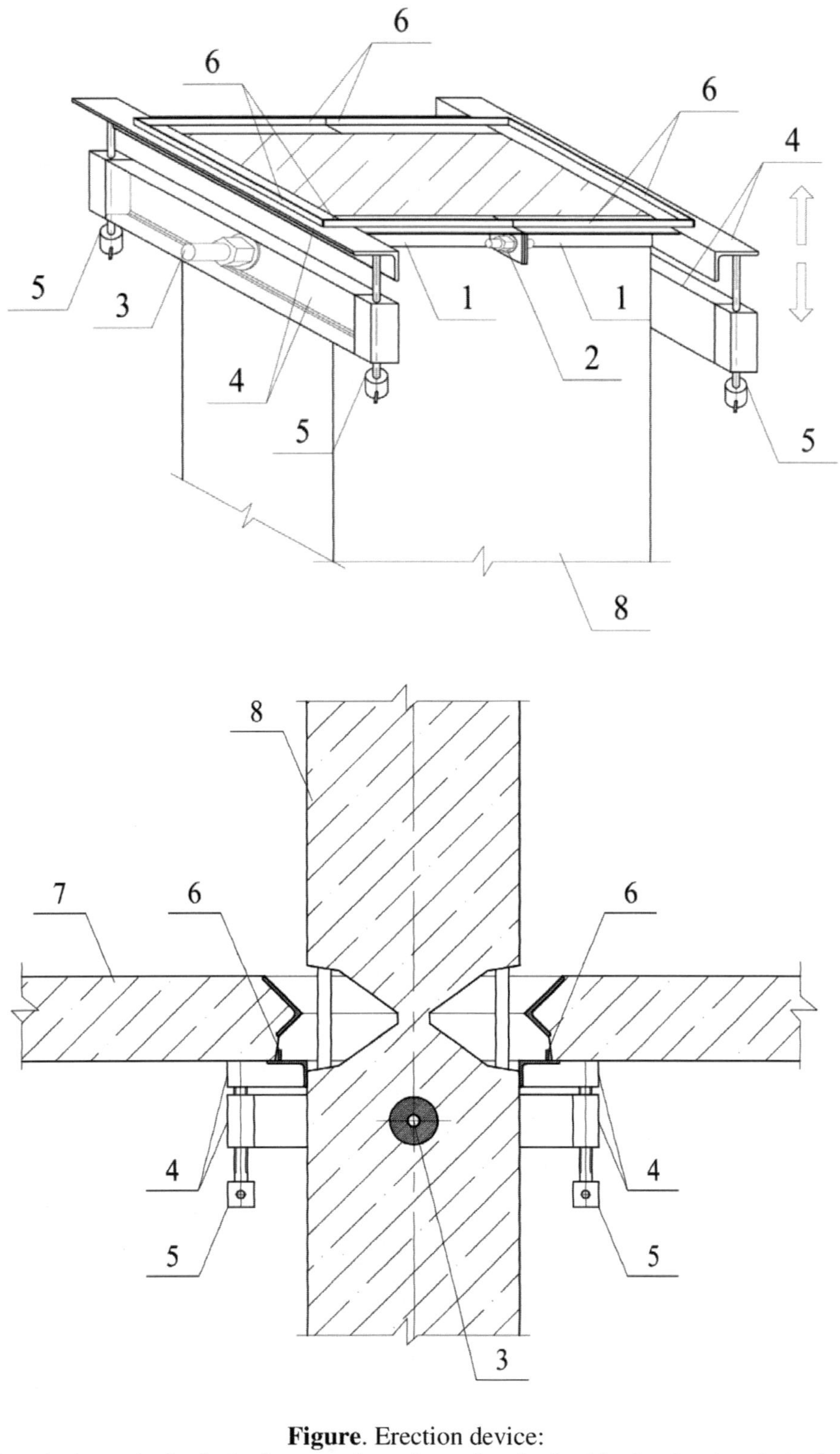

Figure. Erection device:
1 – erection device unit; 2 – bolted connection of erection device blocks; 3 – tierod; 4 – beam; 5 – adjusting screw; 6 – centering plate; 7 – column slab; 8 – column

4. Conclusion

Installation of the column floor slab, provided for a series on jointless ossature without girders and joints KUB-2,5, as well as its modifications (KUB-3V, etc.), is not technologically effective. When using such installation, there is a possibility of technological errors, which can cause a significant deterioration in the performance of load-bearing structures of precast and cast-in-situ building. The use of TRIZ tools allows obtaining new design and technological solutions that eliminate existing shortcomings. The obtained solutions, as a rule, have signs of patentability. According to the results of the research and the solution of the inventive problem with the use of TRIZ tools, a method has been developed: a method for mounting a column floor slab and an erection device for its implementation. The application of the developed method and erection device will allow increasing the technological reliability of the process for assembling a column floor slab of girderless floor construction of the precast and cast-in-situ frame of the KUB system.

References

[1] Elliott K and Jolly C 2013 *Multi-storey Precast Concrete Framed Structures* (London: John Wiley & Sons) 750 p ISBN 978-1-405-10614-6

[2] Elliott K 2016 *Precast Concrete Structures* (London: CRC Press) 732 p ISBN 978-1-498-72399-2

[3] Bachmann H, Steinle A 2011 *Precast Concrete Structures* (Berlin: Ernst & Sohn) 272 p ISBN 978-3-433-02960-2

[4] Chechurin L 2016 *Reseach and Practice of the Theory of Inventive Problem Solving (TRIZ): Linking Creativity, Engineering and Innovation* (New York: Springer) 281 p ISBN 978-3-319-31782-3

[5] Hua Z, Yang J and Coulibaly S 2006 Integration TRIZ with Problem-solving Tools: A Literature Review from 1995 to 2006. 1 (1-2) *International Journal of Business Innovation and Research* pp 111 – 128

[6] Gadd K 2011 *TRIZ for Engineers: Enabling Inventive Problem Solving* (Chichester: John Wiley & Sons. 504 p ISBN 978-0-470-74188-7

[7] Orloff M 2012 *Modern TRIZ: A Practical Course with EASyTRIZ* (Berlin: Springer) 449 p ISBN 978-3-64-225717-4

[8] Orloff M 2006 *Inventive Thinking through TRIZ: A Practical Guide* (Berlin: Springer) 352 p ISBN 978-3-54-033222-0

[9] Pressman D 2015 *How to Make Patent Drawings* (Berkeley: NOLO) 256 p. ISBN 978-1-413-32156-2

[10] Henderson M 2016 *Patent Drawing Rules: Patent Drawing Rules of the United States Patent and Trademark Office and the World Intellectual Property Organization* (Chagrin Falls: Studio 94) 248 p. ISBN 978-0-982-82702-4

[11] Kennedy J and Watkins W 2012 *How to Invent and Protect Your Invention: A Guide to Patents for Scientists and Engineers* (New Jersey: John Willey & Sons) 226 p ISBN 978-1-118-36937-1

[12] Yonover H and Crowe E 2014 *Hardcore Inventing: Invent, Protect, Promote, and Profit from Your Ideas* (New York: Skyhorse) 288 p ISBN 978-1-629-14426-9.

[13] Kannankeril C 2018 *The Inventor in You: A Step-by-Step Guide to Your First Invention* (Bloomington: BalboaPress) 192 p ISBN 978-1-982-20265-1.

Application of effective concrete composites to increase radiation protection of residential areas

V S Rudnov[1], **V A Belyakov**[1]

[1]Ural Federal University, 19 Mira street, Yekaterinburg 620002, Russia

E-mail: v.s.rudnov@urfu.ru, belyakow@rambler.ru

Abstract. The scientific article presents the analysis of studies of radiation safety of residential buildings in the city of Yekaterinburg, Sverdlovsk region and the results of the development of new building materials with protective properties. The review of available statistical data on the specified subject is carried out. It is revealed that the most simple and economically justified method of increasing the protection of the population from the radiation impact of man-made or natural environment is the construction of buildings with the use of innovative building materials or reconstruction of existing ones with the use of heavy concrete with effective radiation-protective aggregates.

1. Introduction

Currently, on the entire surface of the Earth, including the territory of residence of people can meet both small areas and significant areas in which there is an increased radiation background. The causes of this phenomenon are often natural and may be associated with the release of radioactive groundwater, zones of geological faults or halos scattering of radioactive and rare metal deposits. The most common natural source of radioactive radiation is radon gas, which together with its radioactive decay products make the main contribution to the natural radiation background of residential areas, which is about 1 mSv per year when inhaled into the human body.

However, in the residential areas of modern megacities are increasingly hotbeds of such phenomena of man-made origin, due to the progressive global transition of mankind to the use of nuclear energy and the expansion of applications of nuclear fusion. It is possible that at the end of the XXI century in each apartment in the kitchen will be a compact household nuclear power plant to provide electricity and heat needs of residents of one apartment. That is why the problem of protecting the population and the environment from dangerous neutron and γ-radiation arising in the process of nuclear fission is becoming increasingly important.

2. Resultsand discussion

The team of authors of the Institute of new materials and technologies of the Ural Federal University named after the first President of Russia B. N. Yeltsin studied building materials for protection against potential radioactive danger, which can be made on the basis of local raw materials of the Ural region [1-3].

The studies of the potential danger of the territory of Yekaterinburg conducted earlier by ecologists do not always correlate with each other. Yekaterinburg is a 1.5-million-strong metropolis located in the center of the mining and industrial region of the Urals – the Central part of the Russian Federation.

The radiation background of the residential area of the Yekaterinburg agglomeration is influenced by the South Ural radioactive trace, the close location of the Beloyarsk nuclear power plant and other industrial facilities. A number of authors believe that about half of the territory of Yekaterinburg has an increased background radiation mainly due to radon, which is further superimposed on local areas of man-made radioactive contamination [4-6]. Other authors note a slightly increased radon approximately uniform background throughout the metropolis. Also, in different years, epidemiological studies have been conducted to determine the dependence of lung cancer incidence and radiation due to inhalation of radon in the human body, but without proper systematization and without taking into account a significant number of other harmful factors that have carcinogenic effects [7-9].

In the process of nuclear decay, particles with different penetration into other substances are formed: neutrons, electrons, protons, fragments of heavy nuclei, mesons, α-particles and γ-radiation quanta. Charged particles (e.g. α-particles or electrons) are relatively quickly inhibited by matter. Therefore, the main danger for biological beings and humans are neutral particles (neutrons and γ-radiation), which are taken into account in the calculation of biological protection.

The interaction of these types of radiation is the ionization of the latter. In this case, there are special chemical processes that lead to disruption of biochemical processes in cells with the formation of physiological pathological changes. From the point of view of the state of the whole body, more dangerous General radiation. Sensitivity to tissue radiation in the body is different, as well as the sensitivity of individuals. When inhalation enters the human body, the lungs are most likely to suffer, in other cases, the neural tissues of the bone and brain are most exposed to radiation.

In world practice, it is generally accepted that the most optimal material for protection against most types of ion radiation is superheavy concrete, containing in its structure aggregates from a number of rocks, which is the basis for effective protection, including as part of the complex element [10].

For γ-radiation is characterized by the law of attenuation of the flow passing through the substance, which is that with an increase in the thickness of the layer of matter, the intensity of the transmitted radiation decreases in inverse proportion to the thickness of the layer under study. For the characterization of radiation protection it is common to make the thickness of the layer of the substance, in passing through which the intensity of radiation is reduced in two times. Accordingly, increasing the thickness two times, will receive a reduction of radiation intensity by 4 times, and so on. Based on this characteristic (taking into account some other factors), the thickness of the protective fence necessary to attenuate the radiation to the permissible intensity norms is calculated. The thickness of the half-attenuation layer depends directly on the density of the substance (for Portland cement concretes to a greater extent on the filler): the heavier the material, the smaller the required thickness of the fence.

Neutron radiation has a different nature and to weaken it in the structure of the material must be present chemical elements with a small atomic mass, for example, hydrogen. Concrete as a material is the most effective material for protection against these radiation, because only it can combine two types of protection against two types of radiation (fillers reduce γ-radiation, and hydrates slow neutrons). In this case, the cost remains at a relatively low level compared to the single-layer protection of heavy metal (such as lead), as well as a complex multi-layer. To obtain extra-heavy concrete is most effective use of such rocks of high density and with a significant content in the structure of chemically bound water as limonite, serpentinite and others. It is also possible to use artificial aggregates – waste metallurgy: metal scrap, lead shot, etc.

As a binder, Portland cement and its varieties are most often used, but the use of alumina and gypsum alumina cements, when hardening binding more than Portland cement, the amount of water in the crystal hydrates of tumors is also very justified for increasing neutron protection. To further improve the protective properties of the concrete is also introduced additives containing light elements such as carbide, boron, lithium chloride and others.

There are different opinions on the contribution of components to the operational, including radiation-protective properties of concrete. Such scientists as A. P. Veselkin, E. V. Voskresensky, V.

A. Egorov and other researchers believe that the protective properties are determined by fillers. Other scientists, for example V. P. Mashkovich, A. V. Kudryavtseva, B. N. Vinogradov, believe otherwise and emphasize that the type of binder has a significant impact on the protective characteristics of concrete. Therefore, when designing new protective composites, it is necessary to take into account the chemical and physical characteristics of all components [11,12].

The size of aggregates for protective concretes is recommended to determine the massiveness of the concreted structure and choose the largest, and the particle size composition should be chosen so that the content of aggregates in the structure of concrete was as much as possible. In this case, the density of such concrete will increase, and the thickness required for protection will decrease. This is possible with the use of intermittent grain aggregates, as well as the use of three fractions.

The authors conducted research in the laboratory of the Institute of new materials and technologies of the Ural Federal University named after the first President of Russia B. N. Yeltsin. As the object of study were selected rocks, which presumably have the structure of the crystal lattice to retain ionizing radiation. Two types of inert materials were obtained from rock samples: crushed stone of 10-20 mm fraction and sand from crushing screenings. With the help of a binder Portland cement were made samples of concrete, in which a large aggregate is made of the test rock, and in the second case – a small aggregate (table.1).

Table 1. Variants of combination of types of aggregates for protective concrete

Number of sample's	Coarse aggregate	Fine aggregate
1	barite	granite
2	magnetite	granite
3	hematite	granite
4	limonite	granite
5	dunite	granite
6	granite	barite
7	granite	magnetite
8	granite	hematite
9	granite	limonite
10	granite	dunite

Currently, many authors have investigated the inhibition of fast monoenergetic neutrons, which simplifies the understanding of the physical nature of the neutron deceleration and is used in the calculation of protection against this type of impact. The spatial-energy distribution of γ-quantum fluxes in media is described by the transport equation, for which it is difficult to find an exact solution. In this regard, it has recently become accepted to be based on empirical dependences obtained as a result of experiments with radiation sources.

After hardening of the samples under normal conditions up to 28 days, the samples were examined to determine their radiation-protective properties. Studied rocks: barite, magnetite, hematite, limonite and dunite. In the course of experiments, the degree of radiation absorption by concrete samples (in % of the intensity without samples) was determined using a dosimeter with a remote unit (scintillation gamma dosimeter). Isotopic sources were used in experiments to obtain γ-radiation. The studied concrete samples were placed directly in front of the detector at a distance of 420 mm to fix the dose rate of μsv/h radiation obtained in 60 seconds. In the study of the degree of absorption of neutron radiation, the source was a Pu-Be fast neutron reactor with an energy of up to 12 MeV. The distance to the sample and the irradiation time were the same. The results of determining the dependence of the degree of absorption of γ-radiation on the type of rock and (crushed stone on the right, sand – on the left) are shown in figure 1.

Then the degree of neutron radiation absorption from the same parameters (Figure 2).

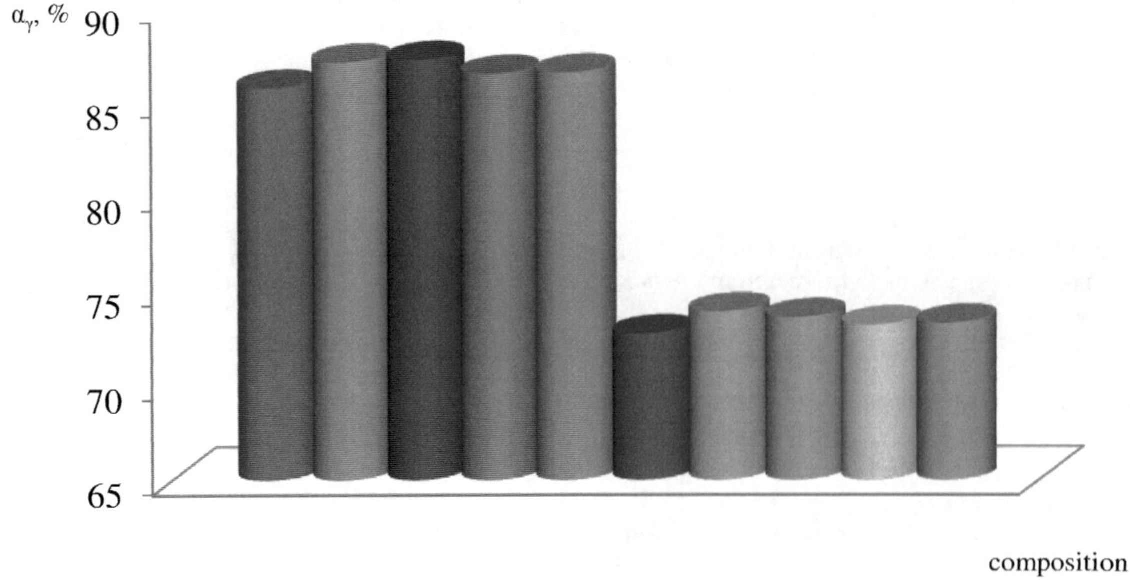

Figure 1. The dependence of the degree of absorption of γ-radiation from the type of filler.

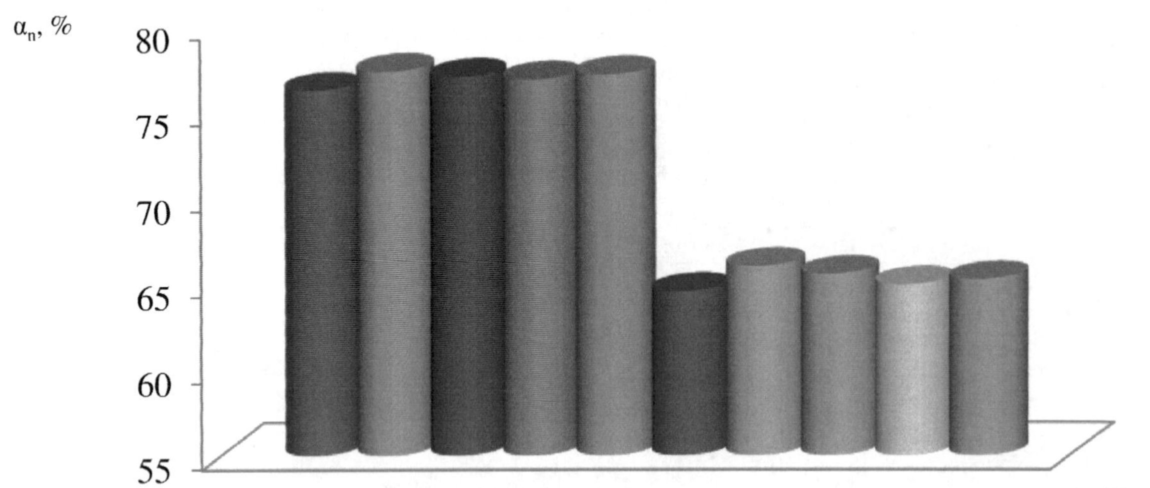

Figure 2. The dependence of the degree of neutron radiation absorption on the type of filler.

3. Conclusion
As a result of the research we can draw conclusions:

- the most effective use of rocks as a large aggregate, because the grinding is destroyed part of the crystallization bonds;
- the studied rocks have comparable efficiency in terms of protection against radiation, which is shown in the study of gravel and sand and does not depend on the size of the material used;

- it is recommended to develop maps of residential areas with an indication of increased radiation hazard and during the reconstruction of buildings in high-risk centers, to work with the use of radiation-protective concrete.
- in areas with high radon content it is recommended to use effective radiation-protective concretes for the construction of basements and basement structures of residential buildings, which will significantly reduce the impact of the main factor of natural origin;
- new construction of buildings or parts thereof that require increased protection against ionizing exposure, the use of concretes with large aggregates of recommended rocks;
- the size of the fillers for the protective concrete is recommended to be taken as much as possible in order to reduce the thickness of the half attenuation.

References

[1] Khoroshavin L B, Medvedev O A and Belyakov V A 2014 Tecnologia smaltimento dei rifiuti radioattivi (RRA) *Italian science review* **10** 175-179

[2] Belyakov V A, Rudnov V S and Akhtyamova V A 2016 Application of Dunite Aggregates to Increase Radiation-Protective Properties of Heavy Concretes *Science and Business: Ways of Development* **12** 46-49

[3] Noskov A S, Rudnov V S and Belyakov V A 2014 Research and development of rational compositions effective for radiation-shielding concrete *Concrete technologies* **10** 20-23

[4] Nikolaev A V 1961 Protection against radioactive radiation (Moskow: Metallurgizdat) p 420

[5] Batlutskaya I V, Bolkhovitina E A, Malanina O A and Khorolskaya E I 2010 Assessment of the environment using information of important indicators of the type of bioindicator *Scientific sheets. Series Natural Sciences* **9** 80-85

[6] Grachev V A 2008 Essays on the ecology of the industrial districts of the Sverdlovsk region Uncomfortable zone habitats of the population of the Middle Ural

[7] Koltover V K 1994 Radiological problem of radon *Radiation biology. Radioecology* Vol. **34** (**2**) 257-264

[8] Poceciun V A, Rudnov V S and Shepel S V 2014 Radioactive safety of residential areas on the example of Yekaterinburg *Ecological safety of mining regions: Proceedings of the II International scientific-practical conference* 156-159

[9] Yarmoshenko I V, Onishchenko D A and Zhukovsky M V 2009 A survey of the levels of accumulation of radon in residential buildings in the city of Yekaterinburg (Yekaterinburg: Institute of industrial ecology) pp 62-69

[10] Korolev E V, Samoshin A P, Smirnov V A, Korolev O V and Grishina A N 2009 Methods and algorithm of synthesis of radiation-protective materials of new generation (Penza: PGWC) p 130

[11] Dvorkin L I and Dvorkin O L 2012 Special concretes (Moskow: Infra-Engineering) p 368

[12] Ruengsri S 2014 Radiation shielding properties comparison of Pb-based silicate, borate and phosphate glass matrices *Science and Technology of Nuclear Installations* **2014** 218-241

IOP Conf. Series: Materials Science and Engineering **972** (2020) 012045 doi:10.1088/1757-899X/972/1/012045

Bochkarihinskoe clay as raw material to building ceramics production

A Yu Pavlovets[1], I A Pavlova[1] and E P Farafontova[1]

[1]Ural Federal University, Mira Str. 19, Ekaterinburg 620002, Russia

E-mail: i.a.pavlova@urfu.ru

Abstract. The possibility of using clay from Bochkarikhinskoe deposit (Sverdlovskaya oblast, Russia) in the ceramic building production has been considered. The chemical, mineral and granulometric composition of clay was determined. The basic technological properties of clay were studied: plasticity, sensitivity to drying, sintering. The frost resistance and mechanical properties of ceramic brick samples were determined. The clay studied with introduction of organic additives also may be used in ceramic production.

1. Introduction

The present work is devoted to expanding the raw materials resources for the production of building materials, specifically an investigation of Bochkarikhinskoe clay in order to determine the possibility of using it in the production of ceramics. Along with refractory clays in the area of the Bogdanovich town in the Sverdlovsk region there are types of red-burning clay [1]. These clays are not utilized in the production of refractories, but they can be used in the production of other, cheaper types of ceramics. The Bochkarikhinskoe clay is one of the varieties of such clays [2].

Red-burning clay is most often used in the manufacture of ceramic bricks [3–12]. In connection with this, it is of interest to study the properties of Bochkarikhinskoe clay in the production of ceramic bricks and other types of building ceramics [13-14].

2. Experimental procedure

The chemical composition and the mineral composition of the clays was studied by means of differential scanning calorimetry using an STA 449 F3 Jupiter calorimeter (Netzsch G. GmbH, Germany) and by means of XPA using a Miniflex 600 diffractometer (rotating anode; CuKα radiation; λ = 1.541862 Å; measurement interval 3.00-90.00°; scan step 0.02°) Rigaku – Carl Zeiss (Japan) with MiniFlex guidance software for control and data collection and a PDXL Basic package for data processing. The JSPDS database was used to identify the diffraction peaks. To determine the chemical analysis, we used emission spectral analysis with inductively coupled plasma on the Optima 4300 DV (PerkinElmer, USA).

The plasticity number of the clays according to the Vasil'ev method [15], coefficient of sensitivity of the clays to drying according to Nosova's method are tested in this study. The water absorption, porosity and apparent density were determined in accordance with [16], the density – in accordance with [17], the compressive strength – [18]. Frost resistance was determined by an indirect method using the formula: $G = B_1/B_2$, where B_1 – water absorption of samples determined according to [19];

IOP Conf. Series: Materials Science and Engineering **972** (2020) 012045 doi:10.1088/1757-899X/972/1/012045

B_2 – water absorption determined after boiling for 4 hours with further cooling to room temperature. Frost-resistant brick is when the value of the coefficient G is less than 0.85.

3. Results and discussion

The chemical composition of clay is presented in Table 1.

Table 1. Chemical composition of clay of Bochkarikhinskoe deposit.

Material	Weight content (%)								
	LOI	SiO_2	Al_2O_3	Fe_2O_3	CaO	MgO	Na_2O	K_2O	TiO_2
Dry matter	9.03	64.84	11.65	7.16	2.70	1.23	0.82	1.54	0.72
Calcined matter	–	71.30	12.81	7.87	2.97	1.35	0.90	1.69	0.79

The Al_2O_3 content of the clay shows the Bochkarikhinskoe clay to be acidic raw material. In terms of the content of coloring oxides the clay is raw material with high content of coloring oxides.

The granulometric composition of clay of the Bochkarikhinskoe deposit is presented in Table 2. According to the fracture content of size < 0.01 mm the clay studied is a medium-disperse raw material, and according to the fracture content of size < 0.001 mm it is a low-disperse raw material.

Table 2. Granulometric composition of clay of the Bochkarikhinskoe deposit.

Parameter	Particle size (mm)				
	0.25–0.06	0.06–0.01	0.01–0.005	0.005–0.001	< 0.001
Weight content of fraction (%)	15.96	22.59	12.73	14.54	34.18

According to the content of coarse-grained (> 0.5 mm) inclusions (14.9%) the clay from Bochkarikhinskoe deposit is a raw material with a high content of coarse-grain inclusions (Table 3). In terms of the size of the inclusions the clay belongs to a class of raw materials with small inclusions (62.4% particles of size 1–0.5 mm).

The inclusions in clay are represented predominately by quartz sand with rare carbonate inclusions of dolomite.

Table 3. Distribution of coarse-grain fractions.

Parameter	Weight content of fraction (%) of size (mm)					
	1–0.5	2–1	3–2	5–3	> 5	Total
with particles of size < 0.5 mm	9.3	2.2	1.8	0.5	1.1	14.9
without particles of size < 0.5 mm	62.4	14.8	12.1	3.4	7.4	100.0

According to the DTA data the clay contains montmorillonite, dolomite and quartz. Total weight loss under firing up to 1200°C is 16.3 wt%. X-ray phase analysis was conducted mainly to determine the composition of the impurity mineral phases. The presence of calcium-containing minerals – dolomite and anorthite is established. However, both mineral phases are crystalline. They form highly porous granules within which other impurities are located, for example, quartzite. Both calcium-containing phases are approximately equal.

According to the plasticity number (P = 13), air shrinkage (S_a = 9.8%) and coefficient of sensitivity to drying (K_s = 3.5), the clay under study is medium plastic, poorly drying and low-sensitive to drying.

IOP Conf. Series: Materials Science and Engineering **972** (2020) 012045 doi:10.1088/1757-899X/972/1/012045

The relation of Bochkarikhinskoe clay to sintering is studied under firing in temperature range of 850–1050°C. The clay was ground to a grain size of less than 3 mm. The body with optimal molding moisture (W = 27%) has been prepared and 35×30 mm 15 samples were molded. The properties of fired samples are presented in Table 4.

Table 4. Post-firing properties of the samples.

Firing temperature (°C)	Shrinkage (%)		Water absorption (%)	Open porosity (%)	Apparent density ($kg \cdot m^{-3}$)	Frost resistance	Compressive strength (M)Pa
	total	fire					
850	7.4	0.8	10.5	20.9	1990	1.17	41.2
900	8.2	1.6	10.5	21.1	2020	1.15	47.2
950	9.0	2.5	7.2	14.9	2150	1.05	52.1
1000	9.0	2.5	8.1	16.7	2080	0.96	26.9
1050	8.9	2.4	7.4	15.2	2050	0.97	21.4

As the temperature increases up to 950°C the density of the samples increases from 1.99 to 2.15 g/cm^3, the compressive strength increases from 41.2 to 52.1 MPa, at the same time water absorption decreases from 10.5 to 7.2% and porosity decreases from 20.9 to 14.9%.

A further increase in the firing temperature leads to structure loosening, which is confirmed by a decrease in density from 2.15 to 2.05 g/cm^3 during firing in the temperature range of 950–1050°C. It also decreases the strength from 52.1 to 21.4 MPa. The shrinkage of the samples increases with firing in the range of 850–950°C, with a further increase in temperature no shrinkage of the samples is observed. All samples calcined in the range of 850–950°C are non-frost-resistant, although an increase in the frost resistance of the samples with an increase in the calcination temperature is observed.

All clay samples have white inclusions – "blowing" associated with dolomite decomposition and formation of both magnesium oxide and calcium oxide. "Blowing" because of interaction with water, increases in volume and destroy the structure of the ceramic product. All samples after firing have a bright orange color.

All samples have a black core. As the temperature increases the black core becomes more pronounced. This is associated with the burning of organic impurities from clay, as well as the decomposition of carbonate inclusions. The access of oxygen to the central part of the samples during firing becomes complicated with the appearance on the surface of the samples of an impermeable layer resulting from the sintering process. In this regard, a reducing atmosphere and carbon deposition in the pores of the samples occurs inside the sample. Even though sufficiently durable samples were obtained, in accordance with [20] grade M 200 and higher, the production of ceramic bricks based on clay from the Bochkarikhinskoe deposit is complicated by a high content of carbonates and the formation of "blowing", as well as black core.

To eliminate such a defect, it was proposed to carry out more fine grinding of clay to a size of < 0.5 mm. Samples of this clay were molded and fired in the temperature range of 900–1000°C. The properties of Bochkarikhinskoe clay samples are presented in Table 5.

As the temperature increases in temperature range of 900–1000°C the fire shrinkage increases from 1.5 to 3.2%, total shrinkage increases from 8.1 to 9.6%, the density increases from 2.00 to 2.06 g/cm^3, the compressive strength increases from 19.0 to 42.7 MPa, at the same time the water absorption decreases from 10.3 to 8.5% and open porosity decreases from 20.6 to 17.5%. All samples are not frost-resistant, although as the firing temperature increases the frost resistance increases.

Table 5. Post-firing properties of the samples of Bochkarikhinskoe clay.

Firing temperature (°C)	Shrinkage (%)		Water absorption (%)	Open porosity (%)	Apparent density (kg·m^{-3})	Frost resistance	Compressive strength (MPa)
	total	fire					
900	8.1	1.5	10.3	20.6	2000	0.99	19.0
950	8.9	2.4	10.4	17.1	2040	0.95	34.7
1000	9.6	3.2	8.5	17.5	2060	0.92	42.7

No "blowing" defects of samples is observed. Small grains of calcium and magnesium oxides after dolomite decomposition are transferred to the glass phase, which improves the quality of the samples. All samples have a black core.

In addition, the clay studied can be recommended to produce expanded clay since when it is fired, a black core is formed. To increase the coefficient of expansion, it is necessary to introduce organic materials in the form of, for example, technical oil [1].

Conclusion

In summary, clay from Bochkarikhinskoe deposit is suitable for obtaining ceramic brick with a compressive strength of more than 10 up to 40 MPa. To eliminate the appearance of "blowing", the clay must be thoroughly ground to a grain size of < 0.5 mm to turn calcium and magnesium oxides after the dolomite decomposition into the glass phase. Even though the black core does not affect the strength and appearance of the brick, its presence is still undesirable in finished products. To eliminate it, it is necessary to introduce additional components into the composition of masses to produce bricks. The obtained brick samples have low frost resistance, and therefore it is necessary to conduct additional studies on the compliance of frost resistance with the requirements of national standard.

References

[1] Pavlova I A 2017 *Glass and Ceramics* **74** 104–106
[2] *Mineral deposits Bochakrihinskoe* [in Russian] http://www.catalogmineralov.ru/deposit/bochkarihinskoe_mestorozhdenie
[3] Adylov G T, Menosmanova G S, Riskiev T T, Rumi M Kh and Faiziev Sh A 2010 *Glass and Ceramics* **67** 63–65
[4] Nestertsov A I 2009 *Glass and Ceramics* **66** 253–254
[5] Nastro V, Vuono D, Guzzo M, Niceforo G, Bruno I and De Luca P 2006 *Journal of Thermal Analysis and Calorimetry* **84** 181–184
[6] Guryeva V A, Doroshin V A and Dubineckij V V 2019 *Materials Science Forum* **945** 1036–1042
[7] Guryeva V A and Doroshin V A 2018 *Solid State Phenomena* **284** 910–915
[8] Guryeva V A, Doroshin V A and Andreeva Yu 2018 *MATEC Web of Conferences* **143** 02014
[9] Vlasov V A, Skripnikova N K, Yuriev I Yu, Kosmachev P, Litvinova V A and Nikonenko E L 2016 *Key Engineering Materials* **683** 156–161
[10] Salman G A, Salih Sh A 2019 *International Journal on Advanced Science, Engineering and Information Technology* **9** 321–328
[11] Bozhko Y A and Lapunova K A 2018 *Journal of Construction and Architecture* **1** 160–167
[12] Trepalina Yu N and Kirillova N K 2019 *Bulletin of Belgorod State Technological University named after V. G. Shukhov* **4** 138–143
[13] Galán-Arboledas R J, Cotes-Palomino M T, Bueno S, Martínez-García C 2017 *Construction and Building Materials* **144** 327–337
[14] Sokolár R, Vodová L, Grygarová S, Stubna I and Sin P 2012 *Ceramics International* **38** 6607–6612

IOP Conf. Series: Materials Science and Engineering **972** (2020) 012045 doi:10.1088/1757-899X/972/1/012045

[15] *GOST 21216–93: Clay Raw Materials: Methods of Analysis* [in Russian], replaces GOST 21216–81, introduced October 21, 1993: Izd. Standartov, Moscow (1993) (international standard)
[16] *GOST 2409–95 Refractories. Method for determination of bulk density, apparent and true porosity, water absorption* [in Russian], replaces GOST 2409–80, introduced January 01, 1997: Izd. Standartov, Moscow (1995) (international standard)
[17] *GOST 2211–65 Refractories and refractory raw materials. Method of true density determination* [in Russian], introduced July 01, 1966: Izd. Standartov, Moscow (1966) (international standard)
[18] *GOST* 473.6–81 *Chemically resistant and heat resistant ceramic wears. The method for determination of compression strength* [in Russian], replaces GOST 473.6–72, introduced January 01, 1982: Izd. Standartov, Moscow (1981) (international standard)
[19] *GOST 13993–78 Chemically and thermal resistant ceramic products. Method for determination of water permeability* [in Russian], replaces GOST 13993–68, introduced January 01, 1979: Izd. Standartov, Moscow (1978) (international standard)
[20] *GOST 530–2012: Ceramic brick and stone. General specifications* [in Russian], replaces GOST 530–2007, introduced July 01, 2013: Standardinform, Moscow (2009) (international standard)

Deformation of composite plates with a layer of aging materials

Y E Yakubovskiy, V I Kolosov, I A Donkova

[1]Department of Applied Mechanics, Industrial University of Tyumen, Lunacharskogo st., 2, Tyumen, 625001, Russia

Email: jakubovskijje@tyuiu.ru, kolosova_l_v@mail.ru, irina_donkova@mail.ru

Abstract. The paper presents a calculation of the stress-strain state of composite plates. The mathematical model obtained by the authors earlier has been applied. In contrast to the previous results, the viscous-elastic properties of the material are taken into account in describing the behavior of the intermediate layer. The developed creep kernel of the layer material contains the expression that serves to describe the properties of the aging material. The compliance of the proposed formulas with actual processes has been proved by comparison with experimental data. In the numerical implementation decomposition of the expressions under integral sign into the Maclaurin series has been used. Based on the experimental data, the parameters of creep kernels have been determined. The parameters have been determined by minimizing the calculated data standard deviation from the experimental ones. The algorithm based on the random search method has been implemented. The results of the calculation according to the developed mathematical model have been given. The analysis of time variation in the stress-strain state of the joints between the layers of the composite structure has also been given.

Keywords: composite plate and shallow shell theory, composite structures, composite plate bending, visco-elastic properties, aging material.

1. Introduction

The article covers the solution to the problem of composite plate bending taking into account the creeping and relaxation processes of the material of the most frequently implemented structures – composite shells, plates and beams [1]. Connections between the layers of composite plates can be provided by adhesive bond or anchors [2, 3]. A record of the relations between stresses and strains for aging material has been proposed. On the basis of the elastic-creeping body theory the possibility of analytical description of the high-rate creep movement has been shown. Using the N. Kh. Arutyunyan and A. A. Zevin method new creep kernels have been built with respect to aging material creeping. A numerical implementation of the kernels has been considered on the example of A. A. Ross' experimental data of [4]. Error estimates have been given.

The algorithm for solving the problem of composite plate bending during long-term deformation has been developed. The algorithm is based on an introduction of a discrete time scale and creep calculation by areas in accordance with the P. I. Vasilyev stepping-stone method.

IOP Conf. Series: Materials Science and Engineering **972** (2020) 012046 doi:10.1088/1757-899X/972/1/012046

2. Main points

We assume that the material of the construction has creep and aging properties. Then, in the elastic-creeping body linear theory the relationship between stresses and strains in such a construction with a uniaxial stress state has the form [5].

Using the method [5], new kernels to describe aging materials creep were determined. The possibility of calculating the obtained functions using power series for a sufficiently wide time interval was shown [6].

To describe the relationship between stresses and strains in accordance with the N. Kh. Arutyunyan and A. A. Zevin method a development of new kernels was implemented taking into account material aging:

$$K_1(t,\tau) = Q(t,\tau) + B(t-\tau) + \int_\tau^t Q(t,s)B(s-\tau)ds\ ;$$
$$K_2(t,\tau) = Q(t,\tau) + B(t-\tau) + \int_\tau^t Q(s,\tau)B(t-s)ds\ . \tag{1}$$

where $Q(t,\tau)$ is a regular part of the creep kernel; $B(t-\tau)$ is a differential part of the creep kernel.

As a differential component, we use one of the most meeting the entire necessary requirements weakly singular kernel [7].

As a regular component, we use a kernel based on the creep measure of the form [5].

For a numerical implementation the exponent in the expression under integral sign [5] is decomposed into a Maclaurin series.

The error of such a transition is determined by the accuracy of the integral count.

The possibility of using the obtained relations to describe the creep of actual materials was tested on the A. A. Ross' experimental data [4] (Figure 1).

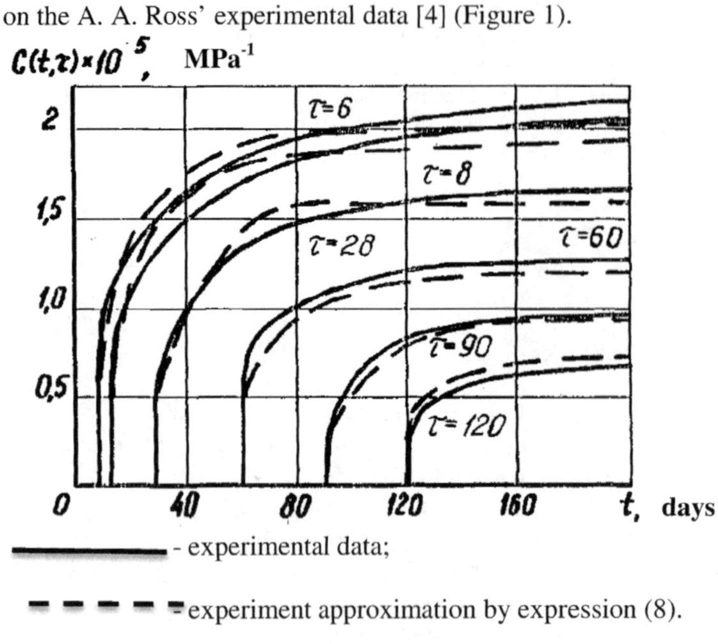

Figure 1. Experimental and theoretical creep measure curves of concrete samples of different ageτ

Creep measure was calculated by formula:

$$C(t,\tau) = \frac{V_r(t,\tau)}{E(t)} - \frac{1}{E(\tau)} + \frac{1}{E(t)}, \tag{2}$$

where, E(t) is the modulus of elasticity, $E(\tau)$ is the stiffness ratio of the concrete at different age.

The kernel parameters were determined according to these data by minimizing the standard deviation from the experiment [6]. The results were used to estimate the accuracy of the transition from the exact relation to approximate [5]. Here, the algorithm was implemented based on the random search method [4]. In the presence of four or more parameters instability in experimental data processing was observed [6]. Therefore, the value of α (the exponent of the kernel expression degree) varied with a step of 0.25 from 0 to 1, and the remaining parameters were determined at fixed α. The parameter search intervals were set based on the minimum error condition when comparing the numerical implementation of the obtained expressions with the experiment.

The approximation of the experimental curves taking into account the obtained parameters is shown in figure 1 (dashed line). The discrepancy between the theoretical and experimental values was not more than 10 %.

The results of the numerical implementation show that the error of the approximate relations is not big with the appropriate set of parameters. These relations can be used both for short times and fairly long ones without using special approximation techniques.

As an example, the problem of bending a three-layer composite plate taking into account the visco-elastic properties of the aging material in the second layer was solved (Figure 2). The dimensions of the plate are: in plan a = b = 1200 mm, thickness h = 120 mm. The outer layers are taken as equal: $E^{(1)} = E^{(3)} = 2.1*10^5$ MPa; $\nu^{(1)} = \nu^{(3)} = 0.3$; $h^{(1)} = h^{(3)} = 1$ mm. The parameters of the intermediate layer are: $E^{(2)} = 8*10^3$ MPa; $\nu^{(2)} = 0.2$; $h^{(2)} = 118$ mm. The stiffness for connection shear between the layers is described by linear relations. The stiffness ratio of the connections varies within: $\eta = 0 \div 20$ N/mm^3.

Resting the pack edge on a hinge has been selected as boundary conditions. The ends are bound with a tape, absolutely rigid in its plane and absolutely flexible from the plane. On the edges X = 0; and we have the following conditions:

$$W = 0;\ M_x = 0;\ T^i = 0;\ N_x^i = 0;\ V^i = 0;$$

$$\frac{\partial^2 W}{\partial x^2} = \frac{\partial^2 W}{\partial y^2} = \frac{\partial^2 \varphi^i}{\partial x^2} = \frac{\partial^2 \varphi^i}{\partial y^2} = 0.$$

The load is evenly distributed in the center of the plate surface on the pad with dimensions of 200 x 200 mm. In the calculations the maximum load intensity was as q = 8 MPa.

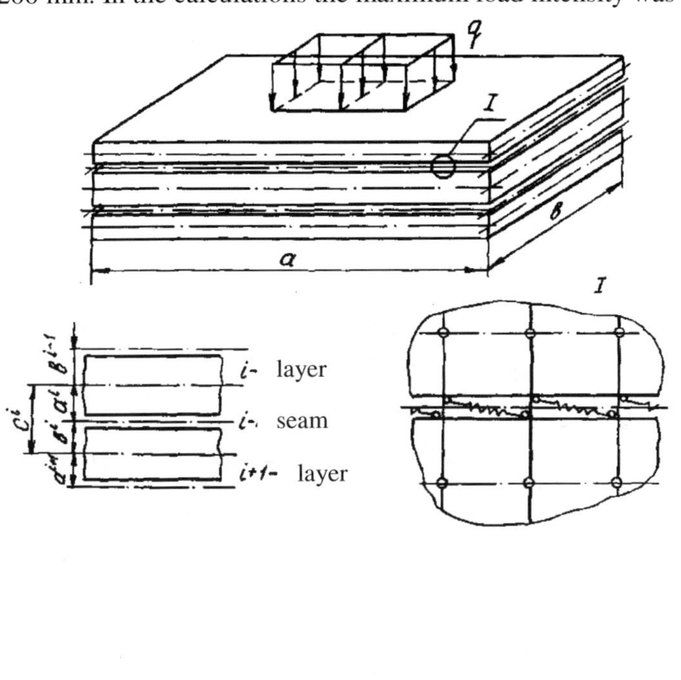

Figure 2. Composite three-layer plate

The calculations were performed for linear deformation of the structure.

The problem, taking into account all the factors, was solved with the step-by-step method [8]. In the initial moment of time t_1 a calculation of instantaneous bending was carried out. In the second calculation cycle, with an underestimated value of the secant module, the visco-elastic properties of the material in the intermediate layer were taken into account. The creep strain was determined by the moment of time t^2. Then the secant modulus and the coefficient of transverse deformation were calculated. The integral of the creep kernel had the following form [5].

Thus, the solution to the problem of bending a three-layer composite plate taking into account the visco-elastic properties of the aging material of the intermediate layer was brought to the required moment of time t_n.

The calculation results are reflected in figures 3-5. The following designations are accepted: the solid line is the solution of the linear creep problem; the dashed line is the solution of the problem of time deformation, taking into account the physically nonlinear properties of the material; and the dash-dot line is an additional account of the material destruction in the stretched zone of the intermediate layer. The calculation was carried out at the local load of q = 5 MPa and joint stiffness $\eta = 1$ N/mm^3. The curves of change of the determined values are built for the loading moments: $\tau = 8$ days, 28 days and 300 days. The description of the intermediate layer material time deformation is used in the form [5].

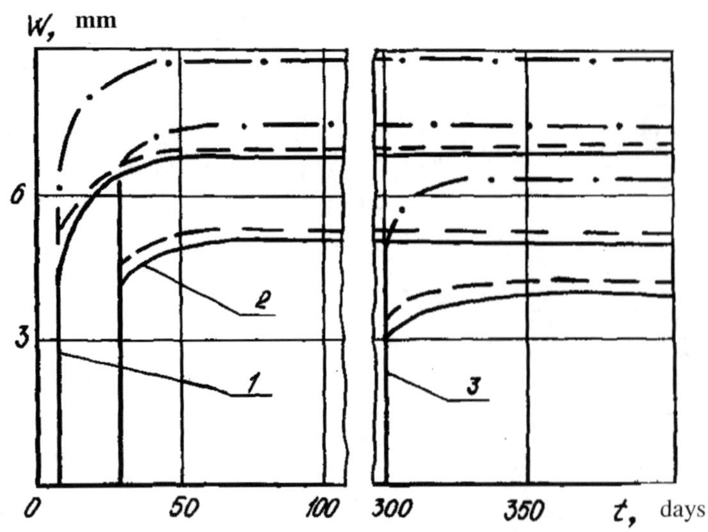

1 – τ = 8 days; 2 – τ = 28 days; 3 – τ = 300 days.

——— - linear bending problem of a composite plate;

– – – – - solution with account of the physically nonlinear properties of the intermediate layer material;

— · — - solution with account of the material destruction in the stretched zone of the intermediate layer.

Figure 3. Time variation of the three-layer composite plate deflection at loading it at different ages of the intermediate layer aging material.

The analysis of the results shows that the age of the material of the structure significantly depends on its stress - strain state. If the nonlinearity of the material properties (at some load values) leads to insignificant changes in the deformation of the pack (for example, the deflection changes by no more than 5 %), then the account of the destruction of the stretched zone of the intermediate layer causes a significant change in the stress-strain state (W increases by 30 - 40 %).

The younger the material of the structure when it is loaded, the more pronounced the creep and relaxation processes are. The displacements and stresses in the composite plate vary by one and a half or more times depending on the age at loading.

The affect associated with the change of stresses in the intermediate layer has been revealed. If the layer material has physically nonlinear properties, the stresses in it are reduced in comparison with the linear version of the calculation. If the material destruction of the stretched zone of the layer occurs, then the stresses in the second layer increase due to the decrease in the stiffness of the pack and the increase in its strain (Figure 4 a).

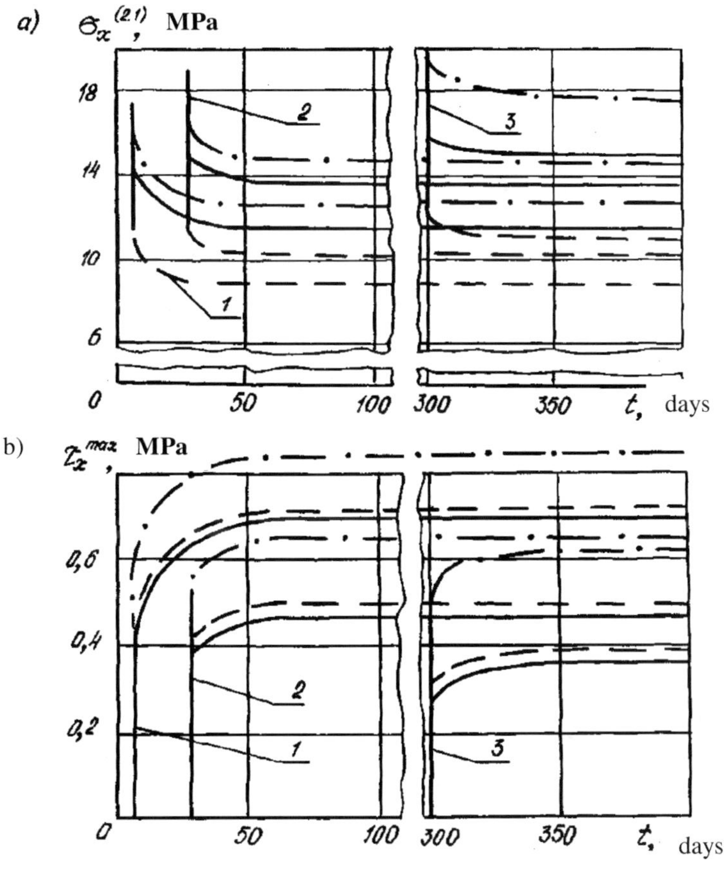

Figure 4. Time variation of normal stresses in the intermediate layer (a) and shear stresses in the joint (b). See designations in figure 3.

Taking into account material destruction also leads to a significant asymmetry of the stress state of the pack along the height of its section. This can be seen from the comparison of the normal stress values in the outer surfaces of steel sheets (Figure 5). The difference in values varies between 40 and 75 %.

During the time deformation not only the value but also the form of stress distribution changes, even in a linear problem. This is most noticeable in the distribution of shear stresses in the joint.

Loading the plate which has the intermediate layer material of an earlier age results in a significant increase in shear stresses when we compare with calculations performed when loading the pack with the second layer material of other ages. This happens for the reason of that the younger material has minor stiffness. Since the stresses are reduced in this case, and the deflection of the pack increases, the joints and outer layers incur a major part of the load.

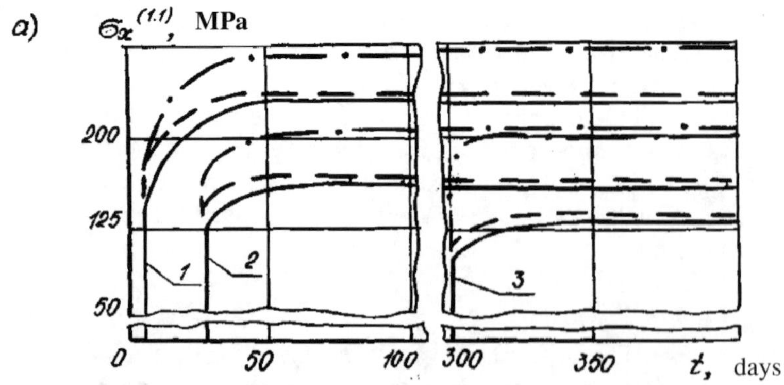

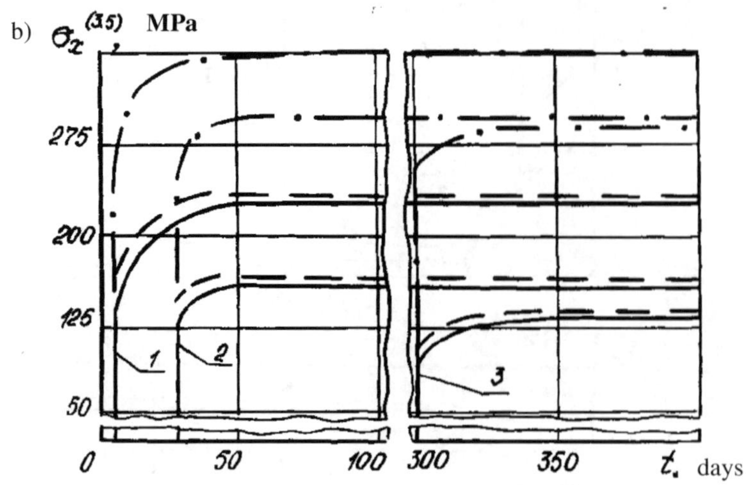

Figure 5. Time variation of normal stresses on the outer surfaces of the first (a) and the third (b) layers. See designations in figure 3.

For the same reason, during time deformation of the plate, the normal stresses in the outer surface of the first layer $\sigma_x^{(1,1)}$ increase by 140 % (8 and 108 days), while the stresses in the upper surface of the second layer $\sigma_x^{(2,1)}$ only decrease by 25 % if the intermediate layer material is loaded at an earlier age. I.e. the overlap of the processes of increasing W and decreasing the stiffness of the intermediate layer changes $\sigma_x^{(2,1)}$ insignificantly. Since the stiffness of the outer layers does not vary in time, the growth of deformation of the plate causes a sharp increase in $\sigma_x^{(1,1)}$.

3. Conclusion

The calculation of the composite plate in time has showed that the age of the structure material at which it is loaded significantly effects its stress - strain state. In time, there is a redistribution of the strain between the layers. Accordingly, the stress state of the layers changes. Shear stresses in the joints are characterized not only by quantitative but also qualitative changes.

References

[1] Yakubovskiy Yu E, Gulyaev B A, Kolosov V I, Krivchun N A and Yakubovskaya S V 2016 Bending of composite plates and shallow shells (Tyumen: Tyumen Industrial University Press) p 185

[2] Yakubovskiy Yu E, Goltsov V S and Kolosov V I 2016 Deformation of working shells of mixing devices in the production of concrete for underground construction *Procedia Engineering* vol **165** pp 1238–45

[3] Yakubovskiy Yu E, Kolosov V I, Gulyaev B A and Goltsov V S 2016 Bending of structurally orthotropic composite structures with anchor connection layers *Procedia Engineering* vol **165** pp 1246–53

[4] Yakubovskiy Yu E 1991 Determination of creep kernel parameters for aging materials *International Applied Mechanics* vol **27** (**6**) pp 37-44

[5] Arutyunyan N Kh, Zevin A A 1982 On a class of creep kernels of aging materials *International Applied Mechanics* vol **8** (**4**) pp 14-21

[6] Yakubovskiy Yu E, Kolosov V I 1991 Creep kernels of aging bodies *Structural Mechanics and Analysis of Constructions* **1** pp 55-61

[7] Rzhanitsyn A R 1968 The creep theory (Moscow: Stroiizdat) p 416

[8] Vasiliev P I, Gavrilin B A and Mal'kevich A B 1983 Issues of the development of the aging medium deformation theory (Research on the theoretical basics of the calculation of building structures) pp 122-5

IOP Conf. Series: Materials Science and Engineering **972** (2020) 012047 doi:10.1088/1757-899X/972/1/012047

Experimental determination of the minimum allowable thickness of the external decorative layer of the silica-granite block under exposure to a radiant heat from the base of the fire

S V Sharhun[1], O A Mokrousova[1], A Y Koshelev[1], E A Ojegov[1], N F Sirina[2]

[1]Ural Institute of State Fire Service of EMERCOM of Russia, Mira str., 22, Yekaterinburg, 620062, Russia

[2]Ural State University of Railway Transport, Kolmogorova st., 66, Yekaterinburg, 620034, Russia

E-mail: S_sharhun@mail.ru

Abstract: The paper presents the results of the full-scale tests that determine the minimum permissible thickness of the protective outer decorative layer of a silica-granite wall block when exposed to radiant heat flux from a fire source, taking into account the location of the constructed object at the standardized distance between buildings and structures. The article presents an analysis of the fire safety regulatory requirements applicable to the Russian Federation, in terms of the location of buildings and structures of various fire resistance levels in the territories of gardeners' and allotment non-profit associations or on land plots intended for individual housing construction in urban and rural settlements.

1. Introduction

Currently the construction industry has been actively developing in the Russian Federation. In large cities there is an active development of both central, already existing areas, and the integrated development of new territories - the creation of new neighborhoods with their own infrastructure. A separate direction has been taken in the construction industry which is the construction of the so-called “nearest suburb” in close proximity to major cities and centers of the constituent entities of the Russian Federation.

It is worth noting that the construction of multi-storey buildings is currently quite well regulated and is subject to constant monitoring from the stage of issuing a design permit to the commissioning of the object, and after that it is also inspected by state fire supervision authorities. The private sector is currently being built often, without any permits (for land plots of certain categories, the construction permits are not provided for at all) and without the proper oversight by the regulatory authorities. What ultimately, as fire statistics show, leads to rather dire consequences.

In the original edition of the Federal Law No. 123 "Technical Regulations on Fire Safety Requirements" dated July 22nd, 2008 [1], there was the article 75 that stated. "Fire safety distances on the territories of garden plots, allotments and household plots", which determined the fire safety distances between buildings and constructions on the territories of garden plots, allotments and household plots. However, later this article was excluded and now in [1] there are no requirements for the placement of buildings and structures in the territories of garden plots, allotments and household plots. These requirements are contained within SP 53.13130.2011 [2].

So according to article 6.5. [2] fire safety distances between residential buildings or residential houses located in adjacent areas, depending on the material of the bearing and enclosing structures, should not be less than the distances illustrated in the table No. 1. At the same time, fire safety distances between the buildings and structures within one garden plot are not standardized.

Table 1. Minimum fire safety distances between the outermost residential buildings (or houses) and groups of residential buildings (or houses) on plots

	The material of bearing and enclosing structures of a building	Distance, m.		
		A	B	C
A	Stone, concrete, reinforced concrete and other non-combustible materials	6	8	10
B	The same, with timber frames and floors, protected by non-combustible and slow-burning materials	8	10	12
C	Timber, framed enclosing structures made of non-combustible, slow-burning and combustible materials	10	12	15

The placement of buildings within the same land plot and the distance between neighboring households in modern gardeners' partnerships and allotment associations, as well as on land plots intended for individual housing, is currently rather chaotic. That being said, we can't speak here about any compliance with the requirements set forth in [2] in regards to the gaps between the buildings at all.

Most studies on this topic, for example [3,4,5], are devoted specifically to issues of preventing the spread of fire to neighboring buildings, and the issues of damage to neighboring objects without the spread of fire are not considered in principle. Figure 1 shows the graphical dependence of the heat flux at a distance from the fire source during the burning of buildings of I-III degree of fire resistance.

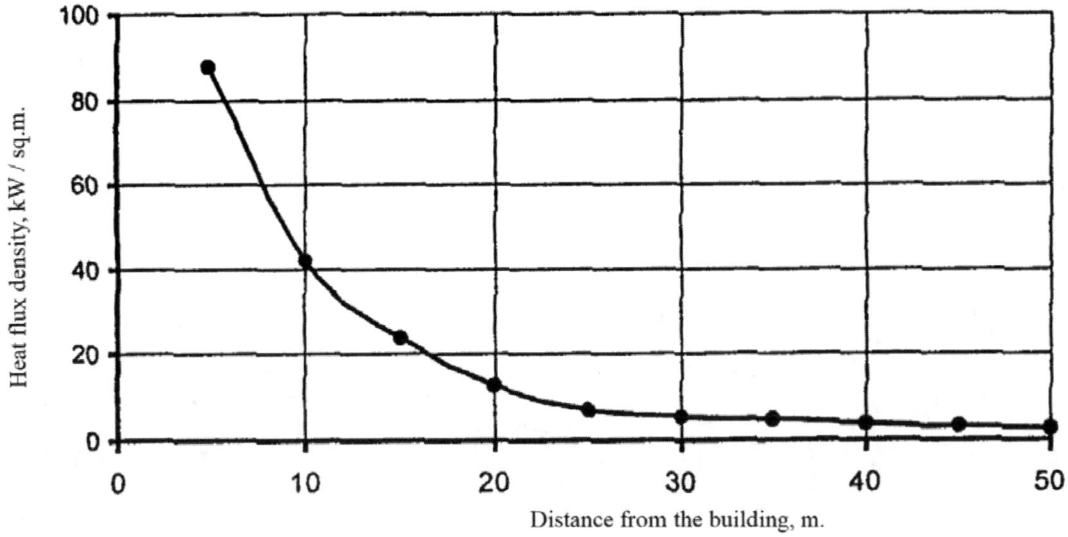

Figure 1. Graphic dependence of the heat flux at a distance from the base of fire during the burning of buildings I-III degree of fire resistance

IOP Conf. Series: Materials Science and Engineering **972** (2020) 012047 doi:10.1088/1757-899X/972/1/012047

Analyzing the graph data, we can conclude that even while maintaining the required distance of 10 meters between the buildings, the value of heat flux that will affect the next building will amount to about 42 kW/m^2. This means the value will rapidly increase with decreasing distance between buildings.

Considering the characteristics of modern building materials (including those used in thermal insulating layers), it can be safely assumed that the effect of heat flux, even if it did not spread the fire, can have a significant impact on the thermal performance of materials used in the construction of buildings.

Currently the so-called multi-layer structures consisting of bearing, thermal insulating and facing layers are widely used. However, if non-combustible thermal insulation is used as a heat-insulating layer in the construction of multistory buildings, then individual residential construction focuses primarily on cost-effective materials, most of which are combustible in their fire performance characteristics. One example of such materials is a silica-granite wall block manufactured according to TU 5835-002-99461491-2008, which are four-layer blocks with an outer textured (silica-granite) layer of heavy and dense concrete, outer undertextured and inner layers of a block of fine-grained dense concrete, connected by plastic, basalt plastic or fiberglass ties. The middle layer is a thermofiller made of expanded polystyrene.

2. Methods and materials

To assess the degree of influence upon the thermophysical properties of an object constructed of such blocks, the heat flux from a fire in a building located in the immediate vicinity, the series of tests were carried out. For this purpose, the device was used, which is a supporting structure housing the following equipment:

- source of thermal radiation (gas infrared burner B64-2 SX SBM);
- the measuring unit moved along the guides with a stand for placing the sample, equipment for measuring the temperature at five points of the sample under study;
- manual drive to move the unit during the experiment;
- equipment control panel;
- working table to accommodate the recording PC;
- gas cylinder with a capacity of 50 liters;
- shut-off and control valves to connect the gas cylinder to the device;

In order to carry out the laboratory tests, a block sample with different thicknesses of the outer layer of 46, 65, 70 and 73 mm. was used.

Figure 2. The visual appearance of the sample for testing

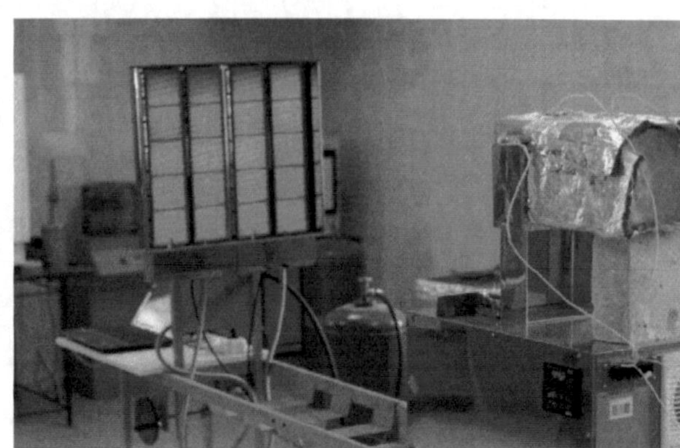

Figure 3. The sample on the laboratory setup

The prepared samples are alternately installed on the device; in order to protect against heat exposure from the sides they are wrapped with thermal insulation material MBOR-16 (Figure 2); to record the data 3 thermocouples are installed: the 1st one on the front surface, the 2nd one behind the facing layer and the 3rd one behind the thermal insulating layer (Figure 3).

After that, the sample was installed at a distance from the heat source at which the value of the heat flux acting on the surface at 42 kW / m^2 was ensured, which, according to the Figure 1, corresponds to the location of buildings at the distance of 10 meters from each other. The time of the test is 60 minutes.

3. Results

The test results, graphical dependences of temperature and the appearance of the sample after the test for block No. 1 (thickness of the outer layer 46 mm) are shown in Pictures 4, 5, for block No. 2 (thickness of the outer layer 65 mm) are shown in Pictures 6, 7, for block No. 3 (outer layer thickness 70 mm) are shown on Pictures 8, 9, for block No. 4 (outer layer thickness 73 mm) are shown on Pictures 10, 11.

Figure 4. The visual appearance of the block No. 1 after the test

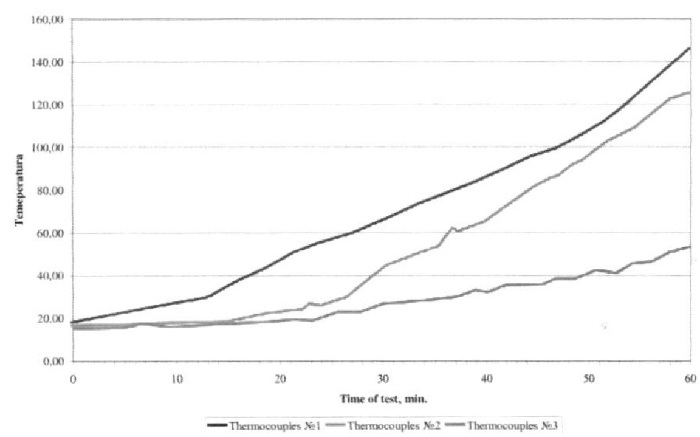

Figure 5. Graphical dependence of temperature according to indicated values of thermocouples No. 1 – No. 3

Figure 6. The visual appearance of the block No. 2 after the test

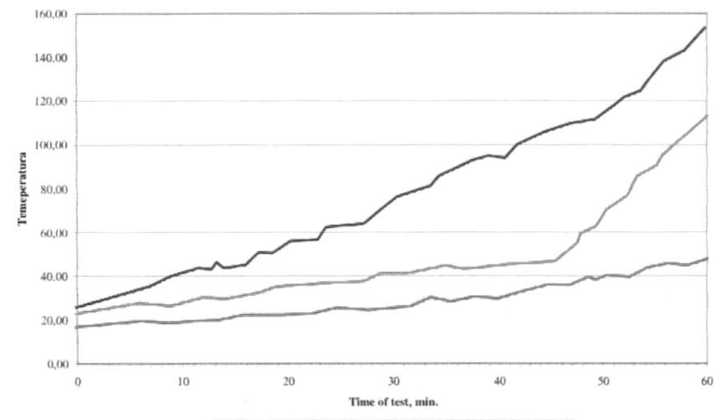

Figure 7. Graphical dependence of temperature according to indicated values of thermocouples No. 1 – No. 3

IOP Conf. Series: Materials Science and Engineering **972** (2020) 012047 doi:10.1088/1757-899X/972/1/012047

Figure 8. The visual appearance of the block No. 3 after the test

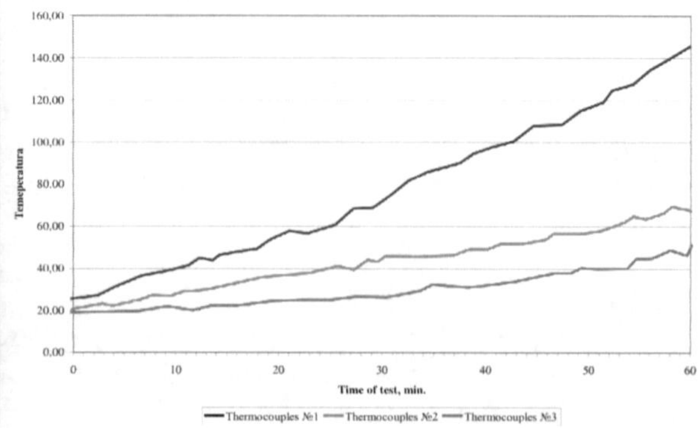

Figure 9. Graphical dependence of temperature according to indicated values of thermocouples No. 1 – No. 3

Figure 10. The visual appearance of the block No. 4 after the test

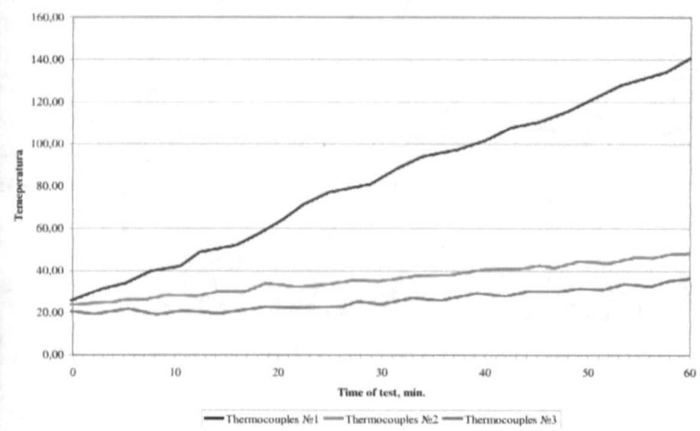

Figure 11. Graphical dependence of temperature according to indicated values of thermocouples No. 1 – No. 3

4. Discussion

Analyzing the results we can draw the following conclusions:

1. When locating the building having a protective facing layer of a thickness 46 mm. and made out of multilayer thermal blocks at a distance equal to 10 meters from the fire source, the maximum temperature (thermocouple No. 2) behind the facing layer will amount to 128 °C within 60 minutes after the start of exposure to radiant heat flux, which leads to the destruction (melting) of 50% of the thermal insulating layer. Consequently, the thickness of the facing coating amounting to 46 mm is not enough.

2. With an increase in the thickness of the protective layer by about 50% to 65 mm, and when locating the building made out of the multilayer thermal blocks at a distance equal to 10 meters from the fire source the maximum temperature (thermocouple No. 2) behind the facing layer will amount to 117 °C within 60 minutes after the start of exposure to radiant heat flux, which leads to the destruction (melting) of 13% of the thermal insulating layer. Therefore, the thickness of the facing coating amounting to 65 mm is also not enough.

3. With an increase in the thickness of the protective layer to 70 mm, and when locating the building at a distance equal to 10 meters from the fire source the maximum temperature (thermocouple No. 2) behind the facing layer will amount to 70 °C within 60 minutes after the start of exposure to radiant heat flux, which leads to the destruction (melting) of 5% of the thermal insulating layer. Therefore, the thickness of the facing coating amounting to 70 mm is also not enough.

4. With an increase in the thickness of the protective layer to 73 mm, and when locating the building at a distance equal to 10 meters from the fire source the maximum temperature (thermocouple No. 2) behind the facing layer will amount to 49 °C within 60 minutes which does not lead to the destruction (melting) of the thermal insulating layer. Consequently, the minimum value of the thickness of the facing coating required for the protection of the thermal insulating layer amounts to 73 mm.

5. Conclusion

During laboratory testing it was discovered that when using expanded polystyrene as a thermal insulating layer and insufficient thickness of the outer (facing) layer of a silica-granite wall block, a concealed damage to the insulating layer is possible even if the buildings are located at a standardized in [2] distance from each other. As a consequence, the thermal conductivity of this construction will be reduced and in the absence of external damage, the thermal performance of the building as a whole will be reduced. The repair of the thermal insulating layer in this case will require significant financial expenses. This can be considered as a hidden damage in the event of a fire, and it may not be discovered during the investigation in the event of a fire, but much later, which in turn in some cases can make it impossible to recover it.

References

[1] Technical regulation on fire safety requirements: Federal Law of the Russian Federation No. 123 – FZ dated July 22nd, 2008 as amended and supplemented 2017 (Yekaterinburg: Kalan)

[2] SP 53.13130.2011Code of practice. Planning and building areas of gardeners' (allotment) associations of citizens, buildings and structures. (Ministry of Regional Development of Russia)

[3] Khasanov I R, Ushakov D V and Sternina O V 2016 *Actual problems of fire safety: materials of the XXVIII Int. Scientific and Practical Conf. Int. Salon of Safety Equipment Integrated Safety – 2016: in 2 parts* (Noginsk) pp 367-375

[4] Bokov S S and Firsova T F 2017 On the need to improve the calculation methods and the definition of fire safety distances *Sci. Mag.* **5(18)** 24-26

[5] Eroshchenko M O and Mamin V S 2015 *Technosphere safety: Proc. of the 2nd Interuniversity Scientific and Technical Conf. with int. participation* (Omsk) pp 105-106

IOP Conf. Series: Materials Science and Engineering **972** (2020) 012048 doi:10.1088/1757-899X/972/1/012048

Fractal analysis of tensile deformation curves of epoxy polymers based on modified epoxy binders

T A Nizina[1], D R Nizin[1], D A Artamonov[1], N S Kanaeva[1]

[1]Department of Building Structures, Ogarev Mordovia State University, Bolshevistskaya str, 68, 430005 Saransk, Russia

E-mail: nizinata@yandex.ru

Abstract. The paper proposes an approach to study the mechanism of deformation of epoxy polymers based on the methods of fractal calculations. The advantages of the method for determining the fractality index of the deformation curves of polymer samples using the minimum coverage method are shown. An algorithm for the quantitative determination of the location of the "critical" points of the deformation curves of polymer samples under tension is proposed. It is shown that the use of the developed methodology for the fractal analysis of time series on the basis of a set of points of deformation curves of epoxy polymer samples under tension, provides valuable information about the processes occurring in the structure of composite materials affected by mechanical loads and various aggressive factors.

1. Introduction

Analysis of the mechanisms of deformation and destruction of composite building materials is an important task of building materials science, its relevance related to the development of new types of composite materials based on polymeric binders is growing year by year. It is known that destruction of composite building materials is a process of multiple origin, development, and aggregation of various types of defects and microcracks up to the appearance of macrocracks [1]. Heterogenous structure of building materials leads to formation of weakened zones, subsequently causing loosening and destruction of composites.

When a microcrack reaches an inclusion (pore or filler), it results in the discharge of the critical energy density at the crack mouth and the system transition to an unstable state (bifurcation point). Branching, a change in the mechanism and direction of the destruction crack development is possible at the bifurcation point. At the same time, the destruction is of a probabilistic nature, and the process of damage accumulation is self-similar which makes it promising to use the fractal analysis methods for its analysis [2–4].

2. Formulation of the problem

The scientific papers provide various methods for determining the fractal dimension of the structure of real composite materials based on cement and polymeric binders [1, 4–10]. In [6, 8], it is shown that using fractal analysis to quantify the structural heterogeneity of polymeric materials and the pore structure of cement composites, we can find a compact way of describing such objects.

In [9, 11, 12], the authors' team proposed a method for determining the fractal dimension of deformation curves based on the minimum coverage method, providing an integral quantitative estimate of

IOP Conf. Series: Materials Science and Engineering **972** (2020) 012048 doi:10.1088/1757-899X/972/1/012048

the process of of building composite destruction under compression, and enabling to determine the position of parametric points of the destruction curve. The proposed method is based on the use of data read using advanced test equipment making it possible to record changes in stresses and strains during the application of high-frequency mechanical loads to a sample. In particular, the WilleGeotechnik® software and hardware complex used in [9, 11] provides readings in increments of a split second.

3. Problem solution

The study object was a polymer based on the modified epoxy resin Etal-247 (Specifications (TU) 2257-247-18826195-07) cured by the hardener Etal-1440N (Specifications (TU) 2257-3570-18826195-03). Technical characteristics of Etal-247 resin: mass fraction of epoxy groups – not less than 21.4-22.8%; Brookfield viscosity at 25 °C – 650-750 centipoise. Compounds hardened by Etal-1440N, have a long viability of 4-5 hours at +20 °C, provide Martens heat resistance of cured compositions up to 140 °C.

The AGS–X series tensile testing machine with TRAPEZIUM X software was used for mechanical tests of polymer composites under tension. The values of stresses and strains were recorded at a frequency of 0.01 sec. We performed tests in accordance with GOST 11262-2017 (ISO 527-2:2012) "Plastics. Tensile Test Method" at a temperature of 23±2 °C and a relative air humidity of 50±5%. The clamps of the tensile testing machine moved at 2 mm/min. At least six 8-shaped samples were tested simultaneously (type 2 according to GOST 11262-2017).

4. Results and discussion

4.1. Analysis of the increase in stresses of polymer composites under tension depending on the step of fixing the readings

Analysis of the results showed that the deformation curve for samples of this composition is characterized by both ascending and descending branches, so we can determine the strength and deformation characteristics of epoxy polymer both at tension and at break (Figure 1, a). The tensile strength and relative elongation at maximum load are 41.05 MPa and 8.53%, respectively; the tensile strength and elongation at break (destruction) are 38.47 MPa and 10.14%. Detailing of the deformation (stress-strain) curve (Figure 1, b), built using the advanced test equipment with a high frequency of readings, shows that the loading process (Figure 1, a) is accompanied by discrete acts of increase and decrease in stress (Figure 1, b). We analyze the change in stress growth in epoxy polymers upon application of tensile forces depending on the reading step increment (from 0.01 to 0.16 s). The analysis of the time dependences built (Figure 2) shows that with an ever greater structuring of the curve is observed when increasing the step. In particular, if at a reading step of 0.01 s (Figure 2, a), the voltage increase variation more resembles noise without a certain tendency (except for the segment corresponding to the maximum destructive loads), then at a step of 0.16 s (Figure 2 b) it is possible to visually trace the formation of three zones with different patterns of change in the values designated by straight lines (red) with different slope angles. The time interval from the beginning of deformation to 40 seconds of loading is characterized by a reduction in the average stress increase from 0.10÷0.11 to 0.6÷0.7 MPa per every 0.16 seconds; at the same time, the characteristic under study in this segment varies from 0.03 to 0.14 MPa.

The second noticeable time interval (from 40 to 90 s) for the sample under study is characterized by a minimal change in the mean values of the stress increase, which makes 0.045÷0.057 MPa (Figure 2, b). For clarity, this segment on the deformation curve (Figure 1, a) is bounded by vertical lines of green (40 s) and blue (90 s). The analysis of data in Figure 2 (b) shows a significant gradient in the studied parameter (from -0.02 to 0.13 MPa) recorded in the time interval of 49.28÷44.94 s, comparable to the stress reduction when reaching maximum tensile force. For the polymer sample under study (red dashed line in Figure 1), this time interval corresponds to the level of relative deformations of 3.28÷3.29%, which is 38.5÷38.6% of the deformation at maximum load, and a stress of 23.08÷23.10 MPa (56.2÷56.3% of the tensile strength limit). Such an abrupt change in stress growth can presumably be explained by the appearance of a "critical" state associated with the formation of defects and cracks in the polymer structure at tensile loads.

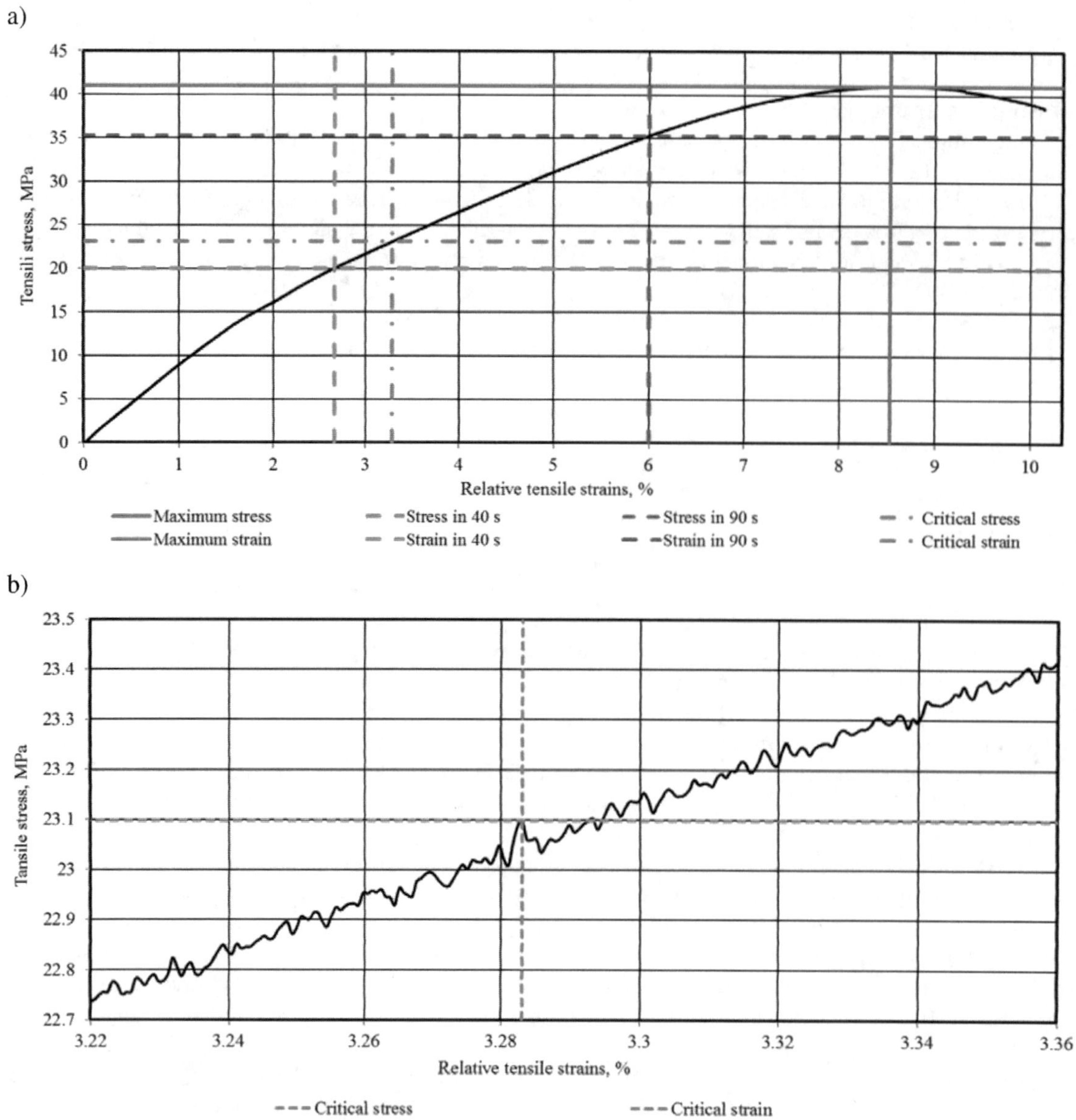

Figure 1. General view (a) and fragment (b) of the strain curve of epoxy polymer samples under tension (solid red line in figure "a" shows the level of maximum stresses and the corresponding strains)

The third time interval is associated with a decrease in the average level of the stress growth which has negative values at a stage close to the achievement of ultimate loads. At the same time, the general tendency for the decrease in this parameter in the time interval from 90 s and further is described by one linear model (Figure 2, b). The point of formation of the main crack leading to the sample destruction under tensile loads, was registered at 139.56÷139.68 seconds from the beginning of loading and corresponds to a stress of 40.48÷40.52 MPa and a relative deformation of 9.30÷9.31%.

The above analysis showed that the study of changes in the growth of stresses built at a small registration step, provides valuable information about the process of deformation of polymer specimens under mechanical loads, but in this case it is extremely important to choose the frequency of the curve building. In addition, certain errors in the analysis can introduce a shift in the beginning and end of the analyzed interval, which requires the search for new analysis techniques, as well as an understanding of which particular stress gradient should be attributed to the "critical" state. At the same time, taking into account that the change in deformations and stresses in the samples under study during mechanical tests was recorded in time with a certain predetermined step (0.01 s), it is feasible to use the theory of fractal analysis of time series to analyze the deformation curves with the identification of characteristic points on it.

a)

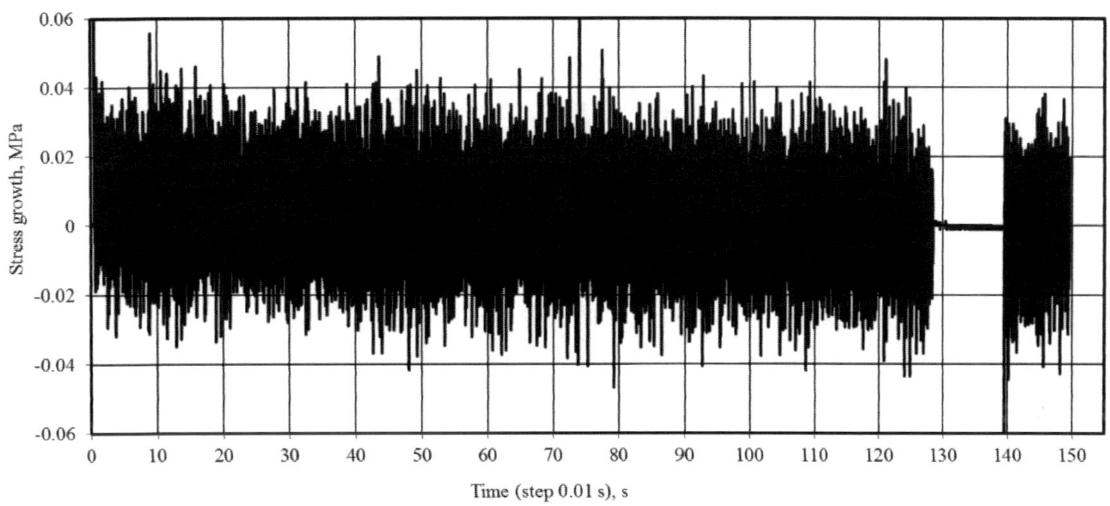

b)

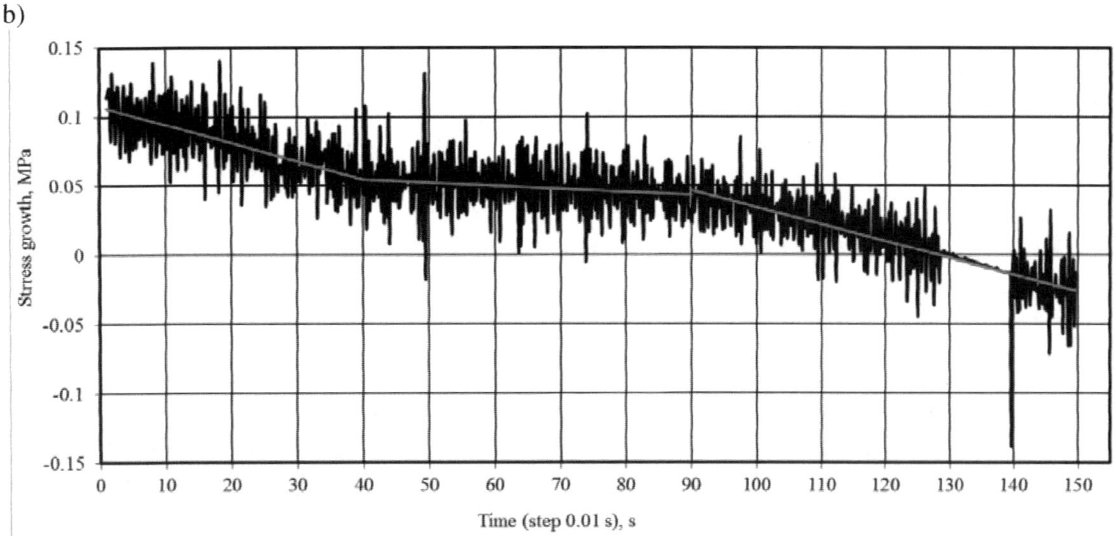

Figure 2. Time series of the growth of tensile stresses in the epoxy polymer sample depending on the reading step: a – 0,01, b – 0,16 s

Consider a time series $y(t)$ defined on a certain segment $[a, b]$. To calculate the fractal dimension, we use the minimum coverage method, which is more accurate than the cell dimension method. Its main aspects are described in [11, 13–15]. The essence of the method consists in uniform breaking down of the segment

$$\omega_m = [a = t_0 < t_1 < \cdots < t_m = b]$$

to m parts and calculating function $y = f(t)$ in the class of coverages consisting of rectangles with base $\delta = \frac{b-a}{m}$ (Figure 3) [15]. Then the height of the rectangle in segment $[t_{i-1}, t_i]$ will be equal to the difference between the maximum and minimum value of function $f(t)$ on this segment – $K_i(\delta)$. By introducing the magnitude of the amplitude variation of function $f(t)$ corresponding to the breakdown scale δ in segment $[a, b]$

$$V_f(\delta) = \sum_{i=1}^{m} K_i(\delta), \tag{1}$$

we obtain the dependence to determine the total coverage area:

$$S_\mu(\delta) = V_f(\delta) \times \delta. \tag{2}$$

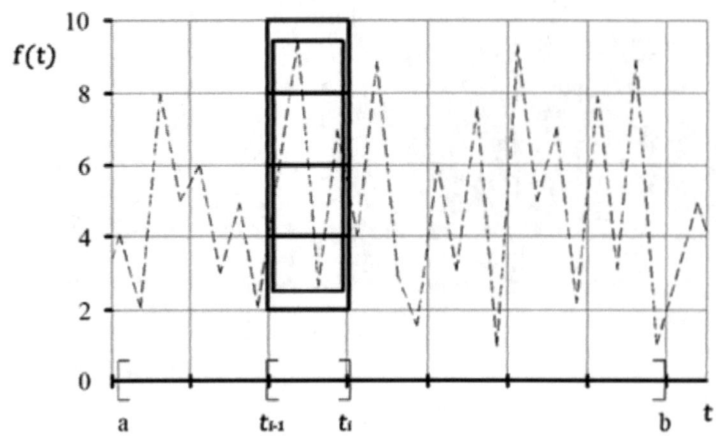

Figure 3. Fragment of cellular (square) and minimal (rectangular) coatings the graph of the fractal function on a segment $[t_{i-1}, t_i]$

Then, according to [2, 15], fractality index μ can be determined from the linearization of dependence

$$V_f(\delta) \sim \delta^{-\mu} \text{ at } \delta \to 0. \tag{3}$$

At the same time, the fractal dimension is related to the fractality index determined by the minimum coverage method, as

$$D_\mu = \mu + 1. \tag{4}$$

To determine fractality index μ, in this work we used sequence m of nested breakdowns, where $m = 2^n$, $n = 0, 1, 2, 3, 4$. Each breakdown consisted of 2^n intervals containing 2^{4-n} experimental points. For each breakdown ω_m, amplitude variation $V_f(\delta)$ was calculated using (1), where $K_i(\delta)$ was determined as the difference between the maximum and minimum tensile strain in time interval $[t_{i-1}, t_i]$. Coefficient β of regression equation $log\left(V_f(\delta)\right) = \alpha_0 + \beta \times log(\delta)$ determined with the least squares method, was used to determine the fractality index and the dimension of the minimum coverage:

$$\mu = -\beta; \; D_\mu = 1 + \mu.$$

4.2. Discussion

Analysis of data reflecting the change in the fractality index depending on the loading duration determined with the shift of the analyzed segment with a step of $2^4 = 16$ points, showed (Figure 4) that the destruction process (to the right of the red line) is characterized by an abrupt decrease in this parameter varying at loading in the interval from 0.33÷0.82 to zero. At the same time, it is known from the analysis of time series [9, 11, 15] that the higher value μ, the more stable the series. If $\mu < 0,5$, then the series is interpreted as a "trend" (a period of sharp upward or downward movement, indicating, as a rule, a "critical" state in the system under study); if $\mu > 0,5$, then the series is interpreted as "flat" (a period of relative calm). At $\mu \approx 0,5$, we can say about the conformity of the changes to the Brownian movement.

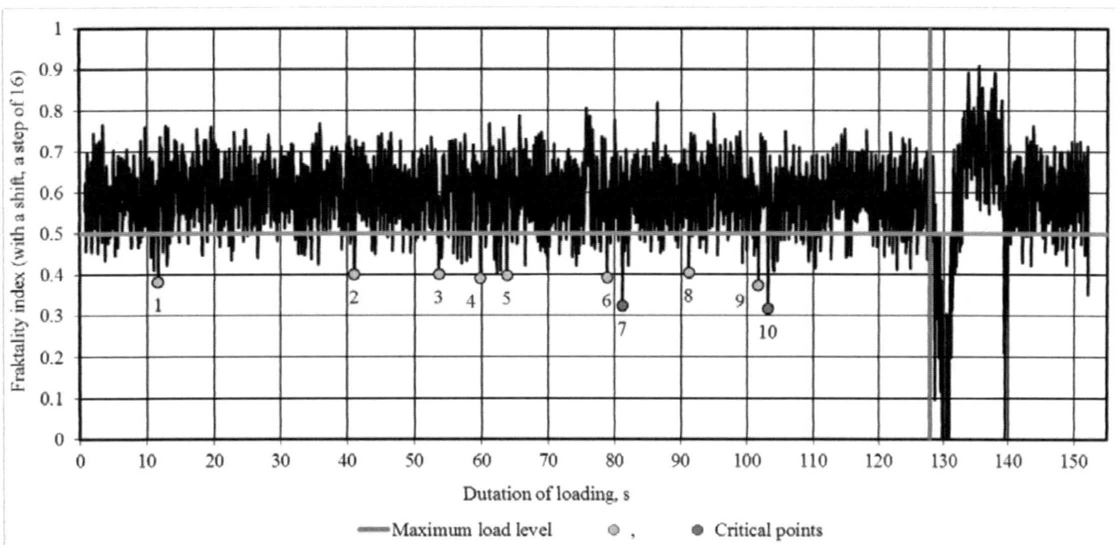

Figure 4. A change in the fractality index of the strain curves of the epoxy composite under tension, depending on the duration of loading

Figure 4 shows that in the process of loading, until the sample achieves maximum tensile stresses after 127.83 seconds from the beginning of the test, there is a systematic abrupt decrease in the fractality index below $\mu = 0,5$, which, in our opinion, is due to the transition of the system to an unstable state associated with the formation of microdamages in the polymer composite structure under tensile loads. In this case, until the composite reaches the state of destruction, the working capacity of the sample is provided mainly by transferring the load to zones without microdefects, expressed on the graph as an increase in μ above 0,5. Therefore, from the analysis of Figure 4, we can identify the "critical" loading points where the fractality index will have the lowest values. In this case, for the analyzed polymer sample, such temporal coordinates corresponding to the bifurcation points were (at $\mu < 0,4$) in: 11.63; 41.02; 53.57; 59.77; 63.79; 78.80; 91.12; 101.71, and 103.14÷103.18 seconds from the beginning of loading (see the points in Figure 4). The lowest values of the fractality index corresponding to $\mu \approx 0,32$ are recorded in 81.17 and 103.15÷103.16 s (highlighted with red dots in Figure 3). Considering that for all the studied points in the interval of 103.14÷103.18 seconds, a significant decrease in the fractality index is observed without its "restoring" to values of 0.5 or more, all of them were combined at number 10. Such a consistent drop in μ in a series of neighboring points indicates the emergence of the growing "difficulties" in the perception of tensile stresses by the sample and the beginning of the process of the structure decompaction, the emergence of foci of local destruction, subsequently leading to its destruction. The numerical values of stress levels and relative deformations in the above “critical” points in absolute and relative values (relative to the point of reaching the maximum tensile stresses) are given in Table 1.

A more detailed analysis of the segments preceding the emergence of “critical” points on the graph of the change in the fractality index versus the loading duration showed that their emergence is observed with a considerable cyclical nature of the change in stresses in the mechanical loading process (Figure 5, a, b). In this case, the emergence of "critical" areas is preceded by the presence of a segment with a small growth of stresses (Figure 5, c, d), which indicates the damage of part of the structure elements and the load transfer to other elements.

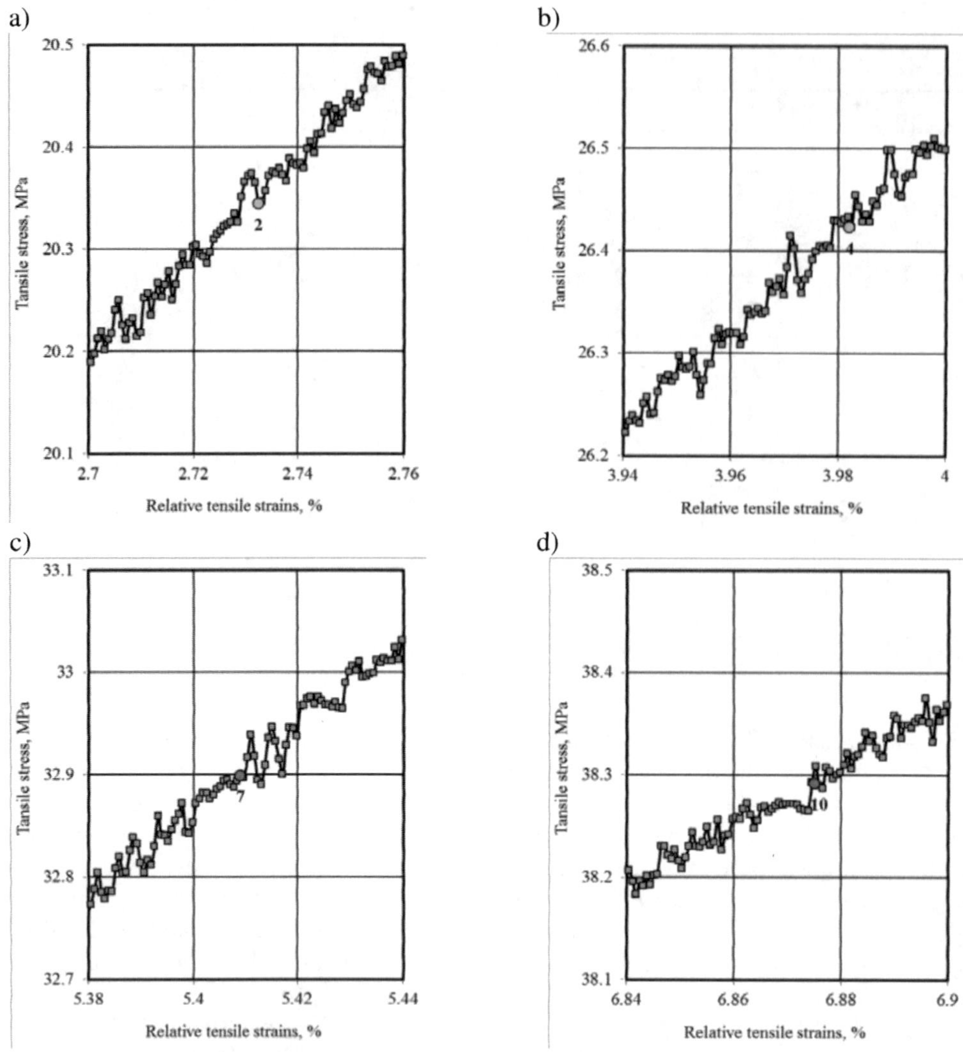

Figure 5. Fragments of the deformation curve of epoxy polymers under tension in the vicinity of "critical" points (the numbers of the points correspond to the values given in Table 1)

Table 1. The coordinates of the "critical" points of the epoxy resin deformation curve under the tensile loads

"Critical" point number	Fractality index μ	Point parameters					
		absolute values			relative values		
		$t_{крит.}$, s	$\sigma_{крит.}$, MPa	$\varepsilon_{крит.}$, %	$\frac{t_{крит.}}{t_{\sigma_{раст.}}}$, %	$\frac{\sigma_{крит.}}{\sigma_{раст.}}$, %	$\frac{\varepsilon_{крит.}}{\varepsilon_{раст.}}$, %
1	0,382	11,63	6,970	0,773	9.10	16.98	9.06
2	0,401	41,02	20,345	2,732	32.09	49.56	32.03
3	0,401	53,57	24,428	3,569	41.91	59.51	41.84
4	0,391	59,77	26,423	3,982	46.76	64.37	46.68
5	0,397	63,79	27,672	4,251	49.90	67.41	49.83
6	0,393	78,8	32,223	5,251	61.64	78.50	61.56
7	**0,324**	**81,17**	**32,899**	**5,409**	**63.50**	**80.14**	**63.41**
8	0,404	91,12	35,575	6,072	71.28	86.66	71.18
9	0,374	101,71	37,96	6,778	79.57	92.47	79.46
	0,375	103,14	38,265	6,873	80.69	93.22	80.57
	0,327	103,15	38,265	6,874	80.69	93.22	80.59
10	**0,317**	**103,16**	**38,292**	**6,875**	**80.70**	**93.28**	**80.60**
	0,355	103,17	38,309	6,875	80.71	93.32	80.60
	0,348	103,18	38,287	6,876	80.72	93.27	80.61

5. Conclusions

The proposed approach for the study of the mechanism of deformation of composite materials under tension, carried out on the basis of the method of fractal analysis of the deformation curves recorded using modern testing equipment with a high frequency, provides valuable information about the process of accumulation in its structure of micro- and macrodefects resulting in the destruction of composites. In our opinion, carrying out such studies on samples of polymer composites of different compositions, as well as after aging under the influence of various aggressive factors, including climatic factors, will provide valuable information about the processes occurring in the structure of composite materials when tensile loads are applied.

Acknowledgments
The reported study was funded by RFBR according to the research project № 18-08-01050.

References

[1] Ivanova V S, Balankin A S, Bunin I Z and Oksogoev A A 1994 Synergetics and Fractals in materials science (Moscow: Science) p 384
[2] Feder E 1991 Fractals (Moscow: Mir) p 254
[3] Mandelbrot B B 1982 The fractal geometry of natur (New York: Freeman) p 480
[4] Selyaev V P, Nizina T A, Lankina Ya A and Tsyganov V V 2006 Determination of the fractal dimension as a structural parameter in the analysis of polymer composites *Achievements, problems and promising directions for development building material science theory and practice: Tenth Academic readings of RAASN* pp 73–76
[5] Bannikov M V, Fedorova A Yu, Terekhina A I and Plekhov O A 2013 Experimental study of the fractal regularities of fatigue crack growth and energy dissipation at its top *Bulletin of PNRPU Mechanics* **2** pp 21–36
[6] Khakhardin A N and Khodykin E I 2007 Fractal dimension of dispersed and porous

materials *Building materials* **8** pp 62–63
[7] Khamidulina D D and Shishkin I V 2015 Application of the fractal geometry theory in building material science *Topical issues of the modern science, technology and education* **2** pp 5–8
[8] Pertsev V T, Kozodaev S P, Ledenev A A and Bobryshev A N 2015 Topological optimization of processes of cement stone and concrete microstructure formation *Scientific bulletin of Voronezh State University of architecture and construction Series: Physical and chemical issues and modern technologies of building material science* **1** pp 21–28
[9] Selyaev V P, Nizina T A, Balykov A S, Nizin D R and Balbalin A V 2016 Fractal analysis of the deformation curves of disperse and reinforced fine-grained concrete at compression *Bulletin of Perm national research polytechnic university Mechanics* **1** pp 129–46
[10] Selyaev V P, Nizina T A and Lankina Ya A 2007 Fractal analysis of the structure filled with polymer composites *News of high schools Construction* **4** pp 43–48
[11] Selyayev V P, Nizina T A, Balykov A S, Nizin D R and Balbalin A V 2018 Fractal analysis of deformation curves of disperse reinforced fine-grained concrete under compression *PNRPU mechanics bulletin* **1-2** pp 84–93
[12] Nizina T A, Nizin D R, Kanaeva N S, Kuznetsov N M and Artamonov D A 2019 Applying the Fractal Analysis Methods for the Study of the Mechanisms of Deformation and Destruction of Polymeric Material Samples Affected by Tensile Stresses *Key engineering materials* vol **799** pp 217-223
[13] Dubovikov M M and Starchenko N S 2003 Variation index and its applications to analysis of fractal structures *Almanac Gordon* **1** pp 1–30
[14] Dubovikov M M, Starchenko N S and Dubovikov M S 2004 Dimension of the minimal cover and fractal analysis of time series *Physica A* **339** pp 591–608
[15] Starchenko N V 2005 Fractality index and local analysis of chaotic time series (Moscow: Degree work of cand. phis. and mat) *Sciences* p 122

IOP Conf. Series: Materials Science and Engineering **972** (2020) 012049 doi:10.1088/1757-899X/972/1/012049

Cement Composition Materials Based on Electrochemically and Electromagnetic-activated Filling Water

V T Erofeev[1], V T Fomichev[2], D V Emelyanov[1]*, A A Matvievskiy[3], K Holschemacher[4], M S Boushi[1] and E Yu Bazeeva[1]

[1]National Research Ogarev Mordovia State University, 68 Bolshevistskaya Str., Saransk 430005, Russia
[2]Volgograd State Technical University, 28 Lenin Avenue, Volgograd 400005, Russia
[3]OJSC «MAKSMIR», Moscow 121099, Russia
[4]Leipzig University of Applied Sciences, 132 Karl Liebknecht Str., Leipzig 04277, Germany

*Email: emelyanoffdv@yandex.ru

Abstract. A technology is proposed for producing electrochemically and electromagnetically activated water and, from the standpoint of resource saving, high-quality cement composite materials. The results of the physicochemical analysis and morphology of the sediment of naturally non-mineralized water activated by different regimes are presented. A quantitative and qualitative X-ray phase analysis, as well as the results of the study of the microstructure of cement stone samples based on mixing water activated in different modes, are given. The strength indicators of cement stone, mortar and concrete samples were established, and the results of a study of the water resistance of cement composites made with the use of activated mixing water were presented. It has been established that electrochemical and electromagnetic activation of mixing water contributes to a change in its physicochemical properties, influences the structure-forming factors of the cement stone and contributes to the improvement of the strength and performance indicators of cement composite materials.

1. Introduction

Currently, more and more attention is paid to improving the quality of concrete mixes and concrete based on them. This is especially true when using concrete products and structures at critical construction sites where high physical, mechanical and performance properties are required (strength and elastic-plastic properties, resistance in aggressive media, durability) [1-7]. At the same time, it is important to reduce the cost of materials produced and products based on them. The modern view of the science of structure formation of concrete from the standpoint of managing material properties and obtaining high-quality products and structures using energy- and resource-saving technologies determines the current direction of research in the field of building materials science.

The growth of the consumption market of monolithic concrete and reinforced concrete requires the solution of important problems in the field of improving the technology for preparing concrete mix. The existing production lines for the production of concrete mixes can be upgraded through the use of installations that allow to improve the technological process of preparing the mixture without significant changes to the used production lines. One of the directions in the area of improving concrete technology is the activation of raw materials, in particular, mixing water [8–13].

Activation of mixing water in the technology of preparing a concrete mixture affects both the processes of dissolution and coagulation, structure and hydrate formation, and the occurrence of condensation-crystallization structure of cement stone, which is formed due to direct chemical interaction of particles with the formation of a rigid bulk structure [14–16].

Treatment of mixing water by external influences, in particular, sequential electrochemical and electromagnetic activation, allows minimizing or eliminating chemical additives and automating the process of preparing a concrete mix [17].

When electrochemical and electromagnetic processing of natural water, along with the influence on the molecular structural characteristics of water, electrochemical processes occurring on the electrodes have a significant effect on the electrodes: the anode, which is under a positive potential, at which oxidation reactions occur ions with negative potentials and a cathode-electrode, which are under a negative potential, at which ion reduction reactions take place, having a positive potential [15]. The resulting compounds, which are in the ultradisperse phase (nanoscale from 1 to 100 nanometers), determine the activity of the water used to mix the mortars and concretes. The dispersed phase of iron oxides and hydroxides (anodic products) and calcium and magnesium hydroxides - due to the cathode reduction of water molecules, form a temporarily stable system of crystallization centers during the processes of transition of cement solutions into the phase of formation of gel structures and the formation of a solid phase with a smaller crystal structure.

The use of sequential electrochemical and electromagnetic activation of mixing water used for the preparation of concrete mixtures, allows to accelerate the setting time of cement paste and increase the strength of cement stone [12, 18]. Regulation of such activation parameters as electromagnetic field intensity, electric current density, water flow rate and processing time makes it possible to control the hardening and physicomechanical properties of binding materials. The optimal parameters of the electromagnetic and electrochemical activation mode and the achieved efficiency of the treatment of the mixing water depend on the properties of the used electrode materials, the physicochemical characteristics of the source water, and the holding time of the treated mixing water before introducing it into the concrete mix.

2. Results and discussion

In order to determine the factors influencing the structure formation of the mineral sediment based on the ionic components of natural water, an experimental study of the dispersed sludge phase, separated during water evaporation before and after its activation, was carried out. To exclude the time factor, the studied water was settled for two days, which led to the deposition of large dispersed particles in the initial water.

The study of the surface morphology of sediments was carried out using an atomic force microscope SPM9600 SHIMADZU with a wide scanning range of 125 μm × 125 μm × 7 μm and a radius of curvature of the probe not exceeding 20 nm in the mode of contact atomic force microscopy at a constant force of interaction of the probe with the surface in air.

For the preparation of electrochemically and electromagnetically activated water was used in the equipment of UPOVS 2-5,0 "Maksmir" [17].

Processing was carried out in three modes, the cipher of which is composed of alphanumeric designations. Mode (1-1) – activated water treated with an electric current with a density of j_{max} = 5.65 A / m^2 in the electrochemical activation chamber and an electromagnetic field of H_{max} = 24 kA / m in the working gap of the electromagnetic activation chamber; (3-3) – j_{max} = 22.58 A / m^2, H_{max} = 75 kA / m; (6-6) – j_{max} = 43.55 A / m^2, H_{max} = 135 kA / m. When treating water in the electrochemical activation chamber, an aluminum anode was used.

The study used tap artesian water of Saransk. The effect of activation treatment on the total and carbonate hardness, oxidability, spare and aggressive carbon dioxide content of the studied natural water is presented in Table 1.

Table 1. Results of physico-chemical analysis of samples of the investigated waters.

Activation mode	Hardness (mmol·eq / l)		Carbon dioxide (mg / l)		Oxidizability ($mg\ O_2 / l$)
	common	carbonate	spare	aggressive	
0	8.21	4.65	—	15.65	1.56
1-1	8.30	5.94	—	15.52	2.03
3-3	7.43	5.21	—	13.31	1.58
6-6	8.33	6.86	—	11.05	1.98

From the results of the analysis of the initial artesian water of Saransk, presented in table 1, it follows that it is characterized by average hardness. Low water oxidability, established in the range of 1.56–2.03 mg·O_2 / l indicates a low content of organic substances and inorganic reducing agents in water. The table shows changes in the hardness of the water. The common hardness index for treated water by the mode (3-3) has the lowest value – 7.43 mmol·eq / l compared to other types. The lowest value of aggressive carbon dioxide was found in water treated by the mode (6-6) – 11.05 mg / l. With an increase in the current density and electromagnetic field under all modes of water activation, the content of aggressive carbon dioxide decreases.

The surface morphology of precipitates and their structure, formed during the evaporation of the water of an artesian well in the city of Saransk before and after activation is shown in Figure 1, 2.

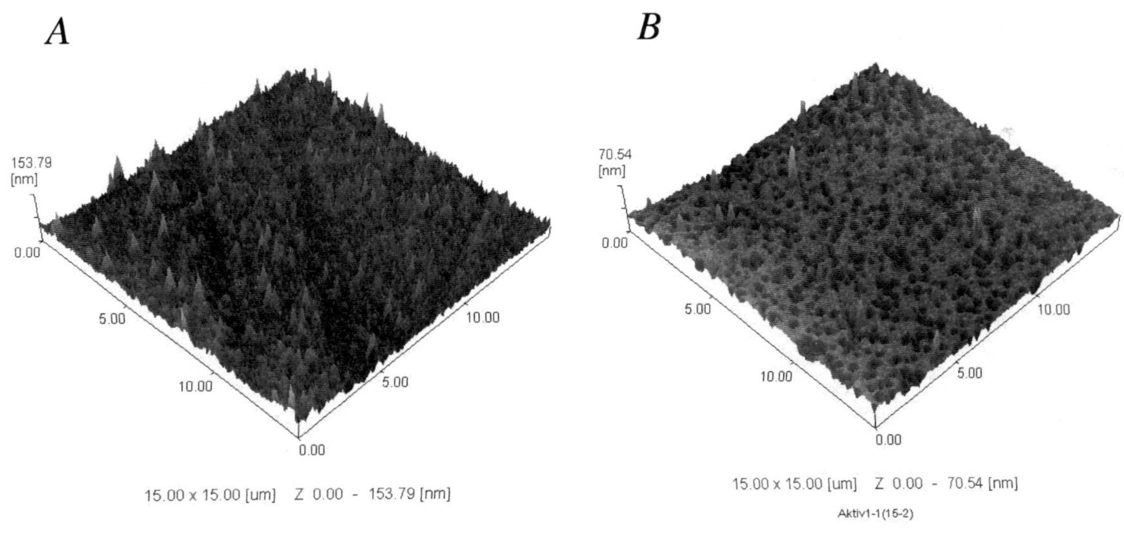

Figure 1. Morphology of artesian water sediment Saransk city: a – not activated water; b – water activated by mode (1-1).

IOP Conf. Series: Materials Science and Engineering **972** (2020) 012049 doi:10.1088/1757-899X/972/1/012049

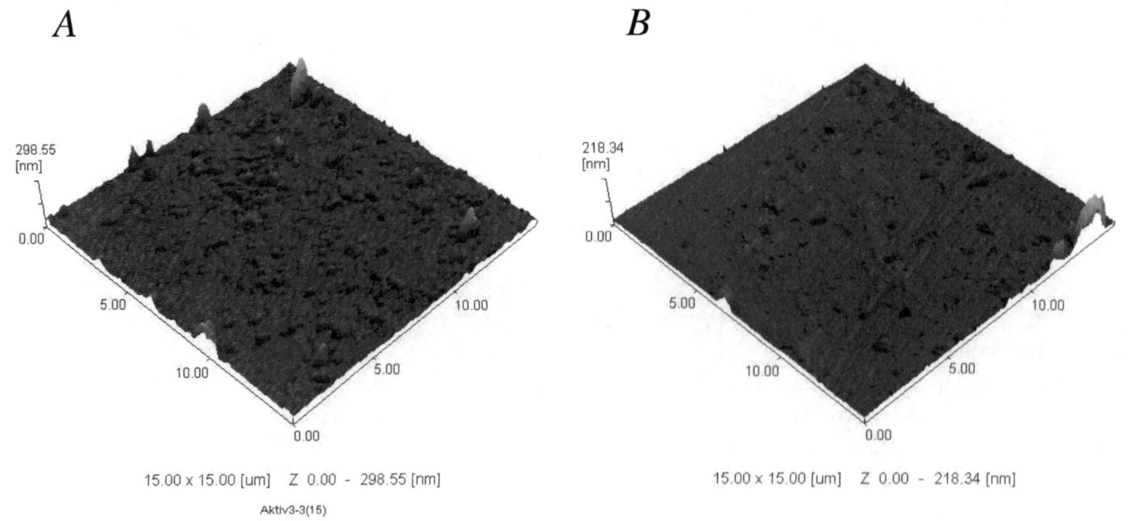

Figure 2. Morphology of artesian water sediment Saransk city: a – water activated by mode (3-3); b – the same, (6-6).

The morphology of sediment from various sources before and after activation is characterized by an average roughness and maximum crystal size (Table 2).

Table 2. Roughness and maximum crystal sizes.

Activation mode	Roughness *R*a (nm)	Maximum size of crystals (nm)
0	11.152	153.79
(1-1)	4.636	70.54
(3-3)	7.96	298.55
(6-6)	4.0	218.34

Analysis of the above data allows us to conclude about the undoubted influence of the parameters of electrochemical and electromagnetic water treatment on the structure formation of mineral sediment and large-scale dimensional levels of crystals. For artesian water, the city of Saransk, a decrease in roughness is observed during the transition from the activation mode (1-1) to the mode (6-6). A definite role in this process is exerted by the ionic composition of water and, in particular, by the level of the total hardness of water. In the course of changing electric and electromagnetic fields, the spatial equilibrium state of nanostructured precipitates is disturbed, which leads either to their consolidation and the formation of fine micrometric systems, or to the formation of nanodispersed systems with excess surface energy. The formation of nanodispersed systems with crystal sizes of 20–100 nm can serve as an additional source of nanometric centers of crystallization of the products of hydration of binders during their hardening.

In order to establish the structural changes occurring in the cement composites based on the activated mixing water, X-ray structural studies were performed, consisting in their analysis. The diffractograms were recorded on an ARL X'tra diffractometer (Switzerland), which is a full-size powder diffractometer q – q geometry with a goniometer radius of 260 mm and a source – a narrow-focus tube with a capacity of 2200 W (Cu anodes).

In the course of the experiment, Portland cement CEM I 42.5 N of the OJSC "Mordovcement" plant was used. Cement stone cubic samples were made of a size of 2 × 2 × 2 cm. The water-cement ratio for all the compositions was determined on the basis of the normal density of the cement paste. Samples hardened in normal conditions for 56 days. Cement stone made on untreated water was considered as a control sample. The obtained samples were ground in an agate mortar with an agate pestle with acetone until all the samples passed through a sieve with an aperture of 90 microns. The attrition samples were placed in round holders with a top loading inner diameter of 25.65 mm and a depth of 1.9 mm made of special steel.

During the qualitative analysis of diffractograms, it was found that the following minerals are present in the composition of all the studied samples of cement stone:

- C_3S (Ca_3SiO_5) with d = [3.044; ... 2.784; ...; 2.617; 2.191; ...; ... 1.77; 1.629; ...];
- C_2S (Ca_2SiO_4) with d = [... 2.886; ... 2.784; 2.755; ... 2.617; ... 2.191 ..];
- C_3A ($Ca_{2.96}Na_{0.08}Al_2O_6$) with d = [2.702; ... 2.206; ...; 1.912; ... 1.3510 ..];
- C_4AF ($Ca_2Fe_{0.28}Al_{1.72}O_5$) with d = [7.321; ... 2.65; ... 2.058; ... 1.923 ...];
- $3CaO \cdot Al_2O_3 \cdot 3CaSO_4 \cdot 32H_2O$ (ettringite) with d = [9.733; ...; 5.619; 3.872; ... 3.49; ...];
- $Ca(OH)_2$ (portlandite) with d = [4.944; 3.04; 2.635; 1.927; 1.8; 1.69 ...].

Also in all samples there is SiO_2 (α-Quartz) with d = [4.261; 3.348; 2.459; 2.283; ...; 2.13; ...; 1.819; 1.543; ...], whose concentration is in the range of 1-3% of the mass. depending on the sample. Further, this phase was not taken into account in the framework of quantitative X-ray phase analysis.

The results of a quantitative X-ray phase analysis of cement stone (Table 3) showed that for cements closed with activated water, on the first day of hardening there is a decrease in the number of phases of all clinker minerals compared to the control composition.

Table 3. Results of quantitative X-ray phase analysis.

Activation mode	The concentration of the phases (% of the mass.)						
	C_3S	C_2S	C_3A	C_4AF	Portlandite	Ettringite	Amorphous phase (C-S-H)
3 days of hardening							
0	22.2	11.50	2.50	10.90	8.00	3.70	41.20
(1-1)	20.8	11.30	1.90	10.10	8.80	3.60	43.50
(3-3)	18.8	11.40	2.10	9.80	9.00	3.70	45.20
(6-6)	17.6	11.30	2.10	9.60	9.20	3.70	46.50
12 days of hardening							
0	10.5	9.40	0.60	8.90	15.30	3.20	52.10
(1-1)	9.5	9.10	0.50	8.50	16.10	3.10	53.20
(3-3)	8.8	9.20	0.50	8.30	16.50	3.20	53.50
(6-6)	8.2	9.20	0.60	8.30	16.50	3.20	54.00
28 days of hardening							
0	8.5	8.50	0.00	7.10	14.20	3.20	58.50
(1-1)	8.2	8.50	0.00	7.00	14.10	3.20	59.00
(3-3)	7.8	8.40	0.00	6.90	14.00	3.10	59.80
(6-6)	7.4	8.60	0.00	6.90	14.30	3.20	59.60
56 days of hardening							
0	7.8	7.10	0.00	6.20	13.50	3.10	62.30
(1-1)	7.2	7.30	0.00	6.10	13.50	3.20	62.70
(3-3)	6.6	7.20	0.00	6.00	13.60	3.10	63.50
(6-6)	6.5	7.30	0.00	6.10	13.40	3.20	63.50

IOP Conf. Series: Materials Science and Engineering **972** (2020) 012049 doi:10.1088/1757-899X/972/1/012049

As a result, the amount of portlandite and calcium hydrosilicate increases. This process can be explained by the effect of electric current and magnetic field on the mixing water. When water is activated, the hardness salts dissolved in it become a different physical state – fine sludge, which forms additional crystallization centers, intensifying the processes of dissolution and hydration of cement [15, 16].

A quantitative analysis of the diffraction patterns of cement stone samples after 28 and 56 days of hardening also indicates a decrease in the number of C_3S in samples based on water treated with a magnetic field and an electric current, which indicates a greater degree of hydration of these samples.

An intensive increase in calcium hydroxide and hydrosilicates in cement stone on activated mixing water slows down by 28 days of hardening, which once again confirms the greater influence of external conditions on the structure formation of cement materials at a later time, compared with mixing water in the initial stages of hardening.

In the samples made on the activated mixing water, an intensification of the transition of C_3A to the solution is observed, which indicates a decrease in the probability of formation of ettringite in the later stages of hardening.

The use of mixing water in cement composites, treated with electric current and magnetic field, allows to intensify the processes of dissolution and hydration of cement, to increase the content of calcium hydrosilicates and, consequently, to increase the strength of cement composites.

In order to identify changes in the microstructure in the early stages of hardening of the cement stone with the use of activated mixing water, a microscopic analysis of the samples of cement stone was carried out. The study of the microstructure of samples at the age of 7 days was carried out using a Quanta 200 scanning electron microscope (Figure 3).

A *B*

Figure 3. Cement stone microstructure: a – on non-activated water; b – on the water activated by the mode of (3-3).

The results of microscopic analysis show that the cement stone with the use of activated mixing water has a more dense structure. The rheology of cement paste with the use of activated mixing water changes in the direction of increasing mobility and, consequently, reducing the volume of micropores of the cement stone.

An important stage of the research is the determination of the influence of the activated mixing water on the persistence of the effects of increasing the strength with the introduction of small and large aggregates. For this, samples were made of cement paste, mortar and concrete mix. Cement produced by the Mordovian cement plant was used as a binder, quartz sand with a fineness modulus 2.0 was used for fine aggregate, 5–10 granite crushed stone was used for coarse aggregate. The mixing water was used activated and non-activated. Water was activated according to the following modes: (1–1), (3–3) and (6–6). The amount of fine aggregate in the composition of the solution was selected from a ratio of 1: 3, and the composition of concrete was adopted from a ratio of 1: 1.053: 1.789 (cement: quartz sand: granite rubble). In each case, equal-moving trains were prepared. The test results are shown in Tables 4–6.

Table 4. Compressive and flexural strength of cement stone at different times of hardening.

Activation mode	3 days of hardening		7 days of hardening		28 days of hardening	
	R_C (MPa)	R_F (MPa)	R_C (MPa)	R_F (MPa)	R_C (MPa)	R_F (MPa)
inactive.	48.50	10.57	61.10	11.95	79.50	15.50
(1-1)	52.38	10.67	69.00	13.15	96.20	18.00
(3-3)	50.90	11.55	67.21	13.63	91.40	18.00
(6-6)	49.00	12.54	70.30	13.03	100.20	15.50

Table 5. Durability at compression and a bend of solutions in various terms of hardening.

Activation mode	3 days of hardening		7 days of hardening		28 days of hardening	
	R_C (MPa)	R_F (MPa)	R_C (MPa)	R_F (MPa)	R_C (MPa)	R_F (MPa)
inactive.	11.00	7.50	24.00	8.57	26.50	9.55
(1-1)	11.00	7.64	25.68	9.00	27.60	10.51
(3-3)	11.22	7.62	26.40	8.83	28.40	10.70
(6-6)	13.75	8.10	30.72	9.17	29.20	10.22

Table 6. Compressive strength and bending of concrete in different periods of hardening.

Activation mode	3 days of hardening		7 days of hardening		28 days of hardening	
	R_C (MPa)	R_F (MPa)	R_C (MPa)	R_F (MPa)	R_C (MPa)	R_F (MPa)
inactive.	32.25	5.83	38.04	6.82	46.42	8.04
(1-1)	35.54	6.35	41.88	7.16	55.40	8.37
(3-3)	32.29	6.05	47.29	7.10	63.25	8.15
(6-6)	32.83	6.24	43.33	7.25	53.17	8.61

From the results of the study, it follows that all materials — cement stone, mortar, and concrete — are characterized by an increase in strength in the case of the use of activated mixing water.

Knowledge of the stability of cement composites when exposed to water is important because the building structures and products of many buildings and structures during operation are in aquatic environments.

Equal-moving cement compositions were considered for the study. The following were used as mixing water: 1) natural non-treated untreated (O), 2) natural, treated with an electromagnetic field with H_{max} = 150 kA / m (M), 3) natural, treated in an electrochemical activation chamber with j_{max} = 43.55 A / m^2 (E6A), 4) natural, sequentially processed by an electromagnetic field with H_{max} = 150 kA / m and electric current with j_{max} = 43.55 A / m^2 (M + E6A), 5) natural, processed sequentially in an electrochemical activation chamber with j_{max} = 43.55 A / m^2 and in the chamber of electromagnetic activation with H_{max} = 150 kA / m (E6A + M). The kinetics of changes in mass content and coefficient of resistance of composites made on the mixing water activated by various modes, depending on the duration of the aging in the medium, is shown in Figure 4, 5.

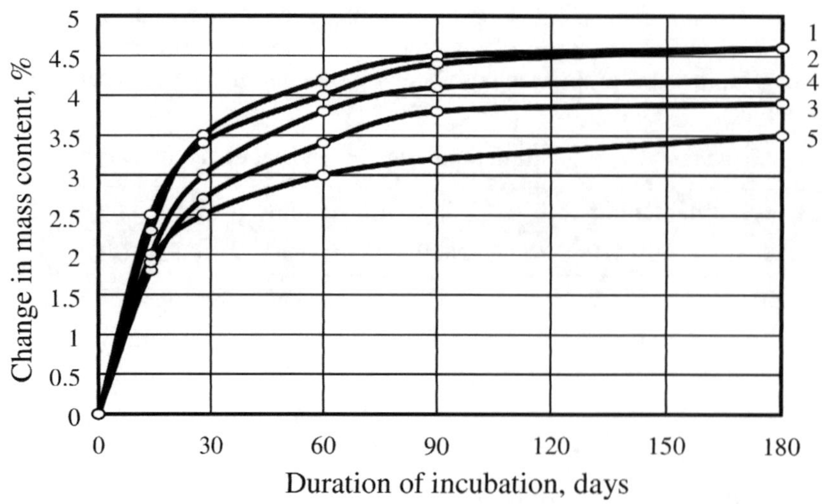

Figure 4. The dependence of the change in mass content of cement composites on the duration of aging in water and the type of activation: 1 – O; 2 – M; 3 – E6A; 4 – M + E6A; 5 – Э6A + M.

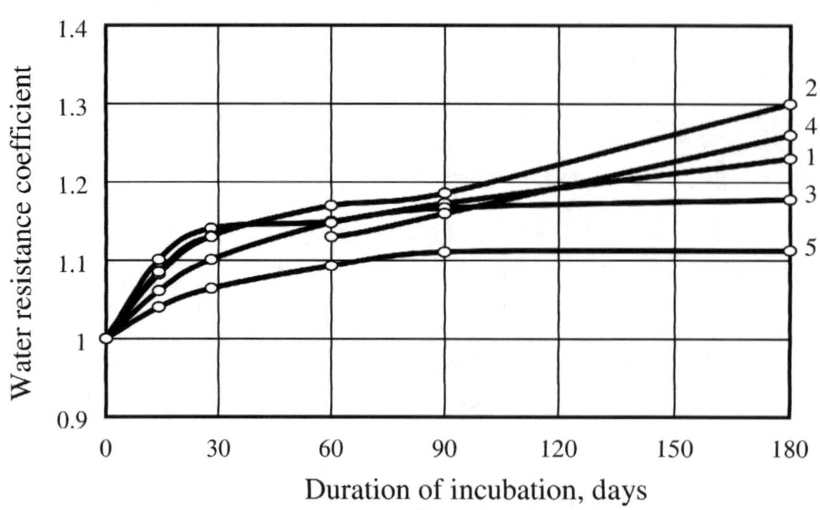

Figure 5. Dependence of change in the coefficient of firmness of cement composites on the duration of aging in water and the type of activation: 1 – O; 2 – M; 3 – E6A; 4 – M + E6A; 5 – Э6A + M.

The kinetics of changes in the mass content of the examined compositions when kept in water has a similar character. In the initial periods of aging in water, an increase in the indicator under consideration occurs. By the three-month period, the samples are completely saturated with water and no further significant changes in the mass content are observed. Of the examined compositions, the most intensive water absorption is characteristic of composites on ordinary water, which have the highest water-cement ratio.

Analyzing the data obtained on the change in the coefficient of water resistance, we can draw the following conclusions. When kept in water for up to 90 days, the strength of all the considered composites increases, which is associated with the ongoing process of cement hydration. After 90 days, the kinetics of change in strength acquires a different character. So, with materials on ordinary water and water treated with a magnetic field, as well as under the combined action of a magnetic field and electric current, strength continues to increase, while materials with water treated by an electric current and under the joint action of electric current and magnetic field do not is changing.

3. Conclusions

Thus, the combined effect of mixing electric and magnetic fields of varying intensity on water significantly affects the structure of the dispersed particles formed and their activity, which allows us to expect an active effect of the treated water on the processes providing the quality characteristics of the concrete products obtained. The use of this technology contributes to the improvement of their operational properties due to the formation of fine-grained, more dense structure of cement stone.

References

[1] Hang M Y, Lan Y J, Zhang B and Qu S Q 2012 Experimental study on high-performance water-reducing agent in C80 concrete *Appl. Mechanics and Materials* **193–194** 435–9 https://doi.org/10.4028/www.scientific.net/AMM.193-194.435

[2] Tuan N V, Ye G, Breugel K and Copuroglu O 2011 Hydration and microstructure of ultra-high performance concrete incorporating rice husk ash *Cement and Concrete Research* **41** 1104–11

[3] Travush V I, Karpenko N I, Erofeev V T, Cherkasov V D, Emelyanov D V and Erofeeva I V 2019 Impact strength of underground cement materials *MATEC Web of Conf.* **265** 01012 https://doi.org/10.1051/matecconf/201926501012

[4] Kalashnikov V I, Erofeev V T, Moroz M N, Troyanov I Yu, Volodin V M and Suzdaltsev O V 2014 Nanohydroxyapatite technology in the manufacture of concrete *Building Materials* **5** 88–91 (In Russian)

[5] Kalashnikov V I, Erofeev V T and Tarakanov O V 2016 Suspension-filled concrete mixtures for powder-activated concrete of new generation *News of Higher Educational Institutions. Construction* **4 (688)** 30–37 (In Russian)

[6] Kaprielov S S, Travush V I, Karpenko N I, Shejnfel`d A V, Kardumyan G S, Kiseleva Yu A and Prigozhenko O V 2008 Modified high-strength concrete classes C80 and C90 in monolithic structures *Building Materials* **3** 9–13 (In Russian)

[7] Bazhenov Yu M, Chernyshov E M and Korotkikh D N 2014 Designing the structures of modern concretes: defining principles and technological platforms *Building Materials* **3** 6–14 (In Russian)

[8] Ryou J S, Song H H, Kim S S and Lee J B 2013 New technology of blastfurnace slag activation with recycling water from ready-mixed concrete production *Cement, Wapno, Beton* **18** 33–38

[9] Dufka Á and Melichar T 2016 The development of materials based on alkali-activated fly ash designed for special applications *Materials Sci. Forum* **865** 114–9 https://doi.org/10.4028/www.scientific.net/MSF.865.114

[10] Boháčová J, Staněk S and Mec P 2014 Preparation and properties of pressed metakaolin and fly ash based alkali-activated binders *Advanced Materials Research* **897** 65–68 https://doi.org/10.4028/www.scientific.net/AMR.897.65

[11] Bulat A D, Filenkov V M, Obrubov V A and Lushkin I A 2018 On the effect of electric fields on cement solutions physico-mechanical properties *Materials Sci. Forum* **931** 614–7 https://doi.org/10.4028/www.scientific.net/MSF.931.614

[12] Erofeev V T, Mitina E A, Matvievskij A A, Osipov A K, Emelyanov D V and Yudin P V 2007 Composite building materials on activated mixing water *Building Materials* **11** 56−57 (In Russian)

[13] Fokin G A, Vilkova N G and Tetyusheva O V 2018 Influence of vortex dynamic field on the properties of building materials *Regional Architecture and Construction* **2 (35)** 36–44 (In Russian)

[14] Judina A and Verstov V 2013 On efficient use of electric treatment methods in the technology of concrete work *World Appl. Sci. J.* **23** 9–12

[15] Bazhenov Yu M, Fomichev V T, Erofeev V T, Fedosov S V, Matvievskij A A, Osipov A K, Emelyanov D V, Mitina E A and Yudin P V 2012 Theoretical justification of production of concrete based on the electrochemical and electromagnetobiology of mixing water *Internet-vestnik VolgGASU. Ser.: Politematicheskaya* **2 (22)** (In Russian) URL:http://vestnik.vgasu.ru/attachments/1_BazhenovFomichev-2012_2(22).pdf

[16] Fomichev V T, Erofeev V T, Emelyanov D V, Matvievskij A A and Mitina E A 2015 The role of anode products in the electromagnetic activation of water *Fundamental Study* **2 (6)** 1194–97 (In Russian)

[17] Patent RU74911U1 RF : IPC C02F 9/00 *Device for reagentless treatment of water systems* Matvievskij A A, Ovchinnikov V G and Chistyakov V Yu ; zayavitel' i patentoobladatel' zakrytoe akcionernoe obshchestvo "Maksmir-M"; publ. 20.07.2008 (In Russian)

[18] Erofeev V T, Mitina E A, Matvievskij A A, Emelyanov D V and Yudin P V 2008 Durability of cement composites on activated water *Industrial and Civil Engineering* **7** 51−53 (In Russian)

IOP Conf. Series: Materials Science and Engineering **972** (2020) 012050 doi:10.1088/1757-899X/972/1/012050

Dynamic Correction of the Construction Projects Scheduling Implementation under Stochastic Influences

M G Dobrosotskikh[1] ,V Ya Mishchenko[1], A V Mishchenko[1]

[1] Department of Technology, Organization of Construction, Examination and Management of the real estate, Voronezh State Technical University, 20-letija Oktjabrja str, 84, Voronezh, 394006, Russia

E-mail: dobrmax@yandex.ru

Abstract. The research results of this paper show that parameters of the catch-up schedule are determined by the dynamics of time-averaged volumes, rates and accelerations of construction jobs only when discrete nature of resources consumed has a weak influence on the rate of project jobs. On the other hand, if the speed of job is determined by the use of nostockable resources, the dynamics of jobs volume demonstrates the weak sensitivity both to the discrete nature of resources and to external stochastic influences. In this, the speed and acceleration of jobs experience jumps at the times of changes in the number of resources used. Thus, a time-averaging operation leads to the loss of information about these features of construction dynamics. In this paper the algorithms for stochastic factors accounting (characterized by different influences on the project implementation) are formulated. Moreover, the algorithm for forming the catch-up jobs schedule is obtained based on the results of monitoring of the jobs pace and averaging only over periods of a constant composition of a number of discrete resources. Furthermore, both empirical methods of separation of stochastic and deterministic impacts on the project implementation, as well as a method based on the qualitative difference in the Fourier spectra of stochastic and deterministic functions, have been developed. The limits of applicability of the empirical and exact methods are analyzed. It is shown that the Fourier analysis of the monitoring results allows to completely recover the deterministic information even against the background of the stochastic signal that coincides with it. It is shown that the proposed algorithm makes the formulation of catch-up jobs scheduling parameters possible, allowing to avoid financial and image losses associated with the untimely end of the project or its main parts.

1. Introduction

In the national [1] and global [2] practice of construction projects implementation the flow line method of scheduling is in widespread, which is determined by a number of inherent advantages. So, in particular, the flow line method ensures the even consumption of resources, the rhythm of finished product production, creating the favorable conditions for allied organizations jobs: contractors, factories, transport enterprises. However, the use of this method is limited by the need to organize equirhythmic or multirhythmic flows components, which is rarely achievable in practice [3]. Besides this, some difficulties were encountered in the practical use of flow line methods, which are determined by the contradictions between the hard labor specialization and the constant change in the ratio of volumes of various types of jobs, related to the dynamic change of construction objects [4].

The practical applicability of this approach is also limited by the instability of the flow processes with the respect to external stochastic influences [5, 6].

A significant drawback of the flow line scheduling method (which does not allow for system optimization) - is the iterative nature of accounting for resource constraints. In this formulation, at the first stage, the resources are considered unlimited and the required resources (in accordance with the project documentation) are calculated by the criterion of the project duration. Next, the construction scheduling is adjusted both in terms of jobs performance and in terms of resources attracted at each construction operation stage. At the same time, as indicated in [2], due to the large range of resources taken into account, the adjustment of this criterion is an extremely difficult task. Therefore, the real process of scheduling design is limited to solving problems taking into account the core resources separately. Correction of individual resources is carried out sequentially in decreasing order of resource importance and scarcity. The order of schedule correction for certain types of resources depends on the specific situation. Most often in construction practice the limiting resource is labor. The consistent approach leads to the situation in which the identification of the impossibility of resource balance maintenance at any scheduling stage forces to return to the previous steps of iteration [2]. Under the given resource constraints, the rate of convergence of the iterative procedure may not be sufficient for practical implementations or the procedure may even diverge.

2. Speed of project implementation

These and other shortcomings of traditional methods of scheduling of construction process stimulated the search for alternative methods of this problem solving [7.]. New prospects for construction scheduling improvement appeared in connection with the introduction of computer technology in the second half of the XXth century and the opportunity to apply methods of operations research (linear and dynamic programming, multicriteria optimization, game theory, etc.) [8]. The use of these methods made it possible to formulate and implement a simple computerized algorithm for describing the project implementation scheduling plan, which became known later as Critical Path Method (CPM) [9]. The development of this approach allowed to create the method of analyzes and assessment of schedules based on the optimization of the process logical scheme (Program (Project) Evaluation and Review Technique - PERT) [10]. Accounting for stochastic external influences, uncertainty and ambiguity of system dynamics is performed by the method of analysis and graphical evaluation (Graphical Evaluation and Review Technique - GERT) [11].

Despite some serious shortcomings inherent in the classical methods of scheduling, they have found the widespread application in world practice. For example, in US, the calendar plan is legally valid and only on its basis relationships between the customer and the contractor are conducted. Thus, the considerable attention is paid to scheduling; it is performed by the individual specialist (scheduler). In the USA scheduling has much more detailed character (than in our national practice) and is based on the widespread introduction of computing technology. [12]. However, the applied over there methods while describing and evaluating schedules do not allow to solve and even set the project optimization problem [13]. In domestic practice, computer technology in solving tasks of scheduling is mostly used in visualization of solutions obtained by empirical methods in the framework of standard software packages [14].

In paper [15] has been shown that the identification of negative trends in the process of project implementation at the early stages of their development is possible not only on the basis of the dynamic analysis of the ratio of planned and actual volumes in physical or price terms, but also in taking into account the planned and real speeds and accelerations of the project jobs (which is the prerequisite for effective and timely correction of project implementation process). Thus, only in this case it is possible to determine the parameters in time and implement a catch-up schedule of jobs, making it possible to minimize the additional project costs and optimize the updated scheduling plan.

The algorithm used in [15] is based on the levelling out the results of dynamic monitoring and on the assumption of a slow change in the averaged speed of the project implementation on the intervals between the monitoring points. This assumption, however, may turn out to be inadequate to the real

IOP Conf. Series: Materials Science and Engineering **972** (2020) 012050 doi:10.1088/1757-899X/972/1/012050

situation as long as that the speed of jobs is determined by a small number of nonstockable resources (hereinafter NR). In this case, the discrete nature of resources qualitatively changes both the scheduling process [16] and the course of the project plan implementation. So, for example, if the construction progress (or its stage) is mainly determined by the performance of NR (personnel, machinery, equipment, etc.), then the change between the monitoring points of the number of resources used from ***n*** to ***n+m*** will lead to the change of project implementation speed in ***(n+m)/n=1+m/n*** times. For the close to absolute values of ***n*** and ***m,*** the jump-like change will be quite significant and the assumption of smooth dependence of the velocity on time is incorrect. Consequently, the result of acceleration calculation will be inadequate to reality. The discrete nature of the NR and the associated stepwise use of power makes the definition of speed (as a derivative of work on time) impossible. Let's define the speed of project implementation at the moment t_i as

$$V(t_i) = \frac{\Delta F(t_i)}{\Delta t}. \quad (1)$$

Here $\Delta F(t_i)$ - the increment of jobs performed for the time period Δt, the value of which (hour, shift, day, week, etc.) is determined by the degree of design detailing and monitoring.

During the levelling out of the monitoring results the initial discrete information will be lost and the forecast of the jobs progress will be impossible. The purpose of this paper is the formulation of the algorithm for dynamic correction of the jobs scheduling performed under the conditions of stochastic influences, taking into account the discrete nature of NR.

The result of the project implementation (with the dominance of discrete resources) is influenced by the two types of stochastic factors. Firstly, by the changes in NR capacity, which is slowly changing during the construction process. For example, throughout the project the personnel qualification, its health condition or the technical condition of equipment may not be in compliance with regulatory requirements. Secondly, the project implementation is influenced by such time-dependent stochastic effects as weather factors, heterogeneity of stored resources, etc. Methods of accounting for these two classes of stochastic factors are different. The influence of these factors slowly varying in time should be dynamically calculated using the monitoring results of construction site and should be taken into account during the forecast update. Moreover, they should be averaged over monitoring intervals.

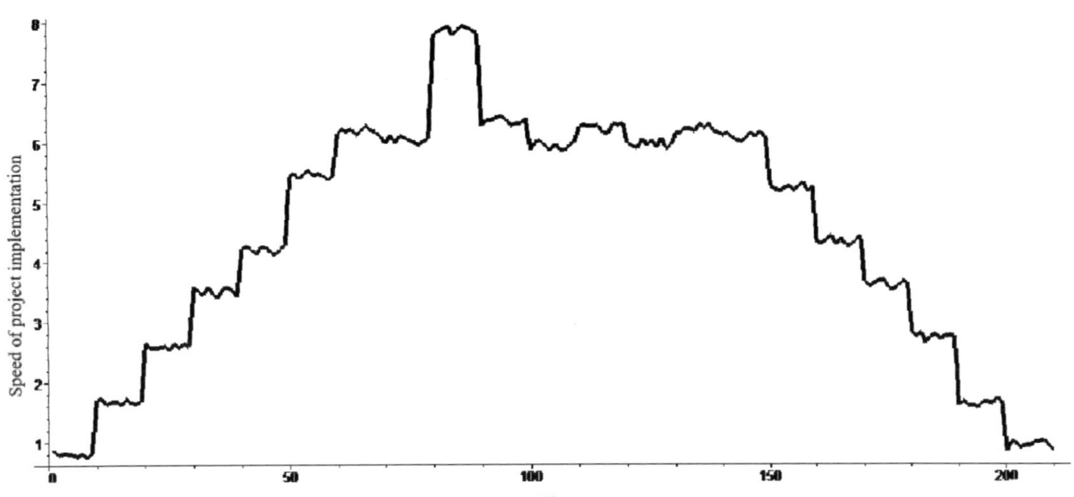

Figure 1. Speed-time relationship of project implementation for the 210 monitoring points.

Let's consider the algorithm for determining the time-independent parameters of NR (using a specific example) and forecasting on the base of the results obtained the progress of construction jobs. The dependence of the jobs speed from time can be seen in Figure 1.

The graph clearly shows the periods of sharp (much higher than random fluctuations) changes in speed associated with changes in the number of NR resources used. There are also periods (e.g. between monitoring points with numbers lying in the range from 130 to 150) in which the modulation fluctuations are close to the deterministic change in velocity. At the same time, both fluctuations and changes in the capacity of resources have practically no effect on the dependence of the total amount of jobs performed $F(t_i)$ on time

$$F(t_i)=\sum_{j=1}^{i}V(t_j)\cdot\Delta t=F(t_{i-1})+V(t_i)\cdot\Delta t \tag{2}$$

at the start value $F(t_0)=0$. Figure 2 shows the dependence $F(t_i)$ throughout the planned time of project implementation. The figure clearly shows the backlog of jobs from the scheduling plan (to the moment of planned project completion) due to the discrepancy between the NR capacities and their passport values.

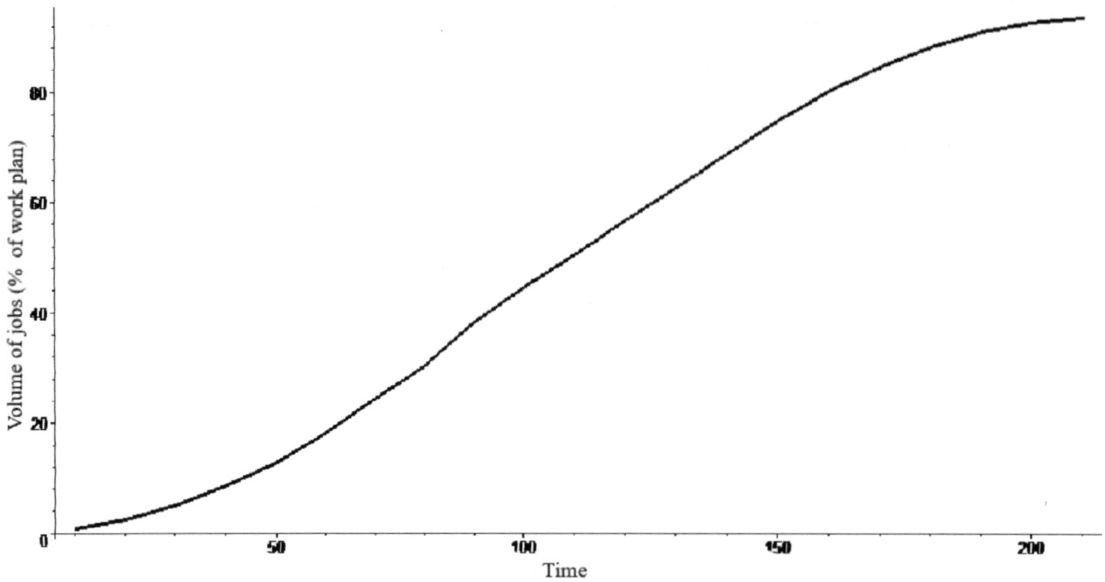

Figure 2. Dependence of the fulfilled jobs on time (in % from the total volume of project jobs) with increment Δt along the X-axis. On both axes variables take on discrete values, but in real scale the discrete nature of the dependence $F(t_i)$ does not affect the graph form.

The planned capacity of all NRs depends on their distribution in time $N_\kappa(t_i)$ and capacity of the separate resource p_κ, where the index defines the types of resources used (e.g. personnel, possessing the relevant competencies; machines or equipment, staffed by personnel, etc.). The assignment of distribution $N_\kappa(t_i)$ determines the resource component of the schedule. If the resource of one type (for example, the main equipment) basically determines the course of the project, the distribution $N_\kappa(t_i)$ determines the planned project speed $\tilde{V}(t_i)$. And this is just the case that will be considered further. The profile of this distribution (corresponding to the construction project considered), is graphically presented in Figure 3.

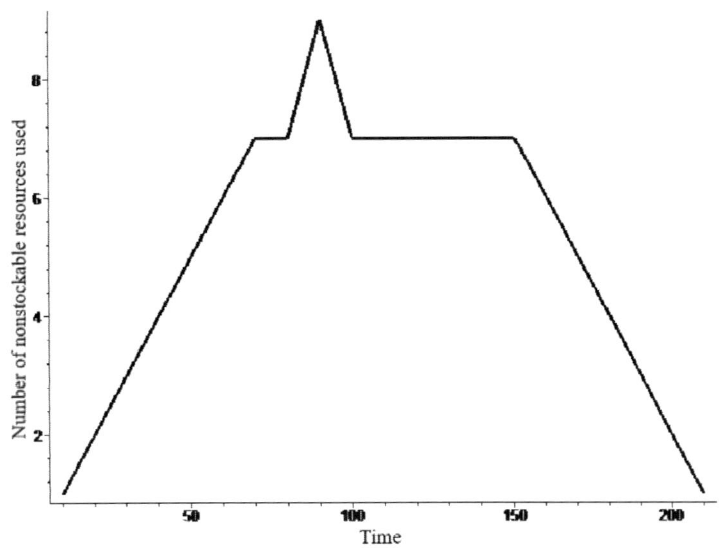

Figure 3. Plan for the use of nonstockable recourse in the course of project implementation.

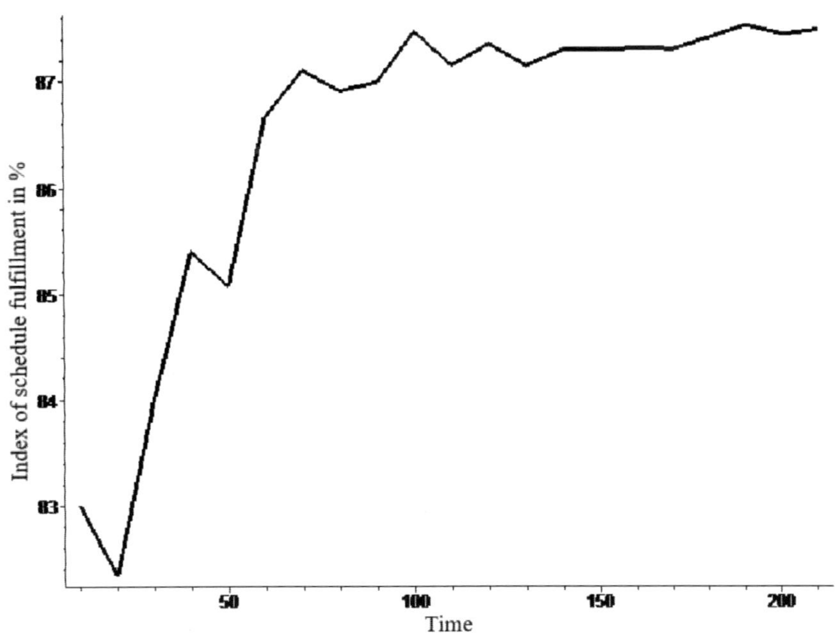

Figure 4. Dependence of the schedule fulfillment index on time. Here one can clearly see the constant backlog from the schedule, associated with the decrease in the capacity of NR in comparison with the design values at the initial stage of construction. At the final stage the backlog became more or less stable.

The schedule of work performed $\tilde{F}(t_i)$ is determined by formula (2) with the replacement of the actual speed of work with the planned one: $V(t_i) \to \tilde{V}(t_i)$

The dependence of schedule fulfillment on time is shown in Figure 4.

IOP Conf. Series: Materials Science and Engineering **972** (2020) 012050 doi:10.1088/1757-899X/972/1/012050

As can be seen from 'Figure 2' and 'Figure 4', the index of jobs fulfillment is much more sensitive to the discrete nature of NR use, than to the amount of jobs performed. The jumps on figure 4 are determined by the change in the amount of resource used and its performance. Monitoring results of the speed of project fulfillment allow to determine the number j_{max} and full set of points in time at which the change in NR composition occurred. Since at the points T_j the sharp change in the capacity of NR used is taking place, the average acceleration

$$a(t_j) = \frac{\Delta V(t_j)}{\Delta t}, \quad j = 1, 2, \dots, j_{max} \tag{3}$$

takes abnormally large values. This fact is reflected in Figure 5.

Abnormally large values of the average acceleration at the moments of NR composition change are evident. The cut-off zone with the width $2a_{max}$ corresponds to the speed fluctuations.

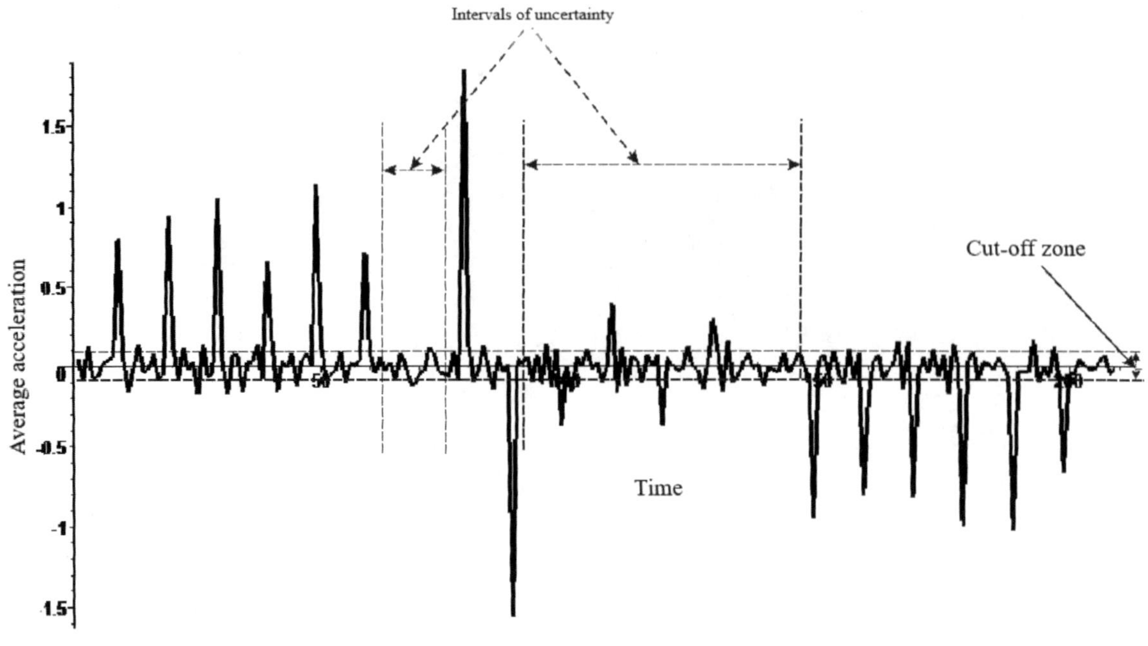

Figure 5. Dependence of the average acceleration $a(t_j)$ of the project implementation on time.

3. Analysis of project implementation

Analysis of the average acceleration on time indicates the presence of uncertainty intervals in which the extremums of average acceleration are present. The formation mechanism of these extremums cannot be clearly established: they can describe both the speed fluctuations and the deterministic processes 'Figure 5'. Within the framework of the model used, fluctuations cannot be clearly separated from the deterministic processes. However, the separation can be qualitatively performed by introducing the upper limit of the acceleration module a_{max} for stochastic processes [17]. The result of the contribution of such is shown in Figure 6.

The result of speed averaging in each interval of zero value $\tilde{a}(t_j)$ is shown in 'Figure 7'.

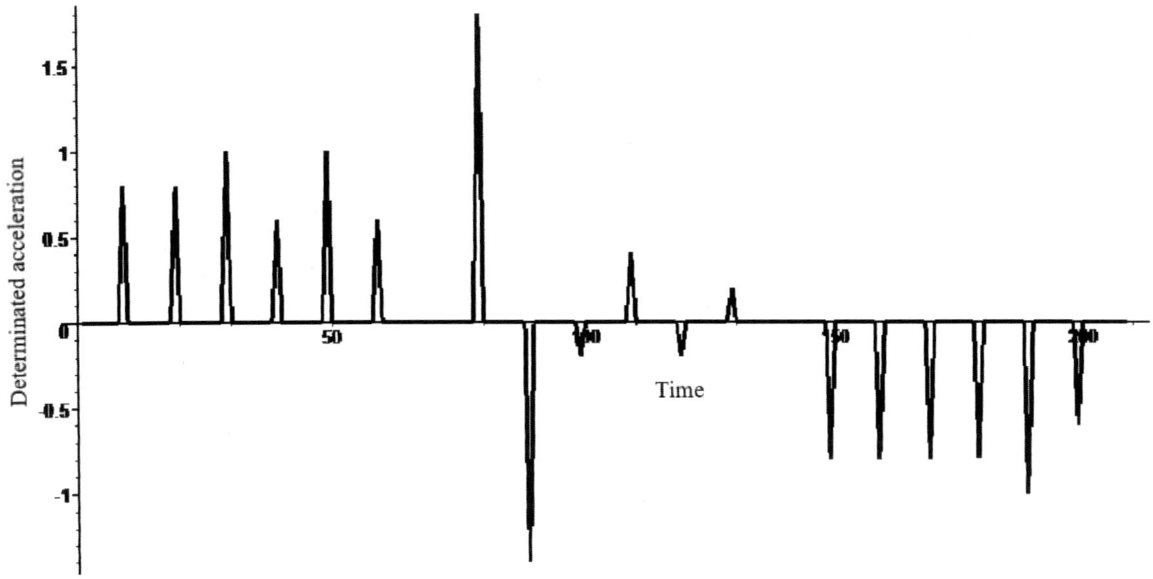

Figure 6. Dependence of the average acceleration $\tilde{a}(t_j)$ determined by the deterministic processes on time with the cut-off parameter $a_{max} = 0.2$.

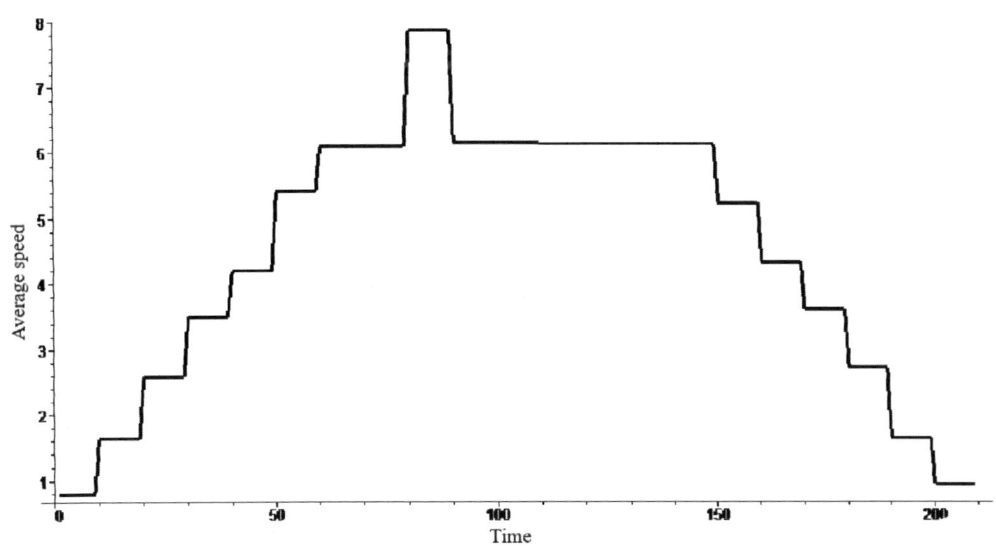

Figure 7. Dependence of average speed on time.

A comparison of 'Figure 1' and 'Figure 7' shows that in the intervals of uncertainty the monitoring performed cannot reliably define the impact of stochastic influences on the process of project implementation. This fact imposes restrictions on the accuracy of forecast of the project completion dates. However, the forecast can be refined by using the alternative method of separating a smooth trend from the stochastic changes, based on the global analysis of empirical array. The principle of separation of the investigated dependencies into deterministic and stochastic parts is based on the decomposition of studied function in a Fourier series [18]

$$V(t)=\sum_{m=1}^{\infty} b_i \sin(\pi\, m\, t)+\sum_{m=0}^{\infty} c_i \cos(\pi\, m\, t) \tag{4}$$

Analysis of the rate decomposition convergence (4) allows to solve the problem of extracting the stochastic information component. This algorithm is based on the qualitatively different behavior of the Fourier decomposition of dynamical and stochastic functions [19]. It allows not only to estimate, but also to isolate the deterministic part of the empirical data. Various modifications of this algorithm are widely used in many areas of science and technology [20]. If there is sufficient empirical information, it allows to select the deterministic part of the information even against the background of stochastic component that coincides with it in order [21]. As a result of using this algorithm, it is possible to fully recover deterministic information and adjust the resource profile on time. The result of the implementation of the catch-up schedule is depicted in Figure 9. The graph clearly shows that the increase in the term of NRs used (without changing their planned volume) allows to finish the construction project in the planned period, even if resources of non-standard capacity are available.

Figure 8. Profile of NR use in the process of catch-up schedule implementation.

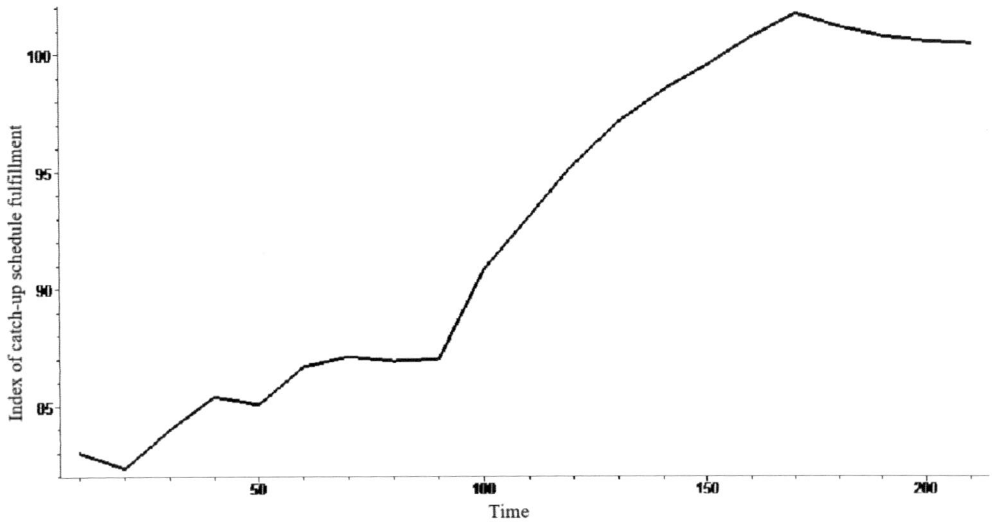

Figure 9. The result of catch-up schedule introduction in the first half of the planned period.

Conclusions

Dynamic adjustment of the schedule of construction projects implementation is highly effective in terms of cost and return. Such adjustment should be based on the jobs monitoring in physical or monetary terms. In the presence of external stochastic influences, the averaging of monitoring results over time is necessary. At the same time, only the timely determination of the parameters of catch-up schedule of jobs allows to minimize additional costs and optimize the updated schedule.

If the discrete nature of the resources consumed has little effect on the pace of project jobs, then the parameters of the catch-up schedule are determined by the dynamics of time averaged volumes, rates and accelerations of jobs. In this case, the time dependence of all these parameters is described by smooth functions. In contrast, if the speed of work is determined by the dynamics of nonstockable resources, the speed of work experiences jumps at times of changes in the amount of resources used, which causes a sharp change and derivatives at these points in time. At the same time, the dynamics of the project jobs volume demonstrates weak sensitivity to both the discrete nature of resources and external stochastic influences. In addition, a complete averaging over time leads to the loss of information about these features of construction dynamics.

The project implementation rate is influenced by two types of stochastic factors, which leads respectively to the quick and slow change in construction speed. The influence of factors rapidly changing in time should be averaged over monitoring intervals. In contrast, the contribution of slowly changing factors should be updated according to the results of construction process monitoring and should be taken into account in forecast creation. Due to these features, the algorithm for formation of the catch-up schedule of jobs cannot be based on the full averaging of monitoring results over time, not on the analyzing the volumes and accelerating the project process. The proposed in this paper algorithm is based on the results of monitoring the jobs pace and averaging them only over the periods of constant composition of discrete resources number.

External stochastic effects lead to the emergence of periods of sharp jobs pace changes (associated with changes in the amount of resources used), as well as to the periods in which the modules of fluctuations have the same order as the deterministic speeds change. The method of separation of stochastic and deterministic effects is based on the analysis of average (for the time between monitoring points) accelerations. The algorithm of extracting of the deterministic component can be based both on the empirical rule of the cut-off zone and on the qualitative difference in the Fourier spectrum of stochastic and deterministic functions. In the first approach the part of the original information is lost. The Fourier analysis of monitoring results allows to fully restore all the deterministic information even against the background of stochastic signal that coincides with it. In any case, the proposed algorithm allows us to formulate the parameters of the catch-up schedule of jobs, which allows to avoid some financial and image losses associated with the late completion of the project or its main parts.

References

[1] Dikman L G 2006 Organization of construction production (Moscow: ASV) [In Russ]
[2] Levy Sidney M 2007 Project management in construction (New York-Chicago –San Francisco: McGraw-Hill)
[3] Vasilev V M 2001 Management in construction (Moscow: ASV) [In Russ]
[4] Kievskiy L V 1987 Complexity and flow (Moscow: Stroizdat) [In Russ]
[5] Afanasiev A V 1982 Improving the organization and management in construction (Leningrad: LECI) pp 13–22 [In Russ]
[6] Gusakov A A 1994 Organizational and technological reliability (Moscow: SVR-Argus) [In Russ]
[7] Larichev O and Sternin M 1992 Knowledge-based approach for solving the multicriteria assignment problem (Models of problem solving. Arbeitspapiere der GMD 630 Sisyphus) ed Linster M
[8] Uskov V V 2016 Innovations in construction: organization and management (Moscow: Infra-Ingeneriya) [In Russ]
[9] Krüger, Wilfried 2006 Excellence in Change - Wege zur strategischen Erneuerung, 3 Auflage, (Wiesbaden: Gabler Verlag) pp 212-213
[10] Project Management Institute 2013 A Guide to the Project Management Body of Knowledge (5th ed.) (Project Management Institute)
[11] MacCrimmon K R Ryavec C A 1964 An Analytical Study of the PERT Assumtion (Opt. Res. vol 12 No 1) pp 16–38
[12] Dikman L G, Dikman D L 2004 Organization of construction in America (Moskow: ASV) [In Russ]
[13] Larichev O I, Pavlova L I and Osipova E A 1990 Problems and methods of making unique and repetitive decisions *Sbornik trudov* № **10** (Moscow: Institut Sistemnogo Analiza) pp 66–74 [In Russ]
[14] Uskov V V 2011 Computer technologies in the preparation and management of construction projects (Vologda: Infra-Ingeneriya) [In Russ]
[15] Mishchenko V YA, Dobrosotskikh M G, Elena El Earthbourne 2019 Optimization of the Construction Operation Scheduling Through Redeployment of Nonstockable Resources. *Nedvizhimost: ekonomika. Upravleniye* vol.**1** pp 83-87
[16] Mishchenko V YA, Dobrosotskikh M G 2016 NP Solvable task of scheduling of construction, reconstruction and repair of objects *Proceedings of higher educational institutions. Textile industry technology* vol. **6** (**366**) (Ivanovo: IVSPU) pp 13-20
[17] Smith L P 2003 Mathematical Methods for Scientists and Engineers (NewYork: Prentice Hall Inc. Englewood Cliff)
[18] Anders V 2003 Fourier Analysis and Its Applications. *Series: Graduate Texts in Mathematics* vol **223** (New York: Springer-Verlag) p 272
[19] Schoenberg I J 1998 Some Analytical Aspects of the Problem of Smoothing, Courant Anniversary Volute (New York: Interscience Publishers)
[20] Preobrazhenskii M, Rudakov O, Popova M and Tran Hai Dang 2017 *Journal of Science and Technology Natural science – engineering – technology* vol **169** **№9** pp 89-92
[21] Pierre Brémaud 2014 Fourier Analysis and Stochastic Processes (Springer International Publishing AG)

IOP Conf. Series: Materials Science and Engineering **972** (2020) 012051 doi:10.1088/1757-899X/972/1/012051

Fortification of building's foundations with eco-friendly materials as means to safer and more comfortable urban environment

L V Zakrevskaya[1], P A Lubin [1], I A Gandelsman [1], K A Andreeva[1]

[1]Department of Construction Industry, Vladimir State University, Gorkogo str., 87, Vladimir, 600000, Russia

E-mail: lvzak@mail.ru

Abstract. In this paper we tested mechanical aspects of chemical reactions as means to fortification of clay soil. We also tested properties of soil-concrete using waste of dolomite and chrysotile asbestos. The composition and properties of the composites are presented along with the comparison with traditional reinforced cementing materials. Main testing means of properties were method of x-ray phase analysis, electric scanning microscopy, soil strength changes and soil internal friction angle changes.

1. Introduction

Traditional methods of foundation fortification are based on the use of cement, lime and organic binders, bitumen is a perfect example. However, it's quite expensive, energy intensive and environmentally dangerous. Production of cement and lime are very energy consuming processes (firing temperatures can vary between 1100 and 1500°C) this accompanied by CO_2 emissions. [1-4].

We attempted to find untraditional fortification additives, which gathered from waste of mining sphere. In particular we looked into waste from dolomite and chrysotile field developments.

In recent years Russia experienced shortages of raw materials that are directly used for production of the building materials. Thus, production of building materials from industry waste has both economic and ecologic upsides.

This article aimed to develop the technology for material production from clay soil, which is suitable for foundation building and masonry material [5-8].

2. Materials and methods

Used soil composition is the following: reddish brown loam, brown loam with interbedded seams, coarse sand, with interlayers of sand, with the inclusion of gravel and rubble up to 7%, glacial.

Soil properties are shown in the table №1.

Table №2 contains comparison of physico-mechanical soil properties

Table 3 shows chemical composition of dolomite waste.

Waste of dolomite production have the following properties: finely dispersed powder with colour range from white to beige. Magnesium binding is created by firing of dolomite waste in the temperature range between 720 and 750°C. In this case, magnesium carbonate dissociates to form magnesium oxide $\rightarrow MgO + CO_2$. Magnesia binders include caustic magnesite (MgO) and caustic dolomite ($MgO + CaCO_3$) - these are finely dispersed powders, the active part of which is magnesium oxide. Calcium carbonate $CaCO_3$ (in dolomite) remains unchanged and is the ballast part of the binder [9-14].

Table 1. Normative and calculated soil property values.

Soil classification based on GOST 25100-2011	Normative values													Calculated values					
	Moisture										Shear resistance					Shear resistance			
	Natural W	At the fluidity stage, Wl	At the lamination stage, Wp	Plasticity index, Ip	Fluidity index, Il	Soil density, ρ, g/sm^3	Soil particle density, ρ, g/sm^3	Porosity coefficient, e	Water saturation coefficient, S_r	Filtration coefficient, K_φ, m/day	Internal friction angle, φ, degree	MAX friction c, MPa	Deformation modulus, E, MPa	Soil density, ρ g/sm^3		Internal friction angle φ degree		MAX friction,c,MPa	
														ρ_{II}	ρ_I	φ_{II}	φ_I	c_{II}	c_I
Stiff loam	0.175	0.261	0.142	0.119	0.28	2.07	2.71	0.536	0.99		15	0.0213	14	2.06	2.06	13	12	0.0154	0.0118

Table 2. Results of physico-mechanical soil properties.

Stiff loam Characteristic	Laboratory tests	Static Probe	Stamp	SP 22.13330.2011	Recommended Normative Values
Internal Friction angle, φ0	15	-	-	23	15
Deformation modulus, E, MPa	-	-	14	47	4
MAX Friction, c, MPa	0.0213	0.0258	-	0.0333	0.0213

Table 3. Chemical composition of dolomite waste.

SiO_2%	Al_2O_3 %	Fe_2O_3 %	FeO %	Ca0 %	MgO %	LOI %	K_20 %	TiO_2%
2.80	0.47	0.21	<0.05	36.17	15.64	44.22	0.16	0.03

The chemical composition of the chrysotile used is: SiO_2 41-43%, MgO 38-41%, Al_2O_3 1-1.5%, FeO and Fe_2O_3 0.3-4%, H_2O 13-14%. Chrysotile asbestos is an aqueous magnesium silicate in its mineralogical composition (3MgO $2SiO_2$ $2H_2O$) with a water content of up to 15%, and has the following dimensions: in cross section less than 0.0001 mm, in length from fractions of a millimeter to several centimeters). It has a very high tensile strength, which provides 20000 MPa, which is 2.57-2.66 g / cm^3.

Traditionally clay soils are stabilized with cement and lime, in which case lime is used to quench humus and create an alkaline environment as the most favorable for cement hardening.

In case of semi-fired dolomite need to use lime is eliminated as lime has an alkaline environment and is itself a type of magnesia cement.

The chemical postulate of "dissolving the similar in the similar" is illustrated in calculation of the modulus of the basicity of magnesian binder based on dolomite and clay soil. It is known that astringent properties appear in substances with Mos. > 1.5.

The modulus of basicity (hydraulicity) M_0.

$$M_0 = \frac{CaO + MgO + Na_2O + K_2O}{SiO_2 + Al_2O_3}, \qquad M_{active.} = \frac{Al_2O_3}{SiO_2}$$

Each binder is characterized by its unique modulus of basicity. The hydraulic properties of the binder are more pronounced with a decrease in M_0 and an increase in $M_{active.}$.

The modulus of basicity of a composite of clay soil and semi-fired dolomite and chrysotile is $M_0 = 1.9$, $M_{active.} = 0.36$, which characterizes it as a quite suitable binder [14-18].

Building composition "binder-soil" including binders (Portland cement) or semi-fired waste of dolomite, clay soil, to create water-resistant structures of phosphates and chrysotile (hydrosilicates) are shown in Table 4.

Table 4. Composition of fortified soils.

	GBD-1	GBD-2	GBD-3	GBD-4	GBD -5	GBC-1	GBC-2	GBC-3
Clay soil	63	63	63	63	63	83	85	90
Cement	-	-	-	-	-	17	15	10
Dolomite	20	18	16	14	12	-	-	-
Crysotile	10	12	14	16	18	-	-	-
KH_2PO_4	7	7	7	7	7	-	-	-

Tests of compressive and tensile strength of the synthesized material were carried out in accordance with DIN EN 206-1 / DIN 1045-2.

Compressive and tensile strength tests were performed on the 28th day.

3. Results and discussion

Comparativ Table 5 shows physico-mechanical properties of the fortified soils.

Table 5. Physico-mechanical properties of fortified soils.

	GBD-1	GBD-2	GBD-3	GBD-4	GBD -5	GBC-1	GBC-2	GBC-3
Compressive strength, MPa (28th day)	10.5	11.2	11.9	12.3	12.8	20	15	10
Water absorption, mass%	6	5.8	5.9	5.4	5.4	7.2	7.6	8
Frost resistance	16	17	20	20	21	17	14	7
Thermal conductivity, W/m^2K	0.55	0.59	0.65	0.71	0.72	0.78	0.70	0.64

Comparison of 2 soil samples before and after fortification is shown in the Figure 1.

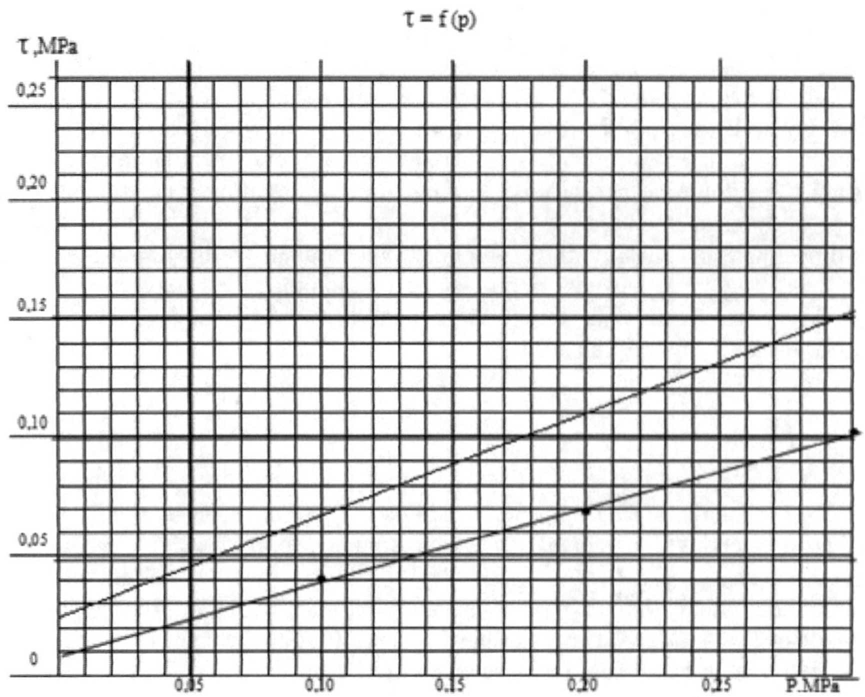

$\tau_{0,10}$	$\tau_{0,20}$	$\tau_{0,30}$
0,0425	0,0700	0,1025
$\varphi°$= 17 C = 0,012 MPa		

Figure 1. Sample tests before and after fortification

Comparison of the soil properties before and after fortification are shown in the Table 6

Table 6. Test results.

	ρ	ρ_σ	ρ_s	W	e	W_L	W_ρ	I_ρ	I_L	S_r	R_c, MPa
Before fortification	2.07	1.76	2.71	0.174	0.540	0.259	0.540	0.540	0.23	0.87	0.87
After fortification	2.2	2.05	2.83	0.07	0.260	-	0.260	0.260	-	-	10.5-12.8

The results of the study of the microstructure of fortified soil are presented in Figure 2, 3

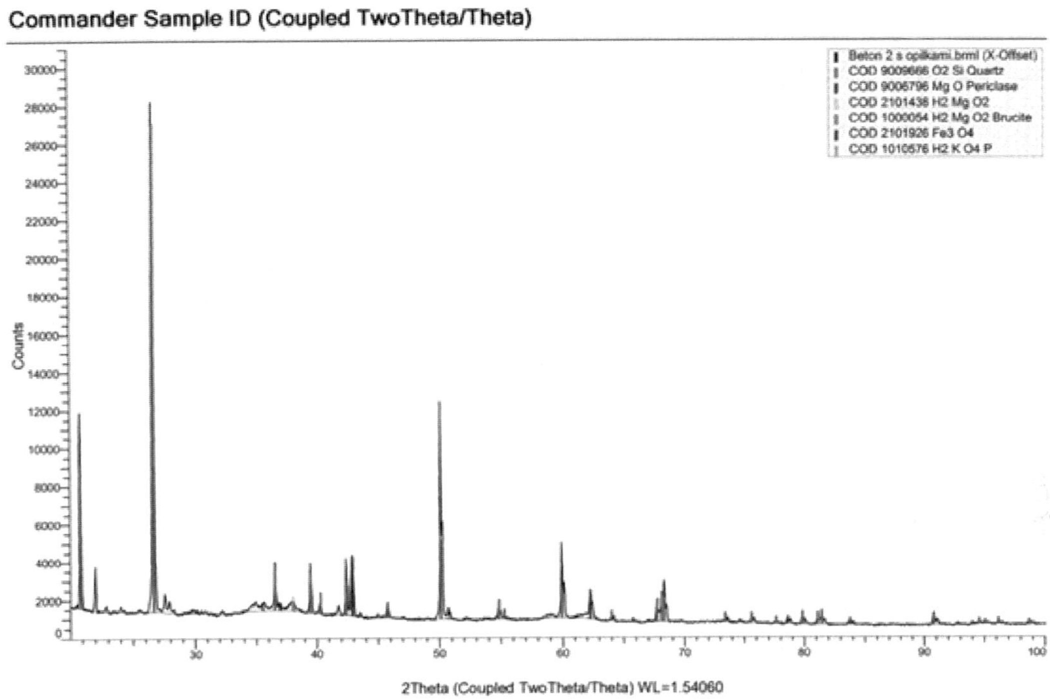

Figure 2. XRF of the fortified soil.

Figure 3. SEM of the fortified soil.

It is clearly seen from the 3rd figure that soil-concrete with dolomite and chrysotile contains strong brucite crystals, which provide high strength properties of this composite.

Wide variety of today's technological solutions to integration of fortified additives into the soil depends on the end usage of the newly obtained material. In case of interlayer fortification (thickness up to 40 cm) horizontal augers are used, which play role in mixing of soil with additives. In the next stage ЗАТВОРИТЕЛЬ is applied and the rolling of the fixed layer is performed. In the process of soil-pile creation, soil from the future pile body is extracted, then the extracted soil is mixed with additives and grouting fluid and then it's placed back in.

In the core of zonal pressing in technology "Russian swing" [19, 20] lies imitation of the effect of the natural process called “Flowing wedge” which is usually used for thixotropic soils capable of creating an extremely dense structure. Excess of moisture and air can be removed by means of special equipment to create and stabilize the achieved sealing.

Technology of “Russian swing” is in the core of this research and is vital in creation of fortified clayey foundations and development of technology for masonry materials from the same clay substances without the need for firing of these materials.

The advantages of the proposed technology include:

- Ability to work in weak water-saturated soils
- High economic performance due to the extensive use of mining waste
- High productivity of the method due to the rapid setting of the complex binder
- Low requirements for compliance with the regulated calcining temperature of dolomite (700°C), since if it is exceeded, the resulting lime does not interfere with the strengthening of the soil, as in the case of production of magnesia binder.

Due to semi-fired dolomite and phosphates, water absorption occurs and a change in the structure and properties of the soil leads to an increase in its carrying capacity due to the formation of hydroaluminates and aluminosilicates of calcium and magnesium.

4. Conclusion

The importance of this research lies in solve environmental problems:

- impervious soil screens for harmful chemicals;
- liberation of land from mining dumps. Of the total dolomite mining, only 30–40% is used in construction, the rest goes to dumps;
- utilization of soils which are left over after digging of pits.

Nanodisperse particles (1-100nm) of chrysotile and semi-fired dolomite possess high chemical activity towards the solid phase of dispersed mineral components of clayey soil. In the process of treating clayey soil with a fortifying composition, nanodispersed alumina, iron and alkali hydroxides are extracted from the crystal lattice of the clayey soil. This contributes to the synthesis of binders, accompanied by the synthesis of hydrosilicates, hydroaluminate hydroferrites [22].

References

[1] Consoli N C, Rotta G V, Prietto P D M 2000 *Influence of curing under stress on the triaxial response of cemented soil* Geotechnique **50(1)** 99–105.

[2] Kasama K, Ochiai H, Yasufuku N 2000 *On the stress–strain behaviour of lightly cemented clay based on an extended critical state concept* Soils Found **40(5)** 37–47.

[3] Miura N, Horpibulsuk S, Nagaraj T 2001 *Engineering behavior of cement stabilized clay at high water content* Soils Found **41(5)** 33–45.

[4] Horpibulsuk S, Miura N, Nagaraj T S 2005 *Clay–water/cement ratio identity of cement admixed soft clay* J Geotechn Geoenviron Eng, ASCE **131(2)** 187–92.

[5] Horpibulsuk S, Katkan W, Sirilerdwattana W, Rachan R 2006 *Strength development in cement stabilized low plasticity and coarse grained soils: laboratory and field study* Soils Found **46(3)** 351–66.

[6] Horpibulsuk S, Liu M D, Liyanapathirana D S, Suebsuk J 2010 *Behavior of cemented clay*

simulated via the theoretical framework of the Structured Cam Clay model Comput Geotechn **37** 1–9.
[7] Horpibulsuk S, Rachan R, Raksachon Y 2009 *Role of fly ash on strength and microstructure development in blended cement stabilized silty clay* Soils Found **49(1)** 85–98.
[8] Prakash K, Sridharan A 2004 *Free swell ratio and clay mineralogy of fine-grained soils* Geotechn Test J, ASTM **27(2)** 220–5.
[9] Gurtug Y, Sridharan A 2004 *Compaction behaviour and prediction of its characteristics of fine grained soils with particular reference to compaction energy* Soils Found;**44(5)**:27–36.
[10] Nagaraj T S, Lutenegger A J, Pandian N S, Manoj M 2006 *Rapid estimation of compaction parameters for field control* Geotechn Test J, ASTM **29(6)** 1–10.
[11] Horpibulsuk S, Katkan W, Apichatvullop A 2008 *An approach for assessment of compaction curves of fine-grained soils at various energies using a one point test* Soils Foundations **48(1)** 115–25.
[12] Ling T and Poon C 2013 *Use of phase change materials for thermal energy storage in concrete: an overview* Constr. Build. Mater. **46** 55–62
[13] Cao L, Tang Y and Fang G 2015 *Preparation and properties of shape-stabilized phase change materials based on fatty acid eutectics and cellulose composites for thermal energy storage Energy* **80** 98-103
[14] Sharma A, Tyagi V, Chen C and Buddhi D 2009 *Review on thermal energy storage with phase change materials and applications* Renew. Sustain. Energy Rev. **13** 318-345
[15] Fang Y, Wei H, Liang X, Wang S, Liu X, Gao X and Zhang Z 2016 *Preparation and thermal performance of silica/n-Tetradecane microencapsulated phase change material for cold energy storage* Energy Fuels **30** 9652-9657
[16] Horpibulsuk S, Kumpala A, Katkan W 2008 *A case history on underpinning for a distressed building on hard residual soil underneath non-uniform loose sand.* Soils and Found **48(2)** 267–86.
[17] Horpibulsuk S, Katkan W, Naramitkornburee A 2009 *Modified Ohio's curves: a rapid estimation of compaction curves for coarse- and fine-grained soils* Geotechn Test J, ASTM **32(1)** 64–75.
[18] Chindaprasirt P, Jaturapitakkul C, Sinsiri T 2005 *Effect of fly ash fineness on compressive strength and pore size of blended cement paste* Cem Concr Compos **27** 425–8.
[19] Xu B and Li Z 2013 *Paraffin/diatomite composite phase change material incorporated cement-based composite for thermal energy storage* Appl. Energy **105** 229–237
[20] Wang K S, Lin K L, Lee T Y, Tzeng B Y 2004 *The hydration characteristics when C2S is present in MSWI fly ash slag* Cem Concr Res **26** 323–30.
[21] Chew SH, Kamaruzzaman AHM, Lee F 2004 *Physicochemical and engineering behavior of cement treated clays* J Geotechn Geoenviron Eng, ASCE **130(7)** 696–706.
[22] Horpibulsuk S, Shibuya S, Fuenkajorn K, Katkan W 2007 *Assessment of engineering properties of Bangkok clay* Can Geotechn J **44(2)** 173–87.

IOP Conf. Series: Materials Science and Engineering **972** (2020) 012052 doi:10.1088/1757-899X/972/1/012052

Study of the Behavior of Cement Composites in the Conditions of Cyclic Exposure Positive and Negative Temperatures

I V Erofeeva[1], V V Afonin[2], D V Emelyanov[1]*, V A Fedortsov[1], V V Moiseev[1], N Yu Podzhivotov[3], M M Zotkina[1]

[1]Department of Building Materials and Technologies, National Research Ogarev Mordovia State University, 68 Bolshevistskaya Str., Saransk 430005, Russia

[2]Department of Automated Information Processing and Management Systems, National Research Ogarev Mordovia State University, 68 Bolshevistskaya Str., Saransk 430005, Russia

[3]Laboratory 30, All-Russian Scientific Research Institte of Aviation Materials, 17 Radio Str., Moscow 105055, Russia

*Email: emelyanoffdv@yandex.ru

Abstract. The paper presents the results of tests of cement composites under the conditions of cyclic exposure to negative and positive temperatures for 45 days. The stability factor (durability) compares the compositions with various fillers and additives. Resistance (long-term preservation of the structure and properties) of composites was estimated from the change in the relative hardness of the material, characterized by the area of the polygon, obtained as a result of a piecewise linear approximation of the exposure points of the test materials.

1. Introduction

One of the most important properties of building materials and, in particular, cement composites is durability under the influence of various operational factors. Quite a lot of work has been devoted to the study of the resistance of building composites to various environments (physical, chemical, biological), including works [1–21]. From literary sources, it follows that for composites with various additives in their composition, the values of resistance coefficients, as a rule, differ markedly.

2. Materials and methods

This article analyzes the results of exhibiting cement composites under the cyclic effects of positive and negative temperatures. As objects of research, cement composites of the following type were considered: cement stone based on a test of normal density (Composition No. 1); cement stone with high water content (Composition No. 2); normal density cement stone with a Melflux hyperplasticizer (Composition No. 3); Cement composite with Melflux hyperplasticizer filled with sand of fractions 0–0.63 and 0.63–2.5 mm with the addition of microquartz (Composition No. 4); Cement composite with Melflux hyperplasticizer filled with sand of fractions 0–0.63 and 0.63–2.5 mm with the addition of microquartz and silica fume (Composition No. 5); Cement composite with Melflux hyperplasticizer filled with sand of fractions 0–0.63 and 0.63–2.5 mm (Composition No. 6). Samples before testing hardness for several days hardened in normal-humidity conditions. Samples were tested according to the following mode.

1. Cooling samples from room temperature (+ 23 °C) to minus 50 °C for about an hour (50–55 minutes).
2. Exposure of samples at a temperature of minus 50 °C – 9 hours.
3. With the camera turned off, the natural heating of the samples to room temperature (+ 23 °C) is at least 5 hours.
4. Exposure of samples at room temperature (+ 23 °C) – 9 hours.

The coefficient of stability (durability) was established by changing the hardness on the surface. As is known, hardness is directly related to the strength index [11, 14, 17]. Composite materials initially have different hardness, because they contain components with initially sharply differing hardness, which does not allow to judge about their resistance to cyclic temperature variation in terms of absolute hardness. Determination of the hardness of composite porous materials has its own characteristics as compared with dense mono-materials. The hardness of porous materials is greatly influenced by the structure of their pore space, which is characterized by the value of the total porosity and the specific surface area of the pore space of the material. The resistance of composite materials to cyclic temperature changes (thermal resistance) compared to mono-material is also influenced by the difference in coefficients of linear temperature expansion of the material components, their Poisson's ratio and elastic modulus. The test points for the time changes in the properties of the composites were recorded after 15 and 45 days of testing. The starting point zero corresponds to the benchmark property. In this regard, all the results obtained are reduced to relative values of hardness relative to the control sample before the start of exposure. As in [16], the concept of a polygon is introduced in the paper, which is formed through piecewise linear interpolation of test data. In addition, assuming that the strength of the cement composites may decrease with the exposure time with cyclical changes in negative and positive temperatures, the assessment of such a change can be characterized by the ratio of the area of the polygon to the area of the rectangle. A rectangle characterizes the immutability of a particular property of cement composites. The explanation for the input ratio (K) of the ratio is shown in Figure 1.

$$K = \frac{S_{AEFCD}}{S_{ABCD}}$$

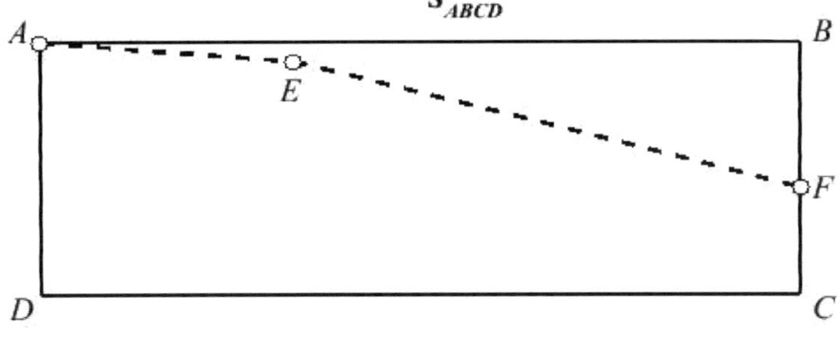

Figure 1. Example of calculating the ratio.

In Figure 1 points A, E, F correspond to the characteristic values of the tested property at the points of exposure. The area of the rectangle ABCD corresponds to an idealized composition, the properties of which do not change during exposure in any adverse conditions. The ratio is a dimensionless quantity, which remains unchanged both for the absolute values of the property under investigation and for their relative values.

For a larger number of composites under study, the composites can be sorted by the method given in [18], when either the area value or the ratio is taken as the key, and the name of the composite is taken as the value. The case of identical keys is also considered in [18].

IOP Conf. Series: Materials Science and Engineering **972** (2020) 012052 doi:10.1088/1757-899X/972/1/012052

3. Results and discussion

The control sample of the composite is designated as composition 1. Comparable samples will be designated as composition 2, composition 3 and elc. The studies were carried out in 4 stages: the results of the composition of cement stone with different water content and with the introduction of a hyperplasticizer were compared (stage No. 1); cement composites, including sandy fractions, as well as the addition of silica fume with stone flour (stage number 2); cement composites incorporating pigments (stage number 3); cement composites, including biocidal additives in their composition, as well as biocidal additives in combination with a hyperplasticizer (stage No. 4). The compositions and test results are given in Table 1.

Table 1. The investigated compositions of cement composites and their initial indicators of hardness.

№ compositi on	Initial components of the composite material	The amount of substance in relation to the cement content (the ratio of the number)	Hardness index (MPa)
1	Cement Ulyanovsk CEM I 42.5 N Water	1 0.267	4010.17
2	Cement Ulyanovsk CEM I 42.5 N Water	1 0.35	2065.24
3	Cement Ulyanovsk CEM I 42.5 N GP "Melflux 1641 F" Water	1 0.009 0.171	7016.08
4	Cement Ulyanovsk CEM I 42.5 N Ground sand (microquartz) Sand fr. 0-0.63 mm Sand fr. 0.63-2.5 mm GP "Melflux 1641 F" Water	1 1.1 2.753 2.347 0.009 0.6	5476.90
5	Cement Ulyanovsk CEM I 42.5 N Ground sand (microquartz) Silica fume Sand fr. 0-0.63 mm Sand fr. 0.63-2.5 mm GP "Melflux 1641 F" Water	1 0.75 0.1 1.775 1.975 0.009 0.475	9746.86
6	Cement Ulyanovsk CEM I 42.5 N Sand fr. 0-0.63 mm Sand fr. 0.63-2.5 mm GP "Melflux 1641 F" Water	1 2.065 1.76 0.009 0.525	4089.57
M3	Cement Ulyanovsk CEM I 42.5 N Activated water on mode (3-3)	1 0.26	5168.04
T3	Cement Ulyanovsk CEM I 42.5 N Biocidal additive "Teflex Universal" Water	1 0.03 0.267	1174.46
P1	Cement Ulyanovsk CEM I 42.5 N Pigment "iron oxide red" Water	1 0.0333 0.292	3600.11
P3	Cement Ulyanovsk CEM I 42.5 N Pigment "iron oxide yellow" Water	1 0.0571 0.331	3074.22

Consider first the composites, which are based on cement stone. The results in absolute and relative values of the coefficient of resistance with different water content are given in Table 2–3, which also includes the values of the areas of polygons and the coefficients of the ratio. The values of the ratio factors are enclosed in brackets. The diagram of the change in the hardness coefficient of the compositions 1, 2 is shown in Figure 1.

Table 2. The results of the exposure of cement composites of compositions 1, 2 at positive and negative temperatures.

Duration of exposure (day)	Hardness indicators (MPa)		Relative hardness values		Polygon area (ratio)	
	1	2	1	2	1	2
0	4010.17	2065.24	1.00000	1.00000	23.9767 (0.5328)	29.1196 (0.6471)
15	2037.07	1274.53	0.50798	0.61713		
45	1349.36	1064.85	0.33648	0.51561		

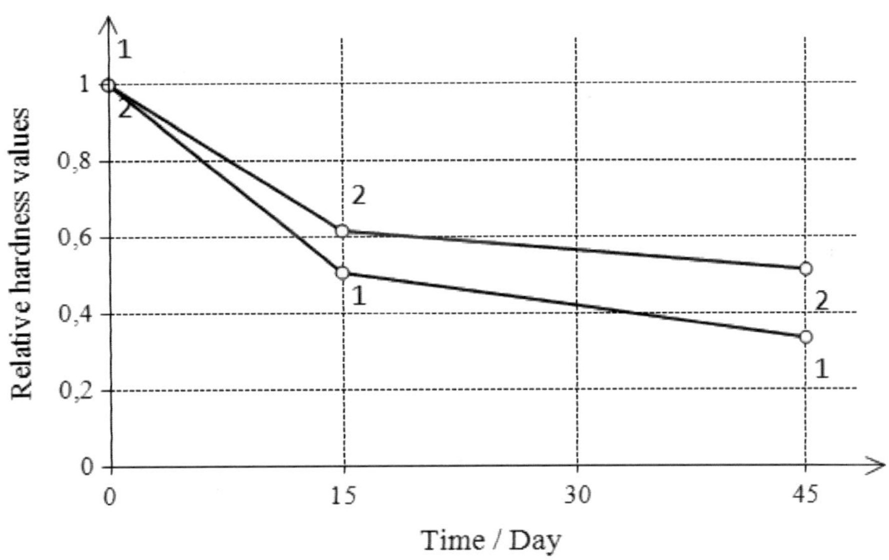

Figure 2. The change in the hardness coefficient of the compositions 1, 2 depending on the duration of exposure.

The thermal stability of porous composite materials, to which all studied compositions belong, essentially depends on their total porosity and average pore diameter, since they are hygroscopic and there is always moisture in their pore space under conditions of natural humidity. Freezing of this moisture leads to the formation of crystallization pressure of the formed ice, hydrostatic and hydrodynamic pressure of the moving fluid on the pore walls, which leads to the destruction of the structure of materials [17]. According to our data (see Table 1), the cement stone of composition 1 has, compared to stone 2, 1.94 times the initial hardness index, and judging by W / C, the total porosity is much less. At the same time, the more preferable initial indicators of the material did not lead to an increase in its durability under the conditions of cyclic exposure to positive temperatures.

It is known that the frost resistance of cement composites decreases with an increase in their porosity. However, in this case, the effect of a different type of cement composites takes place and an increase in the number of large pores is not yet a factor of a more intensive decrease in the hardness of the material.

In general, the following factors can affect the change in surface hardness under conditions of cyclic exposure to positive and negative temperatures: material homogeneity, coefficients of linear and volumetric expansion of components, structural components (total porosity, pore sizes and their nature). Since the first two factors in the cement stone of compositions 1 and 2 have no differences, it is obvious in this case that the intensity of the decrease in the hardness coefficients was affected by the difference in structural components.

From Figure 2 shows that the hardness indicator of cement stone composition 1, under the conditions of cyclic impact of positive and negative temperatures, decreases more intensively than the same characteristic of stone composition 2, although its water-cement ratio is 1.3 times greater, and, accordingly, more porosity than material of the first composition.

According to many studies, in these cases the pores of the stone will differ in their size and character. It can be assumed that a material with a smaller W / C has in its volume, including on the surface, more micropores, in which moisture will condense when the material is cooled.

Filling the pores with moisture followed by freezing will contribute to the reduction of surface hardness. Since the filling of pores with condensate is greater for a stone that has smaller pores, i.e. in this case, for a material of composition 1, then it should be expected to have a more pronounced loss of property index. In addition, the rate of decrease in the property index is influenced by its initial values [18]. The greater the initial indicator of the property, with equal resistance of the composites to aggressive media, the greater its decline.

The study of compositions 1, 3, M3 and T3 are given in Table. 3. Composition 3 with a cement kam-nem of self-compacting cement slurry including a hyperplasticizer. Composition M3 is a cement stone on the mixing water activated in the E + M mode (3-3). Composition T3 is a biostable composite with the additive Teflex Universal. The diagram with changes in hardness coefficients with time for compounds 1, 3, M3 and T3 is shown in Figure 3.

Table 3. The results of exposure of cement composites of compositions 1, 3, M3, T3 at positive and negative temperatures.

Duration of exposure (day)	Hardness indicators (MPa)				Relative hardness values				Polygon area (ratio)			
	1	3	M3	T3	1	3	M3	T3	1	3	M3	T3
0	4010.17	7016.08	5168.04	1174.46	1.00000	1.00000	1.00000	1.00000	(0.5328) 23.9767	(0.5282) 23.7697	(0.6714) 30.2148	(0.6602) 29.7082
15	2037.07	3629.11	3366.30	714.39	0.50798	0.51726	0.65137	0.60827				
45	1349.36	2166.32	2776.63	667.26	0.33648	0.30877	0.53727	0.56814				

The change in the relative value of the hardness ratio of composition 3 almost coincides with the change in the relative coefficient of hardness of composition 1. It is obvious that the resistance of cement stone obtained from the mixture with the lowest water-cement ratio (see part 3, Table 1) with a stone of composition 1. From the research results it follows that the use of activated mixing water (the technology is described in [20]) in obtaining a test of normal thickness contributed to an increase in hardness on the surface of the cement stone and its more High resistance to the environment. The introduction of a biocidal additive led to the same effect. However, compositions of such a composition have the lowest initial hardness (hardness index is 1174.46 MPa, see Table. 1) and the lowest property index after cyclic exposure to positive and negative temperatures (hardness index 667. 26 MPa, see Table 3). Since the composite has a water / cement ratio equal to that of composite 1, the porosity of the materials can be assumed to be equal. Then, with a uniform resistance of composites to aggressive effects, the smaller the initial property index, the slower its decrease [19]. The results of an experimental study applied to composites of compositions 1 and T3, shown in Figure 3, confirm this pattern.

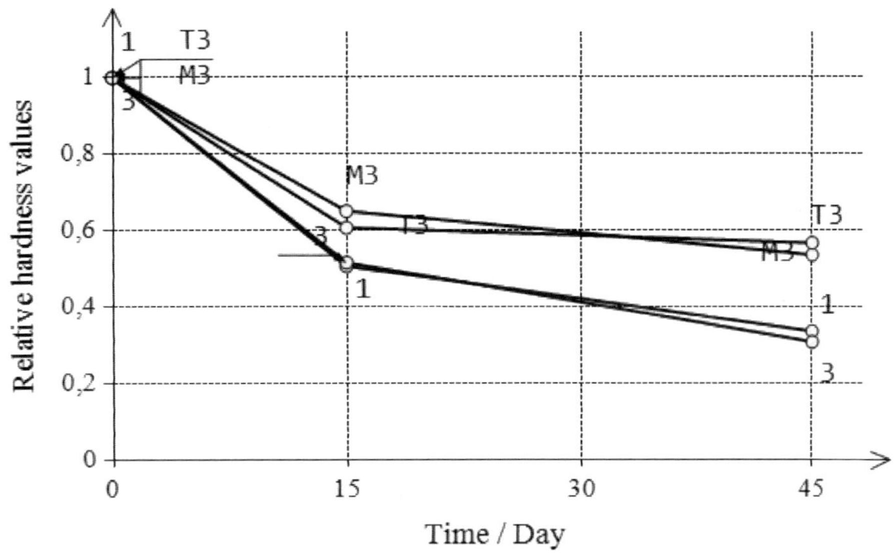

Figure 3. The change in the coefficient of hardness of compositions 1, 3, M3, T3 depending on the duration of exposure.

The study of compositions 1, 3, 4, 5 and 6 are given in Table 4. Composition 4 is a sandy be-tone of a new generation with HP and KM (stone flour) with low cement consumption. Composition 5 is a sandy concrete of the new generation with HP and CM and silica fume (MK). The technology for obtaining concrete of a new generation is given in [20, 21]. Composition 6 is a sandy concrete of transitional generation with GP. The diagram with changes in hardness coefficients over time for compounds 1, 3, 4, 5 and 6 is shown in Figure 4.

Table 4. The results of exposure of cement composites of compositions 1, 3, 4, 5, 6 with positive and negative temperatures.

Duration of exposure (day)	Hardness indicators (MPa)					Relative hardness values					Polygon area (ratio)				
	1	3	4	5	6	1	3	4	5	6	1	3	4	5	6
0	4010.17	7016.08	5476.90	9746.86	4089.57	1.00000	1.00000	1.00000	1.00000	1.00000	(0.5328) 23.9767	(0.5282) 23.7697	(0.7232) 32.5418	(0.4460) 20.0714	(0.7513) 33.8093
15	2037.07	3629.11	3966.10	3786.73	1893.79	0.50798	0.51726	0.72415	0.38851	0.46308					
45	1349.36	2166.32	3194.28	2488.71	4332.25	0.33648	0.30877	0.58323	0.25533	1.05934					

Of all the compositions of composites listed in Table 1 composite of composition 5 has the largest number of components and the highest initial hardness index (9746.86 MPa) and a low cement ratio (0.475) compared to other filled composites, i.e. according to the reasoning below, there are more intensive factors decrease in property, as evidenced by Figure 4. The final value of the hardness index is 0.255, which is the lowest value we have obtained in the experiment. At the same time, a high initial hardness coefficient indicates that the presence of finely ground components leads to filling the pore

IOP Conf. Series: Materials Science and Engineering **972** (2020) 012052 doi:10.1088/1757-899X/972/1/012052

space of the composite with their constituent substances and interaction products and reduces the average pore size, but does not lead to their total clogging, since the material's resistance to aggressive action is negligible.

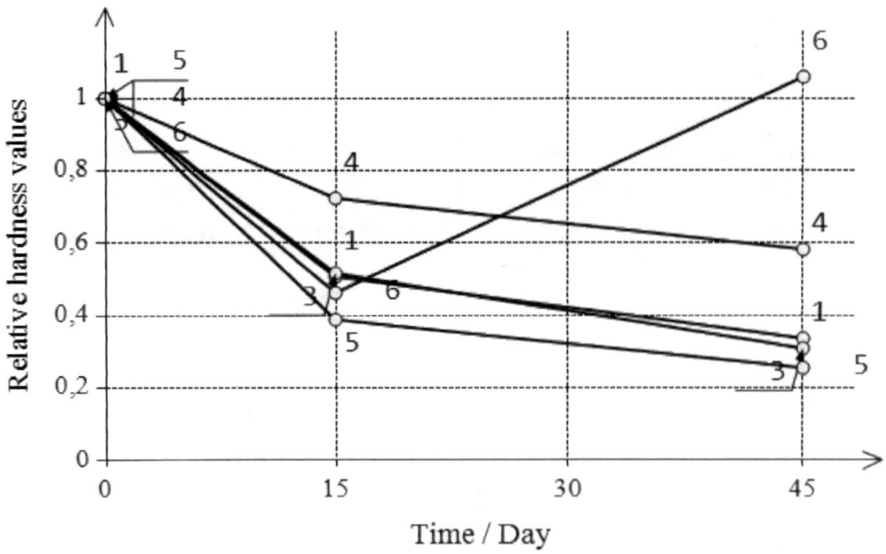

Figure 4. The change in the coefficient of hardness of compositions 1, 3, 4, 5, 6, depending on the duration of exposure.

The composite of composition 4 also contains a microfiller and can therefore be compacted. However, it was obtained from the mixture with the highest water-cement ratio in the experiment (w / c = 0.6). High water / cement ratio leads to a decrease in the number of micropores. In addition, the initial indicator of its hardness is about 1.8 times less than a similar indicator of composition 5. As a result, we have the characteristics of the material, leading to a slower decrease in the indicator of hardness in comparison with the composite of composition 5. The final value of the hardness coefficient is 0.58, which is significantly higher than the composite of composition 5, which is equal to 0.255. The sandy composite of composition 6 does not contain microfiller, its water-cement ratio is intermediate between the same indicators of composites of compositions 4 and 5. The initial microhardness index is about the same as that of the composite (see Table 1). It is obvious that the change in the hardness index of the composite during the cyclic effects of positive and negative temperatures should correspond to these factors. After 15 cycles of exposure to the environment is observed. The composite loses its hardness by about 54% (see Figure 4). However, as further testing is underway, a recovery and even a 6% excess of the initial hardness indicator occurs. Since the regularity of changes in the hardness of this composite under the conditions of cyclic exposure to positive and negative temperatures is an exception (see Table 2–5), it (the regularity) requires independent consideration and confirmation, if necessary, with additional studies. In any case, it is necessary to find out which factor after 15 cycles of exposure to a variable environment has affected the additional solidification of the composite.

The study of compounds 1, P1, P3 are given in Table 5. Composition P1 – decorative composite with iron oxide red pigment. Composition P3 is a decorative composite with the iron oxide yellow pigment. The diagram with changes in the hardness ratio of the compositions 1, P1, P3 is shown in Figure 5.

Table 5. The results of exposure of cement composites of compositions 1, P1, P3 with positive and negative temperatures.

Duration of exposure (day)	Hardness indicators (MPa)			Relative hardness values			Polygon area (ratio)		
	1	P1	P3	1	P1	P3	1	P1	P3
0	4010.17	3600.11	3074.22	1.00000	1.00000	1.00000	(0.5328) 23.9767	(0.9796) 44.0817	(1.0412) 46.8531
15	2037.07	4117.03	3630.86	0.50798	1.14358	1.18107			
45	1349.36	2604.36	2619.05	0.33648	0.72341	0.85194			

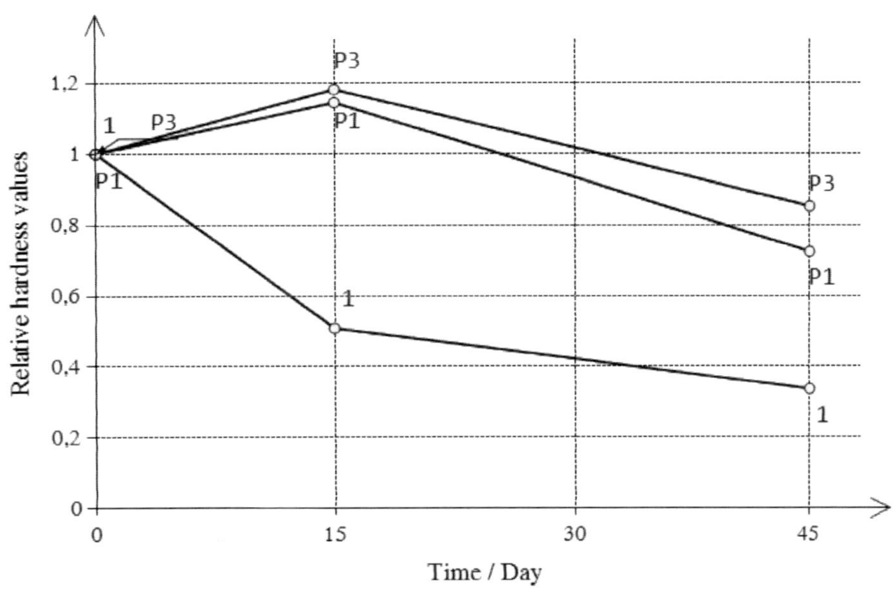

Figure 5. The change in the coefficient of hardness of the compositions 1, P1, P3, depending on the duration of exposure.

It is seen that composites of compositions P1 and P3 under the conditions of cyclic exposure of positive and negative temperatures for 45 days retain high hardness coefficients, which are respectively equal to 0.72 and 0.85 from the initial indicators of 3600.11 MPa and 3074.22 MPa. Water-cement ratios were 0.292 and 0.331. The indicators indicate that the change in the hardness of the composites with the impact should be slow. It can also be assumed that part of the pores that condense moisture when the composite is cooled will be clogged with pigments. It is also impossible to exclude the flow, along with destructive processes, of positive effects, which can lead to the temporary strengthening of composites [19]. In this regard, changes in the hardness indices of composites of compositions P1 and P3 are quite understandable.

4. Conclusions

The conducted studies confirm the regularity given in [18]. Under impacts, the greater the possible change in the characteristic of a property and the less resistance to this change, the greater it is in any suitable period of time, and vice versa. Studies have also established that homogeneous in composition and coefficients of linear and volumetric expansion of cement composites, in which during cooling as a result of condensation of moisture does not occur critical filling of micropores are more stable under the conditions of cyclic exposure to positive and negative temperatures. In addition, research suggests that using the areas of polygons that are under the lines of change under the influence of the characteristics of the property environment, you can estimate the remaining life of the durability of materials and products.

References

[1] Erofeev V T, Fedortsov A P, Karpushin S N, Voronov P V, Rodin A I and Boldyna I V 2017 Corrosion resistance of cement composites made on the basis of biocidal Portland cement with an active mineral additive *Fundamental Research* **4-2** 247–56 (In Russian)

[2] Erofeev V T, Smirnov V F, Lazarev A V, Bogatov A D, Kaznacheev S V, Rodin A I, Smirnova O N and Smirnov I V 2017 Biological and climatic resistance of polymer composites *Academia. Architecture and Construction* **1** 112–19 (In Russian)

[3] Erofeev V T, Afonin V V, Cherushova N V, Zotkina M M, Mitina E A, Zotkin V B and Erofeeva I V 2016 Methods and algorithms for assessing the quality of the surface of building products and structures *Fundamental Research* **4–1** 33–40 (In Russian)

[4] Erofeev V T, Smirnov I V, Voronov P V, Afonin V V, Kablov E N, Startsev O V, Startsev V O and Medvedev I M 2016 Study of the resistance of polymer coatings under the influence of climatic factors of the Black Sea coast *Fundamental Research* **11-5** 911–23 (In Russian)

[5] Smirnov I V, Voronov P V, Svetlov D A, Afonin V V and Kharitonov Ya A 2016 Climatic resistance of epoxy composites depending on the content of the hardener *In The Collection: Composite Construction Materials. Theory and Practice. Collection of Articles of the Int. Sci. and Technical Conf.* 82–84 (In Russian)

[6] Smirnov I V, Voronov P V, Afonin V V, Svetlov D A and Kharitonov Ya A 2016 Climatic resistance of epoxy composites depending on the content of the diluent – butyl alcohol *In The Collection: Composite Construction Materials. Theory and Practice. Collection of Articles of the Int. Sci. and Technical Conf.* 85–88 (In Russian)

[7] Erofeev V T, Rodin A I, Kalashnikov V I, Erofeeva I V and Smirnov V F 2016 Bio-resistance of decorative cement composites *Bulletin of the Volga Regional Branch of the Russian Academy of Architecture and Construction Sci.* **19** 304–8 (In Russian)

[8] Erofeev V T, Bogatov A D, Fedortsov A P and Pronkin S P 2015 Investigation of mechanisms of damage to bitumen composites under the influence of biological aggressive media *Fundamental Research* **2-13** 2787–800 (In Russian)

[9] Zotkina M M, Zotkin V B, Emelyanov D V, Zakharova E A, Cherushova N V, Erofeeva I V and Afonin V V 2015 Changes in the decorative properties of pigmented cement composites as a result of exposure to biological aggressive media *In the Collection: Actual Problems of Architecture and Construction, Materials of the Fourth Int. Scientific and Technical Conf.* 221–4 (In Russian)

[10] Turusov R A, Kuperman A M and Yakhontova E R 2014 Regular composite *Polymer Sci.* D **7** 9–13

[11] Bobryshev A N, Erofeev V T and Kozomazov V N 2012 *Physics and Synergy of Dispersion-Disordered Condensed Composite Systems* (SPb.: Science) 476 p (In Russian)

[12] Sutyagin V M and Lyapkov A A 2010 *Physico-chemical Methods for the Study of Polymers: Studies Manual* (Tomsk: Tomsk Polytechnic University Press) 140 p (In Russian)

[13] Turusov R A, Egorov V I, Schastlivtsev I V, Baranov A O and Naumenko V Yu 2009 The results of tests of tracheal samples and the type of connecting sutures *Mechanics of Composite Materials and Structures* **15** 289–305 (In Russian)

[14] Erofeev V T, Mitina E A, Matvievsky A A, Osipov A K, Emelyanov D V and Yudin P V 2007 Composite building materials on activated water for the construction *Building Materials* **11** 56–57 (In Russian)

[15] Erofeev V T, Afonin V V, Erofeeva I V, Merkulov D A, Smirnov I V and Myshkin A V 2017 The program for assessing the quality of polyester and other polymer composites on the physicomechanical properties and color Characteristics *Certificate of State Registration of a Computer Program* 2017660064 (In Russian)

[16] Erofeeva I, Afonin V, Fedortsov V, Emelyanov D, Podzhivotov N and Zotkina M 2017 Study of the behavior of cement composites in conditions of increased humidity and variable positive temperature *Int. J. for Computational Civil and Structural Engineering* **13** 66–81 https://doi.org/10.22337/2587-9618-2017-13-4-66-81.

[17] Dobshits L M 2017 Physico-mathematical model of the destruction of concrete with alternate freezing and thawing *Housing Construction* **12** 30–36 (In Russian)

[18] Fedortsov A P and Fedortsov V A 2013 On the potential of the material properties and its change *Regional Architecture and Construction* **13 (17)** 39–45 (In Russian)

[19] Fedortsov A P 2015 *Physico-chemical resistance of building composites and ways to increase it: monograph* (Saransk: Mordovian University Press) 464 p (In Russian)

[20] Kalashnikov V I and Erofeeva I V 2016 High-strength concretes of a new generation *Materials of the XII Int. Sci. and Practical Conf., Sci. without Borders* 82–84 (In Russian)

[21] Kalashnikov V I, Erofeev V T and Tarakanov O V 2016 Suspension-filled concrete mixes for powder-activated concrete of a new generation *Proc. of Higher Educational Institutions "Construction"* **4** 37–38 (In Russian)

Construction Materials with Tailings of Quartzite Enrichment

F L Kapustin[1], A A Ponomarenko[1], A M Gorohovskiy[2] and Z G Ponomarenko[2]
[1]Ural Federal University, 19 Mira street, Yekaterinburg 620002, Russia

[2]JSC "Pervouralskiy dinasoviy zabod" silica plant, 1 Il'iycha street, Pervouralsk 623103, Russia

E-mail: f.l.kapustin@urfu.ru

Abstract. The composition and properties of the tailings of quartzite enrichment of "Gora Karaul'naya" deposit, the possibility of its using as an aggregate in the composition of heavy and fine- aggregate concrete and dry construction mixture were determined. The introduction of quartzite tailings increases the water demand of the mixture, reduces the compressive strength of concrete and mortar, increases the bending strength compared to quartz sand.

1. Introduction

The quartzite of "Gora Karaul'naya" deposit is used in the production of Dinas refractories on JSC "Pervouralskiy dinasoviy zabod" silica plant. After crushing it is subjected to enrichment with getting standard fraction more than 5 mm and fine fraction of the raw material (enrichment tailings of current output) in the amount of 30-40 thousand tons per year are placed in slurry storage (sludge pond). To date, more than 0.6 million tons of quartzite tailings have been stored in the dump which pollute the natural environment.

The mineral composition of the tailings of quartzite enrichment (TQE) of the current output and TQE selected from the dump is practically similar and is represented mainly by minerals previously described in quartzites and host rocks of the deposit [1, 2]. Tailings are presented mainly by quartz and muscovite impurities in an amount of up to 2 %. As secondary impurities they contain hematite, magnetite, chlorite, shungite, zircon, rutile, iron hydroxides and pyrite with a total content of up to 1 %. Clay lumps in tailings is absent.

Previously performed studies have been established that taking into account the chemical and mineral composition and dispersion the tailings of ferruginous quartzite can be used as coarse and fine aggregate in the composition of structural concretes and backfilling mixtures, as a filler in the production of composite binders and colorful compositions, and iron-poor calcite tailings – in the production of dinas refractories and other materials [3-7]. However the possibility of JSC "Pervouralsky Dinas plant" TQE using as a part of construction concretes was not considered.

2. Raw materials and methods

In the selection of the structural concrete and mortars compositions the following materials were used: Portland cement CEM II-A/S 32,5 N (Russian standard 31108-2016) of JSC "Sukholozhskcement" production; crushed granite fractions 5-10 and 10-20 mm according to Russian standard 8267-93 with the crushability brand equal 1000; sand from screening of granite crushing according to Russian standard 31424-2010; TQE; acid fly ash of Reftinskaya GRES according to Technical Conditions

5717-004-79935691-2009 and superplasticizer SP-1 according to Technical Conditions 5870-005-58042865-2005 produced by JSC "PoliplastUralSib".

TQE is characterized by the following composition and properties:

- humidity is 3 %, the average density – 2640 kg/m^3; bulk density – 1510 kg/m^3, voidness – 42.8 %, the water demand – 10.1 %;
- the content of particles larger than 0.63 mm – 52,3 %, less than 0.16 mm – 15,0 %, quantity of grains of lamellar and needle forms – 24.3 %, quantity of dust and clay particles – 6.2 %; the fineness modulus – 2.8. The grain composition of the TQE meets the requirements of Russian standard 31424-2010 and it belongs to the II class of sands from the screenings of dense rocks crushing.

In the composition selection of the dry construction flooring mix the following materials are used: Portland cement CEM I 42,5 N (Russian standard 31108-2016) of JSC "Sukholozhskcement" production; fractionated quartz sand of 0.16-2.5 mm fraction; the ground granulated blast furnace slag (Technical Conditions 0799-001-99126491-2013) of JSC "Mechel-Materials" production; superplasticizer Melfux 1641 F of "BASF Constraction Polymers"; cellulose ether Mecellose PMS 50UF of "Samsung Fine Chemicals", antifoam Agitan P813 of "Munzing Chemie".

Humidity, grain composition, bulk and true density of TQE, crushed stone and fractionated sand were determined according to Russian standards 8269.0-97, 8735-88 and 51641-2000 respectively. Calculation and selection of compositions of fine-aggregate and ordinary concretes were carried out according to Russian standard 27006-86, and determination of their density – Russian standard 12730.1-78, concrete strength – Russian standard 10180-2012.

Concrete properties were compared with the requirements of Russian standard 26633-2012. The mobility of the concrete fine-aggregate mix was determined according to Russian standard 310.4-81, the strength of fine- aggregate concrete was determined on samples of 40x40x160 mm at the age of 28 days and after heat and humiditycuring at a temperature of 85 ° C carried out according to the regime: 3 hours heating to a predetermined temperature, 6 hours of isothermal aging and 3 hours of cooling The mortar properties were determined according to Russian standard 5802-86, they were compared with the requirements of Russian standards 31357-2007 and 28013-98.

3. Results and discussion

The composition of fine-aggregate concrete of B25 strength class with the addition of superplasticizer SP-1 in the amount of 1.0 % by weight of cement was calculated to test the possibility of TQE using as a aggregate. The mobility of the concrete mix was 14.7 cm. Composition and physico-mechanical properties of the concrete are presented in table 1. It is established that the actual strength of fine-aggregate concrete based on TQE with a coefficient of variation equal to 9 % (V_R=0.09) corresponds to the specified strength class of B25. The strength of concrete after steaming is 69.8 % of the strength of concrete of normal hardening after 28 days. Thus the TQE can be used as a aggregate in the composition of fine-aggregate concrete. Specific consumption of TQE depends on the concrete strength class, cement and water consumption and can be 1500-1800 kg/m^3.

Table 1. Composition and properties of fine-grained concrete based on quartzite tails.

Material consumption (kg/m^3)				W/C	Compressive strength (MPa) after		Concrete density (kg/m^3)
Cement	TQE	SP-1	Water		28 days	Heat and humidity treatment	
432	1626	4	237	0.54	30.5	21.3	2185

To assess the possibility of using TQE and fractionated quartzite sand (0.16-2.5 mm) on their basis as a fine aggregate in the composition of ordinary concrete raw materials and concrete mixture compositions used at the reinforced concrete plant of JSC "LSR.Stroitelstvo-Ural" (Ekaterinburg sity) were used. The mobility brand of concrete mix was P4. The compositions of concrete and the results of determining the properties of concrete mix and concrete are presented in tables 2 and 3 respectively.

Table 2. The composition of the concrete of B30 class strength.

Composition №	Material consumption (kg/m^3)								
	Cement	Crushed stone fraction (mm)		Sand			Fly ash	Dry SP-1	Water
		5-10	10-20	from screening of crushing granite	TQE	fractionated quartzite			
1	410	435	435	820	-	-	30	1.64	230
2	410	425	425	-	835	-	30	1.64	235
3	400	415	415	-	-	830	30	1.64	237

Table 3. Properties of concrete mix and concrete.

Composition № according Table 2	W/C	Mobility (cm)	Concrete mix density (kg/m^3)	Concrete density (kg/m^3)	Compressive strength (MPa)
1	0.56	21	2360	2335	27.3
2	0.57	20	2360	2307	27.7
3	0.59	20	2330	2332	23.8

It is established that the use of TQE and fractionated quartzite sand does not complicate the process of preparing a concrete mix but to preserve the mobility of the mix it is necessary a greater water consumption (on 5-7 l/m^3) compared to the sand from the granite crushing screenings. The density of the concrete mix with granite sand and TQE is the same and the use of fractionated quartzite sand reduces it by 30 kg/m^3. Concrete based on TQE and sand from the screenings have a close strength and the use of fractionated quartzite sand reduces it by 14.7 %, which may be due to lower consumption of cement (on 10 kg/m^3) and increased water demand of sand. Higher water consumption is associated with the presence in the quartz sand of a significant number of particles of plate and needle shape and the presence of dust-like clay particles on the surface of its grains.

Thus the TQE can be used as a fine aggregate in the ordinary concrete composition. Consumption of TQE in the concrete depending on the quality and consumption of large aggregate can be 750-850 kg/m^3. The use of fractionated quartz sand does not worsen the rheological properties of the concrete mix but reduces the compressive strength of concrete compared to concrete on sand from granite crushing screenings.

Verification of the ability to use fractionated quartz sand in the composition of dry construction mixtures was carried out on the example of the dry mixture for the floor screed in accordance with Russian standard 31358-2007. The control composition of the mix was prepared using fractionated quartz sand having the same grain composition with quartzite sand (table 4). The following composition of the dry mixture, %: 24.6 Portland cement CEM I 42.5 N; 73.9 sand, 1.29 granulated blast furnace slag; 0.085 superplasticizer Malfux 1641 F; 0.025 cellulose ether Mecellose PMS 50UF; 0.1 defoamer Agitan P813 is used.

The amount of water for the preparation of the mortar mix from the dry mix was selected at the same mobility of Pk2 (the actual cone immersion value is 4.2 cm). It is established that using of quartz sand the water-cement ratio was 0.43; quartzite sand – 0.66, i.e. the dry mix on its basis is characterized by increased water demand. The actual composition of the mortar based on quartz and quartzite sands is presented in table 5, the results of determining of the physical and mechanical properties of the mortar – in table 6.

Table 4. Granulometric composition of fractionated sand.

Sand	Quantity of fractions (%)			
	1.25-2.5	0.63-1.25	0.315-0.63	0.16-0.315
Quartzite Quartz	28.91	35.39	20.68	15.02

Table 5. Composition of mortar.

Composition №	Material consumption (kg/m^3)							
	Cement	Sand		Crushed blast furnace slag	Superplasticizer Malfux 1641 F	Cellulose ester Mecellose PMS 50UF	Defoamer Agitan P813	Water
		quartz	quartzite					
1	520	1570	-	6	0.44	0.130	0.52	225
2	500	-	1500	6	0.425	0.125	0.50	330

Table 6. Physical and mechanical properties of mortar.

Composition № according Table 5	Density (kg/m^3)		Bending / compressive strength (MPa) after (days)	
	mix	mortar	3	28
1	2322	2243	5.6 / 39.4	7.4 / 52.5
2	2337	2209	6.1 / 33.7	8.7 / 48.8

It was found that in the early stages of hardening in air-wet conditions the mortar on quartz sand is characterized by lower bending strength (8.9 %) and greater compressive strength (16.9 %) compared with the mortar on quartzite sand. Further hardening up to 28 days retains the predominant bending strength and reduced compressive strength of the mortar on quartzite sand. In this case the difference in compressive strength of the mortar on quartz sand compared with quartzite does not exceed 7.6 %. The investigated composition of flooring dry mix on the quartz sand has a class on the compressive strength of B40, bending class – Btb5,2, and on quartzite sand – Btb5 and B35,2 respectively. Thus quartzite fractionated sand is recommended to be used in the composition of dry construction mixes for which the bending strength and crack resistance is of primary importance for example for the manufacture of flooring concrete or mortar coatings, plaster mortars, etc.

4. Conclusion

The tailings of quartzite enrichment of JSC "Pervouralsky dinas plant" can be used as a aggregate in composition of heavy and fine-aggregate concrete, dry construction mix for a flooring covering. However the introduction of TQE reduces the compressive strength of concrete and mortar and increases the bending strength compared to quartz sand.

References

[1] Perepelitsyn V A, Karpets L A, Rechneva L A and all 1997 Mineral composition of the quartzites and enclosing rocks of "Gora Karaul'naya" quartzite deposit. *Refractories and technical ceramics* **5** 27-32

[2] Perepelitsyn V A, Karpets L A, Kormina I V and etc. 2008 Geochemical features of the quartzites of "Gora Karaul'naya" quartzite deposit. *New Refractories* **3** 103-106

[3] Lesovik R V and Bazhenov Yu M 2013 Fine-aggregate concretes based on composite binders and technogenic sands, (Belgorod: Publishing house of Technical university)

[4] Dvorkin L I and Dvorkin O L 2007 Building materials from industrial waste, (Rostov-on-Don: Phoenix press)

[5] Lesovik R V 2004 Comprehensive use of tailings of wet magnetic separation of ferruginous quartzite. *Mining J* **1** 76-77

[6] Tarasova G I, Sverguzova S V and Starostina I V 2014 The use of ferruginous quartzite enrichment tailings for the production of oil and silicate paints, (Belgorod: Publishing house of Technical university)

[7] Perepelitsyn V A, Kapustin F L, Ponomarenko A A, Rechneva L A and Kolobov A Y 2017 Secondary Mineral Resources for Refractory Manufacture. Part 1. Silica Technogenic Materials. *Refractories and Industrial Ceramics*. **58(3)** 259-268.

IOP Conf. Series: Materials Science and Engineering **972** (2020) 012054 doi:10.1088/1757-899X/972/1/012054

Creep mathematical model on the example of early age concrete

Y E Yakubovskiy[1], V I Kolosov[1], I A Donkova[1] and S O Kruglov[1]

[1]Department of Applied Mechanics, Industrial University of Tyumen, Lunacharskogo st., 2, Tyumen, 625001, Russia

Email: jakubovskijje@tyuiu.ru

Abstract. The objective of the study is development of a creep kernel recording form, which allows obtaining representations for creep curves calculation. Based on the elastic-creeping body theory a possibility of the high-rate creep movement analytical study has been shown. A creep kernel, which contains a formula to describe the aging material properties, has been built. Essential stages of the proposed creep kernel formation have been given. The correspondence of the proposed calculations to real processes has been proved by a comparison with experimental data. In the numerical implementation, decomposition of integration elements into power series has been used. A possibility of calculating functions for a long time interval has been shown. Creep kernel parameters have been determined on the basis of experimental data of early age concrete samples by minimizing the standard deviation of theoretical and empirical values.

Keywords: composite structures, creep kernel, viscoelastic properties, aging material.

1. Introduction

The modern stage of the industry development is characterized by the improvement of technical parameters of building constructions, increasing their reliability and durability. Under market conditions, there is a fast substitution of structural materials. This is due to a reduction of material consumption while increasing strength, durability and efficiency. The solution to the problem of designing facilities is to improve general structural schemes and their calculation methods, as well as to develop effective structural elements and to use new high-strength materials. One of the types of modern efficient structures is composite shells, plates and beams [1, 2]. Individual elements of such structures are made of different materials: polymer, composite, etc. Concrete is used as the cheapest material. Connection between the layers of composite plates can be provided with adhesive joint or anchors [3, 4].

Practice shows that plastic deformations and creep occur when using composite systems. In viscoelastic materials deformation increases (accumulates) in time under a constant load. This leads to a qualitative and significantly quantitative change in the stress-strain state (SSS): stiffness reduction, redistribution of forces between structural elements, etc. In contrast to elasticity problem solutions [5, 6], where the SSS only depends on spatial coordinates, the viscoelastic material SSS components change in time [5, 7-9]. To describe the above phenomena building special physical and mathematical models and methods is required.

In the article a notation of the relationships between the aging material stresses and strains has been proposed. On the basis of the elastic-creeping body theory a possibility of a high-rate creep movement analytical description has been shown. Using the Arutyunyan - Zevin method [9, 10] new creep kernels with respect to aging material have been built. Kernel numerical implementation has been considered on the example of experimental A. A. Ross data [2]. Error estimation has been given.

2. Materials and Methods

It is assumed that structure material has creeping and aging properties. Then, in the linear theory of the elastic - creeping body at a uniaxial stress state, the relationship between stresses $\sigma(t)$ and strains $\varepsilon(t)$ in such a structure is as follows [2]:

$$\varepsilon(t) = \frac{\sigma(t)}{E(t)} + \frac{1}{E(t)} \int_{\tau_0}^{t} K(t,\tau)\sigma(\tau)d\tau\ , \tag{1}$$

where $\varepsilon(t)$ – strain components at an instant of time t, $\sigma(t)$ – stress components at an instant of time t, $E(t)$ – modulus of elasticity of instantaneous strain; $K(t,\tau)$ – creep kernel, t – an instant of time under consideration, τ – an intermediate instant of time, τ_0 – an instant of stress application $\sigma(t)$.

The objective of the study is development of a creep kernel recording form, which allows obtaining "formulization" for creep curves calculation. Using method [2], new kernels for a description of aging material creep have been determined. The possibility of calculating the obtained functions using power series for a sufficiently long time interval has been shown [2, 8].

To describe the relations between stresses and strains in accordance with the Arutyunyan - Zevin method formation of new kernels has been carried out taking into account material aging [2, 10, 11]:

$$K_1(t,\tau) = Q(t,\tau) + B(t-\tau) + \int_{\tau}^{t} Q(s)B(s-\tau)ds\,;$$
$$K_2(t,\tau) = Q(t,\tau) + B(t-\tau) + Q(\tau)\int_{\tau}^{t} B(t-s)ds\,. \tag{2}$$

In formulas (2) the following notations have been adopted: $K_1(t,\tau), K_2(t,\tau)$ – aging material creep kernels, $Q(t,\tau)$ – regular part of the creep kernel; $B(t-\tau)$ – differential part of the creep kernel, $t-\tau$ – difference argument, t – an instant of time under consideration, τ – an intermediate instant of time, s – intermediate argument.

As a difference component, a weakly singular kernel meeting all the necessary requirements is used [2, 12]:

$$B(t-\tau) = \chi\, e^{-\rho(t-\tau)^{\alpha}}\,(t-\tau)^{\vartheta-1}\,\frac{1}{\Gamma(\vartheta)}, \tag{3}$$

where $B(t-\tau)$ – differential part of the creep kernel, $\Gamma(\vartheta)$ – gamma-function, χ, ρ, ϑ – kernel characteristics, $\chi > 0$; $\rho > 0$; $\vartheta \in (0;1)$; $\alpha \in (0;1)$, $t-\tau$ – difference argument, t – an instant of time under consideration, τ – an intermediate instant of time.

The introduction of α parameter extends the possibility of calculating the function using power series in a longer time interval in comparison with the variant when $\alpha = 1$ (the Rzhanitsyn kernel) [12].

As a regular component, the kernel, based on the creep measure of the following form is used [2, 11]:

$$Q(t,\tau) = -\frac{\partial}{\partial\tau}\left[C_0 + Ae^{-\beta\tau}\right]\left[1 - e^{-\gamma(t-\tau)}\right] = \sum_{k=1}^{3} a_k e^{-\gamma_k t} e^{\alpha_k \tau},$$
$$a_1 = \beta A,\quad a_2 = C_0\gamma,\quad a_3 = A(\gamma - \beta),$$
$$\alpha_1 = -\beta,\quad \alpha_2 = \gamma,\quad \alpha_3 = \gamma - \beta,\quad \gamma_1 = 0,\quad \gamma_2 = \gamma_3 = \gamma \tag{4}$$

where $Q(t,\tau)$ – regular part of the creep kernel, $t-\tau$ – difference argument, t – an instant of time under consideration, τ – an intermediate instant of time.
According to paper [10], the resolvent for the regular kernel has the form:

$$G(t,\tau) = Q(t,\tau)e^{\left(-\int_\tau^t Q(t,s)ds\right)} = \sum_{k=1}^{3} a_k e^{\alpha_k\tau - \gamma_k t} e^{-\left[C_0 + Ae^{-\beta\tau}\right]\left[1 - e^{-\gamma(t-\tau)}\right]}. \tag{5}$$

Here $G(t,\tau)$ – regular kernel resolvent, $Q(t,\tau)$ – regular part of the creep kernel, $t-\tau$ – difference argument, t – an instant of time under consideration, τ – an intermediate instant of time, s – intermediate argument.

Substituting formulas (3) and (4) into the formulas for the kernel (2), we obtain:

$$K_r(t,\tau) = Q(t,\tau) + B(t-\tau) + \frac{\chi}{\Gamma(\vartheta)}\sum_{k=1}^{3}\left[a_k\, e^{\alpha_k\tau - \gamma_k t}\int_0^{t-\tau} e^{\eta_{rk}z - \rho z^\alpha} z^{\vartheta-1}dz\right., \tag{6}$$

where $r = 1, 2$; $\eta_{1k} = \alpha_k$; $\eta_{2k} = \gamma_k$, a_k, χ – required parameters, $Q(t,\tau)$ – regular part of the creep kernel, $B(t-\tau)$ – differential part of the creep kernel, $\Gamma(\vartheta)$ – gamma-function, $t-\tau$ – difference argument, t – an instant of time under consideration, τ – an intermediate instant of time, z – intermediate argument.
As a result of integrating formula (6) we obtain:

$$\begin{aligned}
V_r(t,\tau) &= \int_\tau^t K_r(t,s)ds = -\frac{\partial}{\partial s}\int_\tau^t [C_0 + Ae^{-\beta s}][1 - e^{-\gamma(t-s)}]ds + \\
&+ \frac{\chi}{\Gamma(\vartheta)}\int_\tau^t e^{-\rho(t-s)^\alpha}(t-s)^{\vartheta-1}ds + \\
&+ \frac{\chi}{\Gamma(\vartheta)}\sum_{k=1}^{3} a_k e^{-\gamma_k t}\int_\tau^t [e^{\alpha_k s}\int_0^{t-s} e^{\eta_{rk}z - \rho z^\alpha} z^{\vartheta-1}dz]ds = \\
&= \left[C_0 + Ae^{-\beta\tau}\right]\left[1 - e^{-\gamma(t-\tau)}\right] + + \frac{\chi}{\Gamma(\vartheta)}\left(\int_0^{t-\tau} e^{-\rho z^\alpha} z^{\vartheta-1}dz + \right. \\
&+ \sum_{k=1}^{3}\frac{a_k}{\alpha_k}e^{-\gamma_k t}[e^{\alpha_k t}\int_0^{t-\tau} e^{\aleph_{rk}z - \rho z^\alpha} z^{\vartheta-1}dz - -e^{\alpha_k\tau}\int_0^{t-\tau} e^{\eta_{rk}z - \rho z^\alpha} z^{\vartheta-1}dz]),
\end{aligned} \tag{7}$$
$$\aleph_{1k} = 0,\quad \aleph_{2k} = \gamma_k - \alpha_k,$$

where $V_r(t,\tau)$ – creep kernel integral, $\Gamma(\vartheta)$ – gamma-function, $K_r(t,s)$ – aging material creep kernels (r = 1, 2), $t-\tau$ – difference argument, t – an instant of time under consideration, τ – an intermediate instant of time, z – intermediate argument.

3. Results

For numerical implementation of (4) the exponential functions in integration elements (7) are expanded in the Maclaurin power series. When keeping the first three terms of the series, the following result is obtained:

$$\begin{gathered} V_r(t,\tau) = \left[C_0 + Ae^{-\beta\tau}\right]\left[1 - e^{-\gamma(t-\tau)}\right] + \frac{\chi}{\Gamma(\vartheta)}(t-\tau)^{\vartheta}\Big(B\Big[\frac{1}{\vartheta} - \\ - \frac{\rho}{\alpha+\vartheta}(t-\tau)^{\alpha} + \frac{\rho^2(t-\tau)^{2\alpha}}{4\alpha+2\vartheta}\Big] + D_r\Big(\frac{t-\tau}{\vartheta+1} - \frac{\rho(t-\tau)^{\alpha+1}}{\vartheta+\alpha+1} + \\ + \frac{(t-\tau)^2}{2+\vartheta}\sum_{k=1}^{3}\frac{a_k}{\alpha_k}e^{-\gamma_k t}\left[e^{\alpha_k t}\aleph_{2k}^2 - e^{\alpha_k \tau}\eta_{2k}^2\right]\Big)\Big), \end{gathered} \tag{8}$$

where

$$B = 1 + \sum_{k=1}^{3}\frac{a_k}{\alpha_k}e^{-\gamma_k t}\left(e^{\alpha_k t} - e^{\alpha_k \tau}\right); \; D_r = \sum_{k=1}^{3}\frac{a_k}{\alpha_k}e^{-\gamma_k t}\left(e^{\alpha_k t}\aleph_{rk} - e^{\alpha_k \tau}\eta_{rk}\right).$$

4. Discussion

The error of such a transition is determined by the accuracy of the integral count. The numerical implementation of the task showed that with the increase in the number of terms of the Maclaurin power series, the results of the calculations do not go beyond the change of the result by 5 %.

The possibility of using the obtained relations to describe the creep of real materials was tested on experimental data of A. A. Ross [2]. Creep curves obtained from the experimental data and their approximation are shown in figure 1.

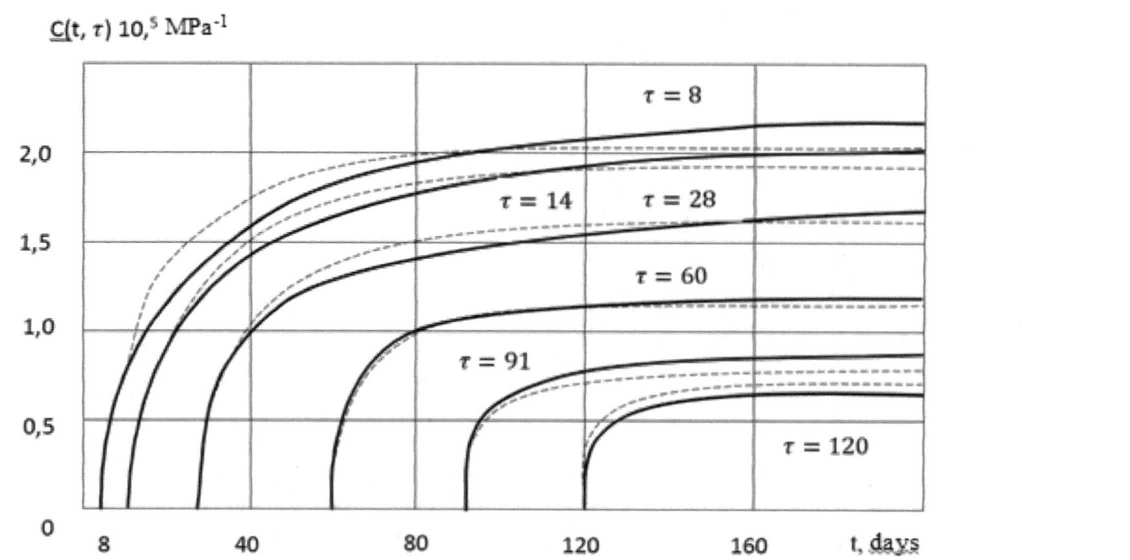

experimental data;
approximation of the experiment with formula (8).

Figure 1. Experimental and theoretical curves of creep measures of concrete samples of different age τ

IOP Conf. Series: Materials Science and Engineering **972** (2020) 012054 doi:10.1088/1757-899X/972/1/012054

The creep measure was calculated by formula:

$$C(t,\tau) = \frac{V_r(t,\tau)}{E(t)} - \frac{1}{E(\tau)} + \frac{1}{E(t)}, \tag{9}$$

where $C(t,\tau)$ – creep measure, $V_r(t,\tau)$ – aging material creep kernel integral, $E(t)$ – modulus of elasticity, $E(\tau)$ – stiffness factor of concrete at different age τ.

On the time interval (Figure 1) experimental curves $\boldsymbol{C(t,\tau)}$ are shown. The transition from creep measure $\boldsymbol{C(t,\tau)}$ to integral $\boldsymbol{V_r(t,\tau)}$ was made using method [10]. The calculations have shown that the stress-strain state of composite structures depends significantly on the age of material $\boldsymbol{\tau}$ at the instant of loading. There are both quantitative and qualitative changes in the distribution of stresses, forces and moments.

5. Conclusions

Comparison of the results obtained during the implementation of the proposed mathematical model and experimental data showed a discrepancy of no more than 10 %, which confirms the validity of the studies. The obtained results are theoretical justification of solutions of composite plates and shells nonlinear strain problems with account for creep, which allows obtaining predictive estimates of structures behavior under real operating conditions.

References

[1] Yakubovskiy Yu, Kolosov V, Krivchun N and Yakubovskaya S 2016 Izgib sostavnykh plastin i pologikh obolochek [Bending of composite plates and flat shells] (Tyumen: TIU) 185 p

[2] Yakubovskiy Yu 1991 *Prikladnaya mekhanika* **27** pp 37 - 44

[3] Yakubovskiy Yu, Kolosov V and Goltsov V_2016 *Procedia Engineering* **165** pp 1238–45

[4] Yakubovskiy Yu, Kolosov V, Gulyaev B and Goltsov V 2016 *Procedia Engineering* **165**, pp 1246–53

[5] Novozhilov V 1991 Osnovy nelineynoy teoriya uprugosti [Fundamentals of nonlinear elasticity theory] (Moskow: OGIZ) 211 p

[6] Novozhilov V 2010 Teoriya tonkikh obolochek [The theory of thin shells] (Leningrad: Politechnika) 380 p

[7] Vasil'ev P, Gavrilin B, Mal'kevich 1983 Voprosy razvitiya teorii deformirovaniya stareyushchikh sred [Questions of development of the theory of deformation of aging media] *Research on theoretical fundamentals of calculation of building structures: Intercollegiate thematic collection works* p 122-128

[8] Mal'tsev L, Karpenko Yu 1999 Teoriya vyazkouprugosti dlya inzhenerov-stroiteley [The theory of visco-elasticity for engineers] (Tyumen: VektorBook) 240 p

[9] Arutyunyan N, Zevin A 1982 Ob odnom klasse yader polzuchesti stareyushchikh materialov [*About one class of creep cores of aging materials*] *Prikladnaya mekhanika* **4** pp 4 - 21

[10] Arutyunyan N and Zevin A 1988 Raschet stroitel'nykh konstruktsiy s uchetom polzuchesti [Calculation of building structures taking creep into account] (Moskow: Stroyizdat) p 256

[11] Yakubovskiy Yu 1994 Nelineynaya teoriya izgiba i raschet sostavnykh plastin i pologikh obolochek peremennoy zhestkosti [Nonlinear bending theory and calculation of composite plates and flat shells of variable stiffness] (Ekaterinburg) 46 p

[12] Rzhanitsyn A 1968 *Teoriya polzuchesti* [Creep theory] (Moskow: Stroyizdat) p 416

[13] Yakubovskiy Yu, Kolosov V 1991 *Stroitel'naya mekhanika i raschet sooruzheniy* [*Construction mechanics and calculation of structures*] **1** pp 55 – 61

The Quality of Heat-Insulating Masonry Mortar with Perlite

E S Gerasimova [1] and A D Il'yina [1]

[1]Ural Federal University, 19 Mira street, Yekaterinburg 620002, Russia

E-mail: e.s.gerasimova@urfu.ru

Abstract. A quality for any building material including the dry construction mixtures is very important. This is a complex concept that includes a set of properties that determines their suitability for intended use. So the heat-insulating masonry mortar should simultaneously have a low coefficient of thermal conductivity, sufficient strength, sufficient frost resistance, etc. And the material not always possesses the characteristics that the manufacturer claims. In this regard it is of interest to conduct tests of finished products for their compliance both Russian Standard and presented characteristics. This work is devoted to main properties determination of heat-insulating masonry mortar produced by JSC "Quick–mix".

1. Introduction

The term "mortar" combines the concepts of "mortar mix", "dry mortar mix" and "mortar". Construction mortar is a material obtained by solidification of a mixture of binder, fine aggregate, water and special additives. Before solidification this mixture is called a mortar mix. Dry mortar mix is a mixture of dry components: binder, aggregate and additives which are batched and mixed as a rule in the factory and mixed with the water before use. The binder in the mortar envelops the aggregate grains reducing friction between them, as a result of which the mortar mixture acquires the necessary mobility to work with it, and in the process of hardening the binder firmly binds the individual particles of the aggregate and the filler. Cement, clay, gypsum, lime or their mixtures are used as a binder, and sand of different origin is used as an aggregate [1-3].

According to Russian Standard 28013-98 mortar can be classified by the main purpose: it is possible to distinguish separate group of masonry mortars. Masonry mortar is used in the construction of masonry of bricks, concrete and stone blocks [4].

In accordance with modern standards of update construction, both wall materials and masonry mortars are subject to increasingly high requirements for heat and energy saving. Therefore, to meet these requirements it is necessary to create innovative mortars that optimally suites to wall materials with high thermal insulation properties [5]. Such thermal insulation mortars are necessary for energy-saving construction, as they can significantly reduce the amount of energy required for heating the buildings. Due to the optimal combination of special light mineral aggregates in the warm mortars composition along with improved mobility and ease of processing, high strength and durability of the masonry is achieved especially in the construction of masonry from porous brick (blocks) and similar materials.

Thermal insulation masonry mortar is a mixture of cement or cement–lime binder with special mineral inclusions that improve its insulating properties. Most often it is a perlite or keramzit (expanded clay aggregates). There are also thermal insulation masonry mortars with expanded polystyrene granules. Due to the introduction of these materials the mortars have a reduced density, reduced thermal conductivity which varies from 0.18 to 0.24 W/m·deg (depending on the brand and manufacturer) which is almost 4 times less than that of ordinary cement mortar (its $\lambda = 0.90$ W/m·deg) [1-2].

The use of a warm mortar for enclosing structures eliminates heat losses through the mortar seams and improves the thermal properties of the masonry by 10-15 % with a seam thickness of about 12 mm while the usual mortar due to the insufficient coefficient of heat transfer creates a cold bridges in the masonry reducing the thermal insulation properties of the wall.

At the moment in the market of construction materials there are many manufacturers of dry mixes, including warm masonry mortars. Table 1 presents a comparative description of these mixtures of the main manufacturers. They are the most popular and common [6].

Table 1. Comparative table of main warm masonry mortars.

Trade mark	The mass of the bag (kg)	λ (W/m·°C)	Bulk density (kg/m^3)	Strength (MPa)
Quick–mix	17	0.190	540	5
Knauf	20	0.240	800–1000	5
Terta	25	0.270	1000	8
Kreisel	20	0.166	640	4
Murexin	20	0.190	650	5

A quality for any building material including the mixtures mentioned above is very important. This is a complex concept that includes a set of properties that determines their suitability for intended use. So the heat-insulating masonry mortar should simultaneously have a low coefficient of thermal conductivity, sufficient strength, frost resistance, etc. And the material not always possesses the characteristics that the manufacturer claims. In this regard it is of interest to conduct tests of finished products for their compliance both Russian Standard and presented characteristics.

2. Results and discussion

The material under study is a heat-insulating masonry mortar designed for laying of wall materials with high thermal insulation properties as well as used to fill cracks and voids. As a heat-insulating filler in its composition perlite is used – crushed rock of volcanic origin mainly consisting of SiO_2. Table 2 presents the technical characteristics of the material according to the manufacturer [7].

Table 2. Technical properties of mortar at + 20 °C relative humidity 60 %.

Mix grade	M50 (according to Russian Standard 31357-2007)
Density of set mortar	$\leq$ 700 (kg/m^3)
Thermal conductivity	$\leq$ 0,18 (W/m·deg)
Compressive strength	$\geq$ 5 (MPa)
Aggregate particles size	0–4 (mm)
Application temperature	up +5 °C to +30 °C
Usage time	~ 2 hours
Mobility grade	Pk2
Frost resistance	F50
Water quantity	~ 10–11 liters per 17,5 kg
Mortar yield	~ 25–27 liters from 17,5 kg of dry mix

In accordance with Russian Standards for dry mixes the properties of the initial dry mix, mortar mixture with water and solidified mortar are determined. For the dry mix humidity and particle size distribution are controlled which was determined using a standard set of sieves according to Russian Standards 8735-88. The result of determining the grain composition of the mix is shown in table 3. Figure 1 shows the dispersion curve from the test results. It was found that the grain content of a certain dispersion is within acceptable limits.

Table 3. Grain composition of the dry mix.

Residue	Residues (% by mass) on sieves						Passage through a sieve with mesh N 0.16 (% by mass)
	5.0	2.5	1.25	0.63	0.315	0.16	
Partial	0.05	5.16	4.24	4.90	52.36	26.34	7.05
Full	0.05	5.21	9.45	14.35	66.70	92.95	100.00

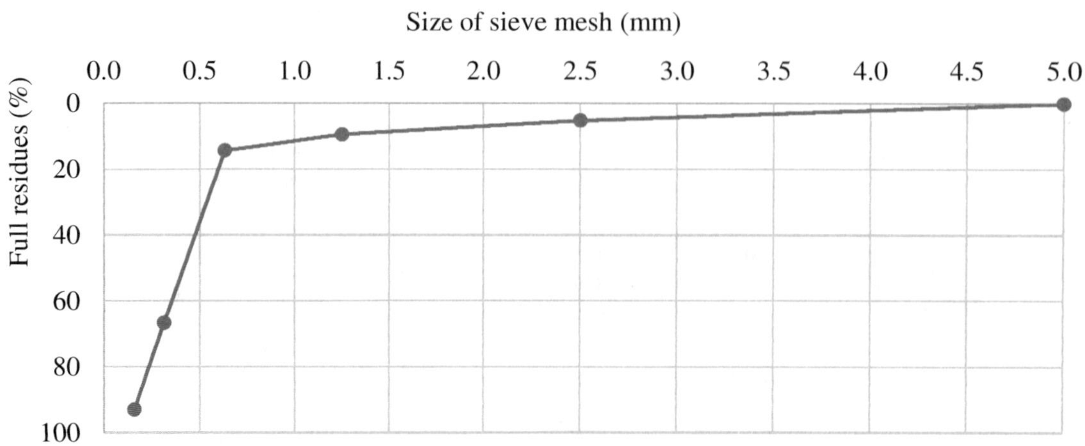

Figure 1. The curve of size dispersion of the dry mix.

The bulk density is 535 kg/m^3 therefore the mix is lightweight. The humidity of the tested dry mix was $\Delta W = 0$ % which corresponds to the requirements of Russian Standards [8-9].

According to the technical data 10-11 liters of mixing water per 17.5 kg of dry mix is required for the selected warm masonry mortar [7]. For the tests a sample equal to 2 kg was taken. Taking into account the recommendations of the manufacturer a water-solid ratio is 0.6. Eventually the volume of the solution mix was increased, its change was ΔV=5 %.

An image of a fresh mortar mixture is shown in figure 2, on the left. As one can see the resulting consistency of the mix does not meet the requirements for masonry mortars: the mixture was rigid, consisting of separate pieces so it was decided to increase the water-solid ratio up to 0.85. The resulting consistency is shown in figure 2, on the right: with increased water-solid ratio the mix is similar to a homogeneous paste without signs of water separation and stratification.

Further the main characteristics of the finished mortar mixture were determined, the results are shown in table 4.

In accordance with Russian Standards 31357-2007 the persistence of the initial mobility of mixtures ready for use is determined by the time of preservation of the initial mobility in minutes [8]. The retention of the initial mobility of the mix shall not be less than the time during which the mixture is produced at the construction or repair site.

According to the technical documentation the time of tested mixture use is 2 hours which means that the mixture must retain its original mobility for 2 hours. To test this indicator an experiment was carried out to measure the mobility of the mix from the moment of its preparation every 15 minutes for

3 hours. The experiment was conducted on a single sample to completely bring the experimental conditions to the real: at the place of use the mixture prepared for using is in the container all the time until it is completely developed. The results of the determination of the mix mobility are shown in figure 3 [10-11].

W/S = 0,6 W/S = 0,85

Figure 2. General view of the mixture at different water-solid ratio.

Table 4. Properties of mortar mix.

Property	Value
The density of the prepared mortar mix (kg/m^3)	1030
Water-holding capacity (%)	98.8
Mobility of mortar mix (cm)	10.7

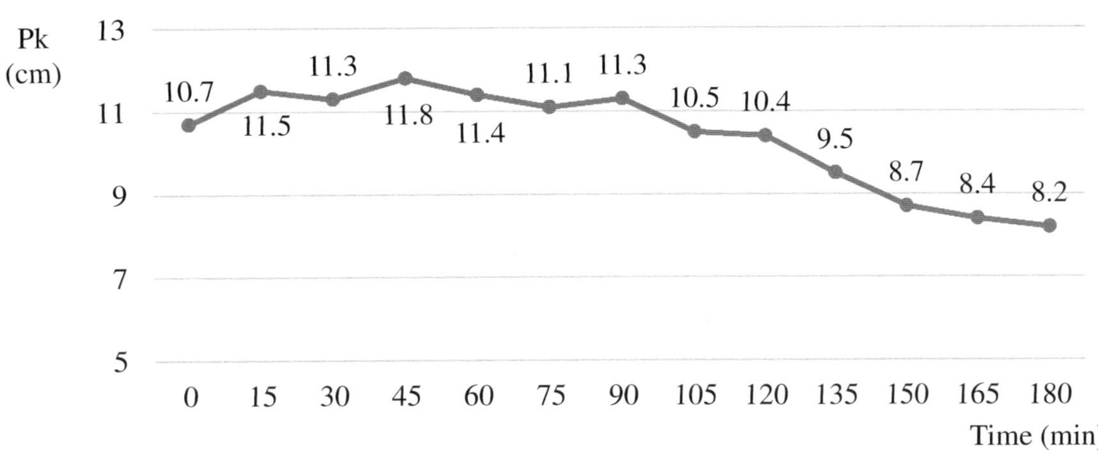

Figure 3. Results of mix mobility kepping measurement.

After conducting all tests with a freshly prepared mixture standard specimens-beams (4x4x16 cm) were molded [10-11]. At the same time it was interesting to study how the holding time of the mix before molding affects the strength of the hardened mortar. For this purpose three batches of samples were formed: the first batch was formed 10 minutes after mixing and preparation of the mix, the second – after 1 hour and the third – after 3 hours. This time was also chosen on the basis of data close to the real conditions of use of such a mixes. The results of measurements of compressive strength of all batches of the mortar are shown in figure 4.

It was found that long-term exposure has no negative effect on the strength index and even when forming samples after an hour the strength of the mortar is greater than the strength of the batch molded immediately after preparation of the mix. This may be due to a violation of the original structure of the cement binder formed and recrystallization of crystallohydrates that are part of Portland cement.

The density of the solidified mortar was equal to 977 kg/m^3 which is 39 % higher than the declared manufacturer. The coefficient of thermal conductivity was equal to 0.2 W/m deg which is 11 % higher than stated.

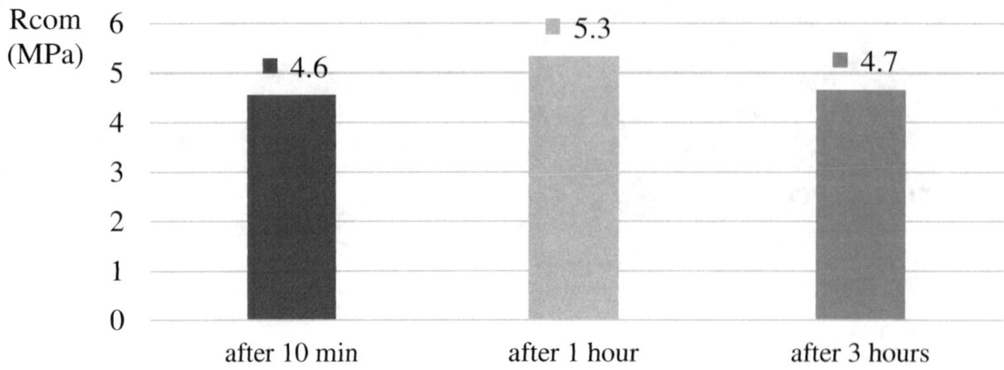

Figure 4. Compressive strength of the mortar depending on the holding time of the mix up to the samples molding.

3. Conclusion

The main characteristics of the dry building mix, freshly prepared mix and solidified mortar in accordance with the requirements of Russian Standards were determined (table 5). The size of the aggregate and humidity correspond to the manufacturer's data and the requirements of Russian Standards.

Table 5. Results of mix testing.

Indicator	Actual value	Manufacturer details	Russian Standards 28013–98
Water-solid ratio	0.85	0.6	–
The size of aggregate (mm)	0-5	0-4	–
Humidity (%)	0	-	corresponds
Water-holding capacity (%)	98.8	-	corresponds
Mobility grade (cm)	Pk 3	Pk2	corresponds
Using time (hours)	2	2	–
Density of set mortar (kg/m$^{3)}$	977	$\leq$ 700	–
Coefficient of thermal conductivity (W/(m·°))	0.20	$\leq$ 0.18	–
Strength grade	M25	M50	corresponds

The water-solid ratio does not correspond to the manufacturer's data. The tested mixture according to Russian Standards 28013-98 has a brand of mobility Pk3 whereas according to the manufacturer –

Pk2. To create the necessary consistency the amount of mixing water was increased by 42 %. Water-holding capacity of the mortar corresponds to Russian Standards 28013-98. The mortar mixture retains mobility for 2 hours which corresponds to the manufacturer. The average density of the solution in the solidified state is higher than indicated by the manufacturer. The solution in accordance with Russian Standards 28013-98 is the brand for the compressive strength of M25 which is not consistent with the data of the manufacturer [4, 8-9].

References

[1] Bazhenov Yu M, V F Korovyakov and Denisov G A 2003 Technology of dry building mixes, (Moscow: Publishing house ASV)
[2] Kozlov V V 2000 Dry building mixes, (Moscow: Publishing house ASV)
[3] Jornet A, Mosca C, Cavallo G and Corredig G. 2012 Comparison Between Traditional, Lime Based, and Industrial, Dry Mortars. In: Válek J., Hughes J., Groot C. (eds) Historic Mortars. RILEM Bookseries (Springer, Dordrecht. vol 7)
[4] Russian Standard 28013–98: Construction mortars. Technical conditions. Russian Standard Institution, Moscow
[5] Miletic M 2014 Use of Transparent Insulation Materials as One of the Measures of Improving Energy of Structures. In: Dincer I., Midilli A., Kucuk H. (eds) Progress in Exergy, Energy, and the Environment (Springer, Cham)
[6] Khomich V A and Emralieva S A 2010 Dry construction mix. *Concrete and reinforced concrete* **2** 18-20
[7] 2019 Official site of Quick-mix. URL: http://katalog.quickmix.ru/lm-21-p-teplyj-kladochnyj-rastvor-s-perlitom (Date 25.05.2019, Electronic Materials)
[8] Russian Standard 31357–2007: Dry building mixtures based on cement binder. Technical conditions. Russian Standard Institution, Moscow
[9] Russian Standard P 57337–2016/EN 998–2:2010: Construction masonry mortars. Technical conditions. Russian Standard Institution, Moscow
[10] Russian Standard 31356–2007: Dry building mixtures based on cement binder. Test methods. Russian Standard Institution, Moscow
[11] Russian Standard 5802–86: Construction mortars. Test methods. Russian Standard Institution, Moscow

IOP Conf. Series: Materials Science and Engineering **972** (2020) 012056 doi:10.1088/1757-899X/972/1/012056

Optimization of the Strength Characteristics of the Cellular Structure in Samples of Thermoplastic Polyester

V T Erofeev[1,2], T F Elchishcheva[3], E M Preobrazhenskaya[3], M V Makarchuk[3], V V Afonin[1]

[1]Faculty of Architecture and Civil Engineering, National Research Mordovian State University named after N.P. Ogarev, Bolshevistskaya str, 68, 430005 Saransk, Russia

[2]Research Institute "Materials Science"

[3]Tambov State Technical University, Sovetskaya str 106, 392000, Tambov, Russia

Email: fac-build@adm.mrsu.ru, elschevat@mail.ru,

Abstract. When making a prototype during design in industrial design, specialists use the data on the strength characteristics of filaments for extrusion provided by the manufacturer of equipment for FDM printing. It is likely that the strength characteristics of the product may deviate from the characteristics of the threads. The authors suggested that the samples of the cellular structure made on 3D printers from polylactide with an internal volume filling of 10% with the filament laying direction according to the standard settings of the equipment manufacturer during the experimental determination of strength characteristics will demonstrate values several times lower than the samples with filling the cell volume with polylactide by 100% and having in their geometry the same direction of laying the extrudable material.To confirm the hypothesis put forward by the authors, a study was carried out of the characteristics of products made of polylactide using the FDM printing method.A comparison is made of the relative change in some parameters of the samples (mass, elongation, and the force used to break the samples), the two devices for growing prototypes – 3D printers MakerBotReplicator 5 Gen and MakerBotReplicator Z18 in percentage terms. Recommendations are given on the optimal operation of equipment for 3D printing, depending on the goals set by the user. The results can be used to select equipment for printing and sample production technology, cell filling density and amount of material when creating a prototype using the additive method in industrial design.

1. Introduction

In the construction industry, various polymer composite materials based on thermosetting and thermoplastic binders are used.There are also products distinguished by pressing, vacuum, extrusion, and centrifugation technologies.A large number of works have been devoted to the experimental and theoretical substantiation of their preparation, the study of physical and mechanical properties, and durability.Under construction conditions, polymeric materials and products are manufactured using injection molding technology, formed in the factory [1-29, 31].The use of additive technologies in prototyping in industrial design has become widespread.This technology involves the use of 3D printing by the FDM (FusedDepositionModeling) method - layer-by-layer filling with an extrudable polymer melt.Manufacturers offer a wide variety of materials for use in 3D printing, their overview is given in table 1. The materials presented differ in consumer and physico-mechanical properties used depending on the requirements for the prototype.The interconnection of technologies and strength characteristics of materials are scrutinized in some publications [30, 32, 33, 34].

Table 1. Popular materials used in 3D printing.

Internationalmat erialdesignation	TensileStrength, MPa	Extrudertemperature, °C
ABS	22.00	220-260
ASA	35.80	270-280
HIPS	62.00	220-260
NYLON	65.99	250-270
PP	25.50	254
PET	50.00	120-160
PLA	57.80	210-240
POM	75.00	210-250
SBS	24.00	240
TPU	40.00	180-220

In the manufacture of the prototype of the selected material by FDM-printing, users rely on the strength characteristics declared for this plastic by the manufacturer. However, the strength characteristics of the manufactured parts do not always correspond to the material properties of the filaments from which they are made. The structure of the part obtained by layer-by-layer deposition of an extrudable melt may differ in properties from unextruded filament.

Another important factor affecting the strength characteristics of the prototype is a feature of the technology. The volume of the product to save material and reduce printing time is filled in the form of a cellular structure. In the case of minimal filling of the volume of the prototype, you can get the shell of the prototype with an empty (unfilled) volume inside. Such a prototype will have low strength characteristics and may be deformed as a result of internal stresses in its design or a significant mass of the prototype.

The additive technology user has a difficult choice to make. You should choose between saving the amount of material and printing time, however, this causes a decrease in the strength of the prototype. Another option is to manufacture a prototype with greater strength, but it will use up more material and spend more time. It can be assumed that polylactide (PLA) cellular samples printed with the manufacturer's recommended settings on a MakerBotReplicator 5 Gen printer with an internal 10% volume filling during tensile testing will demonstrate strength values several times lower than polylactide samples with cell volume filling 100%. The same statement, obviously, is true for samples having the same cellularity made with the MakerBotReplicator Z18 printer.

The results can be used to create a map showing the dependence of the strength characteristics of the prototype on the degree (percent) of the cellular structure, the type of plastic used and other 3D printing parameters, and in the end, to select the manufacturing technology and type of material when creating the prototype in industrial design.

It is obvious that when creating prototypes it is necessary to individually select the parameters of 3D printing depending on the goals and objectives of designers and engineers, taking into account material savings and reducing the mass of the resulting prototype.

Thus, it is relevant to study the strength properties of the product, made by additive technology with different filling of the cellular structure.

2. Research methods

Strength analysis samples were manufactured using additive technology on two 3D printers: MakerBotReplicator Z18 and MakerBotReplicator 5 Gen.

On each of the printers, samples were made in batches of six pieces in each batch with filling of the internal cellular structure of samples from 10% to 100% with a filling step of 10%. A total of 120 samples were manufactured. When growing samples, print settings recommended by the equipment manufacturer were applied. The printing material used was polylactide (PLA) polymer, a polyester of lactic (2-hydroxypropionic) acid CH3-CH (OH) -COOH.

The geometric dimensions of the samples: total length 150 mm, the distance between the wide parts with parallel sides 108 mm, the length of the narrow part with parallel sides 60 mm, the radius of the head 60 mm, the width of the wide part 20 mm, the width of the narrow part with parallel sides 10 mm, thickness 4 mm (GOST 33693-2015).

Studies of the strength characteristics of all manufactured samples were carried out on an MP-0.5-1 tensile testing machine.

3. Research results

On the tensile testing machine MP-0.5-1, the strength characteristics of the samples made of polylactide with various percent filling of the cellular structure were tested.

Using optical microscopy, the fractures of all samples subjected to tensile tests were studied. A typical view of fractures of samples with filling of the cellular structure by 10%, 50%, and 100% is shown in Fig.1.

a *b* *c*

Figure 1. Fracture of samples with 10% (a), 50% (b) and 100% (c) filling the mesh structure

The fracture surface of all samples is porous; the threads of the internally obtained samples are clearly visible. According to the degree of plastic deformation, the samples are fragile. The fracture of the samples under the influence of the applied load is flat with bevels.

The cellular structure is formed in the internal volume of the sample, and its external surface is filled with polymer by 100% (Fig. 1, a, b). An analysis of the samples made with the MakerBotReplicator 5 Gen 3D printer showed that the sample without filling the internal cellular structure has an average weight of 5.66 g, and at 100% filling the average weight of the sample is 10 g. Thus, approximately 4.34 g of the total mass of the sample is involved in the change in the percent filling of the cellular structure of samples from 0% to 100%. For samples made with the MakerBotReplicator Z18 printer, the mass value that changes when filling the mesh structure of the sample from 0% to 100% is 4.16 g. The data obtained show that prototypes manufactured using additive technology for each printer will have a minimum threshold strength.

The test results of the strength characteristics of samples made of polylactide with filling the cellular structure of the sample from 0% to 100% on a tensile testing machine MP-0.5-1 are summarized in table 2.

Table 2. Results of testing the strength characteristics of samples made on 3D printers MakerBotReplicator Z18 andMakerBot Replicator 5 Gen

		MakerBot Z18			MakerBot G5		
%	№ sample	weight / g	mm	MPa	weight / g	mm	MPa
10	1	5.56	5	9.69	6.18	9	13.056
	2	5.62	5	10.557	6.18	9.5	12.852
	3	5.64	6	10.812	6.18	7	13.77
	4	5.62	5.5	10.404	6.14	9	13.056
	5	5.24	7	9.69	6.18	7.5	10.71
	6	5.56	6	9.69	6.16	8	12.852
20	1	6.24	5.5	6.63	6.64	7	12.036
	2	6.18	4	5.61	6.67	6	12.648
	3	6.14	6	8.67	6.62	5.5	10.71
	4	6.18	4	5.61	6.62	6	10.098
	5	6.16	4	4.386	6.62	7	8.67
	6	6.16	3.5	4.896	6.6	4	9.18
30	1	6.6	5.5	11.526	7.1	8	14.076
	2	6.6	5.5	11.118	7.1	6.5	10.608
	3	6.6	5	9.69	7.1	7	13.26
	4	6.6	6	11.628	7.1	4	13.26
	5	6.6	4.5	9.18	7.2	5	11.974
	6	6.6	3.5	12.342	7.1	5	12.056
40	1	7.06	6	9.078	7.6	7	11.73
	2	7.06	5.5	7.548	7.55	6	11.016
	3	7.04	4	5.814	7.3	6.5	11.526
	4	7.08	5.5	7.65	7.45	6	10.2
	5	7.08	6.5	8.772	7.31	5	8.67
	6	7.08	5	6.324	7.52	6	9.18
50	1	7.58	8	13.974	8.1	9	13.056
	2	7.53	8	13.056	8.1	8	14.688
	3	7.55	8	14.076	8.2	8	13.872
	4	7.53	8	12.648	8.1	8	16.32
	5	7.56	7	12.24	8.2	8	14.076
	6	7.53	8	12.852	8.1	8	14.28
60	1	8	5	7.446	8.64	7	11.832
	2	8	8	11.22	8.6	7.5	11.628

		MakerBot Z18			MakerBot G5		
%	№ sample	weight / g	mm	MPa	weight / g	mm	MPa
	3	8	5	7.548	8.62	5	8.67
	4	8	6	8.67	8.64	7	11.73
	5	8	5	7.446	8.52	7	12.75
	6	7.96	6	9.996	8.56	7	12.648
70	1	8.46	9	15.912	9.2	8.5	17.238
	2	8.42	8	14.994	9.14	8	18.462
	3	8.44	8.5	15.3	9.22	8.5	16.32
	4	8.46	11	16.83	9.22	10.5	18.564
	5	8.44	8	14.688	9.24	10	18.87
	6	8.44	8	15.912	9.18	8.5	19.89
80	1	8.76	9	16.524	9.42	8	19.482
	2	8.78	8	16.932	9.48	9.5	20.706
	3	8.82	7	17.85	9.44	8	19.074
	4	8.76	7	17.34	9.5	12	16.116
	5	8.76	6.5	18.054	9.48	10.5	18.258
	6	8.78	9	18.513	9.48	9	19.686
90	1	9.35	6	10.914	9.5	9	18.054
	2	9.3	8	16.014	9.72	8.5	18.564
	3	9.3	7	13.77	9.64	9	14.79
	4	9.3	6	11.526	9.58	9	17.34
	5	9.3	7	15.3	9.78	8.5	17.136
	6	9.2	8	16.83	9.74	9	18.36
100	1	9.65	15	21.573	9.7	7	22.95
	2	9.68	14	22.236	9.8	7	21.114
	3	9.65	11	22.185	10.2	8	20.91
	4	10.04	15	21.522	10.06	8	23.46
	5	9.78	9	19.125	10	8	21.93
	6	9.66	12	20.4	9.9	7	22.44

Dependence plots are constructed for average indicators of mass (Fig. 2), elongation (Fig. 3) and strength (Fig. 4) of samples as a percentage of the filling of the cellular structure. The weight of samples made with the MakerBot G5 printer is on average 0.55 g (8%) larger than the mass of samples made with the MakerBot Z18 printer. The change in mass occurs in proportion to the change in the filling of the cellular structure and averages: 0.38 g for samples made with a MakerBot G5 printer, and

0.41 g for samples made with a MakerBot Z18 printer. It should be noted that the change in mass in each batch of samples occurred uniformly without a significant dispersion of values (Table 2).

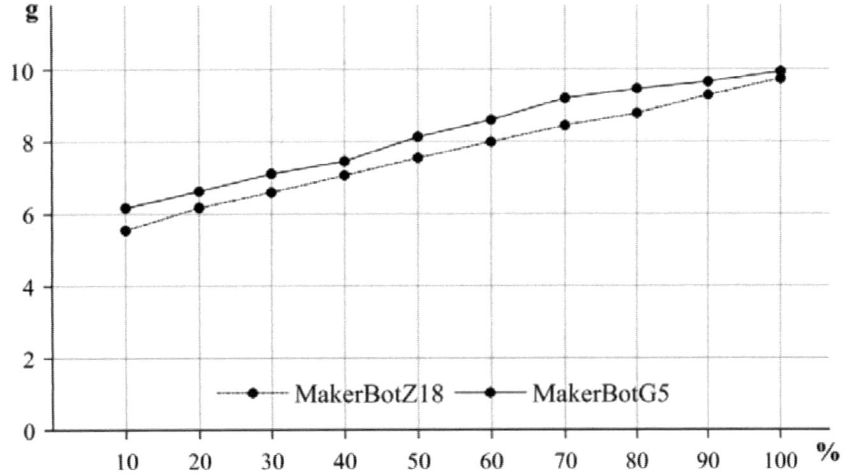

Figure 2. The dependence of the average mass of samples produced on printers Maker Bot Z18 and Maker Bot G5 from the percentage of filling cells

The elongation of samples made with the MakerBotZ18 and MakerBot G5 printers occurs nonlinearly (Fig. 3). The nature of the change in the elongation curves of the samples correlates with the nature of the change in the curves of their tensile strength (Fig. 4). In the batch of samples with the same cell filling, a scatter of the measurement results is observed (Table 2), which may be due to the modes and algorithm of polylactide stacking during printing.

The nature of the change in the tensile strength of samples made on MakerBot Z18 and MakerBot G5 printers is similar (Fig. 4). The characteristic values are the tensile strength for the filling values of the cellular structure of the samples 10%, 50% and 100%.

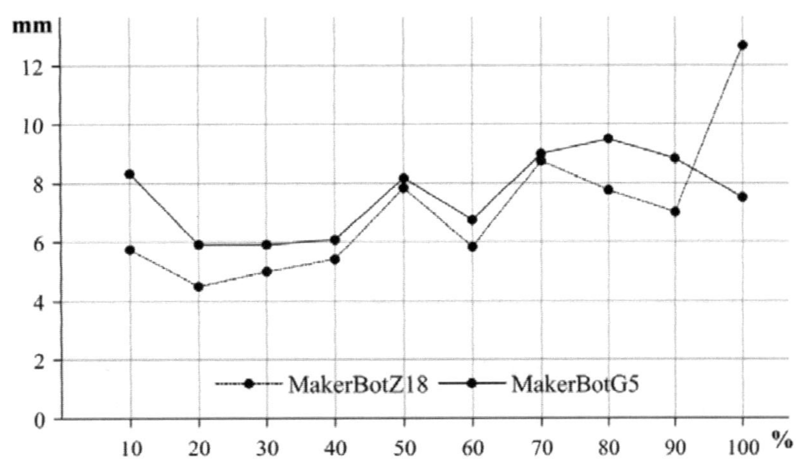

Figure 3. The dependence of the average elongation of samples to the moment of rupture of the percentage of filling cells

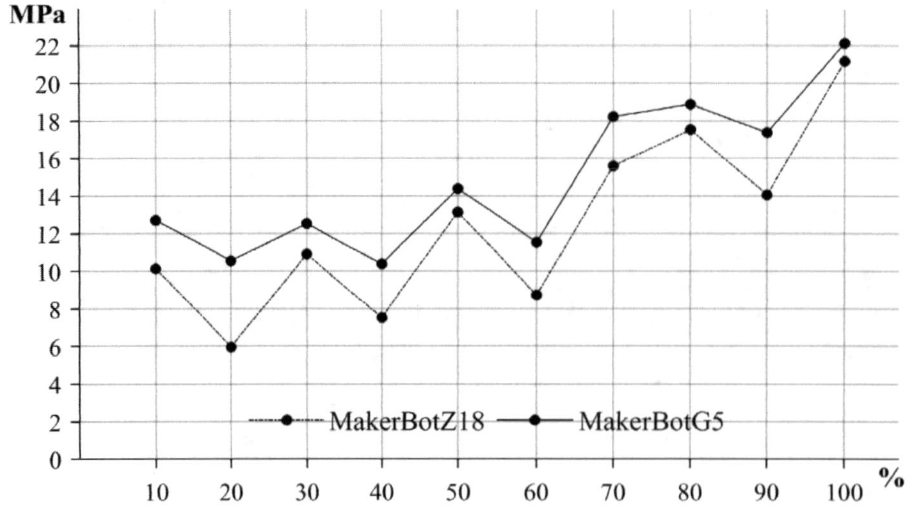

Figure 4. The dependence of the average force applied to break the samples, the percentage of filling cells

For all the samples studied, in the filling interval of the cellular structure above 10% and below 50%, a decrease in the tensile strength is observed in comparison with the strength of the samples with a filling of 10%. Therefore, the manufacture of prototypes with a filling of the cellular structure of 20%, 30% and 40% is irrational.

In the filling interval of the cellular structure above 50% and below 100%, a drop in the tensile strength for a filling value of 60% is observed.The strength value is lower than the strength of the sample with a filling of the cellular structure of 50%. Also, a drop in the tensile strength is observed at a filling value of 90% compared with filling samples by 70%, 80% and 100%. When the cellular structure is filled with 20%, 30%, 40%, 60% and 90%, a significant drop in the tensile strength of the studied samples occurs.

To analyze the operating parameters of printers, we plotted the dependence of changes in indicators (in relative%) —mass, elongation, and the force used to break the samples, samples made on a MakerBotG5 printer compared to samples made on a MakerBotZ18 printer (Fig. 5).Depending on the indicator, which is taken as the basis for the study (mass, elongation, tear strength) and the required level of filling of the samples, a different approach is possible in choosing a printer for 3D printing, as shown below.

When comparing the performance of samples obtained by printing on a MakerBotG5 printer with the performance of samples printed on a MakerBotZ18 printer, it was found that the greatest difference in the mass of samples is observed at 10% cell filling and amounts to + 11.2%, the smallest difference is at 90% cell filling and is 3.9 %. The MakerBotG5 printer at all levels of cell filling shows an increase in sample mass compared to the MakerBotZ18 printer. The average increase in sample mass at all cell fill levels on the MakerBotG5 printer is 6.9%. Therefore, in the manufacture of the prototype without special requirements for its elongation and strength, but with maximum requirements for polymer savings, it is advisable to use the MakerBotZ18 printer at all levels of cell filling.

A comparison of printers in terms of elongation of samples prior to rupture showed that the greatest difference in elongation of samples is observed at 10% cell filling and is + 44.3%, as well as at 100% cell filling and is minus 40.5%.At the same time, the MakerBotG5 printer at cell filling levels from 10% to 90% shows higher elongation rates of the samples compared to the MakerBotZ18 printer.However, from a fill level of 80% to a fill level of 100%, the elongation rate of the MakerBotG5 printer samples is reduced. In this case, the elongation rate of MakerBotZ18 samples increases from a cell filling level of 90% to a cell filling level of 100%.Thus, in the manufacture of the prototype with filling cells over 90%, it is advisable to organize the printing of samples on a MakerBotZ18 printer. The smallest difference in elongation of samples made on different printers is observed at 50% and 70% cell filling and is, respectively, + 2.5% + 2.9%. Therefore, when filling cells of 50% and 70%, it is possible to use Z18 and MakerBotG5 printers for printing samples.

A comparison of printers in terms of the applied force for breaking samples showed that the largest difference in the applied force is observed at 20% cell filling and is + 99.8%, as well as at 40% cell filling and is + 56.4%.The MakerBotG5 printer at all levels of cell filling has higher rates of applied force to break samples compared to the MakerBotZ18 printer.The smallest difference in the magnitude of the applied force for breaking samples made on different printers is observed at 50%, 80% and 100% cell filling and is, respectively, + 9.3%, + 7.6% and + 4.3%. Thus, in the manufacture of a prototype with requirements for filling the cells of 50%, 80% and 100%, it is possible to organize printing of samples on both the MakerBotG5 printer and the MakerBotZ18 printer. At other fill levels, printing on a MakerBotG5 printer is advisable.

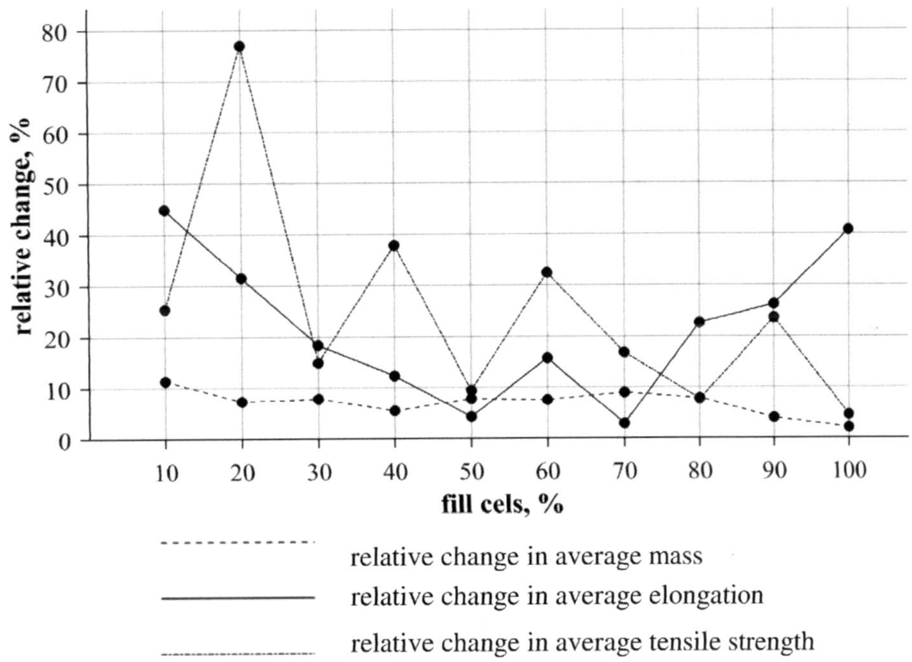

Figure 5. Dependence of the change in the relative parameters of the samples on the percentage cell filling for the MakerBotZ18 and MakerBotG5 printers

4. Conclusions

To determine the strength characteristics of polylactide products grown on 3D printing equipment, samples were prepared with filling the internal cellular structure of samples from 10% to 100%. Samples were weighed and subjected to tensile tests. The results of the experiments are listed in the table, as well as expressed graphically. Analysis of the data showed the following:

- prototypes manufactured using additive technology for each type of printer will have a minimum threshold strength;
- the change in mass in each batch of samples is homogeneous without significant variation in values;
- the nature of the change in the elongation curves of the samples is similar in nature to the curves of the tensile strength of the same images. In batches of samples, significant

differences in measurement results are observed. It was not possible to establish the reason for such differences, this instability can be associated with the modes and algorithm of material stacking. It is a prospect for further research;

- in the filling interval of the cellular structure of more than 10%, but below 50%, a drop in the tensile strength is observed for all studied samples in comparison with the strength of samples with a filling of 10%. Therefore, the manufacture of prototypes with a filling of the cellular structure of 20%, 30% and 40% is irrational;
- when the cellular structure is filled with 20%, 30%, 40%, 60% and 90%, a significant drop in the tensile strength of the studied samples occurs.

The analysis of relative indicators for two printers showed:

- in the manufacture of the prototype without special requirements for its elongation and strength, but with maximum requirements for polymer savings, it is advisable to use the MakerBot Z18 printer at all levels of cell filling;
- the smallest difference in the elongation of samples made on different printers is observed when the cells are filled with 50% and 70%; therefore, at these levels it is possible to use both printers to print samples;
- the smallest difference in the magnitude of the applied force for breaking samples made on different printers is observed when the cells are filled with 50%, 80% and 100%. At these cell fill levels, it is possible to print samples on both printers. At other levels, printing on a MakerBot G5 printer is advisable.

Reference

[1] Bobryshev A N, Erofeev V T, Kozomazov V N 2012 Physics and synergetics of dispersed disordered condensed composite systems (St. Petersburg: Nauka) 476 p
[2] Berlin A A, Wolfson S A, Oshmyan V G, Enikolopov N S 1990 Principles of creating composite polymer materials (Moscow: Chemistry) 240 p
[3] Khozin V G 2004 Strengthening of epoxy polymers (Kazan: Publishing house of PNK "Printing House") 446 p
[4] Askadsky A A et al. 2001 Gradient polymer materials *Russian Chemical Journal* **45** (3) 123-128
[5] Stepanova V F, Stepanov A Yu, Zhirkov E P 2013 Composite polymer fittings (Moscow: Publishing house TsNTP) 200 p
[6] Selyaev V P, IvashchenkoYu G, Nizina T A 2016 Polymer concrete (Saransk: Mordov Publishing House. University) 284 p
[7] Bobryshev A N, Kondratiev E V, Kozitsin V S, Kuznetsov Yu S, Alizaev M A 2004 Alkali-resistant epoxy composites for construction purposes (Penza: PGUAS) 164 p
[8] Vyrovoy V N, Dovgan I V, Semenova S V 2004 Features of structure formation and formation of the properties of polymer composite materials (Odessa: Publishing House TPP) 168 p
[9] LipatovYu S 1996 Physical chemistry of filled polymers (Kiev: NaukovaDumka) 233 p
[10] SokolovaYu. A 1990 Modified epoxy adhesives and coatings in construction (Moscow: Stroyizdat) 174 p
[11] Solomatov V I 1984 Technology of polymer concrete and reinforced concrete products (Moscow: Stroyizdat) 144 p
[12] Bondarev B A, Borkov P V, Komarov P V, Bondare A B 2013 Cyclic durability of polymer composite materials for construction purposes (Lipetsk: LSTU Publishing House) 112 p
[13] Bazhenov Yu M 2011 Concrete technology: a textbook for students of higher educational institutions studying in construction specialties [5th ed.] (Moscow: DIA) 524 p
[14] Lesovik V S, Fedyuk R S 2016 Increasing the impermeability of fiber concrete on a composite binder (Belgorod: BSTU) 165 p
[15] Tynny A N 1990 Strength and destruction of polymers when exposed to liquid media (Kiev: NaukovaDumka) 200 p

[16] Elshin I M 1980 Polymer concrete in hydraulic engineering (Moscow:Stroyzdat) 192 p
[17] Irzhak V I, Rosenberg B A, Enikolopyan N S 1979 Cross-linked polymers - synthesis, structure and properties (Moscow: Stroyizdat) 183 p
[18] Mikhailov K V, Paturoev V V, Chrys R 1989 Polymer concrete and constructions based on them (Moscow: Stroyizdat) 304 p
[19] Kharchevnikov V I 1971 Strength and chemical resistance of fiberglass polymer concrete *Reinforced polymer concrete in building structures: collection. articles.* (Voronezh: Voronezh State University) P 28-30
[20] Vorobyev G Ya 1981 Chemical resistance of polymeric materials (Moscow: Chemistry) 269 p
[21] Zhurkov S K, Kuksenko V S, Slutsekr A I 1972 Micromechanics of polymer destruction *Problems of Strength* **2** 45-50.
[22] Bokshitsky M N 1978 Long-term strength of polymers (Moscow: Chemistry) 309 p
[23] Christensen R M 1982 Introduction to the mechanics of composites (Moscow: Mir) 336 p
[24] Hashin Z 1962 The elastic moduli of heterogeneous materials *J.Appl. Mech* **1** p 143
[25] Startsev V O, Molokov M V, Blaznov A N, Erofeev V T, Smirnov I V 2017 Determination of the heat resistance of polymer construction materials by the dynamic mechanical method *Polymer Science - Series D.* **10** (**4**). Pp. 313-317. doi: 10.1134 / S1995421217040141.
[26] Shafigullin L N, Treschev A A, Romashina A V, Erofeev V T 2017 Concentration of stress on holes in a plate of different resistant material. *Astra Salvensis* **2017** Pp. 213-225
[27] Erofeev V, Bobryshev A, Shafigullin L, Zubarev P, Lakhno A, Darovskikh I, Tretiakov I 2016 Building Heat-insulating Materials Based on the Products of the Transesterification of Polyethylene Terephthalate and DibutyltinDilaurate *Procedia Engineering* **165** Pp. 1455-1459. doi: 10.1016 / j.proeng. 2016.11.879.
[28] Shafigullin L N, Treschev A A, Hodorovich P Y, Erofeev V T 2017 The Stress-Strain State Of Layered Orthotropic Conditional Half-Space Taking Into Account Different Resistance *REVISTA PUBLICANDO* **4** (**13**). Pp. 109-127.
[29] Erofeev V T 2016 Frame Construction Composites for Buildings and Structures in Aggressive Environments *Procedia Engineering* **165** Pp. 1444-1447. doi: 10.1016 / j.proeng.2016.11.11.877.
[30] Chen X, Wang C, Ye X, Xiao Y and Huang S 2001. Direct Slicing from PowerSHAPE Models for Rapid Prototyping *International Journal of Advanced Manufacturing Technology* **17** (**7**) 543-547. DOI: ttps: //doi.org/10.1007/s001700170156.Availablefrom: https://link.springer.com/article/10.1007%2Fs001700170156 [Accessed March 11, 2019].
[31] GOST 33693-2015. Plastics. Samples for testing. (Moscow: Standardinform) Available from: http://docs.cntd.ru/document/1200132500 [Accessed March 5, 2019].
[32] Pham D T and Gault R S 1998. A Comparison of Rapid Prototyping Technologies. International *Journal of Machine Tools and Manufacture* **38** (**10-11**) 1257-1287. DOI: https://doi.org/10.1016/S0890-6955(97)00137-5. Available from: https://www.sciencedirect.com/science/article/pii/S0890695597001375?via%3Dihub [Accessed March 11, 2019].
[33] Pandey P M, Thrimurthulu K and Venkata Reddy N 2004 Optimal part deposition orientation in FDM by using a multi-criteria genetic algorithm *International Journal of Production Research* **42** (**19**) 4069-4089. DOI: https: //doi.org/10.1080/00207540410001708470. Available from: https://www.tandfonline.com/doi/full/10.1080/00207540410001708470 ?scroll=top&needAccess=true [Accessed March 11, 2019].
[34] Santhakumar J, Mohammed Iqbal Uand Prakash M 2017 Investigation on the Effect of Tensile Strength on Fdm Build Parts Using Taguchi-Gray Relational Based Multi-Response Optimization *International Journal of Mechanical Engineering and Technology* (IJMET) **8** (**12**) 53-60. DOI: IJMET_08_12_006. Availablefrom: http://www.iaeme.com/IJMET/ issues.asp?JType=IJMET&VType=8&IType=12 [Accessed March 11, 2019].

IOP Conf. Series: Materials Science and Engineering **972** (2020) 012057 doi:10.1088/1757-899X/972/1/012057

Research of new materials and approaches in additive technologies in the manufacture of prototypes in industrial design

V T Erofeev[1], T F Elchishcheva[2*], E M Preobrazhenskaya [2], M V Makarchuk [2]

[1] Faculty of Architecture and Civil Engineering, National Research Mordovian State University named after N.P. Ogarev, Bolshevistskaya str, 68, 430005 Saransk, Russia

[2]Tambov State Technical University, Sovetskaya str 106, 392000, Tambov, Russia

E-mail: * elschevat@mail.ru

Abstract. In industrial design, in the manufacture of the prototype, the task of choosing the technology and material for its manufacture is taken. At the minimum cost of material, it is necessary to obtain a prototype with high strength indicators, which will guarantee the implementation of the tasks for which the prototype was conceived. The use of additive technologies allows you to get prototypes with a cellular internal structure, which can significantly save the expendable material when prototyping the future product. In this paper, we analyze the approaches that contribute to improving the strength characteristics of a prototype made using additive technology, and also show the promise of research in the direction of finding the shape of cells with the greatest strength. As a hypothesis, it has been suggested that an increase in the strength characteristics of the prototype can be achieved by introducing nanostructured materials into the polymer. Studies were conducted on the synthesis of nanostructured materials in low-temperature plasma and their modification of polylactide. Samples were made from modified polylactide and their strength characteristics were investigated. An analysis of the results showed an increase in the strength characteristics of samples of modified polylactidapo compared with samples made of primary polylactide without nanostructured components.

1. Introduction

Composite building materials based on inorganic and organic binders are widely used in the construction industry. In their manufacture, various technologies are used, including nano- and biotechnologies. Currently, intensive technologies for the production of concrete, polymer concrete and other composites are widely known, which contribute to improving the properties of materials and self-healing in the event of defects. First of all, it is polymer concrete formed by vacuum, pressing and vibrocompression methods, according to frame technology, powder-activated concrete activated by alkaline waste, modified by microbiological environment, etc. [1–13, 16–19, 22–29].

Recently, in industrial design, widespread use of additive technologies. Additive technologies have advantages such as rapid prototype manufacturing, dimensional accuracy, and mobility. However, these technologies have several disadvantages and limitations. For example, a prototype should have strength characteristics close to the designed product, and at the same time, a minimal amount of material should not be used. Additive technology allows you to get a prototype with a cellular internal structure, this significantly saves raw materials when prototyping a future product. Moreover, the smaller the degree of filling of the cells when creating a prototype, the lower strength characteristics it will have. Accordingly, such a prototype will not fulfill the strength requirements for the product being designed. It is necessary to find a balance between the minimum required amount of material used for prototyping and the required strength of the prototype. This problem can be solved in several directions. The most obvious solution is to obtain the dependence of the strength of the prototype on the degree of filling of the cellular structure, expressed as a percentage (%). This dependence will allow for the required strength of the prototype to choose the percentage of filling of the cellular structure.

Another way to solve the problem is to search for the shape of the cells with the greatest strength. As a natural analogue, we consider bionic structures (Fig. 1) [36].

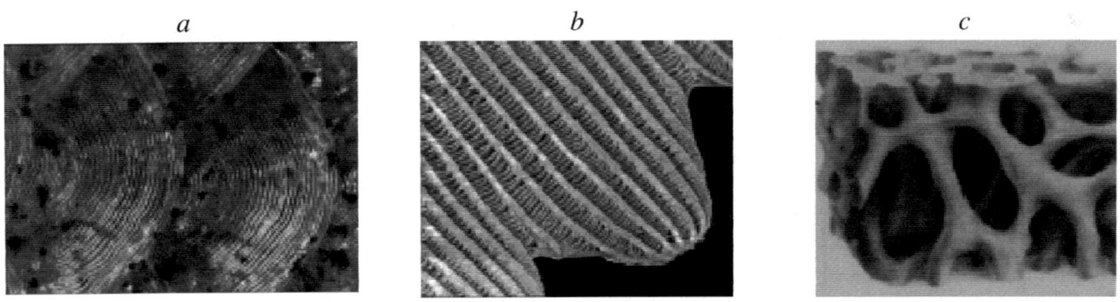

Figure 1. Natural analogues of bionic structures:
a - surface coverings of beetles and butterflies; *b* - lipid binary layers; *c* - block copolymers [36].

Such structures are found in the chloroplast membranes of plants and in cuticular formations on the wings of butterflies. They combine high strength characteristics and low weight [36].

An artificial analogue of such structures is a cellular structure called a "gyroid" with the topology of a triple periodic surface of minimum energy, consisting of repeating elements with the smallest possible area (Fig. 2) [43].

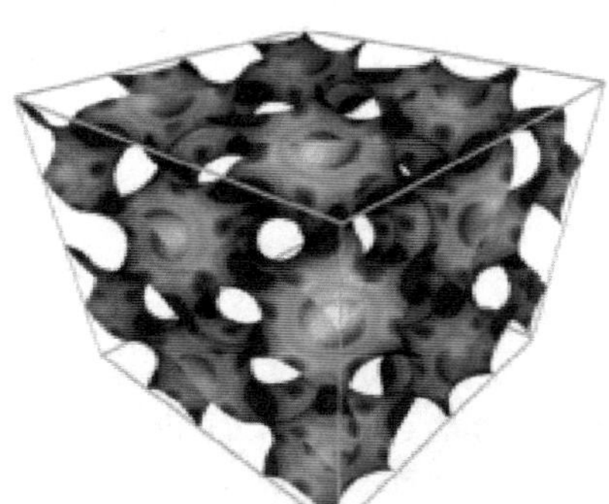

Figure 2. Cellular structure with the topology "gyroid" [43].

Such a structure was made using additive technology from polylactide by the FDM method (Fused Deposition Modeling) - layer-by-layer casting by extruded molten plastic. The physicomechanical properties of the samples were studied. A qualitative improvement of the strength characteristics of the structure using the topology of cells of the "gyroid" type is shown. The authors of the study found that fracture during loading of the sample in different directions relative to the axis of construction occurs, depending on the direction, brittle or plastic [36]. Obviously, further study of this issue is required to describe recommendations on the use of the cell topology of the "gyroid" type in additive technology. This area of research is promising and will allow to qualitatively increase the strength characteristics of the prototype.

No less promising is the direction of using composite materials in additive technology. A large number of studies have been carried out on the introduction of nanostructured additives into polymers, the use of which makes it possible to obtain composites with useful properties [30–33, 35, 37, 42, 44, 45].

For the modification of polymers, the most promising is the modification method under the influence of low-temperature plasma [34, 39,46].

In this work, we studied the strength properties of samples of a modified polymer made by 3D printing. As a modified polymer, a composite of polylactide polymer and nanostructured materials is used.

2. Research methods

To modify the polymer, nanostructured materials were used, which were synthesized by the low-temperature plasma method. In the method used, low-temperature plasma is formed at the site of an electric discharge between two graphite electrodes. The process of polymer modification takes place in one technological volume simultaneously with the synthesis of nanostructured materials during thermal decomposition of graphite by passing polymer strands in the immediate vicinity of the synthesis zone of nanostructured materials. The method allows us to solve several problems in one technological cycle: the direct synthesis and activation of carbon nanostructured material, as well as the modification of polymers in the solid state, synthesized nanostructured materials.

The study of the obtained nanostructured materials was carried out using the methods of derivatography and electron microscopy [40, 41]. A Q1500 D derivatograph and an EMV-100A electron microscope were used for research.

The polymer used was polyamide, a polyester of lactic (2-hydroxypropanoic) acid CH3-CH (OH) - COOH.

The tensile test specimens were manufactured using additive technology on a Maker Bot Replicator Z18 3D printer with 100% fill of the mesh structure according to the settings recommended by the equipment manufacturer. The geometric dimensions of the samples were: total length - 150 mm, the distance between the wide parts with parallel sides - 108 mm, the length of the narrow part with parallel sides - 60 mm, the radius of the head - 60 mm, the width of the wide part - 20 mm, the width of the narrow part with parallel sides - 10 mm, thickness - 4 mm (GOST 33693-2015).

The tensile strength study was carried out by stretching samples of primary and modified polylactide on a tensile testing machine MP-0.5-1 according to the manufacturer's instructions.

3. Research results

The carbon nanostructured material synthesized by thermal decomposition of graphite was subjected to derivatographic studies in order to identify the type of nanostructured formations and their percentage in the total mass of the material. Studies were carried out on the bulk of the synthesized nanostructured material weighing 230 mg. The decomposition of the samples was carried out with a constant increase in temperature in the open sample holder - crucible. The tests were carried out with a change in the heating rate in increments of 10 ° C per minute until a temperature limit of 1000 ° C was reached, followed by the completion of the research.

As can be seen from the graph of the dependence of the sample mass (m, mg) on the annealing temperature (T, °C), the decay of the carbon material began at a temperature of about 530 ° C and ended at a temperature of about 900 ° C (Fig. 3). The nonlinear dependence of the mass decrease with increasing temperature was monitored, since a sharp decrease in the mass of the samples was observed at a temperature of 630 ° C and 846 ° C. Most likely, this is due to the presence in the synthesized nanostructured material of nanofibers with a decay temperature of 630 ° C and nanotubes with a decay temperature of 846 ° C. This assumption was further confirmed by the results of electron-optical studies. Between the indicated points on the temperature scale, the decrease in the mass of the samples occurred close to a linear dependence.

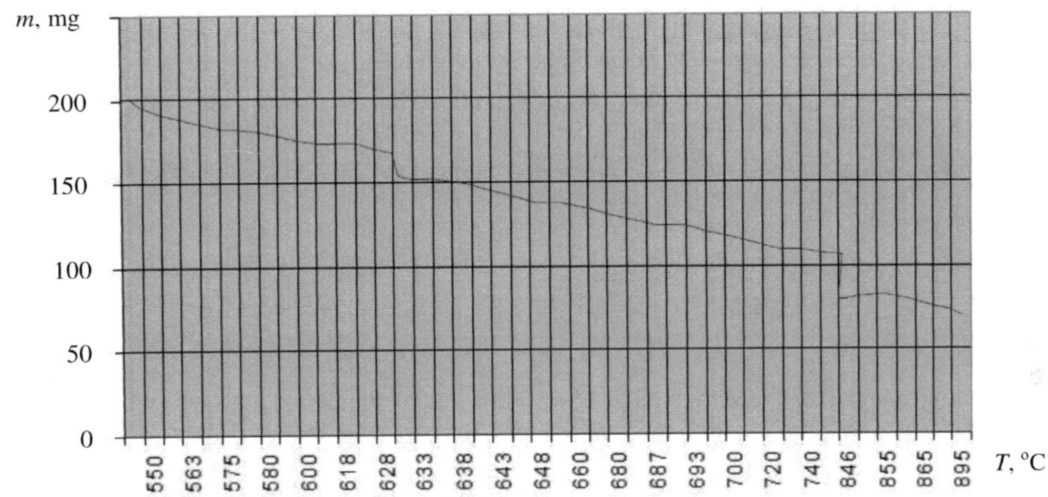

Figure 3. Derivatogram of annealing of carbon nanostructured material

Investigations of the obtained carbon nanostructured materials using an electron microscope made it possible to evaluate the size and morphology of nanoobjects. It was found that during the synthesis of carbon nanomaterials in a low-temperature plasma, carbon nanotubes and nanofibers are formed (Fig. 4, a), as well as needle nanotubes (Fig. 4, b). The method of synthesis of carbon nanostructured materials during thermal decomposition of graphite allows to obtain a sufficient amount of them necessary for polymer modification.

Subsequently, samples of the modified polylactide prepared by the method of exposure to low-temperature plasma were subjected to investigation. An analysis of the data showed that the composition of the modified polylactide contains up to 1% of nanotubes and up to 3% of nanofibers of the total mass of the composite.

It should be noted that carbon nanostructured material is located in the surface layer of polylactide and is evenly distributed over the entire area.

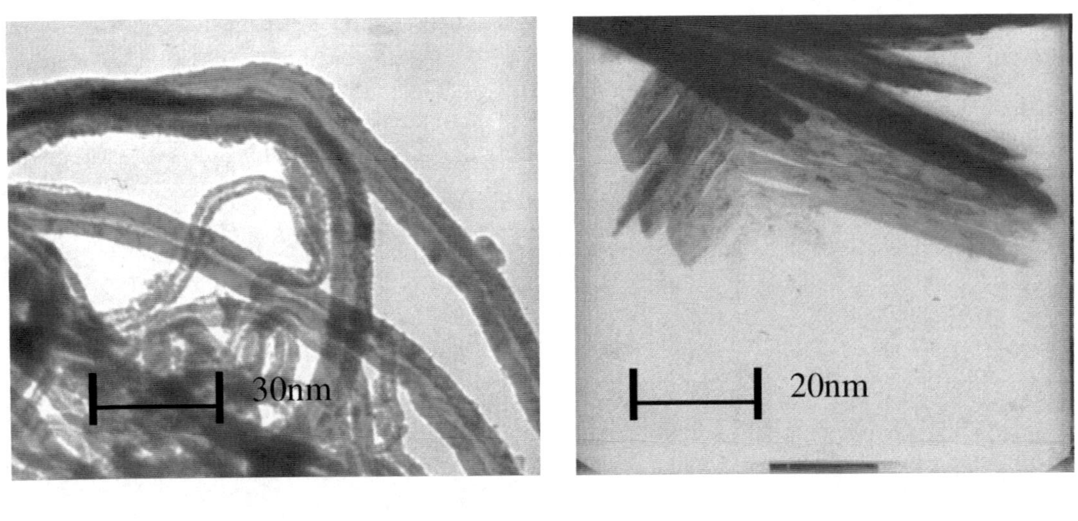

Figure 4. Photos of synthesized nano-objects: a – carbon nanotubes and nanofibres; b – needle nanotubes

Samples were made from modified and primary polylactide by 3D printing. Prior to further research, samples were stored in a package of vapor- and moisture-proof material.

Studies of the ultimate tensile strength of the primary and modified polylactide were carried out on an MP-0.5-1 tensile testing machine.

The first group consisted of control samples from primary polylactid without nanostructure additives - 6 samples. The second group - samples from modified polylactide - 6 samples Using optical microscopy, fractures of samples from primary and modified polylactide were studied. A typical view of the fractures is shown in Fig. 5.

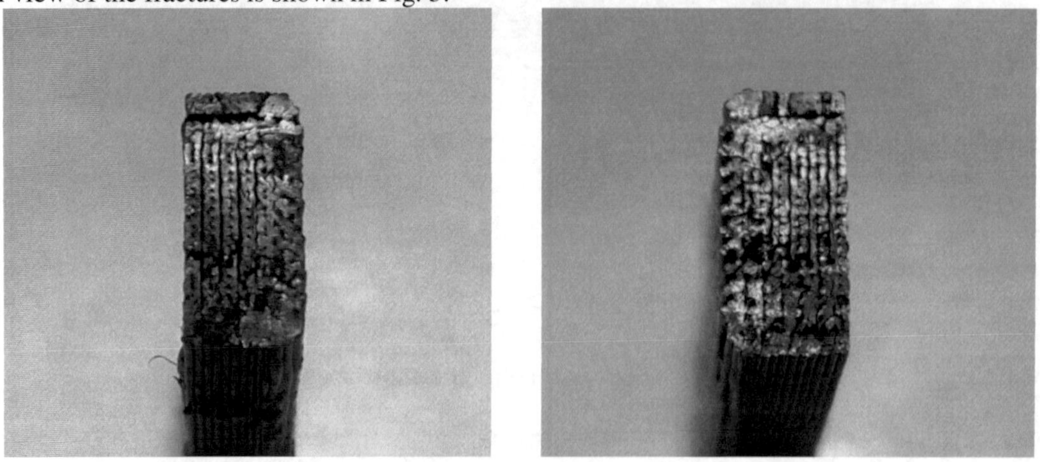

Figure 5. Fracture of samples made of primary (a) and modified (b) polylactide

Visual studies showed that the fracture surface of the samples of each of the two groups was porous. Polymer filaments that formed the samples were clearly visible. By the degree of plastic deformation, the samples of both groups are fragile. The fracture of the samples under the action of the load is flat with bevels. Light reflection by fractures of samples is uneven, matte and shiny areas with yellow and blue tint are observed, occupying up to 30% of the fracture area. Optical studies showed a similarity in the appearance of fractures in all the samples under study. The results of testing two groups of material samples on a tensile testing machine and their statistical analysis are summarized in table 1.

It was found that the values of the mass index within each group of samples are not significantly different. The relationship between the change in the mass of polylactide and its modification is not monitored, since the difference in the average mass of the samples of the two studied groups is less than 1%. A characteristic is the elongation rate of the samples upon rupture of the modified polylactide. Its value decreased by almost 2 times (by 40.8 relative%) compared with the control samples from the primary polylactide. The decrease in the plasticity of the modified polylactide is obviously associated with the presence of nanostructured materials in its structure. The effect obtained requires a more detailed study.

In each group of the studied samples there is a spread in the value of the tensile strength of the material. The tensile strength of samples from modified polylactide, in comparison with control samples, increased by up to 10%.

An analysis of the experimental results showed that it is necessary to conduct additional studies of modified polylactide with varying degrees of filling of the cellular structure of the samples.

The obtained results show the prospect of further research in the direction of polylactide modification and the study of the physicomechanical properties of products made from it.

Table 1. Test results of samples of primary and modified polylactidan rupture

№ sample	Primary polylactide			Modified polylactide		
	weigh, g	elongation, mm	breaking load, MPa	weigh, g	elongation, mm	breaking load, MPa
1	9.65	15	20.74	9.72	7	22.06
2	9.68	14	21.37	9.84	7	22.75
3	9.65	11	21.33	10.22	8	21.57
4	10.04	15	20.7	9.78	6	23.53
5	9.78	9	18.39	9.71	10	22.26
6	9.66	12	21.13	9.73	7	22.75
Average value	9.74	12.7	20.61	9.83	7.5	22.49
Range of variation	0.39	6	2.98	0.51	4	1.96
Dispersion	0.0235	5.8667	1.2640	0.0382	1.9	0.4602
Standard deviation	0.1534	2.4221	1.1243	0.1785	1.3784	0.6784

4. Conclusions

A brief overview of possible approaches to improve the strength characteristics of prototypes made by additive technology of polymers is given.

A method of manufacturing samples from polylactide containing nanostructured materials is proposed. The method made it possible to increase the tensile strength of modified samples by up to 10%. This confirms the assumption about the possibility of increasing the strength characteristics of the prototype when using nanostructured materials. The authors believe that research in the following areas will be promising:

- obtaining the dependence of the strength of the prototype on the degree of filling of its cellular structure;
- conducting research on products made with a cellular structure of the "gyroid" type;
- conducting a study of the strength characteristics of modified polymers depending on the modes of their modification;
- identification of the best option for the dependence of the strength of the product on the above approaches.

Reference

[1] Bobryshev A N, Erofeev V T and Kozomazov V N 2013 Polym. compos. mater. textbook (Publishing house DIA) 480 p.
[2] S A Wolfson V G, Oshmyan N S and Enikolopov 1990 Principles for the creation of compos. polym. mater. (Berlin: Chem) 240 p.
[3] Khozin V G 2004 Strengthening of epoxy polym. (Kazan: Publishing house of the PNK "Printing House") 446 p.
[4] Askadsky A A et al 2001 Gradient polym. mater. *Russian Chem. J.* V**45** No **3** pp123-128
[5] Stepanova V F, Stepanov A Yu and Zhirkov E P 2013 Compos. polym. reinforcement. (Moskow: Publishing house of TsNTP) 200 p.
[6] Selyaev V P, Ivashchenko Yu G and Nizina T A 2016 *Polym. concr.* (Saransk Mordov Publishing House) 284 p.

[7] Paturoev N I 1987 Polym. concr. res. institute of concr. and reinforced concr. (Moskow: Stroyizdat) 286 p.
[8] Vorobyov V A 1967 Fundam. of the techn. of building mater. from plastics (Moskow: Higher school) 280 p.
[9] Gusev B In 1991 Vibration techn. of concr. (Kiev: Alarm clock) 157 p.
[10] Sokolova Yu A 1990 Modified epoxy adhesives and coatings in construction (Moscow: Stroyizdat) 174 p.
[11] Solomatov V I 1984 Techn. of polym. concr. and reinforced polym. concr. prod. (Moscow: Stroyizdat) 144 p.
[12] Solomatov V I 1989 Intensive concr. techn. (Moscow: Stroyizdat) 264 p.
[13] Choshshchiev K H 1983 Techn. of polym. concr. using sand dunes (Ashgabat: Ylym) 231 p.
[14] Bazhenov Yu M 2011 Concr. techn.: a textbook for students of higher educational institutions studying in construction specialties (Moscow: DIA) 524 p.
[15] Lesovik V S 2006 Increasing the efficiency of production of building mater. taking into account the genesis of rocks (Moscow: DIA) 526 p.
[16] Elshin I M 1980 Polym. concr. in hydraulic eng. (Moscow: Stroyizdat) 192 p.
[17] Knyazev V K 1977 *Epoxy Struct. mater. in mechanical eng.* V K Knyazev (Moskow: Eng) 183 p
[18] Zaitsev Yu S, Kochergin Yu S, Pakter M K and Kucher R V 1990 Epoxy oligomers and adhesive compositions (Kiev: Naukova Dumka) 200 p.
[19] Potapov Yu B 1993 Polyester polym. concr. (Voronezh: Voronezh Publishing house) 170 p.
[20] Rebinder P A 1966 Phys. and Chem. Mech. of dispersed structures (Moscow: Science) 400 p
[21] Solomatov V I 1988 Polym.ic compos. mater. in construction (Moscow: Stroyizdat) 312 p.
[22] Suzuki Hiroshi, Hidekiso Kozjo 1972 New epoxyurethane polym.s - curing compositions based on them *Jap Plast* **9** pp 9–12
[23] Erofeev V T, Rodin A I, Yakunin V V, Bogatov A D, Bochkin V S and Chegodajkin A M 2018 Alkali-activated slag binders from rock-wool production wastes *J. of Civil Eng.* **82** (**6**) Pp 219-227 doi: 10.18720 / MCE.82.20.
[24] Yusupova A A, Akhmetova R T, Treshchev A A, Bobrishev A A, Shafigullin L N and Lakhno A A 2016 Production and investigation of properties of sulfide compos. mater. based on technogenic sulfur waste with titanium chloride as an activator *Res. J. of Pharmaceutical, Biological and Chem. Sciences* **7** (**6**) Pp 1614-19
[25] Erofeev V 2016 Frame Construction Compos.s for Buildings and Structures in Aggressive Environments *Procedia Eng* **165** Pp 1444-47 doi: 10.1016 / j.proeng.2016.11.11.877.
[26] Bulgakow A, Emelianov S, Bock T, Sayfeddine D and Erofeev V 2015 Inspection of flyover bridges using quadrotor *32nd Int. Symp. on Autom. and Rob. in Construction and Mining: Connected to the Future, Proceedings* Article number 117997.
[27] Bulgakov A, Erofeev V, Bogatov A, Smirnov V and Schach R 2016 Innovative production techn. of binding and building compos. mater. on the basis of glass wastes *Insights and Innovations in Struct. Eng., Mech. and Comput. - Proceedings of the 6th Int. Conf. on Struct. Eng., Mech. and Comput.* SEMC Pp 1583-86.
[28] Erofeev V, Korotaev S, Bulgakov A, Tretiakov I and Rodin A 2016 Getting Fired Material with Vitreous Binder Using Frame Techn. *Procedia Eng.* **164** Pp 166-171. doi: 10.1016 / j.proeng.2016.11.11.606.
[29] Salman Dawood Salman AL-Dulaimi, Taher AL-Dafafea, Maksimova I N and Erofeev V T 2019 Study of self-healing bio-concr. *Int. J. of Civil Eng. & Techn. (IJCIET)* **10** (**4**) Pp 1354-60.
[30] Breuer O, Sundararaj U 2004. Big returns from small fibers: a review of polym. carbon nanotube compos. *Polym. Compos.* **25** (**5**) pp 630-645.

[31] Badamshina E R, Gafurova M P and Estrin Ya I 2010 Modification of carbon nanotubes and synthesis of polym. compos. involving the nanotubes *Chem. adv.* **79** (**11**) pp 945-979. DOI: https://doi.org/10.1070/RC2010v079n11ABEH004114.

[32] Coleman J N, Khan U and Gun'ko Yu K 2006. Mechanical reinforcement of polym. using carbon nanotubes *Adv. mater* 2006 **18** (**6**) pp 689-706.

[33] Ciardelli F, Coiai S, Passaglia E, Pucci A and Ruggeri G 2008 Nanocompos.s based on polyolefins and functional thermoplastic mater *Polym. Int.* **57** (**6**) pp 805-836. DOI: https://doi.org/ 10.1002 / pi.2415.

[34] Crestinin A V, Kharitonov A P, Shulga Yu M Zhigalin O M, Knerelman E I, Dubois M and others 2009. Obtaining and characterization of fluorinated single-layer carbon nanotubes *Russian nanoTechn.* **4** (**1-2**) 115-131.

[35] Dai L, Mau AWH, 2001 Controlled Synthesis and Modification of Carbon Nanotubes and C60: Carbon Nanostructures for Adv. Polym.ic Compos. Mater.. *Adv. Mater.* **13** (**12-13**), 899-913.DOI: https: // doi .org / 10.1002 / 1521-4095 (200107) 13: 12/13 <899 :: AID-ADMA899> 3.0.CO; 2-G.

[36] Dyachenko S V, Lebedev L A, Sychev M M and Nefedova L A 2018 Phys. and mechanical properties of a model material with a topology of three times periodic surfaces of minimal energy such as a gyroid in the form of a cube *J. of techn. Phys.* **88** (**7**) pp 1014-1017. Available from: https://J.s.ioffe.ru/articles/viewPDF/46169 [Accessed March 5, 2019].

[37] Gonzalez-Guerrero A B, Mendoza E, Pellicer E, Alsina F, Fernandez-Sanchez C and Lechuga L M 2008 Discriminating the carboxylic groups from the total acidic sites in oxidized multi-wall carbon nanotubes by means of acid – base titration *Chem. Phys. Letters* **462** (**4-6**), pp 256-259. DOI: 10.1016 / j.cplett.2008.07.07.071.

[38] GOST 33693-2015 Plastics Samples for testing Moscow Standardinform. Available from: http://docs.cntd.ru/document/1200132500 [Accessed March 5, 2019].

[39 Inagaki N 1996 Plasma Surface Modification and Plasma Polym. (Lancaster: Technomic Publication Co) pp 280.

[40] Korolev A P, Makarchuk M V, Dutov M N and Loskutova A D 2017 The technique of res. of fields of short-range action of cluster nanostructures is developed. *Mech., resource and diagnostics of mater. and structures (MRDMS-2017): Proceedings of the 11th Int. Conf. on Mech., Resource and Diagnostics of Mater. and Structures 1915* 30009-1 -030009-4.DOI: 10.1063 / 1.5017329. Available from: https://www.scilit.net/article/fa6d191492dd89ac37775b74b3abbd84 [date of application: March 5, 2019].

[41] Korolev A P, Makarchuk M V, Dutov M N, Loskutova A D and Firsova A V 2018 Studying the Regimes of Forming Carbonic Nano-Objects on Copper Island Structure *XIV Int. Scientific-Techn. Conf. on Actual Problems of Electronics Instrument Eng.* (APEIE), 36-38 DOI: 10.1109 / APEIE.2018.8545282. Available from: https: //ieeexplore.ieee.org/document/8545282 [Accessed March 5, 2019].

[42] Krueger A 2010 Carbon Mater. and Nanotechn (Weinheim Wiley-VCH Verlag GmbH & Co) pp 474.

[43] Lord E E, McKay A L and Ranganathan S 2010 New geometry for new mater (Moscow: Fizmatlit) pp 264

[44] Morgan P 2005 Carbon Fibers and Their Compos (Boca Raton: CRC Press) 1200 DOI: https://doi.org/10.1201/9781420028744.

[45] Njuguna J, Pielichowski K and Desai S, 2008 Properties of Compos with Carbon Nanotubes *Polym.s for Adv. Techn.* **19** (**8**) pp 947-960.

[46] Syrgiannis Z, Hauke F, Rohl J, Hundhausen M, Graupner R, Elemes Y and Hirsch A 2008 Sideline Functionalization of SWNTs by Nucleophilic Addition of Lithium Amides *Eur. J. of Org. Chem.* **15** 2544- 2550.DOI: 10.1002 / ejoc.200800005.

Carbonate-Quartz Cement Composites

I V Erofeeva[1], D V Emelyanov [2], M A S Bushes [2], E M Balatkhanova[3], S V Nadorov [2], A S Bogdanov [2] and P A Efremov.[2]

[1]Research Institute of Building Physics of the Russian Academy of Architecture and Construction Sciences, 21 Lokomotivny lane, Moscow, 127238, Russia

[2]National Research Mordovian State University named after N.P. Ogarev, Bolshevistskaya str, 68, 430005 Saransk, Russia

[3]Grozny State Oil Technical University named after Academician M.D. Millionshchikov, 100 Huseyn Isaev Avenue, Grozny, Russia

E-mail: emelyanoffdv@yandex.ru

Abstract. The technology of obtaining multicomponent high-quality concrete using limestone powders as fillers is considered. In order to select fillers for powder-activated concrete, a study was made of the water absorption of limestone gravel of various kinds. According to the water absorption index, eight categories of limestone species have been identified. It was revealed that organogenic limestone is one of the materials that is preferred as a filler for high-quality concrete. The experimental results were obtained and graphical dependences of changes in the tensile strength under bending and compression, the dynamic modulus of elasticity and water demand of organogenic cement composites filled with limestone from granulometric composition and filler content were constructed. It has been established that the best strength properties are manifested in materials with optimal indicators of the content and size of limestone and silica powder.

1. Introduction

In world construction practice, the demand for concrete is increasing year after year. According to the International Federation of Structural Concrete, the annual production of concrete currently stands at more than 2.5 billion m3, and in the near future it is projected to reach a volume of up to 4.0 billion m3. In recent years, significant successes have been achieved in the technology for producing high-quality concrete. A significant proportion of research in the field of high-quality concrete is accounted for by the development of multicomponent concrete containing superplasticizers and active dispersed microfillers [1–12]. High results of strength and other properties were obtained on powder-activated sand concrete of a new generation, developed at the Department of TBKiV Penza State University of Architecture and Civil Engineering [13-15]. The term "powder-activated concrete" was proposed by Professor V. Kalashnikov. [16, 17]. Later, a number of works by his followers and students aimed at the study of high-strength and extra-high-strength concrete appeared, in which the basic principles for creating such concrete were formulated [18–23]. Powder-activated concretes of a new generation are plasticized concretes with a high content of suspension component [14]. Despite a large number of studies in the field of creating high-quality concrete, to date, the problems of disclosing the

IOP Conf. Series: Materials Science and Engineering **972** (2020) 012058 doi:10.1088/1757-899X/972/1/012058

mechanism for controlling the surface activity of mineral fillers with optimal particle sizes and the mechanism of interaction in a hydrated multicomponent medium require further study. In this regard, the implementation of scientific research aimed at solving the issue of developing a technology for producing multicomponent high-quality concrete using local mineral fillers of natural and man-made origin is relevant.

2. Research methods

In the manufacture of concrete, materials based on naphthalene and melaminesulfonic acids, polycarboxylates, etc., and finely dispersed fillers: waste from the production of ferrosilicon, slags, ashes, local sands, etc., are used as superplasticizers. The most used are polycarboxylate superplasticizers, quartz and limestone powders. Limestone fillers in cement compositions interact with clinker minerals, reducing the volume of capillary pores. Limestone fillers in their composition contain a large amount of calcium carbonate. Therefore, the consideration of limestone powders as fillers is of considerable interest. It is known that, depending on the amount of water absorption by mass, after 48 hours, limestone is divided into the following categories: limestone of the 1st category ($W < 2\%$) are most suitable for producing concrete of all grades from M200 to M1200; limestone of the 2nd category ($W = 2 \ldots 4\%$) for the production of materials of grades from M200 to M1000; limestone of the 3rd category ($W = 4 \ldots 6\%$) for grades from M200 to M800.

In order to select fillers for powder-activated concrete, a study was made of the water absorption of limestone gravel of various kinds. Eight samples were taken from various deposits: Zhigulevsky quarry (Samara region), Issinsky quarry (Penza region), Yelnikovsky quarry (Republic of Mordovia), limestone crushed stone (quarry of Argun, Chechnya), mountain dolomite (n.a. Duba-Yurt, Chechnya), river limestone (Khul-Khulau river, Chechnya), limestone of the Nizhny Novgorod region (village of Madayevo) and belemnite containing calcite of organic origin. Also for the prototype were taken samples of crushed granite. The chemical composition of some limestone fillers is given in table 1.

Table 1. The chemical composition of limestone fillers

Filler	Content, %											
	other impu-rities	SiO_2	Al_2O_3	Fe_2O_3	CaO	MgO	SO_3^{2-}	K_2O	Na_2O	Cl^-	W	Z
Limestone river	43.77	0.29	0.09	0.09	51.50	4.52	0.18	0	0.04	0.007	0.65	–
Dolomite mountain	41.94	0	0.02	0.06	37.17	20.14	0.04	0	0.18	0.049	0.40	–
Limestone organogenic (belemnite)	45.74	0.116	0.033	0.034	53.100	0.295	0.208	0.004	0.226	–	–	0.003

3. Research results

Before carrying out the research, the samples of crushed stone were dried to a constant weight at a temperature of 105...110°C in an oven. After drying, the samples were weighed and placed in water. After 1 hour, the grains were removed from the water and placed on dry matter. Each grain was individually rubbed with a piece of cloth and weighed on an electronic balance accurate to 0.01 g. The time dependence of the water absorption of limestone rock samples is shown in the diagram (Fig. 1). As can be seen from the results of the study, the water absorption of the samples depends on the porosity of the material and is directly related to its capillary structure. Samples absorb water on the first day of exposure (Fig. 1). After 3 days, the weight of the samples stopped changing, this indicates the complete water saturation of crushed stone. As expected, belemnite and granite rubble have the lowest level of water absorption; in percentage, the level of water absorption after 3 days was 0.27% and 0.63%, respectively. Among the investigated limestones, the lowest water absorption after 3 days has limestone crushed stone (quarry of Argun, Chechnya) - 0.83%. The remaining samples are arranged in the following order: Zhiguli gravel - 1.6%; mountain dolomite (n.a. Duba-Yurt, Chechnya) - 1.63%; river limestone (Khul-Khulau river, Chechnya) - 2.49%; Issinsky quarry (Penza region) - 3.19%, Madaevsky quarry (Nizhny Novgorod region) - 7.77% and Elnikovsky quarry (Republic of Mordovia), which has the highest level of water absorption among the studied samples - 8.87%.

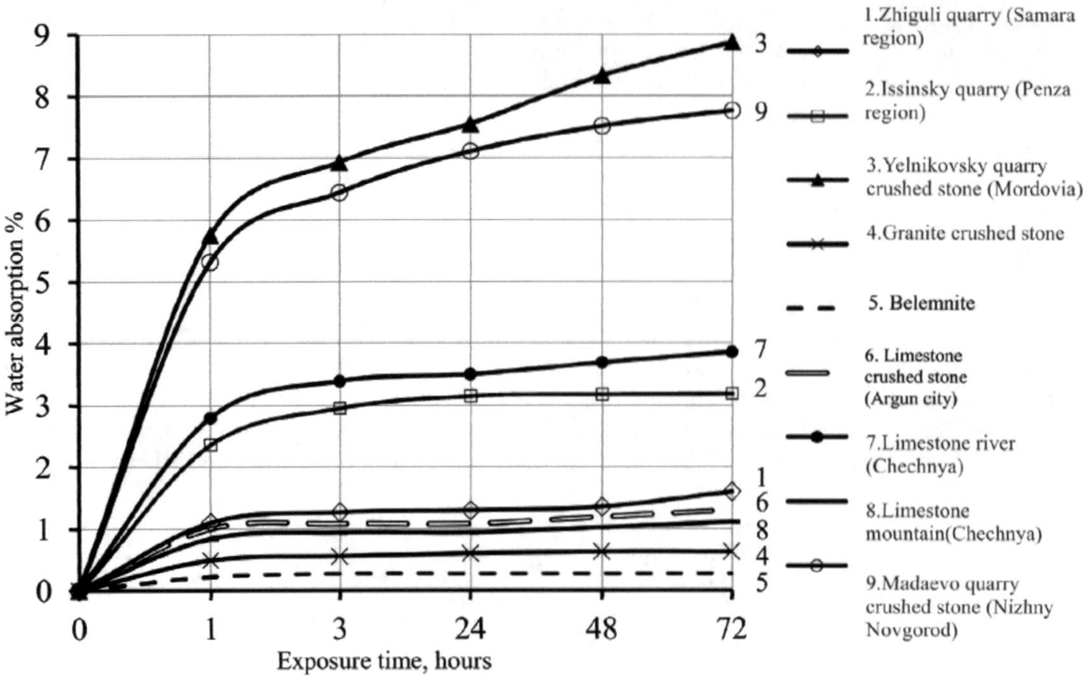

Figure 1. The dependence of changes in water absorption of samples of limestone rocks of various deposits from the exposure time.

Thus, we can attribute to category I (water absorption no more than 2%): limestone crushed stone (quarry in the cities of Argun, Chechnya), crushed stone of the Zhigulevsky quarry (Samara region) and dolomite mountain (e.g., Duba-Yurt, Chechnya). Crushed stone from the Issinsky quarry and river limestone (the Khul-Khulau river, Chechnya), in turn, belongs to the II category (from 2% to 4%), Yelnikovsky crushed stone and crushed stone from the Nizhny Novgorod region (Madayevo) cannot be attributed to any category, so how they have significant water absorption.

Considering that the interaction of limestone with clinker minerals does not lead to the formation of compounds with high strength, it is advisable to use them together with quartz fillers. It is known that the efficiency of filling composite materials is achieved by combining in the structure of the composite fillers not only of a different nature, but also of size. Based on the scientific analysis of domestic and foreign works in the field of creating cement composite materials, it was found that the most preferred are the compositions of composites, in which limestone fillers are used as small components, and quartz powders are larger. The optimization of compositions with carbonate-quartz fillers was carried out using mathematical methods for planning the experiment. To perform the experiment, a matrix in the form of a simplex-lattice Scheffe plan was used, consisting of 10 experiments. The following parameters were considered as optimized parameters: water demand, dynamic modulus of elasticity, bending and compression strength. The studied compositions included Portland cement 500 – D0 produced by the State Unitary Enterprise "Chechencement" (n. Chiri-Yurt), quartz powder (the deposit is located on the Terek River 20 km north of Grozny) and organogenic limestone powders (belemnite) of different fineness . When conducting a scientific experiment, the following factors were chosen as variation factors: X_1 – amount of quartz powder with a fineness of 3 100 ... 3 300 cm^2/g, X_2 – amount of limestone with S_{sp} = 6,000 ... 6,200 cm^2/g, and X_3 – amount of limestone with S_{sp} = 9 000 ... 9 200 cm^2/g. The planning matrix and the experimental results are shown in table 2.

Table 2. Planning matrix and experiment results

№ of composition	Index	The composition of the mixture, wt. hours			Water demand, %	Dynamic modulus of elasticity, E × 10^3 MPa	Strength, MPa	
		X_1	X_2	X_3			when bending	Under compression
1	n_1	1	0	0	47.3	23.18	20.6	53.3
2	n_2	0	1	0	39.7	24.97	23.1	55.7
3	n_3	0	0	1	43.0	22.91	16.1	42.3
4	n_{122}	1/3	2/3	0	41.1	25.12	20.5	49.3
5	n_{133}	1/3	0	2/3	43.8	24.33	19.1	50.5
6	n_{233}	0	1/3	2/3	41.5	23.96	19.4	43.8
7	n_{112}	2/3	1/3	0	42.7	21.27	20.6	43.0
8	n_{113}	2/3	0	1/3	47.7	23.40	19.7	44.6
9	n_{233}	0	2/3	1/3	41.7	25.14	20.3	51.7
10	n_{123}	1/3	1/3	1/3	42.4	22.39	20.2	52.0

Statistical processing of the experimental results made it possible to obtain the dependences characterizing the change in the tensile strength in bending and compression, the dynamic modulus of elasticity and water demand of filled cement composites on the particle size distribution and the content of fillers:

$$R_b = 16,6X_1 + 22,1X_2 + 19,8X_3 + 4,50X_1X_2 - 21,6X_1X_3 - 10,8X_2X_3 + +26,55X_1X_2(X_1\text{-}X_2) + \\ + 54,45X_1X_3(X_1\text{-}X_3) + 95,4X_2X_3(X_2\text{-}X_3) - 218,7\,X_1X_2X_3 \qquad (1)$$

$$R_c = 43,7X_1 + 57,6X_2 + 58,1X_3 + 18,23X_1X_2 - 36,68X_1X_3 - 23,4X_2X_3 + 40,63X_1X_2(X_1\text{-}X_2) + \\ + 207,23X_1X_3(X_1\text{-}X_3) + 151,65X_2X_3(X_2\text{-}X_3) - 472,05\,X_1X_2X_3 \qquad (2)$$

$$E = 20,1X_1 + 25,12X_2 + 24,16X_3 + 9,77X_1X_2 - 7,25X_1X_3 - 2,66X_2X_3 + 36,14X_1X_2(X_1\text{-}X_2) + \\ + 43,7X_1X_3(X_1\text{-}X_3) + 65,07X_2X_3(X_2\text{-}X_3) - 198,22\,X_1X_2X_3 \qquad (3)$$

$$WD = 48,7X_1 + 44,7X_2 + 46,7X_3 - 17,78X_1X_2 - 7,65X_1X_3 + 4,5X_2X_3 - 19,13X_1X_2(X_1\text{-}X_2) - \\ - 13,95X_1X_3(X_1\text{-}X_3) - 76,5X_2X_3(X_2\text{-}X_3) + 170,78\,X_1X_2X_3 \qquad (4)$$

Based on the regression equations, graphical dependencies are constructed (Fig. 2)

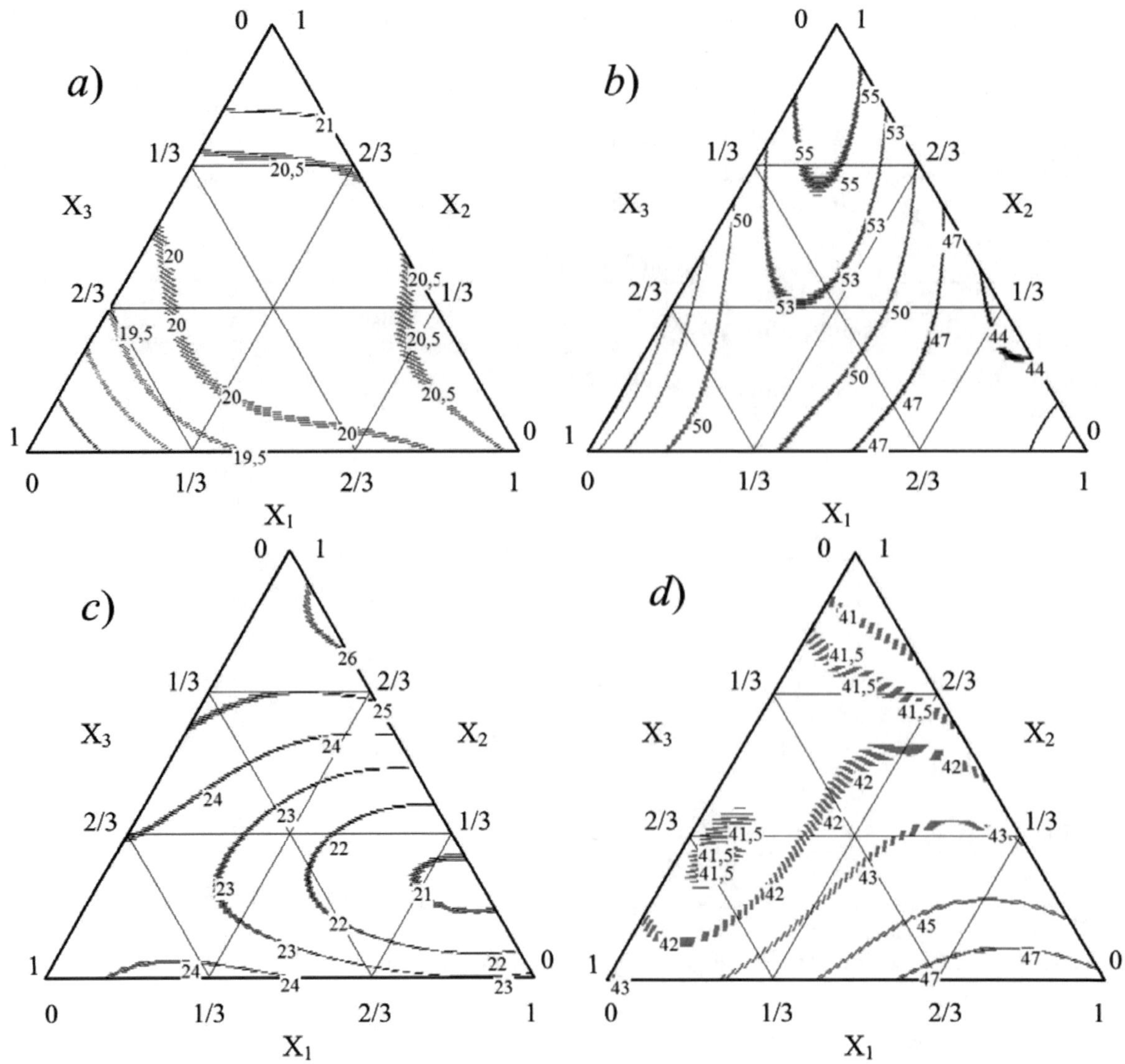

Figure 2. Dependences of changes in bending strength indicators (a), compression (b), dynamic elastic modulus (c) and water demand (d) of cement composites filled with a mixture of limestone of organogenic and quartz sand of different particle size distribution.

4. Conclusions

From the graphic dependences of the change in the physicomechanical properties of the fine-grained compositions of the composites, it follows that the best strength properties are manifested in materials with optimal indicators of the content and size of limestone and quartz powder.

Currently, the authors are developing matrices of various kinds using limestone fillers from deposits in the Volga and North Caucasus regions to create new generation of powder-activated concrete.

References

[1] Bazhenov Yu M 2014 Modern concr. technol. Concr. and reinforced concr. – a look into the future: sci. papers of the III All-Russian (II Int.) conf. on concr. and reinforced concr. (Moscow, May 12-16, 2014): in 7 vol. 7. Plenary reports. (Moscow: MGSU) p. 23-28

[2] Silver Deo 2011 Aspects of the use of non-metallic fiber. Study of the use of fiber for concr. products CPI *Int. Concr. Production* **4** P. 46-56.
[3] Aitchin P C and Neville A 1993 Perfotmance concr. Demnstified, *conct. Yntern* **15** (1) p 21–26
[4] Kaprielov S C, Travush V I and Karpenko N I 2008 Modified high-strength concr. of classes B80 and B90 in monolithic structures *Building mater* **3** p 9-13
[5] Chernyshov E M, Korotkikh D N and Artamonova O V 2010 Nanotechnol. conditions for managing the structure formation of high-strength cem. concr. *Proc. of the Central Regional Branch of RAASN* (Voronezh) p 102-123
[6] Usherov-Morshak A V 2009 Modern concr. and its technol. Coll. *"Concr. and reinforced concr."* (Saint Petersburg: Slavutich) p 20-24
[7] Dvorkin L I, Solomatov V I, Vyrova V N and Chudkovsky S M 1991 Cem. concr. with mineral fillers. (Kiev:. Publishing House Budivelnik) p 135
[8] Erofeev V, Kalashnikov V, Emelyanov D, Balathanova E, Erofeeva I, Smirnova O, Tretiakov I and Matvievskiy A 2016 Biological Resistance of Cem. Compos. Filled with Limestone Powders *Solid State Phenomena* **871** P. 22-27
[9] Erofeev V, Kalashnikov V, Emelyanov D, Balathanova E, Erofeeva I, Smirnov V, Tretiakov I and Matvievskiy A 2016 Biological Resistance of Cem. Compos. Filled with Dolomite Powders *Solid State Phenomena* **871** P. 33-39
[10] Bazhenov Yu M, Chernyshov E M and Korotkikh D N 2014 Design of structures of modern concr. defining principles and technol. platforms *Building mater.* **3** p 6-14
[11] Korotkikh D N 2014 Crack resistance of modern cem. concr. (Voronezh: Voronezh GASU) p. 141
[12] Chernyshov E M and Korotkikh D N 2014 Modification of the structure of cem. stone with micro- and nanosized silica particles (theory and application issues) *Building mater., equipment and technol. of the 21st century* **5** p 30-32
[13] Kalashnikov V I 2008 Through a rational rheology in the future of concr. Part 3. From high-strength and extra-high-strength concr. of the future to superplasticized generalpurpose concr. of the present *Concr. Technol.* **1** p. 22
[14] Kalashnikov V I, Erofeev V T, and Tarakanov O B 2016 Suspension-filled concr. mixes for powder-activated concr. of a new generation *Izvestiya Vysshikh Uchebnykh Zavedenii. Construction* 4 (688) p 30-37
[15] Kalashnikov V I 2012 How to turn old-generation concr. into high-performance new generation concr. *Concr. and reinforced concr.* **1** p. 82
[16] Kalashnikov V I 2011 Terminology of the sci. of new generation concr. *Building mater.* **3** p 103-106
[17] Kalashnikov V I 2012 What is powder-activated concr. of a new generation *Building mater.* **10** p 70–71
[18] Gulyaeva E V, Erofeeva I V, Kalashnikov V I and Petukhov A V 2014 The influence of rationally active additives on the strength properties of plasticized cem. stone *Young scientist* **19**
[19] Moroz M N, Kalashnikov V I and Yerofeyeva I V 2015 Effective new generation concr. with low specific cem. consumption per unit of strength *Young scientist* **6** p 189-191
[20] Kalashnikov V I, Volodin V M, Erofeeva I V and Abramov D A 2015 Highly effective self-compacting powder-activated sand concr. and fiber concr. *Modern problems of sci. and education* **1-2**
[21] Kalashnikov V I, Khvastunov A V and Khvastunov V L 2017 Strength and deformability of high-strength concr. and fiber-reinforced concr. with low specific cem. consumption per unit strength *Compos. building materials. Theory and practice: a collection of articles by the Intern. Sci. and techn. conf. dedicated to the memory of V.I. Kalashnikov*. Edited by V.A. Coastal p 103-113

[22] Kalashnikov V I, Moroz M N, Tarakanov O V, Kalashnikov D V and Suzdaltsev O V 2014 New ideas on the mechanism of action of superplasticizers co-milled with cem. or mineral rocks *Building mater.* **9** p 70
[23] Moroz M N, Kalashnikov V I, Suzdaltsev O V and Yanin V S 2014 High-strength decorative and finishing surface-hydrophobized concr. *Regional architecture and construction* **1** p 18-23

IOP Conf. Series: Materials Science and Engineering **972** (2020) 012059 doi:10.1088/1757-899X/972/1/012059

Nurturing cognitive skills in undergraduates with the help of ontological analysis

A N Gushchin[1], M N Divakova[1]

[1]Department of Urban Development and Landscape Architecture, Ural State University of Architecture and Art, Karl Liebknecht st. 23, Yekaterinburg 620075, Russia

E-mail: alexanderNG@yandex.ru

Abstract. The article deals with the didactic problem that arises in the two-stage system of bachelor's study - master. During the period of study in the master's program, the bachelor needs to make a qualitative leap in the style of thinking and move from the visual-language thinking of the bachelor to the critical conceptual thinking of the master. Why bachelor needs to acquire cognitive skills as soon as possible. The article substantiates the use of the ontological analysis method as a didactic tool for teaching undergraduates. The effectiveness of ontological analysis as a didactic tool with specific examples is demonstrated.

1. Introduction

The transition to a two-stage system of training bachelor - master revealed a didactic problem of the relationship between applied and theoretical. Bachelor - a technical specialist with a fairly wide range of technical skills. As an architect, he participates in the development of project documentation. Master - a specialist with systems thinking skills. As an architect, he can develop concepts for various projects. This can be confirmed by the presence in the standard of the new generation of the Federal State Educational Standards FGOS 3++ competence «UK-1. Able to carry out critical analysis of problem situations on the basis of a systematic approach, to develop a strategy for action» [1]. Thus, the difference in training levels is not quantitative, but qualitative. Therefore, in the opinion of the authors, the main task of training in the magistracy is mastering the skills of systems thinking necessary for the transition to the next level of mastering specialty.

Let us consider in more detail what the critical analysis of problem situations means. Psychologist Diane F Halpern gives the following definition of critical thinking «Critical thinking is the use of those cognitive skills that increase desirable outcome. It used to describe that is purposeful, reasoned, and goal directed – the kind of thinking involved in solving problems, formulating inferences, calculating livelihoods, and making decisions, when the thinker is using skills that are thoughtful and effective for particular context and type of thinking task» [2]. The basis for critical thinking is cognitive skills. Cognitive skills – is skills that use concepts and logical constructions. Cognitive skills in comparison with visual-action skills and graphical-language skills, customary for architects, is a later stage in the development of thinking. In the context of the formulated problem, this is the difference between the visual-language skills of the bachelor and the cognitive skills of the master. This is supported by Le Corbusier's famous statement that architecture is not a profession, but a style of thinking.

However, the nurture of cognitive skills in architects is hampered by the fact that architecture, being a synthetic discipline, comes into contact with many other scientific disciplines, borrowing their conceptual apparatus. As a result, the conceptual apparatus of architectural science itself suffers. The methodological weakness of the conceptual apparatus of architecture is convincingly shown in the work of N.P. Ovchinnikova [3]. This is even more true of landscape architecture, which interacts with even a wider range of scientific disciplines than traditional architecture. Thus, the problem of the nurture of cognitive skills among undergraduates enrolled in the specialty «landscape architecture» is not only relevant, but also complex.

A significant contribution to the study of cognitive skills was made by the Soviet psychologist L.S. Vygotsky [4], who supposed that the study of systemically organized scientific knowledge contributes to the nurture of cognitive skills. Systematically organized scientific knowledge is available in any textbook, and the study of textbooks is an indispensable part of the educational process. But this is a passive way for nurturing the cognitive skills. Today, active educational technologies are needed. Consider modern techniques that can nurture cognitive skills.

One of the current trends is the widespread introduction of information technology. The introduction of information technology is impossible without preparing the environment in which the technology operates. Preparation consists in the formalization and specification of the subject area, i.e. in creating a domain model suitable for use in information technology. In the opinion of the authors, these technologies form a «digital» style of thinking even more than the final result of the introduction of information technologies, which predetermines their significance. One of the technologies for the description and formalization of the subject area is ontological analysis. The modern interpretation of the term «ontology», not a philosophical, applied, is formulated by T. Gruber: «Ontology is the exact specification of conceptualization». «Conceptualization» means «an abstract, simplified view of the world that people use to achieve a certain goal» [5]. The essence of ontological analysis is to highlight the basic concepts of the domain and to establish the relationship between them. Ontological analysis is the modern active technology for the formation of cognitive skills. According to Russian standard GOST R 57297-2016, ontologies can have a hierarchical structure: ontologies of upper levels, ontologies of basic subject areas, ontologies of subject areas, applied ontologies [6]. When using ontological analysis in the educational process, we can talk about applied ontologies within the chosen scientific topic of master's thesis.

Currently, the use of ontologies in the educational process is developing in two directions: the analysis of curricula and programs using ontologies, the introduction of ontologies directly into the educational process. An example of the analysis of the curriculum can be found in the article I.Yu. Shpolyanskaya and I.I. Miroshnichenko [7]. Representatives of the second direction even created a new term "ontological training". Examples of the use of the ontological approach to learning P Buitelaar, P Cimiano, B Magnini [8], as well as A Maedche, S Staab [9]. Ontological learning is based on technology Semantic Web and is designed to train knowledge engineers. Paul Butelara et al. Defines the concept of ontological learning «Ontology learning is inherently multidisciplinary due to strong connection with Semantic Web witch has attached researchers from a very broad variety disciplines: knowledge representation, logic, philosophy, databases, machine learning, natural language processing, image processing, etc.». Previously, the authors analyzed the process of introducing computer technologies into the practice of architectural design and concluded that in the future the architect will need the competence of a system analyst [10].

Another advantage of ontological analysis is that it is a kind of system analysis. The systems approach, about which architects like to talk about, is «the scientific method of knowledge, which is a sequence of actions to establish structural links between variables or permanent elements of the system under study» [11]. Thus, the use of ontological analysis in the educational process will allow us to purposefully form the competence of UK-1 of the standard FGOS 3 ++ about which it was mentioned above.

Thus, the significance of ontological analysis for the training of architects is as follows: ontological analysis is a type of system analysis, ontological analysis is useful for the formation of critical and conceptual thinking. And the development of conceptual thinking is very important for architects who are accustomed to the visual-figurative style of thinking. We demonstrate the prospects of using ontological analysis in the educational process.

2. Method

For the practical implementation of ontological analysis, various standards have been developed, in particular, the IDEF5 ontological analysis standard [12]. According to the standard, the result of the process is an ontological model, which includes two components: a glossary containing the basic concepts (objects) of the subject area with their characteristics and a graphical diagram reflecting the relationship between the concepts (objects). The connections between objects in the IDEF standard are rich and diverse. Authors prefer to use a simpler scheme that includes several types of connections: a connection is «is a subtype», a connection is «consists of», a connection is «converted to» (such connections are called rules or processes). Typical schemes are built on the basis of these connections: a «classification» type scheme – figure 1, a «composition» type scheme – figure 2.

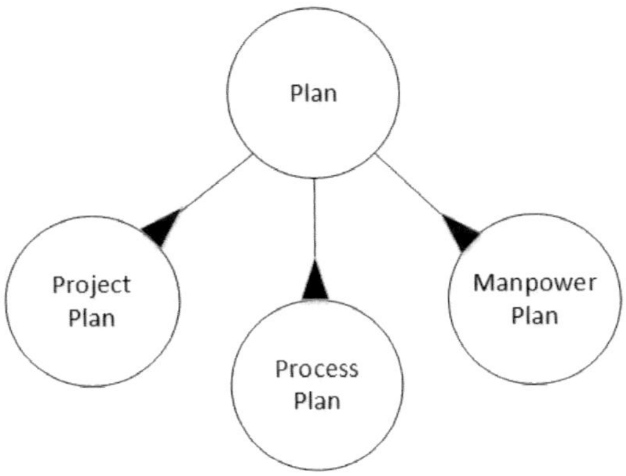

Figure 1. Classification chart

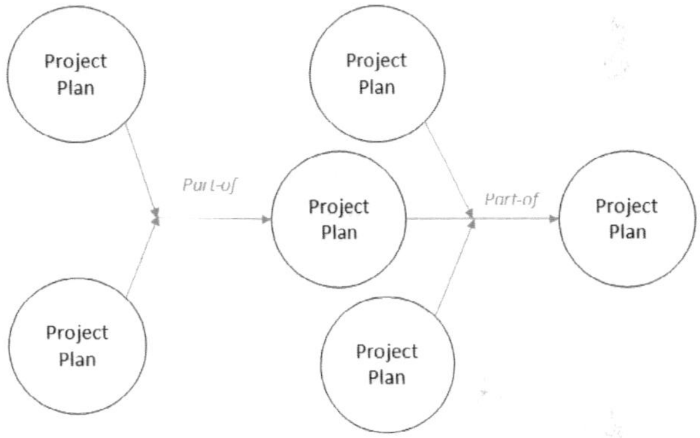

Figure 2. Diagram of the composition (structural)

The third type of charts "is converted to ..."shown in figure 3. Such diagrams are also called rules.

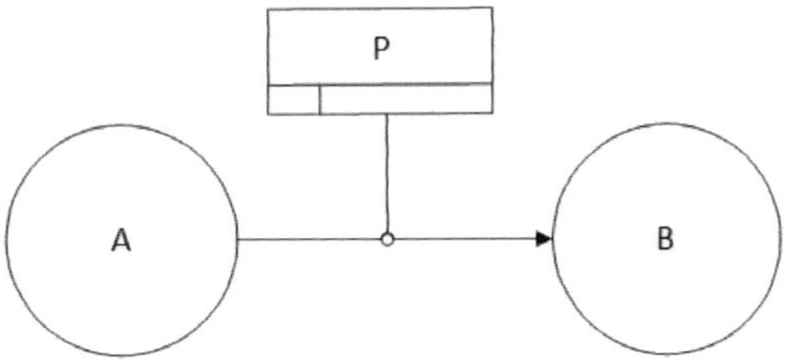

Figure 3. Entity A is converted to entity B using rule P

Using a limited number of schemes is convenient for learning purposes and allows you to quickly form cognitive skills. The presence of a graphic component is valuable for architects who are used to the visual presentation of their ideas.

Further, according to the standard, before starting the study, it is necessary to determine the purpose of the study, its scope and point of view. This is done because the point of view determines the «perspective», under which the researcher analyzes the system. In architecture, they talk a lot about the systems approach, and the realization that the research point of view determines the characteristics of the system under study is very important for the future architect.

The task of performing an ontological analysis of the subject area is given to undergraduates of landscape architects during second semester of the course «Methodology, Methodology and

IOP Conf. Series: Materials Science and Engineering **972** (2020) 012059 doi:10.1088/1757-899X/972/1/012059

Presentation of Research». The task of performing an analysis is formulated as follows: conduct an ontological analysis of the subject area of his dissertation research. The results are presented in the form of ontology in IDEF5 notation. The ontology created by the undergraduate lies in public discussion.

3. Examples

Next, the results of the task on creating the ontology of the domain will be presented (for simplicity, we will give only a graphical relationship between concepts, without a glossary). It is important to note that the diagrams characterize the understanding of the undergraduate at this particular point in time. And work on the correct meaningful interpretation should be made in the course of the educational process.

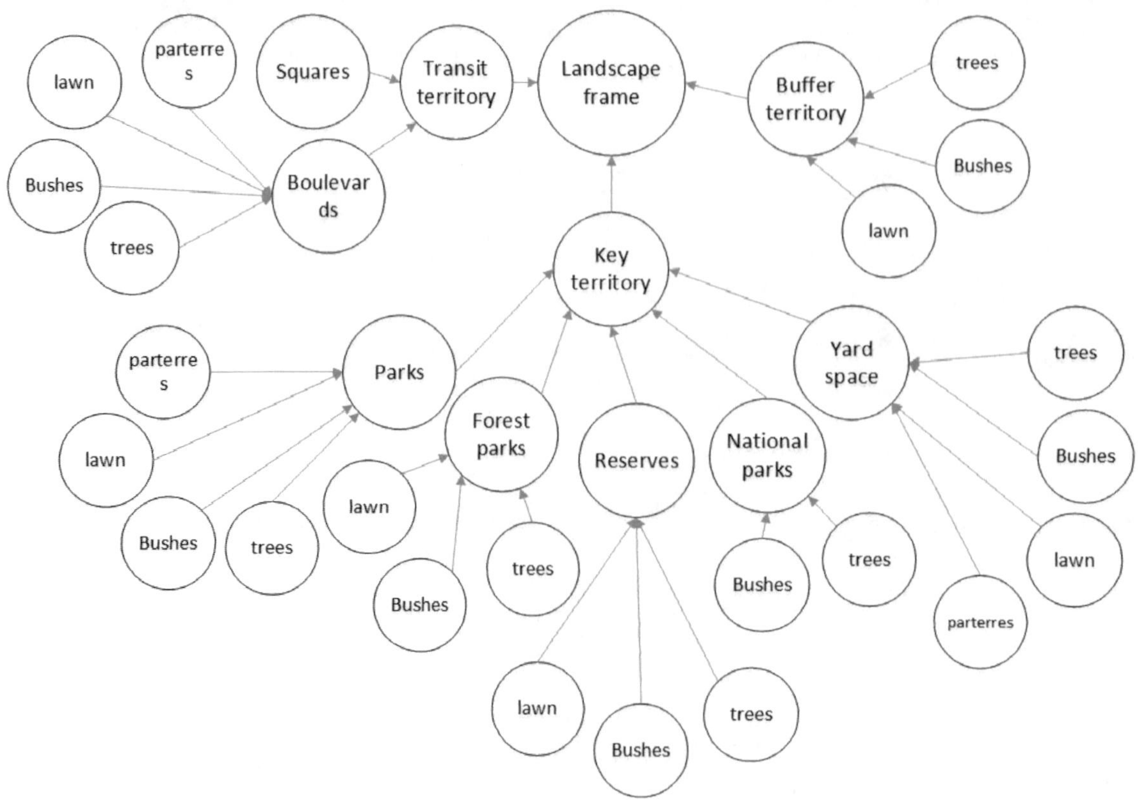

Figure 4. Landscape framework and related concepts. Undergraduate A.K.

Figure 4 shows the graphical part of the ontology, which demonstrates what content the undergraduate invests in the concept of «landscape framework». Relations between concepts allow two interpretations: as a classification diagram and a composition diagram. The undergraduate chooses a composition chart that answers the question «what does it consist of». In the diagram, the undergraduate identified a fairly large number of concepts. At the lower level, the same objects are used: trees, bushes, lawns, but this is due to the possibilities of the graphical representation of concepts. In the glossary, object types are classified according to the definitions [13].

In general, this suggests that the undergraduate prevails practice type of thinking and well-expressed cognitive skill - ability to systematize. At the same time, the model proposed by the undergraduate is not very suitable for the critical analysis of problem situations. Thus, the task performed on ontological analysis allows you to accurately establish thinking skills and ways for further individual work with the undergraduate.

Figure 5 shows the graphical part of the ontology, showing the relationship between concepts in the field of urban identity, beatification and identity. On the scheme, the undergraduate uses all the basic types of connections between concepts allowed by the standard. The urban identity, according to the undergraduate, includes («consists of») the landscape, climate, unique city buildings, a unique urban layout, the history of the city. Branding includes («consists of») style and identity. With the help of style, you can convert a typical beautification in the unique one. The style has the following varieties: constructivism, classicism and other styles.

In general, the ontology built by the undergraduate has a rich internal structure and contains a variety of semantic connections between concepts. In this case, the undergraduate demonstrates not only the developed skills of conceptual thinking, but also builds a model of the subject area, which includes a variety of relationships between entities (objects) of the subject area. There is also no doubt that the dynamic connections identified by the undergraduate can be used for critical analysis. In general, we can conclude that the undergraduate has the competence of UK-1 «It is able to carry out a critical analysis of problem situations on the basis of a systematic approach, to develop a strategy for action».

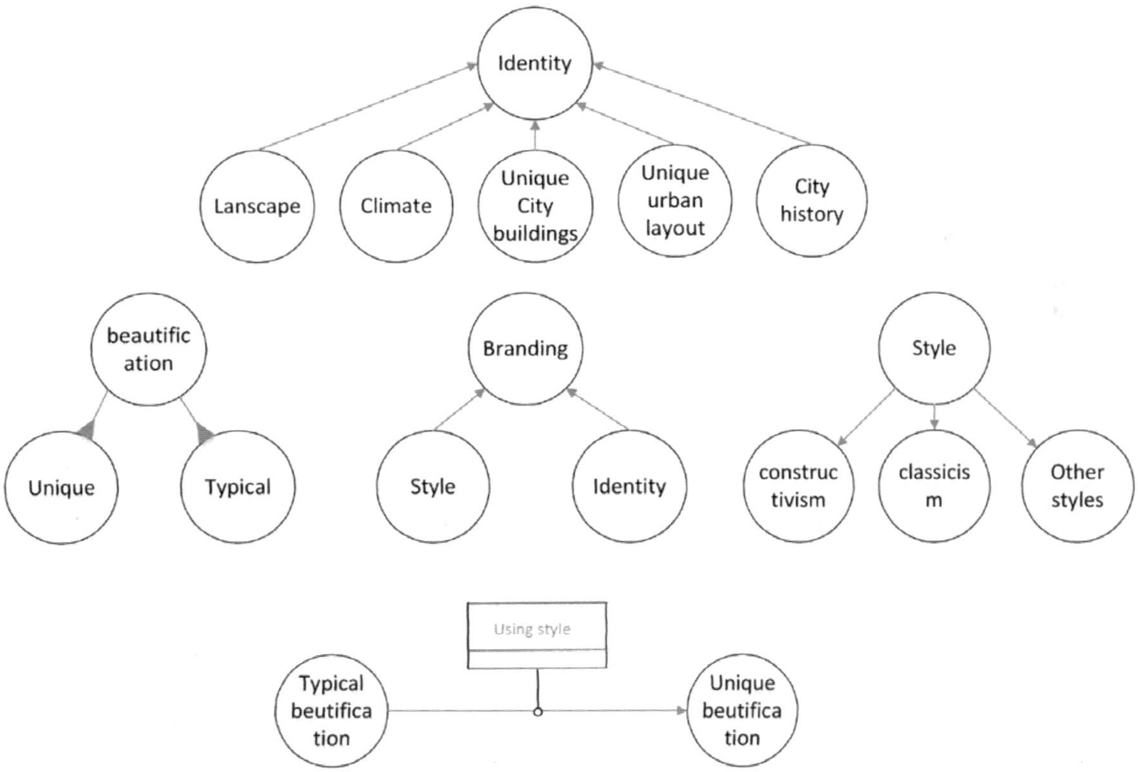

Figure 5. The relationship of concepts in the field of identity. Undergraduate E.T.

4. Conclusion

Earlier (in the introduction to the article) the importance of using ontological analysis in the educational process was justified from a theoretical point of view. The importance of ontological analysis was shown not only for the formation of one or another competence, but also for the formation of critical and conceptual thinking as the basis for the formation of other competencies. Further, using concrete examples, the possibilities of ontological analysis were demonstrated for

IOP Conf. Series: Materials Science and Engineering **972** (2020) 012059 doi:10.1088/1757-899X/972/1/012059

assessing not only the generated competence, but also the assessment of the thinking style of each particular undergraduate.

By their own recognition, undergraduates' task of creating an ontology of the subject area really creates the basis for project and research activities. When grading for the assignment, the number of terms in the ontology is taken into account, first, the number and variety of connections between the terms.

References

[1] On the approval of the federal state educational standard of higher education - master's degree program in the field of preparation 07.04.01 Architecture. 2017 *Ministry of Education and Science of the Russian Federation. Order No. 520 of June 8*

[2] Diane F Halpern 2007 The Nature and Nurture of Critical Thinking *Critical Thinking in Psychology* (Cambridge: University Press) p 6

[3] Ovchinnikova N P 2011 *Osnovy naukovedeniya architecture: studies* (Saint Petersburg: SPSUKU) 288 p

[4] Vygotsky L S 199 *Thinking and speaking* Ed 5 (Moscow: Publishing house «Labyrinth») pp 352

[5] Gruber T R 1993 Translation Approach to Portable Ontology Specifications *Knowledge Acquisition* **5/2** p199-220

[6] GOST R 57297-2016 2016 An integrated approach to the management of information life cycle man-made objects and environments. Libraries of electronic components taking into account the requirements of complex information modeling (Moscow: Standartinform) pp 22

[7] Shpolyanskaya I Yu, Miroshnichenko I I 2010 The ontological model of the information-educational environment of the university subdivision *Bulletin of Rostov State University of Economics (RINH)* **31** p 1-9

[8] Buitelaar P, Cimiano P, Magnini B 2005 Ontology learning from text: methods, evaluation and applications (IOS press)

[9] Maedche A, Staab S 2004 Ontology learning /Handbook on ontologies (Berlin: Springer) p 173-190

[10] Gushchin A N, Divakova M N 2018 System analyst - the future competence of the architect *Architecture and construction of Russia* **3(227)** pp 102-107

[11] Blauberg I V, Yudin E G 1973 *The formation and essence of a systematic approach* Moscow Science pp 270

[12] IDEF5 1994 Method Report Prepared for: Armstrong Laboratory AL/HRGA *Wright-Patterson Air Force Base Ohio 45433 Revision Date: September 21, 1994 Contract Number: F33615-C-90-0012*

[13] Set of Rules 2016 *Beautification of territory* Updated version of SNiP III-10-57 (Moscow)

Theoretical and methodological approaches to management of resource flow processes of development projects on macro-, mezzo-, microlevels

A Platonov[1], V Larionova[1], Y Davy[1], S Bazhenov[1]

[1]Ural Federal University, 19, Mira str., Yekaterinburg, 620002, Russia

Email: a.m.platonov@urfu.ru, naukaservis@rambler.ru, v.a.larionova@urfu.ru, yury.davy@urfu.ru

Abstract. Paper is devoted to the actual problem of managing resource flows of development projects as complex, unique, and open systems, across the different project lifecycle stages that differ in content and are distributed in space and time. The idea was to create a single throughout system for managing resource flow processes of the long-term projects. The research goal included development of theoretical and methodological approaches that can be used in order to manage the combination of development project resource flows using "order from chaos" paradigm. That includes by-stage organization of resource flow processes on of macro-, meso- and microscale, going from sources and sinks of the chaotic project environment resolve to ordered stages of project life cycles. It is proposed to implement resource flow management in form of the stage and resource-based "project-logistic" relay of a triune movement, ordering and transformation of the resource flow processes. This relay is created within a project-logistic field of project management that change by project phases. Synergetic effects of self-organization and self-control appear in the project logistic field due to horizontal interactions, sharing experience, competences, and values among the stakeholders in course of transforming entropic resource-flow processes into the negentropy processes related to project phase goals (attractors). Results can be used to forecast and develop multilevel resource-flow and process systems arranged by resource sources and sinks, and chaotically ordered for projects in different application areas. Further research perspectives include development of self-control and self-organization forms for development projects stakeholders that would comply with international best practices, and fit corresponding project stages.

Keywords: chaotic resource environment, resource sources and sinks, flow ordering and "transformations", entropy, negentropy, attractor stages, project-logistic paradigm, project-logistic relay, project-logistic field.

1. Introduction

The project management is a special kind of management. It is different from business management by socio-economic systems and from management of organizationally cyclic productions and multi reproducible technological operations in different spheres of industry. Project management contents sphere of management of traditional investment-building projects (IBP) for creating objects of estate and development projects (DP) for creating, operation and improving these objects. DP is different from IBP by more significant (in many times) period of life cycle (LC) and much bigger diversity of their project dependent, technologically interconnected phases.

Except preinvestment, investment and less operation phase, realized in traditional or called as "terminal" IBP with named cost and finish periods [1] (in the frames of classical project paradigm), long LC of DP includes phases of repair, reconstruction and modernization of created estate objects, that are changed with their operation phases. Also – finish project phases, liquidation of these objects and forming new project concept [2].

In the practice of project management [3] the phases of LC of DP are detailed by certain, numerous and spaced apart in time and space project-ordered organizationally and technological steps (OTS). In flows of different steps of projects: initialization, concept, marketing and technological and economic project justification; sequential tendering; identify land territories; project, building and operation of creation estate objects as new values; purchase and sale estate objects or their territories and registration of ownership or lease (with providing benefits of investment); production, sale and service in a process of operation of estate objects; reconstruction of buildings, modernization of current production with future operation periods of improving estate objects; finish projects; liquidation of estate objects; creation and development initiatives and concepts of new project development of creation more perfect estate objects etc.

In this context DP is a project of full cycle or developing projects, which have not got named periods of operation and finish goals and can be with open data of their finish [1]. According to this DP is more open and complex space-time and organizationally technological system [3]. This system is formed in flows of different OTS of project LC by interaction and numerous participants (stakeholders) [4], which are like a flow of step by step changing by composition, qualification and quantity of management subjects.

The main goal of long process of development and realization of DP is an achievement of desirable results. Reference points definite by this goal for goal for another levels, for example, for providing OTS of project LC by necessary resources and rational management of all kinds of resource flows.

In this paper resources are that resources of different origin and features which are distributed randomly (chaotic) in space and time of external environment of projects. Energy, labor and financial, informational, service and other kinds of resources are material too. Different processes related to resource flows are formed by numerous stakeholders on the base of need in these resources.

It is obviously that these processes can't be out of space and time period, other words without their beginning and finish. They are from something external and chaotic (to DP) resources on the macro -, mezzo- and micro-resource level and finish in something stocks on these levels and numerous OTS of projects (objects).

Resource flows permeate all OTC of DP LC. According to this their activity is in straight, consistent or parallel directions, for example, activity of material, labor and energy recourses and in counter flows – for example progressive-return activity of information, financial and other kinds of flows. Therefor DP can be like flow systems by sequence of different OTC and their LC and activity of different and step by step conditioned resource flows and by changeable composition of stakeholders on different stages of project.

Numerous and interested in, responsible and interconnected legal entities and individuals with their goals, interests and values are stakeholders of management of resource flows. Namely: project initiatives and customers; investors and developers as the main stakeholders; general and sub-contractor; owners, operated new buildings; realtors and appraisers; sellers, customers and tenants of estate objects; designers, producers and suppliers materials, constructions and equipment; energy organizations; banks, control branches and public organizations; property rights registers, service and production consumers etc. [2,3].

Every stakeholder in the coordination and in interconnection with each other once upon a time makes special functions of realization their OTS that is named by "from a list" without right for mistake and without opportunities of some repeat and of forming and management of step by step recourse and flow processes.

The complex of management of resource and flow processes of DP from one side is sequence and recourse content of project OTS that is strictly ordered. By the way created, operated and modernization property objects for a long time are coordinate set and territorial fixed. From another side geography of excess external resource environment of projects on the macro-, mezzo-, micro levels is no definition and chaotic and globally distributed among sets of different level resources and stocks of the most different resources.

To the peculiarities of the resource management of the processes of DP realization, you can also add the fact that at each of the above-mentioned OTS, flow processes of one nature at subsequent stages resource flows of qualitatively different types arise ("generate"). Therefore, in addition to the problems of managing the processes of movement and the step by step ordering of resource flows, it is also necessary to consider the problem of managing some of their quantitative and qualitative changes ("transformations") occurring at the OTS borders. For example, the "transformation" of information flows of ideas and project concepts into resource flows necessary for the creation of real estate materials, energy and finance. Such resource-transformational processes of the "beginning-finish" of one and the "end-start" of others, the same or different in their origin of resource flows, objectively occur throughout the life cycle of the DP.

All this predetermines the uniqueness of the resource flow management processes in the space and time of their resource-chaotic external environment and before the organization of internal (intra-project, intra-object) resource-ordered flows according to strictly defined OTS life-cycle. And as a result - the need to find a new theoretical and methodological approach (new paradigm) to solving the problems of managing resource flows of DP.

In this regard, the goal of the study is to develop a new project and logistically informative theoretical-methodological approach to the formation of a management system of resource-flow processes of long-term, complex and one-time implemented DP.

The scientific novelty of the research will consist in the development of the theory and methodology of project-logistic management of resource-flow processes of unique DP, which is distinguished by design-informative and step by step detailed (not only in phases) level of their description and structuring. It should include made by stakeholders based on the new order "out of chaos" paradigm, managing the triune process of movements, ordering and "transformations" of resource flows from the chaotic resource environment of projects and towards different, numerous and ordered OTS of their life cycle.

The theoretical significance of the study is to develop, on the basis of the proposed paradigm of new project and logistical approach to managing the realization of the DP as a set of different in their nature and randomly-ordered streaming processes. Practical significance is in the development of legitimate scientific and practical methods for the stepwise and end-to-end management of the resource flows of unique DP. They are proposed to be implemented on the basis of the step by step "project and logistic relay race" of movements, orderliness and qualitative and quantitative "transformations" of the project resource flow processes. This "relay race" can be implemented in the subject-object and management structure that changes in time and space in the form of a globally step by step object - the "project-logistic field" of the flow management of DP resources.

2. The goal and challenges of the management of resource flow processes of DP.

With the acceleration of innovative processes in the technological and organizational and managerial sphere of PD and, consequently, because of increasing the rate of change of OTS and a reduction in their duration over time, the goal of developing a unified theoretical and methodological approach to project and logistics management of a through set of various resource flows is becoming more important. It is especially for large infrastructure projects.

To achieve formulated above goal, the following challenges were set and solved: to investigate the substantive content and possible stepwise changes in the resource-flow processes of DP; based on the analysis of existing logistics concepts to justify the need to form a new theoretical and methodological approach (paradigm) to the stepwise management of various resource flows throughout the life cycle of DP; to substantiate the possibility of creating an end-to-end project-logistic system of managing the triune process of movements, orderliness and "transformations" of various resource flows of DP with a step by step change of project stakeholders; to form a unified and dynamically changing subject-object project and logistic structure for the step by step management of project resource flows throughout their life cycle.

The object of this study is the methodology for the management of DP in the phases of creation and operation, reconstruction and modernization, closing projects and liquidation of estate objects.

The subject of the research is the theoretical and methodological approaches to the organization and management of resource flow processes in the OTS of the life cycle of DP.

A new theoretical and methodological approach to the management of resource-flow processes of DP should provide the overcoming of objectively occurring entropic effects of uncertainty and chaos in the management of resource flows and the formation of positive (non-entropic) organizational and managerial synergistic effects.

3. Existing scientific and practical approaches to the management of resource-flow processes.

Well-known theoretical and methodological approaches to the organization and management of resource and logistics flows [5] are related, as a rule, to the sphere of continuous, mass, serial or multiseries industrial production and to technologically regular and cyclical processes of local production character by place and time.

In recent years, quite extensive scientific research has been conducted in the sphere of logistic industrial production of developing the theory and methodology [6, 7, 8], logistic concepts [9] and strategies [10], methods [11] and models [12, 13] of management of resource flows between suppliers, consumers and carriers and, in particular, along "supply chains" and "logistic channels" [14, 15]. The focus of this work was on warehousing (inventory management), production, procurement, supply and sales, workshop, transport [16] and trade and intermediary logistics at enterprises of various industries.

In a number of foreign works, based on the development of a logistics methodology [17] and strategic approaches to its management [18], challenges were also solved on the formation of concepts, methods and models of logistics management [19, 20] and the project of efficient logistics systems ("supply chains") [21] in the sphere of industrial production.

When analyzing them, it can be noted that the local production and sales logistics system of industrial enterprises is primarily focused on satisfying the demand of numerous consumers (customers) for their products through the supply chain. And its main challenges is to ensure equality in the rate of production of goods to the rate of deliveries on customer orders. At the same time, a high organizational level and cyclical nature of the technological processes of industrial production provides the possibility of multiple repetition and improvement of the "supply chain" management methods. Therefore, the management of cyclically ordered logistic "supply chains" in the industrial mode of production of products remains one of the main resource-flow methods [15].

In contrast, the project-building and step by step method of creating and operating, reconstructing, modernizing and continually improving estate objects with realization of complex and unique DP is characterized by its one-timeness, non-repeatability, and the considerable length of their life cycle, as well as the multiplicity and uncertainty of resource-flow processes, that together requires a completely different management approaches.

Resource and logistics research was also conducted on the problems of resource management and in the sphere of traditional IBP. So, for example, in the work of Kubasova T.I. [22], emphasis was placed on project logistics as an applied direction related to the transaction costs of project participants and the development of the resource concept. S.G. Kiyko [23] devoted his work to the important but narrow issue of modeling the processes of movement of financial resources during the realization of the enterprise's project.

In the foreign scientific literature there are no scientific and methodological approaches to the management of resource-flow processes in the sphere of development in the phases of creation, development and improvement of estate objects or their OTS. Existing researches on this subject, as a rule, in the paradigm of "supply chains", are limited to the phases of the creating estate objects in the implementation of construction projects, or to individual aspects of managing their resource flows. [24, 25, 26, 27, 28] For example, a description of the supply system for construction projects; optimization of supply chains in a construction project, etc.

In contrast to construction projects, in the process of realization of long-term DP, there is a significantly greater uncertainty, openness and non-linearity of the system for managing resource-flow processes carried out by different and step by step teams of stakeholders. Certain contours in the approaches to the study of resource flow processes in the realization of DP were outlined in an earlier work of one of the co-authors of these studies [25].

Thus, it can be stated that the existing and most common logistic concepts (paradigms) for managing resource flows, mainly in industrial production, cannot meet the requirements for managing flow processes during the realization of OTS of unique DP. In the new theoretical and methodological approach to the management of resource and flow processes of OTS of DP, all above features of the realization of these projects and their resource flow management processes should be taken into account.

4. Research method

The main difficulty in developing a research method and substantiating a new theoretical and methodological approach to managing project-logistic resource-flow processes of DP is that the theory of project management is not a science in its strict sense. Created, in the process of realization of unique DP, estate objects can not be returned to their original position and re-created in the best way. It is also impossible based on the theory of project management alternative experimentation in the process of creating these objects or returning them to the initial state.

This is directly related to the management of resource-flow processes of DP. It is also impossible to "experiment" on the withdrawal of the property already used resources and replacing them with others, as well as the appropriate formation of return material and energy and other types of flows to the starting point.

It is also known that the theory of logistics is a theory of descriptive type, which generalizes and organizes the best practices of managing resource flow processes in industrial production, investment and construction, and in other activities.

Based on these provisions, the idea of a new theoretical and methodological approach consisted in a subject-specific step by step analysis of the flow and in a step by step description of the entire set of project-logistic processes of managing various resource flows of DP. The processes of movement, streamlining and "transformations" of resource flows were tracked from the excessive resource environment of projects at the macro, mega and micro levels and further in the direction to the project-defined and ordered OTS of life cycle.

According to the results of the analysis, a paradigm of a new project-logistic resource-flow approach to managing a spatio-temporal and randomly ordered variety of flow processes towards OTS projects can be synthesized as stepwise attracting attractors — future desirable states [30].

In this aspect, the management and self-management processes of the various resource and logistic flows of DP will be built up taking into account the formation of a system of some "virtual" resources. Namely, on the basis of goals, interests and responsibility, the effects of interaction and the mutual influence of stepwise changing project stakeholders, as mutually supporting subjects of managing resource flow processes. At the same time, DP stakeholders, overcoming the entropy resource flow chaos of the external resource environment of projects, will have to form certain non-entropic

IOP Conf. Series: Materials Science and Engineering **972** (2020) 012060 doi:10.1088/1757-899X/972/1/012060

impulses[1] to streamline resource flows and systemic effects of self-organization and self-management [34], focusing on their OTS-attractors of their LC [31, 32, 33]. That, in turn, will allow the emergence of organizational and managerial synergistic effects [35, 36, 37] in the project and logistics of DP.

5. Project and logistic approach to management of resource flow process of DP

"Open flow-type systems" [38], in which self-organization processes can occur, belong to the class of complex non-linear systems that have the properties of self-regulation and self-management and are capable of being attracted to certain intended goals. Nonlinear systems are influenced by random small effects generated by their nonequilibrium and instability. This effect is expressed in the accumulation of fluctuations, in the increase of bifurcations and in the spontaneous transitions of these systems from one state to another [34], which impede the achievement of intended results or require effective organizational and managerial decisions. Such systems can be attributed to the management systems of a vast set of DP resources that are different in their nature, subject to numerous random influences in the excessive resource environment of projects, as well as design and logistic processes during the realization of project oriented OTS projects.

As it is known, the paradigm of a general managerial nature adopted in the theory of logistics - "improving the management of resource supplies" was formulated without taking into account the specifics of the main production-technological or project-specified processes. Therefore, in the development of this paradigm for the sphere of highly organized and technologically advanced industrial production, it was developed substantial and widely used production-logistic and process-logistic resource-management concepts (paradigms) - supply chain management, "just in time", "Kanban" and other.

It can be also traced a certain evolution of the management paradigms for traditional IBP - from managing the cost, timing and quality of project realization to managing the complexity, efficiency and values of projects and further to managing regular interaction with customers of projects in Agile models. In this aspect, for the realization of DP it is also required its own, taking into account their substantive specificity, the project and logistic paradigm of managing resource flows in space and time. Such a project-logistic paradigm of managing incremental processes of movement, orderliness and various "transformations" of resource flows is the "orderliness out of chaos" paradigm as ordering flow processes through OTS projects from the "chaos" of their external resource environment[2].

The application of this paradigm to the processes of managing resource flows of DP is due to the following factors: a variety of project-logistic resource-flow processes that are complex and open, non-linear and multi-component, self-organizing and self-regulating system; the formation by stakeholders of DP of resource flow processes from spatio-temporal and resource-chaotic "sources" and "drains", distributed ("scattered") in some random way at macro, mega and microresource levels; objectively necessary step by step goal orderliness of various resource flows in the direction to the project-specific OTS of projects as goal attractors and their completion at each stage; multiple, objectively and constantly occurring quantitative and qualitative "transformations" of resource flows at the boundaries of the transition from one OTS of projects to another; the interaction of numerous and variable in composition and size, experience and qualifications of project stakeholders to achieve their goals on specific OTS and the overall results of the realization of DP based on trust, mutual exchange of competences and values of activities.

The need to develop a project-logistic paradigm "ordering out of chaos" for an open system of organization and management of resource flow processes of DP can be justified by the following

[1] Hazen A.M. Incorrectness of the non-entropic principle of L. Brillouin // [www.kizsoft.com.ru/intell/KSNews_62.htm / (the data of issue: 10.11.2018).

[2] It is required also for the sphere of for example military (destroyed) projects the development of specific military-logistic paradigm of management of resource flow of war purpose from random and resource environment and to direction of random fronts of war activities also for the sphere of for example military (destroyed) projects.

arguments: the objective complexity of implementing rational management of numerous flow processes of projects from a single center; excessive diversity of the external resource environment of the projects compared to the resources necessary for their realization; the presence of elements of uncertainty and chaos in the management system of flow processes; the emergence and disappearance in time and space of the realization of the OTS of a number of initiative "centers" of self-management and self-organization of stakeholders, complementing the "necessary diversity" of methods, mechanisms and tools for managing the entire set of DP resource flows.

In this case, the project-logistic management paradigm acts as the prevailing conceptual position and the main trend of targeted actions by the participants in the management of resource flow processes of DP. Its application to the field of motion, orderliness and "transformations" of the resource flows of DP that is open and mutually agreed between the stakeholders of the projects is reasonable. At the same time, the end-to-end management of this triune process will represent, in space-time coordinates, a global flow and project-object management mechanism in the form of a step by step resource and randomly ordered "project-logistic relay race (PLRR)" of resource flow processes through LC of OTS projects.

The key point in the proposed mechanism is that the PLRR of the resource flow processes is objectively determined and supported, in turn, by the "relay race" of goals, interests and value attitudes of numerous and step by step changing stakeholders of DP who are interested, interacting and responsible for the results of realization of individual OTS and projects in a whole. The above-described step by step interactions and changes in both subjects and DP objects in the process of managing resource flows is allowed to form, based on the project and logistic management paradigm, "orderliness from chaos" and on the basis of PLRR, a subject-objective, spatial-temporal and step by step changing project management structure - "project-logistic field (PLF)" of resource flow project management during their long-term life cycle.

Thanks to a new theoretical and methodological approach, the activities of stakeholders in managing multiple resource flow processes of DP acquires, taking into account the project and logistics paradigm, a substantive and semantic focus, leading to organizational and managerial synergies in managing resource flows and, ultimately, to step by step planned project results.

6. Forming project logistic structure of resource flow process management of DP

How do they arise, where do they originate from and how are future DP formed? Apparently, from some transcendental spheres, tangible or intangible by people at the level of knowledge and sensations, insights or faith, often outside their consciousness, cognitive abilities and skills. These areas act as global informational unordered "chaos sources" and "chaos-flows" of ideas, concepts and theories developed somewhere, once and by someone. By analogy with this, ideas and concepts of their resource provision and some methods of project and logistics management of various resource flows of projects can also formed by the participants (stakeholders) in the realization of future DP.

Based on the need to form a specific set of resources for future DP, project stakeholders identify and analyze information spheres about external and randomly scattered in space and time "sources" and "flows" of material, energy, financial and other types of resources and possible resource flow processes with a high level of uncertainty and chaos (entropy).

Resource flows generated in such a way from the externally chaotic environment of projects under the influence of project-defined goals and taking into account interests and interactions of stakeholders are then transformed into intraproject (intra-object) and step by step (non-entropy) processes in time and space. The sequence and interaction of these processes sets the step by step PLRR of the various resource flows in all OTS of LC of DP.

Figure 1 shows one of the numerous stepwise fragments of the PLRR in the formation by the stakeholders-contractors of the resource flows of DP from the territorial-object "flows" and "sources" of resources for the realization of a specific and project-specific OTS of their life cycle. In particular, between the territorial logistics terminals and the property for certain resource-flow blocks and areas of PLRR: from the "source" of territorial-terminal flows to the "stock" and "source" of information

IOP Conf. Series: Materials Science and Engineering **972** (2020) 012060 doi:10.1088/1757-899X/972/1/012060

and terminal-object flows and further to the ordered area of object flows and to the object "stock" of resource flows.

As it can be seen in Figure 1, from the “source” of various resource territorial-terminal flows (Block IV, “source”), resource “sink” of mesological information flows (with a high level of entropy) with various resources for realization specific OTS project (Block V, “stock”) is formed in the territorial logistic terminals by stakeholders. In the future, based on it, the “source” resource (with a high level of entropy) of various terminal-object flows (Block V, “Source”) is “generated”.

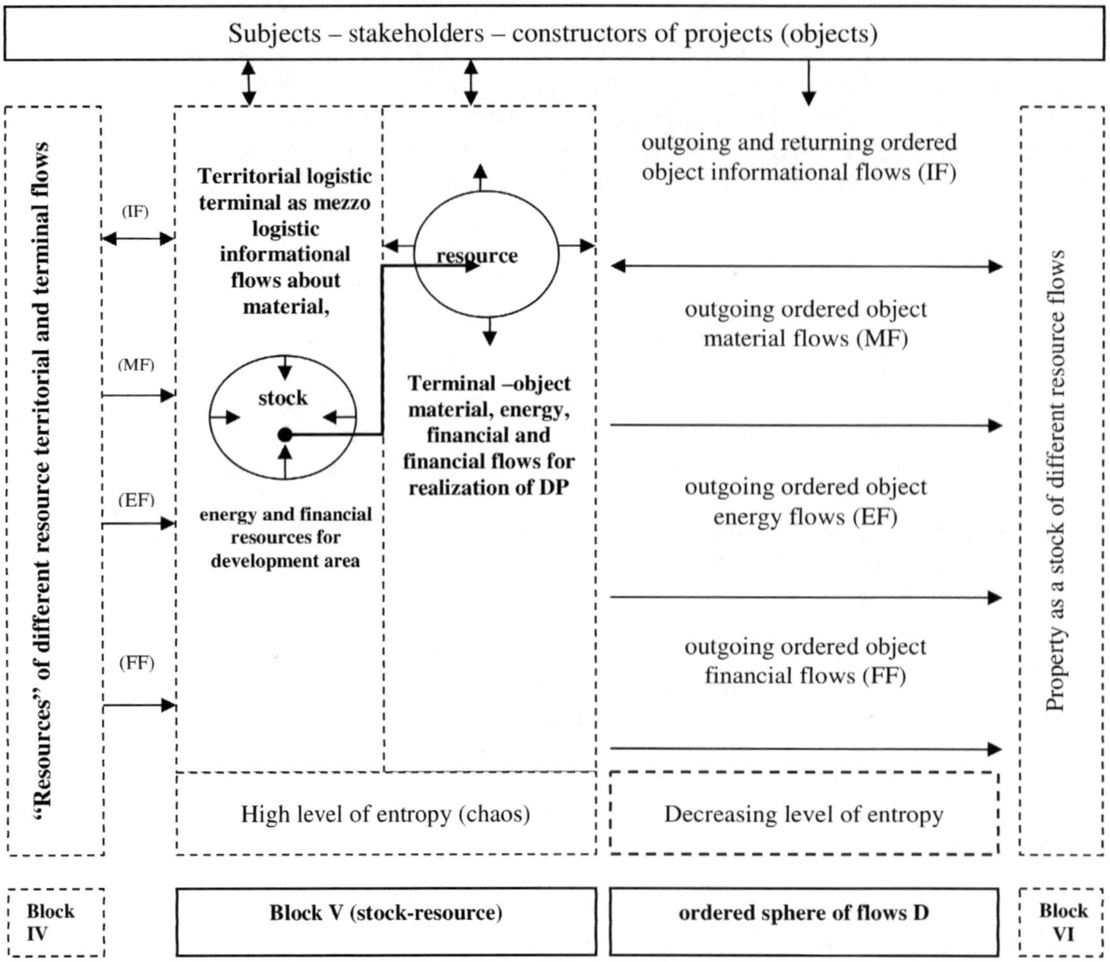

Figure 1. Fragment of the “project and logistics relay race” as a sequence of processes of management of the movement, orderliness and “transformations” of resource flows along OTS of LC of DP caused by interests and the purposes stakeholders – contractors

Stakeholders-contractors carry out at this OTS of DP the processes of ordering various object (outgoing and returning) flows with a corresponding decreasing the level of entropy (ordered sphere of flows D). The resource flows that are ordered in this way can form further on the newly created real estate object, an on-site ordered resource “stock” (Block VI, “stock”) of various resources.

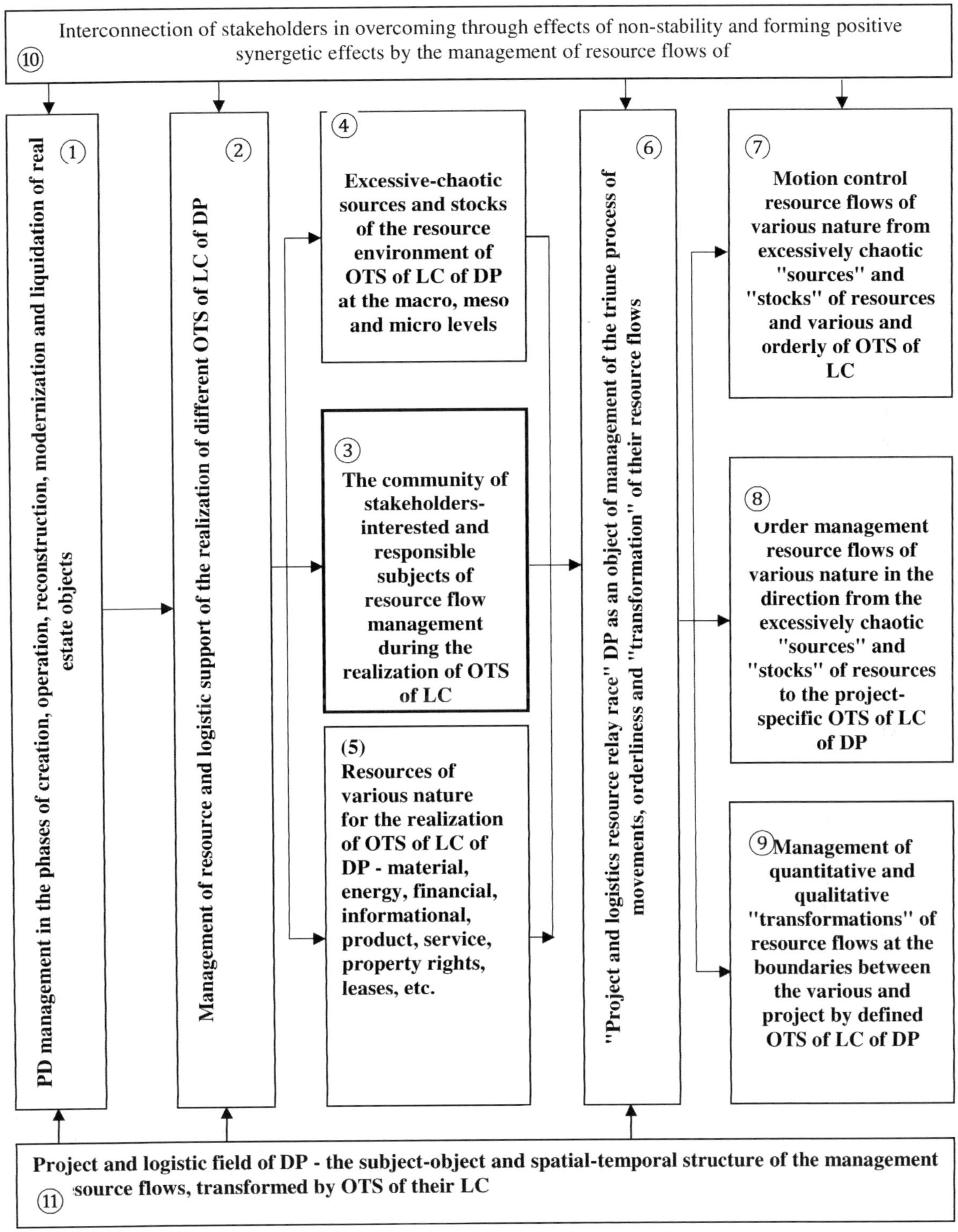

Figure 2. Theoretical and methodological approach to the management of DP resource flows in the creation, operation and improvement of real estate based on the design and logistic paradigm "orderliness out of chaos"

These flows complete the process of creating by stakeholders-constructors property owners (Block VI) as a "stock" of external resource flows of DP. They will be transformed on subsequent OTS projects into a multitude of "sources" and "stocks" of qualitatively different flow processes in the sphere of estate facilities operation. Among the transformable resource-operational flows can be called: flows of rights to the created real estate object; marketing-product, operational-commercial and product-service flows; material, energy, information and financial flows of production of products and services; economic and legal flows of purchase and sale of the property and its individual parts, etc.

Resource and operational flow processes can, in turn, re-generate at the macro, meso, and micro levels a wide variety of investment and construction resource flows for reconstruction and modernization, for capital and current repairs, and again for resource and operational flows on updated real estate objects. And in the future, and a variety of resource flows on the completion of projects, the liquidation of created real estate and the formation of ideas and concepts of new DP.

The above project and logistical aspects of the management of resource-flow processes of DP allow, taking into account their features, to form (Figure 2) a subject-object, space-time and their resource structure that changes in OTS their LC of structure of management of resource flows - PLF of resource-flow management of DP (11). This structure can be represented in the system of interaction between step by step changing subjects of management - stakeholders (3) and management objects – DP (1) – OTS of LC (2) - "sources-stocks" (4) - resources (5) - "project and logistics relay" (6) - movement of flows (7), orderliness of flows (8) - "transformations" of flows (9) - synergistic effects (10).

The main driving force behind the formation of the processes of project-logistic management of resource flows of DP are numerous and interacting with each other, stakeholders, ensuring, within their goals and interests, the realization of a specific development paradigm of managing flow processes - "orderliness from chaos".

At the same way, for the first time, the model of managing the resource-flow processes of DP is presented in the study in the form of a "project-logistic relay race" of resource flows as a single triad of their movements, orderliness and quantitative and qualitative "transformations" from the randomly disordered external resource environment of projects and in the direction of their numerous and project named OTS.

7. Analysis of the results

Analysis of the results of the study of theoretical and methodological problems of project and logistic management of resource-flow processes in the realization of OTS of LC of DP shows the following.

Each DP is one-time and unique in terms of a variety of different flow processes, in a set of interrelated and mutually supporting goals and sub-goals, in non-recurring actions and interactions of stakeholders and in terms of value results, as well as in terms of project realization (often with an open date), used resources and numerous OTS of long-term LC of projects. By these features of DP it is different from cyclically functioning production and technological systems in the industrial sphere and significantly expands the boundaries of traditional IBP.

The management of resource flow processes by the OTS of DP plays a subordinate role in relation to the main project processes. At the theoretical and methodological interface of project management and their logistics, a specific project and logistic paradigm for managing resource-flow processes of long-term and unique DP - orderliness of project-logistic resource flow processes for OTS projects from the chaos of their redundant resource environment is firstly formulated in this paper. The realization of this managerial paradigm is carried out through the step by step, meaningful, and spatial-temporal "project and logistic relay race" of foreign-made resource flows of various natures, transformed in the direction to the project by named (ordered) OTS of DP.

The resource "relay race" also serves here as an organizational and managerial mechanism for the step by step transfer of initiative, goals, and interests in managing resource flow processes from one stakeholder to another. Some stakeholders "leave" the project after completing their activities at a specific OTS, others "enter" the next stage. Thus, in the process of realization of OTS of DP at the moments of the "start" and "finish" of various resource flows there is a change in both the facilities and the subjects of management of the project and logistic flow processes. It necessitates the initiative of horizontal interaction between project stakeholders based on the exchange of their experience and competencies.

Part of the resource "relay race" in the sphere of cyclical industrial production (at the stage of operation of real estate) remains as the "supply chain" as a locally stable and sustainable process-logistic or production-logistic structures with fairly rigidly set "links" and an almost constant composition of stakeholders.

Overcoming the uncertainty, multiplicity and chaos of the resource flow environment of DP and the formation of ordered resource flow processes is carried out on the basis of the specific subject-object, space-time and OTS-changing structure – project flow field- for project management of the resource flow control projects.

In this frames of this "field" and on the basis of a new paradigm, the concept of managing the end-to-end constantly changing flow triad of interconnected processes of step by step movement, orderliness and quantitative and qualitative "transformations" of resource flows of OTS of various nature in time and space is firstly formed in this article.

The project and logistics relay race of resource flows of DP operates in its project and logistics field and is the real embodiment of the continued management paradigm.

The emergence of positive synergistic effects in the management of resource flow processes is ensured by self-organization and self-management, interest, responsibility and interaction of numerous and alternating teams of project stakeholders based on their goals, interests and categorical imperatives of purposeful behavior.

On the basis of the new approach, it can be built: systems of analysis, classification, structuring and forecasting of step by step resource flows of DP; models of a triune process through step by step movements, orderliness and quantitative and qualitative "transformations" of resource flows; forms and methods of interaction, interdependence and mutual influence of project stakeholders, generating the effects of self-organization and self-management of resource-flow processes.

The proposed theoretical and methodological approach opens new opportunities in the study and practical solution of the problems of managing project-logistical processes with the development of appropriate managerial paradigms in various fields. For example, in military and environmental projects or in projects for the realization of online commerce, etc.

References

[1] Razu M L 2011 Managing project. Fundamentals of project management // ed. by. (Moscow: Cambridge University press) pp 96–104

[2] Mazur I I, Shapiro V D 2009 Project management (Moscow: Omega-L) pp 118–189

[3] Mazur I I, Shapiro V D 2009 Real estate development. Guide for professionals (Moscow: Omega-L) 1036 p

[4] Tatarkin A I, Krivorotov V V 2014Competitiveness of socio-economic systems: challenges of the new time (Moscow: Economy) pp 437-464

[5] Afanasiev N B, Shmatko A D 2007 Methods of organization and improvement of logistics systems of the enterprise *Problems of modern science and practice* **1** (**7**) 110-114

[6] Sidorov I I 2006 Logistic model of the processes of production of material goods and their distribution - the basis of new economic theory and control theory *Vestnik of INZHEKON. Ser. Economy* **2** (**11**) 264-273

[7] Sidorov I I, Smirnov D V, Vanichev I A 2004 Fundamentals of logistic flow theory of material production *Digest of 3rd international scientific and practical conference logistics: modern development trends*

[8] Sidorov I I 2001 Logistic concept of enterprise management (Saint Petersburg: BPT Society «Knowledge») 168 p

[9] Ivanov D A 2006 Conceptual model of operational planning and management of logistics chains based on a multi-agent approach *Proceedings of Universities. Instrument making* **11** 9-14
[10] Stock J R, Lambert D M 2005 Strategic logistics management (Moscow: Infra) 757 p
[11] Fedorova M A 2003 Methods of management of the integrated system of distribution logistics. Thesis abstract (Saint Petersburg: Spbgieu)
[12] Zaitsev E I 2007 Modeling of Complex production and logistics networks *Vestnik of Inzhekon. Ser. Economy* **4** (**17**) 276-281
[13] Lukinskiy V V 2007 Design of optimal logistic systems based on EOQ model *Vestnik of Inzhekon. Ser. «Economy»* **4** (**17**) 281-285
[14] Brom A E 2007 Supply chain Management and global logistics *Izv. Universities «Mechanical Engineering»* **4**. P. 68-76.
[15] Kuznetsova O V 2018 Supply Chains and logistics functions in their management at enterprises URL: https://revolution.allbest.ru/marketing/00549402_0.html
[16] Lukinskiy V V, Plestenjak I A 2006 Modern logistics technologies of cargo delivery in mixed traffic *Vestnik of Inzhekon. Ser. «Economy»* **2** (**11**) P 256-263
[17] Hansen T, Voller C, Stensballe B 1992 Design for Logistics: Methodology proposal and Case Study (Morocco: Marrakech) 384 p
[18] Gattorna J, Day A 1986 Strategic Issues in logistics *International Journal of Physical Distribution and Materials Management* **16** (2) 3-42
[19] Langley C J 1986 The Evolution of the Logistics Concept *Journal of Business Logistics* **7** (2) 1-13
[20] Karvinen T 1995 Regional Development of Logistics as a Success Factor of Companies in the Outer-Circle of EU (Helsinki: University of Technology) 209
[21] Gatorna J, Day A, Hargreaves J 1991 Effective Logistics Management *Logistics information Management* **4** (**2**) 2-86
[22] Kubasova T I 2009 Logistic support of investment and construction projects *Problems of management theory and practice* **2** (**64**) 65-68
[23] Kiyko S G 2014 Modeling of processes of management of resource flows of projects *Visnik NTU «XPI»* **2** (**1045**) 96-100
[24] Osipova E, Eriksson E 2011 How procedure options influence risk management in construction projects *Construction Management and Economics* **29** 1149-1158
[25] Pan N-H, Lee M-L, Chen S-Q 2011 Construction Material Supply Chain Process Analysis and Optimization *Journal of Civil Engineering and Management* **17** (**3**) 357-370
[26] Wang L-C, Lin Y-C, Lin P-H 2007 Dynamic mobile RFID-based supply chain control and management system in construction *Advanced Engineering Informatics* **21** 377-390
[27] Irizarri J, Karan E P, Jalaei F 2013 Integrating BIM and GIS to improve the monitoring of construction *Automation in Construction* **31** 241-254
[28] Abduh M, Soemardi B W, Wirahadikusumah R D 2012 Indonesian Construction Supply Chains Cost Structure and Factors: a Case Study of Two Projects *Journal of Civil Engineering and Management*. **18** (**2**) 209-216
[29] Platonov A M, Soldatova Y V 2012 Transformation of logistic flows in the real estate development *Vestnik ENGECON. Ser. Economy* **6** (**57**) 139-146
[30] Gubarev V V 2009 Perspective approaches in management *Management in Russia and abroad* **3** 3-17
[31] Sedov E A 1986 The Relationship of information, energy and physical entropy in the processes of management and self-organization *Information and management* (Moscow: Science) 31-35
[32] Prigogine I, Stengers I 1986 Order from chaos: a new dialogue between man and nature (Moscow: Progress) 432 p
[33] Afanasieva V V 2018 To the philosophical justification of deterministic chaos URL: http://sbiblio.com/biblio/archive/afanasev_k/00.aspx

[34] Nikolis G, Prigogine I 1979 Self-Organization in nonequilibrium systems (Moscow: Mir) 512 p

[35] Kolesnikov A A 2001 Synergetic management theory: concepts, methods, development trends *Proceedings of the southern Federal University. Technical science* 56-63

[36] Selitsky V S 2009 Synergetics and practical management of systems *Bulletin of p. O. Sukhoi state technical University: scientific and practical journal* **4** 109-116

[37] Knyazeva E N, Kurdyumov S P 2018 Bases of synergetics. The synergetic world-view *URSS* 254 p.

[38] Polak L S, Mikhailov A S 1983 Self-Organization in non-equilibrium physical and chemical systems (Moscow: Science) 286 p

Building of effective management structures based on the digital technologies in the mining industry. A case study of UC RUSAL

E N Prokofeva[1*], A V Vostrikov[1], A Y Stebenev[1], D M Volchkov[1] and B L Agasandyan[1]

[1]National Research University Higher School of Economics, 20 Myasnitskaya str, 101000, Moscow, Russia

[*]E-mail: eprokofyeva@hse.ru

Abstract. The most competitive in the current global market and defining the economy of the country persists products of extractive industries. According to this reason precisely this direction requires effective management, continuously connected with advanced technologies. Developing enterprises not only are able to impact the situation in the mining industry but to reform the economy situation in general. Changing external environment produces up-to-date demands on planning based on calculated predictive values, and creates the possibility of increasing the competitiveness of the organization. One can't help underestimating the importance of integrating quality management system to any modern company, as it determines further ways of its prospects.

Keywords: mining industry; development trends; external environment; long-term tasks

1. Introduction

The beginning of the 90s was a very difficult time for the economy of the USSR. In connection with the collapse of the country, almost all industries have lost their former strength. One of the most promising industries was metallurgy, namely the aluminum industry. Its revival has been going on for more than 10 years and was extremely important for the entire infrastructure of the country. The role of aluminum is so great, and the application is so extensive that this metal is not without reason called the "metal of the 20th century." Currently one of the fastest growing and economical industries are minerals extraction and construction-mining functioning of the enterprise. There is not a single sector of the country's economy that would do without non-ferrous metals. The leader in the field of effective development and organization of mining and blasting operations are the companies Alrosa and Sibur, and in the Russian metallurgical sector NorNickel and RUSAL. Digital design of development organization is actively developing in RUSAL. In this work we want to identify the main characteristics of this company, the device, the goals, and the IT technologies that are used in it.

2. Company review

United Company RUSAL (UC Rusal) is one of the largest aluminum and alumina producers in the world and one of the largest private russian companies. The company was established in 2007 as a result of the combination of RUSAL and SUAL with alumina assets of the Swiss Glencore. The capacity of the alumina refinery corporation is 12.3 million tons, and aluminum 4.4 million tons. The main consumer sectors are packaging, transport and construction, and the main sales markets are Russia, the CIS countries, Southeast Asia, Korea, Japan, Europe and North America. RUSAL itself (its assets) is located in 13 countries on five continents [1-5].

The fact that UC RUSAL is one of the largest companies in the world imposes additional responsibility on it. RUSAL survived more than one global economic crisis and intends to work on improving efficiency, expanding the range of products at the expense of value-added products and strengthening its position in key markets, primarily in Russia. In addition, customer needs are becoming a significant reference point for development. The corporation is always ready to work together with customers, jointly develop new technological solutions, invent new ways to create alloys and various materials. This approach contributes to the development of relations with Russian and foreign investors, suppliers and customers.

It is also necessary to note that RUSAL has its own engineering and technological base, which allows for full-scale research in the required areas and industrialize scientific ideas.

3. The emergence and history of development

The first company, called Russian Aluminum, appeared in 2000 through the merger of the assets of Sibneft and Sibirsky Aluminum. The share of this new company in global aluminum production was about 10%. In connection with the formation of RUSAL, the situation in the Russian aluminum industry began to change. Russian businessmen Oleg Deripaska and Roman Abramovich (the owners of the assets of RUSAL) began to unite Russian enterprises producing aluminum, thereby creating a platform for the subsequent modernization and strengthening of the aluminum industry. In addition, the corporation paid significant attention to the development of new methods for the production of aluminum, the development of the technological component. In 2002, the company laid the foundation for the development of scientific and technical base, namely, the Engineering and Technological Center was created, which begins to monitor the research and development and research and development work of the corporation [6-8]. A year later, RUSAL's specialists in record time were developing a new aluminum production method, the RA-300, which, to a large extent, allows for a reduction in electrical energy consumption and an increase in the operating time of the electrolysers. In 2003, new electrolyzers come into operation at the Sayanogorsk Aluminum Smelter. The Krasnoyarsk and Nikolaev combines and the reconstructed ARMENAL foil-rolling plant in Armenia, which was bought by RUSAL in early 2000, also underwent modernization.

It is necessary to note that this corporation throughout its existence led activities to increase its presence in Russia and abroad. In 2002, under the auspices of RUSAL, two Guinean enterprises became - the bauxite-alumina plant Friguia and the mining complex BCC. The company's production base is also expanding through the purchase of a controlling stake in the Novokuznetsk aluminum plant. Closer to 2004, the company's operations are gaining a transatlantic scale. RUSAL begins development of bauxite deposits in Latin America in the Guyana region. The government of Guyana and the management of the corporation create a new subsidiary, the Bauxite Company of Guyana. In addition to operations in Latin America, RUSAL has acquired a significant amount of assets around the world. The largest acquisitions include aluminum and cathode plants in China, Nigeria, Italy, and the Australian alumina refinery Queensland Alumina Limited (the second largest alumina enterprise in the world), the purchase of which contributed to an increase in the self-sufficiency of raw materials by almost 800 thousand tons per year. In the spring of 2007, the consolidation of Russian aluminum manufacturing enterprises was completed. At that time, in addition to RUSAL, there was another company that had a significant role in non-ferrous metallurgy, namely, the Siberian-Ural Aluminum Company. Shareholders of both corporations have decided to merge assets, as well as the purchase of

IOP Conf. Series: Materials Science and Engineering **972** (2020) 012061 doi:10.1088/1757-899X/972/1/012061

the Swiss commodity trader Glencore. The result of these actions was the emergence of RUSAL in the form in which it is known throughout the world today.

It is important to pay attention to the fact that RUSAL and its management, in order to achieve their goals, had to solve many problems and survive global financial crises. So 2009 was the year of the global economic crisis, which became one of the most difficult in the history of the combined company. The cost of aluminum and its demand fell to a record minimum, which could not but affect the financial support of the corporation and relations with creditors. But during the year, the company actively implemented a set of measures that was aimed at reducing costs, minimizing costs, optimizing production, and which eventually allowed to survive the crisis with minimal losses and lay the foundation for further development. This financial crisis has made significant changes in the development of global industry. After analyzing the situation on the global market, RUSAL developed a ten-year program to introduce new technological methods for aluminum production, optimize labor resources management, reduce inefficient capacities, and standardize business processes. In short, the program that will ultimately help strengthen the company's position in the global aluminum industry. The measures applied significantly contributed to the strengthening of the corporation in the international arena, and at the moment RUSAL is one of the most technically advanced companies producing primary aluminum and alumina with a huge raw material base and highly skilled ordinary employees and managers able to effectively manage the company and lead it along the way.

4. Analysis of the internal environment

The internal environment of the organization is part of the external environment. The internal environment of the organization includes many factors that from the inside affect the overall performance of the organization as a whole. In the internal organization of the organization there are many subsystems, among which the following take a special place:

- Marketing;
- Social;
- Financial;
- Organizational.

4.1. Marketing subsystem

Marketing takes an important place in the life of a corporation. RUSAL, being one of the largest companies in Russia and the world, is obliged to conduct high-quality and in-depth studies of the state of the world market. The marketing service in RUSAL works as a buffer between the external environment - the market and the internal environment - the company's resources. One of the main achievements of RUSAL in marketing is the creation of a favorable image of the company, the successful development of new markets and territories, a wide range of products, and relatively low prices. All of them contribute to strengthening the position of the corporation.

4.2. Social subsystem

Personnel policy of the enterprise according to many experts is the basis for the successful functioning of the entire organization. The same opinion is shared by RUSAL. People in the company create its product, are responsible for the internal climate, each employee has his own task, the execution of which affects various business processes. In this regard, the company's management regularly provides funding for this area.

4.3. Financial subsystem

Financial processes are a complex and responsible part of the company. Due to its size, hundreds of thousands of various financial transactions occur every month in an organization, an enterprise has billions of dollars that need to be managed and properly planned for their distribution. For controlling such powerful cash flows, RUSAL attracts highly qualified specialists and purchases the appropriate software.

4.4. Organizational system

For effective management of a transcontinental enterprise, it is necessary to have a developed management structure. RUSAL is actively modernizing its corporate governance system. The highest governing body and control is the General Meeting of Shareholders, which makes decisions on all important issues of the company. The quality of the activities of this body contributes to the attractiveness of the corporation to investors, and also gives additional guarantees to customers and partners in the reliability of the organization.

5. Analysis of the external environment

The external environment of any organization is a source of resources that are necessary for its functioning. The external environment, as well as the internal, includes many components and systems. It is customary to divide it into two large parts: the macroenvironment and the immediate environment of the enterprise.

The macroenvironment of RUSAL

- 1. The political and economic component.

These factors can have an influence on all activities of the organization. RUSAL has always sought to maintain trusting relationships with the authorities and various administrative bodies, since they create the "climate" in a particular region. In addition, the growing pressure from the European Union and the United States, which, through sanctions, influences on the work of the organization. In addition, the management and various experts of the company have to constantly control the level of inflation, the level of employment, business lending rates, etc., as through careful analysis of these factors it is possible to turn the future threat to the organization into good and avoid unnecessary costs.

- 2. Social and cultural components.

External sociocultural factors to a large extent influence the perception of employees of their company. Thus, RUSAL, like almost all enterprises in Russia and the CIS, developed an "individualistic" approach (as opposed to a "family-based" approach in the east), according to which the constant desire for career growth and frequent change of work is the norm of behavior. This approach has a rather negative impact, since there is a high probability of losing frames.

- 3. Scientific and technical progress.
- A scientific and technical environment, if properly analyzed and used, can provide excellent means for the modernization of all systems and processes occurring in an organization. It is worth noting that RUSAL devotes a large amount of time and resources to this factor. Procurement of various software, equipment is actively introduced, and research is being conducted to update the methods of producing aluminum and other substances [9-10].

RUSAL stakeholders

- 1. Competitors

For effective development in the face of tough competition, the company constantly has to modernize and upgrade equipment, develop a marketing subsystem, monitor the situation in competing corporations, and analyze the market. Among the competitors countries in the world can be identified US, European Union, Australia, as well as dynamically developing China. Among competing enterprises, two organizations are worth noting, namely the Chinese Chalco and the British-Australian concern Rio Tinto.

- 2. Consumers

RUSAL is in constant dialogue with its customers, takes into account their needs and wishes when releasing new alloys, and participates with them in creating products for the final consumer. The corporation also actively conducts various advertising campaigns. One of the key factors to increase customer base is the quality of the final product. The company's products meet the best international standards. A significant part of the plants of RUSAL is certified for compliance with the ISO: 9001 quality management system standard. Among the major consumer countries can be identified the United States and European countries, as these regions are highly developed industrially.

- 3. Suppliers

The importance of supplier organizations is due to the fact that such organizations are able to influence the activity of the company-buyer by changing the supply volumes, prices for raw materials, thereby creating the most dependence on resources. In this regard, RUSAL with great attention to the choice of suppliers and uses the following guidelines when choosing: quality, timeliness of supply, cost and efficiency. When working directly with partners, the corporation adheres to the principles of mutual trust and respect, and also monitors the improvement of technology and business processes.

- 4. Investors

Investors are private individuals or organizations who invest their own funds in projects for profit. In this regard, in order to ensure the stability of the investment cash flows, RUSAL maintains a policy of transparency and informational openness. The corporation seeks to publicize the investor community about the results of its operating and financial activities, as well as plans for further development.

6. Determination of the most significant tasks for the company and business processes

UC RUSAL is one of the most efficient aluminum producers in the world. For a corporation to be first means to be first in everything, and in this connection the company has set itself a number of long-term tasks, namely:

- Actively develop scientific and technical potential and develop new technologies in order to subsequently introduce them to enterprises.
- Demonstrate to the investment community the prospects of their enterprise and the ability to effectively use the corporation's benefits and resources.
- Optimize all business processes occurring in the company and increase the speed of interaction between different subsystems, through the introduction of new software and attracting highly qualified specialists.
- Increasing the flexibility of the production process to quickly adapt to rapidly changing market conditions.
- Optimization of the personnel segment of the enterprise.
- Improved accounting and pricing process.
- To provide its customers with products of the highest quality, as well as in a timely manner.

According to many independent experts, the implementation of these very goals will affect the improvement of the quality of work and optimization of processes, as well as the reduction of various costs.

RUSAL builds a further development strategy based on building a tree of goals, identifying top priorities. After conducting this analysis, management decides to update one or another segment of the organization, the purchase of new equipment and software. In connection with the foregoing, in the next part of the analysis of RUSAL, we will discuss the company's IT solutions and the introduction of various software products that ultimately contributed to the achievement of the desired.

7. Analysis of IT-solutions in the corporation

The company RUSAL is a huge company with a rich history and highly developed infrastructure. Managing such a company, analyzing its various subsystems and parts, analyzing the processes occurring in it, is an extremely complex and costly task, requiring, in addition to highly qualified personnel, powerful software. Having set itself the goal of comprehensively improving and optimizing the work of the corporation, RUSAL throughout its history came to various IT solutions and the integration of various software products. Thus, in 2006, at the dawn of the emergence of the modern version of RUSAL, the question arose of a more efficient allocation of company resources, of operational control over the company's activities and the formation of a reliable base for making optimal and timely decisions at all levels of management. On this basis, the leadership of RUSAL made one of the most important decisions, namely, the decision to purchase the SAP ERP product,

IOP Conf. Series: Materials Science and Engineering **972** (2020) 012061 doi:10.1088/1757-899X/972/1/012061

which works according to the ERP system technology and allows to solve most of the issues and tasks faced by modern medium and large enterprises. The introduction of this software into the infrastructure of RUSAL was divided into several stages:

- 1. The first stage included the organizational documentation, namely the order to introduce the new system, the creation of a calendar plan, project charter and a risk management plan.
- 2. The second stage consisted in familiarizing consultants with the peculiarities of the corporate business processes, analyzing the internal structure and communicating directly with the company's employees.
- 3. The third phase is the implementation of conceptual design. The essence of this phase is to develop the rules according to which business processes and system configuration will be built.
- 4. The fourth and final stage was the implementation of the project (integration test and project start).

Three aluminum enterprises of RUSAL in Bratsk, Sayanogorsk and Novokuznetsk felt the impact of this software package. It should be noted that at the factories in Sayanogorsk and Novokuznetsk product introduction took a record short time - 3.5 months before pilot production and 5.5 months before disconnecting inherited systems.

As a result of the introduction of this software, the company has improved the quality of its management in the following number of indicators:

- Improved information management, organization managers began to monitor the processes in the company in real time.
- The quality of management of material and financial flows has increased.
- Updated project management.
- Improved quality of personnel management.

In general, RUSAL, thanks to this software product, has increased the overall efficiency of enterprises. 5 previously existing systems were replaced by a single integrated resource management system. This system allowed to make up to 350 thousand financial operations and up to 520 thousand logistic. In 2009, RUSAL acquired an updated version of this program.

It is worth paying attention to 2010. During this period, the company managed to implement the integration of 2 fundamentally new software packages in a relatively short time. The first was a program called PayDox. PayDox is an electronic document management and business process management system. This software product is a corporate portal that uses electronic document management as a base with a wide range of functions and opportunities for expansion and integration. RUSAL has managed to implement the strengths of this platform in its work and improve the performance of the following processes:

- Support geographically distributed workflow.
- Keeping the entire history of work with documents and projects.
- Management of documents responsible for payments, contract management, OSA.
- Coordination, approval and optimization of document control.
- Interaction with remote units.

The next software product in the queue was BOSS-Kadrovik, which is an automated personnel management system that facilitates simpler and more effective human resource management. This software allows you to plan personnel policy, to conduct activities to structure the organization, recruiting new staff, managing internal processes of the personnel segment, etc. Thanks to this program, it was possible to optimize the work of management and ordinary employees, increase the speed of work between subordinates and improve material support.

Over the next 7 years of the organization's life, only 3 significant software products were introduced, namely the Microsoft .NET Framework, Bpm'online and K2 Blackpearl. Of these, in my opinion, it is worth highlighting the Bpm'online program, since it was this product that brought the most noticeable changes to the company's activities. Bmp'online is a system for managing the entire

IOP Conf. Series: Materials Science and Engineering **972** (2020) 012061 doi:10.1088/1757-899X/972/1/012061

sales cycle from the need analysis to the conclusion of a transaction, the control of payments and the execution of an order. This platform is designed to create flexible information systems that can adapt to the company's business processes. RUSAL Corporation, having introduced this software product into its infrastructure, gained the support of one of the most modern and powerful software packages on the CRM market.

This innovation significantly contributed to the optimization of the work processes related to sales and customer interactions. With the help of this system, the company has been able to better adjust the tactics of transactions, manage the sales funnel and track the "pulse of sales." In addition, it is worth noting the ease of use of this software product compared to others. Bpm'online presents a simple and flexible interface that allows you to place orders and create sales with just a few clicks. At each stage of working with the interface, only the information that is needed at the moment is shown, in addition, the speed of searching for the necessary data is increased. All of these factors contributed to reducing the cost of introducing this product and simplifying employee training [13-16].

8. IT solutions and software products which must be implemented

Up to this point, we conducted a descriptive analysis, studied the history, disassembled the device of the corporation, its various segments and embedded products. Now, on the basis of the data we have studied, we want to pay attention to the fact that despite all the efforts of the corporation, due to its size, it is constantly required to optimize the activities of the employees, improve the accounting work, and also facilitate the work of the staff. Based on the foregoing, we consider it appropriate to implement a specific software package, namely the "System for Automating Pricing Processes in an Enterprise" by Force [11-12].

This system is designed for the effective preparation of planned costing estimates of goods, assistance in the formation of various price offers, price lists and price lists. The introduction of a unified pricing system will help employees of the planning and economic and related departments to conduct a coherent and clear work. Using this platform, RUSAL will be able to form a unified methodology of costing, maintain a unified database of market and procurement prices, reduce the time for collecting baseline data, and also more easily monitor regulatory information (labor intensity, rates, composition of products, etc.). In addition, there are functions for maintaining a database of unstructured information and forming a database of directories of competing companies.

An approximate simplified schedule for the implementation of this product may be as follows:

- Install the solution on the server.
- Analysis of requirements.
- Approval of requirements by the customer.
- Customize the solution in accordance with the requirements.
- User training.
- Preparation of various documentation.
- Transfer to commercial operation.

The total implementation time depends on the size of the enterprise and takes from several months to a year; in the case of RUSAL, we think the process will take at least 6-7 months.

9. Conclusion

In this paper we analyzed the activities of the transatlantic corporation RUSAL, conducted a detailed analysis of most of the company's systems, its infrastructure, main goals and development trends. Indicated what innovations in the IT-sphere were made by the organization. Based on what we now know about this organization, we can say with confidence that this is a dynamically developing, gigantic, complex enterprise in its structure, which is able to overcome various difficulties, act in conditions of fierce competition and ultimately achieve its goals.

References

[1] Potekhin I, Mischenko V, Mottaeva A and Zheltenkov A 2018 Evaluation of possibility to increasing sustainability of high-rise buildings through use university intellectual property E3S Web of Conferences (Eds.). P.03020. https://doi.org/10.1051/e3sconf/20183303020

[2] Goncharenko S N, Duong L B, Petrov M V, Stoyanova I A 2014 Modeling of parameters of innovation water-protection measures on the basis of industrial-technological indices of coal mining at Vietnam enterprises *Gornyi Zhurnal* **9** 143-146

[3] Prokofeva E N, Vostrikov A V, Fernandez E, Borisov N 2017 Navigation satellite systems as the audit foundation for mining companies *Eurasian Mining* **1** 30-32

[4] Prokofeva E N, Vostrikov A V, Shapovalenko G N, Alvarez A 2017 The development of effective geomonitoring for mining area with industrial review *Eurasian Mining* **2** 61-63

[5] Aleskerov F T, Karabekyan D, Ivanov A, Yakuba V I 2018 Individual manipulability of majoritarian rules for one-dimensional preferences *Procedia Computer Science* **139** 212-220

[6] Vasin S, Gamidullaeva L, Shkarupeta E, Finogeev A and Palatkin I 2018 Emerging trends and opportunities for industry 4.0 development in Russia *Eur. Res. Stud. J. XXI* **3** 63-76

[7] Artamonov S 2016 Nonperiodic Modulus of Smoothness Corresponding to the Riesz Derivative // *Mathematical notes* **99** (6) 928-931

[8] Zreckiy A D, Ivanova T E 2018 Industrial technologies and innovations. The standard of the third generation (Saint Petersburg: Peter) 480 p

[9] Michurina O Yu 2010 Theory and practice of integration processes in industry (Moscow: INFRA-M) 287 p

[10] Chernysheva A M, Yakubova T N 2016 Idustrial (B2B) marketing (Moscow: Yurait Publishing House) 433 p

[11] Rusal directorate of new projects 2019 (URL http://rysslandshandel.se/rus/uploads/events/forum2017/Rusal.pdf)

[12] Organization of production aluminum alloys in factories 2019 (URL https://rusal.ru/press-center/files/Belsky%20Alloys.pdf)

[13] About company 2019 (URL https://rusal.ru/about/)

[14] Informational technologies in RUSAL 2019 (URL http://www.tadviser.ru/index.php/Статья:Информационные_технологии_в_РУСАЛ)

[15] System of automation of pricing processes at the company 2019 (URL http://www.fors.ru/industry-solutions/)

[16] Economic information 2019 (URL http://economican.ru/v_man.php?id=7)

Assessment of uncertainties and risks related to implementation of infrastructure projects

D O Reznikov[1]

[1] Mechnical Engineering Research Institute of the Russian Academy of Sciences?
4 Maly Kharitonievsky lane, Moscow, 101990 Russia

E-mail: mibsts@mail.ru

Abstract. The paper considers principles of assessing uncertainties and risks related to realization of infrastructure projects. The procedure for the examination of the proposed infrastructure project should use the methodological toolkit of risk theory, which was developed to assess safety of complex technical systems. A risk-based approach to the assessment of infrastructure projects involves a scenario assessment of the project, taking into account the influence of potential hazardous natural, technological, economic and socio-political factors that can significantly complicate the project implementation. A set of indexes allowing one to assess and compare different variants of the infrastructure project implementation is presented.

Key words: infrastructure project, uncertainty, risk, utility

1. Introduction

The concept of infrastructure project refers to long-term large-scale project that is focused on the development of infrastructure facilities and lifelines and is being carried out in accordance with the strategic plan of the social and economic development of the community, district or territory.

Planning and implementation of infrastructure projects is carried out in uncertain environment. Project uncertainties are related to: (a) random nature of various events and processes that develop at various stages of the project life cycle, and (b) lack of knowledge about many key phenomena that took place in the course of the project development. In some cases it is convenient to distinguish a certain group of uncertainties related to the possibility of occurrence of rare unexpected extreme events that cannot be foreseen at the design (panning) phase of the project. These uncertainties are actually a combination of uncertainties that belong to the first two groups but they are aggregated into the specific group in the view of tremendous influence that they have on the project implementation [1-4].

Due to the above listed uncertainties infrastructure projects should be described in the probabilistic formulation using branched scenario trees. These project uncertainties also cause a set of project risks that are inevitable when infrastructure projects are being developed. Project risks can be classified in accordance with the stage of the project as design, construction, maintenances and decommissioning risks. The following groups of project risks may be distinguished:

- Economic and financial risks include risks related to instant changes in taxes, interest rates, commodity prices, etc.
- Social and political risks may be caused by political conflicts, public unrest, etc.
- Organizational risks referred to human arrows made at various stages of the project development.
- Risks of natural and manmade catastrophes include risks of extreme natural events, technological accidents, unauthorized impacts, etc.

One should note that the presence of uncertainties related to realization of infrastructure projects produce not only risks of various negative factors but also may create chances of some positive influences on the project (appearance of new technologies, improvement of business (economic) conditions, favorable environmental conditions, etc.).

2. Scenario assessment of infrastructure projects

Assessment of uncertainties is an important step of the process of making decisions regarding launching the infrastructure project and planning the schedule of its implementation. Due to the high level of uncertainty related to the project implementation its propagation is multivariate and should be characterized by a branched scenario trees that could be described by graphological models used for assessment of risks in complex engineering systems [4-8]. Each of the scenario $S^{(i)}$ $(i=1,2,...,k)$ terminates with end state (outcome) $ES^{(i)}$ $(i=1,2,...,k)$ of the project. Here each outcome $ES^{(i)}$ can be defined by a set of generalized parameters:

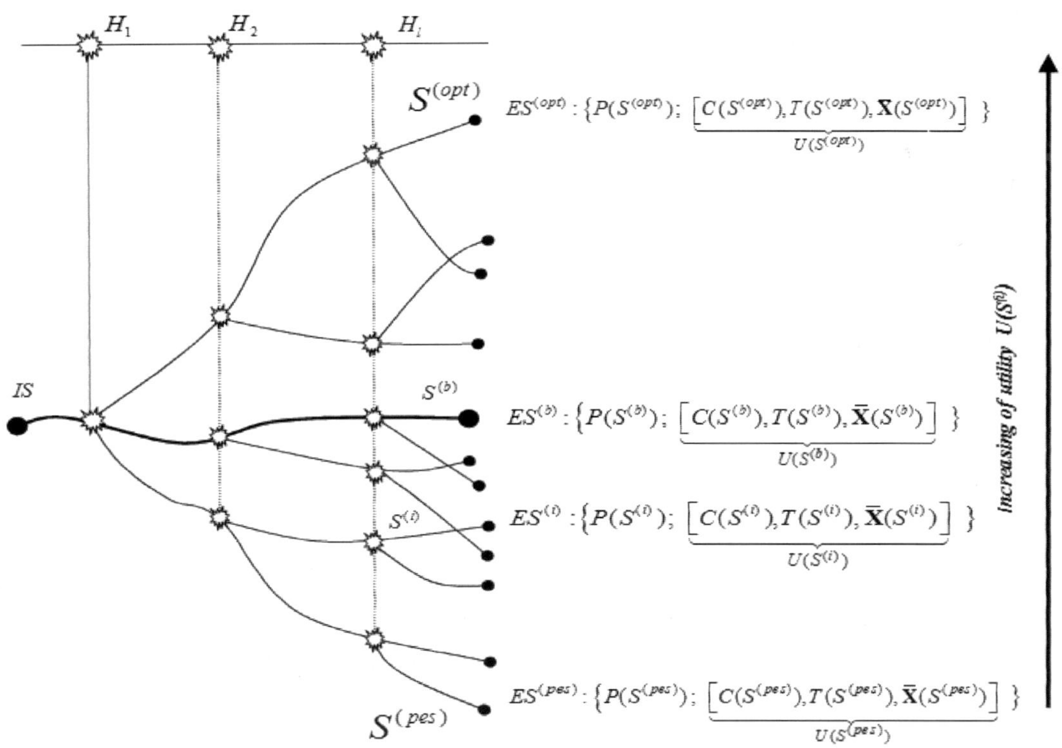

Figure 1. Scenario tree of the project
$S^{(b)}$ - base project scenario; $S^{(opt)}$ -optimistic project scenario; $S^{(pes)}$ -pessimistic project scenario; H_i $(i=1,2,...,l)$ - random factors (risk-factors) influencing the project.

- adjusted cost of the project implementation along scenario $S^{(i)}$ that is calculated through adjusting the flow of payments to a specific date (e.g. to the start of the project), $NPV^{(i)}$;

- time of the project implementation along the specific scenario, $T^{(i)}$;
- an assemblage (aggregate) of the characteristics of the project outcome $X=(x_1, x_2,..., x_n)$ that includes such parameters as cost of a unit of goods or services $Nc^{(i)}$, production rate $E^{(i)}$of the facility (asset) constructed in the frame of the project, its reliability $Re^{(i)}$, service risk $R^{(i)}$, etc.

Each of the project outcomes (end states of the project) can thus be represented by a point in the project space of states Ω: $ES^{(i)}(C^{(i)},T^{(i)}, Nc^{(i)}, E^{(i)}, Re^{(i)}, Z^{(i)}, R^{(i)})\in \Omega$ (Figure 1) and this outcome can occur with a specific probability $P(ES^{(i)})$ that can be estimated using graphological models (event trees, fault trees, Bayesian nets, diagrams of influence, etc.).

Considering $m+2$ dimensional space of the project states and $\Omega(y_1,y_2,..., y_{m+2})$, where $y_1=C$, $y_2=T$, $y_3=x_1$, $y_4=x_2$, …, $y_{m+2}=x_m$. A multiattribute utility function is introduced to compare various project outcomes that are characterized by a set of various parameters $(y_1,y_2,..., y_{m+2})$. This function maps the points $ES^{(i)}$ of the project functional states Ω to the points $u^{(i)}$of the axes R^N of natural numbers:

$$(y_1^{(i)}; y_2^{(i)};..., y_{m+2}^{(i)}) \Rightarrow u^{(i)}, u^{(i)} \in R^N . \qquad (1)$$

This mapping allows describing the preferences of the stakeholders regarding the outcome of the project. In other words the form of the function $u=u(y_1,y_2,..., y_{m+2})$ is determined by the preferences of the stakeholders. It may form the basis for comparison of various scenarios and used in the criteria of the project success. When the whole scenario tree is assessed and the probabilistic distribution of the utility function for this specific variant of the project is obtained one may make judgments estimating the success of the project and compare different variants of the project implementation.

In the simplest formulation the linear utility function can be taken:

$$U = \sum_{j=1}^{m+2} w_j y_j , \qquad (2)$$

where w_j are weighting coefficients that describe the importance of the factors y_j in the eyes of the stakeholders. Weighting factors are constrained by a condition:

$$\sum_{j=1}^{m+2} w_j = 1.$$

Then a specific value of the utility function will correspond to each project scenario $S^{(i)}$:

$$U(S^{(i)}) = \sum_{j=1}^{m+2} w_j \cdot y_j(S^{(i)}) = w_1 \cdot C(S^{(i)}) + w_2 \cdot T(S^{(i)}) + \sum_{j=3}^{m+2} w_j \cdot x_{j-2}(S^{(i)}) .$$

It is clear that the linear form of the utility function is oversimplified and does not allow assessing the interference between various factors and the preferences of the stakeholders to avoid scenarios with strong negative utility. Higher order polynomials should therefore be used for approximation of utility functions for real infrastructure projects.

Then using the project scenario graph all identified project scenarios are considered, and the values of probabilities $P(S^{(i)})$ and utilities $U(S^{(i)})$ for each of the scenario are estimated. Thus in the probabilistic formulation the infrastructure project can be described by a set of triplets:

$$IP = \left\langle S^{(i)}; P(S^{(i)}); U(S^{(i)}) \right\rangle_C . \qquad (3)$$

The subscript C here refers to the complete set of project scenarios. This description of the project is quite similar to classical definition of risk introduced by Sten Caplan in [5]. The structure of triplets determines: (a) possible project scenarios, (b) probability of occurrence of the scenarios, and (c) values of the utility function that correspond to each of the scenario. The results of the scenario assessment are presented in a tabular form (Table 1).

Table1. Listing of the project scenarios.

Scenario	Probability	Utility
$S^{(b)}$	$P(S^{(b)})$	$U(S^{(b)})$
$S^{(1)}$	$P(S^{(1)})$	$U(S^{(1)})$
$S^{(2)}$	$P(S^{(2)})$	$U(S^{(2)})$
.	.	.
.	.	.
.	.	.
$S^{(k)}$	$P(S^{(k)})$	$U(S^{(k)})$

These data for the infrastructure project can be also presented in a graphical form by a set of points $\{P(S^{(i)});\ U(S^{(i)})\}$ at the "probability-utility" plane (Figure 2).

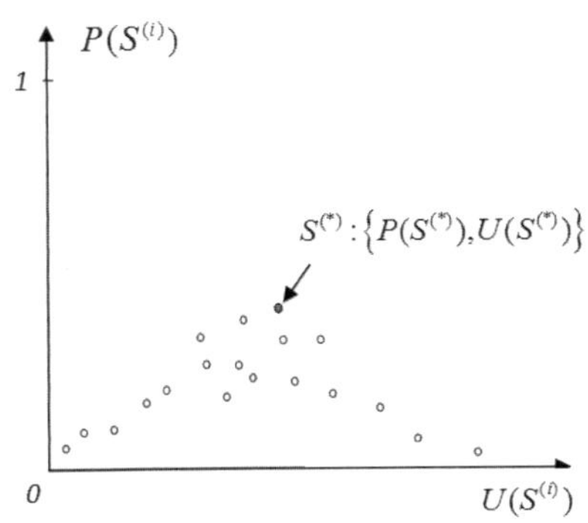

Figure 2. Presentation of the project outcomes by a set of points, $\{P(S^{(*)});U(S^{(*)})\}$ that correspond to various project scenarios.

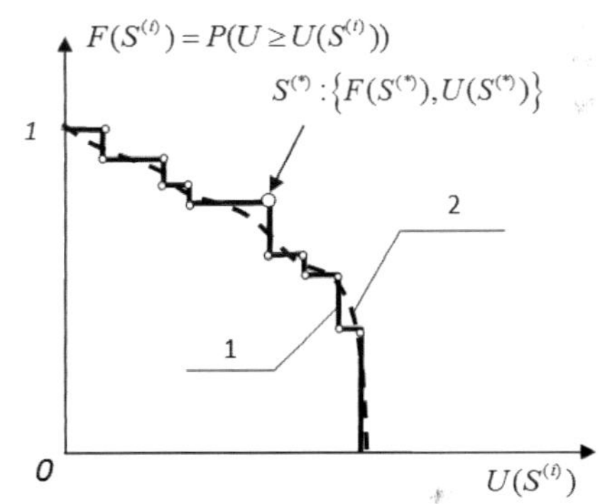

Figure 3. Presentation of the probabilistic assessment of the project in the form of the utility profile.

For the further consideration it will be more convinient to chang the form of peresenting data given in table 1. Here the scenarious $S^{(i)}$ $(i=1,2,\ldots.k)$ that were initally numbered in an arbitrarilly order should be renumbered in such a way that the following conditions hold:

$$U(S^{(0)})\leq U(S^{(1)})\leq U(S^{(2)})\leq\ldots\leq U(S^{(i)})\leq\cdots\leq U(S^{(k)}).$$

Besides the scenario table will be supplemented with an additional collumn containing the values of comulative probabilities $F(S^{(i)})=P(U>U(S^{(i)}))$ for each of the scenario $S^{(i)}$ that equel to the probability that the random value of the project utility U will be higher than the value of utility $U(S^{(i)})$ corresponding to the scenario $S^{(i)}$.

Then the outcome of the scenario assessment of the infrastructure project can be presented in the following triplet form:

$$IP=\{<S^{(i)};F(S^{(i)});U(S^{(i)})>\}_C \quad (i=0,1,2,\ldots,k) \tag{4}$$

Thus each of the scenario can be represented by a point $\{F(S^{(i)});U(S^{(i)})\}$ on the plane "cumulative probability-utility". Connecting these points sequentially one can get a stepwise line $F(U)$ (Figure 3, line 1). This probabilistic distribution of the random value U is called a utility profile of the infrastructure project and can be considered as a general characteristic of the project uncertainties [2].

It should be noted that if a more detailed assessment of the project scenario tree is carried out, each of the scenarios $S^{(i)}$ should be considered as a union of several very similar to each other subscenarios the difference between which were previously considered to be negligible. It means that if the scenario assessment was conducted in a more detailed way the stepwise line of the project utility profile would tend to a smooth curve (Figure 3, curve 2). In this case the discrete random variable U can be considered as a continuous variable u and described by a probability density function $f_U(u)$, probability distribution function $\Phi_U(u)=P(U<u)$ and the project utility profile $F(u)=\text{P}(U>u)=1-\Phi_U(u)$.

Table2. Supplemented listing of the project scenarios.

Scenario	Probability	Utility	Comulative probability
$S^{(0)}$	$P(S^{(0)})$	$U(S^{(0)})$	$F(S^{(0)})=1$
$S^{(1)}$	$P(S^{(1)})$	$U(S^{(1)})$	$F(S^{(1)})=F(S^{(2)})+P(S^{(1)})$
$S^{(2)}$	$P(S^{(2)})$	$U(S^{(2)})$	$F(S^{(2)})=F(S^{(3)})+P(S^{(2)})$
.	.	.	.
.	.	.	.
.	.	.	.
$S^{(i)}$	$P(S^{(i)})$	$U(S^{(i)})$	$F(S^{(i)})=F(S^{(i+1)})+P(S^{(i)})$
.	.	.	.
.	.	.	.
.	.	.	.
$S^{(k-1)}$	$P(S^{(k-1)})$	$U(S^{(k-1)})$	$F(S^{(k-1)})=F(S^{(k)})+P(S^{(k-1)})$
$S^{(k)}$	$P(S^{(k)})$	$U(S^{(k)})$	$F(S^{(k)})=P(S^{(k)})$

3. Indexes of the infrastructure project utility

The procedure described in the p.2 allows one to create of utility profile for the infrastructure project. This profile is the most general characteristics of uncertainties, risks and chances related to implementation of the infrastructure project [2]. However when the comparison assessment of different variants of the project should be carried out it is desirable to have a set of quantitative measures (indexes) of the project utility. These indexes can then be compared with some references or alternative variants of investments. To this end it is desirable to develop some indexes of the infrastructure project utility that would allow one to conduct comparative assessments of different variants of the project and to select the optimal one with accounting for the utility function and the preferences of the stakeholders.

The mathematical mean of the random variable u is one of its most simple and understandable characteristics. It can be referred to as integral index of the project utility:

$$I_E = E\{U\} = \sum_{j=1}^{m} P(S^{(i)})U(S^{(i)}) = \int_0^{\infty} u \cdot f_U(u) \cdot du \cdot \qquad (5)$$

It can be showed that the area S_Q under the utility profile $F(u)$ equals to the mathematical mean of the random variable U:

$$S_Q = \int_0^{\infty} F(u)du = \int_0^{\infty} (1-\Phi_U(u))du = \int_0^{\infty}\int_n^{\infty} f_N(x)dxdu = \int_0^{\infty}\int_0^{u} f_U(x)dudx = \int_0^{\infty} xf_U(x)dx = E\{U\}.$$

In the ideal deterministic case when there is no uncertainties in the course of the project realization the project can be described by a single stepwise line (line 1, Figure 4) that bounds the rectangle and corresponds to baseline scenario of the project realization that occurs with the probability $P(S^{(b)})=1$ and whose utility $U(S^{(b)})=u_b$. When the project is assessed in a probabilistic formulation the project utility profile would be presented by a curve 2 (Figure 4). The position of the curve 2 with respect to line1 is determined by the following conditions:

- The influence of the negative random factors causes the occurrence of a number of pessimistic scenarios whose utility is lower than the utility of the baseline scenario $u<u_b$. Therefore the utility profile 2 will go below the stepwise line 1.
- The influence of the positive random factors may trigger the development of some optimistic project scenarios with utility $u>u_b$. That is why the maximum values of the utility profile u_{max} will exceed the value of u_b.

The area of the curvilinear triangle *ABC* (Figure 5) characterizes the decrease of the project utility (ΔU_-) due to the influence of negative (unfavorable) random factors. The area of the curvilinear triangle *CDE* will on the contrarily define the increase of the expected utility of the project (ΔU_+) due to the influence of the positive (favorable) random factors. Comparing the areas under the line 1 (A_{OABD}) and line 2 (A_{OAE}) one can estimate the decrease (in some cases increase) of the expected project utility when the project is considered in a probabilistic formulation and project uncertainties are taken into account. This value can be considered as another index of the project utility:

$$R_{pr} = A_{OABD} - A_{OAE}\,.$$

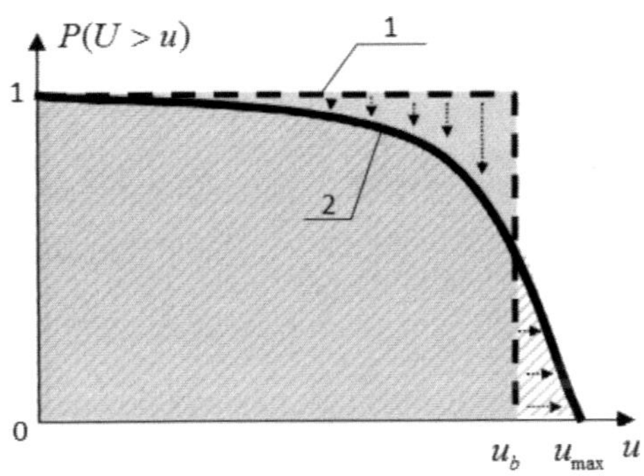

Figure 4. Project utility profiles in case of deterministic (line 1) and probabilistic (line 2) assessment of the infrastructure project utility.

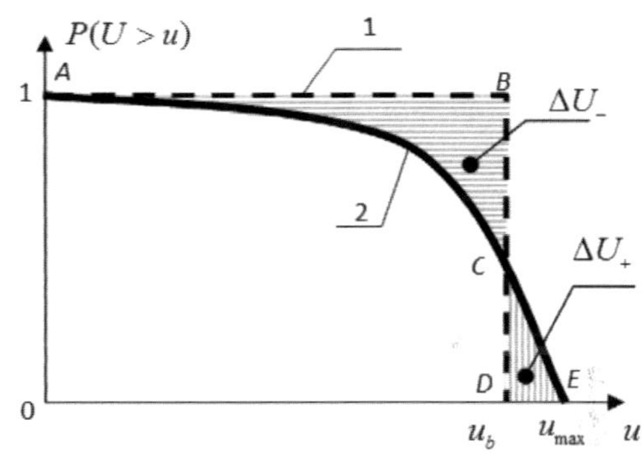

Figure 5. The influence of positive and negative random factors.

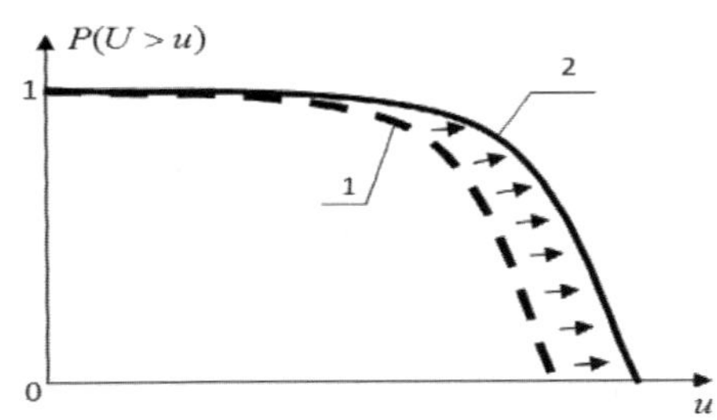

Figure 6. Assessment of the effectiveness of additional measures 1- initial utility profile of the project, 2- utility profile of the project after realization of additional measures.

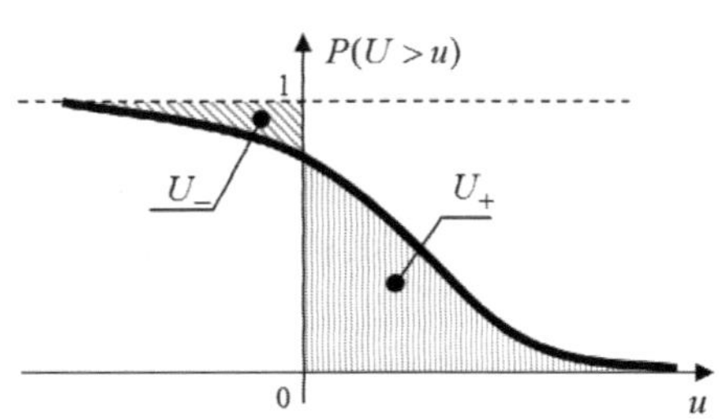

Figure 7. Determination of the integral utility of the project in the presence of scenarios with negative utility.

Assessment of utility profiles allows assessing the efficiency of measures to reduce the cost of project, cut the time of construction or increase safety and security of the project. If the area under the

IOP Conf. Series: Materials Science and Engineering **972** (2020) 012062 doi:10.1088/1757-899X/972/1/012062

utility profile increases after realization of additional measures than these measures should be considered effective.

It should be noted that in the general case given a certain method of choosing a utility function, unfavorable project scenarios may have a negative utility (Figure 7). In this case, the integral utility of the project will be the difference between the absolute values of positive utility U_+ and negative one U_-.

$$U = \|U_+\| - \|U_-\|. \tag{6}$$

The utility profile built for the infrastructure project under consideration is a fairly complete representation of the uncertainties and risks that may occur at different stages of the project life cycle. The utility profile corresponds to the triplet representation of the project uncertainties (4) and makes it possible to assess the feasibility of the project.

Through utility profiles one can judge the effectiveness of various kinds of measures aimed at reducing the cost of work, reducing the project implementation timeframe, increasing safety of the facility being created, and reducing the risks of its operation. If, taking into account the additional costs of the protection measures, the area under the utility curve increases (Figure 6), then the additional measures should be considered justified in terms of the utility criteria of the stakeholders. The advantage of this approach is that it allows one to conduct quantitative estimates and take into account the non-monetary characteristics of the project.

The use of utility profiles also makes it possible to compare different projects. Evaluating the area under the utility profiles of three projects, which are presented in Figure 7, it can be concluded that the project variant 1 is the most preferable.

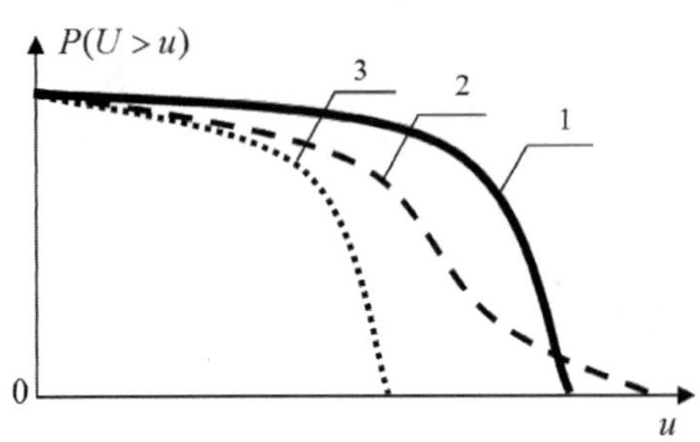

Figure 8. Comparison of various projects $E\{U_1\} > E\{U_2\} > E\{U_3\}$.

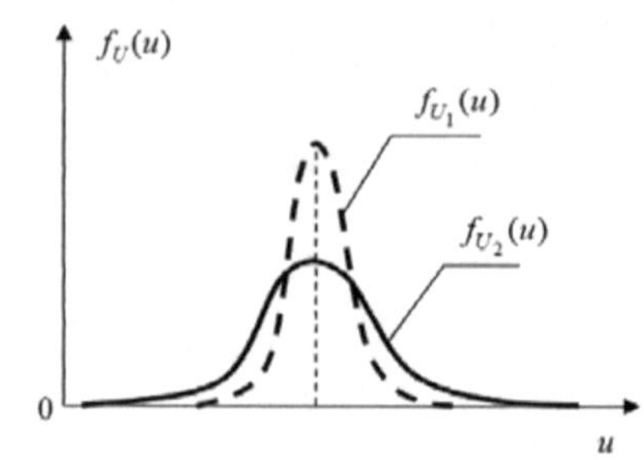

Figure 9. Probability density functions of utilities of various projects $E\{U_1\}=E\{U_2\}$, $\sigma\{U_1\} < \sigma\{U_2\}$.

However a comparison of the mathematical expectations of the project utility for different variants of the project often does not allow choosing the more preferable one (for example, in cases when the utility profiles overlap and have approximately equal areas). A simple comparison of the mathematical expectations of the project utility for cases 1 and 2 (Figure 8) can be insufficient to make a decision which variant is better. Mathematical mean of the project utility is not a complete and comprehensive characteristic of the project as it does not allow describing the aversion of project variants with large scatter of utility values for various scenarios (Figure 9). Bearing this in mind in addition to integral project risk index a ratio of the mathematical mean $E\{U\}$ and standard deviation $\sigma\{U\}$ of the random utility value U can be used as an additional project risk index:

$$I_\beta = E\{U\} / \sigma\{U\}. \tag{7}$$

Another utility index that allows assessing the utility of the project on the bases of its probability distribution is its fractile index:

$$I_p = u_p = \Phi^{-1}(P), \tag{8}$$

where $\Phi^{-1}(P)$ is the so-called quantile function which is an inverse function of the function $\Phi_U(u)$.

This index determines the value of utility u_p which with a given probability P will not be exceeded in the case of implementation of the project.

In particular, when assessing the utility of projects, 5% and 95% quantiles can be used: $I_{0,05}$ and $I_{0,95}$ that represent levels U that will not be exceeded with a probability of 5% and 95%, respectively.

Derivative indices of the project utility (with respect to I_P) are the mathematical expectations of the left and right tails of the distribution of the magnitude U:

$$I_{LT} = E(U \mid U < I_{0,05}), \tag{9}$$

$$I_{RT} = E(U \mid U > I_{0,95}). \tag{10}$$

The issue of development normative values for utility indexes for projects of various types as well as the type of utility function remains largely open now and requires further consideration.

4. Conclusions

1. When planning and implementing infrastructure projects, one encounters two types of uncertainties: (a) uncertainties stemming from the natural variability of conditions and factors affecting the progress of the project; and (b) uncertainties caused by limited knowledge about the processes occurring at different stages of the project development.

2. Scenario analysis of the infrastructure project allows obtaining the so-called profile of the project utility which is the most complete characteristic of uncertainties, risks and chances that accompany its implementation.

3. The proposed set of infrastructure project utility indexes can be used as a basis for making a comparative assessment of various project implementation options.

Acknowledgments

The work was supported by the Russian Foundation for Basic Research (grant № 16-29-09575 "ofi_m")

References

[1] Makhutov N, Reznikov D and Petrov V 2013 Assessment of the success of infrastructure projects implementation, taking into account the factors of uncertainty and multi-criteria decision making *Problems of Risk Analysis* **10** **3** 32-45 (in Russian)

[2] Smith C 2011 Integrated scenario-based methodology for project risk management Ph.D. Thesis. University of Maryland USA

[3] Carbone T and Tippet D Project 2004 Risk Management Using the Project Risk *FMEA Engineering Management Journal* **16(4)** 28-35

[4] Reznikov D 2013 Ways to compensate uncertainties to make complex technical systems secure and optimize the costs of their life cycle *Problems of Mechanical Engineering and Automation* **3** 57-64 (in Russian)

[5] Kaplan S and Garrick J 1981 On the quantitative definition of risk *Risk Analysis* **1** **(1)** 11–27

[6] Makhutov N and Reznikov D 2015 Application of scenario analysis in the assessment of structural reliability of complex technical systems *Journal of Machinery Manufacture and Reliability* **44** **8** pp 1-12

[7] Makhutov N Petrov V Akhmetkhanov R Dubinin E and Dvoretskaya T 2007 Features of scenario analysis of the occurrence and development of man-made disasters *Problems of Safety in Emergency Situations* **3** 3-28

[8] Makhutov N and Reznikov D 2012 Evaluation and regulation of risks associated with the operation of complex technical systems. *Safety in Technosphere* **5** 3-9

Selection of methods of inspection of building structures to prevent damage

V Ya Mishchenko[1], Yu D Sergeev[1], A Yu Sergeeva[1*], Yu V Myasischev[1] and R Yu Myasischev[1]

[1]Department of Technology, Organization of Construction, Examination and Management of the real estate, Voronezh State Technical University, 20-letia Oktjabrja str, 84, Voronezh, 394006, Russia

[*]E-mail: 933947@mail.ru

Abstract. This article discusses the mechanism studies of the functional provisions of the building construction, and identification of symptoms, in the case of which it's possible to forecast the appearance of accidents. The main purpose of this work is to find what kind of next examination should be performed otherwise, what sign that affects better than the others on the breakdown of the construction, you need to find to prevent its collapse and have time to take preventive measures to restore. In the work developed methods to determine the correct symptom, affecting more than any other on the disruption of the structure. An algorithm for the control of the system has been determined, Algorithm of monitoring when the life cycle of construction is maximum. It is proved that the issue of monitoring the priority or determination of the sign influencing the building breakdown can be defined by methods of nonlinear programming (parametric optimization). An algorithm which searches the main technical position of construction when the risk of disruption of the structure is minimal is developed.

1. Introduction

Violation of the building structures, leading to accidents, brings huge financial and property damage. Often predict the failure of the building structure is possible, in a timely manner after identifying signs of affecting her destruction. Therefore, it is required to develop an algorithm for checking the operation of the building structure, with the possibility of recreation, when the length of trouble-free operation becomes maximum [1-3].

The problem can be defined by methods of nonlinear programming (parametric optimization). To solve the formulated problem, the function of the reference point (the life cycle of the structure) S - value as well as the property (for the construction structure) operating time without a fatal failure is recommended. It is assumed that the maximum value of this value – S_{max} for any construction structure is achieved by absolutely accurate determination of its distinctive feature that affects its destruction $\gamma = \gamma$ *(valid)* and subsequent proactive methods of their regeneration [4, 5]. An analogy of the functional purpose of the reference point is ΔS, and a similar goal of improvement is max ΔS, where ΔS – an increase in S as a result of the next survey. Effectively define both functions of the reference point S and ΔS.

IOP Conf. Series: Materials Science and Engineering **972** (2020) 012063 doi:10.1088/1757-899X/972/1/012063

2. Calculation of the duration of the construction of structure without destruction

We calculate the function of the problem S [7, 9]. Perhaps three fundamental principles of calculation *S*. For a good example, imagine that there are three signs that affect the destruction of the building structure, their probabilities are d_1, d_2, d_3, and $d_2 > d_1$ and $d_2 > d_3$. We present conclusion *A* as the most probable sign of structural failure $A = \gamma\{\max(d_\gamma \gamma = 1,2,3,\ldots,n)\}$. In this example, *A* is a sign of failure № 2. Suppose there are three groups of building structures. The first - with a certain set of signs of destruction d_1, the second with a certain set of signs of destruction d_2, the third with a certain set of signs of destruction d_3 [6]. The first idea of calculus S is contained in the following. Suppose that all three groups of building structures are restored according to the definition of the degree of destruction *A*. That is, it is possible to restore the building structure of group d_2 and it is impossible to group d_1 and d_3. The value of S for all types is different (the types are considered to consist of the same structures). S for all building structures as a result of the analysis a_0 is calculated as:

$$S(a_0) = \sum_\gamma d_\gamma(a_o) S_{A,\gamma}, \quad (1)$$

where, $S_{A,\gamma=2} > S_{A\gamma=1}$ $S_{A\gamma=2} > S_{A,\gamma=3}$.

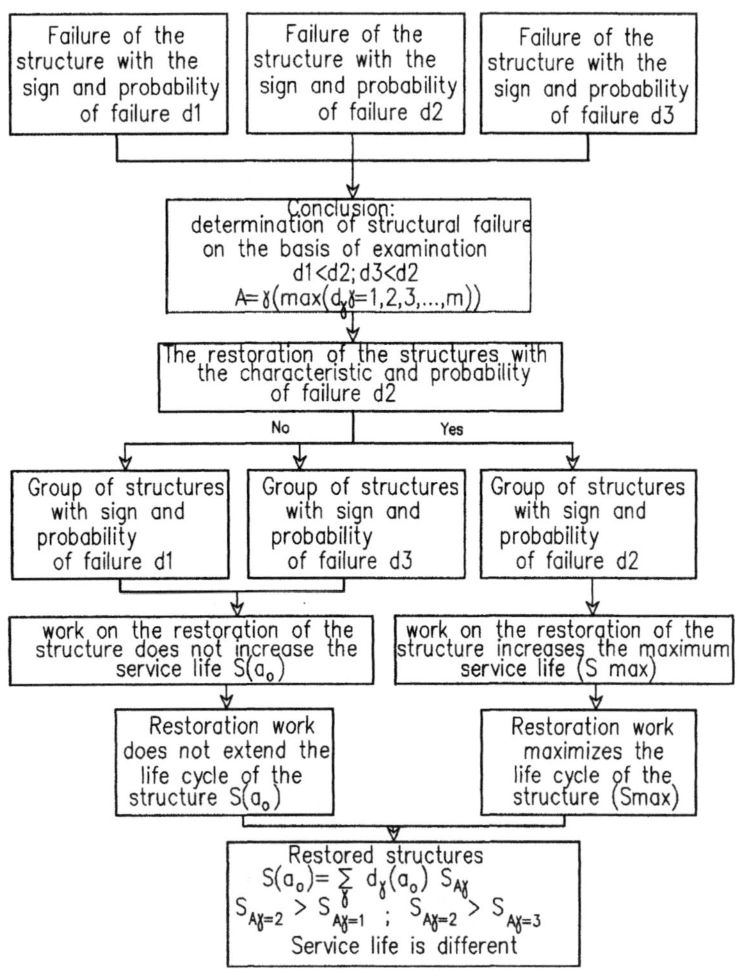

Figure 1. The algorithm for calculating the life cycle of the structure on the principle of 1.

IOP Conf. Series: Materials Science and Engineering **972** (2020) 012063 doi:10.1088/1757-899X/972/1/012063

This is used in practice: the construction structure is restored by determining the degree of destruction, understood as the most likely sign of destruction (Figure 1).

The known methods of probability theory reflect the second principle under consideration [8]. So the systems consist of designs taken in free order. Relative number of constructions in groups d_1, d_2, d_3. The first group is restored from signs of destruction 1, the second - from signs of destruction 2, the third - from signs of destruction 3 (Figure 2).

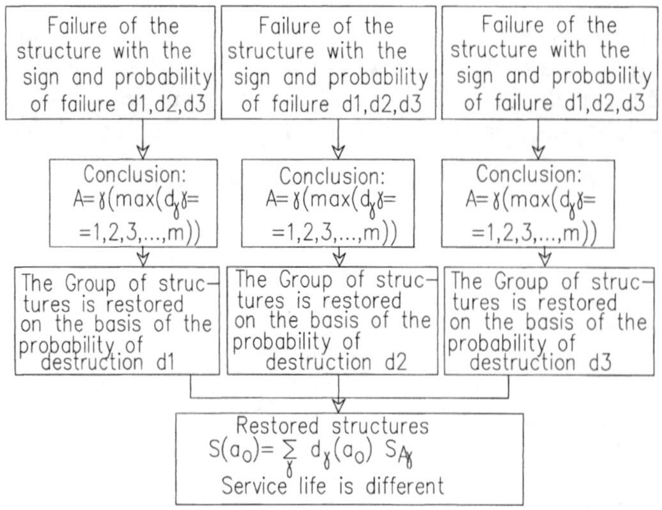

Figure 2. The algorithm for calculating the life cycle of the structure on the principle of 2.

The third thesis is theoretical. For example, if the first type includes structures on the basis of destruction 1, the second - on the basis of destruction 2, then the third group will be construction structures on the basis of destruction 3 (Figure 3).

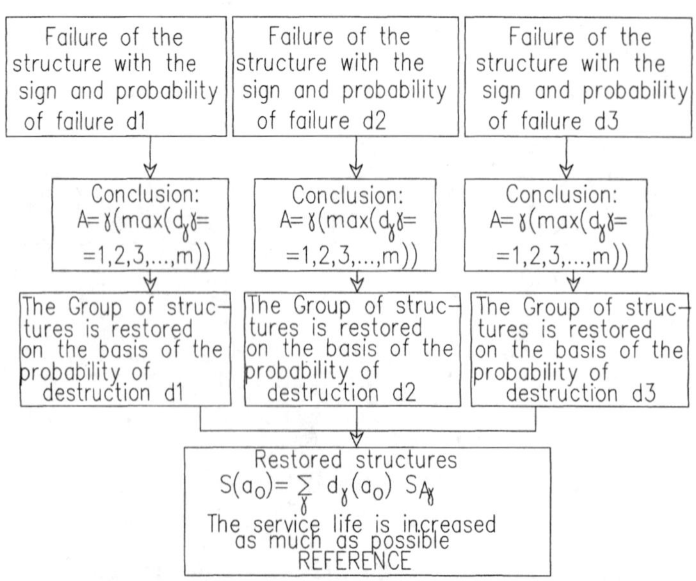

Figure 3. The algorithm for calculating the life cycle of the structure on the principle of 3.

IOP Conf. Series: Materials Science and Engineering **972** (2020) 012063 doi:10.1088/1757-899X/972/1/012063

All building structures, from which species are formed, contain completely identical features that affect the destruction of the structure. Therefore, to specifically form types, each with building constructions one characteristic, influencing the destruction, it is impossible. The third principle of restoration is that the building structures of each group is restored from the sign that destroys the structure that destroyed them, i.e. the restoration of building structures is carried out according to three estimates.

We use the calculus S when the constructions are reconstructed by the first principle. Let us use the matrix-column B with elements $S_{\gamma, A}$. They show the format S of the building structure (the age of the building structure, uneven sediment of the Foundation, the aggressiveness of the medium and other features) of the sign of destruction $\gamma(\gamma = 1,2,3,...,n)$ at restoration, according to the conclusion A. Before the m-th test is carried out, conclusion A is based on the characteristic that is most likely to subject the destruction of the structure $d_{\gamma max} = \max\{d_\gamma \gamma = 1,2,3,\dots,n\}$, before the m-th test is performed:

$$S = \sum_\gamma d_\gamma^{new} S_{\gamma,A}. \tag{2}$$

In formula (2), probabilities are applied d_γ^{new} not d_γ since the distribution will be reliable d_γ^{new}. We can say that the building structures of the additional examination b were reconstructed incorrectly - on the basis of determining the degree of deterioration of A. In the formula (2) S, it is calculated from the old definition of the degree of destruction A. That is, we assume that before determining the degree of destruction of the m-th survey, both the distribution of d_γ and the restoration of the structure are, in General, incorrect. After the survey to determine the degree of destruction of A^{new} will be considered a sign that allows the construction structure, which has the maximum probability:

$$\max\{d_\gamma^{new}(a_{om}), \gamma = 1,2,3,\dots,n)\}. \tag{3}$$

Value S^{new} for examination a_{om}:

$$S_m^{new} = \sum_{C_{om\,min}}^{C_{om\,max}} Q_ь(a_{оь}) \sum_\gamma d_\gamma^{new}(a_{om}) S_{\gamma,A}. \tag{4}$$

where S_m^{new} - the value of S after its examination was carried out (after finding the m-th sign acting on the collapse of the structure).

Consider $S_{\gamma,A}$ – elements of the matrix b column. Its numbering is more equal to the possible symptom of collapse - finding the level of destruction of A^{new}. The expression (4) means that a_{om} passes discrete values from minimum to maximum (from $a_{om\,min}$ to $a_{om\,max}$).

For continuous signs affecting structural failure:

$$S_m^{new} = \int_{C_{om\,min}}^{C_{om\,max}} Q_m(a_{om}) \sum_\gamma d_\gamma^{new}(a_{om}) S_{\gamma,A^{new}} f a_{om}. \tag{4a}$$

In the formula (4a) d_γ, $\gamma = 1,2,3,...,n$ is better to show by vector $\overline{d}(a_{om})$.

In this vector, all coordinates are reset to zero, except for the largest coordinate, which is the conclusion.

This vector will be called $\overline{d}_Д^0(a_{om})$. Then the expression (4a) will be represented as:

$$S_m^{new} = \int_{C_{om\,min}}^{C_{om\,max}} Q_m(a_{om})\ B \times \overline{d}_Д^0(a_{om}) \cdot \overline{E} f a_{om}, \tag{4b}$$

where $\overline{E}$ is a unit vector of dimension n.

By analogy, we rewrite the expression (4). Then according to the 3rd principle of the perfect reanimation of fix as:

$$S_m^{new} = \int_{C_{om\,min}}^{C_{om\,max}} Q_m(a_{om})\ B^d \times \overline{d}(a_{om}) \cdot \overline{E} f a_{om}, \tag{4c}$$

Where B^d is the diagonal matrix obtained from matrix B by zeroing nondiagonal elements (they have $j \neq \gamma$).

3. Determination of the necessary survey, in which the work of the building structure will have the maximum duration

After you define S_m^{new} calculate the effectiveness of implementation of the conclusion of the m:

$$\Delta S(m) = S_m^{new} - S = f(m). \tag{5}$$

After that, by analogy, calculations are carried out for absolutely all k conclusions, features, information of instrumental and visual surveys. In the end, the sought trait (evaluation, instrumental or visual examination) k is characterized by the number of trait acting on structural failure at which ΔS_m is maximal (Figure 4).

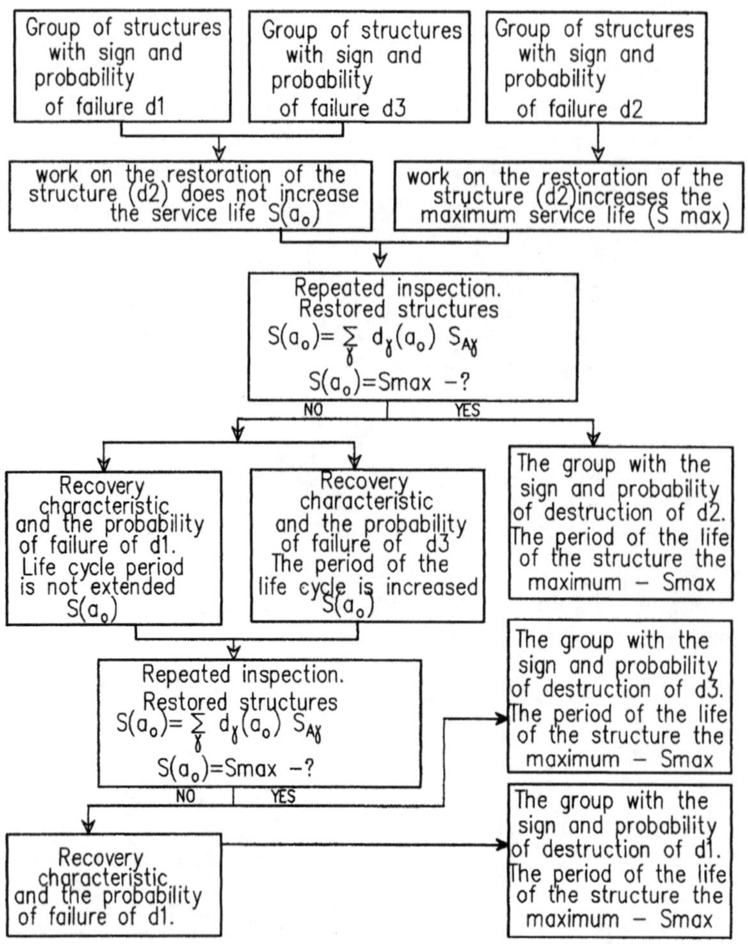

Figure 4. The calculation algorithm in which the maximum life cycle of the structure.

$$k = m\{\max[\Delta S_m m = 1,2,3,...,\delta]\}. \tag{6}$$

The number k is obtained in the reconstruction of the building structure on the first principle.

The k rating is recommended. After the calculation of this estimate k, there is its transition to the estimates of the already known j, $1 \leq j \leq b^{new}$, where $b^{new} = b + 1$.

4. Conclusion

Thus it is proved that the problem of priority of works on inspection of building structures for the prevention of destruction can be determined by methods of nonlinear programming. Methods for determining the characteristic that most affects the destruction of the structure have been developed. An algorithm for monitoring the system is determined, in which the maximum life cycle of the structure.

References

[1] Mishchenko V Ya, Gorbaneva E P, Shishkina O S and Jon R 2015 Monitoring of technical condition of buildings and structures *Sci. J. Eng. S. and Str.* **3**(**20**) P 20-29

[2] Mishchenko V Ya, Golovinsky P A and Drapaluk D A 2009 Predicting rate of deterioration of the housing stock on the basis of monitoring defects of building constructions *Sci. Bul. of V. St. Un. of Arc. and C.* **4** P 111-117

[3] Mishchenko V Ya, Sergeev Yu D, Myasischev R Yu and Sotnikova O A 2015 Systematic approach to improving the performance of industrial and civil buildings *Sci. Bul. of V. St. Un. of Arc. and C.* **4**(**40**) P 49-57

[4] Myasischev Yu V, Sergeeva A Yu, Sergeev Yu D and Myasischev R Yu 2018 Development of model of monitoring of industrial and environmental safety, for an objective assessment of loads and bearing capacity of structures *Con. and R. Es.* **1-1**(**2**) P 63-67

[5] Myasischev Yu V, Sergeeva A Yu, Sergeev Yu D and Myasischev R Yu 2018 Factors affecting the technical and operational condition of building constructions *Con. and R. Es.* **1-1**(**2**) P 67-74

[6] Sergeeva A Yu, Myasischev Yu V, Myasischev R Yu and Sergeev Yu D 2017 Study of the signs of an emergency condition of bearing structures of buildings and constructions *M. Tr. in the Con. and Op. of R. Es., C. of Sci. Art. of M. of Sci-Pr. Con.* P 218-223

[7] Wu J S, Apostolakis G E and Okrent D 1990 Uncertainties in system analysis: probabilistic versus non probabilistic theories *Rel. Eng. and S. Saf.* **30 P** 163–181

[8] Ferson S, Ginzburg L, Kreinovich V, Nguyen H and Starks S 2002 Uncertainty in risk analysis: Towards a general second-order approach combining interval, probabilistic, and fuzzy techniques *Pr. of F-IE Hon. H.* P 1342–47

[9] Feth Th and Oberguggenberger M 2004 Propagation of uncertainty through multivariate functions in the framework of sets of probability measures *Rel. En. and S. S.* Vol 85 **1– 3** P 73–87

IOP Conf. Series: Materials Science and Engineering **972** (2020) 012064 doi:10.1088/1757-899X/972/1/012064

Analysis of industrial areas of Ekaterinburg

A V Alekseeva[1], S V Pridvizhkin[1], G V Mazaev[2]

[1]Ural Federal University, 19 Mira str, 620002 Ekaterinburg, Russia,

[2]Branch of the FSBI "CRDI Minstroy Russia" UralNIIproekt

E-mail: toni.pereskokova@gmail.com

Abstract. The article views the relationship between the change in the purpose of the industrial area and the planning structure of the city. Moreover it includes the proposed gradation of enterprises depending on the area of their industrial sites. Also the text point out the uniqueness of the industrial site of PJSC "Uralmashzavod" among other enterprises of the heavy industry of the city of Ekaterinburg.

1. Introduction

At present, the progressive growth of the Sverdlovsk region continues; it combines industrial, social, cultural, educational, scientific and managerial functions. Old-developed city-factories, such as Ekaterinburg, today bring developed trade and social, cultural and management options into their structure. Nevertheless, industry still has a decisive influence on the economic component of cities.

According to the investment portal of the Sverdlovsk region in 2018, the growth of profits of large industrial enterprises amounted to 41.9% compared with 2017 (Pic. 1). The largest growth in production has been in the engineering sector: the production of machinery and equipment has been 1.5 times higher than the level of January-October 2017, vehicles, trailers and semi-trailers has been 1.3 times; vehicles and equipment has changed by 12.9% [1]. Thus, industrial enterprises provided 41% of investments in the fixed capital of large and medium-sized organizations, which suggests that almost half of the profits in the economy of the Sverdlovsk region are brought by industrial sites. Renovation of industrial areas in order to give the site a different economic function may lead to a loss of production capacity and thereby reduce the economic growth of the region and, consequently, lead to a decline in the economy of the Sverdlovsk region.

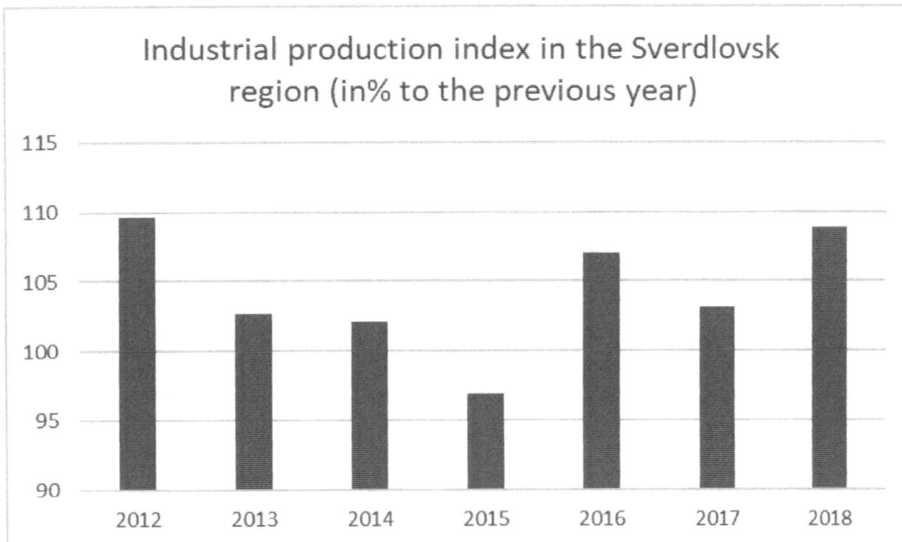

Figure 1. The index of industrial production in the Sverdlovsk region (in % to the previous year).

2. Changes of industrial sites of the city of Ekaterinburg

Despite the fact that production plays a significant role in the development of cities, currently the most popular in Ekaterinburg is the territory redevelopment. Redevelopment is the transfer of an industrial enterprise outside the city limits, and there are a public and residential building on the vacated area. Examples include the residential complex "Universitetsky", built on the site of the Uralobuv Plant, the Malevich residential complex, the Makarovsky Quarter residential complex on the site of the Ekaterinburg Flour Mill. A similar fate awaits the factory of V.V. Vorovskogo, Zavokzalny district, the territory of the former JSC "Uraltransmash" and there is a discussion of a partial change of the territory of PJSC "Uralmashzavod". This method has a number of advantages and disadvantages that affect the urban planning and economic components of the city. Positive and negative points of redevelopment is presented in Table 1 below.

Table 1. advantages and negative points of redevelopment

Advantages	Disadvantages
• Ability to create a new architectural image of the city; • Improving city planning; • Getting the economic benefits in the sale or rental of the premises (apartments) newly constructed buildings; • Elimination of sanitary protection zones, allowing the development of adjacent territories;[4] • Increasing environmental areas	• Violation of existing transport links and the need to build new one; • Overload of existing engineering networks, the need for building new sets; • Loss of jobs in the city; • Creation of additional transport traffic when moving "housing-work"

The pros and cons that arising from a change in the purpose of the industrial site confirm the need for a serious study of project of this type and its impact on the development plan of the city as a whole. The lack of consistency and complementarity of the building, as well as the hasty and ill-conceived disbanding of industry, can lead the city to the loss of existing advantages and a serious lag in development from competitors [2].

3. Analysis of industrial sites

Due to the fact that in the city of Ekaterinburg there is an active redevelopment of territories, the number of production zones decreases. The question arises: how many lands for industrial needs is used for its intended purpose. Since the purpose of the study is PJSC "Uralmashzavod", the data were collected specifically for machine-building enterprises. Information on the number of engineering plants in the city of Ekaterinburg are taken from the official portal of the city; and the area of territories is determined according to the Public cadastral map.

According to official sources, on September 3, 2018 there are about 50 industrial enterprises, which can be divided into 3 groups depending on the area of their industrial zone.

3.1. Enterprises of the first group up to 10 hectares

Basically, these are branches of larger enterprises, or narrowly focused production, such as LLC "ABB - Electrical Engineering" that is the Ural branch of the enterprise (about 35 thousand square meters) or OJSC "Sverdlovsk plant of current transformers" (about 3 hectares). Enterprises of the first group is presented in Table 2 and picture 2 below.

Table 2. Enterprises of the first group up to 10 hectares

Enterprise	Area, hectares
LLC "ABB - Electrical Engineering" is the Ural branch of the enterprise	3.17
CJSC "Yekaterinburg forest machines"	5.67
Ekaterinburg Electric Locomotive Repair Plant is a branch of OAO "Zheldorremmash"	1.11
LLC "Plant" Promavtomatika"	3.64
CJSC "Research and Production Association" League-7"	5.38
LLC "Medin-N"	8.36
CJSC "Reltek" is Russian Electrotechnical Company	2.24
OJSC "Sverdlovsk plant of current transformers"	2.86
JSC "Uralcable"	9.07
LLC "Fotek"	1.17
OJSC "Russian Electrotechnical Company"	1.66

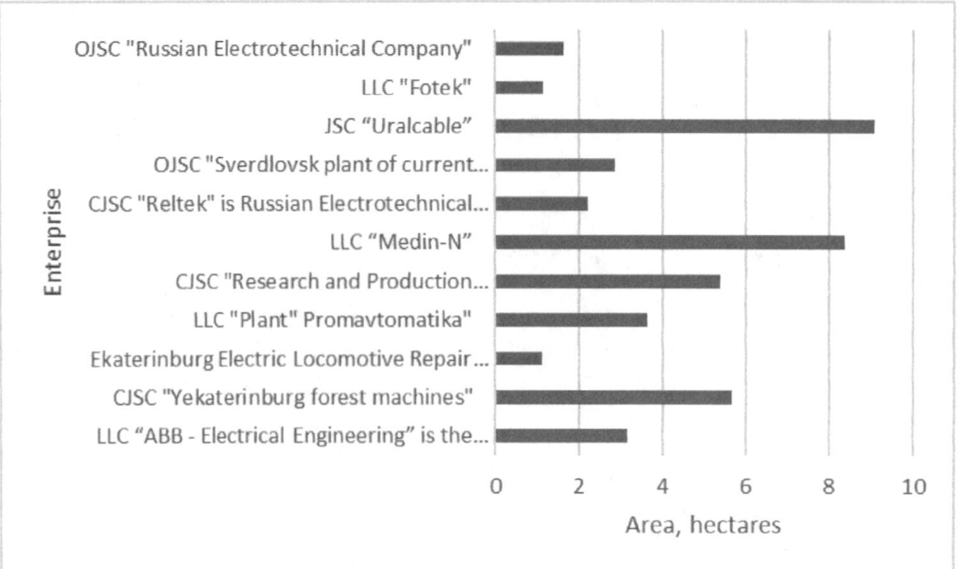

Figure 2. Diagram. Enterprises of the first group.

3.2. Enterprises of the second group from 10 to 50 hectares

Plants that have a larger turnover of industry than in the factories of the first group. For example, OJSC "Drilling and Metallurgical Equipment Plant" is based on about 11 hectares and JSC "Ural Production Enterprise" Vector " is located on about 23 hectares. Enterprises of the second group is presented in Table 3 and picture 3 below.

Table 3. Enterprises of the second group from 10 to 50 hectares

Enterprise	Area, hectares
OJSC "Drilling and Metallurgical Equipment Plant"	10.61
ZAO PO "Complex"	23.39
SC "Ural Transport Engineering Plant"	26.77
JSC "Ural Production Enterprise" Vector "	22.34
JSC "Ural Compressor Plant"	14.38
JSC "Production Association" Ural Optical-Mechanical Plant named after E.S. Yalamova	40.02
Federal State Unitary Enterprise "Ural Electromechanical Plant"	15.91
OJSC "5th Central Automobile Repair Plant"	30.91

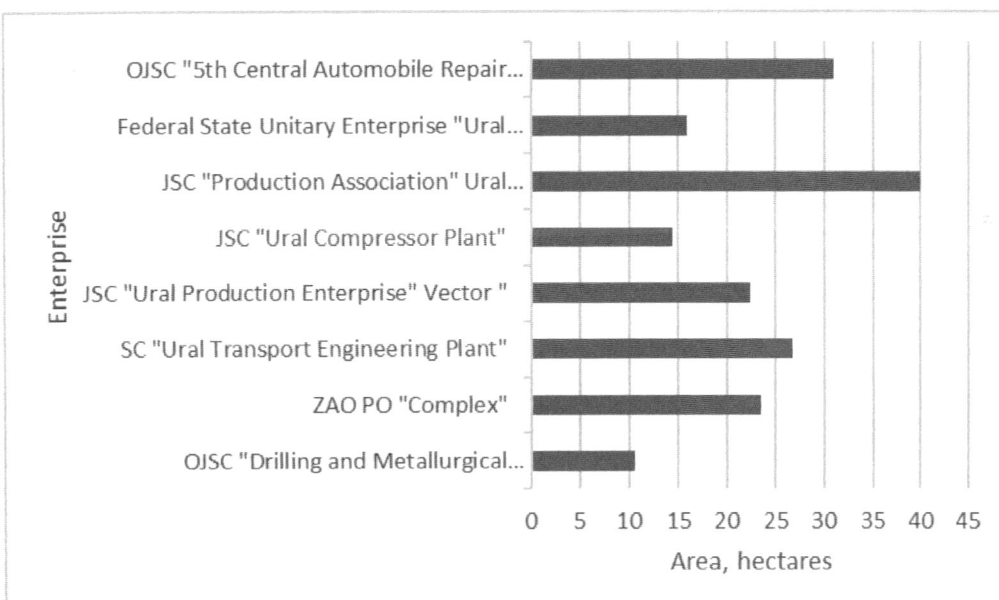

Figure 3. Diagram. Enterprises of the second group.

3.3. Enterprises of the third group of 50 hectares or more

An example of such enterprises is PJSC "MI Engineering Plant Kalinina "(about 70 hectares) and JSC" Ural Turbine Plant "(about 90 hectares). Enterprises of the third group is presented in Table 4 and picture 4 below.

Table 3. Enterprises of the third group of 50 hectares or more

Enterprise	Area, hectares
PJSC "MI Engineering Plant Kalinina "	64.54
JSC "Ural Turbine Plant"	89.04
PJSC "Uralmashzavod"	**2 43.94**

Enterprise	Area, hectares
JSC "Ural Chemical Engineering Plant"	55.75
JSC "Uralelectrotyazhmash"	77.08
LLC "Elmash"	

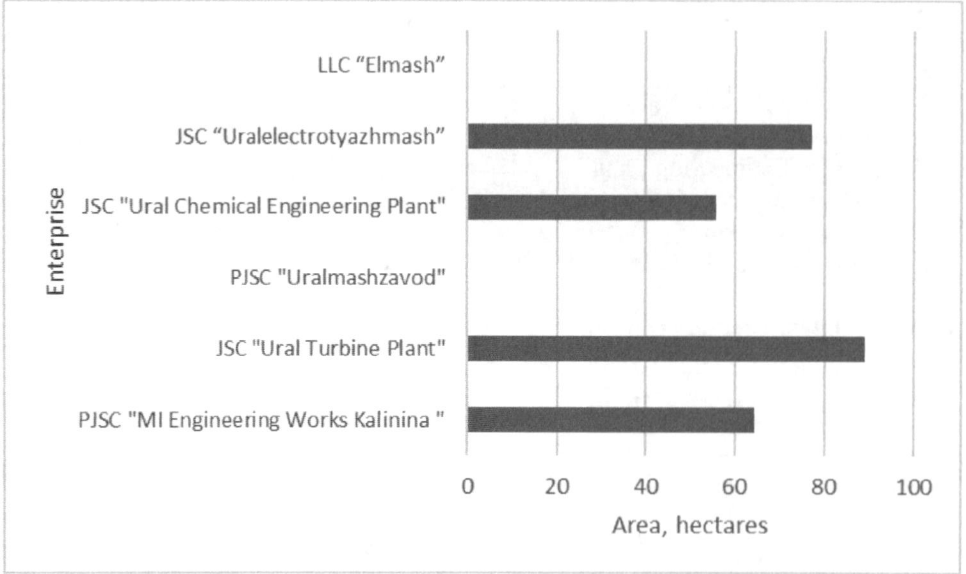

Figure 4. Diagram. Enterprises of the third group.

Based on the proposed classification, PJSC Uralmashzavod is in the third group, but at the same time it has a significantly larger area than other enterprises — more than 200 hectares. Thus, this area is unique among others in terms of area, but this is not the only advantage of the plant. The company is located in the city, which allows workers to travel by public transport and do not have to use factory buses or private cars. At the same time, competent logistics inside and outside the enterprise has been carried out and tested, which allows the delivery of manufactured mining equipment to the installation site in various places of the country and the world.

The draft master plan of the city assumes the division of the territory of PJSC Uralmashzavod by two highways of urban importance in the north-south direction in order to increase the connectivity of the northern part of the city with the center. At the same time, the division of the industrial-communal area of Zavokzalny concern also occurs. Those changes are due to the need to solve the transport problem on two existing highways, that are Cosmonauts Avenue and Donbasskaya Street. Such a solution to the transportation problem will result in restructuring as a result of the fragmentation of its industrial site [2]. Thus, solving one town-planning issue can give rise to an even more serious economic problem, since industry is one of the key advantages of the city of Ekaterinburg and the Sverdlovsk region as a whole. Following the above, we can conclude that PJSC "Uralmashzavod" is a unique enterprise of the city with a unique industrial territory, the loss of integrity of which will cause the dramatic decrease of not only the existing plant, but will not allow developing heavy engineering in the future, without which it is impossible to imagine positive growth of the economy of the city of Ekaterinburg.

4. Conclusion

The city of Yekaterinburg, as one of the old-developed city-factories, must certainly maintain its industrial potential, the security of which is the industrial territories in its planning structure. It is

IOP Conf. Series: Materials Science and Engineering **972** (2020) 012064 doi:10.1088/1757-899X/972/1/012064

necessary to think over the new general plan of the city and calculate all the economic components, so that the residents have the opportunity not only to live in this city, but also to be provided with work, including industrial production, taking into account the development of the economy. Special attention should be paid to the industrial heritage of the city and to preserve the industrial potential of the city of Ekaterinburg.

References

[1] Economy. Statistics. Investment portal of the Sverdlovsk region.. URL: *http://invest.midural.ru/region/obshhie-svedeniya/information/promyishlennost/* (appeal date: 04/07/2019) [In Russ]

[2] Chueva T I 2016 Redevelopment of urban industrial zones of Ekaterinburg: analysis and prospects (Ekaterinburg: Publishing house of the Ural State University of Economics) pp 221-224 [In Russ]

[3] Mazayev G V 2018 Change in the purpose of industrial territories. Theory, practice, methodology (Ekaterinburg: UralNIIproekt) p 120 [In Russ]

[4] Pridvizhkin S V 2007 Formation of local housing markets in the space of the largest cities (Ekaterinburg) p 17 [In Russ]

[5] Spatial development strategy of the city for the period of 2030 (Municipality "City of Yekaterinburg" S) URL: *https: //Екатеринбург.рф* (request date: 12/15/2018). [In Russ]

Some Economical Methods of Assessment of Ecological Products and Service

V B Vlasov[1], V Ya Mishchenko[1] and A I Makeev[2]

[1]Department of Technology, Construction Management, Expertise and Property Management, Voronezh State Technical University, 14 Moskovskiy Avenue, Voronezh 394026, Russia

[2]Department of Building Materials, Voronezh State Technical University, 14 Moskovskiy Avenue, Voronezh 394026, Russia

E-mail: vla-valerij@yandex.ru

Abstract. In this work, authors try to lay the methodological foundation of cost assessment of ecological products and service, which were free in Russian Federation. Transferring more and more products and services to market appeal as market objects means more effective and rational using them with different market mechanisms complexes. Current economic practice does not provide effective using of nature recourses such as water for economical subjects, who control it, and for recourse's consumers. On the basis of research in ten villages of Voronezh region authors reveal some significant reserves in improving effectiveness of the system. The essence of the method is necessity of meticulous research of all characters of resource's demand and revealing its extreme utility. On the basis of research's results were revealed that customer always wants to have more qualify product of service and pay higher price of them. In the job you can find specific example of using the method and some practical recommendations of its rational using. Authors suggest and substantiate the necessity of including the characters of ecological and economical effect and damage in the assessment of effectiveness of different vectors of development.

1. Introduction

One if the main trend of changeable social status is object expansion in our social life, which needs cost assessment, market approach to many things and processes essence. Previous status is probably related to the character of social and state own almost on all that surround us. General nationalization of economical, domestic, nature environment's objects did not suppose their assessment in product form. The scope of provided products and services in the form of social goods was almost not limited. The scope extended to having all types of educational and medical services and nature resources using.

Some serious changings follow the transition to market relationship. Fixing own on most social life's objects and requirement of effective economic activity identified the necessity of using market mechanisms in its assessment. Completely changed legislation, a rigid requirement to account costs, an assessment of the object status demand objective necessity of costs assessment and effective using social goods such as visiting nature reserves, monuments of culture and nature, a possibility to use qualitative water resources, when it is impossible to evaluate it in production form. Entering these objects into market relationship scope, economical subjects, which control it in new economical scope in current system, cannot provide effective economic activity.

2. Estimation method based on marginal utility theory

A feature of using social good is that consuming it by one person do not effect on a magnificent for others. Analyzing the situation using the theory of extreme utility, it is possible to conclude that unlimited possibility of using social good for every extra person who uses it, extra consumption is zero. It means that for defining general readiness to pay for using social good, it is necessary to sum up readiness for all customers to pay for it. It means to sum up individual curve of demand.

Authors consider that the instrument to define the cost of social good is using the method of assessment extreme utility of the good.

2.1. A specific example of using the method

Water system as a communal establishment is given as an example. Water from all sources was almost free in the nearly past. Ineffective work of communal services gives minor proceeds and it lays low qualitative services. In this situation, the method can be used as a variant to analyze markets for management and design systems and establishment tariffs.

The degree of communal services from municipal system of water supply cause many censures. A system of payment for water supply for household in country side is imperfect, so the value of the accumulated sum for water submission cannot provide normal functioning of the system.

The mission that we set is if the system has a reserve for solving the current problem. Using the method of assessment some researches in ten localities of Voronezh region was made.

Was revealed an important factor that it is a desire to pay for water supply according to higher tariff subject to more qualitative service, and the negative moment is higher prepayment for connection to the system.

The curves of demand were counted by received information. In figure 1, we demonstrate the value of connections to the system, the revenue and the size of customer's surplus on current tariff.

The modeled changing in customer's surplus in the situation of increased value of connections increased the size of revenue and customer's revenue on current tariff and doubled tariff, are shown in figures 2 and 3.

Analyses of the results of the research encouragingly admit a possibility to solve the water supply system's problem. It is meant that it is important to concentrate the efforts on municipal companies of more qualitative service, increasing the value of customers and some changings in financial legislation. People rear that the cash they collect for laying aqueduct is not always fit actual outlay. It means that the cost of connecting should be included into the rate of monthly payment for using water, investing and guide it into improving and maintaining the quality of customer's service.

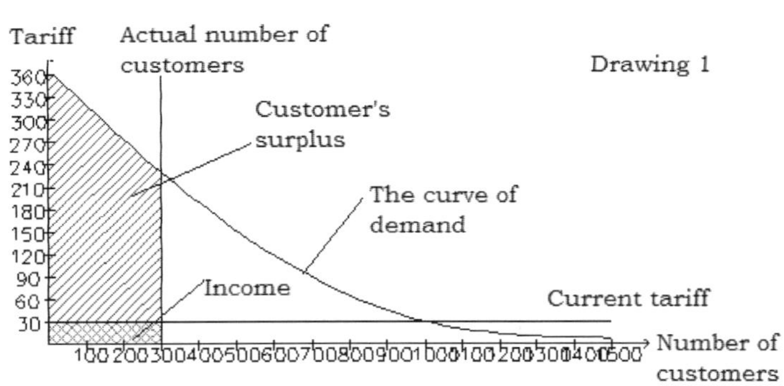

Figure 1. Tariff – number of customers.

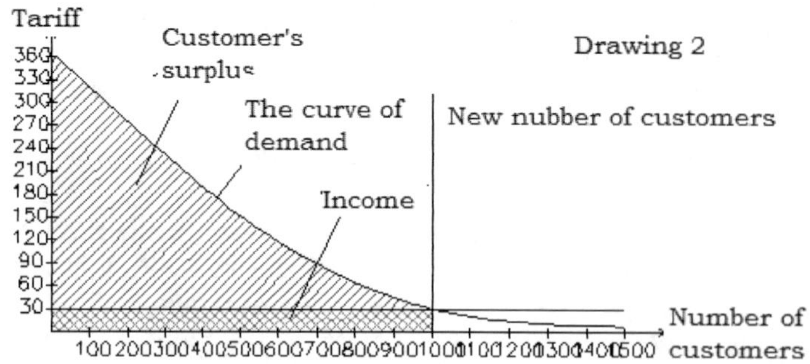

Figure 2. Tariff – number of customers.

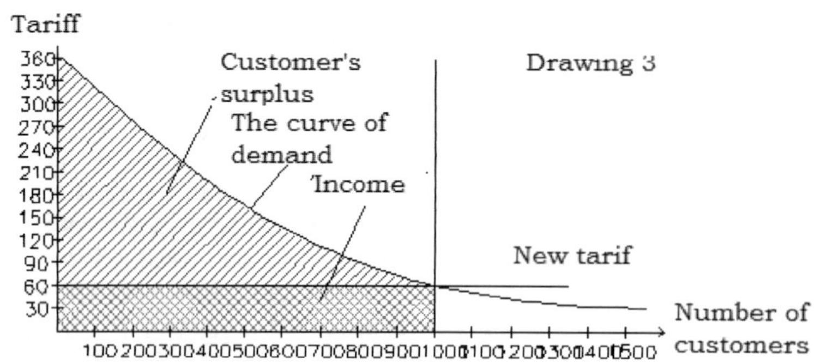

Figure 3. Tariff – number of customers.

3. Analysis of traditional approaches to the assessment of environmental goods

We consider the situation of underestimation of social good, conditioned by the absence of economical mechanism and psychological inhibition in realistic assessment of customer and water resource.

We can find some methods of ecological product's and service's assessment in economical literature in case of solving social or economic problem. Methods are quite popular:

- the method including personal capital's assessment;
- the method including real diseases cost assessment;
- the method with changing of performance;
- the method of preventive and softening cost assessment;
- the method of transport and track costs.

Some methods set a task to find cause-effect connection to measure the damage in different situations. Some methods connected to market's analysis and market's development factors. All methods are in demand in practice of assessment of ecological, nature and social goods' influence.

However, in general, looking from the economical point, the factors cannot be fully the economic factors of social product's and service's assessment.

Without any doubt, common and synthesis character of all method's assessment can be the character of supposed act's effectiveness, classically matching the income and the cost. But right definition of economic value of ecological, social and nature good is very important. The assessment from economic effectiveness point can lead out the good a factor of competitor ability. Accounting of ecological factor of different levels vectors of development can finally increase of reduce economical effect of the vector. If the reduced assessment is chosen of economic, nature or social good, the accounting of ecological damage is wrong, the damage of the territory is not accounted, the reduce of biodiversity, the vector cannot be economically attractive. Inversely - the vector of development including safety of nature, reducing the pollution, saving health and increasing the quality of life - increases economic effectiveness of the vector.

Establishment or characters of economic and ecological damage and economic and ecological effect using the definition of economic effectiveness of different vectors of development will let us make the vectors really effective of not effective, and make the ecological characters the factors of competitor ability.

4. Real valuation of natural resources is a fundamental factor in increasing the efficiency of social production

The main, central issue of solving the problems discussed is the real assessment of environmentally significant resources, including natural ones. The roots of the problem, in our opinion, are as follows. First of all, the realization of state ownership of natural resources in favor of the actual users of it by means of an unfair misappropriation of rent is exaggerated. The functions of state bodies in the regulation of natural resources as objects of ownership, and above all, the determination of their actual economic evaluation, are very blurred. As a result, the actual income from the use of these resources is much less. In addition, this state of affairs creates an economic environment for the extremely inefficient use of resources.

Another major problem that does not allow a real economic assessment of natural resources is the contradiction between the federal laws regulating the activities of users of natural resources, the laws of the constituent entities of the Russian Federation, other regulatory documents and the established environmental and economic practices of resource users, who, naturally, favor their own interests. In this regard, serious cadastral work is required with a thorough economic assessment of all natural resources, which should become the basis for their effective use.

In 2019, the Ministry of Natural Resources and Ecology for the first time considered the value of natural resources in the territory of the Russian Federation. As of the end of 2017, there were 9.04 billion tons of oil and 14.47 trillion cubic meters of natural gas in the territory of the Russian Federation. In rubles it is 39.6 and 11.3 trillion rubles, respectively. The Department estimated the reserves of iron ore at 808 billion rubles, diamonds at 505 billion, and gold at 480 billion. The total cost of all resources amounted to 55.2 trillion rubles [9].

It is imperative to make such an assessment regular. Russian statistics do not analyze natural resources as part of national wealth. Russia is a unique country. Only 2% of the world's population lives here, accounting for 40% of the world's natural resources. However, as shown by the results of a study of the World Bank [10], Russia is at the bottom of the list of countries with average per capita income. For example, we are 17 times poorer than the Swiss and 13 times than the Americans. At the same time, an average of $38,709 of the country's total wealth falls on every Russian. This impressive figure is provided, first of all, by the share of natural resources, which is $17,217 (a little less than half of the total amount) and by production capital of $15,593. For comparison, among the top ten countries, the share of natural resources is not more than 2% of wealth distributed per capita, production capital ranges from 11 to 30%. However, intangible assets in developed countries (technology, science, social institutions, human capital) range from 63 to 87%, while in Russia this figure does not exceed 15% [11].

The ratio of available natural resources to other assets of national wealth - engineering structures, equipment, transport, enterprises and other material values is very important. If we bear in mind that a significant part of fixed assets is worn out, and the objective dynamics has a predisposition to an increase in the value of natural resources, their value may be higher than the production part of national wealth. Despite the fact that there are prerequisites for an increase in the cost of natural resources, there is also anxiety about the decrease in their value as a result of degradation and depletion, including due to the depreciation of equipment and their operating mechanisms. An increase in losses over an increase in the value of natural resources would mean a loss of the natural basis for the growth of national wealth. Unfortunately, today, the rate of use of natural resources exceeds their reproduction.

Another important factor determining the need to increase the role of the valuation of natural resources is the cost of labor in the industries and enterprises that exploit these resources. High capital and operating costs associated with the extraction and processing of natural raw materials, as well as high world prices for natural resources, suggest a high technical level of equipment maintenance, and, consequently, a high level of personnel wages. This is an objective factor in the transfer of resource-processing enterprises from state to private ownership. And here the state is faced with the enormous task of building such a system that would direct privatized resource and processing enterprises to work effectively solely through fair competition, the use of advanced equipment and technology, and the search for intensive factors of increasing productivity, especially considering the negative experience of previous privatizations. On the other hand, this system would assume the interest of mining and exploiting enterprises, which take advantage of their high competitiveness, to share their income with other sectors of the national economy. The task of the state is to create a favorable climate when it will be profitable for extractive industries to invest in the development of other industries.

5. Conclusion

Raw materials should be the most important lever for reforming our economy, and the determination of prices for natural resources is the most important factor in the implementation of market reforms. The current cost of consumed natural resources is much lower than their real economic estimate. This in turn leads to their irrational use and wastefulness of the economy as a whole.

In our opinion, it would be possible not only to set high prices, but to create a system of taxes and payments regulating the prices of natural resources in the domestic market below world prices, thereby creating a more favorable investment climate for other industries.

References

[1] Vlasov V B 2004 Some aspects of effective environmental management. Solving social and economic problems - new approaches *Coll. of Sci. Papers of the Regional Interuniversity Scientific-Practical Conf.* Voronezh: Istoki p 293

[2] Vlasov V B 2004 Economic mechanisms for regulating the environmental process *Coll. of Sci. Papers Economical Sciences* Moscow: MHEI **4** p 42

[3] Umivakin V M, Vlasov V B and Klimov A V 2010 Two-criteria model of sustainable development of territories *Economical Sciences* **71** 269-272

[4] Mihailov V V, Tomilov A A, Umivakin V M, Vlasov V B and Fedorova Ya A 2010 Morphological analysis and synthesis of ecological and economic decisions in the design of natural-anthropogenic systems *Proceedings of the Institute of System Analysis of the Russian Academy of Sciences. System analysis of the problem of sustainable development* **54** 172-184

[5] Pearce D and Moran D 1994 *The economic value of Biodiversi* London: Earthscan Publication Ltd

[6] Dixon J A, Scura L F, Carpenter R A and Sherman P B 1998 *Economic analysis of environmental impacts* London: Earthscan Publications Ltd

[7] Lancaster K J 1966 A new approach to consumer theory *J. of Political Economy* **75** 2

[8] Rowe R D and Chestnut L G 1982 *The value of visibiliti: theory and applications* Cambridge MA: Abt Books.

[9] https://natworld.info/raznoe-o-prirode/prirodnye-resursy-na-territorii-rossii

[10] https://lenta.ru/ Accessed on 28 April 2019

[11] https://ekb.rbc.ru/ Accessed on 13 March 2019

Energy management strategy in the construction industry

S Prokhorov[1]

[1] Vladimir State University named after Alexander and Nikolay Stoletovs, Gorkiy street, 87, 600000 Vladimir, Russia

E-mail: oc204@bk.ru

Abstract. Saving energy resources is the first priority of the world community. In modern society, construction industry has a huge impact on the economic, social and technological environment of human life. New architectural forms and concepts are emerging, building materials and technologies are improving, environmental friendliness and friendliness of buildings to the surrounding nature increases. Currently, issues related to improvement of the economy's energy efficiency are attracting everyone's attention. Projects of energy-saving and energy-active houses are becoming increasingly popular. Existing construction regulations originate in the 70-90s of the last century, and do not reflect modern reality. Another important aspect of the introduction of modern technology is the environmental pressure during construction and installation works. It raises a question about the need to develop economically and technically sound standards in the design and organization of construction, compliance with which would be monitored not only at the stage of project documentation, but also during the building process. The study addressed the issues of energy consumption reduction at the construction design and maintenance stages, during building and installation works, as well as in sanitary, technical and material support of construction production

1. Introduction

The problem of energy efficiency improvement in our country, is regulated by Federal Law No. 261 "On energy saving and on energy efficiency improvement and amendments of certain legislative acts of the Russian Federation". Therefore, the quality and high manufacturability of all components of construction, on which the achievement of high energy efficiency depends, are of particular importance

The life cycle of a building is quite diverse, but it can be divided into several main stages: design, implementation of the project (construction of a building or structure), operation (maintenance) and dismantling (reconstruction). At all these stages, it is necessary to consider the opportunity of minimization of energy costs and the resource saving possibility. And if the approaches to the design and operation of real estate objects are well studied, from the energy saving point of view, then, in modern Russian reality, not enough attention is paid to energy efficiency issues at the stage of project implementation. [1, 2] However, if to analyze the accumulated experience and requirements for passive and energy-active structures, we can build a strategy which will help to improve resource saving and energy efficiency in the performance of installation and construction works. [3,4] Modern construction industry - is a great interconnected system of various departments, the coordinated work of which influences the efficiency of construction in general: starting from the development of method statement, preparation of the building area and the immediate construction works, up to quality control

IOP Conf. Series: Materials Science and Engineering **972** (2020) 012066 doi:10.1088/1757-899X/972/1/012066

and processing of as-built documentation. And there is no need to forget about the logistical and utility support of the construction operations. [5].

2. Methods

BIM-technologies are increasingly used in the design of objects. The result of BIM modeling is a comprehensive computer model that describes both, the object itself, and the process of its construction. [6, 7] Compared with the traditional approach, they have several advantages:

- reduction of time for preparation of project documentation;
- ability to detect 2D design errors;
- reduction of construction costs;
- cost control;
- reduction of time for project implementation.

At the design stage of construction organization plan and the plan for the construction works, a digital model of the construction site allows to avoid undesirable effects of negative and harmful factors, to optimize time, material and human resources. In particular, by analyzing the 3D model of the designed building, it is possible to choose the most rational methods of work, to organize the territory of the construction site and to monitor the processes of technological operations, as well as to carry out the operational management of the financial resources. [8,9]

In addition to the obvious advantages, the use of a complete digital model allows to increase the automation and mechanization of construction works, and therefore to save energy for domestic needs (by reducing the construction period), and also to save energy for the operation of construction equipment (by optimizing the performed work and by reducing unnecessary technological operations), etc. [10]

Currently, the geodesic survey of the site and the geodesic control over the construction process is mainly carried out by hand. The development of computer technology and control systems allows us to speed up this work, without reducing the required accuracy of the measurements. Commercially available unmanned vehicles supplied with high-resolution cameras, lidars and other equipment, can easily explore the construction area and promptly display information on digital maps and three-dimensional models, which is a very effective solution compared to traditional methods. [11] In particular, "Brasfield & Gorrie", one of the largest private construction companies, uses a 3D model obtained with help of drones to compare actual excavation works on a construction site with planned targets. By using unmanned aerial vehicles, they were able to evaluate the results of earthworks much faster than if traditional methods were used, and at the same time didn't lose accuracy. [12]

Another aspect of energy saving during the property construction is the use of modern construction equipment and modern methods of the mechanized works production. This task can be divided into two subtasks: the first is the direct reduction of energy consumption by using modern machines and mechanisms, the second is the minimization of technological operations and errors caused by the human factor. Modern models of construction equipment produced by leading manufacturers, are equipped with hybrid propulsion systems, which allow not only to reduce fuel costs, but also to reduce noise and emissions to the environment during their work. [13] One of the most effective energy saving means of the construction machines with hydraulic drive equipment, is recuperative system with an energy storage device. [14].

For electric propulsions, energy recovery systems are becoming more and more relevant. The regenerative braking system allows to stop the car without using conventional brakes, thereby increasing their service life, and in the meanwhile turning the kinetic braking energy into electric current. [15]

VirtualSite Solutions" software provides real-time monitoring of technical conditions, as well as performance of the equipment, and can quickly identify the underutilized machines. [16] The integration of these products with 2D and 3D positioning systems allows to more carefully monitor the effectiveness of the use of machines and produce their maintenance. The use of modern diagnostic remote control systems, aimed to maintain equipment in good condition, reduces the risk of sudden failure and reduces energy consumption during the operation of machines with defective equipment [17].

The next important aspect which can increase energy saving during the production of construction and installation works, is the household supply.

Building companies already aspire to maximize the use of the thermal energy released inside the residential and public buildings, and maximally protect them from heat loss through external surfaces and ventilation. Unfortunately, this method is marginally used in construction camps and auxiliary facilities. The basic principles of designing an energy efficient facility - are the maximum use of the thermal energy released inside the object, maximum protection against heat loss through external surfaces and ventilation, and the use of alternative energy sources. [18]

These buildings tend to use heating technology, ventilation, lighting, water supply, sewerage which require minimal energy consumption. For this purpose renewable energy sources are used (solar, wind, etc.). Along with this, attention is drawn to reduction of heat loss, reduction of air leakage and its infiltration through gaps and joints, as well as to the increase of the resistance of heat transfer to the outside, in the winter season. [19] In summer, natural ventilation is provided by airing, and for cooling there used radiators or evaporation method. Furthermore, the flow of solar energy reduces by shielding. The above-mentioned measures in the complex with local climatic conditions, make it possible to provide good heat transfer regulation in domestic premises, and to reduce energy costs.

In general, all activities can be divided into:

- Space-planning solutions, which facilitate energy savings;
- The use of energy-efficient thermal insulation of exterior walls, in order to reduce the transfer of heat to the outside of the building;
- The use of energy-saving windows, vents, blinds;
- Increasing the entry of daylight into the room, in order to reduce artificial lighting costs;
- Heat recovery and recovery of warm air which removes from the building.
- The use of energy-saving lighting.

In particular, the transition from stand-alone construction trailers to modular and interlocked premises reduces the area of the cooled surface, and thus improves the thermal performance of buildings. In integrated buildings it is possible to use a heated vestibule, which can significantly reduce the cost for the maintenance of internal temperature.

The use of self-powered heat supply sources with a quantitative and qualitative method of heat output regulation, as well as automated heating units with weather-dependent regulation, can significantly reduce heating costs.

Electric heaters must be replaced with air heating systems, which allows to quickly warm up the room and to maintain the required temperature. It is advisable to combine heating systems with ventilation and air conditioning systems, so this will allow to use recuperation systems and reduce energy consumption for heating and dehumidification.

It is also necessary to foresee the possibility of moving away from traditional types of boilers towards alternative sources of heating (for example, solar panels and collectors), heating with air (which implies electric heating of the air supplied by means of a controlled ventilation system).

As practice has shown, for the summer daylight hours in the Moscow region, one flat-plate collector is able to heat 250-300 liters of water up to 90°C for 7 hours. At the construction site, where the roof area is small, solar collectors can be used as primary water heaters. [20]

To illuminate the construction site, it is advisable to use solar panels with batteries. In particular, the Tvilight system tracks the movement of workers and cars, and automatically adjusts the level of lighting. This will cause the initial costs, but will pay off during constant use. In addition, the use of lamps based on LEDs can significantly reduce energy consumption and regulate the luminous flux.

The studies were carried out basing on the calculation procedure of construction organization plan - Guidance Documents in Construction 12-46.2008 "Methodical recommendations for the development and design of construction organization project, the project of demolition (dismantling), and the method statement"

IOP Conf. Series: Materials Science and Engineering **972** (2020) 012066 doi:10.1088/1757-899X/972/1/012066

The effectiveness of the decisions taken was evaluated using the ratio of the average price per unit, determined by the standard method (P_{av}), to the average price per unit in case of using perspective technologies and perspective methods of work organization (P_{out}) formula (1):

The effectiveness of the decisions was evaluated by determining for each option the amount of construction and operating costs, using the dependence (1):

$$C_{red} = \sum_{t=0}^{t=t_c} \frac{K_t}{(1+E_n)^t} + \sum_{t=1}^{t=t_c} \frac{P_t}{(1+E_n)^t} \quad (1)$$

where K_t and P_t are investments and operating costs in the corresponding year t; $1/(1 + E_n)^t$ is the cost distance factor that takes into account the decrease in the significance of costs incurred in t years; tc is the year that limits the period of summation of costs for the options under consideration:

The use of automation systems helps to improve the use of working time, reduces its loss, and ultimately increases the growth of labor productivity. The growth of labor productivity, while reducing internal labor regime time loss, can be determined by the formula:

$$W = \left(\frac{100-q_{n2}}{100-q_{n1}} - 1\right) * 100 \quad (2)$$

where W - is the growth of labor productivity,%; q_{n1} and q_{n2} - is the loss of working time before and after the implementation of measures,%.

3. Results

"Multifunctional sports complex with a swimming pool, universally applicable hall and SPA" in Reutov, Moscow region, was selected for the study. The built-up area is 7500.00 m^2, the building volume is 78421.12 m^3, the total area of the building is 13094.40 m^2.

Table 1. The results of the calculation

Indicator	Standard (Guidance Document in Construction 12-46.2008)	Calculation (considering energy efficiency measures)
Number of workers per year, people	102	88
The need for utility rooms, m2	354.68	239.36
The need for construction machinery, units	45	30
The operation cost of construction equipment, million rubles	45.568	40.101
The need for energy resources:		
- electric energy, kVA	897.9	688.8
- water m3/day	51.9	38.8
- fuel, tons	89.3	75.5
- heating, MW	12.24	10.11
Estimated cost of construction, million rubles	340.568	368.651
Duration of construction, months	30	24
The effect of duration reduction, million rubles	-	21.337
Profit for early commissioning, million rubles	-	13.563

The project demand for personnel was determined on the basis of performance rate per worker per year, the cost of annual volumes of work and the percentage of the number of employees by their categories.

The need for basic construction machines, mechanisms and vehicles was determined on the basis of the physical volumes of work and the operational performance of machines and vehicles, taking into account the accepted organizational and technological construction schemes.

The need for energy resources is determined by direct counting.

The building period is determined by the calculation based on the construction norms and regulations 1.04.03-85 * Norms of the duration of construction and backlog in the construction of buildings, structures, and industrial facilities.

Over and above the calculated indicators of the construction organization project, it is necessary to take into account the time for the development of technological documentation using modern design systems.

The use of AUTODESK® software products based on BIM 360 modules has greatly simplified the as-built and design documentation, despite the fact, that at first there were certain difficulties in the cooperative interaction between the customer, the contractor and the general designer. But in view of the fact, that the general designer had prior experience of working with BIM modules in the field of architectural and structural design, it was not a big deal to introduce the control and construction management module. The financial costs on the software products are planned to be recouped by the reduction of time for design and construction, and by the use of more technological solutions during the organization of the construction production. The ability to directly monitor the execution of design and construction and installation work in real time via mobile applications, became an extra incentive of the implementation.

4. Discussion

Studies have revealed both positive and negative points of the proposed solutions:

- The use of modern software products that are based on the creation of a unified database, and providing the possibility of joint management, allows to significantly reduce the time for the design work and the coordination of building technological concepts, especially in case of a possible change in the delivery schedule of building materials, or in case of additional restrictions which were not detected at the design stage (in this project, during construction and installation works, it have been discovered that some service lines were not mentioned in the as-built drawing, so it was required to make changes in the earthworks project and the project of communications disassembling);
- The use of modern systems aimed at automation of production works, and in particular production of earthworks, allows to reduce the number of process operations, and as a consequence to increase the productivity, to reduce operating time of the machine, and thus to increase its lifetime before the service and maintenance;
- The implementation of modern methods of work execution, as well as their proper planning, reduces the need for human and material resources, which ultimately affects the energy savings;
- The use of blocked welfare facilities reduces construction needs for electricity, heat energy and air conditioning;
- The use of modern technologies allows not only to reduce the need for material, technical, human and energy resources, but also to reduce the cost of construction, and to increase the competitiveness of the construction organization.

The negative points include:

- BIM-design requires more time and a higher skill level of the staff, as well as additional software costs. However, the initial time spent on building of a full-fledged 3D model is compensated by the reduced duration of coordination, approval and correction of project documentation between the contractor, the customer and other stakeholders, and for large developers, it is easier to manage projects through mobile applications and the Internet;
- The transition to modern systems of automated production in the field of construction and installation works management, allows to improve the quality of work performed, and to

implement integrated control of construction work on the site. However, in this case, the machine parks of construction organizations must be fitted with modern equipment, and staff must be retrained, to be able to work with it, which will lead to the rise of construction costs; this is especially important for regional organizations;

- Automation leads to the reduction of employees engaged in the production of works, and can cause social tensions. On the other hand, it stimulates people to master promising technologies and methods of work, and also frees up personnel and resources for the implementation of a larger number of projects.

References

[1] Tsygankov V 2015. Real estate: Economics, management. **2**. 45-48
[2] Emelyanova Ya. 2018. *Theory. Practice. Innovations*. **11 (35)**. 150-154
[3] Larionov A 2018. *Construction economics*. **1 (49)**. 18-26
[4] Vasilenkoi N 2018 The System Principle of Compactness as a Defining One in the Formation of Functional and Planning Structure and in the Forming of Power-Saving, Power-Active, Environmentally-Friendly and Hi-Tech Buildings. *IOP Conference Series: Materials Science and Engineering(2018) 032086*
[5] Luchkina V 2018. *System technology*. **3 (28)**. 5-13
[6] Park J, Cai H 2017. *Automation in Construction*. **77.** 15-23.
[7] Iyina O 2017. Real estate: Economics, management. **2**. 72-75
[8] Vitasek S, Matejka P. 2017.Utilization of BIM for automation of quantity takeoffs and cost estimation in transport infrastructure construction projects in the Czech Republic. *IOP Conference Series: Materials Science and Engineering. (2017)* **Vol. 236.**
[9] Garyaev N 2018. Analysis of risks arising in the implementation of BIM - technologies in construction organizations. // *MATEC Web of Conferences. (2018)* **Vol. 251.**
[10] Shadyzhev Z, Malsagov M, Evloev Sh, Palunkoev M., Merzhoev I, Mangushev N 2018. *Economics and entrepreneurship*. **8 (97)**. 608-612
[11] Pogorelov V 2016.*Engineering Herald of the Don*. **1 (40)** 58
[12] Korenev V, Orlova N, Ulybin A, Fedotov S 2018. *Construction of unique buildings and structures*. **2 (65)** 40-58
[13] Kristal M, Kisel T 2018. Economic efficiency of innovations in construction: The use of the pilotless equipment. *MATEC Web of Conferences (2018)* **Vol.256.**
[14] Savinkin V, Kuznetsova V, Yakovlev V 2016. *Bulletin of the Siberian state automobile and road Academy*. **2 (48)** 18-25
[15] Mukhametshina R, Petrov A 2017. Topical issues and bases of international cooperation in the field of high technologies. **2** 111-113
[16] Shcherbakov V, Buntsev I , Shcherbakov I, Astrakhantsev V 2018..*Interekspo Geo-Sibir*. **1** 57-63.
[17] Vershinin S 2018. Various aspects of the use of automated control systems in the production of construction works and as part of a complex of precision agriculture (Orel) pp. 78-89
[18] Albuquerque P, Ohi D, Pereira N, Prata B, Barroso G 2018. *Journal of Control, Automation and Electrical Systems*. **29(6)** 718-730
[19] Samarin O 2018. *Reliability and safety of energy*. **Vol. 11**. **No.2**. 149-153
[20] Weeratunge H, Hoog J, Dunstall S, Narsilio G, Halgamuge S 2018. Life Cycle Cost Optimization of a Solar Assisted Ground Source Heat Pump System. *IEEE Power and Energy Society General Meeting (2018)* PESGM.2018.8586063

IOP Conf. Series: Materials Science and Engineering **972** (2020) 012067 doi:10.1088/1757-899X/972/1/012067

Experimental and calculated testing of the efficiency of cyclone filtering devices

G I Belyaeva[1], A T Zamalieva[1], M G Ziganshin[3]

[1]Gazprom transgaz Kazan LLC, 1 Zelenaya street, 420073, Kazan, Russia

[3]Thermal power engineering, Gas supply and Ventilation Department, Kazan State University of Architecture and Engineering, 1 Zelenaya street, 420043, Kazan, Russia

E-mail: gulnazкa16@mail.ru

Abstract. The aim of the study is to identify ways to improve the efficiency, reliability of the dust collection process and improve the aerodynamic properties of cyclone apparatuses in the gas cleaning unit. The analysis of existing methods of cyclone filtration of fuel gases is carried out. Bench tests of a cyclone filter based on the serial cyclone CN-11-200 were made, and the aerodynamic characteristics of its operation were found. The efficiency of the use of multi-cyclone apparatus, the elements of which combine both stages of cleaning, are also analyzed. The problem of finding the optimal location in the apparatus of cyclone elements with semi-tangential inlet nozzles are solved. A numerical model of a multi-cyclone is created and several options for the arrangement of elements are considered. The most effective location is determined taking into account the overall hydraulic resistance of the device. The numerical model also made it possible to evaluate the influence of various factors on the dust sedimentation efficiency in the developed cleaning device.

1. Introduction

Requirements for the efficiency of dust collection processes are constantly increasing with the tightening of regulatory requirements for the purity of atmospheric air and air in the spaces of industrial and civil buildings, as well as with the advent of new technologies, the use of new materials and, therefore, with the release of new emissions [1].

Of practical interest is the combination of the dust trapping by the inertial mechanism with the filtering of the gas flow by porous walls, carried out in the same housing and implemented in the apparatus, known as filter-cyclones. They achieve two-stage dust sedimentation in one unit, facilitating the process of regeneration of the filter sleeves by high concentrations of dust. Such a gas cleaning scheme is especially adapted to local aspiration systems.

Literature analysis [3-10] showed, that the existing devices either provide a high degree of deposition of fine particles of classes PM_{10},$PM_{2,5}$ with high energy consumption, or have a low degree of cleaning. The cyclone filters have higher effectiveness at not high energy costs.

The cyclone filters can also be used as elements of multi-cyclones for increase fine of cleaning. The wide distribution of multi-cyclones in dust cleaning equipment due to the simplicity of their device, reliable in operation, small capital and operating costs. They are effective in the deposition of suspended particles of mid (from 10 microns) and large sizes, which is why they are used in systems as primary processing of emissions. For finishing processing, behind them it is necessary to install fine

IOP Conf. Series: Materials Science and Engineering **972** (2020) 012067 doi:10.1088/1757-899X/972/1/012067

cleaning devices. In this case, small particles are most efficiently deposited by porous filters. However, this leads to higher prices, drop reliability and complicates the cleaning system as a whole.

2. Methods

2.1. Creating of cyclone devices with filter elements

The desire to optimize the processes of dust cleaning is aimed at obtaining a harmonious ratio of the degree of trapping, hydraulic losses and energy costs of gas cleaning equipment. Therefore, it is currently important to create devices with a high degree of deposition of fine particles PM_{10} and $PM_{2.5}$ with a low coefficient of hydraulic resistance.

One of the technical solutions we offer is a cyclone apparatus combining two stages of cleaning for efficient deposition of fine particles of classes PM_{10}, $PM_{2.5}$ with minimal energy and material costs.

To achieve this result, bench tests of the cyclone filter were carried out, the design of which was developed by the authors based on the serial cyclone CN-11 with a diameter of 200 mm. The tests were carried out on a specially prepared dispersion material - chamotte dust with a particle size of up to 63 microns.

A sample of material for testing was taken in accordance with good laboratory practice according to GOST R 51568-99 (ISO 3310-1-90). The selected sample was subjected to sedimentation analysis in distilled water to determine the median particle diameter and dispersion. The change in the mass of crushed material deposited in the dispersion liquid was recorded on a VT-500 torsion scales.

For the sedimentation analysis, a quantity of material was taken from the sifted sample, sufficient for preparing a suspension with a volume fraction of suspension to 0.5 ... 1%. In the future, according to the results of experimental measurements on the basis of the Stokes law the particle sizes were counted:

$$r = \sqrt{\frac{9\eta \cdot H}{2(\rho_p - \rho_G)g \cdot \tau}} \quad (1)$$

where H – particle sedimentation height, τ – settling time, $v=H/\tau$ – sedimentation rate, r – particle radius, ρ_p – particle density, ρ_L , η – density and viscosity of the dispersion medium.

The design of the cyclone filter is shown in Figure 1. The input velocity into the cyclone was measured before dusting with a thermal anemometer (Testo 425 model) installed in the round access port. The dust laden stream passes through the duct into the cyclone inlet and enters the annular space inside it between the body and the exhaust pipe. During such a flow, a number of particles due to inertia come down from curved current lines, collide with the wall of the cyclone and slip into a bunker with a hermetic shutter. The smallest fractions of dust are trapped by a fabric filter, where the deposition of dust of medium and small size along touch and inertial mechanisms. Further, the flow moves down along the helical trajectory, turns up in front of the bunker and exits along the ascending helical trajectory through the exhaust pipe. A micro manometer is installed in the round access port to find the resistance inside the cyclone.

The air flow rate at the cyclone inlet was measured using a collector with an internal diameter of 140 mm and a micro manometer. Adjustment of air flow in the network was performed with a throttle valve in five modes. In to access port of circular section 12, an impactor 20 was installed to determine the dispersed composition of dust after the cyclone-filter.

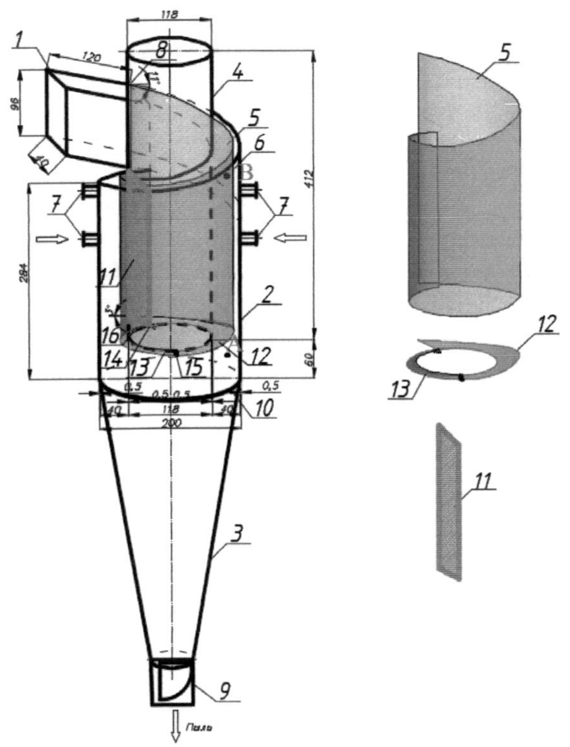

Figure 1. The scheme of the cyclone filter: 1 – inlet duct; 2 - cylindrical body; 3 - conical bottom; 4 - exhaust pipe; 5 - filter material; 6 - metal frame; 7 - purge fittings; 8 - inlet into cyclone; 9 - damper; 10 - fastener parts; 11 - cross filter; 12 - horizontal filter, woven part; 13 - horizontal filter, metal part; 14 - plate with spring; 15 - rod with rings; 16 - folding filter.

Filter material samples were weighed at the beginning and the end of the test and examined under a MIN-8 microscope with 400-fold magnification. This made it possible to determine the particle enumerable concentration of the classes $PM_{2.5}$, PM_{10}, which is necessary for a reliable comparison of the intensity of their sedimentation from the flow to the filter in different modes.

Based on the methods of computational hydrodynamics, a numerical simulation of the aerodynamic parameters of the cyclone filter was performed to observe all the characteristics of the flow, which can be identified in a numerical 3d model. The energy parameters of the rotating flow are comparable with the magnitude of its energy at the entrance to the apparatus.

To ensure the closure of the system in the equations of transport and dissipation of kinetic energy, the Spalart-Allmaras viscosity model is used. [11-14].

2.2. Finding the best location of the cyclone filter elements in a multi cyclone based on numerical studies of the movement of the dispersed flow

With methods of computational fluid dynamics (Computational Fluid Dynamics - CFD) using the RANS turbulence model (Reynolds-Averaged Navier-Stokes equation), was built a 2-d model of the first row of the standard multi-cyclone CN -16 [15-17]. The air flow with dust particles ranging in size from 5 to 45 microns enters at a velocity of 5 m/s, then flows around obstacles. The air with part of the total flow of particles get out of the channel, and part is held on obstacles.

Considered two models of the multi-cyclone , differing in the arrangement of cyclone elements: in the first model, the cyclone elements are located at the same distance from each other, and in the second - the middle cyclone elements are located as close to the extreme side ones as possible. The unstructured mesh based on triangular elements was created using the internal tools of the Gambit preprocessor.

Before starting to solve the problem, were established the following boundary conditions: Velocity Inlet – uniform distribution of velocity at the channel entrance; Pressure-Outlet – atmospheric pressure at the channel exit; Wall – the boundary condition of the wall.

IOP Conf. Series: Materials Science and Engineering **972** (2020) 012067 doi:10.1088/1757-899X/972/1/012067

The problem of finding the best arrangement of the inputs of the elements in the apparatus is further solved by the method of numerical modelling based on the computational fluid dynamics methods in the Fluent processor. In numerical calculations of this construction the two-parameter turbulence model k-ε was adopted, which provides the solution of the problem in close to real conditions at an available computing resource.

3. Results and Discussion

The size of the fractions obtained from the sedimentation analysis is shown in figure 2. The minimum particle size of 1,09 microns, maximum – 60,94 microns, average particle size d_{50} =1,37 microns.

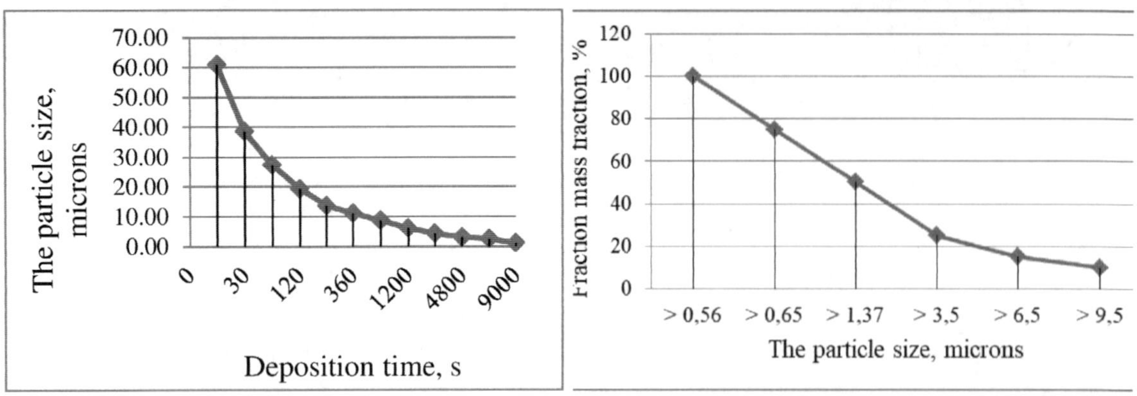

Figure 2. The results of the sedimentation analysis: a – the dependence of the size fractions of particles from the time of deposition; b – dispersed composition of the particles.

To determine the efficiency of the separation of the weighted part of the flow in this cyclone-filter was used dimensionless parameter obtained by methods of similarity theory – the relative Reynolds number Re_r [11]:

$$Re_r = \frac{U_0 \cdot \rho_p^2 \cdot D_p^4}{c \cdot \rho_G \cdot R_2^3 \cdot \eta} \tag{2}$$

where: U_0 -initial velocity, m/s, ρ_P-particle density, kg/m^3, D_p-diameter of particles, m, c - a numerical coefficient, which, depending on the geometric characteristics of the device, ρ_g-gas density, kg/m^3, R_2 -cyclone radius, m, η - coefficient of dynamic viscosity, Pa·s.

The dimension less Re_r parameter is used to determine the efficiency of inertial deposition of particles, as well as the Stokes number St, but unlike it, the Re_r number characterizes the energy costs of the particle to move along a curved trajectory.

The diameter of the particles deposited with an efficiency of 50% under operating conditions at the top B (Figure 4) determined by the formula:

$$D_{50} = D_{50r}\sqrt{\left(\frac{D}{D_r}\right)\left(\frac{\rho_{pr}}{\rho_p}\right)\left(\frac{\eta}{\eta_r}\right)\left(\frac{w_r}{w}\right)} \tag{3}$$

where: D_{50r}=3,65·10^{-6}m, D_r=600mm, ρ_{pr}=1930kg/m^3, η_r=22,2·10^{-6}Pa·s, w_r=3,5 m/s respectively, the average diameter of the deposited in the cyclone particles, the diameter of the cyclone, the particle density, dynamic viscosity and velocity of the gas flow taken according to the reference data; D_{50}, D=200 mm, ρ_p=1990kg/m^3, η=17,96·10^{-6}Pa·s ·, w= 11.29 m/s - the same in the working conditions. Relative Reynolds number Re_r for D_{50}: Re_r =2.5·10^{-9}.

The diameter of the particles captured at 50%, according to the formula (5), is: D_{50}=1,039 microns. Particle size $D_{99.18}$ was determined by the deposition parameter x by the formula:

$$x = \frac{\lg\left(\frac{D_{99.18}}{D_{50}}\right)}{\sqrt{\lg^2 \sigma_\eta + \lg^2 \sigma_p}} \tag{4}$$

where σ_η, D_{50} - respectively, the dispersion and diameter of the particles deposited in the apparatus by 50%. For the cleaning coefficient f (x)=99,18% the value of the deposition parameter at 99.18% x=2,40 [1].

The diameter of the particles captured at 99,18%, according to the formula (2), is: $D_{99,18}$=14,53 microns. Now find the value of the Re_r for $D_{99,18}$=14,53 microns, providing almost complete precipitation of the contaminant from the release: Re_r=of $9{,}55\cdot10^{-5}$.

Thus, The Re_r value for particles captured almost completely (by 99.18%) in the experimental cyclone is $9{,}55\cdot10^{-5}$. Consequently, this cyclone without a filter insert provides complete deposition of particles having an inertial characteristic of motion in a curvilinear flow Re_r above this value. As shown earlier, the size of the particles deposited in the experience on the wall of the cyclone by 50%, amounted to 1,039 microns, indicating the effectiveness of the cyclone with a filter insert, in comparison with the standard value for the CN-11 D_{50}=3,65 microns. Figure 3 shows the flow pattern inside the cyclone in the annular section and inside the exhaust pipe.

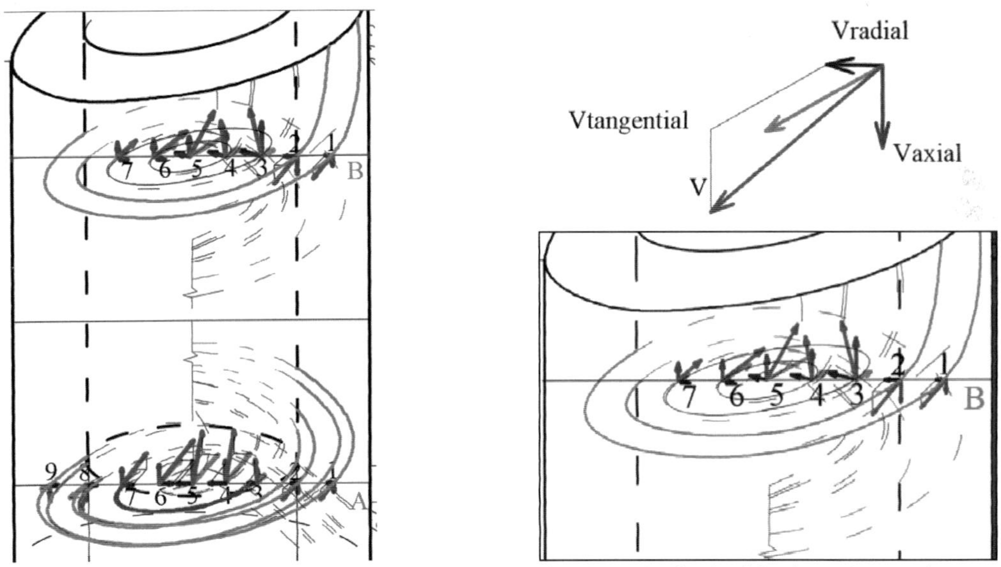

a) velocities in the annular part of the cyclone

b) the velocity inside the exhaust pipe at the top of the measurements *B*

Figure 3. Flow diagram: a) in the annular part of the cyclone at the bottom of the measurements *A* and at the top of the measurements *B*, b) inside the exhaust pipe at the top of the measurements *B*

Figure 4 shows the distribution of axial, tangential and radial flow velocities inside the cyclone at the top point *B*. In the annular section of the cyclone, the maximum velocity was 15 m/s.

The experimental studies have shown the possibility of a constructive additions the return-flowing cyclones the filter insert on the zone where inertial dust precipitation occurs.

IOP Conf. Series: Materials Science and Engineering **972** (2020) 012067 doi:10.1088/1757-899X/972/1/012067

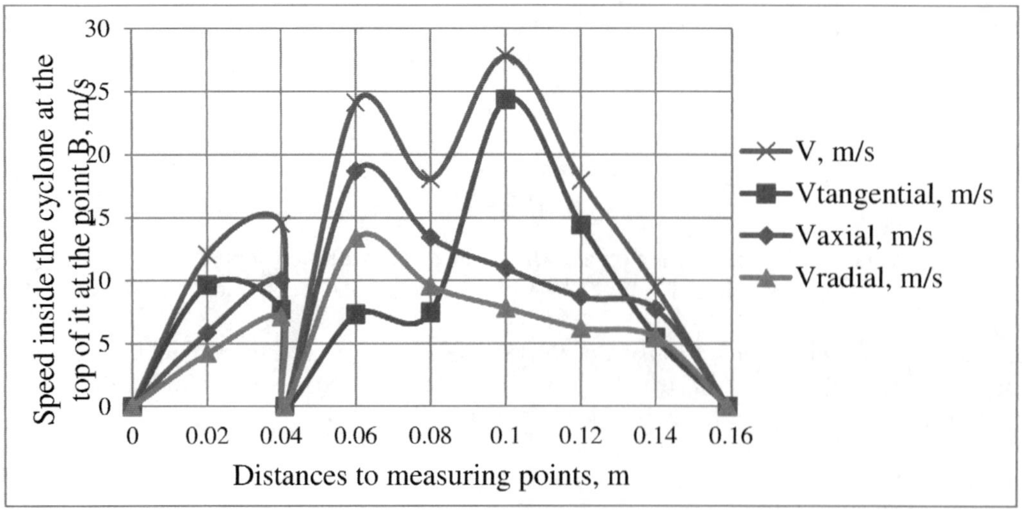

Figure 4. Distribution of tangential, axial and radial flow velocities inside the cyclone at the top point *B*

As can be seen from figures 5 - 6, for the incoming dispersed gas flow, the elements of the multi-cyclone as a whole are represented by an array of obstacles. On the basis of the calculation, taking into account the trajectories of solid particles of different sizes, the optimal variants of the arrangement of filtering cyclone elements are selected. The results of numerical studies show that at the second model in the working space between the elements is higher both the velocity and pressure values, although the flow characteristics at the inlet to the housing are the same for both models. This is due to the flow passing in the second model only in the intervals between the cyclone elements. In the first model, a significant part of the flow passes the area near the walls of the multi-cyclone housing, creating less resistance than the area between the elements.

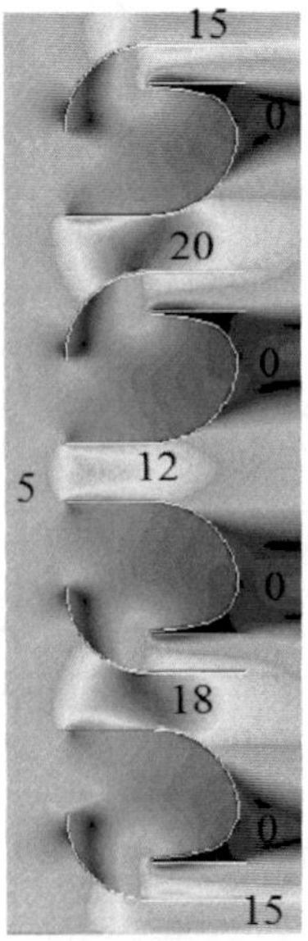

Figure 5. Velocity plots in the first model of the multi-cyclone (velocity values in m/s).

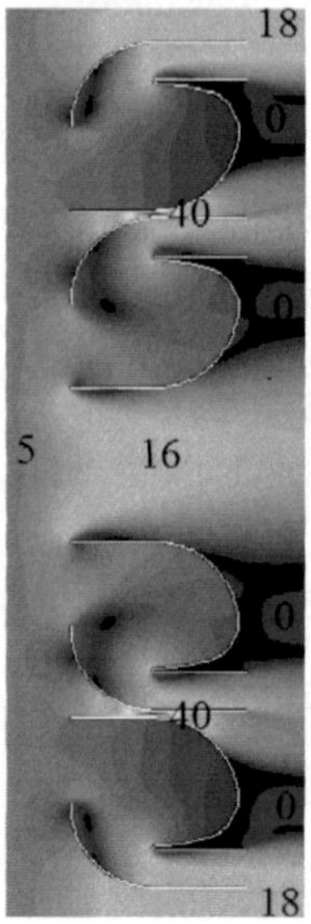

Figure 6. Velocity plots in the second model of the multi-cyclone (velocity values in m/s).

Calculations clearly show that these zones represent local resistances such as unilateral sudden narrowing and sudden expansion. The created numerical model of a multi-cyclone allowed to estimate

IOP Conf. Series: Materials Science and Engineering **972** (2020) 012067 doi:10.1088/1757-899X/972/1/012067

the influence of various factors on efficiency of dust capture in cyclones, and also to create a technique of an assessment of efficiency of a dust collector.

4. Conclusion

The proposed method of equipping a return-flowing cyclones the filter inserts allow you to achieve increased throughput of the treatment apparatus while improving the quality of cleaning, which is expressed in reducing the size of the particles captured at 50 % (diameter of cutting off), from the average for the cyclones values of 5-10 microns to 0,5 microns. This improvement in the quality of cleaning does not require a significant increase in energy costs. It is one of the advantages over conventional cyclonic devices, which for reducing the diameter of cutting off requiring an increase in energy costs up to 15 % every tenth of a micrometer.

The obtained results provide an opportunity to optimize the location of the inputs of semi-continuous nozzles of cyclone elements in the multi-cyclone on the first and second models. They also make it possible to more accurately take into account the hydraulic resistance of the device in determining the most effective location of the cyclone elements.

Thus, the experimental studies have shown the feasibility of constructive additions of return-flow cyclone filter insert in the area where the inertial deposition of dust.

References

[1] Ziganshin M G, Kolesnik A A and Ziganshin A M 2014 Design of devices of a dust and gas cleaning (Saint-Petersburg: Lane) 544 p

[2] Technical information on dust removal units of the PUMA series [Electronic resource: http://www.st-vent.ru/index.php/filter-pyleulovitel/filter-abraziv/filter-puma]

[3] October M L 2016 Equipment for cleaning of the air environment of a working zone *SIBGTU Current problems of a forest complex* № **16** P.14-18

[4] Devisilov V A 2012 Research of hydrodynamic vibration filtering and development of a design of the filter *News of the Samara scientific center of the Russian Academy of Sciences* №**1-3** (Samara) pp 866-876

[5] Plashikhin S V, Beznosik Yu A and Serebryansky D A 2010 Development and modeling of the cyclofilter for cleaning of gas environments *East European journal of advanced technologies* №**4/8** pp 42-44

[6] Serebryansky D A, Zakharov A A and Plashikhin S V 2013 Cyclonic dust collectors. Low-cost modernization *Chemical industry of Ukraine* № **3** pp 70-74

[7] Utikar R 2010 Hydrodynamic Simulation of Cyclone Separators *World's largest Science* (Technology & Medicine Open Access book publisher) 420 p

[8] Potemkina A B 2010 Theoretical and experimental studies of the effect of the depth of immersion of the exhaust pipe on the aerodynamic properties and efficiency of cyclonic devices *Vestnik don. state tech. University* Vol. **10**, № **4** (Rostov-on-don:) P. 526-533

[9] Chistyakov I B 2013 Study of the process of dust collection in centrifugal-inertial devices *Ecology and industry. the youth of Russia* (Moscow) P. 8-11

[10] Odii Ch J 2018 Analysis of fully developed laminar flow between parallel plates with uhf using ANSYS CFD *Global nuclear safety* №**4** (Moscow) P. 31-39

[11] Ziganshin M G 2013 Systems of cleaning of emissions of thermal power plant. Part 1. Efficiency estimates, theoretical bases and evaluation criteria. Monograph (Kazan: Prod. KGEU) 252 p

[12] Smirnov P and Menter F 2018 Sensitization of the SST turbulence model to rotation and curvature by applying the Spalart-Shur correction term *Proc. of ASME Turbo Expo* (Berlin: Power for Land, Sea and Air, GT) p 10

[13] Wilcox D C 1994 Turbulence modeling for CFD (DCW Industries, Inc.) 460 p

IOP Conf. Series: Materials Science and Engineering **972** (2020) 012067 doi:10.1088/1757-899X/972/1/012067

[14] Jagadeesh P and Murali K 2005 Application of Low-Re Turbulence Models for Flow simulations past Underwater Vehicle Hull Forms *Journal of Naval Architecture and Marine Engineering* P. 41-54

[15] Belyaeva G, Zamalieva A and Ziganshin M 2019 Numerical and experimental studies of gas cleaning in multicyclone elements with filter inserts *IOP Conf. Series: Materials Science and Engineering* **481** 0120411-6

[16] By Prosperetti A and Tryggvason G 2009 Computational methods for multiphase flow (Cambridge, New York, Melbourne, Madrid, Cape Town, Singapore, São Paulo, Cambridge University Press) p 484 [available at: www.cambridge.org/9780521847643]

[17] Bagchi P and Balachandar S 2002 Steady planar straining flow past a rigid sphere at moderate Reynolds number J. (Fluid Mech.) P. 365–407

Modern Air-Conditioning Systems without Wiring Duct

N P Shiryaeva[1], E A Malyar[1], P V Ustyantsev[1]
[1]Ural Federal University, 19 Mira Street, Ekaterinburg 620002, Russia

E-mail: n.p.shiriaeva@urfu.ru, e.a.maliar@urfu.ru

Abstract. The calculation involved in the design of integrated heat-hydraulic air-conditioning system in which air is supplied through the raised floor was presented. Calculations were made using the premises of an office building in Yekaterinburg, Russia. Raised floor in the current study was presented in the form of a channel with a concrete wall on one side and chipboard - on the other. It has been successfully established that the loss of pressure during air movement in the raised floor space is minimal, this was suggested as a result of the possible uniformity in air supply to the air distributors. It was also demonstrated that due to transmission losses through the walls of the raised floor, the cooling capacity of the air conditioning system was significantly reduced (up to 15%). Following the results obtained, conclusions were drawn about the main features of such systems.

Keywords: Raised Floors; Displacement Ventilation; Air Distributor.

1. Introduction

The use of air-conditioning systems without ducting in office buildings project have been employed in western countries. This system is gaining prominence in Russia recently, such as the Okhta Center (St. Petersburg), the Moscow-City MIBC, the Waltzing Couple building (Moscow).

The principle of the system relies on the air handling unit supplying air to the ceiling or raised floor space. Air distributors are provided in the raised floor covering, through which air escapes from the raised floor space or suspended ceiling to the room. The structure composition of floor and ceiling are of high density and low air penetration. However, ceilings are more expensive than floors, and such systems are less common. When air is supplied from under the raised floor with a sufficient height of rooms (> 3 m), it becomes possible to use displacement ventilation.

The main distinguishing feature of such systems lies in the lack of channels for air transport from the central unit to the air distributors. The space under the raised floor or behind the suspended ceiling has a sufficiently large cross-sectional area, making air moves in it with relatively low speeds (0.1-0.8 m/s), which results in a minimal pressure loss (0.1 - 0.2 Pa), thereby ensuring uniform air distribution.

2. Subject and methods of research

To estimate the pressure loss, we consider a simplified duct model with a cross section of width $B = 7$ m and height $h = 0.3$ m, channel length $L = 25$ m. The dimensions of the room are taken according to the design of an office building in Yekaterinburg plan of 25 m by 7 m, and the height of the room 4 m. In the upper wall of the duct, which is the floor of the conventional office space, there are holes with air distributors. The number of distributors depends on the number of work places. When calculating, it was assumed that the floor area of the workplace is A_p = 6 m^2. The number of workplaces for the project is n = 30. Specific heat excess for office space with windows oriented to the west and walls of double-glazed windows amount to 90 W/m^2. The air parameters in the room are determined taking into account the air exchange coefficient. Supply air parameters: temperature 18°C, relative humidity 70%; exhaust air parameters: temperature 27°C, relative humidity 40%. In the working area, the air temperature is 23°C; relative humidity of air 50%.

Consumption of air flow is determined by the formula [1]:

$$L = \frac{Q}{c \cdot \rho \cdot (t_s - t_{ex})} \cdot 3600$$

where: Q – excess heat, W;
c – heat capacity of air, equal to 1005 J/(kg °C);
ρ – is the air density equal to 1.2 kg/m^3;
t_s – supply air temperature, °C;
t_{ex} – temperature of exhaust air, °C;

Determine the air flow at 1 workplace:

$$L = \frac{6 \cdot 90}{1005 \cdot 1.2 \cdot (27 - 18)} \cdot 3600 = 180 \frac{m^3}{h}.$$

We accept that the consumption for 1 workplace equal to 200 m^3/h. The total air flow was 30×200 = 6000 m^3/h. According to the manufacturer, we could determine that for 1 workplace, 2 air distributors with a diameter of 200 mm are needed.

Noise level and air pressure drop are within acceptable limits. In order to estimate the pressure loss using standard aerodynamic calculation techniques, we assume that the channel under the raised floor is a duct with branches. In such case, the air distributors are located 6 in a row, and having a total of 10 rows.

The layout of the air distributors is shown in Figure 1.

The distances between the rows are equal and amount to: 25 / (10 + 1) = 2.27 m. A 10-site duct was obtained with air flow rates from 600 to 6000 m^3/h. The area of the side branch of each section is equal to six times the area of the hole with a diameter of 200 mm. In the calculation, it was assumed that the air flow evenly enters the channel from the entire side surface at an average speed v, m / s, determined by [2]:

$$V = \frac{L}{3600 \cdot f},$$

where f – is the cross-sectional area of the channel, m^2, equal to 2.1 m^2.

$$V = \frac{6000}{3600 \times 2.1} = 0{,}794 \frac{m}{s}.$$

In that case, if air is supplied to the underfloor through the hole at an average speed of 3-5 m/s, a jet will be generated, that eventually fades out at some distance from the hole. In this case, it can be assumed that the flow rate will gradually be aligned with the width of the box, and the accepted calculation model of the flow will be executed starting only from the third section.

Furthermore, the average air flow rate, pressure loss for each section was determined according to [2]. The results of the calculations are shown in Table. 1, it highlighted that the pressure loss in the raised floor space was 0.2 Pa. However, in some areas, the pressure loss is from 0.002 Pa to 0.04 Pa. The pressure loss in the air distributors according to the manufacturer's nomograms is 20 Pa, which is hundreds of times greater than the pressure loss in the raised floor space. Pressure losses on the branches are also a small fraction - from 0.4 Pa to 0.85 Pa.

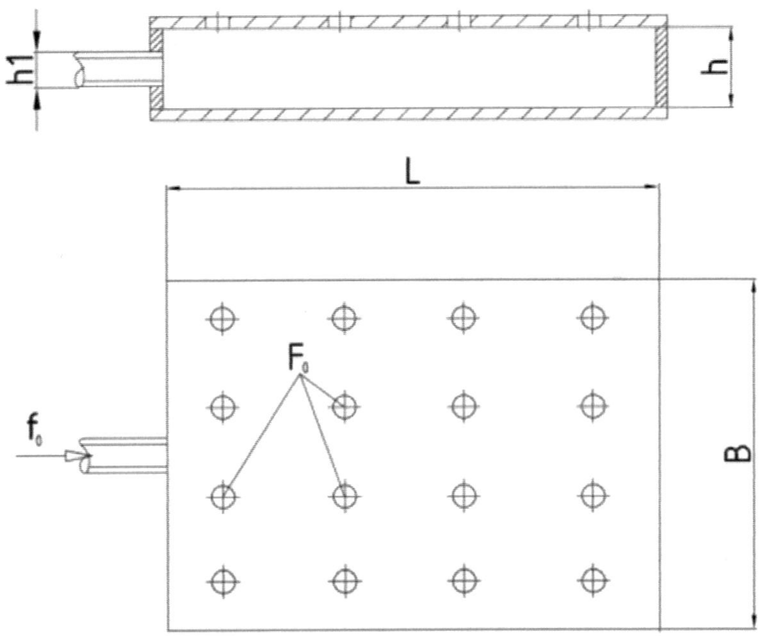

Figure 1. Channel design.

The distances between the rows are equal and amount to: 25 / (10 + 1) = 2.27 m. A 10-site duct was obtained with air flow rates from 600 to 6000 m^3/h. The area of the side branch of each section is equal to six times the area of the hole with a diameter of 200 mm. In the calculation, it was assumed that the air flow evenly enters the channel from the entire side surface at an average speed *v*, m/s, determined by [2]:

$$V = \frac{L}{3600 \cdot f},$$

where *f* – is the cross-sectional area of the channel, m^2, equal to 2.1 m^2.

$$V = \frac{6000}{3600{\times}2.1} = 0{,}794\,\frac{m}{s}.$$

In that case, if air is supplied to the underfloor through the hole at an average speed of 3-5 m/s, a jet will be generated, that eventually fades out at some distance from the hole. In this case, it can be assumed that the flow rate will gradually be aligned with the width of the box, and the accepted calculation model of the flow will be executed starting only from the third section.

Furthermore, the average air flow rate, pressure loss for each section was determined according to [2]. The results of the calculations are shown in Table. 1, it highlighted that the pressure loss in the raised floor space was 0.2 Pa. However, in some areas, the pressure loss is from 0.002 Pa to 0.04 Pa. The pressure loss in the air distributors according to the manufacturer's nomograms is 20 Pa, which is hundreds of times greater than the pressure loss in the raised floor space. Pressure losses on the branches are also a small fraction - from 0.4 Pa to 0.85 Pa.

IOP Conf. Series: Materials Science and Engineering **972** (2020) 012068 doi:10.1088/1757-899X/972/1/012068

Therefore, it should be expected that the uniform distribution of air across all air distributors is provided without additional devices, with the help of which pressure losses in networks are connected with ducts.

Table 1. Hydraulic channel calculation (results)

№ study	Air flow, L, m^3/h	Air velocity, v, m/s	Coefficient of local resistance	Pressure loss at the site, Pa	Total pressure loss, Pa	Air temperature, °C
1	600	0.079	0.4	0.002	0.002	21.36
2	1200	0.159	0.1	0.004	0.006	20.79
3	1800	0.238	0.04	0.006	0.013	20.30
4	2400	0.317	0.03	0.01	0.022	19.93
5	3000	0.397	0.02	0.014	0.036	19.58
6	3600	0.476	0.01	0.019	0.055	19.24
7	4200	0.556	0.01	0.025	0.08	18.92
8	4800	0.635	0.01	0.031	0.112	18.60
9	5400	0.714	0.01	0.039	0.15	18.30
10	6000	0.794	0	0.045	0.195	18.00

In previous studies, M.I. Grimitlin [3] provided the recommended ratios of the dimensions of the distribution chamber, the holes and the inlet nozzle, which ensures uniform distribution of air flow through the holes:

$$F_0/f_0 \leq 7; \qquad L/B = 0.5 - 1; \qquad L/h \leq 5; \;\; h_1/h > 0.6; \qquad F_0/F < 0.09,$$

where: F_0 – hole area, m^2;
h_1, f_0 – height, m, and the cross-sectional area of the inlet pipe, m^2, respectively;
F – the surface area of the raised floor, m^2;
L, B, h – length, width and height of the underground space, m.

For the adopted model of the duct (Fig. 1), we determined the specified ratios values.

We get $F_0 = 60\ \pi\ (0.22/4) = 1.885\ m^2$, hence the area of the inlet should be no more than $1.885/7 = 0.27\ m^2$. Accordingly, the hole width is $b_0 = 0.27/0.3 = 0.9$ m, the average discharge velocity in the hole is $V = 6000/3600 / 0.27 = 6.17$ m/s; $F = 175\ m^2$. The ratio $L/B = 25/7 = 3.6$ is more than the recommended 3.6 times. The ratio $L / h = 25 / 0.3 = 83.3$ is 16.7 times the maximum. Based on the ratio $h_1/h > 0.6$, the nozzle size should be at least $h_1 = 0.6 \times 0.3 = 0.18$ m. The ratio $F_0/F = 1.885/175 = 0.011$. Based on the results of the analysis, we could conclude that a uniform distribution of air is not provided for all air distributors.

For systems with raised floors, air distributors with highly efficient air mixing should be employed to minimize the volume of the zone with unacceptably high air jet parameters. Air distributors are elements of the floor, so they have a design possessing sufficient mechanical strength for the perception of the design loads on the floor. Some foreign manufacturers offer air distributors which in the design mode of operation limits the dimensions of such a zone within (conditionally) cylinder with a diameter of 30-40 cm and a height of 10 cm.

Manufacturers offer a variety of diffusers in different appearance and form. In the raised floor plates, manufacturer provides openings for air distributors, on order. With discomfort associated with excessive jet parameters which could be experienced in almost any office space, the air distributor can be moved a greater distance from the person by changing the location of the raised floor plates with air distributors. This is also relevant for a free office layout, highlighting the possibility of arranging a personal supply air distributor for each workplace.

IOP Conf. Series: Materials Science and Engineering **972** (2020) 012068 doi:10.1088/1757-899X/972/1/012068

When using the air supply from under the raised floor, the basic rules for designing displacement ventilation must be applied [4, 5]. The flow rate of incoming fresh air is not less than the flow rate in convective jets from heat sources in a given room, fresh air consumption - at least person per person [1]. This effect is described theoretically, and the conditions of its existence will be flooding the lower area of the room prepared with relatively cool fresh air. Accordingly, during convective heat transfer, an upward convective air flow arises near the surface of a person, which is fed by fresh air from the lower zone of the room. This will decrease the required amount of outside air while maintaining the same air quality in the human breathing zone and consequently save resources on the treatment of outside air. In this case, the air is removed from the upper zone of the room, where a "heat cushion" will be observed - air with a high temperature and degree of pollution. The need for wiring exhaust air ducts may rely on the room size. Due to the increased temperature difference between the expelled and supply air, air exchange will also decrease due to the apparent warmth. When calculating the air flow, the limitation of the temperature difference over the height of the working area should be considered [6]. There is no simple, unambiguous method for determining the air temperature gradient over the height of a room when air is supplied from the floor. The question of overcooling of the lower area of the room and the possible health consequences remain unclear.

Overlap heat transfer can also result in deterimental effect on the adjustability and controllability of the system, peculiarly on air distributors located in the most distant places from the unit. During condition of low speeds of air movement, the temperature around the concrete slab will be determined by the calculation of [7], using the temperature of the active heat exchange layer of the slab itself.

The stationary mode of heat exchange between the air in the channel and the environment must considered, and the surface density of the heat flux is determined by the formula.

$$q = \frac{t_{in} - t_{ext}}{R_{\text{int}} + \sum_{i=1}^{n} R_i + R_{ext}},$$

where q – is the surface density of the heat flux through the walls of the channel, W;
t_{in}– channel air temperature, °C;
t_{ext} – is the air temperature in the room, °C;
R_{int} – resistance to heat transfer on the inner surface of the channel wall, (m^2•°C)/W;
$\sum_{i=1}^{n} R_i$ – total thermal resistance to conductive heat transfer of the multilayer wall, (m^2•°C)/W;
R_{ext} – resistance to heat transfer on the outer surface of the channel wall, (m^2•°C)/W.

Thermal resistance to heat transfer on the surface of the inner and outer surfaces of the wall, and conductive heat transfer in the channel wall are determined using the formulas, respectively:

$$R_{\text{int}} = \frac{1}{\alpha_{\text{int}}}; \quad R_{ext} = \frac{1}{\alpha_{ext}}; \quad R_i = \frac{\delta_i}{\lambda_i};$$

where α_{int}, α_{ext} – heat transfer coefficients on the inner and outer surface of the channel wall, respectively W/(m^2•°C);
λ_i – coefficient of thermal conductivity of the material of the i-th layer of the wall, W/(m•°C);
δ_i – the thickness of the i-th layer of the channel wall, m.

The determination of the heat transfer coefficient from air to the walls inside the channel α_{int} depends on the flow regime for each section [4]. The main characterization criterion of the flow regime, is the Reynolds number Re.

$$Re = \frac{wl}{\nu},$$

where w – average flow rate of air obtained from aerodynamic calculation;
v – coefficient of kinematic air viscosity, m^2/c;
l – characteristic channel size, m.

The heat transfer coefficient on the inner surface of the channel is determined using the criterial equation

$$Nu = \frac{\alpha_{\text{int}} l}{\lambda},$$

where Nu – Nusselt number,
λ – air thermal conductivity coefficient, W / (m•°C).

Heat transfer coefficient on the outer surface of the channel α_{ext} to the air of the room was determined by the formula:

$$\alpha_{ext} = \alpha_{con} + \alpha_{rad},$$

where $\alpha_{con} + \alpha_{rad}$ – coefficients of radiant and convective heat transfer on the outer surface of the channel, respectively, W/(m^2•°C), defined by [8].

In accordance to the terms of the applied calculation methods, it was assumed that all the walls of the channel are identical and homogeneous, and the surrounding environment has an equal temperature distribution on all sides.

In the considered example, the upper wall of the channel consists of a chipboard 60 mm thick, the lower wall of concrete 250 mm thick, ambient temperature near the upper wall (raised floor coverings) of 21°C, and an ambient temperature near the bottom wall (under the concrete slab) of 27°C. In order to evaluate the combined effect of various conditions, calculations of heat transfer in channels with different composition walls and ambient temperature were performed. The calculation was performed on a computer during the iterations using an algorithm developed in the MS Excel program.

Calculations showed that the heat flux through the concrete wall amounted to 3730 W, through the wall of chipboard – 1040 W. For the selected office space of 175 m^2 with the specific heat excess of 90 W / m^2, the cooling capacity of the air conditioning system will be calculated at 15 750 W. Consequently, the transmission loss of cold will be 15%. In addition, it follows from the calculations that the temperature of the supplied air in the channel will be increased by almost 3°C.

3. Conclusion

From the result obtained from the calculation of aerodynamic of the canal-raised floor model, it was revealed that unlike duct networks, uniform air supply to all air distributors should be provided without installing additional devices to link pressure losses. Additionally, it was shown that the uniform distribution of air when it is supplied through the raised floor is significantly affected by the ratio of the dimensions of the raised floor, the height and cross-sectional area of the inlet pipe and the area of the air outlets. Furthermore, from the calculation of heat transfer between the air in the channel and the environment, it could be established that without isolated floors, and the use of underground distribution of air according to the principle of displacing ventilation, there will be significant transmission losses of cold and an increase in the temperature of the air supplied.

References

[1] SP 60.13330.2012 Heating, ventilation and air conditioning (Moskow: Ministry of Regional Development)
[2] Handbook of calculations of hydraulic and ventilation systems 2001 (Saint-Petersburg: ANO NPO "Peace and Family")
[3] Grimitlin M I and others 1978 Ventilation and heating of machine-building factories (Moskow: Mashinostroenie)

[4] Skiestad H, Mundt E, Nielsen P 2006 Displacing ventilation in non-production buildings (Moskow: NP AVOK)
[5] Belova E M 2006 Central air conditioning systems in buildings(Moskow: Euroclimate)
[6] GOST 30494-96 "Residential and public buildings. Indoor microclimate parameters 1999 (Moskow: Gosstroy of Russia)
[7] Isachenko V P 1975 Heat Transfer: a textbook for universities (Moskow: Energy)
[8] Shumilov R N, Tolstova Yu I 2004 Calculations of heat and mass transfer in the design of ventilation (Ekaterinburg: UGTU-UPI)

IOP Conf. Series: Materials Science and Engineering **972** (2020) 012069 doi:10.1088/1757-899X/972/1/012069

Modelling and monitoring of security unique underground structures of the sewerage system of major towns in difficult ground conditions

N A Perminov[1]

[1]Road construction of the transport complex, St. Peterburg University of Means of Communication, Moscow Avenue 9, St. Peterburg, 190031, Russia

E-mail: perminov-n@mail.ru

Abstract. On the basis of many years of experience in the construction, reconstruction and operation of underground engineering infrastructure of the metropolis developed the concept and philosophy of the creation and operation of Geomonitoring security system of the tunnel collector. The philosophy of construction of space-time structure of Geomonitoring is described. The article presents the experience of modeling , monitoring and taking protective measures aimed at ensuring the safety of tunnel reservoirs located in the area exposed to the geotechnical effects of underground and high-rise construction. The analysis of calculation and experimental data and the results of long-term monitoring aimed at ensuring the safety of underground drainage and treatment facilities of megepolis. The special protection against technogenic influence is required for long-operated tunnel sewage collectors. This paper, based on the extensive data of studying the work of the tunnel collectors in conditions of soft soils and intense technogenic influences, describes new methods of diagnosing, the features of defects, and the results of modeling joint interaction of the "tunnel – soil bulk" as well as experimental data on experimental objects. The geotechnical methods for protecting the long-operated tunnel collectors have been developed, and there has been given the unique experience of their application in the objects of St. Petersburg together with their monitoring support.

1. System of geomonitoring ensuring of construction and maintenance of engineering subsurface structures

Geomonitoring system is intended to ensure resistance of engineering structures of town-planning environment to the exposure, on one hand, and development of preventive measures in order to eliminate negative influence on town-planning environment as a result of breaking of regular operating modes of engineering systems functioning.

Civil engineering infrastructure of Saint-Petersburg develops under increasing requirements to ecology and rational use of land resources. At engineering development of underground space of such a megalopolis, elaboration of complex measures on protection of town-planning environment against negative anthropogenic impact is of special actuality. This elaboration includes geomonitoring insurance of technosphere safety.

Experience of designing, construction and maintenance of a unique treatment facilities complex in Saint-Petersburg pointed to the necessity of creation of such systems. It is known that Saint-Petersburg complex of sewage drainage and disposal consists of more than 110 underground pumping stations. Sewage collectors with diameter from 1.5 to 4.5 m have a developed network of more than 300 km long and 15 to 80 m deep. They are 35-45 years old, mainly laid in one line, and are heavily worn out. Base pumping stations (BPS) together with other engineering structures (at 70 m embedding into the ground and with cross-section up to 2000-3000 thous. sq.m.) have side surface contact area of up to 150-200 thous. sq.m, intersect several (up to five and more) waterbearing formations and have a considerable effect on town-planning environment. As per analysis data [1], almost in 60 per cent of cases deformations of urban development are conditioned by the adjacent construction and influence of engineering systems.

On the instructions of SUE "Vodokanal of St. Petersburg" the institute "Lengiproinzhproekt" together with PGUPS developed system of geomonitoring ensuring of technosphere safety. Currently it is implemented step-by-step at the construction and maintenance of subsurface engineering structures of the city (see Figure 1).

The scientific and production company "Transspetsstroy", being a constant and reliable partner, is a scientific and production base for aprobation and implementation of solutions developed in the institute.

The basis for formation of space-time structure of geomonitoring and conditions of its functioning is as follows: 1 – programs of complex calculations and predictions of change in engineering and geologic conditions and stress-strain state of the geomassif at different modes of structure erection; 2 - technical equipment system of instrumental observation and control over change in the individual elements of "structure-geomassif" system; 3 - information-measuring system of collection, processing, storage and identification of parameters (data) of observation and control; 4 – complex of geotechnological methods of purposeful impact on the ground massif and structure.

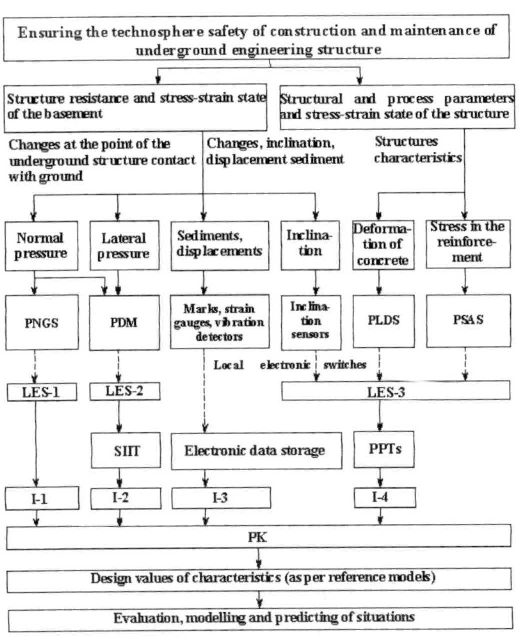

Figure 1. System of geomonitoring ensuring of technosphere safety of construction and maintenance of underground engineering structures

2. Eexperience of geomonitoring use in construction of large-size structures

Elaborated and tested on the basis of results of full-scale and experiment-calculated works was the complex system of geotechnical ensuring of immersion of open caissons with diameter of 50 and 66 m at construction of treatment facilities in Saint-Petersburg. The slotted soil column for monitoring objects is characterized as follows: top part is presented by Quaternary beddings to the depth of 14.0-25.0 meters (middle-density water-saturated dust sand E=11 MPa, C=0 MPa, φ=30°; laminar silt sandy loam E=4 MPa, C=0.01 MPa, φ=15°; laminar silt loam, very soft E=9 MPa, C=0.025 MPa, φ=16; semisolid silt loam with gravel, pebbles E=14 MPa, C=0.028 MPa, φ=28°), lower part is represented by top of positioned Proterozoic bluestone (E=19 MPa, C=0.04÷0.06 MPa, φ=18-21°).

This system includes three complexes: control and measurement complex; complex of signal estimate and transmission for control over engineering procedures; complex of operational influence on engineering procedures.

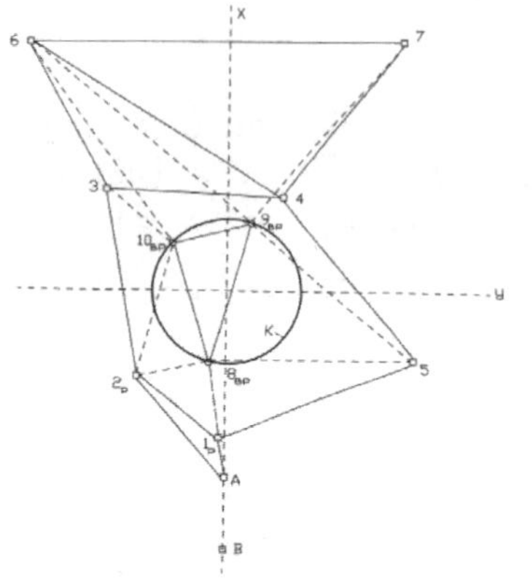

Figure 2. Scheme of estimated network of geomonitoring: K-caisson contour; A, B – trigonometric points; 1p, 2p – origins of trigonometric networking; 3-7 – permanent points of trigonometric network; 8вр, 9вр, 10вр – temporary points at the inner wall of the caisson

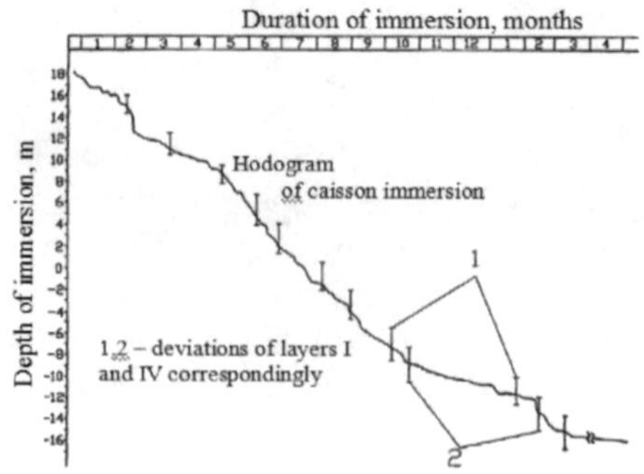

Figure 3. Monitoring at construction of large-scale open caisson KNS in Saint-Petersburg

Measuring complex (see Figure 2-4) provided control over attitude position, displacement of contour of the c aisson cover and ground massif. It included the dimensional orientation devices (optical range finders, reflector marks, roll measuring gauges).

Figure 4. Change of radius R and displacement of center of caisson circle O at its immersion to the depth H: a,b – 6-th and 7-th layers of concrete casting correspondingly, distance from banquette to the top of layer is 34.5 and 40.5 m; I - for H=17 m R=32.93 m, Mx,y=0.067 m (September); II - for H=29 m R=32.85 m, Mx,y=0.05 m (December); III - for H=25 m R=32.75 m, Mx,y=0.048 m (November); IV - for H=29 m R=32.76 m, Mx,y=0.067 m (December); O(I-II)=0.108 m; O(III-IV)=0.234

NDS control subsystem ensured monitoring of parameters characterizing behaviour of the "structure-geomassif" system (see Figure 5), particularly: mode of deformation of ground, which is estimated as per results of measurement of contact pressures, pore pressure and earth handling (sets); mode of deformation of the structure material, including concrete and reinforcement; inclination of structure and sets of ground in the impact area. Sensing devices were installed in design sections along the contour and layers.

Complexes of discrete-continuous registration of sensor readings included digital period meter PTsP-1; stand-alone interface device – K1-20; PTsP controller, controller switch, controller computer, PTsP switch, controller-PDU, switch-sensors).

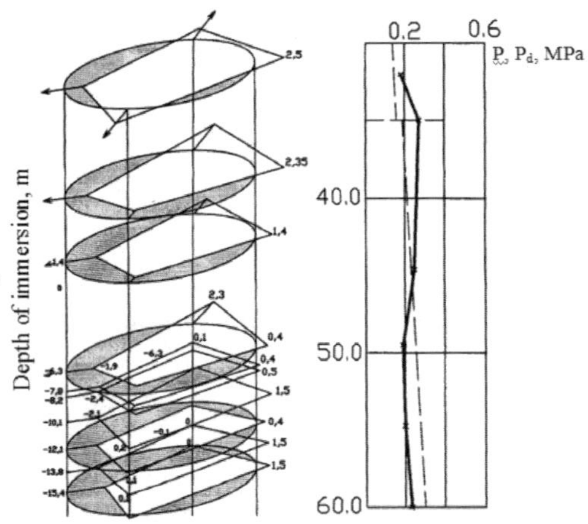

Figure 5. Three-dimensional diagram of "n"-coefficient of horizontal stresses irregularity at immersion of caisson: "n" = Pп / Pг, where Pп – experimental "peak" values of soil pressure, Pг – design values of soil pressure [2]

At the stages of erection, which were predicted to be most difficult (with "inclinations", "strippings", etc.) by way of numeric calculations the technologically feasible geotechnological methods have been chosen: changes in geomassif stress-strain state in the basement of structure or along the side surface, arrangement of leading pores, decrease of friction by means of electroosmosis.

As a result of monitoring the following fact of great importance is determined: peak values of horizontal pressure exceed design values in the moment of "inclination" by 2.3-2.5 times [2]. It may cause microcracks in the structure concrete, which will inevitably entail damage of the structure waterproofing. This fact was noticed after 10-15 years of maintenance of BPS in Saint-Petersburg treatment facilities complex.

3. Experience of geomonitoring ensurance of technosphere safety of tunnel collectors in the area of geotechnical impact of overground construction

In order to evaluate the state of line underground structures – collectors in the course of construction works in the area of geotechnical impact the system of structure deformation and displacement monitoring is developed. The system of monitoring of objects state includes: 1 – subsystem of computerized geodetic testing, by means of GPS connected to the base station; 2 – subsystem of continuous control of mode of deformation (NDS) of tunnel collector structures; 3 – subsystem of assessment of the collector vibrodynamical resistance; 4 – subsystem of control over line structure offsetting at the selected base and crack opening dynamics.

This system was successfully tested on the number of objects of civil engineering infrastructure, such as "Monitoring of sewage collector state at arrangement of pile field and loading of piles in the area of construction of complex "Shkipersky Rynok", "Monitoring of sewage collector state at construction and reconstruction of buildings complex "Birzhevoy Complex", "Monitoring of state of structures of the collector brick vault in Konnogvardeysky boulevard at performance of anti-wreck measures", "Monitoring of sewage collector in the area of geotechnical impact of construction of high-rise buildings of "Okhta-Center".

Collector oscillating process was examined by means of sets of sensors SM-3KV installed in the collector vault. Use of this kind of sensors allows registering oscillations with the amplitude from 0.5 μm to 2000 μm, frequency from 2 to 200 Hz, error of not more than 10%. Sensors are protected from the external field viewing, have a temperature compensation device, waterproofing (for recording

IOP Conf. Series: Materials Science and Engineering **972** (2020) 012069 doi:10.1088/1757-899X/972/1/012069

oscillations in water). They are characterized by minimum mutual impact (up to 5% of orthogonal oscillations). Diagram of oscillations registering is shown in Figure 6.

The set includes three sensors, which allow recording vertical and two horizontal (orthogonal) components of oscillation amplitudes.

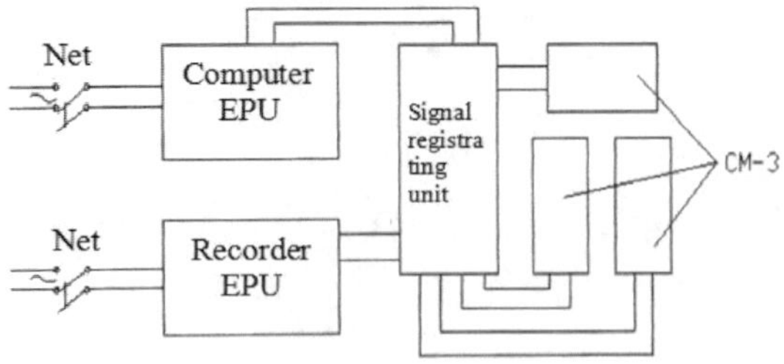

Figure 6. Recording geometry of oscillations in the collector vault

Oscillations monitoring has been performed continuously in the period of construction works related, for example, to pile driving in immediate proximity to the collector or to erection of underground part of the object under construction.

Results were processed as per each component of oscillation amplitude. As a result, average and maximum possible values of oscillation amplitudes were obtained. In all cases probability level was 0.995.

At monitoring of sewage collector state at arrangement of pile field and loading of piles in the area of construction of complex "Shkipersky Rynok" the analysis of obtained results showed that maximum amplitudes of displacements and ground displacement speed at change of distance from the oscillation source are described as per power dependence. Thus, for example, at pile arrangement at a distance of 6 ÷ 8 m from the testing point maximum amplitudes of displacement Amax and displacement speed Vmax were, correspondingly, as follows: 30 ÷70 μm and 0.11 ÷ 0.15 cm/s, in case of moving away to a distance of 20 ÷ 25 m they decreased to the level of 1 ÷ 2 μm and 0.012 ÷ 0.045 sm/s, i.e. decreased by 25 ÷ 30 times.

Data obtained at monitoring indicated that in case of pile arrangement as per low impact technology with arrangement of pilot hole, critical distance of its immersion is considered to be the moving-away from the nearest point of collector in the range L = 6 ÷ 10 m.

In order to ensure this condition, pile footing structure was developed, which takes external action of the multitiered shopping center being constructed above the collector.

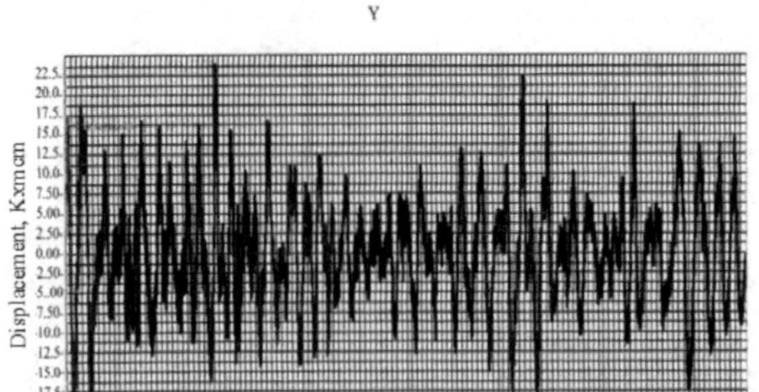

Figure 7. Typical records of oscillating process of the collector lining

Figure 8. Mounting of measuring equipment for collector geomonitoring

It was established, that works performed at the construction site related to the arrangement of pile field in the area of collector road caused definite active background of external dynamic loads, which value is higher than that of the background ones, but lower than it is prescribed by instructions and standards. These loads caused no considerable inertial processes in structures, which could result in oscillations of the "collector structure – ground massif" system as a whole, old cracks opening and generation of new ones.

Figure 9. Vibration detector CM-3KV installed in the collector vault

In order to ensure resistance of emergency collector located at crossing of Konnogvardeysky boulevard and Dekabristov proezd, relieving archy screen structure was developed and collector brick vault was enhanced by structural reinforcing with carbon fiber. As a result of analysis of the obtained monitoring results the following conclusion on the collector vault rigidity increase was made: after arrangement of anti-wreck measures periods of the collector own oscillations along the orthogonal axes changed correspondingly from 0.12 s and 0.54 s to 0.07 s and 0.19 s, i.e. by 35 ÷ 58 %.

Control of the line structure offsetting at the selected base and crack opening dynamics was performed by means of linear displacement transducers of magnetic and striction type BTL5-T110-V0050-P-103 with sensitivity of 1 micron. Subsystem included base controller unit, interface, sensors with IP level of protection, special-purpose software. Structure of the whole subsystem allowed easy enhancing of its facilities, both as per interval of measured displacements and as per number of measuring lines. Information received from sensors with the preset interrogation interval accumulated in the memory block.

4. The general characteristic and degree of wear of the sewage tunneling systems in the largest Russian cities

In cities with an expanded historical center, long-operated tunnel sewage collectors need special protection from anthropogenic influence. As objects referred to the structures of enhanced level of responsibility tunnel collectors should have the required bearing capacity and exploitation reliability in order to meet normative standards of safe use excluding the risk of occurring hazardous technogenic accidents [3].

The authors have collected and analyzed data on the current technical state of sewage collectors of 15 largest cities in Russia including Moscow, St. Petersburg, Ekaterinburg, Samara, Novosibirsk etc. of total length of more than 2300 km. For general sampling there have been used such properties as the length of the tunnel network (from 100 km to 500 km and more); the period of exploitation (from 30 to 70 years); complex geotechnical condi-tions of tunneling (presence of unstable soils, intersecting aquifers including the heading ones).

Practically for the majority of the cities under consideration the network of the tunnel collectors has an average level of physical wear in the range of 55-66% with the dynamics of development of 0.5-1.0% per year. It is registered that in the cities with an expanded historical center and developed transport infrastructure, first and foremost, Moscow and St. Petersburg, but the degree of tunnels wear has higher values of 71% and 83% with the faster dynamics of development up to 1.5-2% per year.

As the system of tunnel sewage in Moscow had been established by the end of the1930s and in St. Petersburg it had happened by the end of 1950s, the degree of the tunnel wear and the dynamics of

IOP Conf. Series: Materials Science and Engineering **972** (2020) 012069 doi:10.1088/1757-899X/972/1/012069

development is faster in St. Petersburg. The tasks of protection of tunnel collectors and identifying the reasons of their high degree of wear are extremely urgent for substantial development of this unique historical city (see Figure 10).

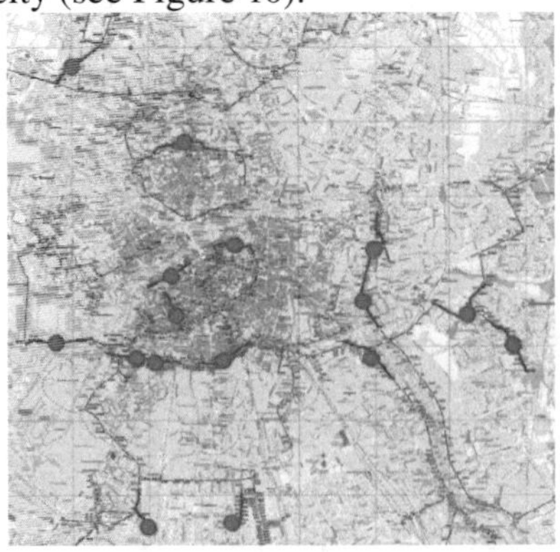

Figure 10. The scheme of location of the tunnel sewage collectors in St. Petersburg.

5. The geotechnical analysis of the influence of anthropogenic factors on reliability of the long-operated sewage tunnels

5.1. The analysis of the defects of the sewage tunnel collectors, their structure and the dynamics of their development.

There have been analyzed the archive data of technical inspections of the tunnel collectors belonging to the "Vodokanal-SPb" by the maintenance service. In the period from the late 1970s to nowadays the authors have conducted the tool inspections of the technical state of the tunnels (the total length of about 15 km) located in the historical city center (the depth of 12-15 m) and beyond it, within the approaches to three main sewage pumping stations collecting all sewage outflows of the city, embedded at the depths of 53-71m.

a b

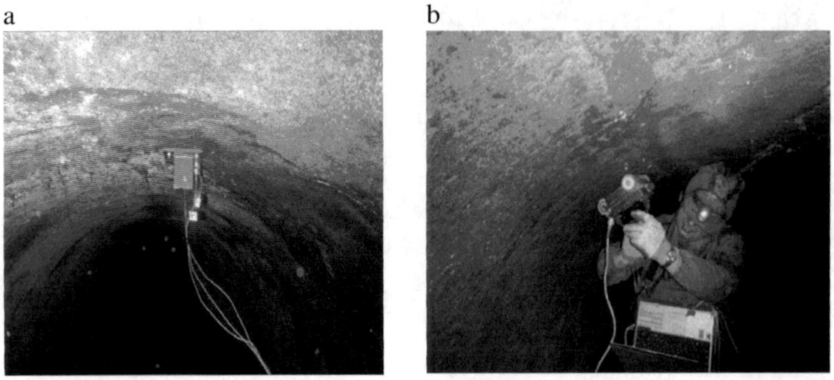

Figure 11. Tests of the structures in the operating tunnel sewage collector: a) oscillation measurements using the ICP accelometers; b) scanning integrity of the structures of the tunnel using a georadar OKO-2.

The technical tool inspections included: in-situ tachometric survey of the spatial location of a tunnel in between the shafts; georadar scanning of the state of the tunnel internal surface evaluating its integrity; core sampling and conducting tests using a tear-off method with scanning to assess strength properties of concrete; sampling for chemical and biochemical analyses; evaluation of a corrosion degree and reinforcement using non-invasive methods; vibro-dynamic tests on oscillations of internal tunnel structures caused by external and construction vibration impacts.

Uniqueness of the data on observations of the state of the tunnels is that the technical inspection of the structures has been conducted during the long period since the late 1970s to nowadays (the end of 2019). During this period the same parts of collectors have been inspected several times. As a rule, each inspection was followed up by the tunnel state monitoring performed for several years. Therefore, there has appeared a possibility to track the dynamics of defects development.

The observation period was divided into three sub-periods: a) the 1970-80s; b) the 1980-2000s; c) 2000-2019. The identified defects which are more typical and influencing the exploitation reliability and bearing capacity of the tunnel were divided into 7 groups: d1- shrinkage cracks in the concrete liner; d2- gas corrosion features; d3- dripping leakages; d4- force cracks in the arch and along the lateral surface of the tunnel; d5- concrete biological corrosion features; d6- reinforcement corrosion, conduit dusting; d7- the presence of heading leakages.

The Figure 12 shows the dynamics and structure of defect manifestation during the operation period of the tunnel sewage collectors from the late 1970s to 2019.

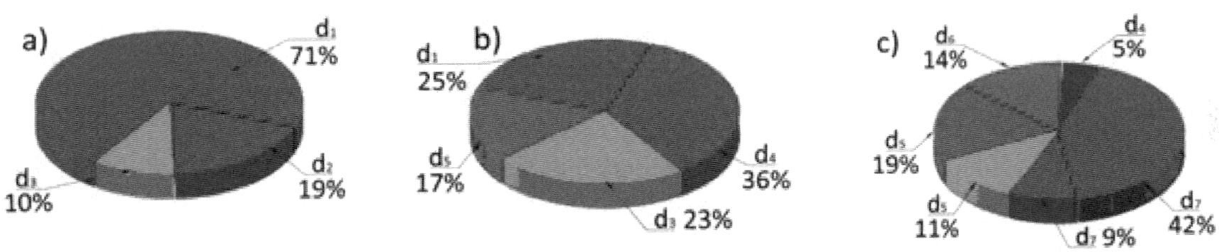

Figure 12. The dynamics and structure of defect manifestation in a long-term operation period of tunnel sewage collectors: a) 1970-80s; b) 1980-2100s; c) 2000-2019. Classification of defects: d1-d7 – primary defects caused by shrinkage cracks and manifestation of gas corrosion features before development of force cracks and occurring heading leakages (see description in the text).

The analysis of the dynamics and structure of defects manifestation shows that during the initial period of operation of the tunnel collectors there have been observed defects in the form of shrinkage cracks in the concrete liner with dripping leakages and gas corrosion features. The nature of the defects prevailing in the first 15-20 years of the tunnel exploitation and their influence on the bearing capacity and exploitation reliability can be considered as negligible; their technical state can be regarded as serviceable according to the existing Russian State Standards and Construction Codes.

The subsequent 20 years of the tunnel operation are characterized by manifestation of force cracks and considerable increase of the defects in concrete due to its gas and biological corrosion. However, their technical state was described as partially serviceable.

The analysis of the defects appeared in the period of 2000-2019 is of particular interest. It is noteworthy that in this period the amount of the defects influencing the tunnel bearing capacity has increased: almost 40% growth in the number of force cracks in the arch and along the lateral surface of the tunnel; the features of reinforcement corrosion and dusting of the conduit; the presence of heading leakages in the tunnel body; the increase in the proportion of the defects associated with biological and gas corrosion. The technical state of the tunnel structure is characterized as partially capable. The tunnel sewage collectors require monitoring of their technical state, taking actions on reconstruction of the bearing capacity and exploitation reliability of the structure.

5.2. The factors conditioning intensive development of the defects of the tunnel sewage collectors

The comprehensive long-term technical inspections of the tunnel sewage collectors allow conducting a retrospective analysis and identifying factors, which condition manifestation and development of the defects. The factors can be divided into internal and external ones according to the nature of their influence.

The nature of the internal factors is explained by the processes linked to transportation of sewage outflows. The main factors are the corrosion of the internal structures (gas-biological), the conduit dusting, sewage pumps etc. These factors lead to rather static development of the defects. The definition of the ultimate acceptable values of their influence on the tunnel sewage collectors pertains to the area of research and activity of the sanitary-technical services [4,5].

The external factors are characterized by the complex interaction of the tunnel structure and the embedding soil bulk, which form a natural-technical system under the influence of the anthropogenic factors.

The evaluation of this interaction and definition of the area of the ultimate acceptable values of mutual influence of the natural-technical system elements is a task of geotechnical predictions. Other papers of the author is dedicated to the issues of optimum management of the natural-technical system elements [6,7].

In the framework of the given inspection we have a task of geotechnical support of the safe level of external anthropogenic impacts on the tunnel structure with account of its residual bearing capacity.

6. Preventive geotechnical measures of protection of the tunnel sewage collectors

Geotechnical and structural measures to ensure reliability and safety of the long-operated tunnel sewage collectors should be selected basing on comparison of competing options, but the main requirement they should meet is a possibility of preventive use substantiated by geotechnical and structural calculations.

The Table 1 presents examples of some geotechnical measures successfully taken in the projects of protection of the tunnel sewage collectors in St. Petersburg in 1998-2019.

Table 1. Geotechnical measures of protection of the tunnel sewage collectors.

The nature of technogenic impacts	Geotechnical and structural measures	The object for protection
Supernormative static and dynamic influences on the brickwork arch of the collector	Construction of an unloading screen and reinforcement of the structures; monitoring.	The brickwork sewage collector XVIII century under Konnogvardeysky boulevard
The increase in static load applied to the tunnel sewage collector due to the weight of the building being constructed above it	Construction of an unloading bridge with load transfer to the soils below the tunnel bottom, monitoring.	The sewage collector in Nalichnaya street
Unloading of a soil bulk during the construction of the deep transport tunnel above the tunnel sewage collector	Construction of a protective screen of reinforced soil to prevent the uplift of the pit bottom, monitoring.	The transport tunnel above the tunnel sewage collector along the Obvodny Canal and the bridge of Alexander Nevsky
Prevention of horizontal displacement of the tunnel sewage collector at unilateral unloading of a soil bulk during reconstruction of the embankment.	Construction of a geotechnical barrier, monitoring.	The tunnel sewage collector along the highway in the embankment of the Fontanka River
Expansion of an area of ultimate acceptable displacements of the tunnel sewage collector at unilateral unloading of a soil bulk during reconstruction of the embankment.	Structural reinforcement and sanation of the tunnel sewage collector using the spiral-coiling method in order to increase the ultimate acceptable tunnel displacements, monitoring.	The tunnel sewage collector along the embankment of the Kryukov Canal and the Moyka River
The increase in static load applied to the tunnel sewage collector due to the impact of heavy road transport and trams.	Structural reinforcement and sanation of the tunnel sewage collector using the coiling method in order to restore the bearing capacity of the tunnel up to the design level, monitoring	The 840-m-long tunnel sewage collector in Tovarischesky prospect

7. Nnumerical modeling and monitoring of geotechnical system "tunnel collector - city planning object - protective measures".

Earlier [6,7,8] the large-sized subsurface object of monitoring was represented from the point of view of system approach and methods of optimum control in the form of the integrated geotechnical system (IGTS) – "subsurface structure and its elements – geological object (the geomassif area containing the structure and interacting with it) – city planning object (the fragment of urban development limited by the area of active influence of subsurface construction). On the basis of numerical modeling and monitoring means a system and methods of protective measures geotechnology optimization were created.

As for the object of buildings of "Cultural and Business Center Okhta", the monitoring and geotechnical substantiation of protective measures in tunnel collector 2,250 mm in diameter located at a depth of 14-15 m in very soft thixotrop loam was carried out. The collector has been in operation for more than 30 years and its condition, as per the technical inspection materials, was considered as restrictedly serviceable.

The collector being in close proximity to the projected complex "Cultural and Business Center Okhta" falls within the area of its influence at the stage of construction and operation. To provide geotechnical safety and monitoring of the collector the influence of "Cultural and Business Center Okhta" complex was evaluated and methods of its geotechnical protection were developed. In the course of solving such a task, use has been made of the finite element method, geotechnical packages "Plaxis 3D Foundation", "Plaxis 3D" Tunnel and design package "Robot 3D".

Construction activities and arrangement of protective measures were simulated in the form of calculation steps which are given in Table 2

Table 2. Numerical modeling steps.

№	Description of calculation steps
1	Original stress and strain state of the system
2	Arrangement of protective measures
3	Construction of walls in the ground
4	Construction of piers and barrettes
5	Tier-wise development of ground with concrete casting of flooring and bottom
6	Application of load from building and foundation loading (700 kPa for tower and 200 kPa for stylobate)

The following was simulated as protective measures (Figure 13): complete ground substitution – a; partial ground substitution – b; application of geotechnical barrier in the form of protective trench made of thixotropic paste with weighting agent ($E=2\div3$ MPa; $\gamma=2.15\div2.20$ t/m^3; $\varphi=5^0$; $c=0.003$ MPa; $\nu=0.48$) – c.

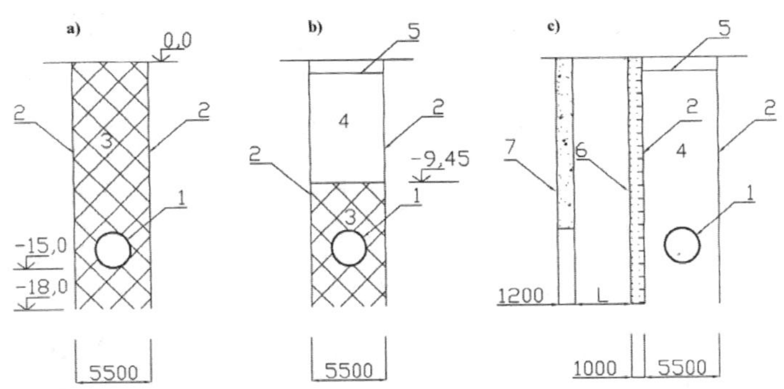

Figure 13. Protective measures for decrease of influence on collector: 1 – reinforcement of tunnel structure as per spiral-wound technology; 2 - brace sheet walls L-5 to the length of the protected section; 3 - stabilized soil E=200 MPa; 4 – existing soils; 5 - metal binding belt; 6 – geotechnical barrier (thixotropic paste with weighting agent - barium); 7 – wall in ground

IOP Conf. Series: Materials Science and Engineering **972** (2020) 012069 doi:10.1088/1757-899X/972/1/012069

The task of numerical modeling in the objective allocation by Plaxis3D Foundation means includes evaluation of polydimensional deformations and strain level in longitudinal cross-sections of the collector liner. Selection of this model was conditioned by that the collector has curved alignment in plan view and the high-rise complex of buildings is represented in the wrong form (figure 14).

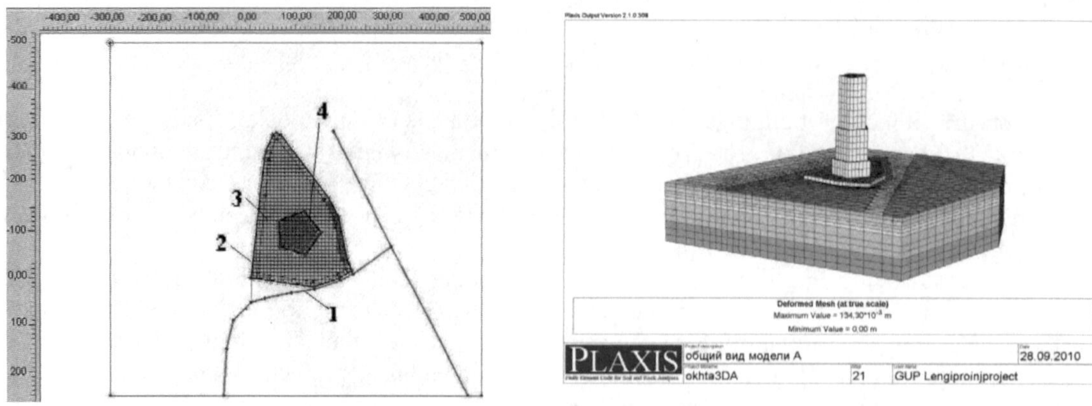

Figure 14. Geometry of a 3D model of the complex (the plan and general view): 1 – the protected collector; 2 – the wall in soil; 3 – the stylobate part of the building; 4 – the high-rise part of the building.

The 3D model is represented as a block with plan view dimensions 800x800 m. The lower boundary of the model is located at an actual elevation of minus 170 m which is conditioned by the depth of mass contractibility according to the geology and loads (see Figure 15).

a)

Number	Name	ν [10^{-3}]	E_{ref} [10^3 kN/m^2]	c_{ref} [kN/m^2]	φ [°]	γ_{unsat} [kN/m^3]	γ_{sat} [kN/m^3]
1	ИГЭ 01 насыпные	300,000	25,000	25,000	25,000	18,000	19,000
2	ИГЭ 01а техногенный	300,000	10,000	1,000	23,000	18,000	19,000
3	ИГЭ 02	340,000	8,000	17,000	18,000	19,900	19,900
4	ИГЭ 03А суглинки	360,000	7,000	8,000	12,000	19,600	19,600
5	ИГЭ 03Б суглинки	360,000	6,000	7,000	9,000	18,600	18,600
6	ИГЭ 04/4а водоупор	360,000	6,000	10,000	11,000	18,600	18,600
7	ИГЭ 02...04	340,000	8,000	10,000	10,000	19,000	19,000
8	ИГЭ 05Б пески	300,000	30,000	6,000	34,000	21,100	21,100
9	ИГЭ 05 супеси	340,000	8,000	12,000	16,000	20,100	20,100
10	ИГЭ 06* супеси	320,000	31,000	33,000	25,000	21,700	21,700
11	ИГЭ 11* глины дислоцир	340,000	40,000	150,000	17,000	21,300	21,300
13	ИГЭ 12* глины твердые	340,000	180,000	340,000	26,000	22,200	22,200
14	ИГЭ 13* глины твердые	340,000	250,000	610,000	32,000	22,300	22,300

b)

ID	Name	ν [-]	E_{ref} [kN/m^2]	γ_{unsat} [kN/m^3]	γ_{sat} [kN/m^3]	R_{inter} [-]
12	B35 стены	0,20	3,45E7	25,0	25,0	0,10
13	барретты	0,35	2E7	24,0	24,0	1,00
15	сваи	0,35	1,5E7	22,5	22,5	1,00
17	бентонит	0,40	1,0	12,0	12,0	1,00
18	maquette	0,35	5E6	2,0	2,0	0,10
19	perm16	0,45	3000,0	16,0	16,0	0,10
20	perm18	0,45	3000,0	18,0	18,0	0,10
21	perm20	0,45	3000,0	20,0	20,0	0,10
[illegible]	[illegible]	0,48	600,0	22,0	22,0	0,10
23	perm14	0,48	800,0	14,0	14,0	0,10
24	жидкий бетон	0,40	1000,0	25,0	25,0	1,00

Figure 15. Geological column – a, and materials parameters –b

At the stage of operating loads application from the high-rise building complex the collector deformations in the area of the maximum approach were 62 mm and settlements – 134 mm; at that, axial forces had variation range from +0.2 MN (elongation) to – 0.5 MN (compression), bending moments – up to 2 NMm[9,10,11].

Flat model (2D) served for analysis of protective measures influence on the collector and represented the schematic cross-sections of the high-rise buildings complex and the adjacent collector. The calculation model is represented as a block with dimensions 160x125 m (figure 16).

IOP Conf. Series: Materials Science and Engineering **972** (2020) 012069 doi:10.1088/1757-899X/972/1/012069

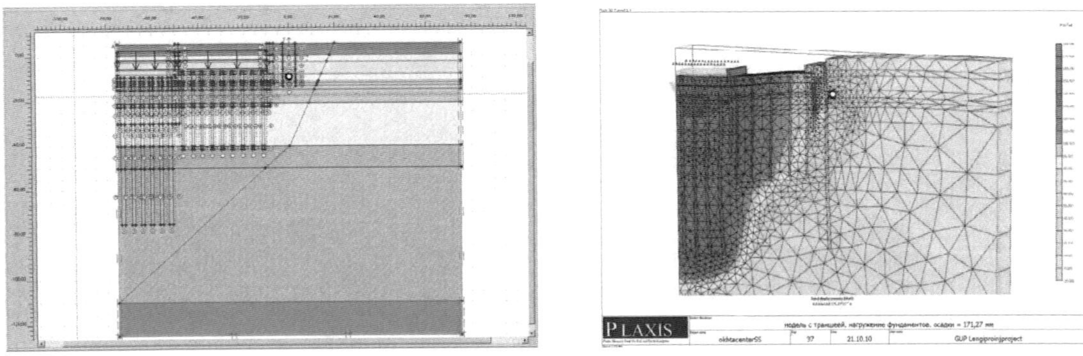

Figure 16. Simulation of the protective measures.

The modeling results are represented in Table 3 and show that the trench protection provides for decrease in not only the vertical but also the horizontal deformations. General decrease in deformations is observed as more than in 3 times.

Table 3. Displacement of collector at construction stages.

Stage of development	Protective measure *	Displacements, mm X	y	total
Development of the first layer with concrete casting of flooring slab	Trench –GTB	-3,99	+2,10	4,46
	no	-5,38	+1,39	5,45
Development of the second layer with concrete casting of flooring slab and with account of piers	Trench –GTB	-2,89	+7,04	7,43
	no	-12,97	+15,34	20,07
Development of a ditch with simultaneous concrete casting of the bottom	Trench –GTB	-7,59	+23,87	2477
	no	-11,17	+35,95	37,57
Loading of building model	Trench –GTB	-6,66	-20,08	23,93
	no	-30	-60,26	67,22

*GTB – Geotechnical barrier

Structure preservation criterion is maximum deformation and maximum oscillation of the collector, which ensure chatter stability of the ground massif, which contains collector. Maximum deformations and values of allowable displacements depending on the length of deformed section of tunnel collector were calculated as per Robot Structural Analysis Professional 2009 program for different sections of collector (see Figure 17).

Monitoring was aimed at provision of that the values of the collector lining displacements do not exceed design values shown at the diagram (see Figure 18).

Reinforcement of collector lining as per the spiral-wound technology (SPR) makes it possible to increase the maximum allowable tunnel displacement up to 3-5 times, depending on the length of the section being deformed.

The results of numerical modeling and monitoring of the geotechnical system "tunnel collector - city planning object - protective measures" used in the area of geotechnical influence of "Okhta-Center" construction can be recommended for other large-scale unique objects under construction in the area of tunnel collectors location[12,13,14].

IOP Conf. Series: Materials Science and Engineering **972** (2020) 012069 doi:10.1088/1757-899X/972/1/012069

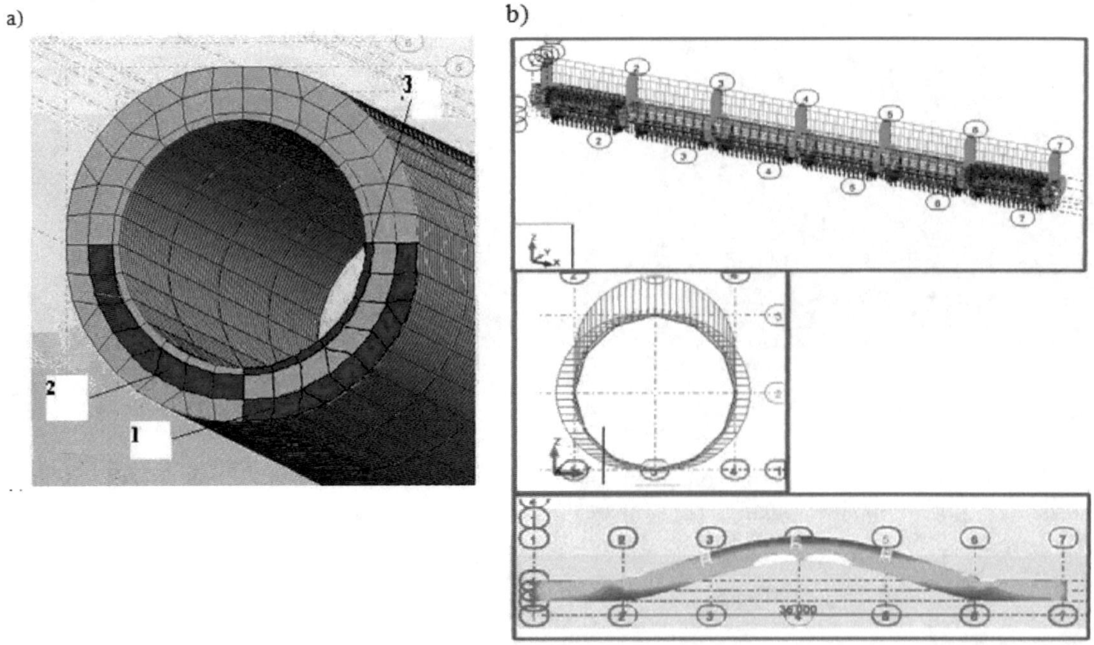

Figure 17. Fragment of design diagram – a, and modeling of maximum deformations of collector – b: 1 – collector tubing; 2 - reinforced-concrete jacket; 3 – reinforcement layer of the tunnel.

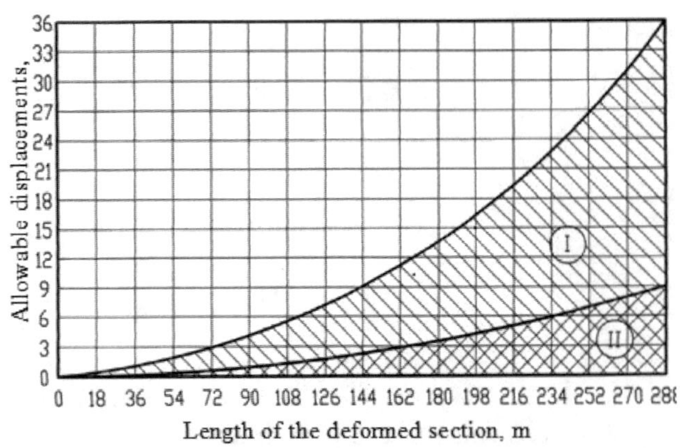

Figure 18. Values of the collector lining allowable displacements depending on the length of deformed section: I-monitoring of "safety corridor" of the collector displacements with lining reinforcement; II-initial state of collector without reinforcement

Concept of the geomonitoring ensuring of technosphere safety of construction and maintenance of tunnel collectors, principles of its formation and functioning should provide reliability, safety and efficiency of underground structures construction. In some cases they can be successfully used at engineering development of underground space of a megalopolis, when it is necessary to provide protection of town-planning environment against negative technical impact. In other cases, vice versa, they can be used to provide preservation of underground spot or line structure being erected against change in condition of exposure of town-planning environment or other polyanthropogenic factors, which disturb conditions of technosphere safety.

References

[1] Perminov N A 1997 Comprehensive geotechnical support for the construction of large edifices as part of St.Petersburg Geotechnical Engineering for the Preservation of Monuments and Historic Sites, Viggiani (ed.)-1997 Balkema Rotterdam ISBN 9054108711

[2] Perminov N A and Lombas S V 2004 Complete geo-technical and monitoring services for the construction of the underground structures in a megapolis. *Geotechnical problems on construction of large –scale and unique projects Proceedings of International Geotechnical Conference dedicated to the Year of Russia in Kazakhstan* 23-25 September 2004 Almaty Kazakhstan. ISBN 9965-25-409-5 Edited by academic of NAN RK, Dr.Sc., Professor Sh.M. Aitalyev and Dr.Sc.,Professor A. Zhusupbekov

[3] The Federal Law 2009 The Technical Regulations for safety of buildings and structures 12.10.2009 #384-FL (the updated version 2016)

[4] Pankova G A and Klementiev M N 2015 The experience of exploitation of sewage tunnels in St. Petersburg *Water supply and sanitary equipment* vol **3** pp 55-61

[5] Vasiliev V M, Pankova G A and Stolbikhin Yu V 2013 Destruction of sewage tunnel and above structures due to the impact of microbiological corrosion *Water supply and sanitary equipment* vol **9** pp 55-61

[6] Perminov N A 2004 Complete geotechnical and monitoring services for the construction of the underground structure in a megapolis *Proceedings of the International Geotechnical Conference Almaty* pp 361-366

[7] Perminov N A 2014 Geotechnical Aspect of Safety Assurance for Long-Used Engineering Infrastructure Facilities in Large Cities in Complicated Ground Conditions. *Geotechnics of Roads and Railways* pp 1195-1201

[8] Perminov N A 2019 Simulation of Defectless Lifecycle of Unique Underground Structures of the Sewage System at the Stage of Their Construction in Difficult Soil Conditions *International Journal for Computational Civil and Structural Engineering* vol **15 #1** pp 119-131

[9] Barabash M 2012 Simulation of beyond-design impacts at studying the lifecycle of buildings and structures *International Journal for Computational Civil and Structural Engineering* vol **12 #3** pp 15-25

[10] Ilichev V A 2008 A flat problem of a plate on elastic subsoil with account of a technology of its construction *Soils, foundations and soil mechanics* vol **3** pp 12-16

[11] Perelmuter A V and Kazantsev O V 2015 The analysis of structures with a changing calculation scheme (Moskow: AS Publishing House) p 148

[12] Ilichev V A and Mangushev R A 2016 The reference book of a geotechnical engineer Soils, foundations and underground structures (The publishing house of the Association of Construction Universities) p 1034

[13] Carmody Y and Sterling R L 1993 Underground space design A guide to subsurface utilization and design for people in underground spaces (New York: VNR) p 328

[14] Ponomaryov A V, Kaloshina S V, Zakharov A V, Bezgodov M A, Shenkman R I and Zolotozubov D G 2015 Results of geotechnical modeling of the influence of construction of the large foundation ditch on the existing historical building Japanese Geotechnical Society *Special Publication: the 15th Asian Regional Conf. on Soil Mechanics and Geotechnical Engineering: Geotechnical Heritage Part 2* (TC 301/ATC 19 Session) Vol **78** pp 2676-2679

IOP Conf. Series: Materials Science and Engineering **972** (2020) 012070 doi:10.1088/1757-899X/972/1/012070

Method for vibration measurement in pipelines of residential buildings

Anan S E Nazzal [1], A A Sekacheva[1] and A S Noskov[1]

[1] Ural Federal University, 19 Mira Street, Yekaterinburg 620002, Russia

E-mail: tonechka_marakulina@mail.ru

Abstract. The article is devoted to the problem of insufficiency and incompleteness of existing methods for measuring vibration in pipelines of residential buildings. Existing methods for measuring vibrations in pipelines are considered and their effectiveness is analyzed. The assessment of complexity, cost efficiency and accuracy of the existing vibration measurement methods is made. An improved method for measuring the vibration of a pipeline section with a tap based on the analysis of the trace of blurring a circular mark and the Doppler Effect is proposed.

1. Introduction

Technological processes need to be diagnosed. One of the main parameters of the diagnostics of engineering systems of buildings is vibration. Increased vibration has a negative effect on process equipment and people. Vibration is measured by vibrometers and vibration analyzers.

Vibration measurement is carried out by contact and contactless methods. The simplest contact method, based on the use of piezoelectric sensors, is highly inertial and not suitable for small mass objects. The open resonator method is also not applicable for objects of small mass, but it has a small inertia with respect to piezoelectric sensors.

Contactless vibration measurement methods are based on the use of sound waves and visible light. The method of ultrasonic phase measurement has the speed of measurement, relative cheapness and high accuracy, but it is sensitive to the measuring environment, it has a strong attenuation of the frequency in the air and reduced accuracy with increasing frequency of vibration of the object under study.

Optical methods have high speed and accuracy of measurement; they do not affect the object under study. The main disadvantages of optical methods are the high cost and complexity of the equipment, high energy consumption.

Technological processes require appropriate monitoring of the state of both the process itself and the state of the process equipment. One of the main controlled parameters of the process being studied is often the amplitude of vibration (vibration displacement) and the frequency of vibration (vibration velocity).

Vibrations are mechanical fluctuations of an object. The full body vibration can be described by a set of six simplest types of motion: translational motion in three mutually perpendicular directions of Cartesian coordinates and rotational with respect to mutually perpendicular axes X, Y, Z. Therefore, it is possible to speak about six degrees of freedom of the object under study.

Object vibration occurs due to various excitation forces acting on it. These forces may occur inside the object (the piping system vibrates due to water pulsations) or may act on the object from the outside (the piping system vibrates due to the distribution of vibration from the process equipment to the pipeline attached to it). To obtain the necessary information about the vibrations of the system under study, it is necessary to analyze the main parameters affecting the occurrence of vibration: the disturbance force, its frequency and direction [1].

Increased vibration of pipelines of residential buildings has a negative impact not only on the state of engineering systems, but also on people nearby. Low-frequency noise can contribute to increased fatigue, the emergence of unreasonable feelings of anxiety, irritation, which in general has a negative impact on human health.

To determine the vibroacoustic characteristics of the piping system, it is necessary to measure the vibration during operation and compare the maximum values of the vibration parameters with the standard values.

Often vibration measurement is performed with low accuracy or has significant costs, for example, for the purchase of expensive vibration measuring equipment, therefore the development of vibration measurement methods is relevant.

2. Vibration diagnostics of pipelines of building engineering systems

The purpose of the study is to find the causes of vibration pipelines. The causes of increased vibrations are [2]:

- the coincidence of the natural frequencies of vibration with the frequencies of the pumped water;
- the pulsation of pumped water arising due to various mechanical and technological reasons;
- suboptimal hydrodynamic characteristics due to deficiencies in the design or installation of pipelines.
- the inadequate condition or location of pipeline support structures.

For a full study of the vibration of the pipeline section, it is necessary:

- to measure the basic vibration parameters: vibration velocity, vibration acceleration, vibration displacement [3];
- to determine permissible vibration values of the piping system. For residential buildings, the permissible vibration values are given in Table 1 [4];
- to calculate the natural frequency of the piping system;
- to determine the effect of measuring equipment on vibration and correct values.

When detecting increased vibrations of the pipeline system that differ in frequency from the natural frequencies of the pipeline system, it is necessary to conduct a full diagnosis. Vibration diagnostics of building engineering systems is a complex and time-consuming task, requiring expensive instrumentation and software products, which is often beyond the capacity of service organizations.

Today, in the field of vibration diagnostics, methods using vibration meters and vibration analyzers are widely used [5-10].

In vibrometers, primarily piezoelectric sensors are used as primary transducers, which are fixed at the site of the pipeline system. Modern vibrometers measure the values of vibration velocity, vibration acceleration and vibration displacement. They are capable of storing and transmitting data for further research. They are also able to measure high frequency vibrations.

Recommended limits for measuring parameters for vibrometers are [2]:

- the RMS value of vibration velocity in the range of 0.5 ... 100.0 mm/s and higher in the working frequency range of 10 ... 1000 Hz;
- the range of vibration displacement in the range of 1.0 ... 1000.0 microns and higher in the working frequency range of 3 ... 300 Hz;
- the peak value of vibration acceleration in the range of 1 ... 200 m/s^2 and higher in the working frequency range of 10 ... 3000 Hz.

One of the drawbacks of a vibrometers is the ability to measure vibration parameters at a specific point in time without the ability to capture changes in parameters over time. This deficiency is absent in vibration analyzers.

A vibration analyzer in comparison with a vibrometer is a more complex measuring instrument. In addition to all the functions of a vibrometer, a vibration analyzer is able to record vibration parameters in the dynamics and output them in a form convenient for further analysis: graphs, tables, etc. Also, a vibration analyzer, unlike a vibrometer, has other functions, for example, the ability to determine the required additional weight for balancing the system [3].

Vibration analyzers provide [2]:

- the measuring of vibration parameters: vibration velocity, vibration acceleration and vibration displacement with a permissible error;
- the calculating of the main coefficients: coefficient of kurtosis, coefficient of peak, etc;
- the determining of natural frequencies of oscillations of the measured pipeline sections;
- the determining of the amplitude-phase-frequency characteristics of vibration when opening / closing valves, equipment start-up, etc.;
- the registering of the change in the shape of the vibration signal in time;
- the analyzing, transmission and storing of measured data in the memory.

Table 1. Maximum permissible vibration levels in residential buildings

Geometric average frequency bands, Hz	Valid values for the X, Y, Z axes			
	Vibration accelerations		Vibration velocities	
	$m/c^2 \cdot 10^{-3}$	dB	$m/c \cdot 10^{-4}$	dB
2	4.0	72	3.2	76
4	4.5	73	1.8	71
8	5.6	75	1.1	67
16	11.0	81	1.1	67
31.5	22.0	87	1.1	67
63	45.0	93	1.1	67
Equivalent corrected values of vibration velocities or vibration accelerations and their logarithmic levels	4.0	72	1.1	67

3. Methods for measuring vibration of technological pipelines

Methods for measuring vibration are divided into two groups:

- the contact methods have a direct mechanical connection of the primary converter with the measured object;
- the contactless methods have not a mechanical connection with the object being measured.

3.1. Contact measurement methods

The simplest contact vibration measuring method is the use of piezoelectric sensors. This method is quite accurate in the range of low frequencies and relatively large amplitudes of vibration. Measurement of high-frequency and low-amplitude vibrations by piezoelectric sensors is impossible due to the high inertia of the transducer, which leads to distortion of the waveform. Also, piezoelectric sensors have a significant impact on the vibration parameters of the studied objects of small mass.

Some disadvantages of using piezoelectric sensors are absent when using the open resonator method. This method is based on measuring the ultra-high frequencies of the resonator, which are changed by the vibration of the object under study. The resonator consists of two mirrors, one of which has a mechanical connection with the object under study, the second is stationary. Measurement of small vibration amplitudes is performed by the amplitude method for changing the reflected or output power. The limit of measurement of this method is 3 microns. With respect to piezoelectric sensors, the open resonator method has low inertia. This method also has drawbacks: the mass of the mirror fixed on the object of study must be significantly less than the mass of the object itself, a constant power supplied to the resonator and a stable frequency of excitation is necessary [11].

3.2. Contactless measurement methods

The most contactless vibration measurement methods are based on the use of sound waves and visible light.

The method of ultrasonic phase measurement is based on measuring the current value of the phase difference between the reference and the reflected from the object of study signals. Signals are given at an ultrasonic frequency. Primary signal transducers are piezoelectric ceramics. The measurement limit of the vibration displacement of the object of study for this method at a frequency of 240 kHz ultrasound is 10 mcm at a distance from the object of 1.5 m. At increase in frequency of the original signal the sensitivity of this method increases. Among the advantages of this method, it is possible to specify a small measurement time, high measurement accuracy, compactness of equipment, relative cheapness, etc. The disadvantages include the dependence of ultrasound on the medium state through which it passes, the strong attenuation of the frequency in the air and the drop in the measurement accuracy with increasing frequency of vibration [12].

Optical methods are divided into holographic methods and methods based on the Doppler Effect.

The Doppler Effect is the basis of laser vibrometry. This is a modern method of measuring the vibration of the studied objects. The advantages of laser methods are:

- the high measurement accuracy;
- the ability to measure the vibration of objects of any mass;
- the remote measurement of research objects;
- the high measurement speed.

The main disadvantages of optical methods are the complexity and high cost of equipment. Also among the drawbacks, it is possible to specify the harmful effects on the staff, high energy consumption, special requirements for the surface of the object of study and the measurement environment.

4. Measurement of vibration parameters by the trace of image blur

Another optical method for measuring vibration is the method of measuring vibration parameters by the trace of a blurred image of a round mark [13, 14, 15]. This method consists of applying a round-shaped retroreflective mark to the object under study, forming a binary image of this mark without vibration and fixing the trace of vibrational blurring. A further comparative analysis of the geometrical parameters of the mark and the trace of the vibratory blur of the mark makes it possible to determine the modulus and direction of the amplitude vectors of the vibration displacement of the object under study.

The advantages of this method are the possibility of measuring the vibration parameters of objects of any mass, short measurement time, high accuracy, etc.

The main disadvantage of these methods [14, 15] is the impossibility of determining some important vibration parameters for analyzing the increased vibrations of pipelines, such as vibration velocity and vibration acceleration.

The improvement of this method for measuring the vibration parameters of a pipeline section with a tap consists of applying three reflective marks in mutually perpendicular planes, as shown in Figure 1. A light beam with a given frequency is fed to the reflectors. The device receiver registers not only the trace of the image blur, but also the frequency of the reflected flow. Registration must occur synchronously with the three tags.

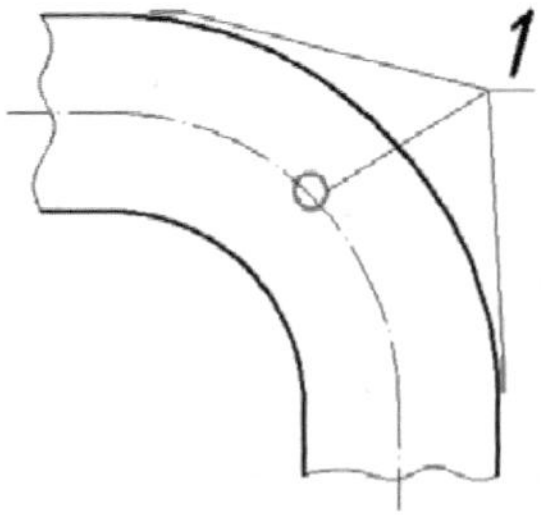

Figure 1. Reflective marks on the pipeline section
1 - reflective marks

The movement of the reflective marks relative to the device-receiver affects the frequency of the recorded signal of the reflected flow due to the Doppler Effect. Using this method, it is possible to calculate the vibration velocity of the object under study in three mutually perpendicular components and construct the amplitude-frequency characteristic of the vibration.

This method will provide all the necessary information about the vibration parameters of the pipeline section. Compared to the method of measuring vibration on the trace of blur of a round mark, it is more complex and expensive to perform.

5. Conclusion

Vibration control of high-rise building pipelines is an urgent task, since increased vibrations lead to accelerated wear of the piping system and process equipment, as well as the noise generated by vibration, have a negative impact on people.

Analyzed vibration measurement methods have several disadvantages. Contact methods affect the vibration parameters of the studied objects, especially low mass, which can lead to significant errors. Contactless methods are expensive and difficult to implement, but more accurate and fast.

The proposed method for measuring of the vibration of a pipeline section with a tap, based on the analysis of the trace of blurring of a round mark image and the Doppler Effect, makes it possible to determine all the pipeline vibration parameters necessary for the analysis. However, like other optical methods, it is quite complex in execution and expensive. This method requires experimental research and further improvement to reduce costs.

References

[1] Metody i sredstva izmereniya vibratsii [Methods and tools for measuring vibration] *Student library online* [Electronic resource] URL: https://studbooks.net/1757020/geografiya/metody_sredstva_izmereniya_vibratsii (the date of application: 10.05.2019)

[2] Mashinnoye oborudovaniye i tekhnologicheskiye truboprovody khimicheskikh, neftekhimicheskikh i neftepererabatyvayushchikh proizvodstv [Machinery and technological pipelines of chemical, petrochemical and oil refining industries] [Electronic resource] URL: https://pandia.ru/text/77/213/96819-2.php (the date of application: 10.05.2019)

[3] Kak vybrat' pribor dlya izmereniya vibratsii [How to choose a device for measuring vibration] 19.09.2018 [Electronic resource] URL: https://tze1.ru/articles/detail/kak-vybrat-pribor-dlya-izmereniya-vibratsii/ (the date of application: 10.05.2019)

[4] SanPiN 2.1.2.2645-10 Sanitary and epidemiological requirements for living conditions in residential buildings and premises (with changes from December 27, 2010)/ approved 10.06.2010 by Chief state health officer of the Russian Federation p. 17

[5] Bochkarev N N, Kurochkin A A and Andreev M I 2010 Diagnostika vibratsii magistral'nykh truboprovodov v sistemakh upravleniya dvizheniem vnutritrubnykh ob"ektov [Diagnostics of trunk pipeline vibrations in control systems of objects movement inside pipeline] *Doklady TUSUR* **2** **(22)** 209-11

[6] RD 153-34.0-20.340-98 Metodicheskiye ukazaniya po kontrolyu za sostoyaniyem metallicheskikh napornykh truboprovodov gidroelektrostantsiy [Guidelines for monitoring the status of metal pressure pipelines of hydroelectric power plants] / *developed by the OAO "ORGRES" and entered into force on 01.06.99* p. 20
[7] Lepesh G V, Kurtov V N, Motylev N G, Krasilnikov A Yu, Telyatnikov S V and Chilipenko A L 2009 Operativnyy kontrol' i diagnostika oborudovaniya [Operational monitoring and diagnostics of equipment] *Technical and technological problems of service* **3 (9)** 8-16
[8] Zaitsev N L and Sukhanov V I 2014 Sovershenstvovaniye metodiki vibrodiagnosticheskogo kontrolya truboprovodnykh obvyazok oborudovaniya [Improvement methods of vibrodiagnostic control of pipeline bindings of gas main equipment] *Bulletin of the South Ural State University Series "Metallurgy"* **14 (1)** 89–92
[9] Petrukhin V V and Petrukhin S V 2010 Osnovy vibrodiagnostiki i sredstva izmereniya vibratsii [Fundamentals of vibration diagnostics and vibration measurement tools] (Vologda: Infra-Engineering) p.168.
[10] Klyuev V V 2005 Nerazrushayushchiy kontrol' [Non-Destructive Testing] Vol. 7 Book 2 (Moscow: Mashinostroenie) p.829
[11] Metody izmereniya vibratsii [Vibration Measurement Methods] *World of knowledge* [Electronic resource] URL: http://mirznanii.com/a/172855/metody-izmereniya-vibratsii (the date of application: 10.05.2019)
[12] Kochegarov I I and Moiseev S A 2015 Beskontaktnoye izmereniye vibratsii [Contactless Vibration Measurement] *Proc. of the Int. Symp. "Reliability and Quality"* **2** 301-3
[13] Grigoriev A V, Danilova E A, Brostilov S A, Naumova I Yu, Lapshin E V and Baranov A A 2015 Struktura metodiki izmereniya parametrov vibratsiy po sledu razmytogo izobrazheniya krugloy metki [Structure of the method for measuring vibration parameters following the blurred image of a round mark] *Proc. of the Int. Symp. "Reliability and Quality"* **2** 13-6
[14] Drzhevetsky A L, Yurkov N K, Grigoriev A V, Zatylkin A V, Kochegarov I I, Kuznetsov S V, Drzhevetsky Yu A and Derkach V A 2013 Patent 2535522 of the Russian Federation IPC G01H9 /00 Sposob izmereniya vibratsiy [Method of measuring vibrations]
[15] Drzhevetsky A L, Yurkov N K, Grigoriev A V, Zatylkin A V, Kochegarov I I, Kuznetsov S V, Drzhevetsky Yu A and Derkach V A 2013 Patent 2535237 of the Russian Federation IPC G01H9/00 Sposob izmereniya vibratsiy [Method of measuring vibrations]

Variable boundary conditions for solving the problem of dispersed flow moving

A Paramonov[1], L Mostovenko[1]

[1]Omsk State Technical University, Power Engineering Institute, 11 Mira Ave., 644050, Omsk, Russia

E-mail: teploblv@mail.ru

Abstract. The paper presents comparing the results of numerical and full-scale experiments which conducted on an inertial-vacuum dust collector (IVDC). In September 2018, a full-scale experiment was held at the Omsk heat power station №4, it was allowed to verify the calculated data in ANSYS CFX. The boundary conditions were corrected after the initial calculations on IVDC of second generation. The results of numerical experiments, that conducted earlier, were corrected. In article it is spoken in detail about comparison of the dependences of the individual elements heights and their influence on the installation collection efficiency

Keywords: laden flow, ivdc, inertial dust collector, ansys cfx, relaminirization.

1. Introduction

There are many ways to capture the ash. Despite this, the question of ash collector creating with a highly collection efficient, a long service life without going into repair remains unresolved. The development of a new method of ash collection is an extremely time-consuming work, complicated by the creation of cumbersome expensive experimental facilities, carrying out field experiments on existing steam generators. The most appropriate creation variant are preliminary studies of newly created devices based on numerical calculations using programs such as ANSYS, which allows for a full-scale numerical experiment.

The inertial vacuum dust collector (IVDC) [1] is an ash collecting device designed for improving collection efficiency from thermal power plants.

According to Figure 1, the inertial-vacuum dust collector contains a vertically positioned two-stage body *1*, the lower stage of which is an ash-collecting chamber (ACC) *6*, and the upper one is intended for ash separation and it is made in the form of two coaxially arranged cylindrical shells *2*,*3*. The inner cylindrical shell *2* serves as an inlet pipe *4* for supplying the flue gases, and the annular space between the two shells is a channel *5* for exhausting the cleaned flue gases. The ash separator also contains a turning camera (TC) *8* in the lower part of the separating stage informing the output from the channel *4* to the channel of the outlet nozzle *5*. TC *8* is made with rings *12* on the crosspiece *13*, above which a cone-shaped divider *9* is installed along the axis of the housing *1*. The lateral surface of the latter, together with the lower part of the inner shell *2*, forms a confused output nozzle *10* of channel *4*. The inlet section *11* of the annular channel *5* is made diffuser. The height h of the divider *9* is 0.5 ... 0.8 of height H of the inner shell *2*, and the angle α of narrowing confused nozzle *10* is equal to the angle of increase efficiency degree β of expansion of the diffuser section *11* of the channel *5* of exhaust gases and is $\alpha = \beta = (15 \ldots 20)°$.

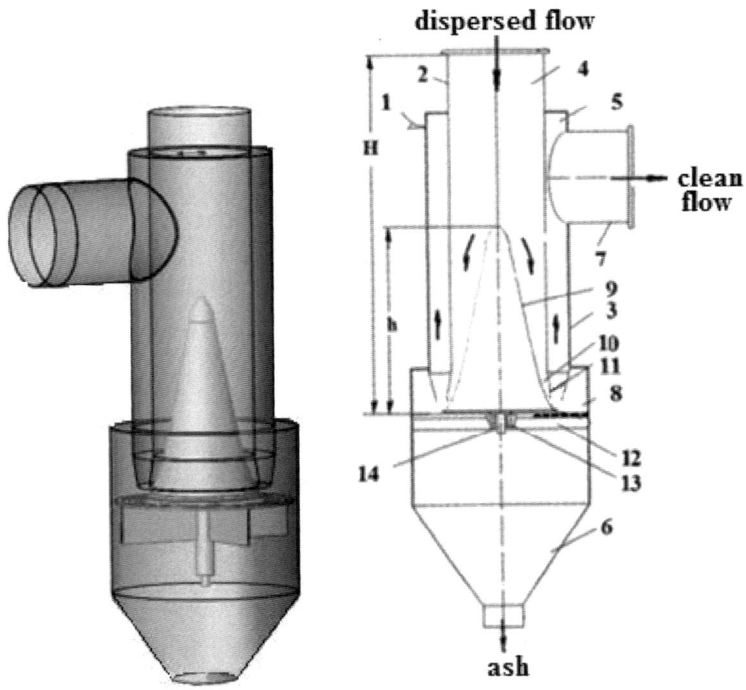

Figure 1. Inertial vacuum ash collector of the second generation. *1* - two-stage case, *2*, *3* - cylindrical shells, *4* - inlet, *5* - channel for purified flue gases, *6* - ash-collecting chamber (ACC), *7* - outlet, *8* - turning camera TC, *9* - divider, *10* - confused output nozzle, *11* - inlet diffuser section, *12* - rings, *13* - retaining rings, cross, *14* - axis.

The operation principle [12] of the IVDC: gas with ash particles through the inlet *4* in the guide arrow enters the ash collector, then moves along the air channel and exits the accelerating nozzle *10*. The particles acquire kinetic energy along the divider *9*, by inertia move into the rotating chamber *8* and due to force gravities are deposited in the bunker *6*. Particles with only the smallest diameters are not deposited. In the TC *8* an aerodynamic trap [11] is formed - particles have the opportunity to get into it, but they cannot exit. In consequence, they are completely deposited in the bunker. Thus, particles can enter the channel, and cannot exit.

2. Methods

In September 2018, in the Omsk region in the JSC "TGC-11" the joint venture "CHP-4" a full-scale experiment was conducted with the IVDC on the boiler BKZ-320-140. In the course of the experiment the measurements were made indicated in Figure 2. The main measurements allow to verify the turbulence mathematical model of using the IVDC: the dust concentration in flow before and after the IVDC; volumetric gas flow before and after IVDC; temperatures on inlet and outlet; pressure drop.

IOP Conf. Series: Materials Science and Engineering **972** (2020) 012071 doi:10.1088/1757-899X/972/1/012071

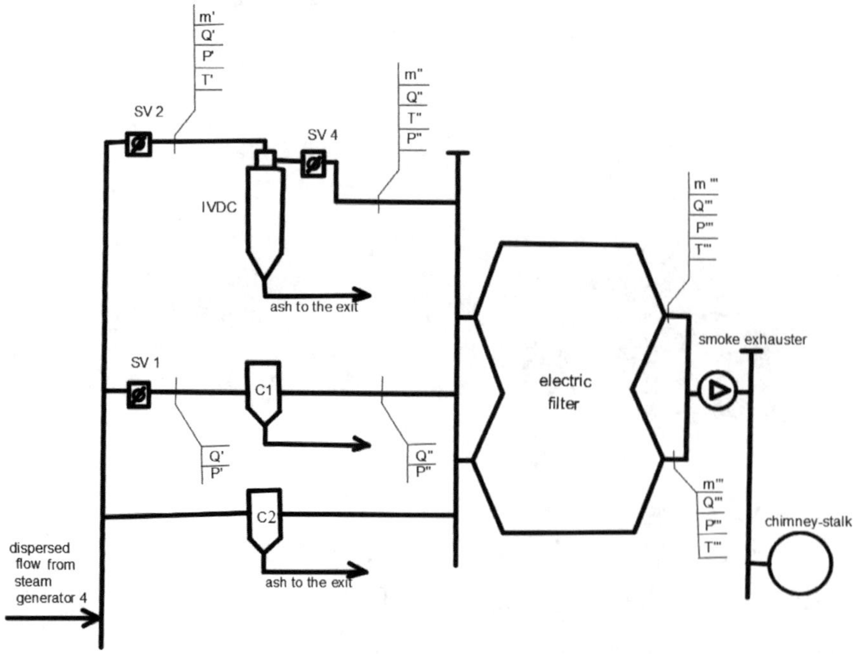

Figure 2. Scheme of ash collector with IVDC. IVDC – inertial vacuum dust collector, C 1,2 – cyclones, SV 1,3,4 – stop valves

The research methodology used in this article was carried out according to Regulatory guide 153-34.1-27.301-2001 «Test method of ash collecting plants of thermal power and boilers plants». This technique can be used for testing various types of dust collectors, provided that the temperature, pressure and polluting components concentration of the cleaned gases are similar to those of flue gases of boilers or atmospheric air. The measuring section in the gas duct is selected in accordance with the requirement of GOST 17.2.4.06, GOST 17.2.4.07, GOST R 50820. Measured sections should be located on straight, preferably vertical, sections of constant-flow ducts before and after ash collectors. In these areas there should be no ash deposits. When zero-type pneumometric tubes are used and dust sampling probes are used for measurement, the length of a straight section of constant cross section must be at least four hydraulic diameters of the flue. Measurements are carried out in a section dividing the plot in a 3: 1 ratio in the direction of gas flow. Sections of circular cross section are preferable to square, and square should be rectangular. The IVDC has a non-standard design, and its installation had a limited space on which the cyclone was previously installed. It was impossible to comply with one of the conditions during the test - to arrange dimensional sections, at a distance of not less than four hydraulic diameters of the flue. In fact, this distance is about two hydraulic diameters. Such a deviation is due to the fact that IVDC has the opposite position of the inlet and outlet, rather than the cyclone.

The ash mass that enters the ash collecting plant with gas, as well as contained in the gas emitted into the atmosphere after cleaning. It can be determined by sampling dusty gas using dust collectors of various types, respectively, before and after the ash collecting plant.

A representative sample of the ash mass of a dusty gas can be taken under the condition that the gas velocities are equal in the sampling channel of the probe (the gas duct at the sampling point). To do this, the sampling sites must be correctly selected according to the "Methodology" and GOST 50820, and the work with the probe is properly organized.

The cross section of the flue is divided into equal elementary. Samples are taken at all scheduled points located in the center of these sites. The gas suction time from each point must be the same (at least 0.5 min).

The total gas sampling time from all the planned points of the cross section of the flue is chosen so that the ash mass collected by the dust extraction probe is at least 0,1 of the mass of the clean filter. Selection is considered complete when samples of dusty gas are taken at all scheduled points.

Such ash sampling was complicated by the fact that the velocity at the exit section of the IVDC was unevenly distributed. It is in the range of 5–20 m/s, which affects the partial sedimentation of particles in the outlet section of the ash collector. The last factor affects the efficiency of the ash collector.

3. Results and Discussion

The data of the field experiment were obtained by conducting experiments [4] with the position of the divider at elevations, as shown in Figure 3. The divider was lifted, the installation resistance was greater [5], the exhaust gas flow rate was smaller, the velocity of the gases in the narrow bore was greater and ash collection efficiency is higher.

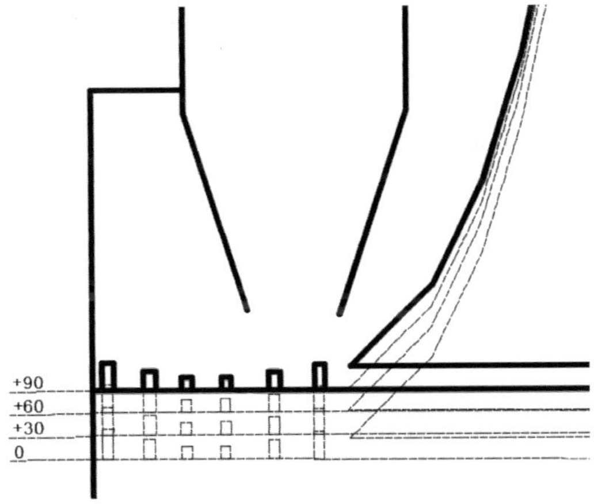

Figure 3. Four marks for a full-scale experiment.

According to the results of the obtained data, the numerical experiments were recalculated, on the basis of which the design of the ash collecting apparatus was carried out. Previously, the boundary conditions were the differential pressure, speed, concentration of ash at the entrance. With the new numerical calculations, the inlet boundary conditions became the mass flow rate of gases 150000 m^3/h, the ash concentration - 70 g / m^3; and the outlet pressure is 0 Pa. The program was calculated the speed of the dusty flow through the flow part. The calculations were performed in the ANSYS CFX software module [6-10] using the control volume method.

The recalculation was made according to the values obtained after changing the height of the ACC (Fig. 4) and the divider height (Fig. 5).

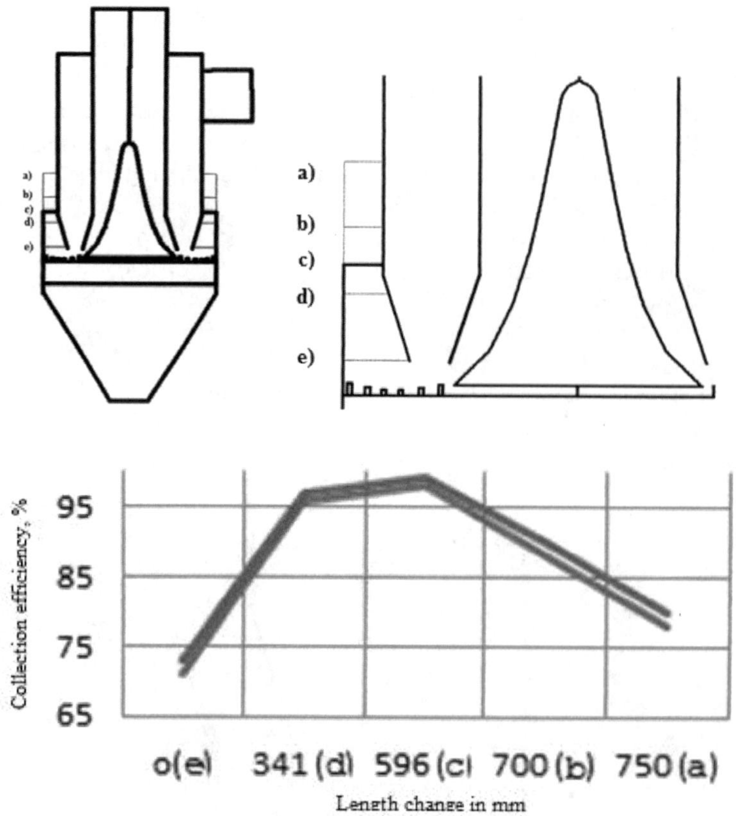

Figure 4. Five positions of the height of the ACC, that the calculations were carried out.

From the graph in Figure 4 it can be seen that a change in the height of the ACC indicates the an extremum presence in the graph after which it is not advisable to increase the height of the camera. Thus, the increase in height is unjustified and carries only the useless consumption of metal.

The good collection efficiency in the case of: it (c) can be explained by the fact that the kinetic energy gained by the passage of the particle along the divider profile, it enough to pass the TC, lose energy in the TC and collapse due to gravity into the ash bunker.

When the results (a) and (e) were unsatisfactory. The lack of TC is resulted in particles moving at a speed of 60.8 m / s directly to the wall of the bunker, in connection with which increased abrasive surface wear could occur, or most of the particles moved towards the exit. TC element is needed to improve the quality of the IVDC, because its presence improves the installation collection efficiency by 30% from option (d).

The divider [3], the different heights of which are depicted in Figure 5, is a very important element of the IVDC. The flow relaminization is achieved because of it [2]. If its height is small, large velocity gradients aren't appeared, ripple smoothing doesn't occurs. Increasing the height of the divider leads to smoothing of possible flow pulsations. The same height is important for an uniform set of speed of the particles, and their subsequent inertial separation from the carrier phase.

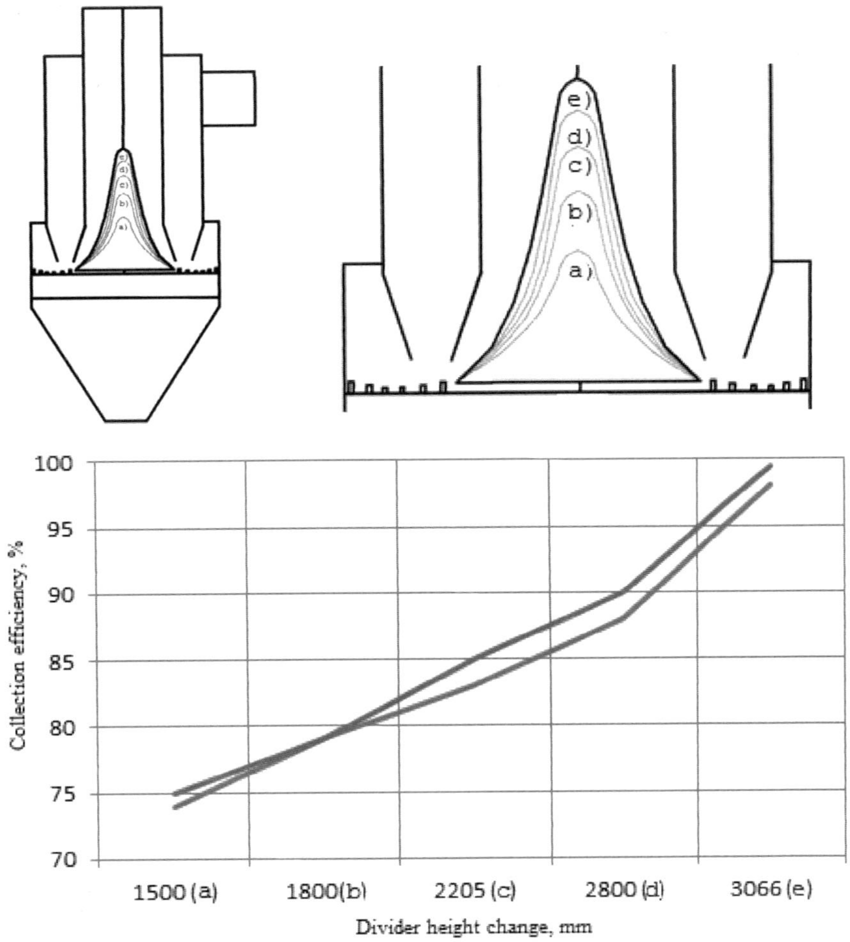

Figure 5. Five marks for changing the height of the divider and its influence on collection efficiency.

It depends on the divider height whether the stream coming from the entrance will gain the necessary speed of about 60 m / s and two independent streams: ash and gas will be divided.

Figures 4 and 5 show 2 dependences: in red and blue colors. Red color dependence is data of a numerical experiment on primary boundary conditions, blue color dependence is data of a numerical experiment with verified secondary boundary conditions. According to the new results, the efficiency in both cases becames lower due to the influence maximum speed in narrow bore.

4. Conclusions

The results of the numerical experiment were verified with respect to the full-scale experiment. Changes have been made to them. Previously obtained capture efficiency values are recalculated based on these data. They became reduced by 1-3%. Maximum installation efficiency 95%. The drop in efficiency is justified by changing the speeds along the flow along the divider, the pressure drop been recalculated.

The data presented in this article are given taking into account the errors arising during the tests.

Research on the ash collector development can be continued further. You can use the additional effect that was obtained on the inertial-vacuum dust collector of the first generation. This ash collector was tested in Omsk in 2008 and was showed efficiency in the range of 83-90%. To date, it has been possible to improve the design of the apparatus flow-through part, so that the efficiency has become 93% without the possible discharge effect.

Additional studies of the movement of the guide vane in the vertical plane can show qualitative and quantitative changes in the flow rate and ash particles collection, allow to establish the real range for the occurrence of relamininization.

It is possible to make holes in the divider to create an additional effect of discharge.

It is also necessary to analyze the possible abrasive wear of the internal surfaces of the installation, their influence on the efficiency of the ash collecting device and to assess the durability of the installation.

Acknowledgments
The authors are grateful to Public Joint Stock Company «Inter RAO UES» «Energy without borders» found for sponsoring contractual theme №15001 «Development, designer supervision of installation and participation in testing a full-scale inertial vacuum dust collector».

References
[1] Beloglazova L V 2017 Inertial-vacuum dust collector for gas cleaning (Moscow) Patent 175570
[2] Beloglazov V P 2018 To the question of «relanimirization» (Omsk: Omsk Scientific Bulletin) **4** 64–67
[3] Beloglazov V P 2016 The influence of the divider height on the capture efficiency of Ekibastuz coal ash in inertial vacuum dust collector (Moscow: Safety & Reliability of Power Industry) **1** 46–49
[4] Regulatory guide 153-34.1-27.301-2001 2003 Test method of ash collecting plants of thermal power and boilers plants (Moscow: Joint Stock Company «All-Russia thermal engineering institute)
[5] Loitsyansky L G 2003 Fluid and gas mechanics (Moscow: Drofa)
[6] Florian Meter 2011 Turbulence Modeling for Engineering Flows: ANSYS CFX.
[7] Durbin P A 2011 Statistical Theory and Modeling for Turbulent Flows (John Wiley & Sons)
[8] Dr. William L O 2004 Verification and Validation in Computational Science and Engineering, (Salt Lake City : Hermosa Publishers)
[9] Spalart P R 2000 Strategies for turbulence modeling and simulations *Int. J. Heat Fluid Flow* **21** 252-263
[10] Keis V M 1972 Convective heat and mass transfer (Moscow: Energiya)
[11] Varaksin A Yu 2003 Turbulent gas flow with solid particles (Moscow: Fizmatlit)
[12] Ziganshin M G 2014 Design of dust removal machines (Moscow: Lan')
[13] Altukhov Yu A 2012 Research of aerodynamics and capture efficiency in labyrinth ash collector *Kuzbass State Technical University Bulletin* **1** 97–101

The Heating System Capacity Calculation of the Cultivation Facilities

N Shiryaeva[1], O Mitroshina[1] and N Kharkova[1]

[1]Ural Federal University, 19 Mira Str., Ekaterinburg 620002, Russia

E-mail: n.p.shiriaeva@urfu.ru

Abstract. The article is devoted to the heating system capacity calculation of the cultivation facility, based on the conditions for ensuring the microclimate parameters regulated by the standards of technological design, depending on the external climatic conditions and the facility design parameters. One of the distinguishing features of the heating system capacity calculation of greenhouses is that the heat balance equation of the structure must be considered together with the equations of heat and mass transfer on the soil and covering constructions. The calculation is made for the greenhouse block № 3 in The Greenhouse Complex with a production area about 12 ha in the village of Sadovy, Ordzhonikidze District, Ekaterinburg. It takes into account convective and radiant heat exchange between the soil and the covering constructions, heat and mass exchange during evaporation and condensation and heat loss for heating infiltrating air.

Keywords. Cultivation Facility; Microclimate; Heat and Mass Transfer; Heating System Capacity.

1. Introduction

The plant's vital activity directly depends on the microclimate parameters. Therefore, it is necessary to maintain a certain level of air and soil temperature, as well as relative humidity in the cultivation facilities work zone.

It is analytically impractical to take into account just transmission heat loss in the heating systems design practice of the cultivation facilities, because the airspace, soil surface and covering constructions heat exchange must be considered.

The below calculation establishes an analytical relationship between the required parameters of the microclimate inside the greenhouse and the required heat flows for their provision.

The main feature of the cultivation facilities is their close connection with the external meteorological conditions, especially with the intensity of solar radiation. However, this calculation is made for the night period, as it is the most unfavorable in terms of heat loss, and, therefore, the load on the heating system is the peak.

2. Object of study

The greenhouse block № 3 in The Greenhouse Complex with a production area about 12 ha in the village of Sadovy, Ordzhonikidze District, Ekaterinburg, was chosen as the object of study. This facility is a block-type greenhouse, shown in Figure 1, and is a combination of several duo-pitch links (sections) under one roof, where the walls between the links are replaced by supporting poles. Covering constructions are made of: walls - double glazing, slopes - single glazing.

Figure 1. The greenhouse block № 3 in The Greenhouse Complex with a production area about 12 ha in the village of Sadovy, Ordzhonikidze District, Ekaterinburg.

The following assumptions were made when drawing up the design scheme for the greenhouse block energy balance, shown in Figure 2:

- Construction night mode;
- Mass-exchanging processes take places only in the work zone;
- The screen (biomass) is not taken into account, due to the decline of radiant heat flow from the soil to the covering constructions;
- The heat and mass-exchanging processes conditions inside the building are stable.

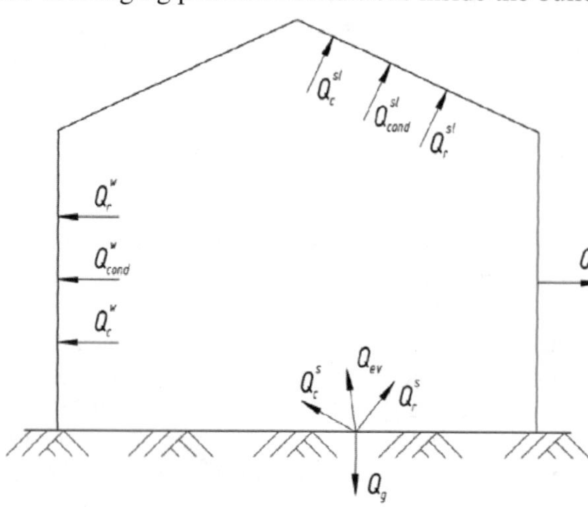

Figure 2. Heat flows scheme in the greenhouse for the night mode

The heat balance equation for the greenhouse airspace is:

$$Q_h = Q^s + Q^{con} + Q_{inf}, \tag{1}$$

where Q_h is the installed capacity of the greenhouse heating systems, *W*; Q^s is the ground mass heat loss, *W*; Q^{con} is the heat loss through the covering constructions, *W*; Q_{inf} is heat loss by the infiltration, *W*.

The heat amount for heating soil is:

$$Q^s = Q_g + Q_c^s + Q_r^s + Q_{ev}, \tag{2}$$

where Q_g is the heat loss through the soil massif, *W*; Q_c^s and Q_r^s are respectively, convective and radiant heat flows from the soil to the air in the work zone, *W*; Q_{ev} is heat flow characterizing the heat expenditure for moisture evaporation from the soil, *W*.

The heat amount for heating covering constructions is:

$$Q^{con} = Q_c^{con} + Q_r^{con} - Q_{cond}^{con} = Q_c^{sl} + Q_r^{sl} - Q_{cond}^{sl} + Q_c^w + Q_r^w - Q_{cond}^w, \tag{3}$$

where Q_c^{sl} and Q_r^{sl} are heat flow as a result of heat exchange by convection of the inner surface, respectively, of slopes and walls with air in the work zone, *W*; Q_r^{sl} and Q_r^w are radiant heat flow from the inner surface, respectively, of the slopes and walls, *W*; Q_{cond}^{sl} and Q_{cond}^w are heat flow characterizing the heat released during the condensation of steam on the inner surface, respectively, of the slopes and walls, *W*.

The equation of the heating system capacity calculating of the greenhouse with regarding formulas (2) and (3) is:

$$Q_h = Q_g + Q_c^s + Q_{ev} + Q_c^{sl} + Q_r^{sl} - Q_{cond}^{sl} + Q_c^w + Q_r^w - Q_{cond}^w + Q_{inf}, \tag{4}$$

2.1. Convective heat exchange

The heat amount received (given away) by covering constructions and soil by convection is determined as:

$$Q_c = \alpha_c \cdot F \cdot \Delta t, \tag{5}$$

where α_c is convective heat transfer coefficient, $W/(m^2 \cdot °C)$; F is the heat exchange surface area, m^2; Δt is the difference between the work zone and the heat exchange surface air temperatures, $°C$.

According to [2], turbulent air movement is observed in greenhouses near all surfaces because of their surfaces large sizes of the soil, walls, slopes, and significant temperature differences between covering constructions and air temperature, therefore, the convective heat transfer coefficient does not depend on the size and looks like:

$$\alpha_c = A \cdot m \cdot \sqrt[3]{\Delta t}, \tag{6}$$

where A is the coefficient for the convective heat transfer calculation (is a function of the heat exchange surface temperature and the ambient air, taken from the data given in [4]; m is the coefficient for the convective heat transfer calculation taken according to [4] for vertical surfaces m = 1, for vertical surfaces facing the heated side up m = 1.3, for slopes it is determined by the formula:

$$m = \frac{F_w + 0.7F_{sl}}{F_w + F_{sl}} \tag{7}$$

2.2. Radiant heat exchange

Heat transfer by the radiation of the soil surface to the covering constructions is determined using the Stefan-Boltzmann law with the replacement of the fourth-degree difference by the value difference in the first degree [3]:

$$Q_r = C_r \left[\left(\frac{\tau_0 + 273}{100} \right)^4 - \left(\frac{\tau_{con} + 273}{100} \right)^4 \right] F_0 \approx 0{,}814 C_r F_0 \left(\tau_0 - \tau_{con} \right), \tag{8}$$

where C_r is reduced emissivity; τ_0 is the soil temperature, $°C$; τ_{con} is the temperature of the greenhouse covering constructions, $°C$.

$$C_r = \frac{5{,}67}{\frac{1}{\varepsilon_0} + \frac{F_0}{F_{con}}\left(\frac{1}{\varepsilon_{con}} - 1\right)}, \tag{9}$$

where ε_0 and ε_{con} are the emissivity factor, respectively, of the soil and covering constructions; F_0 and F_{con} are the surface area, respectively, of the soil and covering constructions, m^2.

2.3. Heat and mass exchange during evaporation and condensation

According to [2], 90% of the water used for irrigation evaporates from the soil. In this case, assuming the evaporation uniformity, the heat amount spent on the evaporation processes from the soil is calculated by the formula:

$$Q_{ev} = 0.25 \cdot G_{w.c.} \cdot r_l, \tag{10}$$

where $G_{w.c.}$ is the irrigation water rate, l/h; r_l is the specific latent heat of evaporation, respectively, taken at the temperature of the soil.

According to [3], the amount of moisture obtained by condensation of water vapor is:

$$Q_{cond} = 0.278 \cdot G_{cond} \cdot r_l, \tag{11}$$

where G_{cond} is the amount of moisture condensed on the covering construction surface; l/h.

$$G_{cond} = 2.2 \cdot 10^{-6} \cdot (t_d + 273)^{0{,}42} \cdot \left(\frac{t_{wz} - \tau_{con}}{v^2}\right)^{0.33} \cdot \Delta\varepsilon \cdot F_{con}, \tag{12}$$

where t_{wz} is the temperature in the greenhouse working zone, $°C$; t_d is the determining temperature, $°C$, equal to $\frac{t_{wz}+\tau_{con}}{2}$; $\Delta\varepsilon$ is the difference in volumetric vapor content in air, %; v is the coefficient of kinematic viscosity of air, m^2/s, accepted at t_d.

$$\Delta\varepsilon = 100 \cdot \left(\frac{\varphi_{wz} \cdot e^s_{t_{wz}} - e^s_{\tau_{con}}}{P_n}\right), \tag{13}$$

where φ_{wz} is relative air humidity in the work zone, %; $e^s_{t_{wz}}$ and $e^s_{\tau_{con}}$ are the elasticity of saturated water vapor at t_{wz} and τ_{con}, kPa; P_n is the atmospheric pressure under normal conditions, kPa.

2.4. Heat loss through the soil massif

Heat loss through the soil massif is determined by the formula:

$$Q_g = \frac{t_g - t_{ext}}{R_0^{av}} F_0, \tag{14}$$

where t_g and t_{ext} are, respectively, generalized and external air temperature, $°C$; R_0^{av} is area-weighted average resistance to the soil heat transfer, $(m^2 \cdot °C)/W$.

$$t_g = \frac{133.9 - C_r\left|55 - \left(\frac{\tau_{con} + 273}{100}\right)^4\right|}{C_r + 8.6} \approx \frac{133.9 - C_r|55 - (0.814\tau_{con} + 55.55)|}{C_r + 8.6}, \tag{15}$$

$$R_0^{av} = \frac{2F_{1,0}2.11 + 2F_{2,0}4.3 + 2F_{3,0}8.6 + F_{4,0}14.2}{F_0}, \tag{16}$$

where $F_{1,0}, F_{2,0}, F_{3,0}, F_{4,0}$ are the greenhouse soil surface, adjacent to the covering constructions and with a width of 2 m or 2, 4 and 6 meters over, respectively, m^2.

IOP Conf. Series: Materials Science and Engineering **972** (2020) 012072 doi:10.1088/1757-899X/972/1/012072

2.5. Heat loss for heating infiltrating air

In cultivation facilities, infiltrating air flows mainly through the rabbet ledge of the windows and bay windows and depends on their length and width, wind speed and the difference between the densities of external and internal air [5]. The heat amount required to heat the infiltrating air is determined by the formula:

$$Q_{inf} = \frac{m_{inf} \cdot c \cdot (t_{wz} - t_{ext})}{3600}, \quad (17)$$

where m_{inf} is the air mass entering through the rabbet ledges, *kg/h*; *c* is air specific heat, $J/(kg \cdot °C)$.

$$m_{inf} = \sum (m'la), \quad (18)$$

where m' is the mass of air entering through the 1 *m* rabbet ledge depending on the design wind speed, $kg/(m \cdot h)$ [5]; l is the length of the rabbet ledge, *m*; a is correction coefficient taken 0.5 for windows and bay windows [5].

2.6. Calculation procedure

Initial data: $t_{ext} = -32°C$; $t_{wz} = 15°C$; $\tau_0 = 20°C$; $\varphi_{wz} = 60\%$; $\varepsilon_0 = 0.96$; $\varepsilon_{con} = 0.92$; $F_0 = 40785\ m^2$; $F_w = 4289\ m^2$; $F_{sl} = 45518$ м^2, the temperature on the walls and slopes surfaces, respectively, $\tau_w = -2{,}5°C$ and $\tau_{sl} = -18°C$.

The calculation process is shown in Table 1:

Table 1. Heating system capacity calculation of the cultivation facility.

Formula number	Symbol	The calculation result	Dimension	Notes
6	α_c^s	3.69	$W/(m^2 \cdot °C)$	$A = 1.66$ by $t_d = 17.5\ °C$, $m = 1.3$; $\Delta t = 5°C$
5	Q_c^s	752510	W	$\Delta t = 5°C$
6	α_c^w	4.36	$W/(m^2 \cdot °C)$	$A = 1.68$ by $t_d = 6.3\ °C$, $m = 1$; $\Delta t = 17.5°C$
5	Q_c^w	327250	W	$\Delta t = 17.5°C$
7	m	0.73		
6	α_c^{sl}	3.98	$W/(m^2 \cdot °C)$	$A = 1.7$ by $t_d = -1.5\ °C$, $m = 0.73$; $\Delta t = 33°C$
5	Q_c^{sl}	5978335	W	$\Delta t = 33°C$
9	C_r^w	3.03		$\varepsilon_0 = 0.96$; $\varepsilon_{con} = 0.92$
8	Q_r^w	2266650	W	
9	C_r^{sl}	5.06		$\varepsilon_0 = 0.96$; $\varepsilon_{con} = 0{,}92$
8	Q_r^{sl}	6389045	W	
10	Q_{ev}	1264840	W	$G_{w.c.} = 2040\ l/h$
13	$\Delta\varepsilon^w$	0.52	%	$e^s_{t_{wz}} = 1.71\ kPa$; $e^s_{\tau_{con}} = 0.50\ kPa$
12	G^w_{cond}	216/5	*l/h*	$\nu = 13.83 \cdot 10^{-6}$ by $t_d = 6.3\ °C$
11	Q^w_{cond}	148420	W	
13	$\Delta\varepsilon^{sl}$	0.89	%	$e^s_{t_{wz}} = 1.71\ kPa$; $e^s_{\tau_{con}} = 0.12\ kPa$
12	G^{sl}_{cond}	4975/0	*l/h*	$\nu = 13.15 \cdot 10^{-6}$ by $t_d = -1.5\ °C$
11	Q^{sl}_{cond}	3410639	W	
15	t_g^w	11.1	°C	$C_r^w = 3.03$
16	R_0^{av}	13.67	$(m^2 \cdot °C)/W$	$F_{1,0} = 1704\ m^2, F_{2,0} = 1672\ m^2,$

IOP Conf. Series: Materials Science and Engineering **972** (2020) 012072 doi:10.1088/1757-899X/972/1/012072

Formula number	Symbol	The calculation result	Dimension	Notes
				$F_{3,0} = 1640\ m^2, F_{4,0} = 35770\ m^2$
14	Q_g^w	128615	W	
15	t_g^{sl}	4.6	$°C$	$C_r^{sl} = 5.06$
14	Q_g^{sl}	109080	W	
18	m_{inf}	123354	kg/h	$m' = 4.95\ kg/(m \cdot \text{h})$; $l = 49840\ m$ $a = 0.5$
17	Q_{inf}	1610455	W	$c = 1000\ J/(kg \cdot °C)$
4	Q_h	15267721	W	

3. Conclusion

Thus, the heating system capacity of the greenhouse block № 3 in The Greenhouse Complex with a production area about 12 ha in the village of Sadovy, Ordzhonikidze District, Ekaterinburg, is 15.3 MW, while the transmission heat loss is 13.9 MW. Therefore, the difference between the capacities is 10%. It shows this calculation allows us to more accurately determine the heating system capacity of the cultivation facility.

Acknowledgments

The authors wish to thank The Greenhouse Complex in the village of Sadovy, Ordzhonikidze District, Ekaterinburg, for its precious contribution and help.

References

[1] Rulebook Document 50.13330.2012 Thermal Performance of the Buildings (Moscow: Ministry of Regional Development of Russia) 100 p

[2] Egiazarov A G 1981 Heating and ventilation of buildings and structures of agricultural complexes (Moscow: Stroiizdat) 239 p

[3] Methodological manual on the thermal calculation of cultivation facilities 1971 (Oryol: Glavselstroyproekt) 148 p

[4] Kutateladze S S and Borishansky V M 1959 Reference book on heat transfer (Moscow: Humanitarian and Ecological Institute) 414 p

[5] Andreevsky A K 1982 Heating e (Minsk: Higher School) 364 p

Simulation gas combustion process in modern heat generators of small and medium power

M Valeev[1], A Dyudina[1], A A Fatikhov[1], M Ziganshin[1,2]

[1]Department Thermal Power Engineering, Gas Supply and Ventilation, Kazan State University of Architecture and Engineering, 1 Zelenaya Street, Kazan, 420043, Russia

[2]Department Thermal Power Plants, Kazan State Power University of Architecture and Engineering, 1 Zelenaya Street, Kazan, 420043, Russia

E-mail: valeyev76@mail.ru

Abstract. The paper presents the problems of numerical modelling of burners and fireboxes the heat generators. The designs household generators placed directly on the premises, in the last decades, for the goal of commercial appeal, tend become to more and more compact, which leads to a reduction of the firebox to a burn zone's volume ratio. The numerical study of combustion processes using computational fluid dynamics (CFD) was carried out on the Unimat UT-L18 "Bosch" boiler, the gas heating convector "FEG" Beata 2, and the Vitodens 100-W "Viessmann" boiler. Computations were based on a one-stage combustion process of methane, both mixes with air and oxygen. Geometric models of fireboxes were developed by corresponding to their constructive features, the necessary boundary and initial conditions of furnace processes were determined, temperature, aerodynamic and concentration characteristics of gas fuel combustion processes were also presented. The models provided an opportunity for the quantitative analysis of furnace and burner devices of this heat generators degree of perfections. Also, the completeness of the oxidation of combustible components of the fuel-air mixture in fireboxes, differing in the degree close walls to burn zone, was compared. The obtained results allow to perform the improvement of the furnace and burner devices of domestic heat generators for using at design energy-efficient structures of decentralized heat systems, including individual heat supply in the residential and industrial building.

Keywords: computational fluid dynamics, domestic heat generators, energy saving, combustion process natural gas.

1. Introduction

For decentralized (including individuals) heating of buildings and individual premises, modern types of low and medium capacity heat generators of foreign and domestic manufacturers are used widely. In each new series of such devices are introduced constructive innovations, sometimes - outstanding. Manufacturers readily disclose only the parameters that give the commercial attractiveness of products, and the specific features of the innovative components responsible for improving the consumer properties of the product and often protected as intellectual property are not given. For example, the efficiency of the devices of the last decade is always high, for condensing boilers it is

IOP Conf. Series: Materials Science and Engineering **972** (2020) 012073 doi:10.1088/1757-899X/972/1/012073

always over 95% (in terms of the highest calorific value). Apparently, according to such information, it will not be possible to compare objectively the innovations and receive the right directions for the constructive improvement of the fireboxes and burner devices of modern heat generators of small and medium power. This problem has become aggravated in the last two decades - during the period of intensive distribution of condensing-type devices and reducing the size of the boilers in the pursuit of commercial attractiveness of products at the compactness direction. To dampen the growth of the constraint by furnace walls of the combustion zone, introduced were different innovations. But the question of whether innovations have helped to avoid the growth of chemical underburning or not, is one of the most pressing challenges of our time. It is directly associated with a real reduction in greenhouse gas emissions due to a decrease in fuel combustion for heating buildings. As is known, about a third of the global energy generation is spent on heating buildings. Therefore, in the decisions of the Paris Agreement on climate, the building sector is equated to the main sectors that directly produce greenhouse gases — production, energy, and transport. Consequently, any innovations that make even small cuts in fuel costs to the heating of the building sector facilities deserve close examination.

2. Methods

The boiler designers establish the effectiveness of innovations based on the results of numerical and experimental studies. Objective test by the consumer of the receive result is really possible only on the basis of numerical methods. At the same time, numerical simulation of fuel combustion processes, even in obviously not constrained fireboxes, presents certain difficulties. One of the principal difficulties is the difficulty of establishing a set of elementary stages of the radical-chain mechanism of thermal oxidation of combustible gases, which is close to the reactions that take place in real furnaces of the boilers. In this direction, quite advanced models have recently been implemented, up to direct modeling with detailed kinetics [1]. However, to perform calculations on a scale larger than the thickness of the combustion zone, much less scrupulous calculations are used. In [2-4], for example, discusses the conditions of the numerical simulation of fuel combustion processes, including waste, in enough large the furnaces and boilers. In our work, we consider methods for constructing numerical models and their study by computational fluid dynamics (Computational Fluid Dynamics-CFD) of three heat generators with different burner devices and geometric characteristics of fireboxes: a Buderus hot water boiler Logano S825L with output 2500 kW, a FEG gas heating Beata 2 with a hermetic combustion chamber with a heating capacity of 2.2 kW and a wall-mounted condensing boiler "Viessmann" Vitodens 100-W with a heating capacity of 19 kW. In addition to other parameters, the presented devices differ in the degree of constraint the flame space by the firebox walls, and, accordingly, its heat stress q_v, kJ / m^3. The studies used Gambit software products with the Exceed emulation environment (preprocessor), ANSYS Fluent (solver) and Tecplot (post processing). The fireboxes and burners of each of the devices have a number of their design features, punctual copying of the geometry of which is possible in the preprocessor, but often not justified, since it does not always have a positive effect on accuracy. Therefore, furnace and burner devices were modeled with certain simplifications, which are discussed further.

2.1. Simulation of geometry and combustion processes in the furnace of the boiler Buderus

The furnace volume of the Buderus boiler has a low heat stress q_v as well as a minimal flame constraint, which was done in order to reduce the formation of NO_x and CO during the combustion of the gas-air mixture. The movement of combustion products is three-pass. The first pass is a combustion chamber in the form of a horizontal flame tube with a length of 3149 mm and a diameter of 776 mm, and part of the turn chamber to the second pass. The second and third passages are composed of three rows of horizontal fire-tube with a diameter of about 40 mm, penetrating the water volume of the boiler. A two-dimensional model of the first pass and the turn chamber into the second pass was created. The fluxes in the second and third passages were not modeled, since their effect on the completeness of fuel combustion is substantially less than in the firebox [5, 6]. The coordinates of the nodes (vertex) are constructed in accordance with the technical characteristics of the Unimat UT-L hot water boiler. (URL: https://www.bosch-climate.ru/bosch-products/promyslennye-kotly/utl.html). The model exported to the solver is adapted to eliminate the grid dependence. At the processor, the type of gas-air mixture "Material Type" (air + methane) is set, the initial values and boundary conditions are established. The initial air velocity is v = 2.18 m/s, the flow rate is G = 0.7 m^3/s, the temperature of the mixture leaving the gas ramp is 292 K. The ratio of the methane and air flows velocities is taken at the rate of 1:10 by volume (air excess factor α about 1.05). This corresponds to a mass ratio of gas and air of 1:18, and an initial mass concentration of methane in a mixture of 5.25%. The boundary conditions for the fluxes are established in accordance with [7]. Flow in the furnace is turbulent, adopted model of turbulence k-epsilon. The energy equation and equation radiation of Rosseland are added to calculate the radiative-convective heat transfer in addition to the differential equations of motion and continuity. In the beginning of the first pass on outlet the gas-air mixture from the ramp, the condition of constant gauge pressure P = 102034 Pa was accepted since furnace operating in real conditions under a small supercharging. The boundary conditions at the exit of the turn chamber into the second pass are taken as "Outflow" - the outflow of flue gases without specifying the parameters, since the movement of the streams on the following passes was not considered. The combustion process was modeled as a component transfer (Species Transport) with a single-stage reaction of methane oxidation with air to final products without predicting the formation of NO_x:

$$CH_4 + 2\ O_2 + 7{,}52\ N_2 = CO_2 + 2\ H_2O + 7{,}52\ N_2 \quad (1)$$

For calculations of combustion at processor solver, the model of the eddy-dissipation was adopted.

2.2. Simulation of geometry and combustion processes in the furnace of the gas-fired FEG Beata unit.
The convector FEG Beata is installed directly in a heated room. It is designed with a fence for burning outdoor air and has a compact combustion chamber with a flame more constrained by its walls than the previous device. According to technical data (FÉG Konvektor Rt Ócsa, Kiss János https://docplayer.com/49705331-Feg-konvektor-rt-ocsa-kiss-janos-u-centralnyy-telefon-36-29-faks-36-29.html), with a growth of power on the line of devices FEG BEATA, the efficiency drops from 92% to 86%, which may be due to the cooling effect of the walls of the heat exchanger due to its proximity to the combustion zone. The latter may affect the radical-chain mechanism for the development of oxidation reactions in the flame zone [8-10].

In the preprocessor, a 3D model of the furnace of the FEG Beata 2 was created. The furnace is vertical, of a simplified form, the output of the gas-air mixture is simulated by holes in the hearthstone. The computational grid is tetrahedral-hybrid, with a linear size of the elements of about 0.005 m. The initial number of grid cells is about 50,000. After exporting to the solver, the model is adapted "by region" to practically eliminate the grid dependence. The final number of cells is 400000. Initial characteristics: air-fuel mixture - methane and air; gas flow consumption G_g = 0.25 m^3/s (0.178 kg/s); the initial velocity V_g = 5 m/s, the temperature of the mixture at the inlet is 292 K. The volume ratio of gas-air is 1:10, the mass ratio is 1:18. Air consumption G_{air} = 2.5 m^3/s (3.219 kg/s), its velocity at the outlet from the bottom holes V_{air} = 1 m/s. For gas, air, and combustion products, the same boundary conditions are set as in the previous case, and the radiant heat transfer was calculated using the discrete ordinate (DO) model.

2.3. Simulation of geometry and combustion processes in the furnace of the boiler Vitodens 100-W
In the Vitodens 100-W series of boilers, a rather original design solution of the burner is provided, excluding the direct contact of the furnace walls with a flame. The gas-air mixture is released through a cylinder from net of heat-resistant steel, and the combustion front is divided into a large number of small flames up to 10 mm high. Therefore, the burner was modeled as a porous body (internal cylinders 2 in Figure 1).

About two dozen coils of the spiral slot for the passage of flue gases with a width of about 1 mm pass through the screen surface of the boiler, which is why the firebox was also modeled as a cylindrical porous zone (external cylinders 1 in Figure 1). The grid of the tetrahedral-hybrid type with a linear size of elements of 0.008 m was generated in the volumes the firebox and burner at the preprocessor. The initial number of cells was 63758.

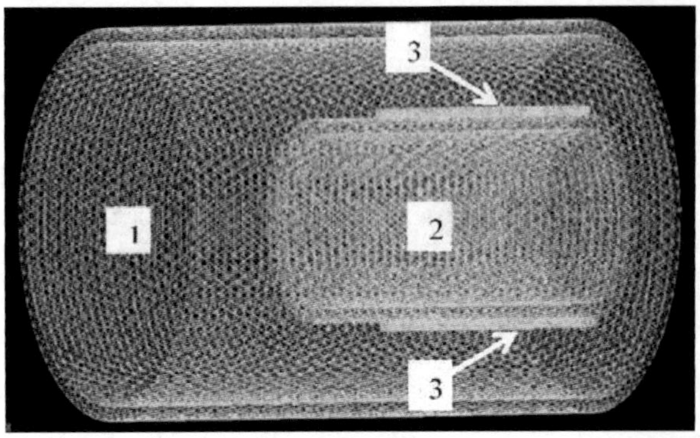

Figure 1. Calculated grid of the 3d model of the firebox (1) and the burner device (2) with ignition electrodes (3) in the Vitodens 100-W boiler.

Unlike previous models, methane and oxygen were taken as components of a combustible mixture. In recent years, such studies have noticeably intensified around the world due to the possibility of intensification of heat exchange and combustion kinetics, while simultaneously achieving a number of environmental benefits, up to a decrease in the yield of greenhouse gases [11–17]. Gas consumption G_g = 2 m^3/s (1,434 kg/s), initial velocity V_g = 12 m/s. The temperature of gas and oxygen at the entrance to the inner space of the burner is 452 K. The initial mass concentration of methane is 20.0%, oxygen is 80.0%, and their mass ratio is 1: 4. Another difference is the introduction of initial values of viscous and inertial resistance in porous zones. Combustion process was simulated as Species Transport with a single-step reaction of methane oxidation in oxygen:

$$CH_4 + 2O_2 = CO_2 + 2H_2O \quad (2)$$

Unlike the previous ones, a combined Eddy-dissipation / Finite Rate model was adopted for the calculations of combustion. This is necessary for the formation the ignite of the fuel-oxygen mixture at the outlet of the burner. In the geometric model, electrodes were installed above and below the burner to ignite the outgoing mixture (position 3, figure 1). The electrode surface temperature of 3000 K was set in a solver.

3. Results and Discussion

3.1. Discussion of the results of numerical studies of furnace process in the boiler Buderus.

Numerical calculations of the 2D model of the Unimat UT-L18 boiler firebox made it possible to work out options for preparing a geometric model in a solver with setting boundary conditions, turbulence and combustion models to ensure physically correct results with minimal computational resources. Distributions of streamlines, temperature, CH_4, O_2, CO_2, H_2O concentrations, pressures and velocities in the firebox and the turning chamber were obtained. The physical adequacy of the streamlines of the combustion products confirmed the correctness of the adopted model and the boundary conditions for hydraulic characteristics. From the velocity field in Figure 2b one can see that the velocity of the combustion products in the radial direction of the furnace falls from the center to the periphery with a velocity gradient of ∇_{vR} = 20 s^{-1}. It can be also seen that the transition from the firebox to the turn chamber has the form of local resistance of the "sudden expansion" type. But the formation of vortex zones in the corners [18], which are common for channels of such a pattern, is not observed here, which can be explained as follows. With the "Outflow" given condition of leaving the combustion products to the left through the upper and lower exits from the chamber, the flow perceives the obstructive effect of the chamber end face, and the configuration of local resistance changes to "180 ° turn". Therefore, the vortex formation passes to the center of the chamber, where the pressure increases, and to the corners adjacent to its end face, where the pressure drops.

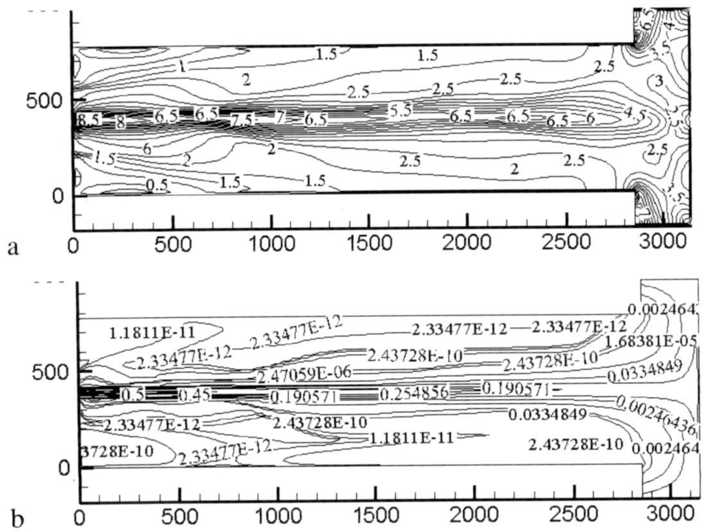

Figure 2. Simulation of the first gas duct and the turning chamber of the furnace of the UT-L18 boiler:

a - velocity field, m / s; b - distribution of methane concentrations, % wt.; dimensions of the firebox on coordinate axes are in mm.

Figure 2b shows the distribution of methane concentration in the combustion space. In the cross section, there is an intensive drop in the concentration of methane from the initial value inlet to trace amounts near the walls ($2.3 \cdot 10^{-10}$% wt.). A less intense CH_4 conversion observed along the flame axis. However, this is can be due to the fact that was simulated separate supply of gas and air. In a real boiler Unimat UT-L18, ensuring complete combustion of methane in the longitudinal direction is not difficult, since gas and air are intensively mixed in the furnace initial part.

3.2. Discussion of the results of numerical studies of furnace process in the FEG Beata heating device

Figure 3 shows the results of calculations performed on the 3D model of the FEG Beata 2 firebox with calculated grid adapted and tested for physical adequacy. Figure 3a shows the temperature distribution in the vertical median plane of the combustion zone. The maximum temperature of 2100 K is reached in the flame core, which is physically correct: the combustion temperature of the methane-air mixture under adiabatic conditions is, as is known, 2200 K. The temperature monotonically decreases from the center to the periphery in all directions of the flame volume, which is also physically correct.

Figure 3b shows the velocity distribution in the same section. The gradient of the speed decrease in the firebox is $\nabla_{vR} = 60$ s^{-1}, which is 3 times higher than the previous case. Despite the smaller than in the previous case, the velocity of the products of combustion in the flame, the value of the gradient increases. This is due to the approach of the walls of the furnace to the combustion zone. Consequently, the calculated value of ∇_{vR} correlates with the flame zone constraint in in hollow fireboxes and, along with the thermal stress, can be used for a comparative assessment of the perfection their designs according to this factor.

Figure 4 shows the distribution of CH_4 concentration in the vertical median section, as well as in the horizontal section at the outlet of the calculated zone. At the periphery of the exit section, the CH4 concentration is mainly 0.4% of the initial amount. Thus, the conversion of methane, after passing through the calculated zone is within 97.9% - 99.6%. Although the percentage of CH4 conversion seems high, the quality side of the process cannot be overlooked. This is especially important when using the single-stage model of thermal oxidation according to the scheme (1). Of the final products, it shows only the products of complete combustion of CO_2 and H_2O. The presence of methane source reagent in them serves as an indicator of the breakage of radical-chain reactions leading to its complete oxidation, already at the initial stages. Therefore, chemical underburning will be higher, and it is advisable to make improvements in the organization of the combustion process.

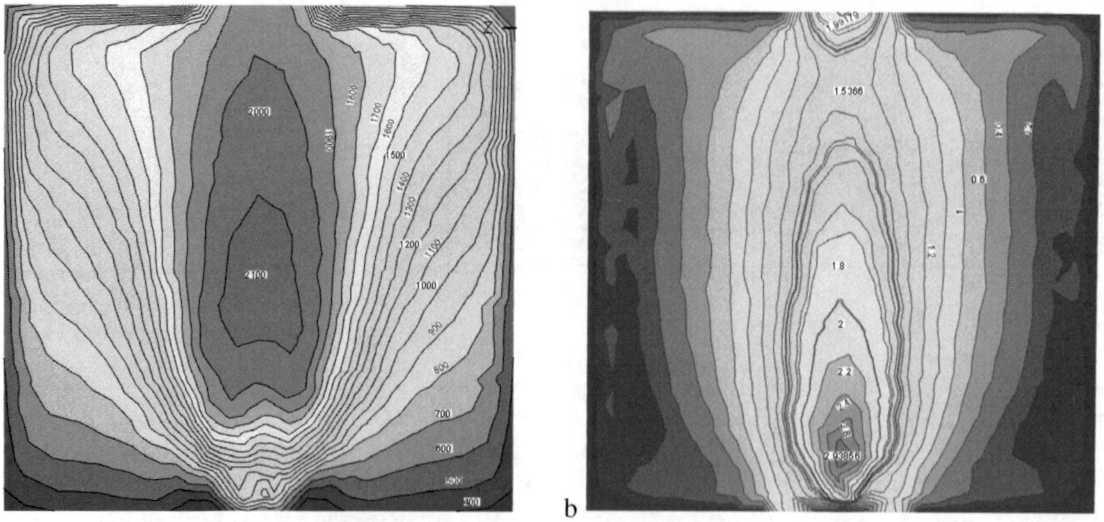

a b

Figure 3. Temperature field (a) and distribution of the flow rate of combustion products (b) in the furnace space of the convector Feg Beata 2.

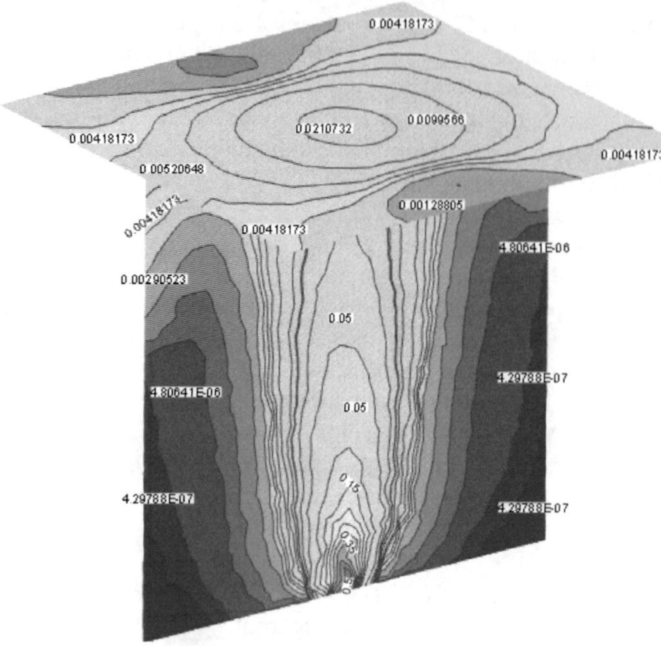

Figure 4. Distribution of methane concentration in the furnace space of the FEG Beata 2.

3.3. Discussion of the results of numerical studies of furnace process in the Vitodens 100-W boiler

The 3D model of the Vitodens 100-W boiler was tested for physical adequacy and adapted "by region" for the achievement of mesh independence of up to 235,000 cells. The results regarding the methane combustion model in oxygen were validated according to well-known empirical data: the maximum temperature of the combustion zone 3370 K obtained in the work exceeds the temperature of the conical combustion front determined by reversing sodium lines [19] by 11%. Therefore, the validation result is satisfactory.

Figure 5a shows the temperature distribution in a longitudinal vertical section along the axis of the firebox. At the beginning of the furnace, the temperature of the mixed methane and oxygen flows is assumed to be 400 K. The mixture of oxygen and methane is ignited by electrodes with a temperature of 3000 K, after which the temperature of the homogeneous phase reaches a maximum of about 3370 K in the flame zone above the burner surface. In the transverse direction, the temperature slowly decreases and at the screen surfaces remains still quite high - almost 2800 K. In the longitudinal direction along the axis in the space behind the burner, the temperature is in the range of 2800-3000 K, and then decreases to the end of the firebox to 1600 K. Thus, manifestations of physical incorrectness of the results of thermal calculations as a whole in the pattern of the temperature distribution in the furnace are not observed.

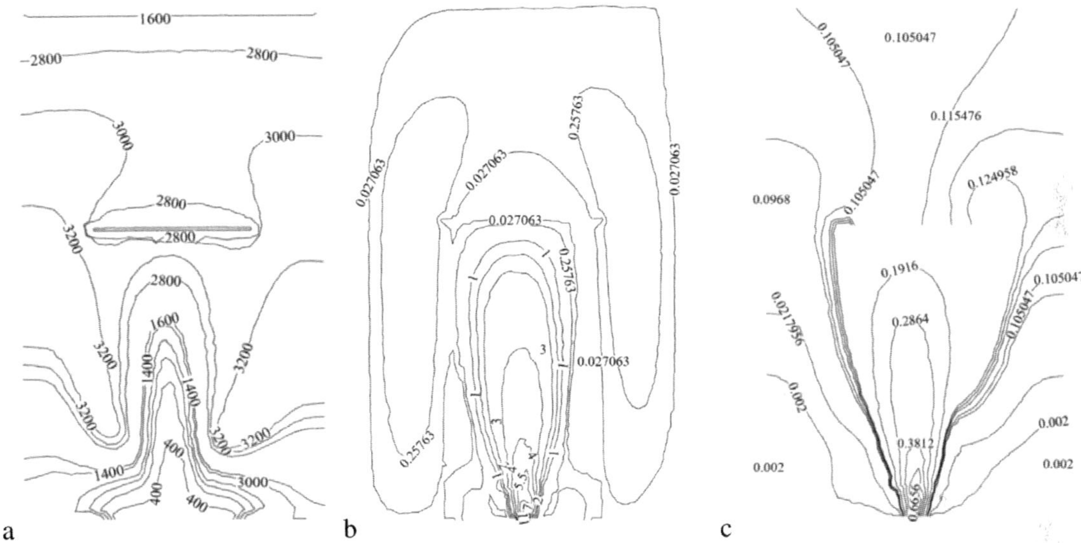

Figure 5. Fields of temperature (a), axial velocity (b) of the flow of combustion products, and methane concentration (c) in longitudinal section along the firebox axis of the Vitodens 100-W boiler.

Figure 5b shows the velocity distribution in the longitudinal section of the furnace. The velocity gradient ∇_{vR} in the transverse direction of the high-temperature zone between the burner and wall the furnace has an order of 12 s^{-1}. We note in this case an additional influence on the velocity of the porous zone, the passage of which the flux spends a significant part of the kinetic energy. As the calculations show (Figure 5b), due to the presence of a porous zone, the flow dramatically slows down and reduces speed by 47.5 times at a distance of 0.005 m. At these space, the gradient ∇_{vR} is close to 200. You can make an estimate of the upper limit ∇_{vR} in the annular space between the burner and the wall of the firebox net the influence of the porous zone. Let us assume that the maximum possible velocity the flux at the surface of the burner does not exceed the initial average flow rate (by mass) of the mixture v_m = 2.64 m / s. Then ∇_{vR} of expanding combustion products in the annular space will be no higher than 80 s^{-1}. Therefore, it can be estimated that the development conditions for the flame in the Vitodens 100-W firebox are similar to that of the Feg Beata 2 convector.

Figure 5c shows the distribution of CH_4 concentration in the longitudinal section of the firebox. It can be seen that after ignition of the gas-oxygen mixture, the CH_4 concentration in the annular space between the burner and the firebox surface decreases to 0.2%, and with such a concentration goes into the convective part of the boiler, where burnout will stop. Not converted methane in the same quantity will be removed with the flue gases. Consequently, the methane conversion in the boiler will be 99.8%, which is close to the results for the FEG Beata 2 convector. The latter was expected as the velocity gradient values are close, which confirms the validity of the ∇_{vR} indicator. The presence of unconverted methane in the flue gases also serves as an indication of the breakage of the radical-chain reactions of methane oxidation in the flame already at the initial stage. Therefore, furnace process is also advisable to improve in order to reduce the proportion of unconverted methane to negligible values.

4. Conclusions

Thus, the methods of modeling the design features of the furnace and burner devices of modern household heat generators used in the work give the correct numerical values of the hydrodynamic, thermal, and concentration parameters of the species. The possibility of using a fairly simple and at the same time fundamental parameter of furnace processes - the radial gradient of the flow velocity in the flame zone ∇_{vR}, for the estimated of the furnace process perfection is illustrated. It is shown that the value of ∇_{vR} correlates with the degree of flame constraint with the firebox walls, which directly affects the completeness of fuel combustion. In the further development of the presented method of modeling through the porous structures the structural features of furnaces and burners with numerous holes and slit, it is necessary to further clarify the values of viscosity and inertial coefficients of hydraulic resistance of specific designs the furnaces and the burners.

References

[1] Aspden A J, Day M S and Bell J B 2016 *Combustion and Flame* **166** pp 266–83
[2] Rahmani A and Dahia A 2009 *Nuclear Technology & Radiation Protection* **24(1)** pp 29-37
[3] Adams B R, Shim H S, Wu S R, Chang W C, Chiao H W and Chen S L 2003 *CFD Proceeding, 4th Asia-Pacific Conference on Combustion,* Nanjing, China, November 2003, pp. 24-26.
[4] Orstnerhof M, Meister G, Jöller M, Dahl J, Braun M and Kleditzsch S 2006 *Progress in Computational Fluid Dynamics* **6** pp. 248-261.
[5] Ganan J, Al-Kassir A, Gonzalez J F, Turegano J and Mirand A B 2005 *Applied Thermal Engineering* **25** pp. 1650-56.
[6] Rahmani A, Trabelsi S 2014 *American Journal of Chemical Engineering* **2** 5 pp. 65-70
[7] Logachev K I, Ziganshin A M, Averkova O A, Logachev A K 2018 *Energy & Buildings* **173** pp. 58–70.
[8] Makarov A N 2015 *Mechanical Engineering Research* **5(1)** 42-55.
[9] Gerlinger P 2003 *Combustion Science and Technology* **175(5)** 841–72.
[10] Kim Jeonglae and Pope S B *Combustion Theory and Modelling* 2014 **18(3)** pp 388–413.
11] Self S J, Rosen M A and Reddy B. V. 2018 *European Journal of Sustainable Development Research* **2(2)** 22 (1-11)
[12] Wu K K, Chang Y C, Chen, C H and Chen Y D 2010 *Fuel* **89(9)** 2455–62.
[13] Yoshiie R, Hikosaka N, Nunome Y, Ueki Y and Naruse I 2014 *Fuel Process Tech* **136** 106-11.
[14] Wall T, Liu Y, Sperob C, Elliott L, Khare S, Rathnam R, Zeenathal F, Moghtaderi B, Buhre B, Sheng C, Gupta R, Yamada T, Makino K and Yu J 2009 *Chem Eng Res Des* **87(8)** 1003–16.
[15] Wall T, Stangera R and Santos S 2011 *Int J Greenh Gas Con* **5(1)** 5–15.
[16] Uchida T, Yamada T, Watanabe S, Kiga T, Gotou T, Misawa N. and Spero C. 2011 *The 7th International Symposium on Coal Combustion (7th1SCC)* Harbin, China, July 17-20
[17] Chen L, Yong S Z and Ghoniem A F 2012 *Prog Energ Combust* **38(2)** 156–214.
[18] Ziganshin A M, Aleshchenko I S, Ziganshin M G and Gimranov I R patent 2604264 of the Rus. Federation IPC F16L 43/00, IPC F16L 25/14
[19] Staskevich N L, Severinets G N, Vigdorchik D Ya 1990 *Handbook of gas supply and use of gas* (Leningrad: Nedra) 762 p.

Comprehensive methods to obtain and process information flows in centralized heat supply systems

I K Sagynganova[1], A I Kvasov[1], A A Kalinin[2]

[1]D. Serikbayev East Kazakhstan State Technical University, 69 Protozanov Street, 070004 Ust-Kamenogorsk, Kazakhstan

[2]Karaganda State Technical University, 56 Nursultan Nazarbayev Ave., 100027 Karaganda, Kazakhstan

E-mail: talgar_45@bk.ru, leokalinin@mail.ru

Abstract. The article discusses the principles of creating the structures of information and control systems for urban heat supply. The necessity to create and develop such systems is due to the significant energy intensity of the country's GDP, which is twice as large as the majority of developed countries. Modern multi-level distributed control systems of the centralized heat supply system facilities, which are built according to the object rule, are described in the paper. The article analyses advantages and disadvantages of various options for information management systems. In the process of thermal automatics development based on software and hardware, including modern intelligent sensors and industrial controllers, modern thermal automation systems began to be intensively introduced. The use of standard techniques of cycling poll of all local objects included in the centralized heating systems, without taking into account the dynamic characteristics of the measured parameters, including the pipeline processing of this information, leads to a significant increase in the time of collection, processing, transmission of information and decision-making in a centralized heating systems. The concept of measuring and processing information in centralized heat supply systems is proposed, based on variable polling cycles of parameters of heat supply systems at heat points taking into account their dynamic characteristics with subsequent conveyor processing of information at the top level of automated centralized heating systems.

Keywords: urban heat supply, information and control systems, measurement and processing, parameter polling, variable cycles, pipeline processing of information

1. Introduction

The cold, sharply continental climate, low average temperature, the long heating period compared to European countries, as well as centralized heat supply, characterized by heavy heating networks, combined with low energy efficiency of production and transfer of thermal energy are a characteristic feature of thermal power systems in Kazakhstan.

Thermal energy consumption according to the Agency of Statistics of the Republic of Kazakhstan and the plans of the Government will increase by 15 thousand Gcal in 2010-2020 and reach 110 thousand Gcal [1].

IOP Conf. Series: Materials Science and Engineering **972** (2020) 012074 doi:10.1088/1757-899X/972/1/012074

Despite the existence of objective reasons for the high energy intensity of GDP of the Republic of Kazakhstan, there is a significant potential for improving energy efficiency and energy saving, since the energy intensity of GDP in Kazakhstan is twice as high as that of most developed countries [2].

Low energy efficiency is characterized by the production and transfer of thermal energy. According to expert data, fuel consumption for power generation is 10-15%, for heat supply - 15-20%. The cost of introducing energy savings is about 5 times lower than for new energy production [1]. The challenge is to determine the heat losses, to personalize which is some difficult due to the branching of the heat distribution networks, the number of boiler houses and pumping and transfer stations, and the irrational configuration of the house heating networks.

In this regard, the necessity to create and implement modern information and control systems for urban heat supply complexes in Kazakhstan is not in doubt.

It is known that the introduction and industrial operation of information management systems for heat supply will tentatively provide [3]:

- saving of heat generation and pumping capacity;
- decrease in accident rate in the megacity heat supply systems;
- reducing maintenance costs for electricity.

2. Research methodology

We now consider the main areas of research and technical solutions for the collection, transmission and processing of information at the base heating facility, i.e. thermal point.

Many researches and developments by V.K. Averyanova and S.I. Bykov [4, 5], B.N. Gromova [6], G.V. Monakhov [7, 8], N.I. Karasyova [9, 10], etc., had an important role in developing this direction. Their works decided to informatize the heat supply systems in megacities, which were considered as objects of control Thus, the works [11, 12] are aimed at the system integration of information and mathematical models that form databases and knowledge for the decision support system to adjust heat supply systems in megacities. When calculating their thermal conditions, the problems of analyzing the steady-state hydraulic regimes of the heat-supplying system were solved, the mathematical model of the decision support system, which contains models of the motion of the heat carrier in technological elements of the heat-supplying system; models of steady-state hydraulic modes of heat supply systems in various regime situations, etc. was developed.

Figure 1 shows a generalized scheme of an automated process control system for the urban centralized heat supply.

The multi-level distributed system for controlling objects of the centralized heat supply system is built on the object principle.

The object of informatization is represented by the following aggregated technological installations: HS is the heat source represented by the technological structure of the TPP or DBH (district boiler house); CHP - central heat point (used to serve a group of consumers (buildings, industrial facilities); IHP - individual heat point of a consumer (used to serve one consumer (building or part thereof); PS - network pumping station on the supply or return pipelines in the main circuit; CS - controlled station for concentrating measuring and control information on the pipeline system.

Manufacturers of hardware and software solution/complex and devices. The ideology of Fieldbus Foundation, which ensures the transfer and implementation of measurement information processing algorithms (filtering, scaling, linearization, etc.), regulation (stabilization, tracking, cascade control, etc.), logic control (start, stop, blocking, etc.) to the lower control level is becoming one of future-oriented ideology of "intellectualizing" devices and I/O units. To implement this ideology, a new typical Fieldbus H1 field network that implements all the functions of HART protocol and also allows through communication with devices to program specific control and management algorithms implemented in devices using special technology languages [13] has been developed.

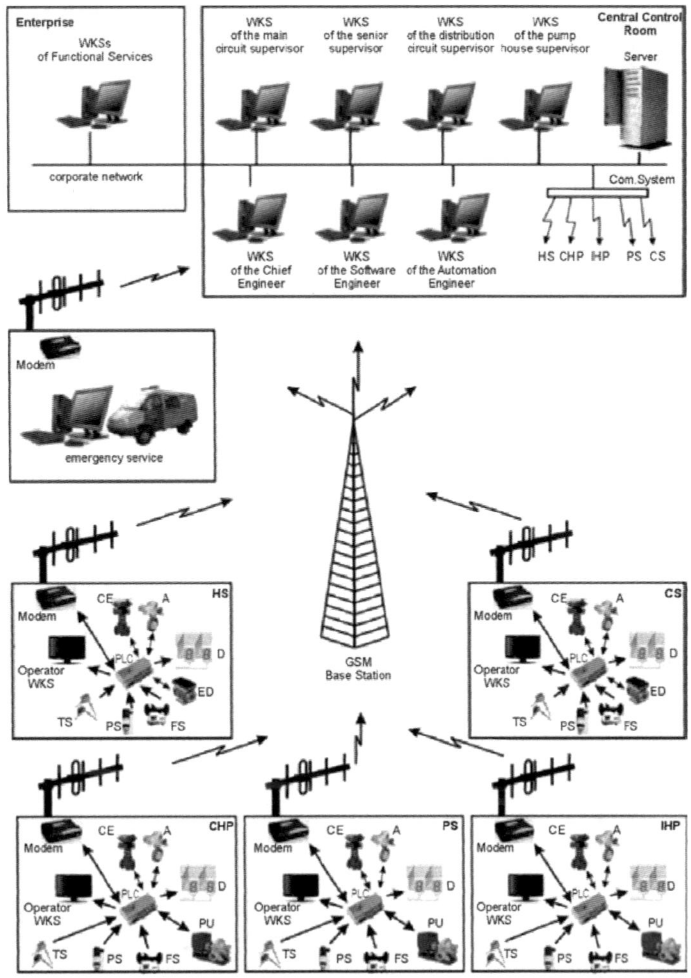

Figure 1. The generalized scheme of an automated process control system for the urban centralized heat supply

The lower control level of all technological installations combines separate controllers with remote I / O units, intelligent instruments and actuators (A - valve actuator, PO - throttling control elements, ДТ - temperature sensors, ДД - pressure sensors, ДР - energy consumption sensors, ЭП - electric drive of aggregates, HA - pumping units, И - control and display means, ИМ - actuators with built-in controllers). Intelligent devices are connected to controllers through a standard digital field bus with a standard HART-protocol, which is supported by all. The ideology of such systems was focused on the insufficient capabilities of thermal automation systems, as well as on mode calculations of systems taking into account various operational and seasonal factors. These are mostly not managing, but information systems.

A pipeline data is effective for processing a significant amount of information [14-16]. This method with regard to the urban centralized heat supply systems are described in [17].

However, the centralized approach, in which the object of control and management for the information management system is the heat supply complex of a large city, requires the organization of two-way functional and redundant information from heat points to servers and from servers to heat points. Cyclic polling of information on the objects of control of each heat station takes large intervals of time, which significantly increases the response time to events and reduces the efficiency of

IOP Conf. Series: Materials Science and Engineering **972** (2020) 012074 doi:10.1088/1757-899X/972/1/012074

managing heat point objects. The whole system becomes inoperative in case of any failure at the top level.

The use of standard technologies for cyclic polling of all local objects included in the centralized heating system, without taking into account the dynamic characteristics of the measured parameters, including conveyor processing of this information, leads to a significant increase in the time of collecting, processing, transmitting information and making decisions in the centralized heating system.

Recently, systems aimed at solving local automation tasks at the level of heat points have been intensively introduced in thermal automation based on software and hardware, including modern intelligent sensors and industrial controllers [18]. The heat point (HP) is one of the main elements in the centralized heat supply systems of buildings that performs functions of receiving the heat carrier, converting (if necessary) its parameters, distribution among consumers of thermal energy and accounting for its costs. Depending on the purpose, conditions of connection of consumers to the heating network, customer requirements and other HPs can be composed of a number of individual functional units.

To simplify the process of designing, assembling and installing, the HP can be manufactured in the factory and delivered to the construction site in the form of ready blocks - a block heat point (BHP). BHP is a separate functional units assembled on a frame into a general structure, as a rule, complete with instruments and devices for control, automatic regulation and management The Danfoss Company offers standard automated block heat points of full factory readiness for use in Kazakhstan that intended for connection to the heat network of various heat consumption systems and executed according to standard technological schemes using water heaters based on soldered or collapsible plate heat exchangers of its own production. The use of Danfoss automated BHPs contributes to solving the most important task in the field of heat supply - raising its quality level, which is to provide comfortable climatic conditions in buildings, the temperatures and hot water consumption required by sanitary standards with the minimum energy consumption [19].

The uniformity of modern technical solutions of BHP and their streamlined production at modern Danfoss plants allow us to: simplify the process of completing a heat supply station with equipment and materials as compared to delivering them to the construction site in bulk, ensure the highest quality of BHP manufacturing, eliminate preparatory and serious installation and commissioning works on site, reducing them to installing the unit in the premises of heat point and connecting it to the building pipelines and power supply networks [19].

In such decentralized systems with their large numbers, and in cities with a significant number of heat points, in the process of work conflict regimes arise when local systems fail to cope with the management of modes, the centralized redistribution of energy flows is necessary.

But an increase in the number of tasks solved by local automation systems at heat points leads to a multiple increase in the measured and processed information. Tens of parameters are subject to measurement with several branch lines at each heat point. A full cycle of polling all urban heat points can take several hours.

At the same time, parameters such as temperature change slowly, and flow rates and pressure have a higher response rate. The protection operation, the identification of emergency situations require even higher speed.

There is proposed the concept of measuring and processing information in centralized heat supply systems, based on variable polling cycles of parameters of heat supply systems at heat points taking into account their dynamic characteristics with subsequent pipelining of information at the top level of automated central heating systems.

There are three groups of parameters to be monitored at the heat point:

- Control of protection operation.
- Control of pressure, flow and power consumption.
- Temperature control.

Polling of parameters of a heat point can be carried out in different periods. Groups of parameters with different dynamic characteristics are distinguished in a heat point.

The heat point receives the setting actions for the drives of pumping stations.

Information on the coolant temperature, averaged pressure and flow data for the heating mains and electricity consumption is periodically transmitted to the central control centers. Information about protection operation is promptly transmitted.

Information processing at central control centers is carried out by pipelined (conveyor) methods.

The solution of such tasks can be provided by a hierarchical structure in which the information necessary for control and management of the entire heat supply system is transmitted to the upper level, and the tasks of managing and controlling each object of the heat supply system are transferred to local subsystems that define variable polling cycles depending on the dynamic characteristics of technological parameters (temperature of heat carriers, pressure, flow, etc.).

The proposed concept combining variable polling cycles at each level of the heat supply system with pipelining of information at central control centers will reduce the load on information transmission channels, significantly reduce traffic when transmitting information from local heat points to central control centers through cellular communication, increase speed, and reduce the time to take solutions in centralized power supply systems.

References

[1] Antonov O 2019 Green Energy of Kazakhstan in the 21st Century: Myths, Reality and Prospects https://polpred.com/?ns=1&ns_id=1030559 (access date 15.02.2019)

[2] The Concept for Developing the Fuel and Energy Complex in the Republic of Kazakhstan until 2030 approved by the Government of the Republic of Kazakhstan 2014

[3] Tomilova N I 2010 Developing an information system to support decision-making on setting up and promising activities in the heat supply systems of megacities *Thesis in support of candidature for a technical degree* (Karaganda) 147 p.

[4] Averyanov V K, Bykov S I 1979 Probabilistic-statistical description of the heating system operation mode *Bulletin of universities: Series: Energy* **11** pp 55-60

[5] Averyanov V K, Bykov S I 1982 Heat transfer in the premises during the programmed heat supply *Engineering and Physics Journal* Vol. **XLIII** No. **3** pp 406-412

[6] Gromov N K 1981 About schemes for automating subscriber installations of large urban centralized heat supply systems *Water supply and sanitary engineering* No. **12** pp 8-9

[7] Monakhov G V, Sibiryakova L P, Karasyov N I, Folgart V I 1985 Software implementation of mathematical models of variable thermal conditions of urban heating networks *Automated systems for scientific research of collective use: collection of research papers of the Scientific Production Association for Automating Urban Management and Computational Work* pp 81-88.

[8] Monakhov G V, Voitinskaya Yu A 1995 Modeling control of heat network modes (Moskow: Energoatomizdat) 221 p

[9] Karasyov N I, Folgart V I, Dirksen V L, et al. 1989 About one model of steady-state flow distribution in large pipeline systems *Energetika: Bulletin of universities* No. **1** pp 113-117

[10] Karasyov N I, Folgart V I, Dirksen V L, et al. 1990 Simulation of steady-state hydraulic regimes of large heat supply systems with throttle control *Electron Modeling* No. **1** pp 72-76

[11] Karasyov N I, Tomilova N I 2008 Synthesis of permissible hydraulic modes of distribution fragments of heat supply systems *Proceedings of the University* Vol. **3** pp 75-77

[12] Karasyov N I, Tomilova N I 2008 Methods of decomposition of hydraulic circuits of heat supply systems *Proceedings of the University* (Karaganda: KSTU) Vol. **2**. pp 72-75

[13] Basics of pumping station automation http://nasosinfo.ru/node/64 (access date 15.03.2019)

[14] Kulba V V, Kosyachenko S A, Lebedev V N 2009 Automated management information systems of socio-economic and organizational structures *Management problems* Vol. **3** No. **1** pp 73-86.

[15] Samofalov K G, Lutsky G M 1989 Fundamentals of the theory of multilevel conveyor computing systems (Moskow: Radio and communication) 271 p.
[16] Conway R V, Maxwell V L, Miller L V 1975 Scheduling theory (Moskow: Science) 360 p.
[17] Rybakova D A, Baklanov A E, Kvass A I 2014 The use of a conveyor data processing system to regulate the operation of heat points *EKSTU Bulletin* No. **3** pp 103-106
[18] Automated pumping station https://www.ngpedia.ru/id478140p3.html (access date 15.02.2019)
[19] Svistunov V A 2013 Automation of a pumping station using a frequency-controlled electric drive *Bulletin of TSU. Engineering science* Vol **12** Part **2** pp 135-139

IOP Conf. Series: Materials Science and Engineering **972** (2020) 012075 doi:10.1088/1757-899X/972/1/012075

Method of assessment of the accidental leakage of trunk the pipeline in the winter

S I Chelombitko[1]

[1]Industrial University of Tyumen, 38, Volodarskogo street, Tyumen, 625000, Russia

E-mail: sjlss@mail.ru

Abstract. Some features of emergency leaks from oil trunk pipelines in winter, when there is snow cover, are considered. The increase in the intensity of accidents and the scale of oil spills in the Northern regions are of concern to both the public and scientists, who are increasingly calling what is happening an environmental disaster. According to various estimates, the rivers of Western and Eastern Siberia annually carry 0.7 to 2 million tons of oil into the sea, and lists of rivers where MPC of pollutants are several times higher than permissible. The same applies to large rivers in their lower reaches, there was a sharp decline in valuable species of fish in their places of historical mass habitat. In rivers and streams that feed them oil products fall primarily with floodwater and sewage. According to the normative documents, the localization (limitation) of the oil slick in winter is carried out by compacting snow, and the first stage of reclamation of oil-contaminated territories (mechanical) involves plowing and harrowing, i.e. the beginning of restoration work is possible after thawing of the soil. This leads to the fact that after the snow melts, the remaining oil after the collection together with flood waters rushes into streams, rivers and other reservoirs. The remaining oil on the surface of the day or seeps into the soil, or after plowing is filtered into the ground with rain and melt water, polluting groundwater. It is established that oil-contaminated mosses and lichens die, vegetation cover with meadow grasses is restored in more than 15-20 years. Therefore, studies that make it possible to realistically assess the consequences of accidental oil spills and to compensate for the damage caused to nature are important and relevant. The proposed method of estimating the amount of spilled oil in winter is based only on the results of a survey of the contaminated area and does not concern either the causes of the accident or the parameters determining the formation of an oil slick, which may be inaccessible or distorted for certain purposes.

1. Introduction

The main explored and developed hydrocarbon deposits of Russia are concentrated in its northern regions, which are characterized by harsh climatic conditions.To transport oil through the territory of Russia, oil trunk pipelines with a total length of 46800 km are used [9], most of which have developed their regulatory resource. It is established that the number of accidents on oil pipelines only increases every year, with an increase in accident rate of 5-9% [13, 14]. Only in Western Siberia through damage to oil pipelines occurs up to 35000 times a year, with oil emissions of more than 10000 tons of about 300 [12]. In Nizhnevartovsk district of Tyumen region about 10 accidents occur on oil pipelines every day [12]. According to various data from the main oil pipelines of the Russian Federation is poured into the environment annually from 10 to 15 million tons of oil, with the largest leaks occur in Western Siberia, where more than 80% of oil is extracted. The total number of accidents on Russia's oil trunk pipelines has increased by almost 20% over the past few years, according to official data [14]. Operation of oil trunk pipelines in the Northern regions of the Russian Federation is carried out in harsh climatic conditions at low ambient temperature, high soil moisture, long-term snow cover (more than six months). This significantly complicates the repair and restoration works, as well as the subsequent remediation of the contaminated area in the event of an accident. Accidental oil spills in the winter are essential features that must be considered when liquidation of consequences and the planning of the rehabilitation works. First of all, in the presence of snow cover, the boundary of the oil slick is mobile and not visible for observation. Therefore, the position of the oil circuit can only be determined by a direct survey of the contaminated area. Snow cover and low load-bearing capacity of swamps create serious problems in the performance of work on the localization of oil spills and collection of spilled oil. For example, in a swampy water-saturated area, soil freezing for the safe movement of road equipment occurs within a few months, and some peat bogs remain impassable all year round. This leads to the fact that the collapse of the oil slick in winter sometimes takes time several times more than the requirement of regulations. During this time, the oil spreads over the frozen surface of the day through a porous snow environment, increasing the scale of pollution. As is known at low temperatures, oil losses from evaporation and infiltration into the frozen soil and vegetation layer are small and neglected in engineering practice. Therefore, it is possible to estimate the amount of emergency leakage from the damaged pipeline in winter by means of experimental determination of the amount of spilled oil located on the day surface and within the snow cover. The use of numerical mathematical models of the process of oil distribution in the snow cover is difficult due to the lack of the necessary reliable information from the leakage and the conditions of formation of the oil spill.

2. Determination of spilled oil volume in case of significant leakage from the main oil pipeline

Significant leakage from the pipeline occurs when the pipe breaks and is accompanied by the release of soil from the trench, possible gushing of oil for a certain time in the process of reducing the pressure in the pipeline. An open stream of "warm" oil is formed around the pipeline damage, which spreads over the day surface and melts some part of the snow cover. In the future, the oil cooled to 0 degrees and below in the form of a semi-pressure flow is filtered in a porous snow environment under the influence of gravity and pressure. The primary task of emergency and recovery services upon arrival at the scene of the accident is to localize the oil slick on the surface of the day by sinking the soil or compaction with snow. As a result of survey of the polluted territory after its collapse it is possible to allocate 3 characteristic sites:

- S_1 - the area of the polluted area around the leak site where the oil has an open surface equal to the area of the oil spill;
- S_2 - the area of the polluted area around the open oil slick where, due to the melting of some part of the snow, the oil is within the snow cover of a smaller thickness than the unpolluted area;
- S_3 - the area of the peripheral part of the contaminated area where the oil cooled to negative temperatures is within the snow cover.

In order to be able to use some of the physical properties of snow to assess its oil capacity, special experimental studies have been conducted in a wide range of changes in its porosity (raft-news). As a result of the data processing it was found that for engineering calculations, the oil capacity of snow can be taken equal to its moisture capacity, while the error does not exceed 5% [17]. This makes it possible to use some of the results of wet snow studies to assess the oil content of snow cover (for example, to Express the free porosity of snow through its density).

Winter in the Siberian regions is characterized by a long period of low temperatures, when there is no film moisture in the snow cover. In this case, the snow is considered to be "dry" and its porosity is unambiguously expressed through its density with the help of a known formula:

$$n = 1 - \frac{\rho}{\rho_{ё}}.$$

To be able to estimate the amount of spilled oil in the snow cover and on the day surface around the site of a significant leak, the following relationship is obtained [7]:

$$V_{н} = \left(h_{жс} - \frac{\rho}{\rho_{в}} h_{c}\right)\cdot S_1 + \left[\left(1-\frac{\rho}{\rho_{л}}\right)\cdot(h_{з}^{(2)} - h_{н,к}) - \frac{\rho}{\rho_{в}}\left(h_{c} - h_{c}^{(2)}\right) + \left(0,5 - 0,55\frac{\rho}{\rho_{в}}\right)\cdot h_{н,к}\right]\cdot S_2 + \left[\left(1-\frac{\rho}{\rho_{л}}\right)\cdot(h_{з}^{(3)} - h_{н,к}) + \left(0,5 - 0,55\frac{\rho}{\rho_{в}}\right)\cdot h_{н,к}\right]\cdot S_3, \quad (1)$$

where: ρ - the average thickness of the density of snow; $\rho_{в}$, $\rho_{л}$ - the density of water and ice respectively; h_{c} -average thickness of snow cover around the contaminated area; $h_{жс}$ - the average on surface-area S_1 thickness of the liquid layer "oil and water"; $h_{c}^{(2)}$ - the average thickness of the snow cover surface S_2; $h_{н,к}$ - the height of capillary rise of oil in the snow; $h_{з}^{(2)}$ - thickness of the contaminated snow layer on the surface S_2; $h_{з}^{(3)}$ - the thickness of the contaminated layer of snow on the surface S_3.

As is known, at a snow temperature greater than -10 ^{0}C, snow contains moisture in the form of a film, and with an increase in temperature, the moisture content increases, which leads to a decrease in the free porosity of snow. When obtaining the formula for determining the amount of spilled oil in the contaminated area, when the snow is "wet", the following known dependence is used [23,25]:

$$n = 1 - \frac{\rho}{\rho_{л}(1-0,083\alpha)},$$

where α - snow humidity.

The formula found for this case is as follows:

$$V_{н} = \left[\left(h_{жс} - \frac{\rho}{\rho_{в}}\left(\frac{1}{(1-0,083\alpha)} + \alpha\right)\cdot h_{c}\right)\cdot S_1 - \frac{\rho\left(h_{c} - h_{c}^{(2)}\right)}{\rho_{в}}\left(\frac{1}{1-0,083\alpha} + \alpha\right)\cdot S_2\right] + \left[\left(1 - \frac{\rho}{\rho_{л}(1-0,083\alpha)}\right)\cdot(h_{з}^{(2)}\cdot S_2 + h_{з}^{(3)}\cdot S_3) + (S_2 + S_3)\cdot\left(0,5 - 0,55\frac{\rho}{\rho_{в}}\right)\cdot h_{н,к}\right]. \quad (2)$$

Thus, the obtained formulas (1) and (2) make it possible to determine the volume of spilled oil as a result of significant leakage /rupture/ on the oil pipeline in winter both at low ambient temperature (less than -10 ^{0}C), when the snow cover does not contain moisture, and in the case of "wet" snow, when the air temperature is more than -10 ^{0}C.

3. Determination of spilled oil volume in case of small leakage from the main oil pipeline

A minor leak in engineering practice are emergency the expiration of oil through the corrosion holes with a small volumetric flow, resulting in emission of soil from the pipeline trench. The oil flowing out of the hole is filtered into the surrounding ground space, sometimes flows along the pipe downhill and comes to the surface in a lowered place. The difficulty of detecting fistula leaks due to the impossibility of their establishment by conventional methods used in practice leads to large and even huge losses of oil. Thus, as a result of fistula (corrosion) oil leaks from the oil pipeline in the Usinsk region of the Komi Republic in August 1994, 79000 to 100000 tons of oil and possibly more [19, 20,

IOP Conf. Series: Materials Science and Engineering **972** (2020) 012075 doi:10.1088/1757-899X/972/1/012075

21] were spilled. Fistula leaks are most often established by visual detection of an oil spot on the day surface during the over flight of the oil pipeline route. The ratio of the amount of spilled oil to the amount of oil in the ground environment depends mainly on the physical properties of the soil (permeability, humidity), the slope of the terrain and the position of the groundwater level [3,4,18].

As a result of heat exchange with the environment facing the surface of the oil prices as a result of the fistula leakage has a lower temperature than the temperature of the oil in the pipeline, so depending on the time of the formation of impurities and conditions of heat exchange (temperature of oil and the environment, snow cover thickness and physical properties of snow and other) possible 3 the following situation.

- Oil is completely within the snow cover and does not have an open surface, while the thickness of the snow cover remains unchanged. This is possible when the output on the daily surface of the cooled oil with a temperature below 0°C.
- In the oil-contaminated area, the snow cover has partially melted, but the oil does not have an open surface.
- On some part of the day surface oil forms an oil slick, on the rest of the contaminated area oil is contained within the snow cover.

For the experimental determination of the amount of spilled oil due to the operation of a small leak for these 3 cases, you can use, respectively, the following dependencies:

- in the first case:

$$V_н = \left[\left(1 - \frac{\rho}{\rho_л} \right) \cdot h_г + \left(0,5 - 0,55 \frac{\rho}{\rho_в} \right) \cdot h_{н,к} \right] \cdot S, \tag{3}$$

where: S - area of contamination of the day surface; $h_г$ - the average thickness of the snow layer containing oil in the form of "gravitational».

- in second case:

$$V_н = \left[\left(1 - \frac{\rho}{\rho_л} \right) \cdot (h_з^{(2)} - h_{н,к}) - \frac{\rho}{\rho_в} \left(h_c - h_c^{(2)} \right) + \left(0,5 - 0,55 \frac{\rho}{\rho_в} \right) \cdot h_{н,к} \right] \cdot S_2 + \\ + \left[\left(1 - \frac{\rho}{\rho_л} \right) \cdot (h_з^{(3)} - h_{н,к}) + \left(0,5 - 0,55 \frac{\rho}{\rho_в} \right) \cdot h_{н,к} \right] \cdot S_3 \tag{4}$$

- in the third case, formula (1).

In the case of "wet" snow and small (corrosive) leakage it is proposed to use the following dependencies:

- in the first case:

$$V_н = \left[\left(1 - \frac{\rho}{\rho_л (1 - 0,083\alpha)} \right) \cdot h_г + \left(0,5 - 0,55 \frac{\rho}{\rho_в} \right) \cdot h_{н,к} \right] \cdot S, \tag{5}$$

- in second case:

$$V_{\text{н}} = \left[\left(1-\frac{\rho}{\rho_{\text{л}}(1-0{,}083\alpha)}\right)\cdot h_{\text{г}}^{(2)} - \frac{\rho\left(h_c-h_c^{(2)}\right)}{\rho_{\text{в}}}\left(\frac{1}{1-0{,}083\alpha}+\alpha\right)\right]\cdot S_2 + \left[\left(0{,}5-0{,}55\frac{\rho}{\rho_{\text{в}}}\right)\cdot h_{\text{н,к}}\right]\cdot S_2 + \left[\left(1-\frac{\rho}{\rho_{\text{л}}(1-0{,}083\alpha)}\right)\cdot h_{\text{г}}^{(3)} + \left(0{,}5-0{,}55\frac{\rho}{\rho_{\text{в}}}\right)\cdot h_{\text{н,к}}\right]\cdot S_3 \quad (6)$$

- in the third case, formula (2).

4. Conclusion

Determination of the volume of spilled oil in winter due to through damage of trunk pipelines is important and relevant for the quality planning of repair and restoration work, remediation of contaminated areas, as well as for the assessment of damage to the environment.

Measurement of characteristic parameters of the snow cover and the oil slick in the course of inspection of the contaminated area allows you to perform estimates using the proposed dependency (1) to(6).

In case of emergency oil spills with an open surface of the oil spill, the formula (1) for "dry" snow and the formula (4) for "wet" snow should be used during the calculations.

When estimating the amount of spilled oil contained on the daily surface in the snow cover in the absence of an oil spill, it is recommended to apply formulas (2)-(3) in the case of "dry" snow and formulas (5)-(6) for "wet" snow.

References

[1] Baranov A V, Grigoriev V Ya, Yakushev N L, Unanyan K L 2010 Degradation and protection in the areas of development of hydrocarbon deposits of the Far North. *Georesources, Geoenergy, geopolitics* **№2** (**2**) [Electronic resource]. Access: http://oilgasjournal.ru/vol_2/articles/7.pdf (date accessed: 10.01.2018).

[2] Resolution of the Government of the Russian Federation of April 15, 2002 №240 "On the procedure for organizing measures to prevent and eliminate oil spills on the territory of the Russian Federation" (as amended on November 14, 2014)

[3] Chelombitko S I, Skavysh S A 2012 Mathematical modeling of the isothermal process of oil distribution in the snow cover in the accident on the main oil pipeline. P*roceedings of the all-Russian scientific and practical conference "Actual problems of pipeline transport in Western Siberia 2012"* (Tyumen: TSOGU) pp 70-72

[4] Chelombitko S I, Skavysh S A 2012 Mathematical model of non-isothermal unsteady filtration of Newtonian oil in snow cover. Materials of the all-Russian scientific-practical conference of students, postgraduates and young scientists «Problems of functioning of transport systems» (Tyumen: TSOGU) pp 389-392.

[5] Gumerov A G, Azmetov A Kh, Gumerov R S, Vekshtein M G 1998 Emergency repair of oil trunk pipelines (Moscow: NEDRA) 271 p

[6] Borodavkin P P, Kim B I 1981 Environmental protection during construction and operation of trunk pipelines (Moscow: NEDRA) 160 p

[7] Terentyev V L, Chelombitko S I 2001 Development of methods for determining the volume of spilled oil in the accident on the main oil pipeline in the winter *Proceedings of Universities. Oil and gas* №**4** pp 87-90.

[8] RD 39-00147105-006-97 instruction on land reclamation, disturbed and contaminated during emergency and major repairs of oil trunk pipelines (date of introduction 01.07.97). 22 p

[9] Bobylev L M 1999 How to avoid the threat of environmental disaster due to oil leaks from oil trunk pipelines. *Oil and Gas technologies* № **6** pp27-29
[10] Resolution of the Government of the Russian Federation of August 21, 2000 № 613 "On urgent measures to prevent and eliminate accidental spills of oil and petroleum products" of 15.04.2002 № 240.
[11] Kutsenko V V 1997 State of raw material base and environmental safety of fuel and energy complex *Green World* №**15** 90 p
[12] Materials of parliamentary hearings "On the impact of the oil and gas industry on the environment" held at the initiative of the Committee on Ecology of the state Duma of the Russian Federation on December 5 2000
[13] Mazur I I 1997 The Economy should be eco-friendly *Oil of Russia* № **10** pp 31-36.
[14] Nizamov H N 1996 Analysis of emergency situations in pipeline systems *Ecology and industry of Russia* № **5** pp 41-43.
[15] Lisanov M V, Pecherkin A S, Sidorov V I, Shvyryaev A A, Safonov V S 1998 Risk Assessment of accidents on the linear part of oil trunk pipelines *Safety in industry* №**9** pp 52-58.
[16] RD 153-39.4-114-01. Rules of elimination of accidents and damages on oil trunk pipelines (date of introduction 20.02.2002)63 p
[17] Terentyev V L, Chelombitko S I 2002 Experimental study of oil saturation of snow. *Proceedings of X-th interdisciplinary seminar "thermal physics, hydro - gasdynamic on, heat engineering"* (Tyumen: TSU) pp 42-43.
[18] Terentyev V L, Chelombitko S I 2002 On the influence of physical and mechanical properties of snow on the calculation of emergency leakage from the main oil pipeline. *Materials of the international seminar "Geotechnical and operational problems of oil and gas industry"* (Tyumen: TSOGU) pp 167-169.
[19] Vladimirov V A 2000 Catastrophes and ecology (Moscow: Contact Culture) 380 p
[20] Vladimirov V A, Dolgin N N, Bogachev V 2001 Catastrophe of the late twentieth century (Moscow: Geopolitics) 424 p.
[21] Bogoslovskaya L S, Bruskin I M, Wilczek G E 1996 The Russian Arctic on the verge of disaster (Moscow: Center of ecological policy of Russia) 206 p
[22] Methods of determining environmental damage in case of accidents on oil trunk pipelines 1996 (Moscow: Transpress) 54 p.
[23] Kuzmin P P 1957 Physical properties of snow cover (Leningrad: Gidrometeoizdat) 183 p
[24] Kuzmin P P 1952 To the method of determining the density of wet snow *Meteorology and hydrology* №**10** (Moscow: Gidrometeoizdat) pp 17-21
[25] Grey D M, E-Mail D H1986 Snow: Directory (Laningrad: Gidrometeoizdat) 751 p

IOP Conf. Series: Materials Science and Engineering **972** (2020) 012076 doi:10.1088/1757-899X/972/1/012076

Reducing water pollution risks by wash water from water treatment plants

G B Brayalovsky[1], O B Naschetnikova[1] and E V Migalatiy[1]

[1]Ural Federal University, 19 Mira Street, Ekaterinburg 620002, Russia,

E-mail:o.b.naschetnikova@urfu.ru

Abstract: The article analyses chemical methods to treat washwater from rapid filters at surface water treatment plants. The review of existing approaches to the application of rapid filter washwater is carried out. During the research work washwater from rapid filters was chemically treated with coagulant "Bopak-E", flocculants "Praestol", and sorbent-coprecipitant reagent "Ecosol-401". The doses of reagents were selected and the efficiency of water purification from turbidity and chromaticity was shown. The proposed technology of washwater treatment allows to obtain drinking quality water after sedimentation and filtering.

1. Introduction

Issues of protection of natural waters from pollution are becoming increasingly urgent nowadays. Currently, as a rule, facilities with two or three-stage technological process are designed for their purification. The operation of such facilities requires a large number of reagents and a significant water consumption for their own needs. Treatment of waste water from sedimentation tanks, filters, clarifiers with a layer of suspended sediment, contact clarifiers and reagent facilities in compliance with the rules of protection of surface sources from pollution can reduce the lost volume and the negative impact on water bodies.

The washwater from the waterworks contains almost all the impurities of the source of water supply and detained during the clarification of natural water. It also includes reagents introduced during the purification of natural water, and with a concentration much greater than the dose in the treatment of water.

Considering the problem of treatment of washwater among the many general issues of water resources protection, it should be noted that it remains one of the least developed, the most technically complex and expensive ones. A successful solution of this problem requires the study of the composition, structure and properties of waste water, the possibility of changing the initial properties of washwater, as well as the determination of optimal parameters of the technological processes that ensure the treatment of washwater of different initial quality [1].

There are several main areas in washwater treatment:

- the supply of the washwater into a pipeline mixer before purifying in the tank-neutralizer, directly into the mixer after precipitating in the mixer after special treatment;
- the discharge of the washwater in the sewage system with subsequent supply to the sewage treatment stations;
- the discharge of the washwater without treatment into the reservoir or to specially designated areas of terrain (without purification) [2].

It is technologically and economically feasible to separate the washwater of the filter facilities from the main water treatment process and use them after purification for another purpose, for example, for washing of filters. However, it is recommended to wash the rapid filters with filtered water. Thus, the quality of washwater used for repeated washing of filters should correspond to the quality of purified natural water [3]. This creates the need to find effective technologies for the treatment of washwater, including reagent.

2. Existing solutions to the recycling of washwater

Taking into account a huge volume of discharges (5-15% of the total capacity of water stations) and consumed reagents, the problem of treatment and reuse of waste water requires an immediate solution [4].

For the purpose of rational use of water and protection of the environment in water treatment plants, it is recommended to reuse water after washing rapid filters and contact purifiers. Clarified water above the sludge can also be reused in technical sedimentation tanks and washwater sedimentation tanks, thickeners, sludge storage, and dewatering sites. It should be noted that there is a need for technological analysis and justification of the re-use of washwater filter facilities in each case. The need for careful justification of reuse of wash water is due to a significant difference in the quality of the source water and, accordingly, the complexity of the technology of water preparation, which can lead to a disruption of natural water purification regimes and deterioration of the quality of treated water [4].

The circulation of wash water from rapid filters, i.e. the establishment of a system of reuse washwater (SRUWW) from rapid filters, both in the past and in the present is developing in several directions.

The first direction is "**the supply to the head of treatment facilities of washwater without water precipitation**" from the intermediate tank (actually performing the function of the neutralizing agent). As a result, the interval between water discharge and the beginning of its supply from the neutralizing tank is reduced to a minimum and the accumulation of the bulk of the sediment is ensured only in sedimentation structures. In addition, the introduction of coagulated suspension into the source water occurs during the supply of undiluted water, which should improve the process of coagulation and sedimentation, especially at low initial values of turbidity.

Studies conducted at the Department of "Water Supply and Sewage Disposal" of Rostov State Construction University (RGCU) indicate that with the introduction of 15% washwater of the main flow rate, there is no deterioration in the quality of purification of the source water, and the increase in doses of the reagents used is not required .

Despite the advantages mentioned, according to the authors, there are some concerns in the application of this method, e.g. the variability of the quality of washwater supplied to the head of the facilities, and the need to maintain impurities in the suspended state [5].

The second direction is "**the supply to the head of treatment facilities with preliminary sedimentation**." The use of sedimentation (reagent-free or with the use of reagents) in the neutralizing tank or in a separate sedimentary basin makes it possible to obtain the discharge water of constant quality, supplied with a constant flow rate. As a result, the main facilities operate in a stable mode, both in terms of consumption and increase in the content of suspended solids due to the addition of waste water. Adding reagents to the waste water and using mixing systems (this requires much less power than in maintaining impurities in suspension) allow to vary the water quality at the outlet of SRUWW and to ensure a narrower range of fluctuations of the quality indicators throughout the seasons of the year. In addition, it is necessary to specify possible variation of a place of supply of the cleared washwater to the basic process depending on the change of its quality and peculiarities of the basic constructions [6].

The third direction is **"the treatment of waste water to the required quality** (drinking, for washing filters, watering, etc.) **with subsequent use"** and can be considered as the development of the second direction. Water after the sedimentation basin (neutralizing tank) is sent for further purification (mainly filters) or to the object of consumption based on the quality of the precipitated water. In this case, the flow rate from water intake facilities is not reduced and there is no impact of waste water on the main flow, but there are additional volumes of water that can be used for appropriate purposes with a certain economic return.

The fourth direction of the development is "**the combination of methods**", i.e. a synthesis of the directions discussed above, which is designed to take into account economic, technological and other factors to optimize the operation of water treatment facilities in general and the system of reuse of washwater from rapid filters, in particular.

The authors have studied the washwater of the Western filtration plant in Ekaterinburg and investigated the treatment methods of water in the second and third directions.

3. Results of the experiment

In the course of the research, the quality indicators of wash water from rapid filters were determined: turbidity is 44.2 mg/dm^3; chromaticity is 23 degrees and hydrogen index is 7.6. This washwater is practically not precipitated, median fall diameter below 0.1 mm/s.To purify this washwater, the reagent aluminium oxychloride of the "BOPAK-E" brand was tested. This coagulant has a number of advantages over other aluminum-containing coagulants and is widely introduced into the waterworks system of the Sverdlovsk region [7].

The results of washwater treatment with this coagulant are presented in Figures 1 and 2.

As shown in Fig.1 and 2, during the experiments it was found that the optimal dose of aluminum oxychloride coagulant required for the treatment of rapid filter washwater is 3 mg/dm^3 of Al_2O_3, thus a further increase in the dose is impractical. To intensify the coagulation process flocculants "Praestol" were tested being introduced after the coagulant.

The experimental results are shown in Figures 3 and 4

Figures 3 and 4 demonstrate that anionic flocculant Pr-2540 proves to be the most effective one. Its efficiency reaches 96% at a dose of 1 mg/dm^3.

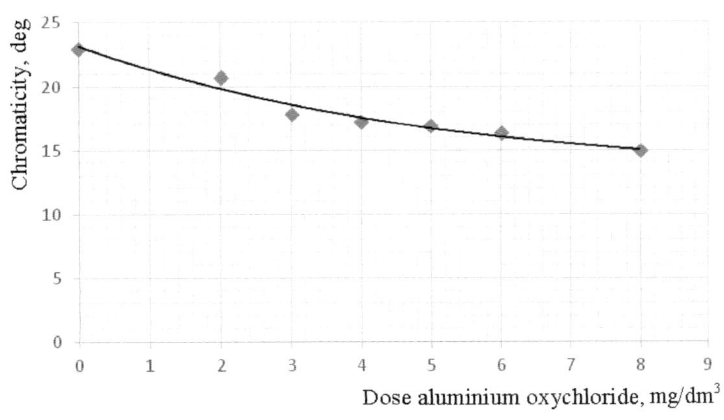

Figure 1. The impact of aluminium oxychloride on chromaticity of cleared water

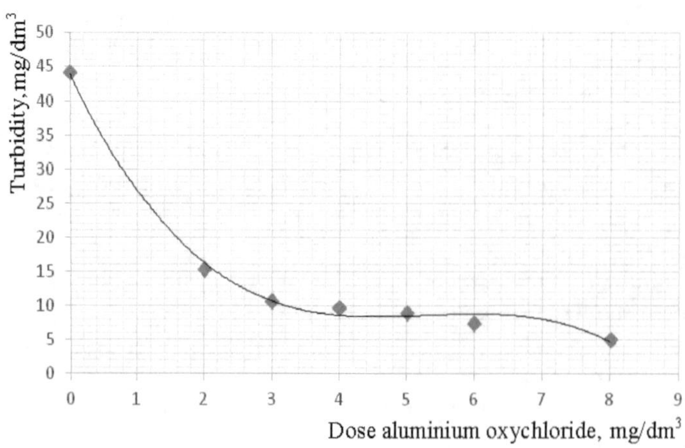

Figure 2. The impact of aluminium oxychloride on the turbidity of the cleared water

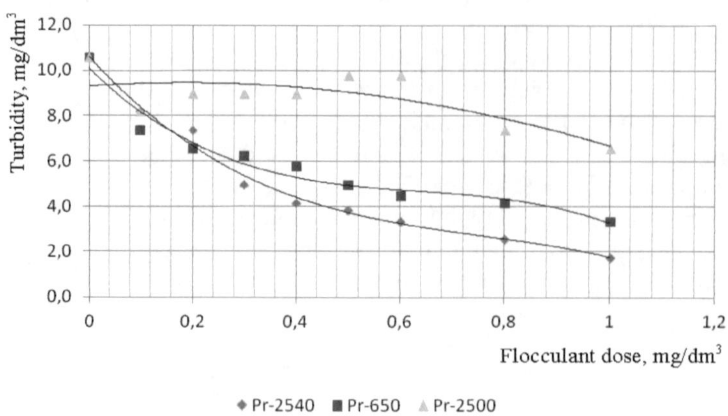

Figure 3. The impact of flocculant dose on turbidity of cleared water (the coagulant dose of 3 mg/dm^3 of Al_2O_3)

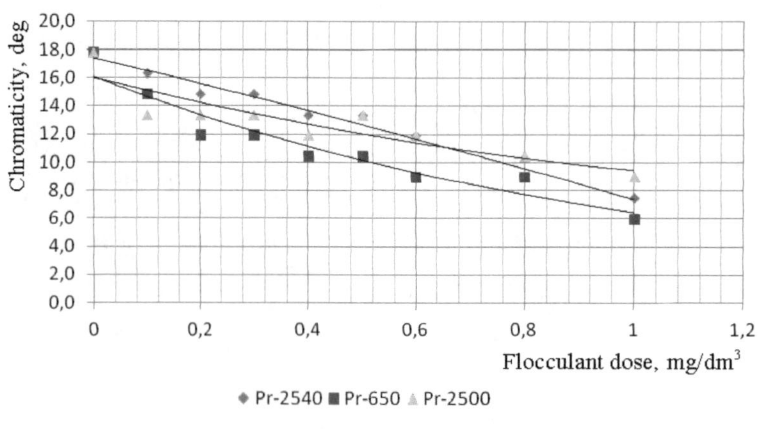

Figure 4. The impact of flocculant dose on the colour of cleared water (the coagulant dose of 3 mg/dm^3 of Al_2O_3)

The introduction of flocculant leads to the intensification of coagulation, the reduction of turbidity, chromaticity and the increase of the effect of clarification, but there is a low rate of deposition of particles. To improve hydraulic size, a decision was made to introduce additional reagent "Ecosol-401" [8]. The dose of this reagent was taken as 50 mg/dm^3. The results of the experiment are shown in Figure5.

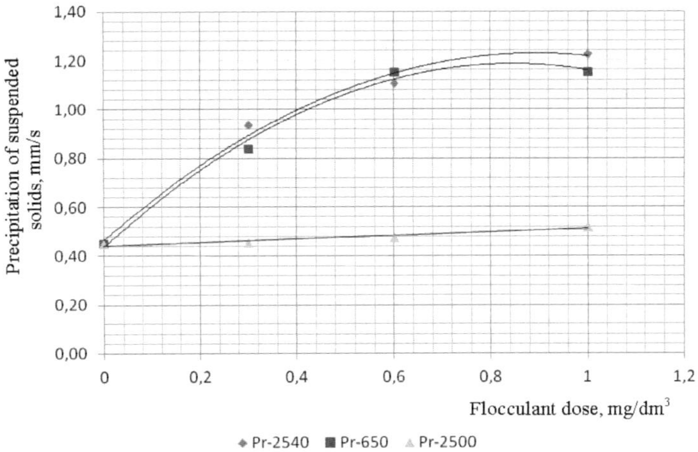

Figure 5. The impact of flocculant dose on precipitation of suspended solids (aluminium oxychloride dose is 3 mg/dm^3; "Ecosol-401" dose is 50 mg/dm^3)

The graph shows that the hydraulic size for the cationic and anionic flocculant is almost the same, and it increases with the increasing dose. The increase of the doses of non-ionic flocculant made little impact on the hydraulic particle size, its change is negligible.

4. Conclusions

Through the research we carried out chemical treatment of the washwater from rapid filters with coagulant "Bopak-E", flocculants "Praestol" reagent and sorbent-coprecipitant "Ecosol-401".

In the first series of experiments, the quality indicators of washwater from rapid filter were determined to be 44.2 mg/dm^3 for turbidity, 23 degrees for chromaticity and 7.6 for hydrogen index, respectively.

In the second series of experiments, we found the optimum dose of coagulant aluminum oxychloride "Bopak-E" required for treatment of washwater from rapid filters, which equals 3мg/dm^3 of Al_2O_3. With the introduction of this dose of coagulant there is a decrease in turbidity to 10 mg/dm^3 and chromaticity to 16. The clarifying effect reaches 76%. The decrease in pH is insignificant.

At the last stage of the work, the type and optimal dose of the flocculant "Praestol" was determined. The best effect of clarification was observed with the introduction of anionic flocculant Pr-2540, the optimal dose of which was 0.6 mg/dm^3. The indicator of the quality of the clarified water turbidity was 3.4 mg/dm^3 and that of chromaticity was 9 degrees.

To increase the speed of flocculation and sedimentation, we introduced the "Ecosol-401" reagent in the amount of 50 mg/dm^3. After its introduction, the hydraulic size increased to 1.11 mm/s.

The proposed technology of washwater treatment allows obtaining drinking quality water after sedimentation and filtering.

References

[1] Butko D, et al 2009 Reagent clarification of washwater from rapid filters *Water supply and sanitary engineering* **9** 53-56

[2] Urvantseva M, Artemenok N 2011 Comprehensive assessment of the clearing processes of washwater from waterworks in Western Siberia *Water supply and sanitary engineering* **2** 25-29
[3] Voitov E, et al 2014 The technology of purification and reuse of washwater *News of universities. Construction* **6** 59-65
[4] The method of determining the basic technological parameters of the structures of water supply and wastewater systems, wastewater treatment and sludge treatment 2014 vol 1 (Moscow: TK Group Ltd) 327p
[5] Butko D 2009 Washwater of rapid filters and their (Rostov-on-Don: Rostov state building university) 122 p
[6] Butko D, Lysov V, Popovyan G 2011 Some aspects of the creation of schemes for a complete working cycle of washwater of water treatment stations *Sanitary engineering and ecology. Bulletin of civil engineers* **1** 114-118
[7] Bogomazov O, Pyankov A Introduction of BOPAK-E aluminum oxychloride coagulant at water treatment plants of Vodokanal Municipal Unitary Enterprise of the city of Yekaterinburg [Retrieved from http://xn--80ab2aof.xn--p1ai/]
[8] Galkin Y, Selitsky G 2009 Deep purification of oil-containing waste water using the sludge flocculators and nano-dispersed reagents *Water supply and sewerage* **7-8** 54- 64

Deceleration of the Cold Flow in the Vortex Tube

N Anoshin[1], A Khait[1], V Bianco[2], A Noskov[1], V Alekhin[1]
[1]Ural Federal University, 19 Mira St., 620002 Ekaterinburg, Russia

[2] DIME/TEC – Division of Thermal Energy and Environmental Conditioning, University of Genoa, Via All'Opera Pia 15A, 16145 Genoa, Italy

E-mail: n.m.anoshin@urfu.ru

Abstract. Enhancement of the cold flow deceleration unit is considered in the paper to improve the energetic efficiency of the vortex tube. It was shown that the gradual expansion of the cross-section in the cold flow diffuser is inefficient for reducing the gas flow velocity in presence of its significant rotation. Installation of a blade grid at the entrance to the cold flow diffuser was suggested for interruption of the flow swirling. A notable reduction of the energy losses has been demonstrated improving the efficiency of the vortex tube.

Keywords: vortex tube, efficiency, deceleration

1. Introduction

The effect of temperature stratification in the vortex flows known also as a Ranque-Hilsch effect is thoroughly investigated both theoretically and experimentally. However, the reasons for the appearance of such a phenomenon are still not well understood [1, 2]. The main difficulty here is the requirement for the velocities of the gas to be extremely high (nearly trans-sonic) to distinguish the effect. The resultant flow appears to be strongly turbulent that significantly complicates any theoretical considerations.

Nevertheless, the Ranque-Hilsch effect has found its application in the devices called vortex tubes that are widely utilized for cooling or heating the pressurized gas. The lack in the complete theoretical understanding of the physical processes involved in the operation of the vortex tubes stipulates a relatively low efficiency of such devices. One of the available ways to improve the performance of the vortex tube is the optimization of its design by performing the numerical or physical experiments.

Following this approach, the influence of the divergence angle of the energy separation chamber was investigated experimentally in [3]. It was shown that the optimal value of the angle lays within the range 2-4°. The number of the inlet nozzle channels and their convergence ratio, length of the energy separation chamber, cold orifice diameter and inlet pressure were studied in [3, 4] and optimized experimentally in [5]. The method of the local entropy generation analysis was implemented in [6] to find the ways for further optimization of the vortex tube performance.

Efficient deceleration of the gas flow before its release from the vortex tube to the ambient environment constitutes also a significant problem. In view of this, the effect of the angle of the hot control valve on the vortex tube efficiency was considered in [7]. On the other hand, the influence of the shape of this valve has been studied in [8]. The spiral motion of the gas at very high velocities makes the deceleration of the flow at the cold outlet of the vortex tube to be a significantly more complicated task. The diffusers having different orifice diameter and divergence angle were tested in [9]. A notable influence of the diffuser angle on the vortex tube efficiency was reported. This points to the insufficient efficiency of the classical design of the cold outlet of the vortex tube. Fins of different shapes were introduced to the geometry of the cold flow diffuser in [10] attempting to reduce the energy losses.

It should be highlighted that in all previous studies the rotation of the gas flow at the cold outlet of the vortex tube has not been taken into account. In this paper we devote a particular attention to interruption of the rotation of the gas flow before its output from the vortex tube. This is expected to allow the reduction of the energy losses associated with the subsequent deceleration process in the cold flow diffuser. Interruption of the gas flow rotation is performed by the specifically designed blade grids having different configurations. The reduction of the energy losses in the cold flow deceleration process will potentially influence the increase of the vortex tube efficiency.

2. Vortex Tube and the cold flow deceleration unit

The domain of the vortex tube is shown in Fig. 1. Pressurized air is supplied to the nozzle inlet 1 consisting of six curved rectangular channels. Reduction of the cross-section of each channel leads to the acceleration of the air to the velocities close to the speed of sound. Accelerated air is then passed to the energy separation chamber 2 having a conical shape. Radial temperature stratification arises in the vortex flow generated in the chamber 2 such that the core region of the vortex has lower total (stagnation) temperature compared to the outer region. The heated air is let out via the left side of the vortex tube, see Fig. 1, whereas the cooled air from the vortex core is exhausted via the opposite side.

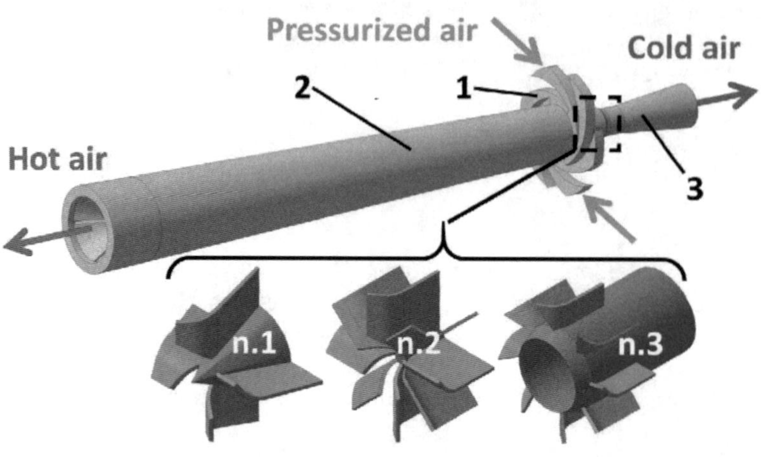

Fig. 1. Schematic of the vortex tube with different blade grids at the cold flow outlet: 1 - nozzle inlet to the vortex tube; 2 - energy separation chamber; 3 - cold flow diffuser

Deceleration of the cold air before its release from the vortex tube to the ambient environment is presently performed by the diffuser 3 (Fig. 1). The reduction of the cold air velocity within the diffuser is expected due to the increase of the cross-section area. Such configuration was proved to be efficient for the unidirectional coaxial flows. Because of the existence of a significant rotational component in the cold flow, application of the simple diffuser is inefficient for the vortex tubes. In this study we attempt to increase the efficiency of the cold flow deceleration by improvement of the configuration of the diffuser 3 (Fig. 1). The suggested method of the vortex flow deceleration consists of two stages. First, we interrupt the rotation of the flow while passing through the blade grids. This will allow to obtain a coaxial flow that may be decelerated in the diffuser.

Different configurations of the blade grids denoted as n.1 – n.3 were considered, see Fig. 1. It is known that the radial distribution of the rotational component of the velocity in the flow under consideration is linear with a negligibly small rotation in the center. Thus, in the first case (n.1), we place the blades only at the outer part of the diffuser and fill the center by the conical body. Second configuration (n.2) incorporates the blades that occupy the entire cross-section. The last option (n.3) is similar to n.1 except for the shape of the body placed in the center. Now we utilize a short tube allowing the core part of the flow to pass the deceleration unit while maintaining the rotation.

3. Numerical model

Numerical model of 3D flow of compressible viscous air was utilized to test the performance of different configurations of the vortex tube shown in Fig. 1. Since the flow under consideration is strongly turbulent, the model was based on Reynolds-Averaged Navier-Stokes (RANS) equations closed with the standard k-ε turbulence model. Ideal gas state equation was used to take into account the dependence of the air density on its static pressure and temperature. The conservation of the energy of the system was ensured by solving the equation for the total enthalpy, which in the case of the ideal gas is $H = c_p T + V^2/2$. The properties of the air were defined by the following constants: gas constant R = 287 $J/(kg\ K)$; dynamic viscosity $\mu = 1.831\times10^{-5}$ $Pa\ s$; thermal conductivity $\lambda = 0.0261$ $J/(m\ s\ K)$; heat capacity c_p = 1004.4 $J/(kg\ K)$. The constants of the k-ε turbulence model were selected as follows: $C_{\varepsilon 1} = 1.44$; $C_{\varepsilon 2} = 1.92$; $C_\mu = 0.09$; $\sigma_k = 1.0$; $\sigma_\varepsilon = 1.3$.

The ANSYS CFX software was used for the numerical solution of the equations using the Finite-Volume Method. High resolution advection scheme was used for estimation of the spatial derivatives, while second-order implicit scheme was used for the temporal integration. Adaptive time step within $\Delta t = 10^{-5}$ - 10^{-7} s was applied. The structured hexahedral mesh consisting of 430 000 cells was utilized to approximate the geometry of the nozzle inlet and the energy separation chamber. On the other hand, the geometry of the cold flow deceleration unit incorporating the blade grid was described by the unstructured tetrahedral mesh consisting of 140 000 – 160 000 cells depending on the particular configuration n.1 – n.3, see Fig. 1. We refer to [6] for the better details on the numerical model.

The dimensions of the domain and the operational parameters of the vortex tube are summarized in Table 1. Operation regime of the vortex tube is usually characterized by the value of the so-called cold flow fraction given as

$$\varphi = G_c / G_i, \qquad (1)$$

where G_c and G_i are the mass flow rates of the cold and the inlet flows respectively. In turn, the efficiency of the vortex tube is estimated by the following coefficient:

$$\eta = \varphi \frac{T_i^* - T_c^*}{T_i^* \left[1 - \left(\dfrac{p_c^*}{p_i^*} \right)^{\frac{\gamma-1}{\gamma}} \right]} \qquad (2)$$

Here superscript * denotes the stagnation value of the quantity; $\gamma = 1.4$ is the adiabatic index.

4. Results and discussion

First, we study the flow observed in the base configuration of the deceleration unit corresponding to the standard conical diffuser without the blades installed. The velocity field in the longitudinal section of such diffuser is shown in Fig. 2(a). It is seen that the velocity magnitudes reach quite significant values around 250 m/s. Because of the centrifugal force caused by the strong rotation of the air, gradual increase of the diameter of the diffuser leads to the reduction of the pressure in the core part where the rotation is almost absent. In the considered case, the pressure drops even below the ambient value leading to the appearance of the large-scale secondary circulation clearly seen in Fig. 2(a). Existence of such a circulation is considered to be one of the main reason of the low efficiency of the deceleration unit.

As seen from Fig. 2(b), deceleration unit n.1 redistributes the velocity in the cross-sections such that the high speed area is now concentrated near the outer walls, whereas the velocity value in the core region is notably reduced. Such a flow configuration shortens the large-scale secondary

IOP Conf. Series: Materials Science and Engineering **972** (2020) 012077 doi:10.1088/1757-899X/972/1/012077

circulation. Implementation of the deceleration unit n.2, see Fig. 2(c), gives even smoother flow with almost constant value of the velocity in every cross-section of the diffuser. The rotational component of the velocity is effectively cancelled, and the secondary circulation is fully suppressed in this case.

Table 1. Dimensions and operation parameters of the vortex tube

	Value	Units
Diameter of the energy separation chamber	16	mm
Length of the energy separation chamber	170	mm
Diameter of the diaphragm	9.5	mm
Cone angle	2	°
Gauge pressure of the supplied air, p_{in}^*	300	kPa
Stagnation temperature of the supplied air, T_{in}^*	300	K
Cold flow fraction	0.1 – 0.8	-

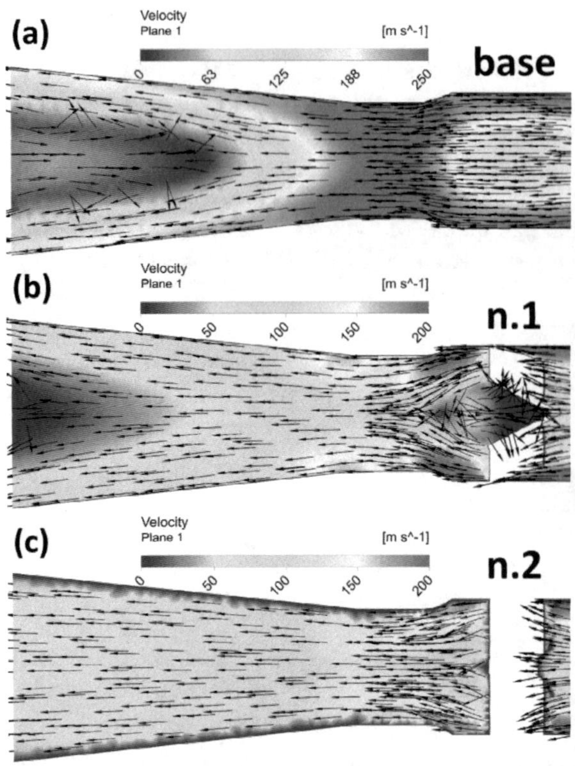

Fig. 2. Distribution of the velocity magnitude [m/s] over the longitudinal section of the cold flow diffuser: (a) base configuration without blades; (b) deceleration unit n.1; (c) deceleration unit n.2

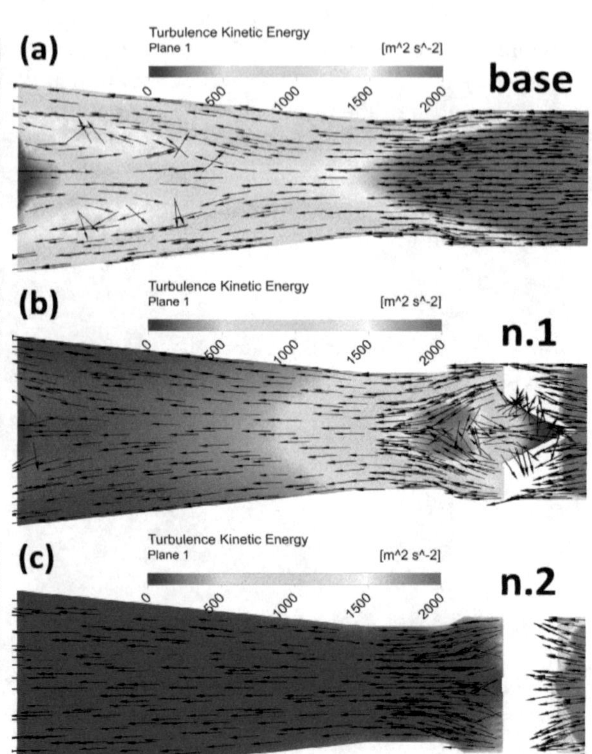

Fig. 3. As in Fig. 2 for the turbulence kinetic energy [m^2/s^2]

In Fig. 3 we study the distribution of the turbulence energy in the longitudinal section of different deceleration units. It is notable that the base configuration shows the strongest turbulization of the flow. In the unit n.1, the turbulence is concentrated only in the proximity of the blade grid, while the turbulence in the unit n.2 is very low in all cross-sections after the blades, cf. Figs. 3(b) and (c).

Taking into account that the irreversible energy losses in k-ε turbulence model are proportional to the square of the turbulence kinetic energy, the utilization of the deceleration unit n.2 is expected to improve the efficiency of the vortex tube as compared to the base design.

Despite the low values of the rotational velocity in the center of the cold flow deceleration unit, application of the blade grid n.3 gave rise to the flow instability. An extremely unstable secondary vortex structure originated from the central flow is observed in Fig. 4. Increase of the turbulence in this case is expected to provoke the additional energy losses and, consequently, the reduction of the efficiency.

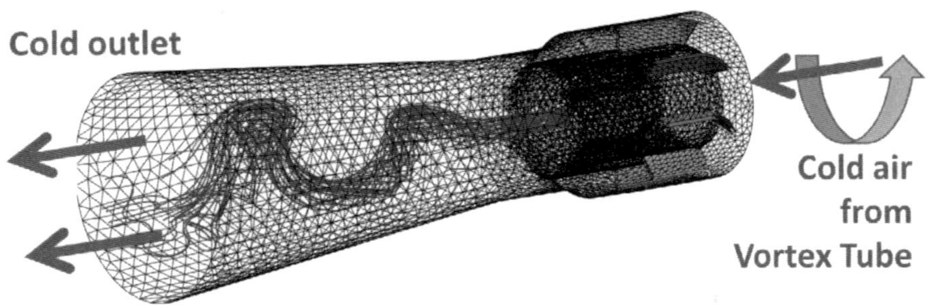

Fig. 4. Secondary vortex structure observed in the deceleration unit n.3

The main energetic characteristic of the vortex tube is plotted in Fig. 5 as function $\eta = \eta(\varphi)$. The highest value of the energy efficiency coefficient (2) is found at the cold flow fraction (1) $\varphi \approx 0.6$ in agreement with the previous investigations. As expected, installation of the deceleration unit n.3 reduced the efficiency as compared to the base design of the vortex tube. On the other hand, unit n.2 (see Fig. 1) allowed a notable improvement of the vortex tube performance. Taking into account that the design of the unit n.2 is not optimal, the efficiency may be increased even further after the corresponding optimization.

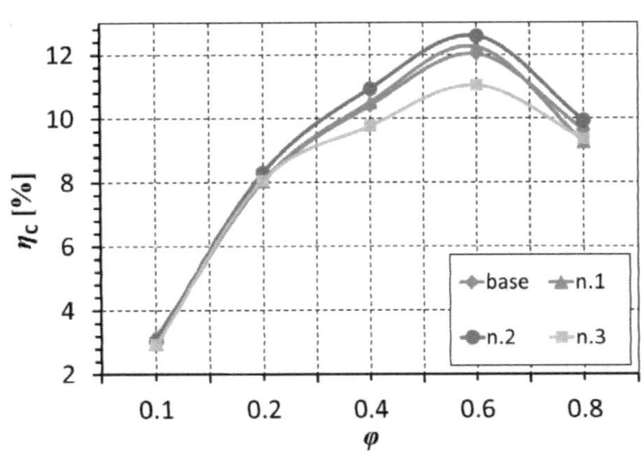

Fig. 5. Energy efficiency of the vortex tube (2) as a function of the cold flow fraction (1)

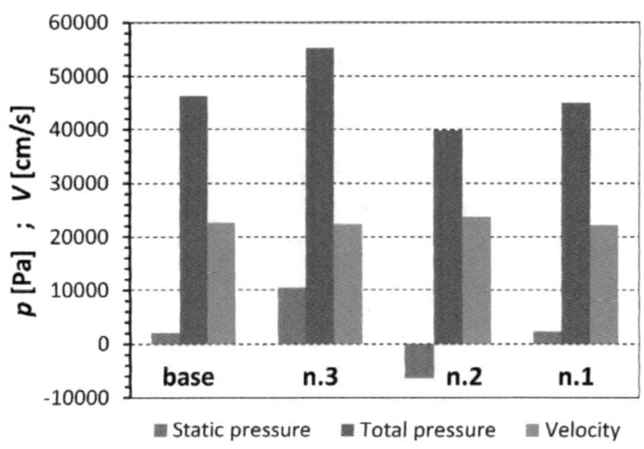

Fig. 6. Mass-flow averaged values of the gauge pressure and the velocity at the entrance to the deceleration units

A physical insight on the reasons of the efficiency improvement may be found from the analysis of the integral value of the mass-flow averaged static gauge pressure at the entrance to the deceleration unit. Fig. 6 shows that the velocity magnitudes at the entrance to the deceleration units are similar in all the considered designs. On the other hand, there is an obvious trend showing that the lower static pressure is observed at the entrance to the deceleration unit the higher efficiency of the vortex tube is obtained. In case the unit n.2 is applied, the static gauge pressure drops even below the ambient value.

Reduction of the given pressure causes increase of the available pressure difference (pressure ratio) between the inlet to the vortex tube and the outlet from the energy separation chamber (2 in Fig. 1).

5. Conclusion

In this study we attempted to reduce the energy losses involved in the deceleration of the gas flow at the cold outlet of the vortex tube. For that purpose, different blade grids were placed before the inlet to the cold flow diffuser. The blades were designed to interrupt the rotation of the air flow. Once the rotation is cancelled, the velocity of the air is efficiently reduced by expanding the cross-section in the diffuser.

It was found that the installation of the blade grids may result in reduction of the flow turbulization. Straightening of the flow allowed avoiding appearance of the secondary recirculation at the cold outlet and notably increased the efficiency of the vortex tube. The physical justification of the improvement of the vortex tube efficiency was found in the analysis of the mass-flow averaged values of the static pressure at the entrance to the deceleration unit. Reduction of this pressure provoked the increase of the available pressure ratio and, consequently, the available work.

It should be emphasized that the geometry of the deceleration units examined in the present paper cannot be considered as optimal. Therefore, the optimization of several constructive parameters of the blade grids and the diffuser may theoretically lead to even more significant increase of the vortex tube efficiency.

References

[1] J. Lagrandeur, S. Poncet, M. Sorin 2019 Review of predictive models for the design of counterflow vortex tubes working with perfect gas *Int. J. Therm. Sci.* **142** 188–204

[2] A. Khait, A. Noskov, A. Lovtsov, V. Alekhin 2014 Semi-empirical turbulence model for numerical simulation of swirled compressible flows observed in Ranque-Hilsch vortex tube *Int. J. Refrig* **48** 132-141

[3] K. Chang, Q. Li, G. Zhou, Q. Li 2011 Experimental investigation of vortex tube refrigerator with a divergent hot tube *Int. J. Refrig.* **34** 322–327

[4] S.E. Rafiee, M. Rahimi 2013 Experimental study and three-dimensional (3D) computational fluid dynamics (CFD) analysis on the effect of the convergence ratio, pressure inlet and number of nozzle intake on vortex tube performance - Validation and CFD optimization *Energy* **63** 195–204

[5] M. Bovand, M.S. Valipour, K. Dincer, S. Eiamsa-Ard 2014 Application of response surface methodology to optimization of a standard Ranque-Hilsch vortex tube refrigerator *Appl. Therm. Eng.* **67** 545–553

[6] A. Khait, A. Noskov, V. Alekhin, V. Bianco 2018 Analysis of the local entropy generation in a double-circuit vortex tube *Appl. Therm. Eng.* **130** 1391–1403

[7] K. Dincer, S. Baskaya, B.Z. Uysal, I. Ucgul 2009 Experimental investigation of the performance of a Ranque-Hilsch vortex tube with regard to a plug located at the hot outlet, *Int. J. Refrig.* **32** 87–94

[8] S.E. Rafiee, M.M. Sadeghiazad 2017 Experimental and 3D CFD investigation on heat transfer and energy separation inside a counter flow vortex tube using different shapes of hot control valves *Appl. Therm. Eng.* **110** 648–664

[9] M. Farzaneh-Gord, M. Sadi 2014 Improving vortex tube performance based on vortex generator design *Energy* **72** 492–500

[10] A. Bazgir, A. Heydari, N. Nabhani 2019 Investigation of the thermal separation in a counter-flow Ranque-Hilsch vortex tube with regard to different fin geometries located inside the cold-tube length, *Int. Commun. Heat Mass* **108** 104273

AUTHOR INDEX